LINEAR NETWORKS AND SYSTEMS

WAI-KAI CHEN
UNIVERSITY OF ILLINOIS AT CHICAGO

 BROOKS/COLE ENGINEERING DIVISION
Monterey, California

To Shiao-Ling and Jerome and Melissa

Sponsoring Editor: Ray Kingman
Production: Brian K. Williams/San Francisco
Manuscript Editor: Sandy Spiker
Interior and Cover Design: Nancy Benedict
Illustrations: Carl Brown
Typesetting: Syntax International
Production Services Manager: Stacey C. Sawyer
Coordinating Designer: Jamie Sue Brooks

BROOKS/COLE ENGINEERING DIVISION
A Division of Wadsworth, Inc.

Library of Congress Cataloging in Publication Data:

Chen, Wai-Kai, 1936–
    Linear networks and systems.

    Includes bibliographies and index.
    1. Electric networks   I. Title.
    TK454.2.C425   1983   621.319′2   82-20647
    ISBN 0-534-01343-0

ISBN 0-534-01343-0

Printed in the United States of America

10 9 8 7 6 5 4 3 2 1

John Senior

LINEAR NETWORKS AND SYSTEMS

PREFACE

Traditionally, two courses in networks are included in the electrical engineering curricula: one given at the sophomore level and followed by another offered at the junior level for a total of four semesters of networks. With the advances in integrated circuits, computers, microprocessors, and solid-state electronics, few curricula can afford the luxury of four semesters of networks. The trend is toward shortening and broadening the second course in networks to a one-semester or two-quarter course on linear networks and systems.

Basic network and system theory is a vast area, and at least two semesters are required to cover the basics. Most books attempt to cover it all, and do a fair job on part of it. What is sorely needed are two books that treat networks and systems in depth: one for basic dc and ac networks, and another for transform and state-variable techniques. There seems to be general agreement to the contents for the first course in networks. However, there is no consensus on the contents of the second course. This suggests a need for innovation. This book is a result of this debate and is designed for the second course in networks with modern systems viewpoint. Its prerequisites are the basic calculus, dc and ac networks, matrix algebra, and some familiarity with linear differential equations which are desirable but not essential. The objective of the book is to select and feature theories and concepts of fundamental importance that are amenable to a broad range of applications. Since juniors have been exposed only to one or two semesters of introductory course in networks at sophomore level, the best way to introduce systems concepts and viewpoint is to use electrical networks exclusively as illustrations. Thus, non-electrical systems are excluded.

With increasing use of computer in undergraduate curricula, I have long believed that the systematic, graph-theoretic methods of writing network equations both in frequency domain and time domain should be taught in a basic course, leaving the numerical computations to computers. Thus, no special techniques or "tricks" are included. This view has been dominant in the book. The average student may find the chapter on graphs a little difficult to relate to, but as a modern text on linear networks and systems where computer-oriented systematic formulation of system equations is necessary, the graph-theoretic concepts are indispensable in this formulation. Furthermore, as graphs are widely used in computer science and engineering, the students should be exposed to this topic more rigorously at this stage.

Recognizing that computers are common tools in modern engineering, canned computer programs are developed throughout the text. The students are asked to use them to solve practical problems, and to add to the illustrative programs. All the programs were written in WATFIV, which can easily be converted to FORTRAN IV language if required. In fact, the name WATFIV stands for *WATerloo Fortran IV*. The WATFIV compiler was developed at the University of Waterloo to satisfy important requirements in both education and research. It is a fast compiler to handle the fast-growing volume of undergraduate student problems in an economical way. Furthermore, it provides error diagnostics which help students catch many obscure programming blunders. This is the primary reason for choosing this language in the text.

The scope of this book should be quite clear from a glance at the table of contents. Chapter 1 introduces many fundamental concepts related to linear, time-invariant systems. Chapter 2 gives a fairly complete exposition of the linear graph concepts and the formulation of the primary network equations. To systematically reduce the number of these equations, Chapter 3 presents procedures for obtaining three secondary systems of equations known as the loop, cutset and nodal systems of equations. The solution of the secondary systems of equations in the time domain is taken up in Chapter 4. The next two chapters study the Laplace transformation and its applications to network analysis. The convolution integral and the representation of a network response in terms of the convolution integral are discussed in Chapter 7. In this chapter, we also study the principle of superposition in the context of the convolution integral.

Chapters 8, 9, and 10 are concerned with Fourier series and transforms. Bilateral and circular convolution, discrete Fourier series (DFS), fast Fourier transform (FFT) algorithm, and the sampling theorem are treated in considerable depth, thus providing basis for other courses in electrical engineering such as digital signal processing, communications, control and signal analysis. The last two chapters deal with the state-variable techniques, which have assumed new importance. The classical transform methods were very popular, but in many quarters there was a movement away from this approach. Underlying this shift of emphasis is of course the availability of computers that can obtain numerical solutions. A more important reason is that the state-variable representation can easily and naturally be extended to time-varying and nonlinear systems. In fact, computer-aided solution of time-varying, nonlinear network problems almost always is accomplished using the state-variable approach. Even if it were not for the above reasons, the new approach provides an alternate view of the physical behavior of a system under study. Therefore, the introduction of linear state-variable concepts to juniors is essential. Before students can learn to solve most nonlinear and time-varying problems, they must understand the general formulation of simpler linear problems.

It is a pleasure to thank the following who have given valuable assistance: Drs. Nasser Jaleeli, Mohammed Jameel, and Robert Lilley of Ohio University; Messrs. Shunquan Gao, Jialong Lau and Zuo-Chen Chi from various parts of China; and Mr. Zhao-Ming Wang of Chengdu Institute of Radio Engineering. Drs. Jaleeli and Jameel read various portions of my manuscript and gave useful suggestions. Mr. Wang, a visiting scholar from China, also read portions of the manuscript and assisted me in preparing some of the computer programs. Special thanks are due to my wife Shiao-Ling, who wrote most of the subroutines, and to my son Jerome for keypunching and running the programs. I am indebted to students who have helped me in fixing the order of presentation and the pattern of emphasis. Finally, I express my appreciation to my wife, Shiao-Ling, and children, Jerome and Melissa, for their patience and understanding during the preparation of the book.

Naperville, Illinois   *Wai-Kai Chen*
May 15, 1982

CONTENTS

CHAPTER THREE

THE SECONDARY SYSTEMS OF NETWORK EQUATIONS   84

CHAPTER FOUR

SIMULTANEOUS LINEAR DIFFERENTIAL EQUATIONS   125

CHAPTER FIVE

THE LAPLACE TRANSFORMATION   173

CHAPTER SIX

NETWORK ANALYSIS   222

CHAPTER SEVEN

INTEGRAL SOLUTION—CONVOLUTION   281

CHAPTER EIGHT

FOURIER SERIES AND SIGNAL SPECTRA   325

CHAPTER NINE

SYSTEM RESPONSE AND DISCRETE FOURIER SERIES   412

CHAPTER TEN

THE FOURIER TRANSFORM AND CONTINUOUS SPECTRA   478

CHAPTER ELEVEN

STATE EQUATIONS   558

CHAPTER TWELVE

SOLUTION OF STATE EQUATIONS   606

APPENDIX A

SUBROUTINE PLOT   A1

APPENDIX B

SUBROUTINE PLOT2   A4

INDEX   A7

LINEAR NETWORKS AND SYSTEMS

CHAPTER ONE

FUNDAMENTAL CONCEPTS

1-1 INTRODUCTION

Over the past two decades, there has been a rapid development of technology in electrical engineering in general, and in solid-state electronics in particular. Solid-state devices such as the transistor, the tunnel diode, and Zener diodes replaced the vacuum tube. Already, integrated circuit technology has emerged to push these relatively recent inventions into obsolescence. In order to cope with the threat of obsolescence and to meet the demand for more broadly educated electrical engineers, it has become necessary to increase the emphasis on both mathematics and physics in the education of electrical engineers. This book's modern application of mathematical methods to the analysis of engineering systems serves both goals.

The term *system*, as generally used, has numerous meanings. One can think of a system as a collection of objects interacting with one another to accomplish a common plan or serve a common purpose. This rather broad definition includes all physical as well as nonphysical systems. Both natural and human-made systems exist in a variety of ways. Electronic systems, mechanical systems, hydraulic systems, and thermal systems are examples of physical (natural) systems. Political systems, economic systems, and social systems are examples of nonphysical (human-made) systems. In a sense, the totality of the universe can be thought of as a single gigantic system, but such a generalization is of no use in the solution of engineering problems. Rather than pursuing such an overwhelmingly complex system, we must narrow the scope of the problem by considering a bounded system consisting of a relatively small ensemble of objects. In this way meaningful and useful solutions can be obtained to analyze the behavior of the system under various conditions.

In this text the term *system* is interpreted as an electrical network composed of a variety of elements such as resistors, capacitors, inductors, transformers, transistors, operational amplifiers, and sources. These physical elements are interpreted as performing certain mathematical functions. For instance, resistors are multipliers, and capacitors and inductors are integrators or differentiators. If we focus our attention on the functions of these

Figure 1.1
Block diagram of a
communication system.

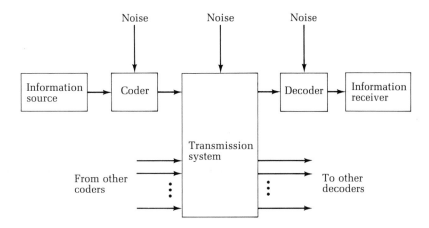

elements rather than on the physical elements themselves, a system may be
viewed as a processor. The processing is carried out with an analog, a
digital, or a hybrid network. If a digital network is used, the system can be
regarded as a digital device with discrete inputs and outputs. These are
three different interpretations of a system.

As engineers we are interested not only in the analysis but also in the
design of systems. However, as in many creative efforts, we must learn how
to analyze a system before we can proceed with synthesis. Generally system
analysis can be divided into three aspects:

1. *Modeling.* This is the technique used to develop a mathematical
    representation that exhibits as many characteristics of the original
    system as appropriate.
2. *Solution.* After a suitable model is obtained, solutions are computed
    in various forms.
3. *Interpretation.* The solutions of the mathematical model are related
    or interpreted in terms of the physical problem.

To obtain a glimpse of the general-system approach, consider a simple
communication system as illustrated by the block diagram of Figure 1.1.
The blocks in this diagram represent functional subsystems, which themselves are studied using the methods of system analysis. In most situations
the transmission system must be used for several signals simultaneously,
each arising from a separate source and each being transmitted to individual
receivers. Radio is an example in which the common transmission system
might be the ionosphere. To distinguish one signal, as defined below, from
another, each signal is coded by a coder. This process is called *modulation*.
When modulation is involved in the coding process, the decoder is designed
not only to select the desired signal for its addressee but also to transform it
back to its original form. This is referred to as *demodulation*.

The system-analysis methods to be discussed in this text are found to
be remarkably effective in a wide variety of applications. In fact, these

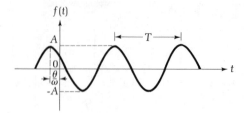

methods constitute the basis of modern engineering systems analysis. Even though the text examples are primarily about electrical signals and systems, the techniques are applicable as well to other systems that can be described by similar mathematical models.

1-2 SIGNALS

A *signal* is a physical embodiment of a message. In most systems a signal can be observed as a varying entity that changes with time. In electrical networks, for example, signals are embodied as various voltages and currents. In a flight control radar system, a light spot on the surface of a cathode-ray display is a signal that might indicate the presence of an aircraft. This light spot derives from a very short pulse, ranging from a fraction of a microsecond to a few microseconds in duration. Thus signals are functions of time. In order to design systems that can handle time-varying signals, such as those described above, more detailed knowledge of their nature is required. It can be shown that a signal, no matter how complicated, may be analyzed in terms of frequency spectrum or frequency content—that is, a collection of frequency components with specified relative amplitudes as defined below. We shall have ample opportunity to discuss and ascertain these frequency components later in the book. For the moment we shall consider certain general ideas.

A *continuous-time signal* is a function $f(t)$ of time t the value of which is specified for every point within a given time interval. For instance, the familiar sinusoidal signal

$$f(t) = A \cos(\omega t + \theta) \tag{1.1}$$

is a continuous-time signal the domain of which is the interval $-\infty \leq t \leq \infty$. The coefficient A is the *amplitude*, θ is the *phase shift*, and ω is the *angular frequency* as given by

$$\omega = \frac{2\pi}{T} \tag{1.2}$$

where T is the period of the sinusoid. This signal is presented graphically in Figure 1.2. We say that $f(t)$ *leads* $B \cos \omega t$ if θ is positive or *lags* $B \cos \omega t$ for negative θ. Note that a continuous-time signal need not be a mathematically continuous function. It may contain discontinuities, and such a function

Figure 1.3
A continuous-time signal
with two
discontinuities.

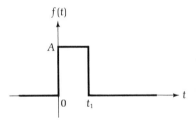

is called a *piecewise continuous function*. A continuous-time signal $f(t)$ is
said to have a *discontinuity* at the point t_0 if for $\epsilon > 0$

$$\lim_{\epsilon \to 0} f(t_0 + \epsilon) \neq \lim_{\epsilon \to 0} f(t_0 - \epsilon) \tag{1.3}$$

To simplify the notation, write

$$f(t_0 +) = \lim_{\epsilon \to 0} f(t_0 + \epsilon) \tag{1.4a}$$

$$f(t_0 -) = \lim_{\epsilon \to 0} f(t_0 - \epsilon) \tag{1.4b}$$

The value of the discontinuity at t_0 is arbitrarily defined to be

$$f(t_0 +) - f(t_0 -) \tag{1.5}$$

The rectangular pulse of Figure 1.3 is an example of a continuous-time
signal containing two discontinuities—namely, at $t = 0$ and $t = t_1$. At these
points the values of the discontinuities are found to be

$$f(0 +) - f(0 -) = A - 0 = A \tag{1.6a}$$

$$f(t_1 +) - f(t_1 -) = 0 - A = -A \tag{1.6b}$$

At other points the signal is defined by the equation

$$f(t) = A, \qquad 0 < t < t_1$$
$$= 0, \qquad t < 0 \quad \text{or} \quad t > t_1 \tag{1.7}$$

Figure 1.4
Samples of the
continuous-time signal
given in Figure 1.3.

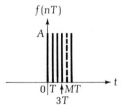

A *discrete-time signal* is a function of t the domain of which is a set of
discrete-time instants and is often generated by *sampling* a continuous-time
signal. For instance, the samples of the continuous-time signal $f(t)$ of Figure
1.3 are shown in Figure 1.4. The samples illustrated were chosen at the
equidistant points even though they can be at any discrete time instants.
For our purposes we shall assume that a discrete-time signal always has its
values at $t = 0, \pm1, \pm2, \ldots$. In fact, the values of t can be taken either to
be the actual time instants at which the values of the signal are known or to
be the indices of successive time instants that may or may not be equally
separated. If the samples are equidistant, then the constant T in the discrete-
time signal $f(nT)$, $n = 0, \pm1, \pm2, \ldots$, is called the *sampling interval*. A
device that converts analog information into digital form is called an *analog-
to-digital* or *A/D converter*. A large number of such devices is available. For
example, one form of digital voltmeter uses the A/D converter. The samples
of Figure 1.3 form a discrete-time signal as illustrated in Figure 1.4, which

can be expressed analytically as

$$f(nT) = A, \qquad n = 0, 1, 2, \ldots, M$$
$$= \text{undefined, otherwise} \tag{1.8}$$

where $MT \leq t_1 < (M + 1)T$.

In networks the voltages, currents, and other physical entities are usually represented by continuous-time signals. However, in numerical processing of these signals, their values are needed only for a discrete set of time instants. It is convenient to express them by discrete-time signals.

1-3 MANIPULATION OF SIGNALS

Signals are frequently processed to yield new signals for various purposes. In the present section, we demonstrate by simple examples how the new signals are generated from known signals.

Sum and Product

The sum of two signals is a signal the value of which at every instant equals the sum of values of the two signals at that instant. For example, the sum of two continuous-time signals

$$f_1(t) = A \cos \omega t \tag{1.9a}$$

$$f_2(t) = B \sin(\omega t + \phi) \tag{1.9b}$$

is a continuous-time signal defined by the equation

$$f(t) = A \cos \omega t + B \sin(\omega t + \phi) \tag{1.10}$$

Likewise the sum of two discrete-time signals [Figure 1.5(a) and (b)]

$$f_1(n) = 0, \qquad n < -2$$
$$= 3^{-n}, \qquad n \geq -2 \tag{1.11}$$

$$f_2(n) = 2^n + 1, \qquad n \leq 0$$
$$= \frac{1}{n}, \qquad n > 0 \tag{1.12}$$

is the discrete-time signal [Figure 1.5(c)]

$$f(n) = 2^n + 1, \qquad n < -2$$
$$= 3^{-n} + 2^n + 1, \qquad -2 \leq n \leq 0$$
$$= 3^{-n} + \frac{1}{n}, \qquad n > 0 \tag{1.13}$$

A system the output signal of which is the sum of its input signals is referred to as an *adder* and is symbolically shown in Figure 1.6 with

$$f(x) = \sum_{i=1}^{k} f_i(x) \tag{1.14}$$

where $x = t$ or n.

Figure 1.5
Sum (c) of two discrete-
time signals (a) and (b).

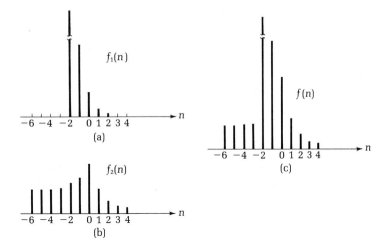

$f_1(n)$

(a)

$f_2(n)$

(b)

$f(n)$

(c)

Figure 1.6
Symbolic representation
of an adder.

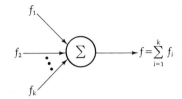

$$f = \sum_{i=1}^{k} f_i$$

The *product* of two signals is a signal the value of which at every instant equals the product of the values of the two signals at that instant. For example, the product of the two continuous-time signals (1.9) is the signal

$$g(t) = AB \cos \omega t \, \sin(\omega t + \phi) \qquad \textbf{(1.15)}$$

whereas the product of the two discrete-time signals (1.11) and (1.12) is another discrete-time signal defined by (Figure 1.7)

$$h(n) = 0, \qquad n < -2$$
$$= 3^{-n}(2^n + 1), \qquad -2 \leq n \leq 0$$
$$= \frac{3^{-n}}{n}, \qquad n > 0 \qquad \textbf{(1.16)}$$

Figure 1.7
Product of the two
discrete-time signals (a)
and (b) of Figure 1.5.

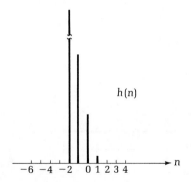

$h(n)$

Figure 1.8
Symbolic representation
of a multiplier.

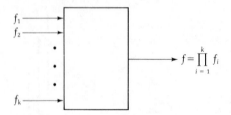

A system whose output signal is the product of its input signals is called a *multiplier* and is depicted symbolically in Figure 1.8 with

$$f(x) = \prod_{i=1}^{k} f_i(x) \tag{1.17}$$

An example of the use of the product signals is the amplitude modulation (AM) in communication.

Shifting

A continuous-time signal $f(t)$ is delayed by T seconds if the independent variable t is replaced by $t - T$, where T is a real positive number. The operation is depicted in Figure 1.9 for the case of a continuous-time signal. Likewise, by replacing t by $t + T$, the signal $f(t)$ is advanced by T seconds, as shown in Figure 1.10. In the case of discrete-time signals, the operations are similar. By replacing the independent variable n in the expression for a

Figure 1.9
Illustration of the shifting operation showing the time delay of a signal.

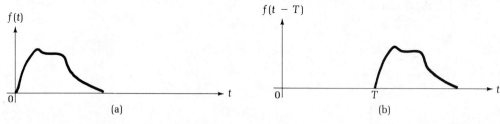

Figure 1.10
The advance of a signal
by the shifting operation.

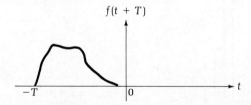

discrete-time signal $f(n)$ by $n + k$ for some integer (positive or negative) k, the signal is advanced or delayed by $|k|$ units, respectively, depending on whether k is positive or negative.

Consider, for example, the discrete-time signal

$$
\begin{aligned}
h(n) &= 0, && n < -2 \\
&= 3^{-n}(2^n + 1), && -2 \leqq n \leqq 0 \\
&= \frac{3^{-n}}{n}, && n > 0
\end{aligned}
\tag{1.18}
$$

as previously given in (1.16) and illustrated in Figure 1.7. For $k = 2$ we have

$$
\begin{aligned}
h(n - 2) &= 0, && n - 2 < -2 \\
&= 3^{-(n-2)}(2^{n-2} + 1), && -2 \leqq n - 2 \leqq 0 \\
&= \frac{3^{-(n-2)}}{n - 2}, && n - 2 > 0
\end{aligned}
\tag{1.19}
$$

which simplifies to

$$
\begin{aligned}
h(n - 2) &= 0, && n < 0 \\
&= 9 \cdot 3^{-n}(\tfrac{1}{4} \cdot 2^n + 1), && 0 \leqq n \leqq 2 \\
&= \frac{9 \cdot 3^{-n}}{n - 2}, && n > 2
\end{aligned}
\tag{1.20}
$$

The delayed signal $h(n - 2)$ is shown in Figure 1.11.

For the continuous-time signal $f(t)$ of Figure 1.3,

$$
\begin{aligned}
f(t) &= A, && 0 < t < t_1 \\
&= 0, && \text{otherwise}
\end{aligned}
\tag{1.21}
$$

we have for $T = t_1$

$$
\begin{aligned}
f(t + t_1) &= A, && 0 < t + t_1 < t_1 \\
&= 0, && \text{otherwise}
\end{aligned}
\tag{1.22}
$$

Figure 1.11
The delay of the discrete-time signal of Figure 1.7 by the shifting operation.

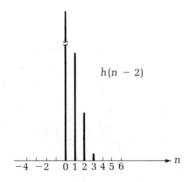

$h(n - 2)$

Figure 1.12
Advance of the
continuous-time signal
of Figure 1.3 by the shifting
operation.

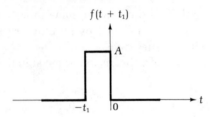

which simplifies to

$$f(t + t_1) = A, \qquad -t_1 < t < 0$$
$$= 0, \qquad \text{otherwise} \tag{1.23}$$

The advanced signal $f(t + t_1)$ is presented in Figure 1.12.

A system the output signal of which is identical with its input signal but is delayed by a constant is called a *delay unit*. On the other hand, a system the output signal of which is identical with its input signal but is advanced by a constant is referred to as a *predictor*. However, as will be indicated in Section 1-5, a predictor system is physically impossible to construct. As a case of the use of the shifting operation, it is employed in phase modulation in communication.

Transposition

A continuous-time signal $f(t)$ is said to be *transposed* if the independent variable t is replaced by $-t$; likewise, a discrete-time signal $f(n)$ is *transposed* if n is replaced by $-n$. The operation is equivalent to folding the signal about the line $t = 0$ or simply interchanging the "past" and "future" of the time signal. For instance, the transposed signals of Figures 1.3 and 1.4 are shown in Figure 1.13, the analytic expressions of which are given by

$$f(-t) = A, \qquad -t_1 < t < 0$$
$$= 0, \qquad \text{otherwise} \tag{1.24}$$

and

$$f(-nT) = A, \qquad n = 0, -1, \ldots, -M$$
$$= 0, \qquad \text{otherwise} \tag{1.25}$$

where, as before, $MT \leqq t_1 < (M + 1)T$. Because transposition means interchanging the "past" and the "future," no physical system can perform such an operation. We introduce it here only for mathematical convenience.

A signal can be simultaneously transposed and delayed. The operations are equivalent to replacing t or n by $-t + T$ or $-n + k$. To see this we consider a continuous-time signal $f(t)$ that is to be transposed and delayed by T seconds. To yield a transposed signal, we replace t by $-t$ in $f(t)$, resulting in $f(-t)$. The transposed signal $f(-t)$ is then delayed by T seconds

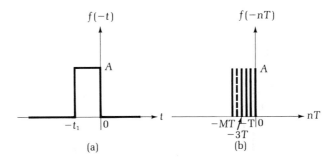

(a)   (b)

to give

$$f[-(t - T)] = f(-t + T) \tag{1.26}$$

as asserted.

As an example we will transpose and delay by 2 seconds the discrete-time signal (1.18). The operations are equivalent to replacing n in $h(n)$ by $-n + 2$, resulting in

$$
\begin{aligned}
h(-n + 2) &= 0, & -n + 2 < -2 \\
&= 3^{n-2}(2^{-n+2} + 1), & -2 \leqq -n + 2 \leqq 0 \\
&= \frac{3^{n-2}}{-n + 2}, & -n + 2 > 0
\end{aligned}
\tag{1.27}
$$

which simplifies to

$$
\begin{aligned}
h(-n + 2) &= 0, & n > 4 \\
&= \frac{3^n(4 \cdot 2^{-n} + 1)}{9}, & 2 \leqq n \leqq 4 \\
&= \frac{3^n}{-9n + 18}, & n < 2
\end{aligned}
\tag{1.28}
$$

A plot of (1.28) is shown in Figure 1.14.

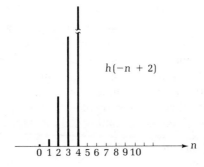

$h(-n + 2)$

0 1 2 3 4 5 6 7 8 9 10

Figure 1.15
(a) The transpose of the discrete-time signal of Figure 1.7, and (b) the transpose of the discrete-time signal of Figure 1.7 delayed by 2.

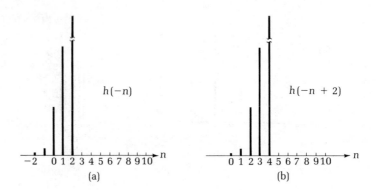

Figure 1.16
The advancing (a) of the discrete-time signal $h(n)$ of Figure 1.7 and then transposing (b).

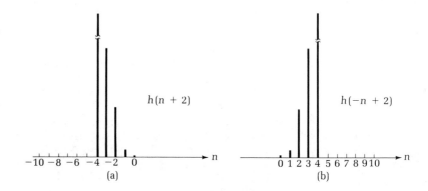

We now demonstrate that the $h(-n + 2)$ plot can also be generated in two other ways: For the signal $h(n)$ as shown in Figure 1.7, we can first determine its transposition as in Figure 1.15(a) and then delay the transposed signal $h(-n)$ by 2 units to obtain $h(-n + 2)$ as in Figure 1.15(b). Or for the signal of Figure 1.7, we can first advance it by 2 units to obtain $h(n + 2)$ as in Figure 1.16(a), and then transpose the advanced signal to yield $h(-n + 2)$ as in Figure 1.16(b). The results are, of course, the same as in Figure 1.14.

Scaling

The magnitude of a signal is scaled by a factor b if the value of the signal at every instant of time is multiplied by the constant b. A system the output signal of which equals the input signal scaled in magnitude by a constant b is known as an *amplifier* if $|b| > 1$, and an *attenuator* if $|b| < 1$. Likewise, the time variable t of a continuous-time signal $f(t)$ is scaled by a positive constant a if t in $f(t)$ is replaced by at.

For instance, consider the damped sinusoidal signal (Figure 1.17)

$$f(t) = Ae^{-\sigma t} \cos \omega t, \qquad \sigma > 0 \tag{1.29}$$

Figure 1.17
Plot of a damped sinusoidal
signal $Ae^{-\sigma t} \cos \omega t$.

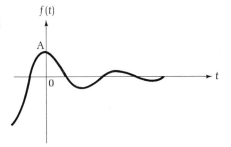

Figure 1.18
Examples of magnitude amplification (a) and attenuation (b) scaling of the damped
sinusoidal function of Figure 1.17.

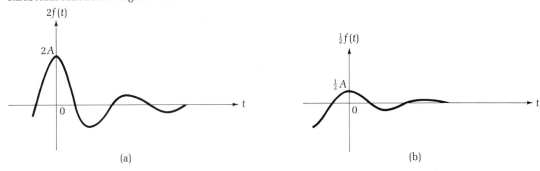

(a)   (b)

Suppose that the signal is magnitude–scaled up by a factor of $b = 2$. The
resulting signal is shown in Figure 1.18(a). Figure 1.18(b) shows the situation where the signal is magnitude–scaled down by a factor of $b = 0.5$.

If we scale the time variable t by $2t$ in the damped sinusoid of (1.29),
we obtain

$$f(2t) = Ae^{-2\sigma t} \cos 2\omega t, \qquad \sigma > 0 \qquad\qquad (1.30)$$

The time variable is said to be scaled by a factor $a = 2$, and the resulting
time response is depicted in Figure 1.19(a). For $a = 0.5$ we have the situation
as depicted in Figure 1.19(b). Finally, if the signal $f(t)$ is simultaneously
magnitude-scaled by a factor $b = 0.5$ and time-scaled by a factor $a = 2$, we
have

$$\tfrac{1}{2}f(2t) = \tfrac{1}{2}Ae^{-2\sigma t} \cos 2\omega t, \qquad \sigma > 0 \qquad\qquad (1.31)$$

The corresponding time response is presented in Figure 1.20.

Intuitively the magnitude scaling by a positive constant b amounts to
stretching or contracting the magnitude axis, depending on whether b is
greater or less than 1. On the other hand, the time scaling by a positive
constant a greater than 1 is equivalent to contracting the time axis and thus

Figure 1.19

Examples of time scaling of the time variable by 2 and by $\frac{1}{2}$ for the damped sinusoidal function of Figure 1.17.

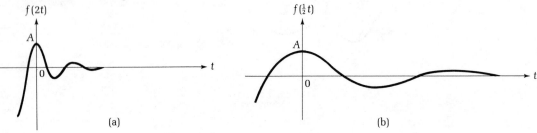

(a)   (b)

Figure 1.20

Illustration of simultaneous magnitude and time scaling of the damped sinusoidal function of Figure 1.17.

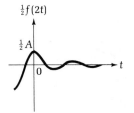

the signal, whereas the time axis and thus the signal are stretched if the scaling constant is less than 1. Such contraction and stretching of a continuous-time signal are illustrated in Figures 1.18–1.20. Note that for our purposes we do not define the operation of time scaling for a discrete-time signal.

1-4 IMPULSE FUNCTIONS

In practical modeling applications, we frequently encounter discontinuities in a continuous-time signal $f(t)$. Such a signal does not possess finite derivatives at its discontinuities. Nonetheless, for conceptual and operational reasons it is desirable to include the derivative of the signal $f(t)$ in our considerations; therefore we introduce the concept of unit impulse function and discuss some of its properties from a nonrigorous approach.

There are various ways of defining the unit impulse function. One of the approaches is as follows: Let $f_n(t)$ be a sequence of pulses defined by

$$f_n(t) = 0, \qquad t < 0$$

$$= n, \qquad 0 < t < \frac{1}{n}$$

$$= 0, \qquad t > \frac{1}{n} \tag{1.32}$$

For $n = 1$, 2, and 3, the pulses $f_1(t)$, $f_2(t)$, and $f_3(t)$ are shown in Figure 1.21. As n increases, the pulse width becomes smaller and the height increases. Consequently, the area under the pulse for each n is unity:

$$\int_0^\epsilon f_n(t)\, dt = 1, \qquad \epsilon > \frac{1}{n} \tag{1.33}$$

In the limit, as n approaches ∞, we have for any positive ϵ

$$\lim_{n \to \infty} \int_0^\epsilon f_n(t)\, dt = 1 \tag{1.34}$$

Figure 1.21
Illustration of unit impulse function as the limit of a pulse the area of which is unity but the width of which approaches zero.

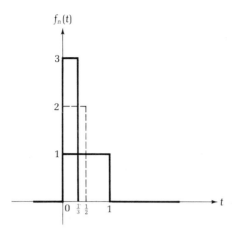

Figure 1.22
Graphical symbol used for the unit impulse.

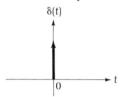

The *unit impulse function*, denoted by $\delta(t)$, is defined by the equation

$$\delta(t) \triangleq \lim_{n \to \infty} f_n(t) \tag{1.35}$$

so that

$$\begin{aligned} \delta(t) &= 0, && t \neq 0 \\ &= \infty, && t = 0 \end{aligned} \tag{1.36}$$

The unit impulse function, represented graphically by a vertical arrow in Figure 1.22, is zero everywhere except at the origin, at which point it is infinite. In practical situations a very short pulse, such as frequently occurred in a radar system, is a good approximation of the impulse function.

A function that is very closely related to the unit impulse is the unit step. The *unit step function*, denoted by $u(t)$, is defined by[†]

$$\begin{aligned} u(t) &\triangleq 0, && t < 0 \\ &= 1, && t > 0 \end{aligned} \tag{1.37}$$

and its value at $t = 0$ may be taken to be 0, $\frac{1}{2}$, or 1. A plot of $u(t)$ is given in Figure 1.23. The unit step function is a continuous-time signal with a discontinuity at $t = 0$, but mathematically it is a piecewise continuous function.

[†] A related function called the *signum function* will be defined in (10.139), the derivative of which is given by (10.140). Compare these with (1.37) and (1.38).

Figure 1.23
Illustration of the unit step function.

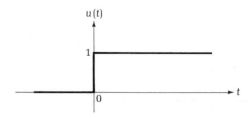

We next show that the derivative of the unit step function is the unit impulse function:

$$\delta(t) = \frac{d}{dt} u(t) \qquad (1.38)$$

Consider the function

$$f(t) = 0, \qquad t < 0$$

$$= \frac{t}{\epsilon}, \qquad 0 < t < \epsilon$$

$$= 1, \qquad t > \epsilon \qquad (1.39)$$

as illustrated in Figure 1.24. It is evident that, as ϵ goes to zero, $f(t)$ approaches the unit step function:

$$\lim_{\epsilon \to 0} f(t) = u(t) \qquad (1.40)$$

The derivative of $f(t)$ is given by the equations:

$$g(t) = \frac{d}{dt} f(t) = 0, \qquad t < 0$$

$$= \frac{1}{\epsilon}, \qquad 0 < t < \epsilon$$

$$= 0, \qquad t > \epsilon \qquad (1.41)$$

Figure 1.24
Plot of a piecewise continuous function $f(t)$ that approaches the unit step function as $\epsilon \to 0$.

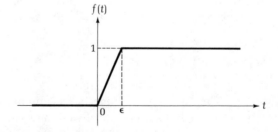

Figure 1.25
The derivative of the function given in Figure 1.24.

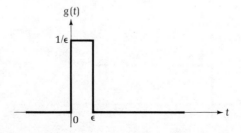

as depicted in Figure 1.25. Observe that (1.32) becomes (1.41) if n is replaced by $1/\epsilon$. Thus as ϵ approaches zero, $g(t)$ becomes the unit impulse function:

$$\delta(t) = \lim_{\epsilon \to 0} g(t) = \lim_{\epsilon \to 0} \frac{d}{dt} f(t) \tag{1.42}$$

After an interchange of the operations of the limit and derivative, we obtain

$$\delta(t) = \frac{d}{dt} \lim_{\epsilon \to 0} f(t) = \frac{d}{dt} u(t) \tag{1.43}$$

Upon integration we have

$$\int_{-\infty}^{t} \delta(t)\, dt = u(t) \tag{1.44}$$

Equation (1.44) indicates that the unit impulse function is not an ordinary point function because the ordinary integral is a continuous function of the upper limit, whereas $u(t)$ is obviously discontinuous. For our system modeling purposes, we shall refer to it as a function as if it were an ordinary point function. In a rigorous mathematical sense, we should interpret the unit impulse function as a *distribution*, an entity invented by the French mathematician L. Schwartz.[†]

EXAMPLE 1.1

We will compute the derivative of the continuous-time signal $f(t)$ illustrated in Figure 1.26 and defined by the equations

$$f(t) = 0, \qquad t \le 0$$
$$= \frac{Bt}{T_1}, \qquad 0 < t < T_1$$
$$= C, \qquad T_1 < t < T_2$$
$$= A, \qquad t > T_2 \tag{1.45}$$

The signal is a sort of a pedestal function used in television.

[†] L. Schwartz, *Théorie des distributions* (Paris: Actualités Scientifiques et Industrielles, Hermann et Cie, 1957 and 1959), Vols. I and II.

Figure 1.26

Continuous-time signal $f(t)$ used in Example 1.1.

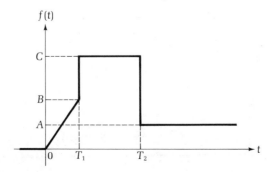

Figure 1.27
Three functions the sum of which comprises the signal of Figure 1.26.

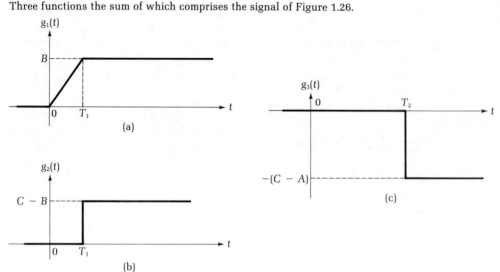

This signal can be viewed as the sum of three functions $g_1(t)$, $g_2(t)$, and $g_3(t)$ of Figure 1.27:

$$f(t) = g_1(t) + g_2(t) + g_3(t) \tag{1.46}$$

where

$$g_1(t) = \frac{B}{T_1}[tu(t) - (t - T_1)u(t - T_1)] \tag{1.47a}$$

$$g_2(t) = (C - B)u(t - T_1) \tag{1.47b}$$

$$g_3(t) = -(C - A)u(t - T_2) \tag{1.47c}$$

Using the derivative relation (1.43)

$$\delta(t - T) = \frac{d}{dt}u(t - T) \tag{1.48}$$

the derivative of $f(t)$ is

$$\frac{d}{dt}f(t) = \frac{B}{T_1}[u(t) - u(t - T_1)] + (C - B)\,\delta(t - T_1) - (C - A)\,\delta(t - T_2) \tag{1.49}$$

as illustrated in Figure 1.28. Observe that two impulses appear in the derivative at the points of discontinuity, $t = T_1$ and $t = T_2$.

Figure 1.28
The derivative of the signal $f(t)$ given in Figure 1.26, the components of which were illustrated in Figure 1.27.

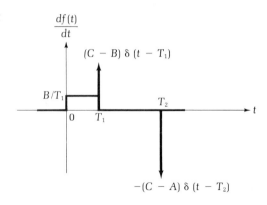

Figure 1.29
Illustration of the delayed pulse function corresponding to that of Figure 1.21 ($n = 1, 2, 3, \ldots$).

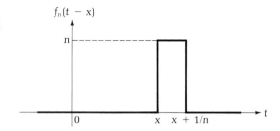

In the impulse function $k\,\delta(t)$, the coefficient k of the unit impulse function $\delta(t)$ is called the *strength* or *area* of the impulse. Accordingly, the two impulses of Figure 1.28 have strengths $(C - B)$ and $-(C - A)$, respectively, located at points T_1 and T_2. The areas of the impulses are equal to the values of the jump discontinuities (Figure 1.26) in $f(t)$—namely,

$$f(T_1 +) - f(T_1 -) = C - B \tag{1.50a}$$

$$f(T_2 +) - f(T_2 -) = A - C = -(C - A) \tag{1.50b}$$

An important utility of the unit impulse function is the *sifting property*, expressed by the relation

$$\int_{-\infty}^{\infty} f(t)\,\delta(t - x)\,dt = f(x) \tag{1.51}$$

for a function $f(t)$ that is continuous at $t = x$. This result can be justified by using the following argument.

Consider the integral

$$\int_{-\infty}^{\infty} f(t) f_n(t - x)\,dt = \int_{x}^{x+1/n} n f(t)\,dt \tag{1.52}$$

where $f_n(t)$ is the sequence of pulse functions defined in (1.32) and $f_n(t - x)$ is the delayed pulse shown in Figure 1.29. Appealing to the continuity of

$f(t)$ at $t = x$, we see that as $n \to \infty$ the function $f(t)$ remains essentially constant and equal to $f(x)$ in the interval $x < t < x + 1/n$. Hence

$$\lim_{n \to \infty} \int_{-\infty}^{\infty} f(t) f_n(t - x)\, dt = f(x) \tag{1.53}$$

Interchanging the operations of the limit and integral in (1.53) and using (1.35) give

$$\int_{-\infty}^{\infty} f(t)\, \delta(t - x)\, dt = f(x) \tag{1.54}$$

EXAMPLE 1.2

Let $f(t) = Ae^{-\sigma t} \cos \omega t$, $\sigma > 0$, be a damped sinusoidal signal of (1.29). Then

$$\int_{-\infty}^{\infty} f(t)\, \delta(t - T)\, dt = \int_{-\infty}^{\infty} Ae^{-\sigma t} \cos \omega t\, \delta(t - T)\, dt$$

$$= Ae^{-\sigma T} \cos \omega T \tag{1.55}$$

If, on the other hand, $f(t) = A \sin(\omega t + \pi/4)$, then

$$\int_{-\infty}^{\infty} f(t)\, \delta(t)\, dt = \int_{-\infty}^{\infty} A \sin\left(\omega t + \frac{\pi}{4}\right) \delta(t)\, dt$$

$$= A \sin \frac{\pi}{4} = \frac{A}{\sqrt{2}} \quad . \tag{1.56}$$

Recall that the sifting property of the unit impulse function as expressed by (1.51) was derived under the assumption that $f(t)$ is continuous at $t = x$. Assume that $f(t)$ is continuous over the interval $-\infty < t < \infty$. Then for a fixed T, we have

$$\int_{-\infty}^{\infty} f(t)\, \delta(t - T)\, dt = f(T) \tag{1.57}$$

If T were to range from $-\infty$ to $+\infty$, $f(t)$ would be reproduced in its entirety by (1.57). The operation is equivalent to moving a sheet of paper with a very thin slit across the plot of $f(t)$, as depicted in Figure 1.30. In other words, the function is *scanned* by the unit impulse function.

Figure 1.30
Illustration of the operation in which a signal $f(t)$ is scanned by the unit impulse function.

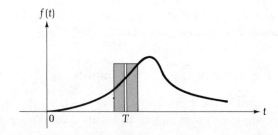

Figure 1.31
The function $g_n(t)$ used to
to illustrate the doublet as
the limit of $g_n(t)$ as n
approaches infinity.

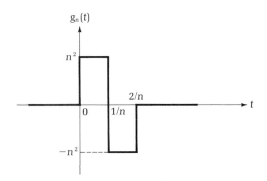

Let us now examine the higher order derivatives of the unit step. The
unit impulse function is obtained by differentiating the unit step function.
If this process is repeated, the resulting function is called a *doublet* or a
second-order impulse function, denoted by the symbol $\delta'(t)$:

$$\delta'(t) = \frac{d}{dt}\,\delta(t) \tag{1.58}$$

$$\int_{-\infty}^{t} \delta'(t)\,dt = \delta(t) \tag{1.59}$$

Alternately, the function $\delta'(t)$ may be defined directly as follows. Consider
a sequence of functions $g_n(t)$ given by

$$g_n(t) = 0 \qquad t < 0$$

$$= n^2, \qquad 0 < t < \frac{1}{n}$$

$$= -n^2, \qquad \frac{1}{n} < t < \frac{2}{n}$$

$$= 0, \qquad t > \frac{2}{n} \tag{1.60}$$

Figure 1.32
Graphical representation
of the doublet function as
the limit of the function
given in Figure 1.31.

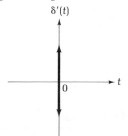

as illustrated in Figure 1.31. A *doublet* is defined by

$$\delta'(t) \triangleq \lim_{n\to\infty} g_n(t) \tag{1.61}$$

giving

$$\delta'(t) = 0, \qquad t \neq 0 \tag{1.62a}$$

$$\delta'(0) = \infty \text{ and } -\infty \text{ simultaneously} \tag{1.62b}$$

The graphical representation of a doublet is shown in Figure 1.32. An
important property of the doublet is the relation

$$\int_{-\infty}^{\infty} f(t)\,\delta'(t-x)\,dt = -f'(x) \tag{1.63}$$

where $f'(x)$ is the derivative of $f(t)$ evaluated at $t = x$.

To verify this relation, we integrate the left-hand side of (1.63) by parts with

$$u = f(t), \qquad dv = \delta'(t - x)\, dt \tag{1.64}$$

Thus

$$\int_{-\infty}^{\infty} f(t)\, \delta'(t - x)\, dt = f(t)\, \delta(t - x)\Big|_{-\infty}^{\infty} - \int_{-\infty}^{\infty} \delta(t - x) f'(t)\, dt$$

$$= 0 - f'(x) = -f'(x) \tag{1.65}$$

In addition to the doublet, higher-order derivatives of $\delta(t)$ can be defined in a similar manner. In fact, if $\delta^{(n)}(t)$ and $f^{(n)}(t)$ denote the nth derivatives of $\delta(t)$ and $f(t)$, respectively, it can be shown that

$$\int_{-\infty}^{\infty} f(t)\, \delta^{(n)}(t - x)\, dt = (-1)^n f^{(n)}(x) \tag{1.66}$$

1-5 SYSTEMS AND THEIR CLASSIFICATIONS

One of the objectives of this text is to help electrical engineers think about electrical problems from the system point of view. Although individual elements of a system will affect its overall behavior, the elements of a complicated system are usually grouped together in such a way that each group of elements performs a certain function. Such a group is often referred to as a *subsystem*. The overall system is then modeled as an interconnection of these subsystems, which interact with one another only at their terminals. For instance, if a digital computer were modeled in terms of all its components, including resistors, diodes, operational amplifiers, capacitors, and so on, the model would be as complicated as the original. Instead, the computer is more conveniently described as a system comprised of separate arithmetic units, control circuits, memory devices, and the input and output terminals. The individual units can be viewed as subsystems of the overall system. Each subsystem itself may be considered as an individual system in its own right and, in turn, can be made up of smaller subsystems. For example, the arithmetic unit of a digital computer is a system made up of subsystems such as adders, counters, and indicators.

As another example, a radio receiver is a system made up of subsystems, including oscillators, amplifiers, demodulators, filters, and so on. If the receiver is a part of a satellite tracking station, it becomes a subsystem of a larger system containing other subsystems—namely, computers, antennas, transmitters, and so forth.

In many situations we are not concerned about the interactions of the various subsystems but only the development of relationships between the outputs and the inputs. The system is represented as in Figure 1.33 with inputs $x_1(t), x_2(t), \ldots, x_n(t)$ and outputs $y_1(t), y_2(t), \ldots, y_m(t)$. A diagrammatic relationship between the outputs and inputs is called the *system model*. Modeling is one of the important steps in studying systems. For instance,

Figure 1.33

A system portrayed as an ensemble of inputs $x_i(t)$ $(i = 1, 2, \ldots, n)$ and outputs $y_j(t)$ $(j = 1, 2, \ldots, m)$.

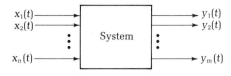

Figure 1.34

Schematic representation of a voltage amplifier.

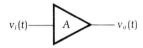

for a voltage amplifier we are not concerned with the input or output currents but only the voltage amplification. Such a relation can be represented schematically as in Figure 1.34 with

$$v_o(t) = Av_i(t) \qquad (1.67)$$

where A is the voltage amplification factor.

One of the major tasks of this text is to present various ways of describing the input-output relationship of systems. Most of our discussion will be limited to systems with a single input and a single output, as depicted in Figure 1.35, although many of the concepts introduced can be readily extended to multiple input-output systems. For the single input-output system of Figure 1.35, the symbol

$$y(t) = H[x(t)] \qquad (1.68)$$

is used to indicate that $y(t)$ is the output of the system corresponding to the input signal $x(t)$. For the ideal voltage amplifier of Figure 1.34, we have $x(t) = v_i(t)$, $y(t) = v_o(t)$, and

$$y(t) = H[x(t)] = Ax(t) \qquad (1.69)$$

Linearity

A system the input-output relation of which is described by (1.68) is said to be *homogeneous* if for any constant α

$$H[\alpha x(t)] = \alpha H[x(t)] = \alpha y(t) \qquad (1.70)$$

In other words, the output of the system to an input signal $\alpha x(t)$ equals α times the response of the system to the signal $x(t)$.

A system is said to be *additive* if for arbitrary input signals $x_1(t)$ and $x_2(t)$

$$y_1(t) = H[x_1(t)] \qquad (1.71a)$$

$$y_2(t) = H[x_2(t)] \qquad (1.71b)$$

Figure 1.35

Depiction of a system with a single input and single output.

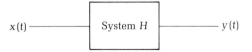

then

$$H[x_1(t) + x_2(t)] = H[x_1(t)] + H[x_2(t)]$$
$$= y_1(t) + y_2(t) \tag{1.72}$$

The additive property, which is also known as the *superposition* property, means that the response of a system to the sum of two arbitrary signals $x_1(t)$ and $x_2(t)$ equals the sum of the responses of the system to the individual signals $x_1(t)$ and $x_2(t)$. A system is *linear* if it is both homogeneous and additive. Although the homogeneous property can be inferred from the additive property if α is a rational number, there are mathematical transformations that are additive but not homogeneous. However, these systems are mostly pathological and would not arise physically.

In a linear system, if for any input signals $x_1(t)$ and $x_2(t)$ and for any constants α_1 and α_2

$$y_1(t) = H[x_1(t)] \tag{1.73a}$$

$$y_2(t) = H[x_2(t)] \tag{1.73b}$$

$$y(t) = H[\alpha_1 x_1(t) + \alpha_2 x_2(t)] \tag{1.73c}$$

then

$$y(t) = \alpha_1 y_1(t) + \alpha_2 y_2(t) \tag{1.74}$$

A system is *nonlinear* if it is not linear.

EXAMPLE 1.3
Consider the network of Figure 1.36 in which it is straightforward to verify that the terminal voltage $v(t)$ and current $i(t)$ are related by the equation

$$v(t) = \frac{3}{2} Ri(t) + \frac{1}{2} E \tag{1.75}$$

Take $i(t)$ to be the input signal and $v(t)$ the output response. Let

$$i_1(t) = i_2(t) = 1A \tag{1.76}$$

be two input signals. The corresponding responses $v_1(t)$ and $v_2(t)$ are found from (1.75) to be

$$v_1(t) = v_2(t) = 1.5R + 0.5E \tag{1.77}$$

Figure 1.36
Electrical network for the system in Example 1.3, which is nonlinear owing to the presence of the independent source E.

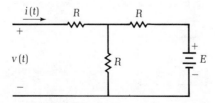

Now apply the input signal that is the sum of $i_1(t)$ and $i_2(t)$—namely,

$$i(t) = i_1(t) + i_2(t) = 2\text{A} \tag{1.78}$$

The corresponding output response is given by

$$v(t) = 3R + 0.5E \tag{1.79}$$

Therefore

$$v(t) \neq v_1(t) + v_2(t) \tag{1.80}$$

unless $E = 0$. Consequently, the single input-output system of Figure 1.36 is nonlinear. The presence of the independent source in the network renders the network nonlinear. On the other hand, if the independent source E is set to zero, we have

$$v(t) = v_1(t) + v_2(t) \tag{1.81}$$

meaning that the system is linear and the superposition principle holds. Our conclusion is that the presence of an independent source can render a network nonlinear.

EXAMPLE 1.4

Consider the RC circuit of Figure 1.37 in which the capacitor is initially charged to a voltage V_0. The terminal voltage $v(t)$ is related to the terminal current $i(t)$ by

$$v(t) = Ri(t) + \frac{1}{C} \int_0^t i(x)\,dx + V_0 \tag{1.82}$$

Take the current $i(t)$ as the input and $v(t)$ the output. Let $i_1(t)$ and $i_2(t)$ be two input signals. The corresponding responses $v_1(t)$ and $v_2(t)$ are found from (1.82) to be

$$v_k(t) = Ri_k(t) + \frac{1}{C} \int_0^t i_k(x)\,dx + V_0, \qquad k = 1, 2 \tag{1.83}$$

If the input signal is taken as

$$i(t) = i_1(t) + i_2(t) \tag{1.84}$$

Figure 1.37
RC circuit used in
Example 1.4.

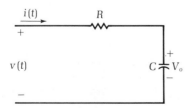

the output becomes

$$v(t) = R[i_1(t) + i_2(t)] + \frac{1}{C} \int_0^t [i_1(x) + i_2(x)]\, dx + V_0$$

$$= v_1(t) + v_2(t) - V_0 \tag{1.85}$$

Thus the system of Figure 1.37 is linear if and only if $V_0 = 0$. We conclude that the presence of any initial condition can render a system nonlinear.

Continuity

A system is said to be *continuous* if it is capable of accepting continuous-time signals as inputs and generating continuous-time signals as outputs. The networks given in Figures 1.36 and 1.37 are examples of continuous-time systems. Note that the input and output signals of continuous systems do not themselves need to be continuous functions. For instance, the input to a continuous system may be the unit step function.

A system is said to be *discrete* if it accepts signals only at discrete times and generates signals only at discrete times. A digital computer is an example of a discrete system. In such a system, the time intervals between signal samples may change. The values of the signals may also change from one sample to the next. Continuous systems are usually modeled using differential equations, whereas the discrete systems are frequently described by the *difference equations*.

For instance, a system takes an input sequence of numbers denoted by $x(1)$, $x(2)$, ..., $x(n)$ and transforms them into an output sequence of numbers $y(1)$, $y(2)$, ..., $y(n)$. The input-output relation of this particular system at time $t = k$ is modeled by the difference equation

$$y(k) = x(k) - 4y(k - 1) \tag{1.86}$$

an equation which expresses the unknown $y(k)$ in terms of $y(k - 1)$ and the known input signal $x(k)$. Assume that the system is connected to the source at $k = 0$. Then (1.86) holds for every $k \geq 1$.

To compute $y(k)$ we set $k = 1, 2, \ldots, n$ in (1.86) and obtain

$$y(1) = x(1) - 4y(0)$$
$$y(2) = x(2) - 4y(1)$$
$$y(3) = x(3) - 4y(2)$$
$$\vdots$$
$$y(n) = x(n) - 4y(n - 1) \tag{1.87}$$

Therefore if we know $y(0)$, we can ascertain $y(1)$ from the first equation since $x(1)$ is the known input. Knowing $y(1)$ we can find $y(2)$ from the second equation, $y(3)$ from the third, and so on. The number $y(0)$ is known as the *initial condition*. Thus as in solving a differential equation, the solution

Figure 1.38
Waveforms for the input
and output of a time-
invariant system.

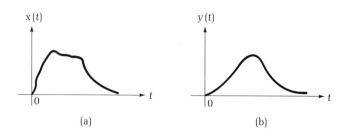

(a)   (b)

Figure 1.39
Waveforms for the input and output of the same time-invariant system used
in Figure 1.38.

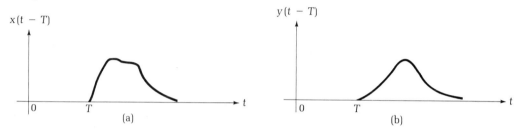

(a)   (b)

of a difference equation requires not only the input source but also the
initial condition.

Finally, a system is said to be *hybrid* if it accepts continuous-time signals
at its inputs and generates discrete-time signals as outputs or vice versa. A
television, for example, is a hybrid system where the input to the antenna
is a continuous-time signal of electromagnetic waves and the outputs are
pictures made up of millions of discrete dots corresponding to the modulation
signals on the RF carrier.

Time Invariance

A system is said to be *time invariant* if a time-shifted input signal will
result in a correspondingly time-shifted output signal. Mathematically,
time invariance can be stated as follows: Let $y(t)$ be the response of a system
to an input excitation $x(t)$. Then for a time-invariant system, the excitation
$x(t - \tau)$ will yield the response $y(t - \tau)$ for any τ. In other words, a time-
invariant system should produce the same response no matter when a
given excitation is applied. Figure 1.38(a) and (b) show the waveforms of
the input and output signals of a time-invariant system, respectively. If
the input signal $x(t)$ is delayed by T seconds, $x(t - T)$, as depicted in
Figure 1.39(a), the corresponding output waveform of Figure 1.39(b) is
identical in shape to that shown in Figure 1.38(b) but is delayed by T seconds.
A system is *time variant* if it is not time invariant.

Strictly speaking, all physical systems are time variant because their
component values change as a result of environmental conditions or aging.

However, in many cases they can be approximated by time-invariant systems. A transistor radio set is such a system. If we assume that the characteristics of the transistors, resistors, capacitors, and other components do not age, the radio can be approximated by a time-invariant system.

Causality

The term *causality* connotes the existence of a *cause-effect* relationship. Intuitively, a causal system cannot yield any response until after the excitation is applied. In other words, a causal system is not anticipative; it cannot predict the future behavior of the excitation. Realizable physical systems must be nonanticipative; however, the systems that we analyze are models composed of elements that are idealizations of actual physical devices. Therefore they may be anticipative. For example, in the study of the effect of bandwidth on pulse duration and rise time to be discussed in Chapter 10, we consider the ideal low-pass filter. The ideal low-pass filter is a system that is not physically realizable and, in fact, is noncausal, but it can be used as a measure with which the performance of other physical systems may be compared. The definition for system causality is formulated as follows.

Consider the system of Figure 1.35, where $y(t)$ is the output due to the input excitation $x(t)$. The system is said to be *causal* if for arbitrary excitations $x_\alpha(t)$ and $x_\beta(t)$ and for any time t_1 when

$$x_\alpha(t) = x_\beta(t), \qquad -\infty < t < t_1 \tag{1.88}$$

then

$$y_\alpha(t) = y_\beta(t), \qquad -\infty < t < t_1 \tag{1.89}$$

where $y_\alpha(t)$ and $y_\beta(t)$ are the responses of $x_\alpha(t)$ and $x_\beta(t)$, respectively. A system is *noncausal* if it is not causal.

The definition states in effect that two identical excitations over any time interval must result in identical responses over the same time interval for a system to be causal. To test for noncausality, it is sufficient to find *some* excitations $x_\alpha(t)$ and $x_\beta(t)$ such that for *some* time t_1, $x_\alpha(t) = x_\beta(t)$ for $t < t_1$ gives rise to $y_\alpha(t) \neq y_\beta(t)$ for some $t < t_1$.

Examples of causal systems are the networks shown in Figures 1.36 and 1.37. For an example of a noncausal system, consider a nonlinear resistor the terminal voltage $v(t)$ and current $i(t)$ of which are related by

$$v(t) = i^2(t) \tag{1.90}$$

Suppose that the voltage is the excitation and the current is the response. Choose two input voltages $v_\alpha(t)$ and $v_\beta(t)$ such that, for $t < t_1$,

$$v_\alpha(t) = v_\beta(t) = e^{-t} \tag{1.91}$$

For this input two different responses correspond,

$$i_\alpha(t) = +e^{-t/2} \qquad\qquad (1.92a)$$

$$i_\beta(t) = -e^{-t/2} \qquad\qquad (1.92b)$$

for $t < t_1$. Hence the system is noncausal under voltage excitation-current response. Now suppose that the excitation is a current source, and the response is the voltage. Then for any two input currents $i_\alpha(t)$ and $i_\beta(t)$ such that

$$i_\alpha(t) = i_\beta(t), \qquad t < t_1 \qquad\qquad (1.93)$$

the corresponding voltages $v_\alpha(t)$ and $v_\beta(t)$ are equal,

$$v_\alpha(t) = i_\alpha^2(t) = i_\beta^2(t) = v_\beta(t), \qquad t < t_1 \qquad\qquad (1.94)$$

showing that the system is causal under current excitation-voltage response.

We emphasize that, in characterizing a system as being causal or noncausal, it is important to specify the input and output variables. A system may be causal under one set of excitation and response and non-causal under a different set. In general, most linear systems are causal, and noncausal linear systems are very rare, occurring only in a few mostly trivial cases.

It has also been suggested by many that causality or nonanticipativeness of a system be defined in such a way that zero excitation over a time interval implies zero response over the same time interval; that is,

$$x(t) = 0, \qquad t \leqq t_1 \qquad\qquad (1.95)$$

implies

$$y(t) = 0, \qquad t \leqq t_1 \qquad\qquad (1.96)$$

This criterion for causality is not satisfactory in that some obviously causal and nonanticipative system may be rendered noncausal and anticipative by this approach. The system of Figure 1.40 comprises a series connection of a resistor and a battery. Assume that the terminal voltage $v(t)$ is the excitation and current $i(t)$ the response. Applying zero excitation $v(t) = 0$

Figure 1.40
A causal network that possesses the property that if the input terminals are shorted there results a nonzero output current even though the input voltage $v(t)$ is zero.

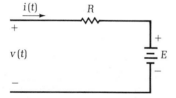

to the network is equivalent to shorting the input terminals. This gives rise to a nonzero current

$$i(t) = -\frac{E}{R} \tag{1.97}$$

indicating that zero excitation $v(t) = 0$ does not imply zero response $i(t) \neq 0$. It would be incorrect to say that this simple network is noncausal and anticipative.

Lumped and Distributed Parameters

A *lumped element* is an element in which the disturbance initiated at any point is propagated instantaneously at every point in the element. A *lumped-parameter system* comprises only of lumped elements. In electrical systems this means that the wavelength of the input signal is large compared with the physical dimensions of the elements. Such systems are modeled by *ordinary differential equations*. Electrical networks are examples of lumped-parameter systems. A *distributed-parameter system* is a system that is not a lumped-parameter system. It can be represented by a *partial differential equation* and generally has dimensions that are not small compared with the wavelength of signals of interest. Transmission lines, antennas, and waveguides are examples of distributed-parameter systems.

Figure 1.41 is a schematic of a lumped-parameter system. The equation for the current in the network is given by the Kirchhoff's voltage law as

$$L\frac{di(t)}{dt} + Ri(t) + \frac{1}{C}\int i(t)\,dt = v_g(t) \tag{1.98}$$

Differentiating and dividing both sides by L give

$$\frac{d^2 i(t)}{dt^2} + \frac{R}{L}\cdot\frac{di(t)}{dt} + \frac{1}{LC}i(t) = \frac{1}{L}\cdot\frac{dv_g(t)}{dt} \tag{1.99}$$

The system, therefore, is described by a second-order ordinary differential equation.

As an example of a distributed-parameter system, consider the current and voltage along a transmission line, as depicted in Figure 1.42. For the transmission line, let

     R = resistance per unit length

     L = inductance per unit length

     G = conductance to ground per unit length

     C = capacitance to ground per unit length

A very small segment of the transmission system can be represented by the equivalent network of Figure 1.43, where $v(x, t)$ is the voltage at time t and at the point that is x units from the source, and $i(x, t)$ is the corresponding

Figure 1.41
Schematic of a lumped-parameter network.

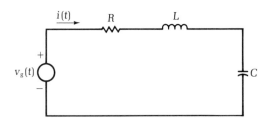

Figure 1.42
An electromagnetic transmission line as an example of a distributed-parameter system.

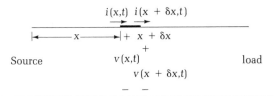

Figure 1.43
Equivalent network model for a segment of the transmission line of Figure 1.42.

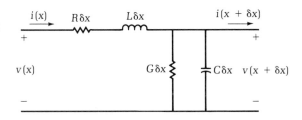

current. To simplify the notation, we write

$$v = v(x) = v(x, t), \qquad i = i(x) = i(x, t) \tag{1.100}$$

Applying Kirchhoff's voltage law to the equivalent network of Figure 1.43, we obtain

$$v(x + \delta x) = v(x) - (R\,\delta x)i - (L\,\delta x)\frac{\partial i}{\partial t} \tag{1.101}$$

or

$$v(x + \delta x) - v(x) \equiv \delta v = -(R\,\delta x)i - (L\,\delta x)\frac{\partial i}{\partial t} \tag{1.102}$$

Dividing both sides by δx and letting δx approach zero yield

$$\frac{\partial v}{\partial x} = -Ri - L\frac{\partial i}{\partial t} \tag{1.103}$$

Similarly, applying Kirchhoff's current law, we obtain

$$\frac{\partial i}{\partial x} = -Gv - C\frac{\partial v}{\partial t} \tag{1.104}$$

To obtain a single equation involving only v, we differentiate (1.103) with respect to x and (1.104) with respect to t. The results are given by

$$\frac{\partial^2 v}{\partial x^2} = -R\frac{\partial i}{\partial x} - L\frac{\partial^2 i}{\partial x \, \partial t} \qquad (1.105)$$

$$\frac{\partial^2 i}{\partial t \, \partial x} = -G\frac{\partial v}{\partial t} - C\frac{\partial^2 v}{\partial t^2} \qquad (1.106)$$

These two equations can be combined by eliminating the term

$$\frac{\partial^2 i}{\partial t \, \partial x} = \frac{\partial^2 i}{\partial x \, \partial t} \qquad (1.107)$$

to yield

$$\frac{\partial^2 v}{\partial x^2} = LC\frac{\partial^2 v}{\partial t^2} + (RC + GL)\frac{\partial v}{\partial t} + RGv \qquad (1.108)$$

Alternatively, differentiating (1.103) with respect to t and (1.104) with respect to x and then eliminating the derivatives of v, we obtain

$$\frac{\partial^2 i}{\partial x^2} = LC\frac{\partial^2 i}{\partial t^2} + (RC + GL)\frac{\partial i}{\partial t} + RGi \qquad (1.109)$$

Equations (1.108) and (1.109) are partial differential equations governing the voltage and current along the transmission line and are sometimes called the *telegraph equations*.

A large complex system may contain both the lumped-parameter subsystems and distributed-parameter subsystems. In a telephone system, for instance, the terminal devices are essentially lumped-parameter, whereas the interconnection among the terminals is distributed-parameter.

Memory

A system is said to be *instantaneous* or *memoryless* if its response at any time t depends only on the excitation at time t, not on any past or future values of the excitation. A resistive network is a typical example of an instantaneous or memoryless system. The voltage amplifier of Figure 1.34 the output $v_o(t)$ of which equals A times the input $v_i(t)$ is another example.

A system that is not instantaneous is said to be *dynamical* and to have *memory*. Thus the response of a dynamical system depends not only on the present excitation but also the values of the past input. A network containing energy storage elements is mostly dynamical. The network of Figure 1.41 is a simple example of a dynamical system.

Unless stated to the contrary, all the systems to be discussed in this text will be assumed to be linear, time invariant, causal, lumped parameter, and dynamical. As a result their basic mathematical models are linear ordinary differential equations with constant coefficients.

1-6   SYSTEM EQUATIONS

The principle problem in system analysis is to find the response of a given system to a specific input. To do so the first step is to develop a suitable mathematical model for the physical problem of interest. Often this is not a simple matter. It involves judgment, experience, experiments, and knowledge of a particular discipline. For example, the electrical engineers will learn how to write equations for a system of electrical networks, civil engineers for a system of structures, and mechanical engineers for a system of mechanisms. The mathematical models that are of primary concern in this text are linear ordinary differential equations with constant coefficients, systems of integrodifferential equations, and systems of first-order differential equations known as the *state equations*. We shall have ample opportunity to study and formulate these equations for networks later in the book. At the moment we shall demonstrate by a very simple example that different mathematical models can be derived for the same physical systems.

The equilibrium equation for the network of Figure 1.44 is given by Kirchhoff's current law as

$$C\frac{dv}{dt} + \frac{1}{R}v + \frac{1}{L}\int v(t)\,dt = i_g \tag{1.110}$$

By differentiation we have

$$C\frac{d^2v}{dt^2} + \frac{1}{R}\cdot\frac{dv}{dt} + \frac{1}{L}v = \frac{di_g}{dt} \tag{1.111}$$

Equation (1.110) is an integrodifferential equation governing the nodal voltage $v(t)$, whereas (1.111) is an ordinary differential equation with constant coefficients that $v(t)$ must satisfy.

To model the network of Figure 1.44 in terms of a system of first-order differential equations, we rewrite (1.110) as

$$C\frac{dv}{dt} + \frac{v}{R} + i = i_g \tag{1.112}$$

Figure 1.44
A lumped-parameter network, which is modeled by a system of first-order differential equations known as the state equations.

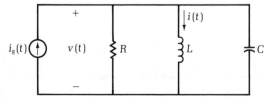

where $i(t)$ is the inductor current, which is related to $v(t)$ by

$$v = L\frac{di}{dt} \tag{1.113}$$

Transposing and rewriting (1.112) and (1.113) give

$$\frac{dv}{dt} = -\frac{1}{RC}v - \frac{1}{C}i + \frac{1}{C}i_g \tag{1.114a}$$

$$\frac{di}{dt} = \frac{1}{L}v \tag{1.114b}$$

or, more compactly, in matrix notation

$$\begin{bmatrix} \dfrac{dv}{dt} \\ \dfrac{di}{dt} \end{bmatrix} = \begin{bmatrix} -\dfrac{1}{RC} & -\dfrac{1}{C} \\ \dfrac{1}{L} & 0 \end{bmatrix}\begin{bmatrix} v \\ i \end{bmatrix} + \begin{bmatrix} \dfrac{1}{C} \\ 0 \end{bmatrix}[i_g] \tag{1.115}$$

Equation (1.115) represents a system of first-order differential equations in the unknown voltage $v(t)$ and current $i(t)$; it is called the *state equation* of the network.

After a suitable model is obtained for a system, we next consider various techniques for its solution and then relate or interpret the solution in terms of the physical problem. If the solution is attained directly from the differential equations without resort to integral or other transformations, the resulting procedure is referred to as the *time-domain analysis* or *solution*. The other approach, in which Laplace or Fourier transformations are used, is called the *frequency-domain analysis* or *solution* to be elaborated shortly.

In carrying out systems analysis in either the time or the frequency domain, it is necessary to specify the conditions of energy storage within the system at the time when the input signal is applied. These specifications are known as the *initial conditions*. There are a number of ways to establish the initial conditions. When a system is modeled by an nth-order differential equation of the form (1.111), an appropriate way to specify the initial conditions is to give the value of the output signal and $n - 1$ values of its derivatives at the time when the input signal is applied. On the other hand, when a system is described by its state equation, the appropriate procedure is to specify the values of the unknown variables at the time when the input signal is applied. For instance, if the input signal is applied to the network of Figure 1.44 at $t = 0+$, an appropriate set of initial conditions for equation (1.111) is $v(0+)$ and

$$v'(0+) = \frac{dv(t)}{dt}\bigg|_{t=0+} \tag{1.116}$$

For state equation (1.115) the required initial conditions are initial inductor current $i(0+)$ and initial capacitor voltage $v(0+)$. It is important to note that as far as system analysis is concerned it makes no difference as to which set of initial conditions is specified. The effect of any set will be exactly the same, regardless of how they came into being. They are different ways of stating the same thing.

An effective technique in solving a system of integrodifferential or differential equations is to use the integral transformations such as the Laplace transformation. As will be shown later in this text, such a transformation will reduce a system of linear simultaneous integrodifferential equations to a simpler system of linear simultaneous *algebraic* equations. A time-domain function is associated with another function, which is defined in terms of a complex variable

$$s = \sigma + j\omega \tag{1.117}$$

called the *complex frequency*. The real part σ describes growth or decay of the amplitudes of signals, and the imaginary part ω is angular frequency in the usual sense. The idea of complex frequency is developed by examining the *cisoidal signal*

$$f(t) = Ae^{j\omega t} \tag{1.118}$$

which can be generalized to yield

$$f(t) = Ae^{st} = Ae^{(\sigma + j\omega)t} \tag{1.119}$$

the real and imaginary parts of which are sinusoids as given by

$$\text{Re } f(t) = Ae^{\sigma t} \cos \omega t \tag{1.120}$$

$$\text{Im } f(t) = Ae^{\sigma t} \sin \omega t \tag{1.121}$$

Depending on the signs of σ, the waveforms of (1.120) and (1.121) are presented in Figure 1.45.

The term angular frequency is often referred to as the *real frequency*, which is the imaginary part of s. The real part σ of s, misleading as it may be, is called the *imaginary frequency* and was in general use before 1930. Another convention is to name ω *radian frequency* and σ *neper frequency*, thus avoiding the nearly metaphysical names. But no matter what we call them, the two components of frequency are conveniently added together to give complex frequency. For our purposes we shall use the term radian or real frequency for ω. Frequency in hertz is given by $f = \omega/2\pi$.

After the integrodifferential equations have been transformed into algebraic equations, the transform variables are then solved algebraically with the insertion of the initial conditions. The process in manipulating the transform variables and the subsequent interpretation of the solutions are referred to as analysis in the *frequency domain*. Finally, an inverse transformation is performed to yield the solution in the time domain.

The above discussion is primarily for motivation, rather than information, and is not intended to dwell at length on the subject. We shall have ample opportunities to discuss the various aspects of the subject in detail later in this book.

Figure 1.45

The waveforms of the signal $f(t) = Ae^{(\sigma + j\omega)t}$ according to whether the real
(a, b, e) or imaginary (c, d, f) part is taken and whether σ is negative (a, c),
positive (b, d), or zero (e, f).

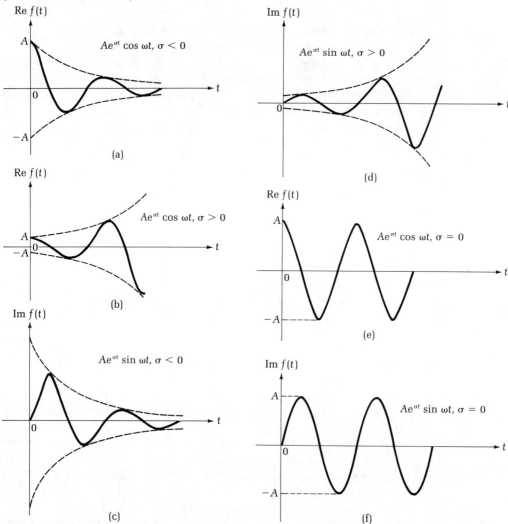

1-7 SUMMARY

We began this chapter by introducing the general concept of systems.
The term *system* has many meanings. The use of the term is best under-
stood from the various examples presented and from the context.

A *signal* is a physical embodiment of a message and can usually be observed as an entity that changes with time. Two types of signal are useful—the *continuous-time signal* and the *discrete-time signal*. The continuous-time signal may be a piecewise continuous function that may contain mathematical discontinuities. Signals are frequently processed to yield new signals for various purposes. For this reason we introduced the following operations on signals: *sum, product, shifting, transposition*, and *scaling*.

A special kind of signal that is widely used in engineering is the *unit impulse*. The unit impulse is a function that is zero everywhere except at the origin, at which its magnitude is infinite. Such a function is not an ordinary point function with which we are familiar and, in a strict mathematical sense, should be interpreted as a *distribution*. An important property of the unit impulse is the *sifting property*, which can be employed to scan a continuous function. The integral of the unit impulse is the *unit step* whereas the derivative of the unit impulse is the *doublet*.

In analysis or design it is advantageous to decompose or compose a large complex system as an interconnection of subsystems. These subsystems interact with one another only at their terminals. In such a system the analysis or design of the various subsystems can usually be handled separately. One of the main objectives of this text is to present various ways of characterizing the input-output relationship of a single input and a single output system. For this we classify systems according to their terminal behaviors. Special techniques can then be developed and important conclusions can be reached for the analysis or design of these various classes of systems.

A system is classified as *linear* if it is both homogeneous and additive. As a result the superposition principle holds for linear systems. A system is classified as (1) *continuous* if it accepts and generates continuous-time signals, (2) *discrete* if it accepts and generates signals only at discrete time instants, or (3) *hybrid* if it is a mix of the other two. A *time-invariant system* is one that will produce the same response no matter when a given excitation is applied. Strictly speaking, all physical systems are time varying; however, some can be approximated by time-invariant systems. A *causal system* connotes the existence of a cause-effect relationship. It is a system that does not anticipate a response until after the excitation is applied.

Depending on the wavelength of the highest significant frequency of the signals of interest, a system is referred to as *lumped parameter* if the wavelength is large compared with the largest physical dimension of the system. Such systems are modeled by ordinary differential equations, whereas partial differential equations are required for the distributed-parameter systems. A system is termed *dynamical* and said to have *memory* if its response depends not only on the present excitation but also on the values of the past input. Otherwise it is *instantaneous* or *memoryless*.

The linear, time-invariant, causal, lumped-parameter, and dynamical systems are represented by the *linear ordinary differential equations* with constant coefficients, the *linear integrodifferential equations* with constant coefficients, or the *linear state equations*. In fact, different mathematical models can be derived for the same physical systems. After a suitable model is obtained, the next problem is to consider various techniques for its solution and then relate the solution in terms of the physical problem. If the solution process involves only the differential equations without resort to integral transformations, the analysis or solution is in the *time domain*. The other approach, in which Laplace or Fourier transformations are used, is called the *frequency-domain analysis* or *solution*.

REFERENCES AND SUGGESTED READING

Gabel, R. A., and R. A. Roberts. *Signals and Linear Systems*. New York: Wiley, 1973, Chapter 1.

Kuo, F. F. *Network Analysis and Synthesis*. 2d ed. New York: Wiley, 1966, Chapters 1 and 2.

Lago, G. V., and L. M. Benningfield. *Circuit and System Theory*. New York: Wiley, 1979, Chapters 1-3.

Lathi, B. P. *Signals, Systems, and Controls*. Scranton, Pa.: Intext Educational Publishers, 1974, Chapter 1.

Liu, C. L., and J. W. S. Liu. *Linear Systems Analysis*. New York: McGraw-Hill, 1975, Chapters 1 and 2.

McGillem, C. D., and G. R. Cooper. *Continuous and Discrete Signal and System Analysis*. New York: Holt, Rinehart & Winston, 1974, Chapters 1, 2, and 3.

Papoulis, A. *Circuits and Systems: A Modern Approach*. New York: Holt, Rinehart & Winston, 1980, Chapter 1.

Van Valkenburg, M. E. *Network Analysis*, 3d ed. Englewood Cliffs, N.J.: Prentice-Hall, 1974, Chapters 1-5.

PROBLEMS

1.1 The response of a system to an input signal $x(t)$ is found to be $y(t) = x(-t)$. Classify the system according to the classifications described in Section 1-5.

1.2 Sketch the functions $(t - 2)u(t - 1)$, $e^{-2t}u(t - 2)$ and $(t - 2)u(t^2 - 1)$.

1.3 Sketch the functions $u(\cos t)$ and $u(\sin^2 t)$.

1.4 A system is characterized by the ordinary differential equation

$$\frac{d^3y}{dt^3} + 2\frac{d^2y}{dt^2} + 3\frac{dy}{dt} + 5y = 3x + 2\frac{dx}{dt} \qquad (1.122)$$

Classify the system according to the classifications of Section 1-5.

1.5 Repeat Problem 1.4 for the system described by the differential equation

$$\frac{d^2y}{dt^2} + 2t\frac{dy}{dt} + 6y = 3x \qquad (1.123)$$

Figure 1.46
Two continuous-time signals that are piecewise continuous functions.

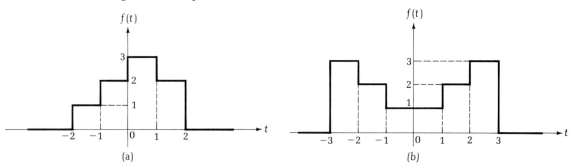

(a)   (b)

Figure 1.47
A series RLC network
with an independent
voltage source.

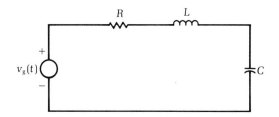

1.6 Evaluate the following integrals:

(a) $\int_{-5}^{5} (t^3 + 1)[\delta(t) - \delta(t - 1)]\, dt$ **(1.124a)**

(b) $\int_{-5}^{5} (3t + 2)[\delta(t + 1) + \delta(t - 1) + \delta(t)]\, dt$ **(1.124b)**

(c) $\int_{-5}^{5} (t^2 + 1)\,\delta'(t)\, dt$ **(1.125a)**

(d) $\int_{-5}^{5} (t^2 + t + 1)\,\delta'(t - 1)\, dt$ **(1.125b)**

1.7 Express the functions shown in Figure 1.46 in terms of the unit step function $u(t)$.

1.8 Let $f(t) = e^{-2t}u(t - 2)$. Find its derivative.

1.9 Prove the following identities:

$$t\delta'(t) = -\delta(t)$$ **(1.126)**

$$\delta'(t) = -\delta'(-t)$$ **(1.127)**

1.10 Find the differential equation and the state equation for the network of Figure 1.47.

1.11 Given two discrete-time signals $f_1(n)$ and $f_2(n)$,

$$f_1(n) = 1, \qquad n < -4$$
$$\quad\ = 2^{-n}, \qquad n \geqq -4$$ **(1.128)**

$$f_2(n) = 2^{-n} + 1, \qquad n \leqq 0$$
$$\quad\ = 1 + \frac{1}{n}, \qquad n > 0$$ **(1.129)**

Find the following:

(a) $f_1(n) + f_2(n)$ (b) $f_1(n)f_2(n)$ (c) $f_1(n-2)$

(d) $f_2(n+2)$ (e) $f_1(-n)$ (f) $f_2(-n+1)$ (g) $f_1(-n)f_2(n+2)$.

1.12 Given two continuous-time signals $f_1(t)$ and $f_2(t)$,

$$f_1(t) = e^{-2t} \sinh t + \cos t \qquad (1.130)$$

$$f_2(t) = \cosh 2t + \sin t \qquad (1.131)$$

Find the following:

(a) $f_1(t) + f_2(t)$ (b) $f_1(t)f_2(t)$ (c) $f_1(t-2)$

(d) $f_2(t+2)$ (e) $f_1(-t)$ (f) $f_2(-t+1)$ (g) $f_1(-t)f_2(t+2)$.

1.13 Given $f_1(t)$ as in (1.130), sketch $2f_1(\frac{1}{2}t)$ and $\frac{1}{2}f_1(2t)$.

1.14 Sketch the functions $\delta(\cos t)$ and $u(\cos t) - 1$.

1.15 A system is characterized by its state equation

$$\begin{bmatrix} \dfrac{dx_1}{dt} \\[2ex] \dfrac{dx_2}{dt} \end{bmatrix} = \begin{bmatrix} -t & 4 \\ -8 & 0 \end{bmatrix}\begin{bmatrix} x_1 \\ x_2 \end{bmatrix} + \begin{bmatrix} 1 \\ 2 \end{bmatrix}, \qquad t \geq 0 \qquad (1.132)$$

Classify the system according to the classifications of Section 1-5.

1.16 Sketch the derivative of the function $f(t)$ shown in Figure 1.48.

1.17 The state equation of a system is found to be

$$\begin{bmatrix} \dfrac{dx_1}{dt} \\[2ex] \dfrac{dx_2}{dt} \end{bmatrix} = \begin{bmatrix} te^{-t} - 5 & 4 \\ -3 & 2 - \dfrac{1}{1+t^2} \end{bmatrix}\begin{bmatrix} x_1 \\ x_2 \end{bmatrix} + \begin{bmatrix} 0 \\ 2 \end{bmatrix}[\sin 2t] \qquad (1.133)$$

Is this system linear? Is it time invariant?

1.18 Give an example of a system that is homogeneous but not additive.

1.19 Solve the difference equation

$$y(k) = x(k) - 4y(k-1) \qquad (1.134)$$

for $k > 0$ with $y(0) = 2$.

The input sequence $x(k)$ is known.

1.20 Verify the following identities:

$$f(t)\,\delta(t-T) = f(T)\,\delta(t-T) \qquad (1.135)$$

$$f(t)\,\delta'(t) = f(0)\,\delta'(t) - f'(0)\,\delta(t) \qquad (1.136)$$

where $f'(0)$ is the derivative of $f(t)$ evaluated at $t = 0$.

Figure 1.48
An even exponential
function.

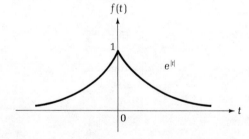

CHAPTER TWO

GRAPHS AND NETWORK EQUATIONS

A *network* is an interconnection of electrical elements such as resistors, capacitors, inductors, transistors, and transformers. The network problem deals with predicting the behavior of the network in terms of the characteristics of the elements and the manner in which these elements are interconnected. The geometrical properties of a network are independent of the constituents of its branches, and so in topological discussions it is usual to replace each branch by a line segment. The configuration thus obtained is called a *linear graph*. The graph is referred to as a *model* of the physical network. In the present chapter we introduce some of the basic properties of a graph and show how they can be applied to the systematic formulation of network equations. The results are indispensable in computer-aided analysis and design of integrated circuits.

2-1 LINEAR GRAPH CONCEPTS AND DEFINITIONS

Figure 2.1(a) is a transistor amplifier, the equivalent network of which is shown in Figure 2.1(b) after ignoring the biasing circuitry. By replacing each element in the equivalent network of Figure 2.1(b) by a line segment, we obtain a geometrical configuration as shown in Figure 2.2. Thus Figure 2.2 is a linear graph model of the transistor amplifier. In general, a line segment may represent a single element such as a resistor, capacitor, inductor, or a dependent or independent source; or it may represent any combination of these elements. In Figure 2.1, for example, we may choose a line segment to represent the series combination of R_1 and L_2 and another line segment to represent the parallel connection of R_6 and C_5. The resulting linear graph is presented in Figure 2.3. Thus a *linear graph* or simply a *graph* consists of a set of elements called *nodes* together with a set of line segments called the *edges*. A node is sometimes also called a *vertex* and an edge a *line*.

A *directed graph* is a graph in which an orientation or direction has been assigned to each of its edges. The orientation is represented geometri-

Figure 2.1

A common-emitter transistor amplifier (a) and its equivalent network (b).

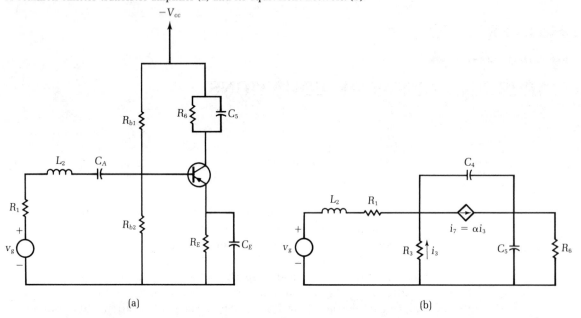

(a)   (b)

Figure 2.2

A linear graph model of the transistor amplifier of Figure 2.1(b).

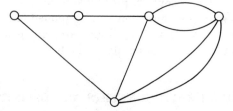

Figure 2.3

Another linear graph model of the transistor amplifier of Figure 2.1(b).

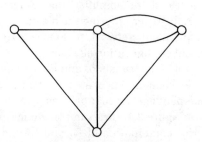

cally by the edge-orientation arrows. An example of the directed graph is shown in Figure 2.4. In some situations the orientation of the edges is a "true" orientation in that a system represented by the graph exhibits some unilateral property. For example, the directions of the one-way streets of a city and the

Figure 2.4
A directed graph obtained
by assigning directions to
the edges of the graph of
Figure 2.3.

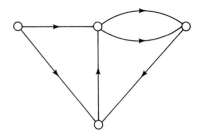

Figure 2.5
The standard symbols
used to indicate the
positive direction of
current in the direction of
the arrow and the
potential drop from the
marked plus sign.

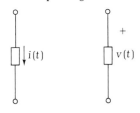

Figure 2.6
The combined references
for voltage and current.

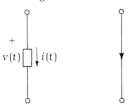

orientations representing the unilateral property of a communication network are true orientations of the physical systems. In other situations, such as in electrical network theory, the orientation is a "pseudo"-orientation used in lieu of an elaborate reference system as briefly discussed below.

In formulating network equations, it is necessary to associate each network element with two variables, a current $i(t)$ and a voltage $v(t)$. The standard symbols are to draw an arrow alongside the element to indicate the positive direction of current and a plus sign to indicate the potential drop from the marked plus sign, as illustrated in Figure 2.5. These two symbols, the arrow and the plus sign, are called the *current* and *voltage references*, respectively. How these references are chosen is immaterial, but it is important that a current and a voltage reference be assigned to each network element. Otherwise the functions $i(t)$ and $v(t)$ for that element are determined only to within a factor of ± 1. Since current and voltage references can be chosen arbitrarily, there is no need to carry two sets of references. Therefore we combine these two references by adopting the convention shown in Figure 2.6 where the voltage + reference is at the tail of the current-reference arrow. In Figure 2.7(a) the amplifier of Figure 2.1(b) is given in the familiar form, with all the current and voltage references shown, including the voltage + reference being kept at the tail of the current-reference arrow. The corresponding directed graph is presented in Figure 2.7(b).

The concepts and definitions introduced below for a directed graph are equally valid for an undirected graph. For our purposes we shall consider only directed graphs. A *subgraph* of a directed graph G is a subset of the edges of G. A node and an edge are *incident* with each other if the node is an endpoint of the edge. For example, in Figure 2.7(b) the edges e_1, e_2, e_3, e_6, and e_8 form a subgraph as exhibited in Figure 2.8(a). Other subgraphs are shown in Figure 2.8(b) and (c). The edge e_1 is incident at nodes 2 and 3 and vice versa. On the other hand, edge e_1 is not incident with node 1, 4, or 5. A directed graph is *connected* if it is in one piece. A maximum connected subgraph of a directed graph is called a *component*. The directed graph in Figure 2.8(a) is connected and thus contains only one component. Figure 2.9 is an example of an unconnected directed graph containing two components $e_1 e_2 e_3 e_4 e_5$ and $e_6 e_7 e_8$. The subgraphs $e_1 e_2 e_3 e_4$ and $e_1 e_2 e_3 e_4 e_6$ are not components, since $e_1 e_2 e_3 e_4$ is not a maximum connected subgraph and $e_1 e_2 e_3 e_4 e_6$ is not connected.

Figure 2.7
The assignment (a) of voltage and current references to the amplifier of Figure 2.1(b) and the corresponding directed graph (b).

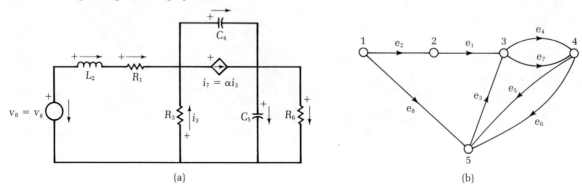

(a)   (b)

Figure 2.8
Subgraphs of the directed graph of Figure 2.7(b).

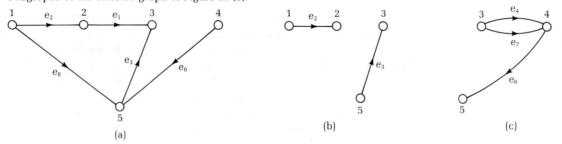

(a)   (b)   (c)

Figure 2.9
An unconnected graph with two components.

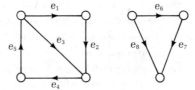

2-1.1   Circuit or Loop and Cutset

The words *circuit* and *network* have been considered synonymous in the language of electrical engineers. Modern electrical engineering terminology tends to designate an electrical network as a *network* and not a circuit since the original meaning of *circuit*, translated from the German *Kreis*, is "circle." We shall use the word circuit in its original meaning. Initial confusion may result for those used to the older terminology but the modern usage is much preferred.

Figure 2.10
Examples of paths in the directed graph of Figure 2.7(b).

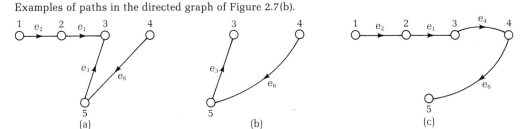

Figure 2.11
Examples of circuits in
the directed graph of
Figure 2.7(b).

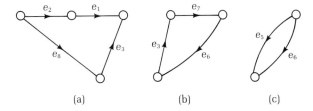

A *path* of a directed graph G is a subgraph consisting of a sequence of edges, with the possible exception of the first and last edges, each edge having one node in common with the preceding edge in the sequence and the other node in common with the succeeding edge; furthermore each node is incident at most with two edges in the sequence. The node of the first edge not shared by the second edge is the *initial node* and the node of the last edge not common to the previous edge is the *final node*. Thus if the initial and final nodes are different, each is incident with one edge and all other nodes are each incident with exactly two edges in the sequence. On the other hand, if they coincide, the path becomes a *closed path* and each node is incident with exactly two edges. A closed path is called a *circuit* or *loop*. In Figure 2.7(b), the subgraph $e_2e_1e_3e_6$ as shown in Figure 2.10(a) is a path with initial node 1 and terminal node 4 or initial node 4 and terminal node 1. Other examples of a path are shown in Figure 2.10(b) and (c). The subgraph $e_2e_1e_3e_8$ as depicted in Figure 2.11(a) is a circuit or loop of Figure 2.7(b). Other circuits or loops are given in Figure 2.11(b) and (c).

The concept of circuit or loop is indispensable in formulating Kirchhoff's voltage law equations for a network. The dual concepts in writing Kirchhoff's current law equations are known as the cutset. A *cutset* of a directed graph G is a minimum collection of edges whose removal will increase the number of components by one. Thus if G has more than one component, a cutset can only be formed from the edges of one of its components. If we cut the edges of a cutset, one of the components will be cut into two pieces. The name *cutset* has its origin in this interpretation. Note that a component of G may consist of an isolated node.

Figure 2.12

A transformer-coupled transistor amplifier (a), its equivalent network (b), and the associated directed graph (c).

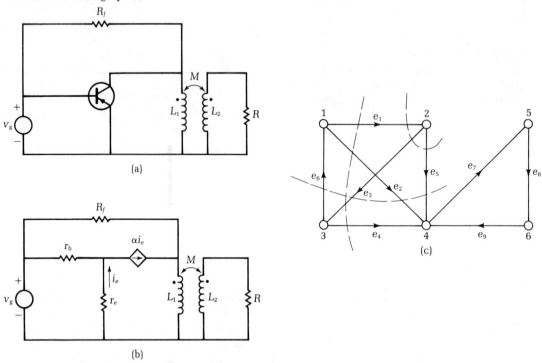

Consider the transistor amplifier, its equivalent network, and the associated directed graph G shown in Figure 2.12. Examples of cutsets of G are the subgraphs $e_1 e_2 e_3 e_4$, $e_2 e_3 e_5 e_6$, and $e_1 e_3 e_5$. The broken lines in Figure 2.12(c) show how these cutsets *cut* G. However, the subgraph $e_1 e_2 e_3 e_4 e_7$ is not a cutset because it is not a minimum collection of edges; the removal of $e_1 e_2 e_3 e_4$ will also increase the number of components by one—from one to two. The subgraph $e_2 e_4 e_5 e_7 e_9$ is not a cutset because the removal of these edges will increase the number of components by two—from one to three, as shown in Figure 2.13. In general, the set of edges incident at a node does not constitute a cutset.

The subgraph formed by the edges incident at a node of G is termed an *incidence cut*. In Figure 2.12(c), the subgraphs $e_1 e_3 e_5$ and $e_2 e_4 e_5 e_7 e_9$ are examples of incidence cuts although $e_1 e_3 e_5$ is also a cutset. As indicated above, a cutset or an incidence cut can be represented by drawing a broken line to show how it cuts the directed graph. However, there are cutsets which cannot be shown by a single line drawn across the directed graph. For example, in Figure 2.12(c) the cutset $e_1 e_4 e_5 e_6$ cannot be represented by

Figure 2.13
The subgraph resulting from the removal of the edges e_2, e_4, e_5, e_7, and e_9 from the directed graph of Figure 2.12(c).

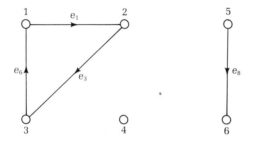

Figure 2.14
Representing the cutset $e_1 e_4 e_5 e_6$ by drawing a line through their edges.

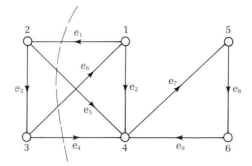

drawing a line through its edges. Only when the directed graph is redrawn as in Figure 2.14 can we draw a line through them.

2-1.2 Tree and Cotree

The tree is perhaps the single most important subgraph in electrical network theory because a number of fundamental results in network theory can be related to it. For example, the number of independent Kirchhoff's equations, the number of state equations, the methods of choosing the independent equations, and the topological formulas for network functions may all be stated in terms of the single concept of a tree.

A *tree* of a directed graph G is a connected subgraph that contains all the nodes of G but does not contain any circuits. For G to contain a tree, it must be connected itself. The word *tree* signifies a treelike structure, a structure in one piece with branches connecting to other branches. A tree has three important characteristics: (1) connectedness, (2) all nodes, and (3) no circuits. If any one of these characteristics is not satisfied, the subgraph is not a tree.

For instance, the directed graph of Figure 2.14 has many trees; four of them are shown in Figure 2.15. The subgraphs $e_1 e_2 e_4 e_9$, $e_1 e_3 e_7 e_9$, and $e_1 e_2 e_3 e_4 e_7 e_9$ shown in Figure 2.16 are not trees because $e_1 e_2 e_4 e_9$ does not contain all the nodes of the directed graph, $e_1 e_3 e_7 e_9$ is not connected, and $e_1 e_2 e_3 e_4 e_7 e_9$ contains a circuit. As another example, the directed graph of

Figure 2.15

Four of the trees of the directed graph of Figure 2.14.

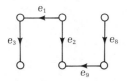 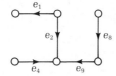 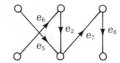 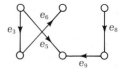

Figure 2.16

Subgraphs that are not trees of the directed graph of Figure 2.14.

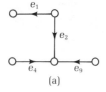 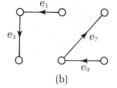 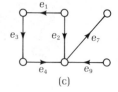

(a)   (b)   (c)

Figure 2.17

A directed graph containing eight trees.

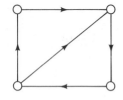

Figure 2.17 contains eight trees which are presented in Figure 2.18. As might be expected from the intuitive interpretation of a tree, an edge of a tree is called a *branch*.[†] Thus each edge in Figure 2.18 is a branch of some tree.

For convenience the complement of a tree in a directed graph is called a *cotree*—namely, the subgraph formed by the edges not contained in the tree. The word *cotree* is an abbreviation of the phrase "the *complement of a tree*." It is defined only with respect to a chosen tree. The edges of a cotree are called *links*. Figure 2.19 shows the eight cotrees of the directed graph in Figure 2.17. Each edge in Figure 2.19 is a link of some cotree. Observe that cotrees need not be connected and may contain a circuit. For instance, in the associated directed graph G of a low-pass filter shown in Figure 2.20(a) and (d), a tree is presented in Figure 2.20(b) and its corresponding cotree is given in Figure 2.20(c). For a tree T, it is convenient to denote the associated cotree by $\overline{T}$.

2-1.3 Fundamental Circuits and Cutsets

The methods of choosing the independent network equations are based on the concepts introduced in this section.

When we speak of links and branches, it is with respect to a chosen tree. Since a tree is connected and contains no circuits, there is a unique path between any two nodes in a tree. If we add one link to the tree, the resulting graph is of course no longer a tree. The link and the unique path in the tree between the two endpoints of the link constitute a circuit or loop.

[†] The currents and voltages of network elements are frequently referred to as the *branch currents* and the *branch voltages*. The word *branch* has nothing to do with the tree branch defined here and should not create any difficulty.

Figure 2.18
All the trees of the
directed graph of Figure
2.17.

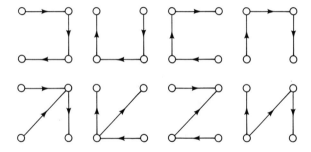

Figure 2.19
All the cotrees of the
directed graph of
Figure 2.17.

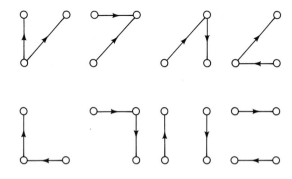

Figure 2.20
The associated directed
graph (a) of a low-pass
filter (d) and a tree (b) and
its corresponding cotree
(c).

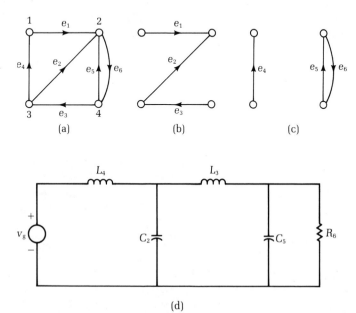

Figure 2.21
Three f-circuits defined with respect to the tree $e_1 e_2 e_3$ of the directed graph of Figure 2.20(a).

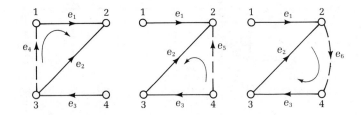

Figure 2.22
The pictorial representations of two different orientations of a circuit.

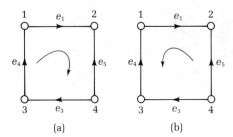

(a)   (b)

Therefore each link of a cotree defines a unique circuit in the directed graph with respect to the chosen tree. Such a circuit is called a *fundamental circuit* or simply *f-circuit*. The number of f-circuits constructed in this way with respect to a chosen tree is clearly equal to the number of links in the corresponding cotree.

As an example consider the directed graph G of Figure 2.20(a). Let $T = e_1 e_2 e_3$ of Figure 2.20(b) be the chosen tree. The corresponding cotree $\overline{T} = e_4 e_5 e_6$ is shown in Figure 2.20(c). Now if we add one link at a time to the tree T, the resulting graphs are shown in Figure 2.21. Each contains a unique circuit formed by the defining link denoted by the broken line and the unique tree path between the two endpoints of the link. The three f-circuits obtained are $e_1 e_2 e_4$, $e_2 e_3 e_5$, and $e_2 e_3 e_6$.

In our application it is useful to orient the circuits. An *oriented circuit* of a directed graph is a circuit with an orientation assigned by cyclic ordering of nodes along the circuit. For example, in Figure 2.20(a) the circuit consisting of the edges e_1, e_5, e_3, and e_4 can be oriented as (1, 2, 4, 3) or (1, 3, 4, 2). This orientation can also be represented pictorially by an arrowhead as depicted in Figure 22(a) or 22(b). For the f-circuits, the orientations are chosen to agree with those of the defining links. For instance, the orientations of the f-circuits in Figure 2.21 are as indicated in the figure. As a more complicated example, choose the tree $T = e_2 e_3 e_6 e_7 e_9$ in Figure 2.14. The tree T together with the cotree $\overline{T} = e_1 e_4 e_5 e_8$ denoted by the broken lines is redrawn in Figure 2.23. The four f-circuits and their orientations are presented in Figure 2.24. One of the important properties of the f-circuits is that in each f-circuit there is an edge—namely, the defining link, which is not contained in any other f-circuits. Thus in using these circuits in writing

Figure 2.23
A tree (solid lines) and its corresponding cotree (broken lines).

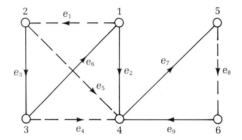

Figure 2.24
The f-circuits defined for the tree shown in Figure 2.23 and their orientations.

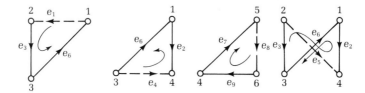

Figure 2.25
Symbolical representation of an f-cutset.

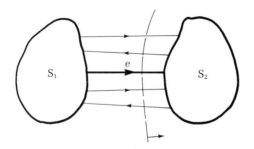

Kirchhoff's voltage equations, we can be sure they are linearly independent; that is, one cannot be obtained from the others by linear combinations.

So far we have been concentrating on circuits. A dual operation can be applied to cutsets. Let T be a tree of a directed graph G and let e be a branch of T. Since T is connected and contains no circuits, the removal of e from T results in a subgraph consisting of two components. If S_1 and S_2 denote the node sets of these two components, then S_1 and S_2 are mutually exclusive and together include all the nodes of G. Therefore the branch e of T defines a partition of the nodes of G in a unique way. The set of edges of G connecting a node in S_1 and a node in S_2 is clearly a cutset of G; as depicted symbolically in Figure 2.25, the branch e is assumed to be directed from a node in S_1 to a node in S_2. Such a cutset is termed a *fundamental cutset* or simply an *f-cutset*. An important property of an f-cutset is that it contains only one branch—namely, the defining tree branch—and some links of the cotree (with respect to the same tree).

To illustrate we consider the directed graph G of Figure 2.23. As before, let $T = e_2 e_3 e_6 e_7 e_9$ be the chosen tree. Suppose that e_6 of T is the branch

Figure 2.26
Subgraph resulting from
the tree $e_2e_3e_6e_7e_9$ of
Figure 2.23 by removing
the branch e_6.

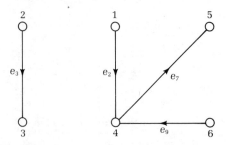

Figure 2.27
The f-cutset $e_1e_4e_5e_6$
defined by the branch e_6
for the tree $e_2e_3e_6e_7e_9$.

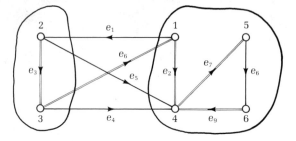

Figure 2.28
The f-cutset $e_2e_4e_5$ defined
by the branch e_2 for the
tree $e_2e_3e_6e_7e_9$.

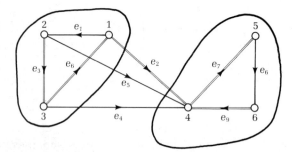

of interest. Removing e_6 from T results in two components e_3 and $e_2e_7e_9$ as
shown in Figure 2.26. The node sets of these two components are respectively
$S_1 = \{2, 3\}$ and $S_2 = \{1, 4, 5, 6\}$. Thus the branch e_6 of T partitions the
nodes of G into two mutually exclusive sets S_1 and S_2. The set of edges of G
connecting a node in S_1 and a node in S_2 defines an f-cutset $e_1e_4e_5e_6$ of G.
The operation is depicted in Figure 2.27. Let us now consider the branch e_2,
the removal of which partitions the node set of G into two mutually exclusive
subsets $S_1 = \{1, 2, 3\}$ and $S_2 = \{4, 5, 6\}$. The f-cutset defined by e_2 is
the set of edges connecting a node in $S_1 = \{1, 2, 3\}$ and a node in $S_2 = \{4, 5, 6\}$ and therefore comprises the edges e_2, e_4, and e_5 as illustrated in
Figure 2.28. The other f-cutsets defined by branches e_3, e_7, and e_9 are
respectively given by $e_1e_3e_5$, e_7e_8, and e_8e_9. The complete set of f-cutsets are
presented pictorially in Figure 2.29 by drawing a broken line across each
f-cutset.

Figure 2.29
Pictorial representation of
the f-cutsets defined for
the tree $e_2 e_3 e_6 e_7 e_9$ by
drawing a broken line
across each f-cutset.

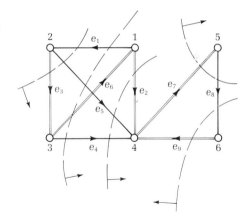

Like the circuits, it is useful to orient the cutsets. To do this we recall that a cutset "cuts" a component of a directed graph into two pieces. Let S_1 and S_2 be the node sets of the two pieces. Then the cutset is said to be *oriented* if the node sets S_1 and S_2 are ordered either as (S_1, S_2) or as (S_2, S_1). In most cases the orientation of a cutset can be represented pictorially as an arrow by placing an arrowhead near the broken line defining the cutset. The orientation of an f-cutset is chosen to coincide with the direction of the defining tree branch. In Figure 2.27 the cutset $e_1 e_4 e_5 e_6$ can be oriented either as $(\{2, 3\}, \{1, 4, 5, 6\})$ from left to right or as $(\{1, 4, 5, 6\}, \{2, 3\})$ from right to left. Since $e_1 e_4 e_5 e_6$ is an f-cutset, we choose its orientation to coincide with the direction of the defining tree branch e_6. Thus it is oriented from left to right. The orientations of other f-cutsets are indicated in Figure 2.29.

2-2 GRAPH MATRICES AND KIRCHHOFF'S EQUATIONS

In this section we define three matrices associated with a directed graph and show how they can be applied to the systematic formulation of Kirchhoff's current and voltage equations.

2-2.1 The Incidence Matrix and Kirchhoff's Current Law Equations

A network must satisfy Kirchhoff's[†] two laws regardless of whether it is linear or nonlinear and time varying or time invariant. Kirchhoff's laws are therefore independent of the constituents of network elements and can be defined in terms of the associated directed graph.

Consider the transistor amplifier of Figure 2.30(a). After the biasing circuitry is removed, its equivalent network is presented in Figure 2.30(b). Using the reference convention adopted in Figure 2.6, the associated directed

[†] Named for German physicist Gustav Kirchhoff (1824–1887), professor of physics in Heidelberg.

Figure 2.30

A transistor amplifier together with its biasing circuitry (a), its equivalent network (b), and the associated directed graph (c).

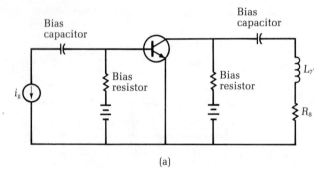

(a)

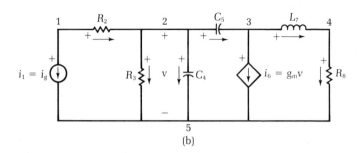

(b)

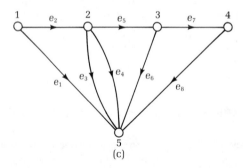

(c)

graph G is shown in Figure 2.30(c). Let $i_k(t)$ and $v_k(t)$ be the current and voltage of the edge e_k ($k = 1, 2, \ldots, 8$). Using Kirchhoff's current law, abbreviated as KCL, equations for this network can be written as

$$
\begin{array}{lll}
\text{node 1:} & i_1(t) + i_2(t) & = 0 \\
\text{node 2:} & -i_2(t) + i_3(t) + i_4(t) + i_5(t) & = 0 \\
\text{node 3:} & -i_5(t) + i_6(t) + i_7(t) & = 0 \\
\text{node 4:} & -i_7(t) + i_8(t) = 0 \\
\text{node 5:} & -i_1(t) \quad -i_3(t) - i_4(t) \quad -i_6(t) \quad -i_8(t) = 0
\end{array}
$$

$$(2.1)$$

which can be rewritten in matrix notation as

$$
\begin{array}{c}
\\
\text{node 1} \\
\text{node 2} \\
\text{node 3} \\
\text{node 4} \\
\text{node 5}
\end{array}
\begin{array}{c}
\begin{array}{cccccccc}
e_1 & e_2 & e_3 & e_4 & e_5 & e_6 & e_7 & e_8
\end{array} \\
\left[\begin{array}{cccccccc}
1 & 1 & 0 & 0 & 0 & 0 & 0 & 0 \\
0 & -1 & 1 & 1 & 1 & 0 & 0 & 0 \\
0 & 0 & 0 & 0 & -1 & 1 & 1 & 0 \\
0 & 0 & 0 & 0 & 0 & 0 & -1 & 1 \\
-1 & 0 & -1 & -1 & 0 & -1 & 0 & -1
\end{array}\right]
\end{array}
\begin{bmatrix}
i_1(t) \\
i_2(t) \\
i_3(t) \\
i_4(t) \\
i_5(t) \\
i_6(t) \\
i_7(t) \\
i_8(t)
\end{bmatrix}
=
\begin{bmatrix}
0 \\
0 \\
0 \\
0 \\
0
\end{bmatrix}
$$

$$(2.2)$$

This equation can be expressed more compactly as

$$\mathbf{A}_a \mathbf{i}(t) = \mathbf{0} \tag{2.3}$$

The coefficient matrix $\mathbf{A}_a$ is called the *complete incidence matrix* of G and $\mathbf{i}(t)$ the *branch-current vector*. In general, if G is a directed graph of n nodes and b edges, the complete incidence matrix of G is a matrix of order $n \times b$ such that if

$$\mathbf{A}_a = [a_{ij}] \tag{2.4}$$

then

$a_{ij} = 1$ if edge e_j is incident at node i and is directed away from node i

$a_{ij} = -1$ if edge e_j is incident at node i and is directed toward node i

$a_{ij} = 0$ if edge e_j is not incident at node i

Each row of the matrix $\mathbf{A}_a$ is associated with a node of G and each column an edge as shown in (2.2). In Figure 2.30(c) the edge e_1 is directed from node 1 to node 5. Hence there is a 1 in the first row position of column 1 and a -1 in the fifth row position. Likewise edges e_3 and e_4 are directed from node 2 to node 5 and there are 1's in second row and -1's in fifth row positions of the columns corresponding to e_3 and e_4—namely, columns 3 and 4. The rows and columns of $\mathbf{A}_a$ need not be arranged in the natural orders of the node and edge labelings of G; they may be arranged in any order whatsoever. The resulting matrix is of course dependent on the order in which they are arranged. For example, a different complete incidence matrix of Figure 2.30(c) is given below.

$$
\begin{array}{c}
\\
\text{node 3} \\
\text{node 4} \\
\text{node 5} \\
\text{node 1} \\
\text{node 2}
\end{array}
\begin{array}{c}
\begin{array}{cccccccc}
e_1 & e_4 & e_5 & e_8 & e_2 & e_3 & e_6 & e_7
\end{array} \\
\left[\begin{array}{cccccccc}
0 & 0 & -1 & 0 & 0 & 0 & 1 & 1 \\
0 & 0 & 0 & 1 & 0 & 0 & 0 & -1 \\
-1 & -1 & 0 & -1 & 0 & -1 & -1 & 0 \\
1 & 0 & 0 & 0 & 1 & 0 & 0 & 0 \\
0 & 1 & 1 & 0 & -1 & 1 & 0 & 0
\end{array}\right]
\end{array}
\tag{2.5}
$$

The first four columns correspond to the branches of the tree $e_1 e_4 e_5 e_8$ and the last four columns to the links of the cotree $e_2 e_3 e_6 e_7$. The branch e_5 is directed from node 2 to node 3, so there is a 1 in row 5 corresponding to node 2 and a -1 in row 1 corresponding to node 3 of column 3, which corresponds to edge e_5. This matrix can be obtained from (2.2) by a series of row and column interchanges.

Kirchhoff's current equations can therefore be written compactly in the form of (2.3). In writing this form, we implicitly assume that the elements of the branch-current vector $\mathbf{i}(t)$ have been arranged in the same order as edges corresponding to the columns of $\mathbf{A}_a$. If, for instance, we use the complete incidence matrix of (2.5), the current vector $\mathbf{i}(t)$ must be arranged as

$$\mathbf{i}(t) = [i_1(t),\, i_4(t),\, i_5(t),\, i_8(t),\, i_2(t),\, i_3(t),\, i_6(t),\, i_7(t)]' \tag{2.6}$$

The prime denotes the transpose of a matrix. This symbol will be used throughout the remainder of the book.

Observe that each column of $\mathbf{A}_a$ contains exactly two nonzero elements, a 1 and a -1. Hence the sum of all the rows of $\mathbf{A}_a$ is a row of zeros, showing that the rows of $\mathbf{A}_a$ are not linearly independent. In other words the rank of the matrix $\mathbf{A}_a$ is at most $n - 1$ and the equations of (2.3) are not linearly independent, one being the linear combinations of the others. It can be shown that the rank of $\mathbf{A}_a$ is also at least $n - 1$ if G is connected. A proof of this can be found in Chen.[†] We conclude as follows.

Property 2.1 *The rank of the complete incidence matrix $\mathbf{A}_a$ of an n-node directed graph G having c components is equal to $r = n - c$.*

The number $r = n - c$ is also called the *rank* of G. As a result of Property 2.1 there is no need for us to consider all the rows of $\mathbf{A}_a$; r linearly independent rows are sufficient. A submatrix of $\mathbf{A}_a$, denoted by $\mathbf{A}$, is called a *basis incidence matrix* if $\mathbf{A}$ is of order $r \times b$ and of rank r. If G is connected, a basis incidence matrix $\mathbf{A}$ can be obtained from the complete incidence matrix $\mathbf{A}_a$ by deleting any one of its rows. Therefore an appropriate set of linearly independent KCL equations may be compactly written as

$$\mathbf{A}\mathbf{i}(t) = \mathbf{0} \tag{2.7}$$

In obtaining $\mathbf{A}$ from $\mathbf{A}_a$, the deleted row in $\mathbf{A}_a$ corresponds to the reference-potential point in the network.

As an illustration, consider the network of Figure 2.30(b) and its associated directed graph G of Figure 2.30(c). The complete incidence matrix $\mathbf{A}_a$ of G is the coefficient matrix of (2.2). Suppose that node 5 is chosen as the reference-potential point for the network. Then a basis incidence matrix $\mathbf{A}$ can be obtained from $\mathbf{A}_a$ by deleting the row corre-

[†] W. K. Chen, *Applied Graph Theory: Graphs and Electrical Networks*, 2d ed. (New York: American Elsevier, 1976), p. 38.

sponding to node 5, which is row 5, giving

$$\mathbf{A} = \begin{bmatrix} 1 & 1 & 0 & 0 & 0 & 0 & 0 & 0 \\ 0 & -1 & 1 & 1 & 1 & 0 & 0 & 0 \\ 0 & 0 & 0 & 0 & -1 & 1 & 1 & 0 \\ 0 & 0 & 0 & 0 & 0 & 0 & -1 & 1 \end{bmatrix} \qquad (2.8)$$

Before we proceed to generalize the above results by using the cutsets, we shall use the symbols n, c, r, $\mathbf{A}_a$, and $\mathbf{A}$ defined above throughout the remainder of the book.

2-2.2 The Cutset Matrix and Kirchhoff's Current Law Equations

A generalized version of Kirchhoff's current law equations written for individual nodes will now be extended to the cutsets. The result is essential in the formulation of state equations and in the cutset system of equations to be discussed later in the book.

Let us again consider the network and its associated directed graph G of Figure 2.30. The current equations of the network are given in (2.1). Suppose that we add some of the equations—say, the equations for nodes 1 and 2. The resulting equation is

$$i_1(t) + i_3(t) + i_4(t) + i_5(t) = 0 \qquad (2.9)$$

We shall look into the physical meaning of this equation. Observe that currents that appear in this equation are exactly those in edges e_1, e_3, e_4, and e_5 of G, which have one node in the node set $\{1, 2\}$ and the other node in the node set $\{3, 4, 5\}$. These edges form a cutset separating the nodes of G into two subsets $\{1, 2\}$ and $\{3, 4\ 5\}$. If instead we add equations for nodes 1, 2, and 5, we obtain the new equation

$$i_5(t) - i_6(t) - i_8(t) = 0 \qquad (2.10)$$

Again the currents that appear in this equation are exactly those in edges e_5, e_6, and e_8 of G, which have one node in the node set $S_1 = \{1, 2, 5\}$ and the other node in the node set $S_2 = \{3, 4\}$. The subgraph $e_5 e_6 e_8$ is thus a cutset separating the node sets S_1 and S_2. The currents of the edges the endpoints of which both appear in S_1 or S_2 do not appear in the resulting equation because of cancellation. In Figure 2.31 the two cutsets together with their orientations are shown by means of the broken lines. In general, the above operation will yield either a cutset or an edge-disjoint union of cutsets. Physically (2.9) and (2.10) can be interpreted as follows: the total current crossing the broken lines in the direction of the arrow in Figure 2.31 is zero at all times, or equivalently the algebraic sum of currents through a cutset is zero at every instance. This may be considered a generalization of KCL.

The directed graph G of Figure 2.31 obviously has many cutsets besides the two shown in the figure. Some of these cutsets may simply be the in-

Figure 2.31
Physical interpretation of
total current crossing the
broken lines in the
direction of the arrow
associated with a cutset.

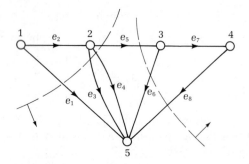

cidence cutsets—that is, the sets of edges connected to a node. For each
of the cutsets, a current equation can be written. For our purposes we
consider the network and with its associated directed graph G of Figure
2.32. All the cutsets of G together with their orientations are shown by the
broken lines in Figure 2.32(b). The KCL equations obtained for these cutsets
are

$$
\begin{aligned}
\text{Cutset 1:} \quad & i_1(t) + i_2(t) & & & = 0 \\
\text{Cutset 2:} \quad & -i_2(t) + i_3(t) & & + i_5(t) &= 0 \\
\text{Cutset 3:} \quad & & -i_4(t) - i_5(t) &= 0 \\
\text{Cutset 4:} \quad -i_1(t) & \quad -i_3(t) + i_4(t) & & &= 0 \\
\text{Cutset 5:} \quad i_1(t) & \quad +i_3(t) & & + i_5(t) &= 0 \\
\text{Cutset 6:} \quad & i_2(t) - i_3(t) + i_4(t) & & &= 0
\end{aligned}
\tag{2.11}
$$

which can be written in matrix notation as

$$
\begin{array}{c}
\begin{array}{ccccc}
e_1 & e_2 & e_3 & e_4 & e_5
\end{array} \\
\begin{array}{l}
\text{Cutset 1} \\
\text{Cutset 2} \\
\text{Cutset 3} \\
\text{Cutset 4} \\
\text{Cutset 5} \\
\text{Cutset 6}
\end{array}
\begin{bmatrix}
1 & 1 & 0 & 0 & 0 \\
0 & -1 & 1 & 0 & 1 \\
0 & 0 & 0 & -1 & -1 \\
-1 & 0 & -1 & 1 & 0 \\
1 & 0 & 1 & 0 & 1 \\
0 & 1 & -1 & 1 & 0
\end{bmatrix}
\begin{bmatrix}
i_1(t) \\
i_2(t) \\
i_3(t) \\
i_4(t) \\
i_5(t)
\end{bmatrix}
=
\begin{bmatrix}
0 \\
0 \\
0 \\
0 \\
0 \\
0
\end{bmatrix}
\end{array}
\tag{2.12}
$$

or, more compactly, as

$$
\mathbf{Q}_a \mathbf{i}(t) = \mathbf{0}
\tag{2.13}
$$

The coefficient matrix $\mathbf{Q}_a$ is called the *complete cutset matrix* of G and
as before $\mathbf{i}(t)$ is called the branch-current vector. In general, if G has q
cutsets and b edges, the complete cutset matrix $\mathbf{Q}_a$ is a matrix of order
$q \times b$ such that if

$$
\mathbf{Q}_a = [q_{ij}]
\tag{2.14}
$$

Figure 2.32

A network (a) and its
associated directed graph
(b).

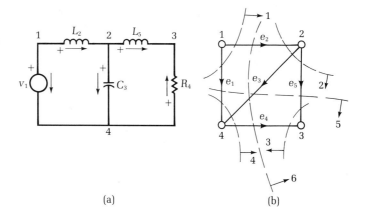

(a)   (b)

then

$q_{ij} = 1$ if edge e_j is in cutset i and the orientations of the cutset and
the edge coincide

$q_{ij} = -1$ if edge e_j is in cutset i and the orientations of the cutset and
the edge are opposite

$q_{ij} = 0$ if edge e_j is not in cutset i

Each row of the matrix $\mathbf{Q}_a$ is associated with a cutset of G and each
column with an edge, as shown in (2.12). In Figure 2.32(b) the edge e_3 is
contained in cutsets 2, 4, 5, and 6, so there are nonzero elements in rows
2, 4, 5, and 6 of column 3 corresponding to the cutsets 2, 4, 5, and 6. The
element in row 5 is 1 because the orientation of cutset 5 coincides with that
of edge e_3, and the element in row 6 is -1 because the orientation of cutset
6 is opposite to that of edge e_3. As in $\mathbf{A}_a$, the rows and columns of $\mathbf{Q}_a$ may
be arranged in any manner whatsoever and need not be arranged in natural
orders as that shown in (2.12). The resulting matrix is, of course, different,
within a permutation of rows and columns.

Observe that (2.12) has six equations in five unknown currents. Evidently
they are not linearly independent since some of these equations are obtainable from the others by linear combinations. A question naturally asked at this
point is then "What is the minimum number of linearly independent equa-
tions?" or, equivalently, "What is the rank of the matrix $\mathbf{Q}_a$?" Since $\mathbf{Q}_a$
contains $\mathbf{A}_a$ as a submatrix,[†] we know right away that the rank of $\mathbf{Q}_a$ is at
least r, the rank of the directed graph G. It turns out that r is also the rank
of $\mathbf{Q}_a$. A proof of this may be found in Chen.[‡] Thus we have

Property 2.2 *The rank of the complete cutset matrix $\mathbf{Q}_a$ of an n-node
directed graph G having c components is equal to $r = n - c$.*

[†] More precisely, $\mathbf{Q}_a$ contains rows of $\mathbf{A}_a$ and/or rows formed by the linear combinations of the
rows of $\mathbf{A}_a$.
[‡] Chen, *Applied Graph Theory*, p. 50.

This shows that we need only consider r linearly independent equations of (2.13). For this we say that a submatrix of $\mathbf{Q}_a$, denoted by $\mathbf{Q}$, is a *basis cutset matrix* if $\mathbf{Q}$ is of order $r \times b$ and of rank r. All the information contained in (2.13) is now contained in

$$\mathbf{Q}\mathbf{i}(t) = \mathbf{0} \tag{2.15}$$

The 4-node directed graph G of Figure 2.32(b) has five edges, giving $n = 4$, $b = 5$, and $c = 1$. Thus the rank of G and the rank of the coefficient matrix $\mathbf{Q}_a$ of (2.12) are both $r = 3$ $(=4 - 1)$. The submatrix formed by the first three rows of $\mathbf{Q}_a$ as given below

$$\mathbf{Q} = \begin{bmatrix} 1 & 1 & 0 & 0 & 0 \\ 0 & -1 & 1 & 0 & 1 \\ 0 & 0 & 0 & -1 & -1 \end{bmatrix} \tag{2.16}$$

is a basis cutset matrix.

One of the systematic ways in writing down a basis cutset matrix is based on the concepts of f-cutsets. Denote by $\mathbf{Q}_f$ the cutset matrix corresponding to a set of f-cutsets. Since an important property of the f-cutsets is that each one contains a branch—namely, the defining tree branch—not contained in any other f-cutsets, $\mathbf{Q}_f$ must contain the identity matrix of order r. Hence $\mathbf{Q}_f$, being of order $r \times b$ and of rank r, is a basis cutset matrix. In doing so we have implicitly assumed that G is connected, and therefore it has a tree. This should not be deemed a restriction because for unconnected graphs we can consider each component individually.

The directed graph of Figure 2.32(b) is redrawn in Figure 2.33. Choosing tree $T = e_2 e_3 e_4$ yields three f-cutsets as shown in Figure 2.33. The f-cutset matrix is given by

$$\mathbf{Q}_f = \begin{array}{c} \text{f-cutset 1} \\ \text{f-cutset 2} \\ \text{f-cutset 3} \end{array} \begin{bmatrix} \overset{e_1}{1} & \overset{e_5}{0} & \overset{e_2}{1} & \overset{e_3}{0} & \overset{e_4}{0} \\ 1 & 1 & 0 & 1 & 0 \\ 0 & 1 & 0 & 0 & 1 \end{bmatrix} \tag{2.17}$$

Figure 2.33
Pictorial representation of the f-cutsets defined for the tree $e_2 e_3 e_4$ by drawing a broken line across each f-cutset.

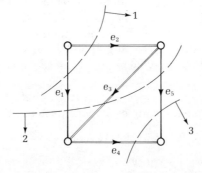

This matrix possesses the identity matrix of order 3 formed by the last three columns corresponding to the branches of the chosen tree and therefore is a basis cutset matrix.

2-2.3 The Circuit Matrix and Kirchhoff's Voltage Law Equations

In this section we turn our attention to the systematic formulation of Kirchhoff's voltage law, abbreviated as KVL, equations. The approach is parallel to those discussed in the preceding section.

Consider the low-pass filter of Figure 2.32(a) the associated directed graph G of which is redrawn in Figure 2.34. As before, let $i_k(t)$ and $v_k(t)$ be the current and voltage of the edge e_k ($k = 1, 2, 3, 4, 5$). Using the three loops as indicated in Figure 2.34, KVL equations can be written as

$$\begin{aligned}
\text{loop 1:} \quad & -v_1(t) + v_2(t) + v_3(t) && = 0 \\
\text{loop 2:} \quad & \qquad\qquad\qquad -v_3(t) - v_4(t) + v_5(t) = 0 \\
\text{loop 3:} \quad & -v_1(t) + v_2(t) \qquad\quad - v_4(t) + v_5(t) = 0
\end{aligned} \tag{2.18}$$

which can be rewritten in matrix notation as

$$
\begin{array}{c}
\text{loop 1} \\
\text{loop 2} \\
\text{loop 3}
\end{array}
\begin{array}{ccccc}
e_1 & e_2 & e_3 & e_4 & e_5
\end{array}
\begin{bmatrix}
-1 & 1 & 1 & 0 & 0 \\
0 & 0 & -1 & -1 & 1 \\
-1 & 1 & 0 & -1 & 1
\end{bmatrix}
\begin{bmatrix}
v_1(t) \\
v_2(t) \\
v_3(t) \\
v_4(t) \\
v_5(t)
\end{bmatrix}
=
\begin{bmatrix}
0 \\
0 \\
0
\end{bmatrix}
\tag{2.19}
$$

or, more compactly, as

$$\mathbf{B}_a \mathbf{v}(t) = \mathbf{0} \tag{2.20}$$

The coefficient matrix $\mathbf{B}_a$ is called the *complete circuit matrix* of G and $\mathbf{v}(t)$ the *branch-voltage vector*. In general, if G is a directed graph of n nodes, b edges, and c components and contains κ circuits, the complete circuit matrix of G is a matrix of order $\kappa \times b$ such that if

$$\mathbf{B}_a = [b_{ij}] \tag{2.21}$$

then

$b_{ij} = 1$ if edge e_j is in circuit i and the orientations of the circuit and the edge coincide

$b_{ij} = -1$ if edge e_j is in circuit i and the orientations of the circuit and the edge are opposite

$b_{ij} = 0$ if edge e_j is not in circuit i

Therefore each row of $\mathbf{B}_a$ is associated with a circuit of G and each column an edge, as shown in (2.19). In Figure 2.34 edge e_3 is contained in circuits

Figure 2.34
Three loops used to write KVL equations.

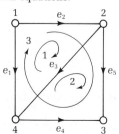

1 and 2, so that nonzero elements in rows 1 and 2 of column 3 correspond to circuits 1 and 2. The element in row 1 is 1 because the orientation of circuit 1 coincides with that of edge e_3; the element in row 2 is -1 because the orientation of circuit 2 is opposite to that of edge e_3. Like $\mathbf{A}_a$ and $\mathbf{Q}_a$, the rows and columns of $\mathbf{B}_a$ may be arranged in any way deemed convenient for the situation at hand.

Observe from (2.18) that the third equation is the sum of the first two. Thus the equations in (2.20) in general are not linearly independent. Then what is the minimum number of linearly independent equations? Or, equivalently, what is the rank of the matrix $\mathbf{B}_a$? Recall that an important property for the f-circuits is that in each f-circuit there is an edge—namely, the defining link—which is not contained in any other f-circuits. Thus equations written for these circuits are linearly independent. The number of f-circuits with respect to a chosen tree is equal to the number of cotree links. It is not difficult to show that the number of branches in a tree is $n - 1$, n being the number of nodes of the tree (Problem 2.4). We conclude that the number of f-circuits with respect to a chosen tree is $b - n + 1$. In other words (2.20) contains at least $b - n + 1$ linearly independent equations, meaning that $\mathbf{B}_a$ is of rank at least $b - n + 1$, provided that G is connected. It can also be shown (for an example, see Chen[†]) that the rank of $\mathbf{B}_a$ is at most $b - n + 1$ for a connected G. In the situation where G is not connected, the above result can be applied to each of its components. Thus we can state the following.

Property 2.3 *The rank of the complete circuit matrix $\mathbf{B}_a$ of an n-node directed graph having b edges and c components is equal to $m = b - n + c$.*

The number $m = b - n + c = b - r$ is referred to as the *nullity* of G, r being its rank. A submatrix of $\mathbf{B}_a$, denoted by $\mathbf{B}$, is called a *basis circuit matrix* if $\mathbf{B}$ is of order $m \times b$ and of rank m. All the information contained in (2.20) is now contained in

$$\mathbf{B}v(t) = \mathbf{0} \tag{2.22}$$

For Figure 2.34 we have $n = 4$, $b = 5$ and $c = 1$, giving nullity $m = 2$ ($= 5 - 4 + 1$). The submatrix formed by the last two rows of $\mathbf{B}_a$ in (2.19), as given below,

$$\mathbf{B} = \begin{bmatrix} 0 & 0 & -1 & -1 & 1 \\ -1 & 1 & 0 & -1 & 1 \end{bmatrix} \tag{2.23}$$

is a basis circuit matrix.

As in the cutset case, a simple and systematic way in writing down a basis circuit matrix uses the f-circuits. Denote by $\mathbf{B}_f$ the circuit matrix

† Chen, *Applied Graph Theory*, p. 44.

Figure 2.35
The f-circuits defined for the tree $e_2 e_3 e_4$.

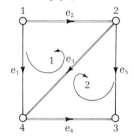

corresponding to a set of f-circuits. Since each f-circuit contains an edge that is not contained in any other f-circuits, $\mathbf{B}_f$ must contain the identity matrix of order m. Hence $\mathbf{B}_f$, being of order $m \times b$ and of rank m, is a basis circuit matrix. The procedure is possible only if G is connected. For unconnected graphs we must treat each component individually.

In Figure 2.34 choosing tree $T = e_2 e_3 e_4$ yields two f-circuits as shown in Figure 2.35. The f-circuit matrix is given by

$$\mathbf{B}_f = \begin{array}{c} \\ \text{loop 1} \\ \text{loop 2} \end{array} \begin{array}{cccccc} e_1 & e_5 & e_2 & e_3 & e_4 \\ \begin{bmatrix} 1 & 0 & -1 & -1 & 0 \\ 0 & 1 & 0 & -1 & -1 \end{bmatrix} \end{array} \qquad (2.24)$$

The matrix contains the identity matrix of order 2 formed by the first two columns corresponding to the links of the cotree $e_1 e_5$, showing that $\mathbf{B}_f$ is indeed a basis circuit matrix.

2-3 INTERRELATIONSHIPS OF GRAPH MATRICES

A basis incidence matrix $\mathbf{A}$ contains all the information about the directed graph G. The reason is that for a given $\mathbf{A}$ we can restore the complete incidence matrix $\mathbf{A}_a$ by appealing to the property that each column of $\mathbf{A}_a$ contains exactly two nonzero elements, a 1 and a -1. Once we have $\mathbf{A}_a$, G can then be drawn in a straightforward manner. For instance, suppose $\mathbf{A}$ of (2.8) is given. The complete incidence matrix $\mathbf{A}_a$ in (2.2) can easily be restored by adding a row at the bottom of $\mathbf{A}$. The elements of the added row are chosen in such a way that each column of the resulting matrix is comprised of exactly one 1 and one -1, yielding the desired $\mathbf{A}_a$ in (2.2). As a result of this observation, it is logical to expect that we should be able to derive formulas relating the matrices $\mathbf{A}$, $\mathbf{B}_f$, and $\mathbf{Q}_f$. Before we proceed, we need the following result, which is one of the fundamental theorems in graph theory. For a proof of this, see Chen.[†]

> **Property 2.4** *If the columns of the complete incidence matrix* $\mathbf{A}_a$, *the complete cutset matrix* $\mathbf{Q}_a$, *and the complete circuit matrix* $\mathbf{B}_a$ *of a directed graph G are arranged in the same edge order, then*
>
> $$\mathbf{A}_a \mathbf{B}_a' = \mathbf{0} \qquad (2.25)$$
>
> $$\mathbf{Q}_a \mathbf{B}_a' = \mathbf{0} \qquad (2.26)$$
>
> *where the prime, as before, denotes the matrix transpose.*

We illustrate this property by the following example.

† Chen, *Applied Graph Theory*, p. 44.

EXAMPLE 2.1

We use Figure 2.34 to verify Property 2.4. From (2.12) and (2.19) and Figure 2.34, the required graph matrices are given by

$$
\mathbf{Q}_a = \begin{matrix} & e_1 & e_2 & e_3 & e_4 & e_5 \\ & \begin{bmatrix} 1 & 1 & 0 & 0 & 0 \\ 0 & -1 & 1 & 0 & 1 \\ 0 & 0 & 0 & -1 & -1 \\ -1 & 0 & -1 & 1 & 0 \\ 1 & 0 & 1 & 0 & 1 \\ 0 & 1 & -1 & 1 & 0 \end{bmatrix} \end{matrix}
\tag{2.27}
$$

$$
\mathbf{B}_a = \begin{bmatrix} -1 & 1 & 1 & 0 & 0 \\ 0 & 0 & -1 & -1 & 1 \\ -1 & 1 & 0 & -1 & 1 \end{bmatrix}
\tag{2.28}
$$

$$
\mathbf{A}_a = \begin{bmatrix} 1 & 1 & 0 & 0 & 0 \\ 0 & -1 & 1 & 0 & 1 \\ 0 & 0 & 0 & -1 & -1 \\ -1 & 0 & -1 & 1 & 0 \end{bmatrix}
\tag{2.29}
$$

It is straightforward to verify that

$$
\mathbf{Q}_a\mathbf{B}'_a = \mathbf{0} \quad \text{and} \quad \mathbf{A}_a\mathbf{B}'_a = \mathbf{0}
\tag{2.30}
$$

Recall that the submatrix of an f-cutset matrix $\mathbf{Q}_f$ formed by the columns corresponding to the defining tree branches is the identity matrix. Likewise the submatrix of an f-circuit matrix $\mathbf{B}_f$ formed by the columns corresponding to the defining links is the identity matrix. If edges of G are labeled so that the first m columns of $\mathbf{Q}_f$, $\mathbf{B}_f$, and $\mathbf{A}$ correspond to the m defining links and the last r columns to the r defining tree branches and if the m f-circuits and r f-cutsets are numbered correspondingly, then $\mathbf{Q}_f$, $\mathbf{B}_f$, and $\mathbf{A}$ can be partitioned as

$$
\mathbf{Q}_f = [\mathbf{Q}_{f11} \quad \mathbf{1}_r]
\tag{2.31}
$$

$$
\mathbf{B}_f = [\mathbf{1}_m \quad \mathbf{B}_{f12}]
\tag{2.32}
$$

$$
\mathbf{A} = [\mathbf{A}_{11} \quad \mathbf{A}_{12}]
\tag{2.33}
$$

where $\mathbf{1}_m$ is the identity matrix of order m. Since $\mathbf{Q}_f$, $\mathbf{B}_f$, and $\mathbf{A}$ are respectively the submatrices of $\mathbf{Q}_a$, $\mathbf{B}_a$, and $\mathbf{A}_a$, identities (2.25) and (2.26) remain valid when $\mathbf{Q}_a$, $\mathbf{B}_a$, and $\mathbf{A}_a$ are replaced by $\mathbf{Q}_f$, $\mathbf{B}_f$, and $\mathbf{A}$, giving

$$
\mathbf{A}\mathbf{B}'_f = \mathbf{0}
\tag{2.34}
$$

$$
\mathbf{Q}_f\mathbf{B}'_f = \mathbf{0}
\tag{2.35}
$$

Substituting (2.31)–(2.33) in (2.34) and (2.35) yields

$$\mathbf{AB}'_f = [\mathbf{A}_{11} \quad \mathbf{A}_{12}] \begin{bmatrix} \mathbf{1}_m \\ \mathbf{B}'_{f12} \end{bmatrix} = \mathbf{A}_{11} + \mathbf{A}_{12}\mathbf{B}'_{f12} = \mathbf{0} \tag{2.36a}$$

$$\mathbf{Q}_f\mathbf{B}'_f = [\mathbf{Q}_{f11} \quad \mathbf{1}_r] \begin{bmatrix} \mathbf{1}_m \\ \mathbf{B}'_{f12} \end{bmatrix} = \mathbf{Q}_{f11} + \mathbf{B}'_{f12} = \mathbf{0} \tag{2.36b}$$

or

$$\mathbf{B}'_{f12} = -\mathbf{A}_{12}^{-1}\mathbf{A}_{11} \tag{2.37}$$

$$\mathbf{Q}_{f11} = -\mathbf{B}'_{f12} \tag{2.38}$$

In deriving (2.37), we have used the fact that the submatrix $\mathbf{A}_{12}$ is non-singular, so its inverse $\mathbf{A}_{12}^{-1}$ exists. This is always the case as long as the columns of $\mathbf{A}_{12}$ correspond to the branches of a tree. A detailed justification may be found in Chen[†].

With (2.37) and (2.38), $\mathbf{Q}_f$ and $\mathbf{B}_f$ can be expressed directly in terms of the submatrices of $\mathbf{A}$:

$$\mathbf{Q}_f = \mathbf{A}_{12}^{-1}\mathbf{A} = [\mathbf{A}_{12}^{-1}\mathbf{A}_{11} \quad \mathbf{1}_r] \tag{2.39}$$

$$\mathbf{B}_f = [\mathbf{1}_m \quad -\mathbf{A}'_{11}\mathbf{A}'^{-1}_{12}] \tag{2.40}$$

The significance of the results is that once $\mathbf{A}$ is given, the matrices $\mathbf{Q}_f$ and $\mathbf{B}_f$ can be computed directly from $\mathbf{A}$ without the necessity of first forming the f-cutsets and f-circuits—a real saving in computer-aided analysis.

EXAMPLE 2.2

We use Figure 2.35 to verify the above results. Choosing tree $T = e_2e_3e_4$ and partitioning the columns of $\mathbf{A}$ in the order of links e_1 and e_5 and branches e_2, e_3, and e_4 yield from (2.29), after deleting the last row,

$$\mathbf{A} = [\mathbf{A}_{11} \quad \mathbf{A}_{12}] = \begin{array}{c} \begin{array}{ccccc} e_1 & e_5 & e_2 & e_3 & e_4 \end{array} \\ \begin{bmatrix} 1 & 0 & 1 & 0 & 0 \\ 0 & 1 & -1 & 1 & 0 \\ 0 & -1 & 0 & 0 & -1 \end{bmatrix} \end{array} \tag{2.41}$$

First we compute

$$\mathbf{A}_{12}^{-1}\mathbf{A}_{11} = \begin{bmatrix} 1 & 0 & 0 \\ -1 & 1 & 0 \\ 0 & 0 & -1 \end{bmatrix}^{-1} \begin{bmatrix} 1 & 0 \\ 0 & 1 \\ 0 & -1 \end{bmatrix}$$

$$= \begin{bmatrix} 1 & 0 & 0 \\ 1 & 1 & 0 \\ 0 & 0 & -1 \end{bmatrix} \begin{bmatrix} 1 & 0 \\ 0 & 1 \\ 0 & -1 \end{bmatrix} = \begin{bmatrix} 1 & 0 \\ 1 & 1 \\ 0 & 1 \end{bmatrix} \tag{2.42}$$

[†] Chen, *Applied Graph Theory*, p. 41.

Hence from (2.39) we have

$$\mathbf{Q}_f = [\mathbf{A}_{12}^{-1}\mathbf{A}_{11} \quad \mathbf{l}_3] = \begin{bmatrix} 1 & 0 & | & 1 & 0 & 0 \\ 1 & 1 & | & 0 & 1 & 0 \\ 0 & 1 & | & 0 & 0 & 1 \end{bmatrix} \tag{2.43}$$

confirming (2.17); and from (2.40)

$$\mathbf{B}_f = [\mathbf{l}_2 \quad -(\mathbf{A}_{12}^{-1}\mathbf{A}_{11})'] = \begin{bmatrix} 1 & 0 & | & -1 & -1 & 0 \\ 0 & 1 & | & 0 & -1 & -1 \end{bmatrix}. \tag{2.44}$$

confirming (2.24).

2-4   TELLEGEN'S THEOREM

Kirchhoff's current and voltage laws are algebraic constraints arising from the interconnection of network elements and are independent of the characteristics of the elements. As a consequence we show that they imply the conservation of energy. In other words, conservation of energy need not be an added postulate for the discipline of network theory.

In the foregoing we demonstrated that a network fulfills KCL and KVL if and only if the equations $\mathbf{Qi}(t) = \mathbf{0}$ and $\mathbf{Bv}(t) = \mathbf{0}$ are satisfied. The most general solutions of $\mathbf{i}(t)$ and $\mathbf{v}(t)$ satisfying $\mathbf{Qi}(t) = \mathbf{0}$ and $\mathbf{Bv}(t) = \mathbf{0}$ can be written explicitly as

$$\mathbf{i}(t) = \mathbf{B}'\mathbf{i}_m(t) \tag{2.45}$$

$$\mathbf{v}(t) = \mathbf{Q}'\mathbf{v}_c(t) \tag{2.46}$$

where $\mathbf{i}_m(t)$ is an arbitrary m-vector, a vector consisting of m elements; and $\mathbf{v}_c(t)$ an arbitrary r-vector. Recall that n, b, c, m, and r denote respectively number of nodes, number of edges, number of components, and the nullity and rank of the associated directed graph of the network. To verify we compute

$$\mathbf{Qi}(t) = \mathbf{QB}'\mathbf{i}_m(t) = \mathbf{0i}_m(t) = \mathbf{0} \tag{2.47}$$

$$\mathbf{Bv}(t) = \mathbf{BQ}'\mathbf{v}_c(t) = (\mathbf{QB}')'\mathbf{v}_c(t) = \mathbf{0v}_c(t) = \mathbf{0} \tag{2.48}$$

since from (2.26) $\mathbf{QB}' = \mathbf{0}$. As a result we immediately have

$$\sum_{k=1}^{n} v_k(t)i_k(t) = \mathbf{v}'(t)\mathbf{i}(t) = \mathbf{v}_c'(t)\mathbf{QB}'\mathbf{i}_m(t) = \mathbf{v}_c'(t)\mathbf{0i}_m(t) = \mathbf{0} \tag{2.49}$$

showing that the sum of instantaneous powers delivered to all the elements, the voltages and currents of which are respectively $v_k(t)$ and $i_k(t)$, is equal to zero. If we integrate (2.49) between any two limits t_0 and t, we obtain the total energy stored in the network from t_0 to t as

$$w(t) = \sum_{k=1}^{n} \int_{t_0}^{t} v_k(x)i_k(x)\,dx = \text{constant} \tag{2.50}$$

Stated differently (2.50) shows that conservation of energy is a direct consequence of Kirchhoff's two laws KCL and KVL, and there is no need to add the postulate that the total energy (electric, mechanical, thermal, or otherwise) is conservative in the discipline of network theory.

Since Kirchhoff's laws are independent of the characteristics of the network elements, an even more general result can be stated. Consider another network $\hat{N}$ having the same associated directed graph G. Let $\hat{i}(t)$ and $\hat{v}(t)$ denote the branch-current and branch-voltage vectors of $\hat{N}$. Then we have $\hat{i}(t) = \mathbf{B}'\hat{i}_m(t)$ and $\hat{v}(t) = \mathbf{Q}'\hat{v}_c(t)$, where $\hat{i}_m(t)$ and $\hat{v}_c(t)$ are arbitrary m-vector and r-vector, respectively. We can now compute

$$\mathbf{v}'(t)\hat{i}(t) = \mathbf{v}_c'(t)\mathbf{QB}'\hat{i}_m(t) = \mathbf{v}_c'(t)\mathbf{0}\hat{i}_m(t) = 0 \tag{2.51}$$

$$\hat{v}'(t)i(t) = \hat{v}_c'(t)\mathbf{QB}'i_m(t) = \hat{v}_c'(t)\mathbf{0}i_m(t) = 0 \tag{2.52}$$

By taking the transposes of these equations and observing that the transpose of a scalar—namely, a 1×1 matrix—is itself, we obtain several variations of (2.51) as follows:

$$\mathbf{v}'(t)\hat{i}(t) = \hat{v}'(t)i(t) = i'(t)\hat{v}(t) = \hat{i}'(t)v(t) = 0 \tag{2.53}$$

Equation (2.53) is referred to as the *Tellegen's theorem.*[†] The entity $\mathbf{v}'(t)\hat{i}(t)$ does not have physical significance as the sum of instantaneous powers because $\mathbf{v}(t)$ and $\hat{i}(t)$ belong to two different networks. We emphasize that Tellegen's theorem is valid under the assumption that two different networks under consideration have the *same* topology—namely, the same associated directed graph. The theorem is valid whether the networks are linear or nonlinear and time invariant or time varying. In addition to its fundamental importance, the theorem provides a way to check the accuracy of the numerical results of a computer solution. After all the voltages and currents are computed, the sum of the products of the corresponding terms must be zero for every value of t. Another important application is its fundamental role in sensitivity analysis through adjoint networks. This aspect, however, will not be discussed here.

We illustrate the above result by the following examples.

EXAMPLE 2.3

Consider the networks N and $\hat{N}$ in Figure 2.36, the associated directed graph of which is shown in Figure 2.37. The branch-current and the branch-

---

[†] B. D. H. Tellegen, "A General Network Theorem, with Applications," *Proc. Inst., Radio Engrs. Australia* 14 (1953): 265–270. Bernard D. H. Tellegen (1900–   ) is with Philips Research Laboratories, Netherlands, and Technological University, Delft.

Figure 2.36
Two networks possessing
the same topology.

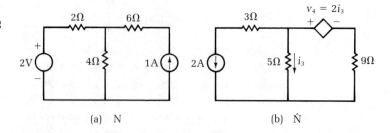

(a) N   (b) N̂

Figure 2.37
The associated directed
graph of the networks of
Figure 2.36.

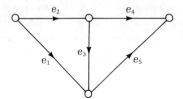

voltage vectors of N and N̂ are easily shown to be

$$
\mathbf{i} = \begin{array}{c} e_1 \\ e_2 \\ e_3 \\ e_4 \\ e_5 \end{array}
\begin{bmatrix} \frac{1}{3} \\ -\frac{1}{3} \\ \frac{2}{3} \\ -1 \\ 1 \end{bmatrix},
\mathbf{v} = \begin{bmatrix} 2 \\ -\frac{2}{3} \\ \frac{8}{3} \\ -6 \\ -\frac{26}{3} \end{bmatrix},
\hat{\mathbf{i}} = \begin{bmatrix} 2 \\ -2 \\ -\frac{3}{2} \\ -\frac{1}{2} \\ \frac{1}{2} \end{bmatrix},
\hat{\mathbf{v}} = \begin{bmatrix} -\frac{27}{2} \\ -6 \\ -\frac{15}{2} \\ -3 \\ \frac{9}{2} \end{bmatrix}
\qquad \textbf{(2.54)}
$$

Since the two networks have the same topology, Tellegen's theorem applies.
It is easy to verify (2.53), as we have

$$
\mathbf{v}\hat{\mathbf{i}} = \hat{\mathbf{i}}'\mathbf{v} = 2 \times 2 + \left(-\frac{2}{3}\right) \times (-2) + \left(\frac{8}{3}\right) \times \left(-\frac{3}{2}\right) + (-6) \times \left(-\frac{1}{2}\right)
$$

$$
+ \left(-\frac{26}{3}\right) \times \left(\frac{1}{2}\right) = 0 \qquad \textbf{(2.55a)}
$$

$$
\hat{\mathbf{v}}'\mathbf{i} = \mathbf{i}'\hat{\mathbf{v}} = \left(-\frac{27}{2}\right) \times \left(\frac{1}{3}\right) + (-6) \times \left(-\frac{1}{3}\right) + \left(-\frac{15}{2}\right) \times \left(\frac{2}{3}\right)
$$

$$
+ (-3) \times (-1) + \left(\frac{9}{2}\right) \times 1 = 0 \qquad \textbf{(2.55b)}
$$

EXAMPLE 2.4
In the network N of Figure 2.38, the current i_1 is known to be $-830/541$ A,
and the current i_2 is $-50/541$ A. Suppose that a voltage source of $v_2 = 20$ V
replaces the shorted wire between terminals c and d, and a 6Ω resistor replaces the 10V generator connected between terminals a and b as indicated

Figure 2.38
A given resistive network.

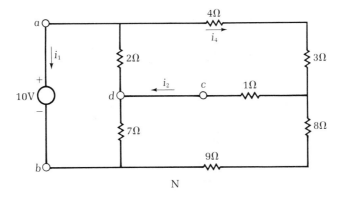

N

Figure 2.39
A resistive network
having the same topology
as that of Figure 2.38.

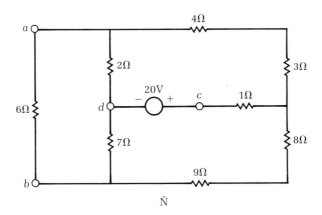

N̂

Figure 2.40
The associated directed
graph of the networks of
Figures 2.38 and 2.39.

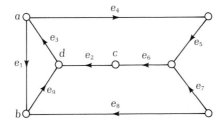

in Figure 2.39. We wish to compute the current in the 6Ω resistor by means
of Tellegen's theorem.

The associated directed graph G of the networks N and $\hat{N}$ of Figures
2.38 and 2.39 is presented in Figure 2.40. The branch voltage and the branch
current associated with the edge e_k are denoted by v_k and i_k for N, and $\hat{v}_k$
and $\hat{i}_k$ for $\hat{N}$. By (2.53) we have $\mathbf{v}\hat{\mathbf{i}} = \mathbf{0}$, giving

$$v_1\hat{i}_1 + v_2\hat{i}_2 + \sum_{k=3}^{9} v_k\hat{i}_k = 0 \qquad\qquad (2.56)$$

Denote by R_k the resistance associated with edge e_k $(k = 3, 4, \ldots, 9)$. Then we have

$$v_1\hat{i}_1 + v_2\hat{i}_2 = -\sum_{k=3}^{9} R_k i_k \hat{i}_k \qquad (2.57)$$

Likewise, using $\hat{\mathbf{v}}'\mathbf{i} = 0$ we obtain

$$\hat{v}_1 i_1 + \hat{v}_2 i_2 = -\sum_{k=3}^{9} R_k \hat{i}_k i_k \qquad (2.58)$$

Equating (2.57) and (2.58) yields

$$v_1\hat{i}_1 + v_2\hat{i}_2 = \hat{v}_1 i_1 + \hat{v}_2 i_2 \qquad (2.59)$$

Substituting the known values in (2.59), we get

$$10\hat{i}_1 + 0\hat{i}_2 = \hat{v}_1\left(-\frac{830}{541}\right) + 20 \times \left(-\frac{50}{541}\right) \qquad (2.60)$$

which in conjunction with $\hat{v}_1 = 6\hat{i}_1$ gives the solution

$$\hat{i}_1 = -\frac{300}{3117} = -0.096 \text{ A} \qquad (2.61)$$

2-5   PRIMARY SYSTEM OF NETWORK EQUATIONS

The purpose of this section is to present the primary system of network equations in matrix form. The secondary systems of network equations will be formulated in Chapters 3 and 11.

Kirchhoff's two laws are concerned with the way the network elements are interconnected. The characteristics of the network elements do not enter into the discussion. On the other hand, the character of the individual network element is independent of where the network element is situated in the network. In solving the network problem these two aspects of a network must both be considered.

Consider a transistor amplifier, its equivalent network, and the associated directed graph G of Figure 2.30. As before, denote by v_k and i_k the voltage and current associated with edge e_k $(k = 1, 2, \ldots, 8)$. The character of the network elements are described by the following system of equations:

$$i_1(t) = i_g(t) \qquad\qquad v_2(t) = R_2 i_2(t)$$

$$v_3(t) = R_3 i_3(t) \qquad\qquad i_6(t) = g_m v_3(t)$$

$$i_4(t) = C_4 \frac{dv_4(t)}{dt} \qquad\qquad i_5(t) = C_5 \frac{dv_5(t)}{dt}$$

$$v_7(t) = L_7 \frac{di_7(t)}{dt} \qquad\qquad v_8(t) = R_8 i_8(t) \qquad (2.62)$$

which can be written in matrix notation as

$$\begin{bmatrix} 1 & 0 & 0 & 0 & 0 & 0 & 0 & 0 \\ 0 & R_2 & 0 & 0 & 0 & 0 & 0 & 0 \\ 0 & 0 & 1/R_3 & 0 & 0 & 0 & 0 & 0 \\ 0 & 0 & 0 & C_4 p & 0 & 0 & 0 & 0 \\ 0 & 0 & 0 & 0 & C_5 p & 0 & 0 & 0 \\ 0 & 0 & g_m & 0 & 0 & 0 & 0 & 0 \\ 0 & 0 & 0 & 0 & 0 & 0 & L_7 p & 0 \\ 0 & 0 & 0 & 0 & 0 & 0 & 0 & R_8 \end{bmatrix} \begin{bmatrix} i_1(t) \\ i_2(t) \\ v_3(t) \\ v_4(t) \\ v_5(t) \\ v_6(t) \\ i_7(t) \\ i_8(t) \end{bmatrix} = \begin{bmatrix} i_g(t) \\ v_2(t) \\ i_3(t) \\ i_4(t) \\ i_5(t) \\ i_6(t) \\ v_7(t) \\ v_8(t) \end{bmatrix} \qquad (2.63)$$

where $p = d/dt$ is the *linear differential operator*. More compactly we have[†]

$$\tilde{\mathbf{H}}(p)\tilde{\mathbf{u}}(t) = \tilde{\mathbf{y}}(t) \qquad (2.64)$$

The coefficient matrix $\tilde{\mathbf{H}}(p)$ of the operator p is called the *branch-immittance matrix operator*, which in general is of order b, the number of network elements, Equation (2.64) is referred to as the *element v-i equations*.

The element v-i equations together with KCL and KVL equations, summarized below as

$$\mathbf{Q}\mathbf{i}(t) = \mathbf{0} \qquad (2.65a)$$

$$\mathbf{B}\mathbf{v}(t) = \mathbf{0} \qquad (2.65b)$$

$$\tilde{\mathbf{H}}(p)\tilde{\mathbf{u}}(t) = \tilde{\mathbf{y}}(t) \qquad (2.65c)$$

constitute the *primary system of network equations*. In the primary system, there are $2b$ unknowns, b unknown currents in $\mathbf{i}(t)$ and b unknown voltages in $\mathbf{v}(t)$; there are exactly $2b$ $(=r + m + b)$ equations, r KCL equations in (2.65a), m KVL equations in (2.65b) and b equations in (2.65c). The problem is then to find the vectors $\mathbf{i}(t)$ and $\mathbf{v}(t)$ such that (2.65) holds true. Theoretically the process is straightforward; it amounts to solving a system of linear algebraic and differential equations. In practice we seldom go through the normal process to get the solutions because the number of equations involved is rather large. For instance, the network of Figure 2.30(b) would involve a system of sixteen linear algebraic and differential equations, some of which are rather trivial. Few electrical engineers would actually solve the problem in this way even with the aid of a digital computer. In Chapters 3 and 11, we shall present techniques that amount to a systematic organization of the variables in the primary system of equations (2.65) so that the number of equations needed can be reduced.

EXAMPLE 2.5

We formulate the primary system of equations for the network of Figure 2.30(b). For this we choose a tree, say $T = e_1 e_4 e_5 e_8$ in Figure 2.30(c). The

---

[†] Alternatively, $\tilde{\mathbf{y}}(t)$ can be decomposed as the sum of two vectors, one corresponding to sources and one to all others.

KCL and KVL equations are obtained as follows:

$$
\mathbf{Q}_f \mathbf{i}(t) =
\begin{array}{cccccccc}
e_2 & e_3 & e_6 & e_7 & e_1 & e_4 & e_5 & e_8
\end{array}
\begin{bmatrix}
1 & 0 & 0 & 0 & 1 & 0 & 0 & 0 \\
-1 & 1 & 1 & 1 & 0 & 1 & 0 & 0 \\
0 & 0 & -1 & -1 & 0 & 0 & 1 & 0 \\
0 & 0 & 0 & -1 & 0 & 0 & 0 & 1
\end{bmatrix}
\begin{bmatrix}
i_2(t) \\ i_3(t) \\ i_6(t) \\ i_7(t) \\ i_1(t) \\ i_4(t) \\ i_5(t) \\ i_8(t)
\end{bmatrix}
$$

$$
=
\begin{bmatrix}
0 \\ 0 \\ 0 \\ 0
\end{bmatrix}
\tag{2.66}
$$

$$
\mathbf{B}_f \mathbf{v}(t) =
\begin{array}{cccccccc}
e_2 & e_3 & e_6 & e_7 & e_1 & e_4 & e_5 & e_8
\end{array}
\begin{bmatrix}
1 & 0 & 0 & 0 & -1 & 1 & 0 & 0 \\
0 & 1 & 0 & 0 & 0 & -1 & 0 & 0 \\
0 & 0 & 1 & 0 & 0 & -1 & 1 & 0 \\
0 & 0 & 0 & 1 & 0 & -1 & 1 & 1
\end{bmatrix}
\begin{bmatrix}
v_2(t) \\ v_3(t) \\ v_6(t) \\ v_7(t) \\ v_1(t) \\ v_4(t) \\ v_5(t) \\ v_8(t)
\end{bmatrix}
$$

$$
=
\begin{bmatrix}
0 \\ 0 \\ 0 \\ 0
\end{bmatrix}
\tag{2.67}
$$

Equations (2.66) and (2.67) together with (2.63) constitute the primary system of equations for the network of Figure 2.30(b). The significance of the primary system is not its use in computing network solutions, rather it is the foundation in formulating the secondary systems of equations such as the state equations, loop equations, and cutset or nodal equations.

2-6   PLANARITY AND DUALITY

A graph is said to be *planar* if it can be drawn on a plane such that no two edges have an intersection that is not a node. Figure 2.40 is an example of a planar graph. Figure 2.41(a) is also planar because it can be redrawn as shown in Figure 2.41(b). Planar graphs are extremely useful to the designers of printed circuits and integrated circuits for reasons of economy as well as

Figure 2.41

Two isomorphic planar graphs.

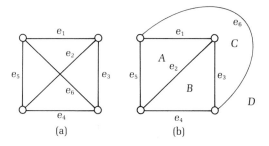

(a)   (b)

Figure 2.42

Examples of nonplanar graphs.

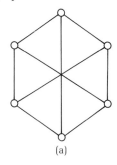

(a)

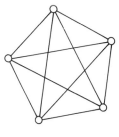

(b)

technology of manufacturing. In the present section, we outline some fundamental properties of a planar graph.

A planar graph, when drawn on a plane without intersections or crossings, divides the plane into *regions*. The unbounded region is called the *outside region* and the other regions *inside regions*. Thus a region is characterized by the edges on its boundary. In network theory the circuit formed by the boundary edges of a region is called a *mesh* because it has the appearance of a mesh of fish net. The graph of Figure 2.41(b) divides the plane into four regions A, B, C, and D, as indicated. Regions A, B, and C are the inside regions and D the outside region. The meshes of these regions are $e_1e_2e_5$, $e_2e_3e_4$, $e_1e_3e_6$, and $e_4e_5e_6$. Figure 2.42 gives two examples of nonplanar graphs because no matter how one redraws the graphs there is at least one crossing as indicated in Figure 2.43. It was first pointed out by Kuratowski[†] that the two nonplanar graphs of Figure 2.42 and their variants are necessary and sufficient to characterize a planar graph.

Kuratowski's Theorem on Planar Graphs *A graph is planar if and only if it does not contain either of the two basic nonplanar graphs of Figure 2.42 as a subgraph, where each edge in Figure 2.42 may represent a single edge or a series connection of edges.*

[†] C. Kuratowski, "Sur le Probléme des Courbes Gauches en Topologie," *Fund. Math.* 15 (1930): 271–283. The theorem is named after Casimir Kuratowski (1896–   ).

Figure 2.43

The nonplanar graphs of Figure 2.42 exhibiting at least one crossing.

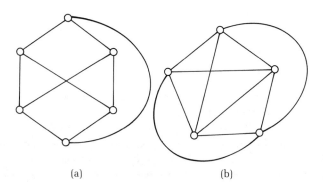

(a)   (b)

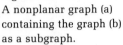

Figure 2.44
A nonplanar graph (a)
containing the graph (b)
as a subgraph.

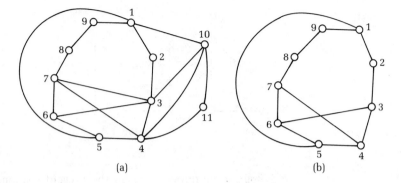

(a)   (b)

Figure 2.45
A planar graph (a) used to
illustrate the construction
of its dual graph (b),
which is redrawn in (c).

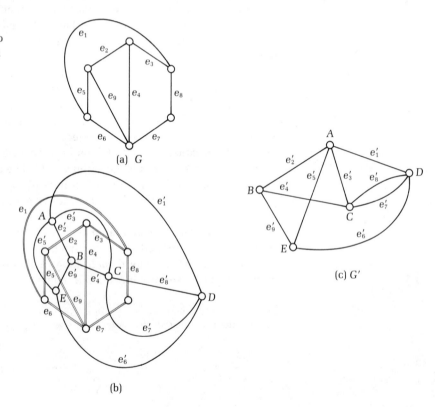

(a) G

(b)

(c) G′

The graph of Figure 2.44(a) is not a planar graph because, according to Kuratowski's theorem, it contains a subgraph shown in Figure 2.44(b). This subgraph can be transformed into that of Figure 2.42(a) by replacing the series connection of three edges between nodes 1 and 7 by a single edge and the series connection of two edges between nodes 1 and 3 by another single edge.

In the planar graph G of Figure 2.45(a), place a dot in each region including the outside region. Connect two dots by a line segment which crosses an edge e_j of G if edge e_j is on the boundary between the two regions associated with the two dots. As illustrated in Figure 2.45(b), the dots are marked as nodes A, B, C, D, and E with the prime denoting the line segment e_j' of e_j. The resulting graph G' is called a *dual graph* of G and is redrawn in Figure 2.45(c) for clarity. It is easily seen that if G' is a dual of G, then G is a dual of G'. Hence we shall say that G and G' are dual graphs. From the construction of a dual graph, we see that a circuit (cutset) in G corresponds to a cutset (circuit) in its dual G', and vice versa. As a result, KVL (KCL) equations formulated for G are identical in form to KCL (KVL) equations for G' and vice versa. Apart from signs, the circuit matrix of the associated directed graph of either graph is the cutset matrix of the other. In Figure 2.45 a circuit, say $e_2 e_3 e_5 e_6 e_7 e_8$ of G, corresponds to a cutset $e_2' e_3' e_5' e_6' e_7' e_8'$ in its dual G'; and a cutset, say $e_1 e_2 e_4 e_8$ of G, corresponds to a circuit $e_1' e_2' e_4' e_8'$ of G'. Conversely, a circuit, say $e_1' e_2' e_6' e_9'$ of G', corresponds to a cutset $e_1 e_2 e_6 e_9$ in its dual G; and a cutset, say $e_1' e_3' e_4' e_5' e_9'$ of G' corresponds to a circuit $e_1 e_3 e_4 e_5 e_9$ of G. In particular, the incidence cuts of G correspond to the meshes of G' and vice versa. Recall that if G is connected and contains n nodes, the KCL equations written for any $n-1$ incidence cuts are linearly independent. Since each mesh of G' corresponds to a node of G, we conclude that the KVL equations written for any $n-1$ meshes of G' are linearly independent. Therefore for a planar network, the circuit matrix corresponding to the meshes of all inside regions is a basis circuit matrix.

We illustrate the above results by the following example.

EXAMPLE 2.6

The associated directed graphs of the graph G and its dual G' of Figure 2.45 are presented in Figure 2.46 as G_1 and G_2. A basis incidence matrix of G_1 obtained from the complete incidence matrix by deleting the row corresponding to node 6 and a basis circuit matrix formed by the meshes as shown in

Figure 2.46
The associated directed graphs G_1 and G_2 of the graph G and its dual G' of Figure 2.45.

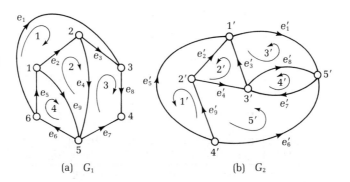

(a) G_1 (b) G_2

Figure 2.46(a) are found to be

$$
\mathbf{A}_1 =
\begin{array}{c}
\\
\text{node 1} \\
\text{node 2} \\
\text{node 3} \\
\text{node 4} \\
\text{node 5}
\end{array}
\begin{array}{c}
e_1 \quad e_2 \quad e_3 \quad e_4 \quad e_5 \quad e_6 \quad e_7 \quad e_8 \quad e_9 \\
\left[
\begin{array}{rrrrrrrrr}
0 & 1 & 0 & 0 & -1 & 0 & 0 & 0 & 1 \\
0 & -1 & 1 & 1 & 0 & 0 & 0 & 0 & 0 \\
-1 & 0 & -1 & 0 & 0 & 0 & 0 & 1 & 0 \\
0 & 0 & 0 & 0 & 0 & 0 & -1 & -1 & 0 \\
0 & 0 & 0 & -1 & 0 & 1 & 1 & 0 & -1
\end{array}
\right]
\end{array}
$$

$$\text{(2.68)}$$

$$
\mathbf{B}_1 =
\begin{array}{c}
\text{mesh 1} \\
\text{mesh 2} \\
\text{mesh 3} \\
\text{mesh 4}
\end{array}
\left[
\begin{array}{rrrrrrrrr}
1 & -1 & -1 & 0 & -1 & 0 & 0 & 0 & 0 \\
0 & 1 & 0 & 1 & 0 & 0 & 0 & 0 & -1 \\
0 & 0 & 1 & -1 & 0 & 0 & -1 & 1 & 0 \\
0 & 0 & 0 & 0 & 1 & 1 & 0 & 0 & 1
\end{array}
\right]
$$

$$\text{(2.69)}$$

Likewise the basis incidence and the basis circuit matrices of G_2 of Figure 2.46(b) are given by

$$
\mathbf{A}_2 =
\begin{array}{c}
\\
\text{node 1}' \\
\text{node 2}' \\
\text{node 3}' \\
\text{node 4}'
\end{array}
\begin{array}{c}
e_1' \quad e_2' \quad e_3' \quad e_4' \quad e_5' \quad e_6' \quad e_7' \quad e_8' \quad e_9' \\
\left[
\begin{array}{rrrrrrrrr}
1 & -1 & -1 & 0 & -1 & 0 & 0 & 0 & 0 \\
0 & 1 & 0 & 1 & 0 & 0 & 0 & 0 & -1 \\
0 & 0 & 1 & -1 & 0 & 0 & -1 & 1 & 0 \\
0 & 0 & 0 & 0 & 1 & 1 & 0 & 0 & 1
\end{array}
\right]
\end{array}
$$

$$\text{(2.70)}$$

$$
\mathbf{B}_2 =
\begin{array}{c}
\text{mesh 1}' \\
\text{mesh 2}' \\
\text{mesh 3}' \\
\text{mesh 4}' \\
\text{mesh 5}'
\end{array}
\left[
\begin{array}{rrrrrrrrr}
0 & 1 & 0 & 0 & -1 & 0 & 0 & 0 & 1 \\
0 & -1 & 1 & 1 & 0 & 0 & 0 & 0 & 0 \\
-1 & 0 & -1 & 0 & 0 & 0 & 0 & 1 & 0 \\
0 & 0 & 0 & 0 & 0 & 0 & -1 & -1 & 0 \\
0 & 0 & 0 & -1 & 0 & 1 & 1 & 0 & -1
\end{array}
\right]
$$

$$\text{(2.71)}$$

Thus a basis incidence matrix of either G_1 or G_2 equals a basis circuit matrix of the other—namely,

$$\mathbf{A}_1 = \mathbf{B}_2 \quad \text{and} \quad \mathbf{A}_2 = \mathbf{B}_1 \qquad\qquad \text{(2.72)}$$

This means that with appropriate changes of variables the KCL equations written for either G_1 or G_2 are identical with the KVL equations written for the other.

One remaining question is whether a graph can have more than one dual. This is why we use the phrase "a dual" instead of "the dual." The answer is affirmative. One can conceive a simple way of constructing an example

Figure 2.47
An unconnected graph (a)
and its dual constructed
in (b) and redrawn in (c).

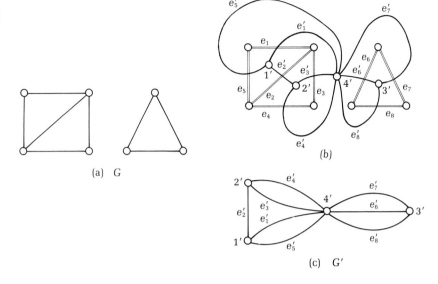

(a)   G

(b)

(c)   G'

Figure 2.48
The construction (a) of a dual of the graph of Figure 2.47(c) with the resulting dual
redrawn in (b).

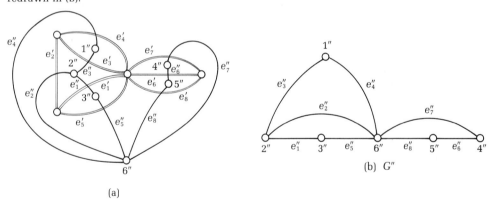

(a)

(b)   G''

where a graph has two different duals. We begin with an unconnected
planar graph G and use the procedure outlined in this section to find its dual
G'. Clearly G' is connected. Next let us take G' as a given planar graph and
construct its dual G'' by the same procedure. Then G'' is also connected.
Since G and G'' are both duals of G' and since G' is connected and G is not,
we obtain two different duals for G'. For instance, the graph G of Figure
2.47(a) is unconnected; its dual G' is constructed in Figure 2.47(b) and is
given in Figure 2.47(c). We next take G' and construct a dual as shown in
Figure 2.48(a). The resulting graph G'' is redrawn in Figure 2.48(b) for clarity.
Thus G and G'' are two different duals of G'.

Figure 2.49

Two isomorphic directed graphs.

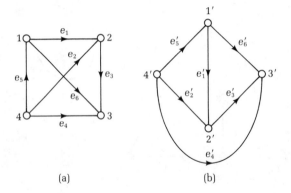

(a)   (b)

We see that in drawing a graph we have great freedom in the choice of the location of nodes and in the form of the line segments joining them. This may make the diagrams of the same graph look entirely different. An example is shown in Figure 2.41. Thus when we say that two graphs are the "same," we really mean in mathematical terms that they are *isomorphic*, which means that there exists a one-to-one correspondence between the nodes and edges of the two graphs such that the incidence relationships are preserved. For directed graphs, in addition to the above requirements, the directions of the corresponding edges must be preserved. For example, the two directed graphs of Figure 2.49 are isomorphic because we can establish a one-to-one correspondence between their nodes and edges, as indicated in the figure, such that the incidence relationships and directions of the corresponding edges are preserved.

2-7 SUMMARY

We began this chapter by defining and formulating network graph theory and introducing the basic concepts of *circuits*, *cutsets*, *trees*, *cotrees*, *f-circuits*, and *f-cutsets* and some of their fundamental properties. Applications of these results to the systematic formulations of *Kirchhoff's equations* were taken up next. Specifically, we showed that there are r linearly independent KCL equations that can be obtained systematically from the r f-cutsets defined for a chosen tree having r branches. Likewise we showed that there are m linearly independent KVL equations that can be written systematically from the m f-circuits defined over m links of a cotree. Together they form a system of $b = m + r$ linearly independent equations, b being the number of network elements. These equations are algebraic constraints arising from the interconnection of network elements and are independent of the characteristics of the elements. On the other hand, the character of the individual network element is independent of where the network element is situated in the network. The characteristics of the network elements are described by a system of b element v-i equations. These equations together with KCL and KVL equations constitute the *primary system of network equations*, which involves $2b$ equations in $2b$ unknowns. The significance of the primary system lies in its use in the formulation of *secondary systems of equations* such as the *state equations*, *loop equations*, and *cutset* or *nodal equations* to be discussed in Chapters 3 and 11.

Some properties of a *directed graph* are most succinctly described by a matrix. For this we introduced three types of matrices: the *incidence matrix*, the *cutset matrix*, and the *circuit matrix*. We showed that the f-cutset matrix and the f-circuit matrix can be computed directly from the incidence matrix without the necessity of first forming the f-cutsets and the f-circuits; this represents a real saving in computer-aided network analysis.

In addition, we introduced *Tellegen's theorem*, which states that, in two different networks having the same topology, the sum of products of branch voltages in one network and the corresponding branch currents in the other is identical to zero at all times. In particular, we demonstrated that Kirchhoff's two laws imply the conservation of energy and that there is no need to add separately the postulate of conservation of energy in network theory. In addition to playing a fundamental role in sensitivity analysis, Tellegen's theorem provides a way by which the accuracy of a computer solution can be checked.

Finally we introduced *planar graphs* and the notion of *duality*. A graph has a dual if and only if it is planar. In general there can be more than one dual graph to a given planar graph. The significance of dual graphs is that the KCL equations written for one are identical in form with KVL equations written for the other.

REFERENCES AND SUGGESTED READING

Blackwell, W. A. *Mathematical Modeling of Physical Networks.* New York: Macmillan, 1968, Chapters 4 and 5.

Chan, S. P. *Introductory Topological Analysis of Electrical Networks.* New York: Holt, Rinehart & Winston, 1969, Chapters 1–3.

Chen, W. K. *Applied Graph Theory: Graphs and Electrical Networks.* 2d ed. New York: American Elsevier, 1976, Chapters 1 and 2.

Chua, L. O., and P. M. Lin. *Computer-Aided Analysis of Electronic Circuits: Algorithms & Computational Techniques.* Englewood Cliffs, N.J.: Prentice-Hall, 1975, Chapter 3.

Koenig, H. E., Y. Tokad, and H. K. Kesavan. *Analysis of Discrete Physical Systems.* New York: McGraw-Hill, 1967, Chapter 4.

Penfield, Jr., P., R. Spence, and S. Duinker. *Tellegen's Theorem and Electrical Networks.* Cambridge, Mass.: The M.I.T. Press, 1970.

Mayeda, W. *Graph Theory.* New York: Wiley, 1972, Chapters 1–4 and 6.

Seshu, S., and M. B. Reed, *Linear Graphs and Electrical Networks.* Reading, Mass.: Addison-Wesley, 1961, Chapters 1–5.

Figure 2.50
An RLC network used to
determine the numbers of
linearly independent KCL
and KVL equations.

Figure 2.51
Graphs used to determine
their planarity.

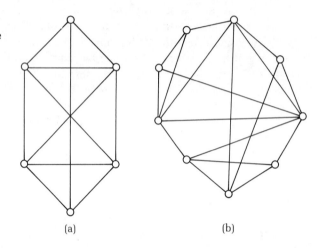

(a)   (b)

PROBLEMS

2.1 List all the circuits and trees of the directed graph in Figure 2.49(a).

2.2 List all the cutsets and cotrees of the directed graph in Figure 2.49(a).

2.3 For the directed graph of Figure 2.30(c), choose the tree $T = e_2e_3e_6e_7$. Determine the f-cutset matrix and the f-circuit matrix with respect to this tree.

2.4 Prove that every tree of a connected n-node graph contains $n - 1$ branches.

2.5 For the network shown in Figure 2.50, determine the numbers of linearly independent KCL and KVL equations. Obtain an associated directed graph of the network and write down a system of linearly independent KCL and KVL equations. Set up the primary system of equations for the network.

2.6 Determine whether each of the graphs of Figure 2.51 is planar. If it is a planar graph, redraw the graph to eliminate all crossings that are not nodes. If not, show it contains one of the basic nonplanar graphs of Figure 2.42 or their variants.

2.7 A directed graph G may be obtained from the graph of Figure 2.51(a) by assigning orientations to its edges. Pick a tree in G and determine the f-cutsets and the f-circuits with respect to this tree. Write the corresponding f-cutset matrix and the f-circuit matrix. Using these two matrices, verify the identity (2.35).

2.8 Is the directed graph of Figure 2.52 isomorphic to that of Figure 2.49(a)? Justify your statement.

2.9 Select a tree in the directed graph of Figure 2.52. Using (2.39) and (2.40), compute the f-cutset matrix $\mathbf{Q}_f$ and the f-circuit matrix $\mathbf{B}_f$ with respect to this tree from a basis incidence matrix obtained from the complete incidence matrix by deleting any one of its rows.

2.10 The equivalent networks of the amplifiers of Figure 2.53(a) and (b) are shown in Figure 2.53(c) and (d), respectively. Obtain directed graphs of the equivalent networks (c) and (d) and determine appropriate sets of linearly independent KCL and KVL equations for these networks.

Figure 2.52
A given directed graph.

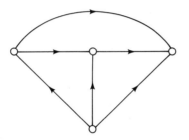

Figure 2.53
Transistor feedback amplifiers (a) and (b) and their equivalent networks (c) and (d).

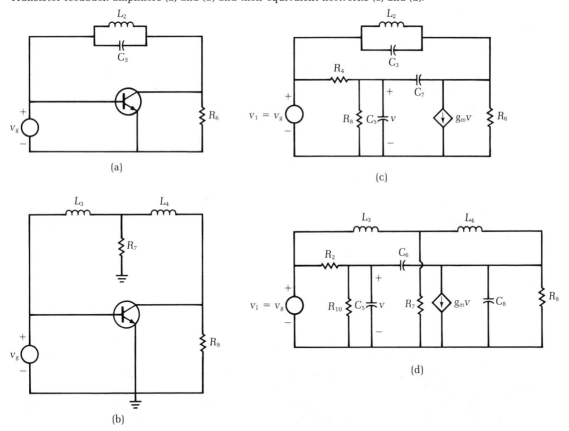

2.11 Determine the primary system of network equations in matrix form describing the equivalent network of Figure 2.53(c) for the amplifier of Figure 2.53(a).

2.12 Construct a dual graph of the directed graph G of Figure 2.52. Is this dual graph isomorphic to G? Is there more than one dual graph of G?

Figure 2.54
A cubic graph.

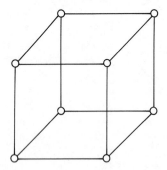

2.13 A basis incidence matrix of a directed graph is given by

$$\mathbf{A} = \begin{bmatrix} 0 & 1 & 0 & 0 & 0 & 1 & 0 & 1 \\ 0 & 0 & -1 & 0 & 0 & -1 & 1 & 0 \\ 0 & 0 & 0 & -1 & 1 & 0 & -1 & -1 \\ -1 & -1 & 1 & 1 & 0 & 0 & 0 & 0 \end{bmatrix} \qquad (2.73)$$

Find the directed graph from **A**.

2.14 The subgraph corresponding to the first four columns of **A** of (2.73) is known to be a tree. Using (2.39) and (2.40), compute the f-cutset matrix and the f-circuit matrix with respect to this tree. (Hint: Permute the columns of **A** first.)

2.15 Find the rank and nullity of the graph of Figure 2.54. Assign orientations to the edges and select a tree. Using this tree determine the f-cutsets, f-circuits, and their corresponding f-cutset and f-circuit matrices.

2.16 Construct a dual graph of the graph in Figure 2.54.

2.17 Assigning orientations to the edges of the graph in Figure 2.54, write a system of linearly independent KVL equations for the meshes of the resulting directed graph. Verify that the coefficient matrix of the system is a basis circuit matrix.

2.18 An f-cutset matrix of a connected directed graph is given below:

$$\mathbf{Q}_f = \begin{bmatrix} 1 & 0 & 0 & 0 & -1 & 0 & 0 & 0 \\ 0 & 1 & 0 & 0 & -1 & 1 & 0 & 0 \\ 0 & 0 & 1 & 0 & 0 & 1 & 0 & 1 \\ 0 & 0 & 0 & 1 & 0 & 0 & -1 & -1 \end{bmatrix} \qquad (2.74)$$

Find the directed graph.

2.19 In the network N of Figure 2.38, the current i_1 is known to be $-830/541$ A and the current i_4 is $190/541$ A. Suppose that a voltage source of 20V is inserted in series with the 4Ω resistor with potential rise from left to right and a 6Ω resistor replaces the 10V generator. Using Tellegen's theorem compute the current in the 6Ω resistor.

2.20 Let **A** be a basis incidence matrix of a connected directed graph G. It can be shown that the number of trees of G is equal to det **AA'**. Using **A** of (2.73), determine the number of trees of G.

2.21 Using the formula given in Problem 2.20, determine the number of trees in the directed graph of Figure 2.7(b).

2.22 Using the formula given in Problem 2.20, determine the number of cotrees in the directed graph in Figure 2.30(c).

2.23 Using the formula given in Problem 2.20, determine the number of trees in the graph of Figure 2.45(c).

2.24 Figure 2.55(a) is an equivalent network of a transformer-coupled transistor amplifier of (b). Obtain an associated directed graph of the equivalent network. Write an appropriate set of linearly independent KCL and KVL equations for the network. Verify that the coefficient matrices of KCL and KVL equations are a basis cutset matrix and a basis circuit matrix. Also determine the element v-i equations in matrix form.

Figure 2.55
An equivalent network (a)
of a transformer-coupled
transistor amplifier (b).

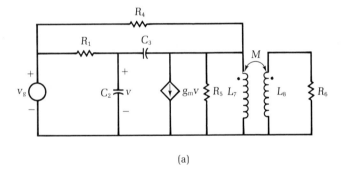

(a)

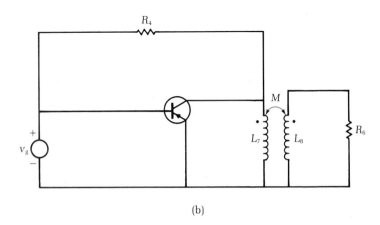

(b)

2.25 Show that a graph contains a tree if and only if it is connected.

2.26 Let $\mathbf{B}_f$ and $\mathbf{Q}_f$ be an f-circuit matrix and an f-cutset matrix of a directed graph G. It can be shown that the number of trees of G is given by the formula

$$\text{number of trees} = \det \mathbf{B}_f \mathbf{B}_f' = \det \mathbf{Q}_f \mathbf{Q}_f' \qquad (2.75)$$

Use this formula to determine the number of trees in the graph of Figure 2.54.

CHAPTER THREE

THE SECONDARY SYSTEMS
OF NETWORK EQUATIONS

The primary system of network equations formulated in the preceding chapter contains the basic information on network theory. The next problem is to calculate the branch voltages and branch currents for given input excitations and initial conditions. In principle the process is straightforward; it amounts to solving the primary system of equations. In practice we seldom go through the normal process of getting the solutions because of the large number of equations involved. In the present chapter, we shall present techniques that systematically organize the variables in the primary system so that the number of equations needed in solving the network problem can be reduced. The resulting equations are referred to as the *secondary systems of network equations*. This systematic treatment is particularly important at present when computers are used to perform the necessary numerical computations.

Three secondary systems of equations will be systematically developed. They are known as the *loop system of equations*, the *nodal system of equations*, and the *cutset system of equations*. The methods are general and can be applied to any linear time-invariant network. The classical method of solution of linear differential equations with constant coefficients and the operational method that uses the Laplace transformation will be presented in succeeding chapters.

In addition to the three secondary systems to be presented here, network equations can also be formulated as a system of first-order differential equations known as the *state equations*. The state equations are very important, and their formulation and solution will be taken up in Chapters 11 and 12.

For a given network, its associated directed graph is assumed to contain b edges and is of rank r and nullity m. The primary system of equations, composed of r linearly independent Kirchhoff's current law (KCL) equations, m linearly independent Kirchhoff's voltage law (KVL) equations, and b element v-i equations

$$Qi(t) = 0 \tag{3.1a}$$

$$Bv(t) = 0 \tag{3.1b}$$

$$\tilde{H}(p)\tilde{u}(t) = \tilde{y}(t) \tag{3.1c}$$

where $p = d/dt$ is the differential operator, constitutes the starting point of the development. Equation (3.1) is a system of $2b$ linear simultaneous algebraic and differential equations in $2b$ variables. Our objective is to reduce this number from $2b$ equations to r or m equations in r unknown voltages or m unknown currents.

3-1 THE LOOP SYSTEM OF EQUATIONS

In this section we demonstrate that the primary system (3.1) can be reduced to a system of m equations in m unknown currents.

Consider the simple transistor amplifier of Figure 3.1. Ignoring the biasing circuitry, the small-signal equivalent network of the amplifier is shown in Figure 3.2(a). The associated directed graph is given in Figure 3.2(b), with v_g and R_1 considered as a single edge and $g_m R_4 v$ and R_4 as another for convenience. As before, denote by v_k and i_k the voltage and current

Figure 3.1
A common-emitter transistor amplifier with biasing circuitry.

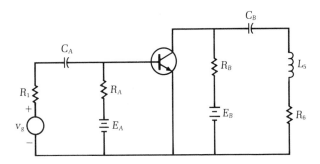

Figure 3.2
The small-signal equivalent network (a) of the transistor amplifier of Figure 3.1 and its associated directed graph (b).

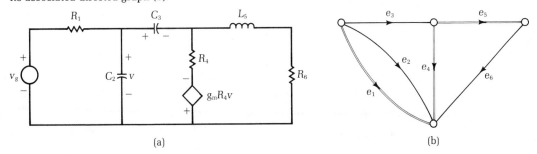

(a)   (b)

of edge e_k ($k = 1, 2, \ldots, 6$). The characteristics of the network elements are described by the following v-i relations:

$$v_1 = v_g + R_1 i_1 \tag{3.2a}$$

$$i_2 = C_2 \frac{dv_2}{dt} = C_2 p v_2 \tag{3.2b}$$

$$i_3 = C_3 \frac{dv_3}{dt} = C_3 p v_3 \tag{3.2c}$$

$$v_4 = R_4 i_4 - g_m R_4 v_2 \tag{3.2d}$$

$$v_5 = L_5 \frac{di_5}{dt} = L_5 p i_5 \tag{3.2e}$$

$$v_6 = R_6 i_6 \tag{3.2f}$$

To express v_2 and v_3 in terms of i_2 and i_3, we integrate (3.2b) and (3.2c) from initial time t_0 to time t to give

$$v_2(t) = \frac{1}{C_2} \int_{t_0}^t i_2(\tau)\, d\tau + v_2(t_0) \tag{3.3a}$$

$$v_3(t) = \frac{1}{C_3} \int_{t_0}^t i_3(\tau)\, d\tau + v_3(t_0) \tag{3.3b}$$

It is convenient to use the symbol p^{-1} or $1/p$ to denote the definite integral

$$p^{-1} i = \frac{1}{p} i = \int_{t_0}^t i(\tau)\, d\tau \tag{3.4}$$

The symbol p^{-1} or $1/p$ is called the *integral operator*. Using it the element v-i relations can be written in matrix form as follows:

$$
\begin{bmatrix} v_1 \\ v_2 \\ v_3 \\ v_4 \\ v_5 \\ v_6 \end{bmatrix}
=
\begin{bmatrix}
R_1 & 0 & 0 & 0 & 0 & 0 \\
0 & \dfrac{1}{C_2 p} & 0 & 0 & 0 & 0 \\
0 & 0 & \dfrac{1}{C_3 p} & 0 & 0 & 0 \\
0 & -\dfrac{g_m R_4}{C_2 p} & 0 & R_4 & 0 & 0 \\
0 & 0 & 0 & 0 & L_5 p & 0 \\
0 & 0 & 0 & 0 & 0 & R_6
\end{bmatrix}
\begin{bmatrix} i_1 \\ i_2 \\ i_3 \\ i_4 \\ i_5 \\ i_6 \end{bmatrix}
+
\begin{bmatrix} v_g \\ v_2(t_0) \\ v_3(t_0) \\ -g_m R_4 v_2(t_0) \\ 0 \\ 0 \end{bmatrix}
\tag{3.5}
$$

or more compactly as

$$\mathbf{v}(t) = \mathbf{Z}(p)\mathbf{i}(t) + \mathbf{e}(t) \tag{3.6}$$

where $\mathbf{Z}(p)$ is the *branch-impedance matrix operator* and $\mathbf{e}(t)$ is the *branch voltage-source vector*, which includes the contribution of the initial conditions. Observe that in converting the hybrid *v-i* equations (3.2) to those of (3.5) or (3.6) with the branch voltages appearing exclusively on the left-hand side and branch currents on the right, the original set of differential equations becomes a set of simultaneous *integrodifferential* equations—that is, equations involving unknown functions, some of their derivatives, and some of their integrals. In the following we shall present a systematic method for obtaining the loop system of integrodifferential equations for any linear time-invariant network. These equations are necessary for the computation of the complete response of a given network to a given input and a given set of initial conditions. Before we do this, we emphasize that $\mathbf{Z}(p)$ is a matrix of the operators p and p^{-1}. Care should be taken to see that variables are not moved across the differentiation or integration sign by a careless interchange of the order of factors containing variable coefficients. However, in many respects the operator p can be handled as though it were a simple algebraic entity. For instance, for any positive integers m and n

$$p^m p^n = p^{m+n} \tag{3.7a}$$

and for any real numbers α_1, α_2, β_1, and β_2

$$(\alpha_1 p^n + \beta_1)(\alpha_2 p^m + \beta_2) = \alpha_1 \alpha_2 p^{m+n} + \alpha_1 \beta_2 p^n + \alpha_2 \beta_1 p^m + \beta_1 \beta_2 \tag{3.7b}$$

On the other hand, $pp^{-1} \neq p^{-1}p$. For example, applying pp^{-1} to a function f yields

$$pp^{-1}f = \frac{d}{dt}\left[\int_{t_0}^{t} f(\tau)\,d\tau \right] = f(t) \tag{3.8a}$$

whereas

$$p^{-1}pf = \int_{t_0}^{t} \left[\frac{df(\tau)}{d\tau} \right] d\tau = f(\tau)\Big|_{t_0}^{t} = f(t) - f(t_0) \tag{3.8b}$$

With these preliminaries, we now proceed to derive the loop equations. Using (3.6) to replace (3.1c), the primary system (3.1) becomes

$$\mathbf{Q}\mathbf{i}(t) = \mathbf{0} \tag{3.9a}$$

$$\mathbf{B}\mathbf{v}(t) = \mathbf{0} \tag{3.9b}$$

$$\mathbf{v}(t) = \mathbf{Z}(p)\mathbf{i}(t) + \mathbf{e}(t) \tag{3.9c}$$

To eliminate $\mathbf{v}(t)$, we substitute (3.9c) in (3.9b) and obtain

$$\mathbf{B}\mathbf{Z}(p)\mathbf{i}(t) + \mathbf{B}\mathbf{e}(t) = \mathbf{0} \tag{3.10}$$

The most general solution satisfying (3.9a) is given by

$$\mathbf{i}(t) = \mathbf{B}'\mathbf{i}_m(t) \tag{3.11}$$

as demonstrated in (2.47), where $\mathbf{i}_m(t)$ is an arbitrary m-vector called the *loop-current vector*. To continue the elimination process, we substitute (3.11) in (3.10), giving the *loop system of equations*:

$$\mathbf{BZ}(p)\mathbf{B}'\mathbf{i}_m(t) = -\mathbf{Be}(t) \tag{3.12}$$

or

$$\mathbf{Z}_m(p)\mathbf{i}_m(t) = \mathbf{e}_m(t) \tag{3.13}$$

where

$$\mathbf{Z}_m(p) = \mathbf{BZ}(p)\mathbf{B}' \tag{3.14a}$$

$$\mathbf{e}_m(t) = -\mathbf{Be}(t) \tag{3.14b}$$

The coefficient matrix $\mathbf{Z}_m(p)$ is called the *loop-impedance matrix operator* and the m-vector $\mathbf{e}_m(t)$, the *loop voltage-source vector*. Equation (3.13) represents a system of m integrodifferential equations in m loop currents as defined by the elements of $\mathbf{i}_m(t)$. The reason for the name *loop current* is that the elements of $\mathbf{i}_m(t)$ can be identified physically as a set of circulating currents around the set of circuits defined by the rows of the basis circuit matrix $\mathbf{B}$. In the particular situation, when the meshes of a planar directed graph are used in formulating the loop equations, the term *mesh current* is frequently used in lieu of loop current, and *loop analysis* becomes *mesh analysis*.

The actual solution of the loop system of equations (3.13) will be deferred to the following chapters. Once the loop-current vector $\mathbf{i}_m(t)$ is known, the branch-current vector $\mathbf{i}(t)$ can be found immediately from (3.11) and the branch-voltage vector $\mathbf{v}(t)$ from (3.9c). Thus all branch voltages and currents are known once the loop currents are ascertained.

We illustrate the above results by the following examples.

EXAMPLE 3.1
Consider the amplifier network and its associated directed graph of Figure 3.2. Choose tree $T = e_1e_4e_5$. The f-circuit matrix with respect to T is found to be

$$
\mathbf{B}_f = \begin{array}{c} \\ \\ \end{array} \begin{array}{cccccc} e_1 & e_2 & e_3 & e_4 & e_5 & e_6 \\ \left[\begin{array}{cccccc} -1 & 1 & 0 & 0 & 0 & 0 \\ -1 & 0 & 1 & 1 & 0 & 0 \\ 0 & 0 & 0 & -1 & 1 & 1 \end{array}\right] \end{array} \tag{3.15}
$$

The branch-impedance matrix operator $\mathbf{Z}(p)$ and the branch voltage-source vector $\mathbf{e}(t)$ were determined earlier and are given in (3.5). Substituting these in (3.14) yields

$$\mathbf{Z}_m(p) = \mathbf{B}_f \mathbf{Z}(p) \mathbf{B}'_f = \begin{bmatrix} \dfrac{1}{C_2 p} + R_1 & R_1 & 0 \\[3mm] R_1 - \dfrac{g_m R_4}{C_2 p} & R_1 + R_4 + \dfrac{1}{C_3 p} & -R_4 \\[3mm] \dfrac{g_m R_4}{C_2 p} & -R_4 & R_4 + R_6 + L_5 p \end{bmatrix}$$

$$(3.16)$$

$$\mathbf{e}_m(t) = -\mathbf{B}\mathbf{e}(t) = \begin{bmatrix} v_g - v_2(t_0) \\[2mm] g_m R_4 v_2(t_0) - v_3(t_0) + v_g \\[2mm] -g_m R_4 v_2(t_0) \end{bmatrix} \qquad (3.17)$$

Writing $\mathbf{i}'_m(t) = [i_{m1}, i_{m2}, i_{m3}]$, the loop system of equations becomes

$$\begin{bmatrix} R_1 + \dfrac{1}{C_2 p} & R_1 & 0 \\[3mm] R_1 - \dfrac{g_m R_4}{C_2 p} & R_1 + R_4 + \dfrac{1}{C_3 p} & -R_4 \\[3mm] \dfrac{g_m R_4}{C_2 p} & -R_4 & R_4 + R_6 + L_5 p \end{bmatrix} \begin{bmatrix} i_{m1} \\[2mm] i_{m2} \\[2mm] i_{m3} \end{bmatrix} = \begin{bmatrix} v_g - v_2(t_0) \\[2mm] v_g + g_m R_4 v_2(t_0) - v_3(t_0) \\[2mm] -g_m R_4 v_2(t_0) \end{bmatrix}$$

$$(3.18)$$

The loop currents i_{m1}, i_{m2}, and i_{m3} can be interpreted physically as the circulating currents around the three f-circuits $e_1 e_2$, $e_1 e_3 e_4$, and $e_4 e_5 e_6$ defined by the tree $T = e_1 e_4 e_5$ as depicted in Figure 3.3. Once the loop currents are determined, the branch currents and voltages are found immediately from (3.11) and (3.9c), as follows:

$$\mathbf{i}(t) = \mathbf{B}'_f \mathbf{i}_m(t) = \begin{bmatrix} -1 & -1 & 0 \\ 1 & 0 & 0 \\ 0 & 1 & 0 \\ 0 & 1 & -1 \\ 0 & 0 & 1 \\ 0 & 0 & 1 \end{bmatrix} \begin{bmatrix} i_{m1} \\ i_{m2} \\ i_{m3} \end{bmatrix} = \begin{bmatrix} -i_{m1} - i_{m2} \\ i_{m1} \\ i_{m2} \\ i_{m2} - i_{m3} \\ i_{m3} \\ i_{m3} \end{bmatrix} \qquad (3.19)$$

Figure 3.3
Physical interpretation of
the loop currents as the
circulating currents
around the f-circuits
defined by the tree $e_1 e_4 e_5$.

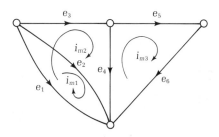

$$\mathbf{v}(t) = \mathbf{Z}(p)\mathbf{i}(t) + \mathbf{e}(t) = \begin{bmatrix} -R_1(i_{m1} + i_{m2}) + v_g \\[2mm] \dfrac{1}{C_2} \displaystyle\int_{t_0}^{t} i_{m1}(\tau)\,d\tau + v_2(t_0) \\[2mm] \dfrac{1}{C_3} \displaystyle\int_{t_0}^{t} i_{m2}(\tau)\,d\tau + v_3(t_0) \\[2mm] R_4(i_{m2} - i_{m3}) - g_m R_4\left[v_2(t_0) + \dfrac{1}{C_2} \displaystyle\int_{t_0}^{t} i_{m1}(\tau)\,d\tau \right] \\[2mm] L_5 \dfrac{di_{m3}}{dt} \\[2mm] R_6 i_{m3} \end{bmatrix}$$

$$(3.20)$$

EXAMPLE 3.2

We determine the mesh system of equations for the same transistor amplifier network of Figure 3.2(a). The meshes of the associated directed graph of Figure 3.2(b) are illustrated in Figure 3.4. The basis circuit matrix of these meshes is found to be

$$\mathbf{B} = \begin{bmatrix} -1 & 1 & 0 & 0 & 0 & 0 \\ 0 & -1 & 1 & 1 & 0 & 0 \\ 0 & 0 & 0 & -1 & 1 & 1 \end{bmatrix} \qquad (3.21)$$

Using this in conjunction with (3.5), we obtain the mesh system of equations from (3.13) and (3.14):

$$\begin{bmatrix} R_1 + \dfrac{1}{C_2 p} & -\dfrac{1}{C_2 p} & 0 \\[3mm] -(g_m R_4 + 1)\dfrac{1}{C_2 p} & R_4 + (g_m R_4 + 1)\dfrac{1}{C_2 p} + \dfrac{1}{C_3 p} & -R_4 \\[3mm] g_m R_4 \dfrac{1}{C_2 p} & -R_4 - g_m R_4 \dfrac{1}{C_2 p} & R_4 + R_6 + L_5 p \end{bmatrix} \begin{bmatrix} i_{m1} \\ i_{m2} \\ i_{m3} \end{bmatrix}$$

$$= \begin{bmatrix} v_g - v_2(t_0) \\ (1 + g_m R_4)v_2(t_0) - v_3(t_0) \\ -g_m R_4 v_2(t_0) \end{bmatrix} \qquad (3.22)$$

3-1.1   Writing Loop Equations by Inspection

The general formulation of loop equations as discussed in the foregoing is very important for two reasons. First, it is completely general and works in all cases, and, second, it is readily amenable to automatic computations.

Figure 3.4
The meshes associated with the directed graph of Figure 3.2(b).

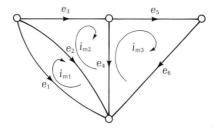

Figure 3.5
The replacement of a capacitor with initial voltage $v_C(t_0)$ by a capacitor of the same capacitance without initial voltage in series with a constant voltage source $v_C(t_0)$.

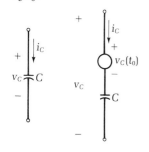

In the case of networks made up only of resistors, capacitors, self-inductors, and independent and dependent sources, the loop equations can be written by inspection. For a self-inductor of inductance L, the term Lp is called the *impedance operator*. Likewise, the term $1/Cp$ is the *impedance operator* for a capacitor of capacitance C. To avoid getting bogged down in notation, we shall demonstrate the procedure by way of an example.

EXAMPLE 3.3

Let us write the loop system of integrodifferential equations for the network of Figure 3.2(a)

Step 1. Replace each controlled or dependent source by an independent source, and replace each capacitor with initial voltage $v_C(t_0)$ by a capacitor of the same capacitance without initial voltage in series with a constant voltage source $v_C(t_0)$, as shown in Figure 3.5, the justification being that

$$v_C(t) = \frac{1}{C} \int_{t_0}^{t} i_C(\tau)\, d\tau + v_C(t_0) \tag{3.23}$$

For the transistor amplifier network of Figure 3.2(a), the resulting network is presented in Figure 3.6, where the controlled voltage source $g_m R_4 v$ is replaced by the independent voltage source v_α.

Step 2. Select a set of f-circuits or the meshes of a planar network and write the loop equations for the modified network obtained in Step 1 as follows:

$$\tilde{\mathbf{Z}}_m(p)\tilde{\mathbf{i}}_m(t) = \tilde{\mathbf{e}}_m(t) \tag{3.24}$$

Figure 3.6
The modified network used to write the mesh equations (3.25).

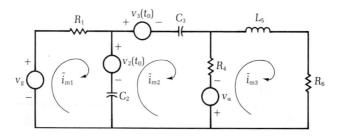

a. Call $\tilde{z}_{ii}$ the diagonal element of $\tilde{\mathbf{Z}}_m(p)$ in the ith row and ith column, and define the *impedance operator* as Lp for a self-inductor of inductance L, $1/Cp$ for a capacitor of capacitance C, and R for a resistor of resistance R. Then $\tilde{z}_{ii}$ is the sum of all impedance operators for the inductors, capacitors, and resistors in the ith circuit.

b. Call $\tilde{z}_{ij}$ the element of $\tilde{\mathbf{Z}}_m(p)$ in the ith row and jth column, $i \neq j$. Then $\tilde{z}_{ij}$ is equal to the sum of all impedance operators common to circuits i and j and traversed by these oriented circuits in the same direction minus the sum of all impedance operators common to circuits i and j and traversed by these circuits in the opposite directions.

c. Call $\tilde{e}_{mk}$ the kth row element of $\tilde{\mathbf{e}}_m(t)$. Then $\tilde{e}_{mk}$ is the algebraic sum of all the source voltages in circuit k with the positive sign associated with voltage rise in the direction of the oriented circuit k.

For the network of Figure 3.6 with the meshes chosen as indicated, the mesh equations are found to be

$$
\begin{bmatrix}
R_1 + \dfrac{1}{C_2 p} & -\dfrac{1}{C_2 p} & 0 \\[2ex]
-\dfrac{1}{C_2 p} & R_4 + \dfrac{1}{C_2 p} + \dfrac{1}{C_3 p} & -R_4 \\[2ex]
0 & -R_4 & R_4 + R_6 + L_5 p
\end{bmatrix}
\begin{bmatrix}
\tilde{i}_{m1} \\[2ex]
\tilde{i}_{m2} \\[2ex]
\tilde{i}_{m3}
\end{bmatrix}
=
\begin{bmatrix}
v_g - v_2(t_0) \\[2ex]
v_\alpha + v_2(t_0) - v_3(t_0) \\[2ex]
-v_\alpha
\end{bmatrix}
$$

$$(3.25)$$

Observe that the coefficient matrix for this class of networks is always symmetrical. For example, $\tilde{z}_{11}$ denotes the sum of resistance (impedance operator) R_4 and impedance operator $1/C_2 p$ in circuit i_{m1}. The element common to circuits i_{m1} and i_{m2} is the impedance operator $1/C_2 p$, which is traversed by the oriented circuits i_{m1} and i_{m2} in the opposite directions. Thus $\tilde{z}_{12} = -1/C_2 p$. The algebraic sum of all the source voltages v_g and $v_2(t_0)$ in circuit i_{m1} is $v_g - v_2(t_0)$, giving $\tilde{e}_{m1} = v_g - v_2(t_0)$.

Step 3. Express each controlled source, which was treated as an independent source in the modified network, in terms of the loop currents and independent sources.

For the network of Figure 3.2(a), we have

$$v_\alpha = g_m R_4 v = g_m R_4 \frac{1}{C_2 p} (\tilde{i}_{m1} - \tilde{i}_{m2}) + g_m R_4 v_2(t_0) \qquad (3.26)$$

Step 4. Substitute the expressions obtained in Step 3 in the loop equations of the modified network and rearrange terms.

For the network of Figure 3.2(a), we substitute (3.26) in (3.25) and rearrange terms, yielding (3.22) after removing the tilde.

The following example will demonstrate that for a linear resistive network the loop equations consist of a system of m linearly independent linear algebraic equations in the m unknown loop currents. A simple matrix inversion will yield the desired loop-current vector.

EXAMPLE 3.4

Figure 3.7(a) is a common-emitter transistor feedback amplifier with a resistor R_f connecting from the output to the input to provide external feedback. Assume that the transistor is described by its hybrid parameters. The low-frequency small-signal equivalent network of the amplifier is presented in Figure 3.7(b). The element values are listed below:

$$
\begin{aligned}
h_{fe} &= 50 & h_{ie} &= 1\text{ k}\Omega \\
h_{re} &= 2.5 \times 10^{-4} & h_{oe} &= 25\ \mu\text{mho} \\
R_1 &= 10\text{ k}\Omega & R_2 &= 4\text{ k}\Omega \\
R_f &= 40\text{ k}\Omega
\end{aligned}
\tag{3.27}
$$

Replace the source combination $h_{fe}i_b$ and $1/h_{oe}$ by the Thévenin equivalent. The resulting network is shown in Figure 3.8. We follow the four steps outlined in the foregoing to write down the loop equations by inspection.

Step 1. The modified network of Figure 3.8 is illustrated in Figure 3.9 where the two controlled voltage sources $h_{re}v_2$ and $h_{fe}i_b/h_{oe}$ are now treated as independent voltage sources v_α and v_β.

Figure 3.7
A common-emitter transistor feedback amplifier (a) and its low-frequency small-signal equivalent network (b).

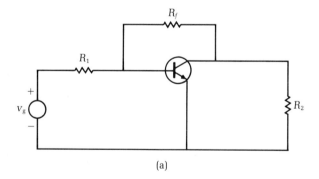

(a)

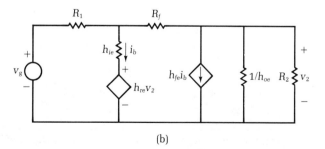

(b)

Figure 3.8
An equivalent network of
the network of Figure
3.7(b).

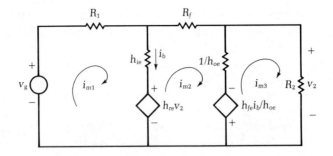

Figure 3.9
The modified network
used to write the loop
equations (3.28).

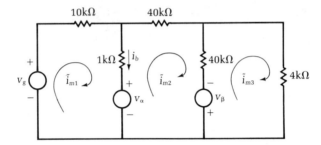

Step 2. Choose the three meshes as indicated in Figure 3.9. The associated loop equations can be written down by inspection as follows:

$$10^3 \begin{bmatrix} 11 & -1 & 0 \\ -1 & 81 & -40 \\ 0 & -40 & 44 \end{bmatrix} \begin{bmatrix} \tilde{i}_{m1} \\ \tilde{i}_{m2} \\ \tilde{i}_{m3} \end{bmatrix} = \begin{bmatrix} v_g - v_\alpha \\ v_\alpha + v_\beta \\ -v_\beta \end{bmatrix} \qquad \textbf{(3.28)}$$

Observe that the coefficient matrix is again symmetrical.

Step 3. We express v_α and v_β in terms of the loop currents and v_g. From Figure 3.8 we have

$$v_\alpha = h_{re}v_2 = h_{re}R_2\tilde{i}_{m3} = 2.5 \times 10^{-4} \times 4 \times 10^3 \, \tilde{i}_{m3} = \tilde{i}_{m3} \qquad \textbf{(3.29a)}$$

$$v_\beta = h_{fe}i_b/h_{oe} = \frac{h_{fe}(\tilde{i}_{m1} - \tilde{i}_{m2})}{h_{oe}} = 2 \times 10^6(\tilde{i}_{m1} - \tilde{i}_{m2}) \qquad \textbf{(3.29b)}$$

Step 4. Substituting (3.29) in (3.28), rearranging terms, and omitting the tilde yield the desired loop equations of Figure 3.8:

$$10^3 \begin{bmatrix} 11 & -1 & 0.001 \\ -2001 & 2081 & -40.001 \\ 2000 & -2040 & 44 \end{bmatrix} \begin{bmatrix} i_{m1} \\ i_{m2} \\ i_{m3} \end{bmatrix} = \begin{bmatrix} v_g \\ 0 \\ 0 \end{bmatrix} \qquad \textbf{(3.30)}$$

To determine the loop currents, we invert the coefficient matrix and obtain

$$\begin{bmatrix} i_{m1} \\ i_{m2} \\ i_{m3} \end{bmatrix} = 10^{-6} \begin{bmatrix} 98.18647 & 0.413564 & 0.373745 \\ 79.26308 & 4.750659 & 4.317088 \\ -788.0969 & 201.4595 & 205.8948 \end{bmatrix} \begin{bmatrix} v_g \\ 0 \\ 0 \end{bmatrix}$$

$$= 10^{-4} v_g \begin{bmatrix} 0.981865 \\ 0.792631 \\ -7.880969 \end{bmatrix} \qquad (3.31)$$

The coefficient matrix in (3.30), being devoid of the operators p and $1/p$, is called the *loop-impedance matrix* or more precisely the *loop resistance matrix*, which is independent of the sources. If there are other sources, they will be included on the right-hand side of (3.30).

To compute the branch voltages and currents, we return to Figure 3.8. Its associated directed graph is shown in Figure 3.10, in which each of the voltage sources and its series resistance is considered as a single edge. The meshes or loop currents are also as indicated. The corresponding basis circuit matrix is found to be

$$\mathbf{B} = \begin{bmatrix} -1 & 0 & 0 & 1 & 0 \\ 0 & 0 & 1 & -1 & 1 \\ 0 & 1 & 0 & 0 & -1 \end{bmatrix} \qquad (3.32)$$

The element v-i relations are described by

$$\begin{bmatrix} v_1 \\ v_2 \\ v_3 \\ v_4 \\ v_5 \end{bmatrix} = 10^3 \begin{bmatrix} 10 & 0 & 0 & 0 & 0 \\ 0 & 4 & 0 & 0 & 0 \\ 0 & 0 & 40 & 0 & 0 \\ 0 & 10^{-3} & 0 & 1 & 0 \\ 0 & 0 & 0 & -2 \cdot 10^3 & 40 \end{bmatrix} \begin{bmatrix} i_1 \\ i_2 \\ i_3 \\ i_4 \\ i_5 \end{bmatrix} + \begin{bmatrix} v_g \\ 0 \\ 0 \\ 0 \\ 0 \end{bmatrix} \qquad (3.33)$$

The branch currents are obtained from (3.11), as follows:

$$\begin{bmatrix} i_1 \\ i_2 \\ i_3 \\ i_4 \\ i_5 \end{bmatrix} = \begin{bmatrix} -1 & 0 & 0 \\ 0 & 0 & 1 \\ 0 & 1 & 0 \\ 1 & -1 & 0 \\ 0 & 1 & -1 \end{bmatrix} \begin{bmatrix} 0.981865 \\ 0.792631 \\ -7.880969 \end{bmatrix} 10^{-4} v_g = 10^{-4} v_g \begin{bmatrix} -0.981865 \\ -7.880969 \\ 0.792631 \\ 0.189234 \\ 8.673600 \end{bmatrix}$$

$$(3.34)$$

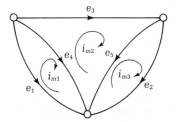

Figure 3.10
The associated directed graph of the network of Figure 3.8.

Finally, to obtain the branch voltages, we substitute (3.34) in (3.33), giving

$$
\begin{bmatrix} v_1 \\ v_2 \\ v_3 \\ v_4 \\ v_5 \end{bmatrix} = v_g \begin{bmatrix} 0.01814 \\ -3.15238 \\ 3.17052 \\ 0.01814 \\ -3.15238 \end{bmatrix}
\tag{3.35}
$$

Once the loop currents are known, the branch currents can be written directly from Figure 3.10 by inspection without the necessity of first obtaining the basis circuit matrix. The loop currents can be interpreted physically as the circulating currents along the circuits used in formulating the loop equations. The branch current of an edge is equal to the algebraic sum of loop currents passing through that edge. We choose the positive sign if the orientations of the loop current and the edge coincide, and the negative sign if they are opposite. In Figure 3.10, for example, $i_4 = i_{m1} - i_{m2}$ and $i_5 = i_{m2} - i_{m3}$. The others can easily be verified from (3.34). Also note that in carrying out the numerical computation, six significant digits were retained in order to get consistent sets of branch currents and voltages. For practical purposes only two or three significant figures are needed.

3-1.2   Voltage Source Transformation

The illustrative examples in the preceding sections all have one common feature in that all the voltage sources, dependent or independent, are connected in series with other passive branches. Quite often we find situations in which a voltage source is connected between two nodes without a series passive branch as shown in Figure 3.11(a). We shall be in some difficulty in writing the v-i equation for this source since its current is unknown. This difficulty may be overcome by performing an operation called the *e-shift* or the *Blakesley transformation*. The procedure is valid for both independent and dependent sources.

Consider the voltage source v_k of Figure 3.11(a). This source can be shifted forward through node b into each of the other branches connected to the node, preserving its reference orientation as shown in Figure 3.11(b). Alternatively, the voltage source v_k can be shifted backward through node a into each of the other branches connected to node a as shown in Figure 3.11(c). We observe the validity of the above transformations by examining the loop equations of the network before and after the change. As can be seen from (3.14b), the branch-voltage sources of $\mathbf{e}(t)$ are included in the loop equations only as the product $-\mathbf{Be}(t)$ $[=\mathbf{e}_m(t)]$. This means that the actual branch distribution of the voltage sources is not important; only their circuit distribution $\mathbf{e}_m(t)$ matters. Consequently, the loop equations remain

Figure 3.11
Portions of a network illustrating the operations of the e-shift.

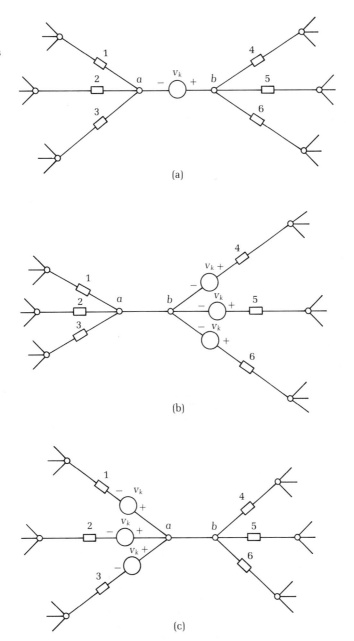

(a)

(b)

(c)

unaltered. The e-shift is really a current-invariant transformation, meaning that the branch currents are invariant under the transformation.

We illustrate the above procedure by the following example.

Figure 3.12

A voltage-shunt feedback amplifier (a) and its low-frequency small-signal
equivalent network (b).

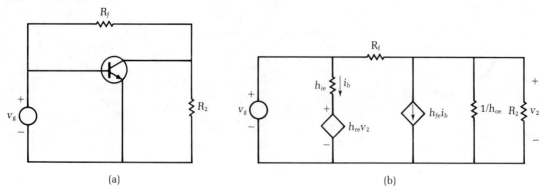

(a)   (b)

Figure 3.13

The networks obtained
from that of Figure 3.12(b)
by performing the
operation of shifting the
voltage source v_g forward
(a) or backward (b).

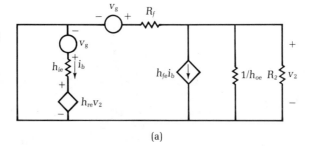

(a)

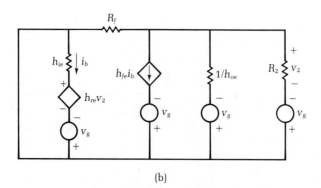

(b)

EXAMPLE 3.5

Consider the voltage-shunt feedback amplifier of Figure 3.12(a), the equivalent network of which is shown in Figure 3.12(b). The voltage source v_g
can be shifted forward as shown in Figure 3.13(a) or backward as in Figure

Figure 3.14
The networks resulting
from the redrawing of
those of Figure 3.13.

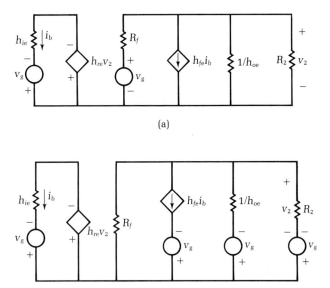

(a)

(b)

3.13(b). With the exception of the currents of the shifted voltage sources
v_g, the branch currents remain unaltered for all three networks because
they all have same loop equations provided that corresponding sets of
basis circuits are used. The networks of Figure 3.13 can be redrawn as
those shown in Figure 3.14.

The following example will demonstrate that in loop analysis not all
the current sources need be converted to equivalent voltage sources.

EXAMPLE 3.6
Using the element values of (3.27), the network of Figure 3.7(b) may be
simplified to that of Figure 3.15. Selecting the loop currents as indicated and

Figure 3.15
The simplified network of
Figure 3.7(b) with explicit
element values.

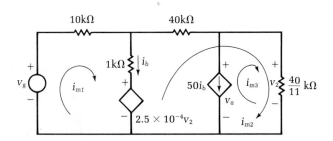

following the steps outlined in the preceding section, the loop equations are found to be

$$
10^3 \begin{bmatrix} 11 & -1 + 1/1100 & 1/1100 \\ -1 & 41 + 40/11 - 1/1100 & 40/11 - 1/1100 \\ 0 & 40/11 & 40/11 \end{bmatrix} \begin{bmatrix} i_{m1} \\ i_{m2} \\ i_{m3} \end{bmatrix} = \begin{bmatrix} v_g \\ 0 \\ v_a \end{bmatrix} \quad \textbf{(3.36)}
$$

where the voltage v_a across the controlled current source $50i_b$ is unknown. However, in this case the loop current i_{m3} is related to other two loop currents by the relation

$$
i_{m3} = -50i_b = -50(i_{m1} - i_{m2}) \quad \textbf{(3.37)}
$$

Substituting this in (3.36) and collecting the terms, the first two equations become

$$
\frac{10^4}{11} \begin{bmatrix} 12.05 & -1.049 \\ -201.05 & 249.049 \end{bmatrix} \begin{bmatrix} i_{m1} \\ i_{m2} \end{bmatrix} = \begin{bmatrix} v_g \\ 0 \end{bmatrix} \quad \textbf{(3.38)}
$$

yielding

$$
\begin{bmatrix} i_{m1} \\ i_{m2} \end{bmatrix} = 10^{-4} \begin{bmatrix} 0.981865 & 0.004136 \\ 0.792631 & 0.047507 \end{bmatrix} \begin{bmatrix} v_g \\ 0 \end{bmatrix} = 10^{-4} v_g \begin{bmatrix} 0.981865 \\ 0.792631 \end{bmatrix} \quad \textbf{(3.39)}
$$

and from (3.37)

$$
i_{m3} = -50(i_{m1} - i_{m2}) = -9.461697 v_g \cdot 10^{-4} \quad \textbf{(3.40)}
$$

Once the loop currents are known, the branch currents and voltages can be determined directly from Figure 3.15; the details are left as an exercise (Problem 3.3). As before, we retain six significant digits in the computation so that the results can be compared with those in (3.34). For practical purposes only two or three significant figures are sufficient.

3-2 THE CUTSET SYSTEM OF EQUATIONS

In the present section we show that the primary system of equations (3.1) can be reduced to a system of r equations in r unknowns. The development is exactly parallel to the loop system, and therefore we can afford to keep our discussion shorter than that of the preceding section.

Again we start our development by considering the simple transistor feedback amplifier of Figure 3.1, the equivalent network of which is shown in Figure 3.2(a). For our purposes we convert the two voltage sources to Norton equivalent current sources. The resulting network and its associated directed graph are presented in Figure 3.16. Our first step is to express the

Figure 3.16
An equivalent network (a)
of Figure 3.2(a) and its
associated directed graph
(b).

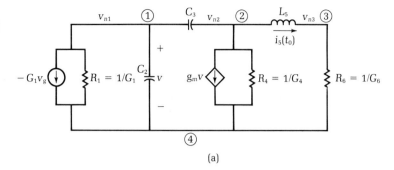

(a)

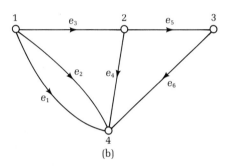

(b)

branch currents in terms of the branch voltages and independent sources
and initial conditions. From Figure 3.16(a) we have

$$i_1 = G_1 v_1 - G_1 v_g$$
$$i_2 = C_2 p v_2$$
$$i_3 = C_3 p v_3$$
$$i_4 = G_4 v_4 + g_m v_2 \tag{3.41}$$
$$i_5 = \frac{1}{L_5 p} v_5 + i_5(t_0)$$
$$i_6 = G_6 v_6$$

which can be written in matrix notation as

$$
\begin{bmatrix} i_1 \\ i_2 \\ i_3 \\ i_4 \\ i_5 \\ i_6 \end{bmatrix}
=
\begin{bmatrix}
G_1 & 0 & 0 & 0 & 0 & 0 \\
0 & C_2 p & 0 & 0 & 0 & 0 \\
0 & 0 & C_3 p & 0 & 0 & 0 \\
0 & g_m & 0 & G_4 & 0 & 0 \\
0 & 0 & 0 & 0 & \dfrac{1}{L_5 p} & 0 \\
0 & 0 & 0 & 0 & 0 & G_6
\end{bmatrix}
\begin{bmatrix} v_1 \\ v_2 \\ v_3 \\ v_4 \\ v_5 \\ v_6 \end{bmatrix}
+
\begin{bmatrix} -G_1 v_g \\ 0 \\ 0 \\ 0 \\ i_5(t_0) \\ 0 \end{bmatrix}
\tag{3.42}
$$

or more compactly

$$\mathbf{i}(t) = \mathbf{Y}(p)\mathbf{v}(t) + \mathbf{j}(t) \qquad (3.43)$$

where $\mathbf{Y}(p)$ is the *branch-admittance matrix operator* and $\mathbf{j}(t)$ is the *branch current-source vector* containing the contribution of the initial conditions. Thus with (3.43) replacing (3.1c), the primary system (3.1) becomes

$$\mathbf{Qi}(t) = \mathbf{0} \qquad (3.44a)$$

$$\mathbf{Bv}(t) = \mathbf{0} \qquad (3.44b)$$

$$\mathbf{i}(t) = \mathbf{Y}(p)\mathbf{v}(t) + \mathbf{j}(t) \qquad (3.44c)$$

To eliminate $\mathbf{i}(t)$ we substitute (3.44c) in (3.44a) and obtain

$$\mathbf{QY}(p)\mathbf{v}(t) + \mathbf{Qj}(t) = \mathbf{0} \qquad (3.45)$$

The most general solution satisfying (3.44b) is from (2.46)

$$\mathbf{v}(t) = \mathbf{Q}'\mathbf{v}_c(t) \qquad (3.46)$$

where $\mathbf{v}_c(t)$ is an arbitrary r-vector called the *cutset-voltage vector*. The reason for the name *cutset voltage* is that the elements of $\mathbf{v}_c(t)$ can be identified or interpreted physically as a set of voltages associated with the cutsets defined by the rows of $\mathbf{Q}$. For some choices of the cutsets, the cutset voltages become the voltages between some pairs of nodes in the network and are known as the *node-pair voltages*.

To continue the elimination process, we substitute (3.46) in (3.45) and and obtain the *cutset system of equations*:

$$\mathbf{QY}(p)\mathbf{Q}'\mathbf{v}_c(t) = -\mathbf{Qj}(t) \qquad (3.47)$$

or

$$\mathbf{Y}_c(p)\mathbf{v}_c(t) = \mathbf{j}_c(t) \qquad (3.48)$$

where

$$\mathbf{Y}_c(p) = \mathbf{QY}(p)\mathbf{Q}' \qquad (3.49a)$$

$$\mathbf{j}_c(t) = -\mathbf{Qj}(t) \qquad (3.49b)$$

The coefficient matrix $\mathbf{Y}_c(p)$ is known as the *cutset-admittance matrix operator*, and the r-vector $\mathbf{j}_c(t)$ the *cutset current-source vector*. Equation (3.48) represents a system of r integrodifferential equations in r cutset voltages.

As before, once the cutset voltage vector $\mathbf{v}_c(t)$ is known, the branch-voltage vector $\mathbf{v}(t)$ can be determined immediately from (3.46) and the branch-current vector $\mathbf{i}(t)$ from (3.44c). The technique is general and is applicable to any linear time-invariant network.

Two special situations of the cutset system will be discussed because of their specific forms and their wide use in solving practical problems: the *nodal analysis* and the *f-cutset analysis*. The methods are also general and can be applied to any network.

3-2.1 The Nodal System of Equations

The matrix $\mathbf{Q}$ in (3.47) is any basis cutset matrix. In particular, if we choose $\mathbf{Q} = \mathbf{A}$, the basis incidence matrix, (3.47) can be rewritten as

$$\mathbf{Y}_n(p)\mathbf{v}_n(t) = \mathbf{j}_n(t) \tag{3.50}$$

where

$$\mathbf{Y}_n(p) = \mathbf{A}\mathbf{Y}(p)\mathbf{A}' \tag{3.51a}$$

$$\mathbf{j}_n(t) = -\mathbf{A}\mathbf{j}(t) \tag{3.51b}$$

The coefficient matrix $\mathbf{Y}_n(p)$ is referred to as the *node-admittance matrix operator*, and $\mathbf{j}_n(t)$ the *nodal current-source vector*. The elements of $\mathbf{v}_n(t)$ can now be identified as the *node-to-datum voltages* in the network.

Once the *node-to-datum voltage* vector $\mathbf{v}_n(t)$ is known, the branch voltages and currents are determined immediately by

$$\mathbf{v}(t) = \mathbf{A}'\mathbf{v}_n(t) \tag{3.52a}$$

$$\mathbf{i}(t) = \mathbf{Y}(p)\mathbf{v}(t) + \mathbf{j}(t) \tag{3.52b}$$

The above results are illustrated by the following examples.

EXAMPLE 3.7

We formulate the nodal system of equations for the network of Figure 3.16(a). From Figure 3.16(b) the basis incidence matrix with node 4 chosen as the reference node is given by

$$\mathbf{A} = \begin{bmatrix} 1 & 1 & 1 & 0 & 0 & 0 \\ 0 & 0 & -1 & 1 & 1 & 0 \\ 0 & 0 & 0 & 0 & -1 & 1 \end{bmatrix} \tag{3.53}$$

The branch-admittance matrix operator $\mathbf{Y}(p)$ and the branch current-source vector $\mathbf{j}(t)$ were computed earlier and are given in (3.42). Substituting these into the equations (3.51) yields

$$\mathbf{Y}_n(p) = \mathbf{A}\mathbf{Y}(p)\mathbf{A}' = \begin{bmatrix} G_1 + (C_2 + C_3)p & -C_3 p & 0 \\ g_m - C_3 p & G_4 + C_3 p + \dfrac{1}{L_5 p} & -\dfrac{1}{L_5 p} \\ 0 & -\dfrac{1}{L_5 p} & G_6 + \dfrac{1}{L_5 p} \end{bmatrix} \tag{3.54}$$

$$\mathbf{j}_n(t) = -\mathbf{A}\mathbf{j}(t) = \begin{bmatrix} G_1 V_g \\ -i_5(t_0) \\ i_5(t_0) \end{bmatrix} \tag{3.55}$$

Writing $\mathbf{v}'_n(t) = [v_{n1}, v_{n2}, v_{n3}]$, the nodal system of equations becomes

$$\begin{bmatrix} G_1 + (C_2 + C_3)p & -C_3 p & 0 \\ g_m - C_3 p & G_4 + C_3 p + \dfrac{1}{L_5 p} & -\dfrac{1}{L_5 p} \\ 0 & -\dfrac{1}{L_5 p} & G_6 + \dfrac{1}{L_5 p} \end{bmatrix} \begin{bmatrix} v_{n1} \\ v_{n2} \\ v_{n3} \end{bmatrix} = \begin{bmatrix} G_1 V_g \\ -i_5(t_0) \\ i_5(t_0) \end{bmatrix} \tag{3.56}$$

The node-to-datum voltages v_{n1}, v_{n2}, and v_{n3} can be identified as the voltages from nodes 1, 2, and 3 to the reference node 4, respectively, and hence the name. Once the nodal voltages are known, the branch voltages and currents are determined immediately from the branch equations (3.52) or directly from Figure 3.16 by inspection:

$$\mathbf{v}(t) = \mathbf{A}'\mathbf{v}_n(t) = \begin{bmatrix} 1 & 0 & 0 \\ 1 & 0 & 0 \\ 1 & -1 & 0 \\ 0 & 1 & 0 \\ 0 & 1 & -1 \\ 0 & 0 & 1 \end{bmatrix} \begin{bmatrix} v_{n1} \\ v_{n2} \\ v_{n3} \end{bmatrix} = \begin{bmatrix} v_{n1} \\ v_{n1} \\ v_{n1} - v_{n2} \\ v_{n2} \\ v_{n2} - v_{n3} \\ v_{n3} \end{bmatrix} \tag{3.57}$$

$$\mathbf{i}(t) = \mathbf{Y}(p)\mathbf{v}(t) + \mathbf{j}(t) = \begin{bmatrix} G_1(v_{n1} - v_g) \\ C_2 \dfrac{dv_{n1}}{dt} \\ C_3 \dfrac{d}{dt}(v_{n1} - v_{n2}) \\ g_m v_{n1} + G_4 v_{n2} \\ \dfrac{1}{L_5} \displaystyle\int_{t_0}^{t} [v_{n2}(\tau) - v_{n3}(\tau)]\, d\tau + i_5(t_0) \\ G_6 v_{n3} \end{bmatrix} \tag{3.58}$$

Thus the central problem is to determine the nodal voltages from the three integrodifferential nodal equations of (3.56).

As in the loop case, if a network is comprised only of resistors, capacitors, self-inductors, and independent and dependent sources, the nodal equations

Figure 3.17
The replacement of an
inductor with initial
current $i_L(t_0)$ by an inductor of the same inductance
without initial current in
parallel with a constant
current source $i_L(t_0)$.

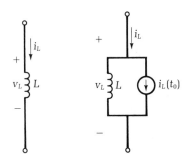

can be written by inspection. For a capacitor of capacitance C, the term
Cp is called the *admittance operator*. Likewise, the term $1/Lp$ is the *admittance operator* for a self-inductor of inductance L, and G is the *admittance
operator* for a resistor of resistance $1/G$. The procedure for writing the nodal
equations by inspection is as follows:

Step 1.   Replace each controlled source by an independent source, and
replace each inductor with initial current $i_L(t_0)$ by an inductor of the same
inductance without initial current in parallel with a constant current
source $i_L(t_0)$, as shown in Figure 3.17. The justification is that

$$i_L(t) = \frac{1}{L} \int_{t_0}^{t} v_L(\tau)\, d\tau + i_L(t_0) \tag{3.59}$$

Step 2.   Write the nodal equations for the modified network obtained in
Step 1 as follows:

$$\mathbf{Y}_n(p)\tilde{\mathbf{v}}_n(t) = \tilde{\mathbf{j}}_n(t) \tag{3.60}$$

  a.   Call $\tilde{y}_{ii}$ the diagonal element of $\tilde{\mathbf{Y}}_n(p)$ in the ith row and ith column.
       Then $\tilde{y}_{ii}$ is the sum of all admittance operators for the resistors,
       capacitors, and inductors connected to node i.
  b.   Call $\tilde{y}_{ij}$, $i \neq j$, the element of $\tilde{\mathbf{Y}}_n(p)$ in the ith row and jth column.
       Then $\tilde{y}_{ij}$ is the negative of the sum of all admittance operators
       connecting node i and node j.
  c.   Call $\tilde{j}_{nk}$ the kth row element of $\tilde{\mathbf{j}}_n(t)$. Then $\tilde{j}_{nk}$ is the algebraic sum of
       all source currents entering node k.

Step 3.   Express each controlled source, which was treated as an independent source in the modified network, in terms of the nodal voltages and
independent sources.

Step 4.   Substitute the expressions obtained in Step 3 in the nodal equations
of the modified network and rearrange terms.

We illustrate these steps by the following examples.

EXAMPLE 3.8

We write the nodal equations for the network of Figure 3.16(a) by inspection following the steps outlined above.

Step 1. Replace the controlled current source $g_m v$ by an independent source i_α and connect in parallel with L_5 a constant current source $i_5(t_0)$ to obtain the modified network as illustrated in Figure 3.18.

Step 2. Apply the rules outlined above and write the nodal equations for the network of Figure 3.18 by inspection as follows:

$$
\begin{bmatrix}
G_1+(C_2+C_3)p & -C_3 p & 0 \\
-C_3 p & G_4+C_3 p+\dfrac{1}{L_5 p} & -\dfrac{1}{L_5 p} \\
0 & -\dfrac{1}{L_5 p} & G_6+\dfrac{1}{L_5 p}
\end{bmatrix}
\begin{bmatrix}
\tilde{v}_{n1} \\ \tilde{v}_{n2} \\ \tilde{v}_{n3}
\end{bmatrix}
=
\begin{bmatrix}
G_1 v_g \\ -i_\alpha - i_5(t_0) \\ i_5(t_0)
\end{bmatrix}
$$

(3.61)

Observe that the coefficient matrix for this class of networks is always symmetrical. For example, $\tilde{y}_{11}$ is the sum of admittance operator or conductance G_1 and admittance operators $C_2 p$ and $C_3 p$ connected to node 1. The element connecting nodes 1 and 2 is the capacitor C_3. Thus $\tilde{y}_{12} = -C_3 p$. The algebraic sum of source currents entering node 1 is $G_1 v_g$, giving $\tilde{j}_{n1} = G_1 v_g$.

Step 3. The controlled current source i_α can be expressed in terms of the nodal voltage $\tilde{v}_{n1}$:

$$
i_\alpha = g_m v = g_m \tilde{v}_{n1}
$$

(3.62)

Step 4. Substitute (3.62) in (3.61) and rearrange terms to obtain the nodal equations of (3.56) after removing the tilde.

Figure 3.18
The modified network used to write the nodal equations (3.61).

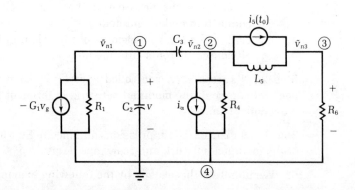

The following example will demonstrate that for a linear resistive network the nodal equations consist of a system of r independent *linear algebraic* equations in the r unknown nodal voltages, which can be solved by a simple matrix inversion. Also we indicate that in nodal analysis not all the voltage sources need be converted to equivalent current sources.

EXAMPLE 3.9
Consider the common-emitter feedback amplifier and its equivalent network of Figure 3.7. We use nodal analysis to solve this network problem. For our purposes we convert the source combination v_g and R_1 to its Norton equivalent, as shown in Figure 3.19(a). Using the element values of (3.27) and choosing node 4 as the reference potential, we obtain the network of Figure 3.19(b) with all conductances given in μmhos. Following the four steps outlined above yields the nodal equation

$$10^{-3}\begin{bmatrix} 1.125 & -0.025 & -1 \\ 50 - 0.025 & 0.3 & -50 \\ -1 & 0 & 1 \end{bmatrix}\begin{bmatrix} v_{n1} \\ v_{n2} \\ v_{n3} \end{bmatrix} = \begin{bmatrix} v_g \cdot 10^{-4} \\ 0 \\ -i_x \end{bmatrix} \quad (3.63)$$

where i_x is the unknown current of the controlled voltage source as indicated in Figure 3.19(b). However, the nodal voltage v_{n3} in the present situation is related to nodal voltage v_{n2} by

$$v_{n3} = 2.5 \cdot 10^{-4} v_{n2} \quad (3.64)$$

Figure 3.19
An equivalent network (a) of Figure 3.7(b) with element values shown in (b) where all conductances are given in μmhos.

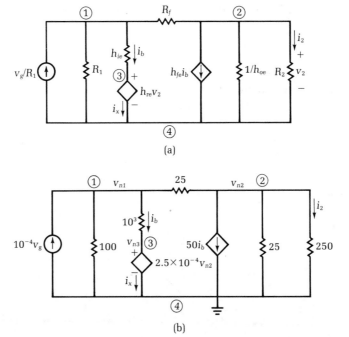

(a)

(b)

Substituting the nodal voltage (3.64) in (3.63) and rearranging terms, the first two equations become

$$\begin{bmatrix} 11.25 & -0.2525 \\ 499.75 & 2.8750 \end{bmatrix} \begin{bmatrix} v_{n1} \\ v_{n2} \end{bmatrix} = \begin{bmatrix} v_g \\ 0 \end{bmatrix} \qquad (3.65)$$

giving

$$\begin{bmatrix} v_{n1} \\ v_{n2} \end{bmatrix} = \begin{bmatrix} 0.01814 & 0.00159 \\ -3.15238 & 0.07096 \end{bmatrix} \begin{bmatrix} v_g \\ 0 \end{bmatrix} = \begin{bmatrix} 0.01814 \\ -3.15238 \end{bmatrix} v_g \qquad (3.66)$$

Once the nodal voltages are known, the branch voltages and currents can be determined by inspection directly from Figure 3.19(b). The results are the same as given in (3.34) and (3.35). Note that the source combination of v_g and R_1 in Figure 3.7(b) has been converted to equivalent Norton current source in Figure 3.19(b).

3-2.2 The f-cutset System of Equations

The f-cutset system is a generalization of the nodal system discussed in the foregoing. In the present case we choose $\mathbf{Q} = \mathbf{Q}_f$, a basis f-cutset matrix. Recall that in writing the basis f-cutset matrix $\mathbf{Q}_f$ we already have chosen a tree for which the r f-cutsets are defined. Using this $\mathbf{Q}_f$ in (3.47) gives the *f-cutset system of equations*. One of the important advantages of this choice is that the elements of the resulting $\mathbf{j}_c(t) [= - \mathbf{Q}_f \mathbf{j}(t)]$ can be interpreted physically as the voltages of the tree branches of the defining tree for the f-cutsets.

EXAMPLE 3.10
We write the f-cutset system for the network of Figure. 3.16(a), the directed graph of which is shown in Figure. 3.16(b). Choosing the tree $T = e_2 e_4 e_5$, the three f-cutsets $e_1 e_2 e_3$, $e_3 e_4 e_6$, and $e_5 e_6$ defined with respect to T are depicted in Figure 3.20. The corresponding f-cutset matrix is found to be

Figure 3.20
Representation of the f-cutsets defined for the tree $e_2 e_4 e_5$ by drawing a broken line across each f-cutset.

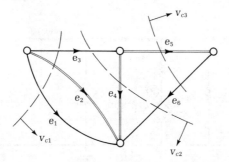

$$
\mathbf{Q}_f = \begin{array}{cccccc}
 & e_1 & e_2 & e_3 & e_4 & e_5 & e_6 \\
\end{array}
\begin{bmatrix}
1 & 1 & 1 & 0 & 0 & 0 \\
0 & 0 & -1 & 1 & 0 & 1 \\
0 & 0 & 0 & 0 & 1 & -1
\end{bmatrix}
\tag{3.67}
$$

The branch-admittance matrix operator $\mathbf{Y}(p)$ and the branch current-source vector $\mathbf{j}(t)$ for this network were given in (3.42). Substitutïng these in (3.49) yields

$$
\mathbf{Y}_c(p) = \mathbf{Q}_f\mathbf{Y}(p)\mathbf{Q}_f' =
\begin{bmatrix}
G_1 + (C_2 + C_3)p & -C_3 p & 0 \\[2mm]
g_m - C_3 p & G_4 + G_6 + C_3 p & -G_6 \\[2mm]
0 & -G_6 & G_6 + \dfrac{1}{L_5 p}
\end{bmatrix}
\tag{3.68}
$$

$$
\mathbf{j}_c(t) = -\mathbf{Q}_f\mathbf{j}(t) =
\begin{bmatrix}
G_1 v_g \\
0 \\
-i_5(t_0)
\end{bmatrix}
\tag{3.69}
$$

Writing $\mathbf{v}_c'(t) = [v_{c1}, v_{c2}, v_{c3}]$, the f-cutset system of equations becomes

$$
\begin{bmatrix}
G_1 + (C_2 + C_3)p & -C_3 p & 0 \\[2mm]
g_m - C_3 p & G_4 + G_6 + C_3 p & -G_6 \\[2mm]
0 & -G_6 & G_6 + \dfrac{1}{L_5 p}
\end{bmatrix}
\begin{bmatrix}
v_{c1} \\
v_{c2} \\
v_{c3}
\end{bmatrix}
=
\begin{bmatrix}
G_1 v_g \\
0 \\
-i_5(t_0)
\end{bmatrix}
\tag{3.70}
$$

The cutset voltages c_{c1}, v_{c2}, and, v_{c3} can be interpreted physically as the voltages associated with the f-cutsets $e_1 e_2 e_3$, $e_3 e_4 e_6$, and $e_5 e_6$, respectively. The branch voltage of an edge is equal to the algebraic sum of the cutset voltages containing this edge. We choose the positive sign if the orientations of the f-cutset and the edge coincide and the negative sign if they are opposite. Thus v_{c1}, v_{c2}, and v_{c3} are also the branch voltages of the defining tree branches e_2, e_4, and e_5 of the f-cutsets, respectively. In Figure 3.20 edge e_3 is contained in f-cutsets v_{c1} and v_{c2}. Since the orientations of f-cutset v_{c1} and e_3 coincide and those of v_{c2} and e_3 are opposite, we have $v_3 = v_{c1} - v_{c2}$. Likewise edge e_6, which is contained in f-cutsets v_{c2} and v_{c3}, has the same orientation as v_{c2} and is opposite to that of v_{c3}. This gives $v_6 = v_{c2} - v_{c3}$.

Alternatively the branch voltages can be determined from (3.46) and are given by

$$
\mathbf{v}(t) = \mathbf{Q}'_f \mathbf{v}_c(t) =
\begin{bmatrix}
1 & 0 & 0 \\
1 & 0 & 0 \\
1 & -1 & 0 \\
0 & 1 & 0 \\
0 & 0 & 1 \\
0 & 1 & -1
\end{bmatrix}
\begin{bmatrix}
v_{c1} \\
v_{c2} \\
v_{c3}
\end{bmatrix}
=
\begin{bmatrix}
v_{c1} \\
v_{c1} \\
v_{c1} - v_{c2} \\
v_{c2} \\
v_{c3} \\
v_{c2} - v_{c3}
\end{bmatrix}
\tag{3.71}
$$

confirming that $v_6 = v_{c2} - v_{c3}$. Thus the tree branch voltages v_2, v_4, and v_5 are the cutset voltages v_{c1}, v_{c2}, and v_{c3}, respectively. Once the branch voltages are known, the branch currents can either be computed directly from (3.42) or by inspection from Figure 3.16:

$$
\mathbf{i}(t) = \mathbf{Y}(p)\mathbf{v}(t) + \mathbf{j}(t) =
\begin{bmatrix}
G_1 v_{c1} - G_1 v_g \\[6pt]
C_2 \dfrac{dv_{c1}}{dt} \\[6pt]
C_3 \dfrac{d}{dt}(v_{c1} - v_{c2}) \\[6pt]
g_m v_{c1} + G_4 v_{c2} \\[6pt]
\dfrac{1}{L_5} \displaystyle\int_{t_0}^{t} v_{c3}(\tau)\, d\tau + i_5(t_0) \\[6pt]
G_6(v_{c2} - v_{c3})
\end{bmatrix}
\tag{3.72}
$$

Observe that in Figure 3.20 if we choose tree $T = e_2 e_4 e_6$, the corresponding f-cutset matrix becomes the incidence matrix of (3.53) and the cutset voltages become the nodal voltages of Figure 3.16. At any rate, both the nodal analysis and the f-cutset analysis are special cases of the more general cutset analysis presented in (3.48). The special cases are important in that, apart from the resulting simplicity, the methods are general and can be applied to any linear time-invariant network.

As in the nodal analysis, if a network is comprised only of resistors, capacitors, self-inductors, and independent and dependent sources, the cutset equations can be written by inspection, as follows:

Step 1. This step is the same as in Step 1 of nodal analysis.

Step 2. Write the cutset equations for the modified network obtained in Step 1, as follows:

$$
\tilde{\mathbf{Y}}_c(p)\tilde{\mathbf{v}}_c(t) = \tilde{\mathbf{j}}_c(t)
\tag{3.73}
$$

    a.   Call $\tilde{y}_{cii}$ the diagonal element of $\tilde{\mathbf{Y}}_c(p)$ in the ith row and ith column. Then $\tilde{y}_{cii}$ is the sum of all conductances and admittance operators for the inductors and capacitors contained in the cutset i.

b. Call $\tilde{y}_{cij}$, $i \neq j$, the element of $\tilde{\mathbf{Y}}_c(p)$ in the ith row and jth column. Then $\tilde{y}_{cij}$ is equal to the sum of all admittance operators common to cutsets i and j and traversed by these oriented cutsets in the same direction minus the sum of all admittance operators common to cutsets i and j and traversed by these cutsets in the opposite directions.

c. Call $\tilde{j}_{ck}$ the kth row element of $\tilde{\mathbf{j}}_c(t)$. Then $\tilde{j}_{ck}$ is the algebraic sum of all source currents contained in cutset k. Choose the positive sign if the orientation of the cutset k and the reference direction of a source current are *opposite* and the negative sign otherwise.

Step 3. Express each controlled source, which was treated as an independent source in the modified network, in terms of the cutset voltages and independent sources.

Step 4. Substitute the expressions obtained in Step 3 in the cutset equations of the modified network and rearrange terms.

The procedure outlined above is valid for any set of r linearly independent cutsets in general, and for the f-cutsets in particular. We illustrate this procedure by the following examples.

EXAMPLE 3.11

We write the cutset equations for the network of Figure 3.16(a), using the directed graph of Figure 3.20.

Step 1. The modified network is given in Figure 3.18.

Step 2. Applying the rules outlined above for the three f-cutsets as illustrated in Figure 3.20, the cutset equations for the modified network of Figure 3.18 can be written by inspection, as follows:

$$
\begin{bmatrix}
G_1 + (C_2 + C_3)p & -C_3 p & 0 \\
-C_3 p & G_4 + G_6 + C_3 p & -G_6 \\
0 & -G_6 & G_6 + \dfrac{1}{L_5 p}
\end{bmatrix}
\begin{bmatrix}
\tilde{v}_{c1} \\ \tilde{v}_{c2} \\ \tilde{v}_{c3}
\end{bmatrix}
=
\begin{bmatrix}
G_1 v_g \\ -i_\alpha \\ -i_5(t_0)
\end{bmatrix}
$$

(3.74)

Observe again that the coefficient matrix is symmetrical. This is valid in general because, with the replacement of the controlled current source $g_m v$ by an independent current source i_α, the modified network is reciprocal. For example, $\tilde{y}_{c22}$ is the sum of admittance operators or conductances G_4 and G_6 and the admittance operator $C_3 p$ contained in the f-cutset $e_3 e_4 e_6$. The element common to f-cutsets $e_1 e_2 e_3$ and $e_3 e_4 e_6$ is the admittance operator $C_3 p$, which is traversed by these cutsets in the opposite directions. Thus $\tilde{y}_{c12} = -C_3 p$. The algebraic sum of the source currents in f-cutset $e_3 e_4 e_6$ is $-i_\alpha$. We pick the negative sign because the f-cutset orientation and the reference direction of i_α coincide. Thus $\tilde{j}_{c2} = -i_\alpha$.

Step 3. The controlled current source i_α can be expressed in terms of the cutset voltage $\tilde{v}_{c1}$ as

$$i_\alpha = g_m v = g_m \tilde{v}_{c1} \tag{3.75}$$

Step 4. Substituting (3.75) in (3.74) and rearranging terms, we obtain the cutset equations of (3.70) after removing the tilde.

EXAMPLE 3.12

We use cutset analysis to solve the network problem of Figure 3.19, which is redrawn in Figure 3.21(a) with all the conductances given in μmhos. The associated directed graph is given in Figure 3.21(b). Because the controlling voltages are v_1 and v_2, it is convenient to choose a tree containing edges e_1 and e_2. One of such choices is depicted in Figure 3.21(b) together with the f-cutsets defined by this tree. The cutset voltages v_{c1}, v_{c2}, and v_{c3}, which are also the branch voltages v_1, v_2, and v_3 associated with the f-cutsets $e_1 e_4 e_5$, $e_3 e_4 e_5$, and $e_2 e_5 e_6$, respectively, are also indicated in the figure—that is, $v_1 = v_{c1}$, $v_2 = v_{c2}$, and $v_3 = v_{c3}$. Following the four steps described in the foregoing, we obtain the cutset equations for the network of Figure 3.21(a):

$$10^{-6} \begin{bmatrix} 25 + 1000 + 100 & -25 & 25 + 100 \\ 50000 - 25 & 25 + 275 & -25 \\ 25 + 100 & -25 & 25 + 100 \end{bmatrix} \begin{bmatrix} v_{c1} \\ v_{c2} \\ v_{c3} \end{bmatrix} = \begin{bmatrix} v_g \\ 0 \\ v_g - i_3 \end{bmatrix} 10^{-4} \tag{3.76}$$

Note that the current i_3 of the controlled voltage source is unknown. However, the cutset voltage v_{c3} is related to v_{c2} by

$$v_{c3} = 2.5 \times 10^{-4} v_2 = 2.5 \times 10^{-4} v_{c2} \tag{3.77}$$

Figure 3.21
An equivalent network (a) of Figure 3.19(b) with all conductances given in μmhos and its associated directed graph (b).

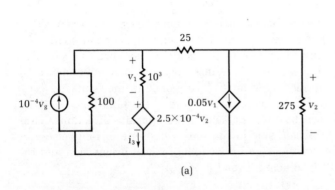

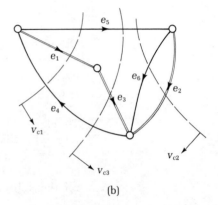

(a)   (b)

Substituting (3.77) in (3.76) and rearranging terms, the first two equations reduce to

$$\begin{bmatrix} 11.25 & -0.24969 \\ 499.75 & 2.99994 \end{bmatrix} \begin{bmatrix} v_{c1} \\ v_{c2} \end{bmatrix} = \begin{bmatrix} v_g \\ 0 \end{bmatrix} \tag{3.78}$$

The coefficient matrix can be inverted to give

$$\begin{bmatrix} v_{c1} \\ v_{c2} \end{bmatrix} = \begin{bmatrix} 0.01892 & 0.00158 \\ -3.15238 & 0.07096 \end{bmatrix} \begin{bmatrix} v_g \\ 0 \end{bmatrix} = v_g \begin{bmatrix} 0.01892 \\ -3.15238 \end{bmatrix} \tag{3.79}$$

obtaining from (3.77)

$$v_{c3} = -0.00079 v_g \tag{3.80}$$

The branch voltages are determined immediately from (3.46) in conjunction with Figure 3.21(b):

$$\mathbf{v}(t) = \mathbf{Q}_f' \mathbf{v}_c(t) = \begin{bmatrix} 1 & 0 & 0 \\ 0 & 1 & 0 \\ 0 & 0 & 1 \\ -1 & 0 & -1 \\ 1 & -1 & 1 \\ 0 & 1 & 0 \end{bmatrix} \begin{bmatrix} 0.01892 \\ -3.15238 \\ -0.00079 \end{bmatrix} v_g = \begin{bmatrix} 0.01892 \\ -3.15238 \\ -0.00079 \\ -0.01813 \\ 3.17052 \\ -3.15238 \end{bmatrix} v_g \tag{3.81}$$

Alternatively we can use the physical interpretation of the cutset voltages to compute the branch voltages. For instance, edge e_5 is contained in all three f-cutsets such that the orientation of the edge and that of the f-cutsets v_{c1} and v_{c3} coincide, whereas the direction of e_5 and that of f-cutset v_{c2} are opposite, giving

$$v_5 = v_{c1} + v_{c3} - v_{c2} = 3.17052 v_g \tag{3.82}$$

confirming the fifth row element of (3.81). Likewise the branch currents can either be determined from (3.44c) or directly from Figure 3.21(a) in conjunction with the directed graph of Figure 3.21(b). Since $i_1 = i_3$, we have

$$\mathbf{i}(t) = \mathbf{Y}(p)\mathbf{v}(t) + \mathbf{j}(t) = \begin{bmatrix} 0.01892 \\ -0.86691 \\ 0.01892 \\ 0.09818 \\ 0.07926 \\ 0.94617 \end{bmatrix} 10^{-3} v_g \tag{3.83}$$

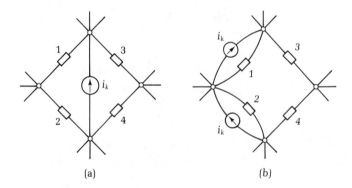

(a)   (b)

3-2.3 Current Source Transformation

In loop analysis we find that it is convenient to consider an independent or
dependent voltage source and its series passive branch as a single edge,
performing the e-shift if necessary. Likewise in cutset analysis it is advantageous to view an independent or dependent current source and a passive
parallel branch as a single edge. Quite often we find situations where a
current source is connected between two nodes without any passive branches
connecting in parallel with the source, such as shown in Figure 3.22(a). This
difficulty can easily be circumvented by an *i-shift*, as illustrated in Figure
3.22(b). The operation effectively places the current source i_k in parallel with
each of the branches that form a circuit with i_k. In this way the net current
at each node remains unaltered. As can be seen from (3.48), the branch-
source currents in $\mathbf{j}(t)$ are included in the cutset equations only as the product $-\mathbf{Q}\mathbf{j}(t)$ $[=\mathbf{j}_c(t)]$. This means that the actual branch distribution of the
current sources is not important; only the cutset distribution $\mathbf{j}_c(t)$ matters.
The current-source transformation depicted in Figure 3.22 keeps the cutset
distribution of current sources invariant. Consequently, the cutset voltages
remain unaltered. With the exception of the voltages of the shifted current
sources, the i-shift is a voltage-invariant transformation.

EXAMPLE 3.13
Consider the voltage-shunt transistor feedback amplifier of Figure 3.7(a).
Instead of using the hybrid parameters for the transistor of Figure 3.7(b),
in this current transformation approach we use the low-frequency small-
signal equivalent T-model. The resulting network is presented in Figure 3.23.
Following the four steps outlined in Section 3-2.1, the nodal equations are
obtained from Figure 3.23 as

$$
\begin{bmatrix}
G_1 + g_e + G_f & -g_e & -G_f \\
(\alpha - 1)g_e & g_b + (1 - \alpha)g_e & 0 \\
-\alpha g_e - G_f & \alpha g_e & G_2 + G_f
\end{bmatrix}
\begin{bmatrix}
v_{n1} \\
v_{n2} \\
v_{n3}
\end{bmatrix}
=
\begin{bmatrix}
G_1 v_g \\
0 \\
0
\end{bmatrix}
\qquad \textbf{(3.84)}
$$

Figure 3.23
An equivalent network of the voltage-shunt transistor feedback amplifier of Figure 3.7(a).

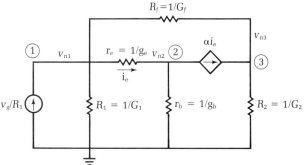

Figure 3.24
Networks resulting from shifting the controlled current source αi_e in Figure 3.23 to (a) r_b and R_2 and (b) r_e anf R_f.

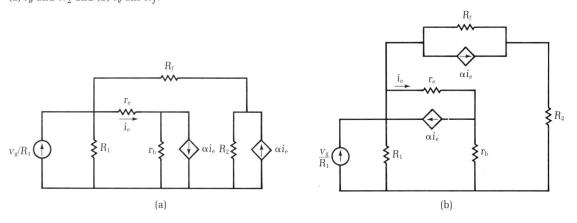

(a)   (b)

The controlled current source αi_e can be shifted to r_b and R_2 as shown in Figure 3.24(a) or to r_e and R_f as in Figure 3.24(b), which are redrawn in Figure 3.25 for clarity. Suppose that we write the nodal equations for the network of Figure 3.25(b) using the simplified procedure. The resulting equations are again given by (3.84), verifying the validity of the i-shift being a voltage-invariant transformation.

EXAMPLE 3.14
An active filter is presented in Figure 3.26(a). Assume that the operational amplifier has infinite input resistance and zero output resistance with finite voltage gain α. The equivalent network of the filter is shown in Figure 3.26(b). We wish to determine the output to input voltage ratio v_o/v_g.
    In order to prepare the network before writing nodal equations, we

Figure 3.25
Networks obtained by
redrawing those shown in
Figure 3.24.

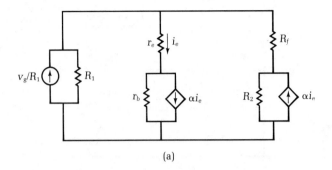

(a)

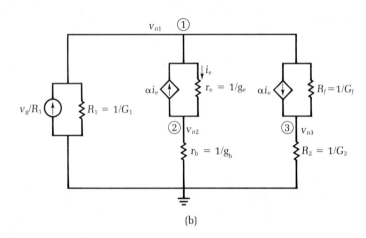

(b)

apply the e-shift to the controlled source αv by shifting it backward to resistors R_3 and R_4. The resulting network is shown in Figure 3.27(a). Now converting the two voltage sources to their Norton equivalents, we obtain the final network of Figure 3.27(b) with R denoting the parallel combination of the resistances R_1, R_3, and R_5. Let $G_i = 1/R_i$ ($i = 1, 2, 3, 4, 5$). The nodal equations can be written by inspection from Figure 3.27(b):

$$\begin{bmatrix} G_1 + G_2 + G_3 + G_5 & \alpha G_3 - G_2 \\ -G_2 & G_2 + (1 + \alpha)G_4 \end{bmatrix} \begin{bmatrix} v_{n1} \\ v_{n2} \end{bmatrix} = \begin{bmatrix} G_1 v_g \\ 0 \end{bmatrix} \tag{3.85}$$

giving

$$\begin{bmatrix} v_{n1} \\ v_{n2} \end{bmatrix} = \frac{G_1 v_g}{\Delta} \begin{bmatrix} G_2 + (1 + \alpha)G_4 \\ G_2 \end{bmatrix} \tag{3.86}$$

where

$$\Delta = (G_1 + G_2 + G_3 + G_5)(G_2 + G_4 + \alpha G_4) + G_2(\alpha G_3 - G_2) \tag{3.87}$$

Since

$$v_o = -\alpha v = -\alpha v_{n2} \tag{3.88}$$

Figure 3.26
An active filter (a) and its equivalent network (b).

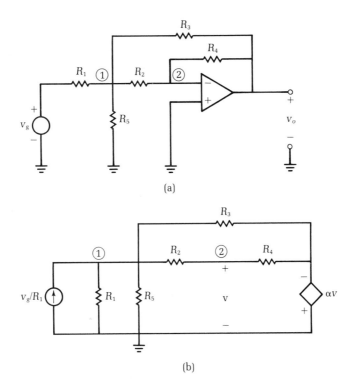

(a)

(b)

Figure 3.27
Network (a) resulting from that of Figure 3.26(b) by shifting the controlled source αv backward to resistors R_3 and R_4, and its equivalent network (b) obtained by converting the two voltage sources to their Norton equivalents.

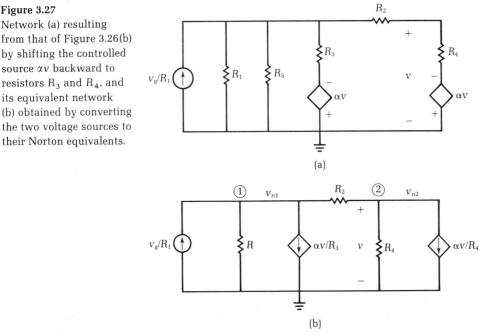

(a)

(b)

from (3.86) we have

$$\frac{v_o}{v_g} = \frac{-\alpha G_1 G_2}{(G_1 + G_2 + G_3 + G_5)(G_2 + G_4 + \alpha G_4) + G_2(\alpha G_3 - G_2)} \tag{3.89}$$

In the limit, as $\alpha \to \infty$, the operational amplifier being ideal, the ratio reduces to

$$\frac{v_o}{v_g} = \frac{-G_1 G_2}{G_4(G_1 + G_2 + G_3 + G_5) + G_2 G_3} \tag{3.90}$$

As in loop analysis, if $\mathbf{Y}_c(p)$ is devoid of the operators p and $1/p$, the matrix $\mathbf{Y}_c$ is known as the *cutset-admittance matrix*. Likewise if $\mathbf{Y}_n(p)$ is devoid of p and $1/p$, it is called the *node-admittance matrix*. The coefficient matrix of (3.76) is an example of the cutset-admittance matrix. The coefficient matrices of (3.84) and (3.85) are the node-admittance matrices of the networks of Figures 3.23 and 3.27(b). More precisely, they should be called the *cutset-conductance matrix* and *node-conductance matrix*.

3-3 SUMMARY

In this chapter we introduced three secondary systems of network equations known as the *loop system*, *nodal system*, and *cutset system*. The equations amount to a systematic organization of the variables in the primary system so that the number of equations needed in solving the network problem can be reduced. The nodal analysis may be considered as a special case of the more general cutset analysis.

In loop analysis we showed that the primary system can be reduced to a system of m integrodifferential equations in m unknown loop currents. The loop currents possess the physical significance of currents circulating around the defining circuits. In preparing a network for loop analysis, we usually manipulate all the sources until equivalent voltage sources are obtained. We found that it was convenient to consider a voltage source and its series passive branch as a single edge. If no such passive branch exists, it is necessary to perform the *e-shift* operation, which is a current-invariant transformation.

In nodal analysis the primary system is reduced to a system of r integro-differential equations in r unknown nodal voltages. The nodal voltages can be identified as the node-to-datum voltages in the network. In preparing a network for nodal analysis, we manipulate network sources so that only current sources are in the resulting network. In writing the nodal equations, it is advantageous to consider a current source and a passive branch connected in parallel with this source as a single edge. If no such parallel passive branch exists, a simple *i-shift* is required to prepare the network in standard form. The i-shift operation is a voltage-invariant transformation.

The cutset analysis is a generalization of the nodal analysis. It involves a system of r integrodifferential equations in r unknown cutset voltages. The cutset voltages have the physical significance of the voltages associated with the defining cutsets. In the particular situation where a set of f-cutsets is chosen, the cutset voltages are the voltages of the tree branches of the tree used to define the f-cutsets.

The three systematic methods are all general and can be applied to any linear time-invariant network. Then for a given network, which of the three techniques shall we choose? Our choice will be made in terms of such factors as these: (1) Which technique will yield the smallest number of equations? (2) Will the computer be used in the solution scheme? (3) What is the objective of our analysis? One or several voltages and/or currents? All the voltages or currents? Consider, for example, the series connection of a resistor, an inductor, a capacitor, and a voltage source. If we use nodal or cutset analysis, we need two equations after converting the voltage source and the resistor to their Norton equivalent. On the other hand, one loop equation will be sufficient. The reverse is true if a resistor, an inductor, a capacitor, and a current source are connected in parallel. It is always better to know two ways of solving a problem rather than one, for then one can choose a particular approach or combination of approaches so as to solve the network problem at hand in the simplest and most satisfying manner.

The solution of the integrodifferential equations formulated in this chapter has not been discussed. The classical method of solution and the operational method that uses the Laplace transformation will be presented in the following chapters.

Finally, we mention that in addition to the three secondary systems discussed above network equations can also be formulated as a system of first-order differential equations known as the *state equations*. State equations are another extremely important secondary system. Their formulation and solution will be the subject of Chapters 11 and 12.

REFERENCES AND SUGGESTED READING

Balabanian, N., and T. A. Bickart. *Electrical Network Theory.* New York: Wiley, 1969, Chapter 2.

Chan, S. P., S. Y. Chan and S. G. Chan. *Analysis of Linear Networks and Systems.* Reading, Mass.: Addison-Wesley, 1972, Chapter 3.

Chen, W. K. *Applied Graph Theory: Graphs and Electrical Networks.* 2d ed. New York: American Elsevier, 1976, Chapter 2.

Chen, W. K. *Active Network and Feedback Amplifier Theory.* New York: McGraw-Hill, 1980, Chapter 2.

Chua, L. O., and P. M. Lin. *Computer-Aided Analysis of Electronic Circuits: Algorithms & Computational Techniques.* Englewood Cliffs, N.J.: Prentice-Hall, 1975, Chapters 4 and 6.

Desoer, C. A., and E. S. Kuh. *Basic Circuit Theory.* New York: McGraw-Hill, 1969, Chapters 10 and 11.

Jensen, R. W., and B. O. Watkins. *Network Analysis: Theory and Computer Methods.* Englewood Cliffs, N.J.: Prentice-Hall, 1974, Chapter 4.

Peikari, B. *Fundamentals of Network Analysis and Synthesis.* Englewood Cliffs, N.J.: Prentice-Hall, 1974, Chapter 4.

Wing, O. *Circuit Theory with Computer Methods.* New York: Holt, Rinehart & Winston, 1972, Chapter 6.

PROBLEMS

3.1 Consider the networks of Figure 3.28 with all initial conditions as indicated. Obtain in matrix form the following integrodifferential equations by the systematic methods:

(a) The loop equations

(b) The nodal equations

(c) The cutset equations other than the nodal equations obtained in (b)

Figure 3.28
Networks with initial
conditions.

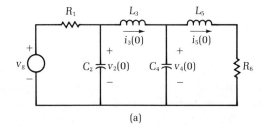

(a)

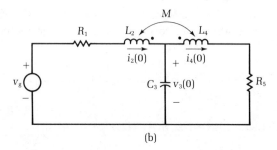

(b)

Figure 3.29
An equivalent network of
a transistor feedback
amplifier in which all
initial conditions are zero.

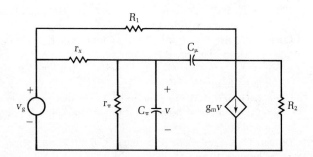

3.2 Write the loop, nodal, and cutset equations by inspection for the network of Figure 3.28(a), making nodal and cutset equations different.

3.3 Using (3.39) and (3.40), compute the branch currents and voltages of the network of Figure 3.15.

3.4 Figure 3.29 is an equivalent network of a transistor feedback amplifier in which all initial conditions are zero. Obtain in matrix form the following integrodifferential equations by the systematic methods:
  (a) The loop equations
  (b) The nodal equations
  (c) The cutset equations formed by a set of f-cutsets

3.5 Write the loop, nodal, and cutset equations of Problem 3.4 by inspection.

3.6 Consider the band-pass active filter of Figure 3.30. Obtain the following integrodifferential equations in matrix form by the systematic methods assuming zero initial conditions. (The operational amplifier is assumed to be ideal.)
  (a) The loop equations
  (b) The nodal equations
  (c) The cutset equations formed by a set of f-cutsets

3.7 Figure 3.31(a) illustrates a current-shunt feedback amplifier. Assume that the two transistors are identical, with the following specifications. (The transistors are described by their hybrid parameters.)

$$h_{ie} = 1.1 \text{ k}\Omega \qquad h_{fe} = 50$$
$$h_{re} = h_{oe} = 0 \qquad R = 3 \text{ k}\Omega$$
$$R_e = 50 \text{ }\Omega \qquad R_f = 1.2 \text{ k}\Omega \qquad \qquad \textbf{(3.91)}$$
$$R_1 = 1.2 \text{ k}\Omega \qquad R_2 = 500 \text{ }\Omega$$

Using these values, a low-frequency small-signal equivalent network for the amplifier is presented in Figure 3.31(b) with all the conductances given in μmhos and $g_m = 455 \times 10^{-4}$ mhos. Determine the branch currents and voltages of the network.

3.8 Obtain the loop integrodifferential equations for the active filter of Figure 3.32 under the following situations, assuming zero initial conditions:
  (a) The operational amplifier has finite input and output resistances with finite gain.
  (b) The operational amplifier has infinite input resistance and zero output resistance with finite gain.
  (c) The operational amplifier is ideal.

3.9 Repeat Problem 3.8 by writing the nodal integrodifferential equations.

3.10 Repeat Problem 3.6 for the network of Figure 2.7(a).

3.11 Repeat Problem 3.6 for the network of Figure 2.30(b).

3.12 Compute the branch currents and voltages for the network of Figure 2.38.

3.13 Repeat Problem 3.6 for the network of Figure 2.53(c).

3.14 Repeat Problem 3.6 for the network of Figure 2.53(d).

Figure 3.30
A band-pass active filter.

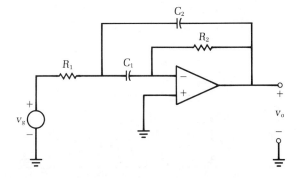

3.15 Repeat Problem 3.6 for the network of Figure 2.55(a).

3.16 A practical version of a field-effect transistor (FET) Colpitts oscillator is shown in Figure 3.33(a). After the biasing circuitry and the radio frequency (RF) choke have been removed, an equivalent network is presented in Figure 3.33(b). Repeat Problem 3.6 for this network.

3.17 A voltage-series feedback amplifier is shown in Figure 3.34(a). Assume that the two transistors are identical and characterized by the hybrid parameters with $h_{ie} = 1.1 \text{ k}\Omega$, $h_{fe} = 50$, and $h_{re} = h_{oe} = 0$. An equivalent network of the amplifier is presented in Figure 3.34(b) with all conductances given in μmhos and $g_m = 455 \cdot 10^{-4}$ mhos. Compute the branch currents and voltages of the network.

Figure 3.31
A current-shunt feedback amplifier (a) and its low-frequency small-signal equivalent network (b) with all conductances given in μmhos.

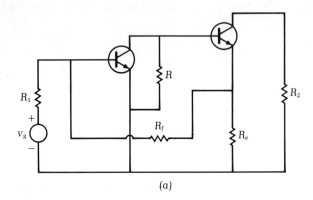

(a)

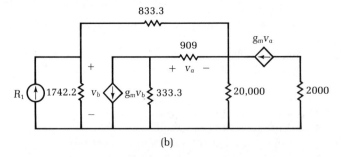

(b)

Figure 3.32
An active filter.

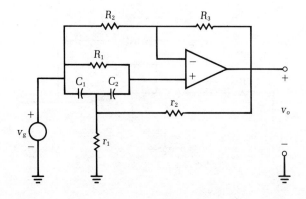

Figure 3.33
A field-effect transistor (FET) Colpitts oscillator (a) and its equivalent network
(b) after the removal of the biasing circuitry and the radio frequency (RF) choke.

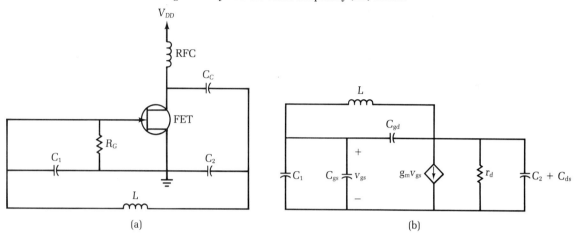

(a)   (b)

Figure 3.34
A voltage-series feedback
amplifier (a) and its
equivalent network (b),
assuming the identical
transistors with all
conductances given in
μmhos.

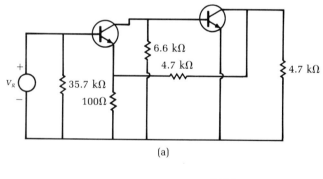

(a)

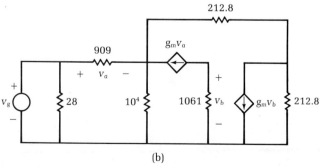

(b)

3.18 Figure 3.35 is a low-pass active filter. Obtain the loop integrodifferential equations for the active filter under the situation that the operational amplifier has infinite input resistance and zero output resistance with finite gain. Assume zero initial conditions.

3.19 Repeat Problem 3.18 by writing the cutset integrodifferential equations with respect to a set of f-cutsets.

3.20 Consider the network of Figure 3.36 with all initial conditions as indicated. Obtain the loop integrodifferential equations by the systematic method with respect to a set of f-circuits.

3.21 Obtain the loop integrodifferential equations for the network of Figure 3.36 by inspection with respect to a set of f-circuits.

3.22 Obtain the nodal integrodifferential equations for the network of Figure 3.36 both by inspection and by the systematic method.

3.23 Determine the cutset integrodifferential equations for the network of Figure 3.36 with respect to a set of f-cutsets both by the systematic method and by inspection.

3.24 Repeat Problems 20 and 21 by choosing a set of meshes.

Figure 3.35
A low-pass active filter.

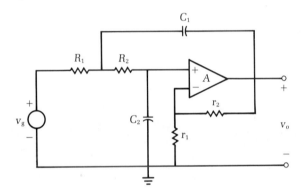

Figure 3.36
A given network with initial conditions.

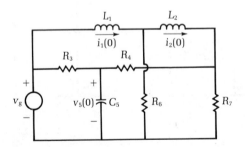

CHAPTER FOUR

SIMULTANEOUS LINEAR DIFFERENTIAL EQUATIONS

The secondary systems of network equations formulated in the preceding chapter are systems of linear simultaneous integrodifferential equations. The classical methods of solution of these equations and the properties of the solutions will be studied in this chapter. The operational method of solution that uses the Laplace transformation will be presented in the following chapter.

We stress the importance of these methods because they are fundamental and effective tools for studying linear time-invariant systems and because so many engineering problems depend on them. However, we shall only develop those properties that serve our present purposes. These properties are important in that they help us understand the system behavior and we must study them carefully.

There are various ways of solving such systems. The most obvious approach is to reduce the equations by successive elimination of the unknowns until a single differential equation in one variable remains. This equation is solved for that variable, and then, working backward, the other variables are determined one by one until the problem is solved. A second method is to convert the system to a system of differential equations and obtain solutions for all the variables at the same time. Finally, the use of the Laplace transformation provides an operational method of solution.

In this chapter we shall attempt through examples to present the second method, leaving the third to Chapter 5, and to study the properties of the solutions.

4-1   SIMULTANEOUS LINEAR DIFFERENTIAL EQUATIONS

In order to develop the theory, we first convert a secondary system to a system of linear simultaneous differential equations, and then study the properties of its solutions.

Consider the following nodal system of integrodifferential equations for the network of Figure 3.16(a), as given by (3.56):

$$
\begin{bmatrix}
G_1 + (C_2 + C_3)p & -C_3 p & 0 \\
g_m - C_3 p & G_4 + C_3 p + \dfrac{1}{L_5 p} & -\dfrac{1}{L_5 p} \\
0 & -\dfrac{1}{L_5 p} & G_6 + \dfrac{1}{L_5 p}
\end{bmatrix}
\begin{bmatrix}
v_{n1} \\ v_{n2} \\ v_{n3}
\end{bmatrix}
=
\begin{bmatrix}
G_1 v_g \\ -i_5(t_0) \\ i_5(t_0)
\end{bmatrix}
$$

(4.1)

If we differentiate the second and third equations of (4.1) and use

$$
\frac{d^k w}{dt^k} = p^k w
$$

(4.2a)

all integral operators $1/p$ will be eliminated and we shall have a system of differential equations:

$$
\begin{bmatrix}
G_1 + (C_2 + C_3)p & -C_3 p & 0 \\
-C_3 p^2 + g_m p & C_3 p^2 + G_4 p + 1/L_5 & -1/L_5 \\
0 & -1/L_5 & G_6 p + 1/L_5
\end{bmatrix}
\begin{bmatrix}
v_{n1} \\ v_{n2} \\ v_{n3}
\end{bmatrix}
=
\begin{bmatrix}
G_1 v_g \\ 0 \\ 0
\end{bmatrix}
$$

(4.2b)

The procedure can be used to convert any secondary system of integrodifferential equations to a system of differential equations. The latter is most conveniently written in matrix notation as

$$
\mathbf{W}(p)\mathbf{x}(t) = \mathbf{f}(t)
$$

(4.3)

We use the neutral symbols $\mathbf{W}(p)$, $\mathbf{x}(t)$ and $\mathbf{f}(t)$ to represent the coefficient matrix of the operator p, the unknown vector $\mathbf{x}(t)$ and the known *forcing* or *excitation vector* $\mathbf{f}(t)$ of the sources in order not to be tilted toward a particular system such as nodal, cutset, or loop equations. Because of the presence of the known source vector $\mathbf{f}(t)$, (4.3) is called *nonhomogeneous*. If $\mathbf{f}(t)$ is identically zero, we have the so-called *homogeneous* equation

$$
\mathbf{W}(p)\mathbf{x}(t) = \mathbf{0}
$$

(4.4)

We shall now establish certain properties of the solutions of a system of differential equations even though the form of those solutions is not known. We shall begin by stating three fundamental properties pertaining to the solutions of the general equations (4.3) and (4.4). These bear a remarkable resemblance to the similar properties of systems of linear algebraic equations, and it may be helpful to keep the analogy in mind.

Property 4.1   *If $\mathbf{x}_1$ and $\mathbf{x}_2$ are any two solutions of (4.4), then*

$$
\mathbf{x}(t) = c_1 \mathbf{x}_1(t) + c_2 \mathbf{x}_2(t)
$$

(4.5)

where c_1 and c_2 are arbitrary constants, is also a solution.

To justify this it is only necessary to substitute the expression (4.5) into the given differential equation and verify that the latter is satisfied:

$$\mathbf{W}(p)\mathbf{x}(t) = \mathbf{W}(p)(c_1\mathbf{x}_1 + c_2\mathbf{x}_2) = c_1\mathbf{W}(p)\mathbf{x}_1 + c_2\mathbf{W}(p)\mathbf{x}_2$$
$$= c_1\mathbf{0} + c_2\mathbf{0} = \mathbf{0} \tag{4.6}$$

Property 4.1 can easily be extended to any number of solutions. It assures us that if we have n solutions of the homogeneous equations (4.4), then we can obtain infinitely many other solutions simply by forming the arbitrary linear combinations of these n solutions. However, it leaves unanswered the important question of whether or not *all* solutions of (4.4) can be obtained in this manner and also the problem of determining the minimal number of such solutions required so that all other solutions can be obtained in this fashion. To answer these questions, we need a few more terms.

By the *order* of a matrix differential equation (4.3) is meant the degree of the determinant of the operator matrix $\mathbf{W}(p)$, regarded as a polynomial in p. A solution of a matrix differential equation of order n containing n arbitrary constants is said to be *complete* if every solution of the differential equation can be obtained from it by assigning suitable values to the constants which appear in it. The determinant of the coefficient matrix of (4.2b), if treated as a polynomial in p, is of degree 4. Therefore (4.2b) is a matrix differential equation of order 4 and its complete solution must contain four arbitrary constants.

The solution of the nonhomogeneous equation (4.3) is based on the following property.

Property 4.2 *If $\mathbf{x}_p$ is any solution whatsoever of the nonhomogeneous equation* (4.3), *and if $\mathbf{x}_h$ is the complete solution of the homogeneous equation* (4.4), *then the complete solution of* (4.3) *is given by*

$$\mathbf{x}(t) = \mathbf{x}_h(t) + \mathbf{x}_p(t) \tag{4.7}$$

To prove this, let $\mathbf{x}_\alpha$ be any solution whatsoever of the nonhomogeneous equation (4.3). Then

$$\mathbf{W}(p)\mathbf{x}_\alpha(t) = \mathbf{f}(t) \tag{4.8}$$

and similarly, since $\mathbf{x}_p$ is also a solution of (4.3),

$$\mathbf{W}(p)\mathbf{x}_p(t) = \mathbf{f}(t) \tag{4.9}$$

Subtracting the last two equations yields

$$\mathbf{W}(p)\mathbf{x}_\alpha(t) - \mathbf{W}(p)\mathbf{x}_p(t) = \mathbf{0} \tag{4.10a}$$

or

$$\mathbf{W}(p)[\mathbf{x}_\alpha(t) - \mathbf{x}_p(t)] = \mathbf{0} \tag{4.10b}$$

Thus the vector $\mathbf{x}_\alpha - \mathbf{x}_p$ satisfies the homogeneous equation (4.4), and hence, by assigning suitable values to the arbitrary constants contained in $\mathbf{x}_h$, we have

$$\mathbf{x}_h(t) = \mathbf{x}_\alpha(t) - \mathbf{x}_p(t) \tag{4.11}$$

or

$$\mathbf{x}_\alpha(t) = \mathbf{x}_h(t) + \mathbf{x}_p(t) \tag{4.12}$$

Thus any solution of (4.3) can be expressed as in (4.7).

The vector $\mathbf{x}_p$, which can be any special solution whatsoever of the nonhomogeneous equation (4.3), is called a *particular integral*[†] of (4.3). The vector $\mathbf{x}_h$, which is the complete solution of the associated homogeneous equation (4.4) obtained from (4.3) by deleting $\mathbf{f}(t)$, is called the *complementary function* of (4.3). In anticipation of these, we have chosen the subscripts p and h in $\mathbf{x}_p$ and $\mathbf{x}_h$ to signify their significance. Thus to carry out the solution of the nonhomogeneous equation (4.3), we follow the steps given below:

Step 1. Delete the forcing function $\mathbf{f}(t)$ from (4.3) and obtain the complete solution $\mathbf{x}_h$ of the resulting homogeneous equation (4.4). The vector $\mathbf{x}_h$ is the complementary function of (4.3).

Step 2. Find one particular solution $\mathbf{x}_p$ of the nonhomogeneous equation (4.3).

Step 3. Add the complementary function $\mathbf{x}_h$ found in Step 1 to the particular solution $\mathbf{x}_p$ found in Step 2 to obtain the complete solution of (4.3) as

$$\mathbf{x}(t) = \mathbf{x}_h(t) + \mathbf{x}_p(t) \tag{4.13}$$

In the following sections, we show how these theoretical steps can be carried out. However, before we do this, we illustrate the above results by the following example.

EXAMPLE 4.1

In the network of Figure 4.1, switch S_1 is open and switch S_2 is closed for $t < 0$. At $t = 0$, switch S_1 is flipped to closed position and switch S_2 is open. We wish to determine the branch voltages and currents of the network.

At $t = 0$, the initial capacitor voltages are evidently $v_1(0-) = 2$ V and $v_2(0-) = 0$. For $t \geq 0$, the equivalent network of Figure 4.1 is shown in Figure 4.2(a). Figure 4.2(b) is the associated directed graph of Figure 4.2(a). Applying the procedure outlined in Section 3-2.1, the nodal differential equation in the form of (4.3) is obtained by inspection as

$$\begin{bmatrix} 3 + 2p & -2 - 2p \\ -2 - 2p & 2 + 3p \end{bmatrix} \begin{bmatrix} v_{n1} \\ v_{n2} \end{bmatrix} = \begin{bmatrix} 1 \\ 0 \end{bmatrix}, \qquad t \geq 0 \tag{4.14}$$

[†] A particular solution is any special solution whatsoever of the nonhomogeneous equation (4.3). A particular integral is also frequently referred to as a *particular solution*.

Figure 4.1
A network with switch S_1 closed and switch S_2 open at $t = 0$.

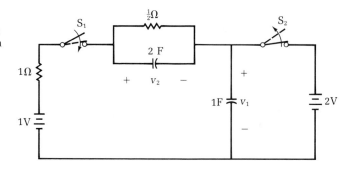

Figure 4.2
An equivalent network (a) of Figure 4.1 for $t \geq 0$ and its associated directed graph (b).

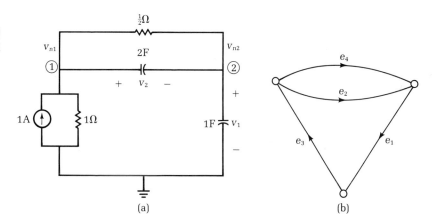

The determinant of the operator matrix, regarded as a polynomial in p, is found to be

$$\det \mathbf{W}(p) = 2p^2 + 5p + 2$$

Thus the matrix differential equation (4.14) is of order 2 because the determinantal polynomial $\det \mathbf{W}(p)$ is of degree 2.

Setting the right-hand side of (4.14) to zero gives the associated homogeneous equation

$$\begin{bmatrix} 3 + 2p & -2 - 2p \\ -2 - 2p & 2 + 3p \end{bmatrix} \begin{bmatrix} v_{n1} \\ v_{n2} \end{bmatrix} = \begin{bmatrix} 0 \\ 0 \end{bmatrix}, \qquad t \geq 0 \tag{4.15}$$

We can show that vectors of the form

$$\mathbf{v}_1(t) = \begin{bmatrix} v_{n1} \\ v_{n2} \end{bmatrix} = c_1 \begin{bmatrix} 1 \\ 2 \end{bmatrix} e^{-0.5t}, \qquad t \geq 0 \tag{4.16a}$$

$$\mathbf{v}_2(t) = \begin{bmatrix} v_{n1} \\ v_{n2} \end{bmatrix} = c_2 \begin{bmatrix} 2 \\ 1 \end{bmatrix} e^{-2t}, \qquad t \geq 0 \tag{4.16b}$$

are solutions of (4.15), where c_1 and c_2 are arbitrary constants. To verify, we substitute (4.16a) in (4.15) and obtain

$$
\begin{bmatrix} 3 + 2p & -2 - 2p \\ -2 - 2p & 2 + 3p \end{bmatrix} \begin{bmatrix} c_1 e^{-0.5t} \\ 2c_1 e^{-0.5t} \end{bmatrix} = \begin{bmatrix} 3c_1 - c_1 - 4c_1 + 2c_1 \\ -2c_1 + c_1 + 4c_1 - 3c_1 \end{bmatrix} e^{-0.5t}
$$

$$
= \begin{bmatrix} 0 \\ 0 \end{bmatrix} \tag{4.17}
$$

Similarly (4.16b) is also easily verified. By Property 4.1 the sum of $\mathbf{v}_1$ and $\mathbf{v}_2$ is also a solution of (4.15):

$$
\mathbf{v}_h(t) = \mathbf{v}_1(t) + \mathbf{v}_2(t) = c_1 \begin{bmatrix} 1 \\ 2 \end{bmatrix} e^{-0.5t} + c_2 \begin{bmatrix} 2 \\ 1 \end{bmatrix} e^{-2t}
$$

$$
= \begin{bmatrix} c_1 e^{-0.5t} + 2c_2 e^{-2t} \\ 2c_1 e^{-0.5t} + c_2 e^{-2t} \end{bmatrix}, \quad t \geq 0 \tag{4.18}
$$

To confirm this, it is necessary only to substitute (4.18) in (4.15) to verify that (4.15) is satisfied. In fact, as will be shown later, (4.18) is the complete solution of (4.15) containing two arbitrary constants c_1 and c_2. Thus (4.18) is the complementary function of the nonhomogeneous equation (4.14).

On the other hand, the vector

$$
\mathbf{v}_p = \begin{bmatrix} 1 \\ 1 \end{bmatrix} \tag{4.19}
$$

is a very special solution of the nonhomogeneous equation (4.14). Therefore it is a particular integral of (4.14). From Property 4.2 the complete solution of the nonhomogeneous equation (4.14) becomes

$$
\begin{bmatrix} v_{n1} \\ v_{n2} \end{bmatrix} = \mathbf{v}_h(t) + \mathbf{v}_p(t) = \begin{bmatrix} 1 + c_1 e^{-0.5t} + 2c_2 e^{-2t} \\ 1 + 2c_1 e^{-0.5t} + c_2 e^{-2t} \end{bmatrix}, \quad t \geq 0 \tag{4.20}
$$

Knowing the nodal voltages v_{n1} and v_{n2}, the capacitor voltages v_1 and v_2 are found to be

$$
\begin{bmatrix} v_1 \\ v_2 \end{bmatrix} = \begin{bmatrix} v_{n2} \\ v_{n1} - v_{n2} \end{bmatrix} = \begin{bmatrix} 1 + 2c_1 e^{-0.5t} + c_2 e^{-2t} \\ -c_1 e^{-0.5t} + c_2 e^{-2t} \end{bmatrix}, \quad t \geq 0 \tag{4.21}
$$

Since voltages across the capacitors cannot change instantaneously, $v_1(0+) = v_1(0-) = 2v$ and $v_2(0+) = v_2(0-) = 0$. Using these in (4.21) gives

$$
2c_1 + c_2 = 2 - 1 = 1 \tag{4.22a}
$$

$$
-c_1 + c_2 = 0 \tag{4.22b}
$$

yielding

$$
c_1 = c_2 = \frac{1}{3} \tag{4.23}
$$

which when combined with (4.21) gives the capacitor voltages

$$
\begin{bmatrix} v_1 \\ v_2 \end{bmatrix} = \begin{bmatrix} 1 + \dfrac{2}{3}\,e^{-0.5t} + \dfrac{1}{3}\,e^{-2t} \\[4mm] -\dfrac{1}{3}\,e^{-0.5t} + \dfrac{1}{3}\,e^{-2t} \end{bmatrix}, \qquad t \geqq 0 \tag{4.24}
$$

Once the capacitor voltages are known, the branch currents and other branch voltages can be determined from Figure 4.2(a) in conjunction with Figure 4.2(b): $v_3 = -v_{n1} = -v_1 - v_2$, $v_4 = v_2$, and

$$
\begin{bmatrix} i_1 \\ i_2 \\ i_3 \\ i_4 \end{bmatrix} = \begin{bmatrix} \dfrac{dv_1}{dt} \\[3mm] 2\dfrac{dv_2}{dt} \\[3mm] 1 - 1v_{n1} \\[3mm] 2v_2 \end{bmatrix} = \begin{bmatrix} -\dfrac{1}{3}\,e^{-0.5t} - \dfrac{2}{3}\,e^{-2t} \\[3mm] \dfrac{1}{3}\,e^{-0.5t} - \dfrac{4}{3}\,e^{-2t} \\[3mm] -\dfrac{1}{3}\,e^{-0.5t} - \dfrac{2}{3}\,e^{-2t} \\[3mm] -\dfrac{2}{3}\,e^{-0.5t} + \dfrac{2}{3}\,e^{-2t} \end{bmatrix}, \qquad t \geqq 0 \tag{4.25}
$$

4-2   HOMOGENEOUS LINEAR DIFFERENTIAL EQUATIONS

In following the theory established in the last section, our task here is to find the complete solution of the matrix homogeneous linear differential equation

$$
\mathbf{W}(p)\mathbf{x}(t) = \mathbf{0} \tag{4.26}
$$

obtained from (4.3) by deleting the forcing function $\mathbf{f}(t)$. In searching for a solution of (4.26), it is natural to try the exponential form

$$
\mathbf{x}(t) = \mathbf{K}e^{\alpha t} \tag{4.27}
$$

where the scalar α and the vector of constants $\mathbf{K}$ are yet to be determined. Since

$$
p^k(e^{\alpha t}) = \alpha^k e^{\alpha t} \tag{4.28}
$$

substituting (4.27) in (4.26) yields

$$
\mathbf{W}(\alpha)\mathbf{K}e^{\alpha t} = \mathbf{0} \tag{4.29}
$$

Dividing out the nonvanishing scalar factor $e^{\alpha t}$ gives

$$
\mathbf{W}(\alpha)\mathbf{K} = \mathbf{0} \tag{4.30}
$$

This is a system of simultaneous linear algebraic equations in the unknowns of **K**. As is well known, this system will have nontrivial solution if and only if the coefficient matrix is identically singular, or

$$\det \mathbf{W}(\alpha) = 0 \tag{4.31}$$

This purely algebraic equation is called the *characteristic* or *auxiliary equation* of (4.3) or (4.26). It is significant to observe that (4.31) is nothing but the determinant of the operator matrix $\mathbf{W}(p)$ of the system equated to zero, with p replaced by α. The roots of the characteristic equation are called the *characteristic roots* or *the eigenvalues*. The characteristic roots are often referred to as the *natural frequencies*. A formal definition on this will be given in Chapter 12.

For each characteristic root α_j there will be a solution vector $\mathbf{K}_j$ of (4.30) determined to within an arbitrary scalar factor c_j. If the characteristic equation (4.31) is of degree n and if its characteristic roots are all distinct—that is, $\alpha_i \neq \alpha_j$ if $i \neq j$, n solutions of (4.26) will be obtained:

$$c_1\mathbf{K}_1 e^{\alpha_1 t}, c_2\mathbf{K}_2 e^{\alpha_2 t}, \ldots, c_n\mathbf{K}_n e^{\alpha_n t} \tag{4.32}$$

According to Property 4.1 an infinite family of solutions can be formed from these n linearly independent solutions

$$\mathbf{x}(t) = c_1\mathbf{K}_1 e^{\alpha_1 t} + c_2\mathbf{K}_2 e^{\alpha_2 t} + \cdots + c_n\mathbf{K}_n e^{\alpha_n t} \tag{4.33}$$

The question arises as to whether or not this constitutes the complete solution of (4.26). This is answered by the following property.[†]

Property 4.3 *The number of arbitrary constants in the complete solution of a system of simultaneous linear differential equations with constant coefficients is equal to the order of the system.*

Our conclusion is that if the characteristic roots are all distinct then (4.33) is the complete solution of the homogeneous equation (4.26) because it contains n arbitrary constants. Any solution of (4.26) can be obtained from (4.33) by a suitable choice of the constants $c_1, c_2, \ldots, c_n$.

We illustrate the above procedure by the following example.

EXAMPLE 4.2
Consider the homogeneous nodal differential equation for the network shown in Figure 4.2(a), valid for $t \geq 0$

$$\begin{bmatrix} 3 + 2p & -2 - 2p \\ -2 - 2p & 2 + 3p \end{bmatrix}\begin{bmatrix} v_{n1} \\ v_{n2} \end{bmatrix} = \begin{bmatrix} 0 \\ 0 \end{bmatrix} \tag{4.34}$$

[†] For a proof of this result, see, for example, E. L. Ince, *Ordinary Differential Equations*, (New York: Dover, 1944), pp. 144–50.

which was discussed in Example 4.1. In searching for a solution, we let

$$
\begin{bmatrix} v_{n1} \\ v_{n2} \end{bmatrix} = \begin{bmatrix} k_1 \\ k_2 \end{bmatrix} e^{\alpha t} \tag{4.35}
$$

where k_1, k_2, and α are unknown scalars to be ascertained. Substituting (4.35) in (4.34) yields

$$
\begin{bmatrix} 3 + 2\alpha & -2 - 2\alpha \\ -2 - 2\alpha & 2 + 3\alpha \end{bmatrix} \begin{bmatrix} k_1 \\ k_2 \end{bmatrix} e^{\alpha t} = \begin{bmatrix} 0 \\ 0 \end{bmatrix} \tag{4.36}
$$

Dividing both sides by the nonvanishing scalar $e^{\alpha t}$ gives

$$
\begin{bmatrix} 3 + 2\alpha & -2 - 2\alpha \\ -2 - 2\alpha & 2 + 3\alpha \end{bmatrix} \begin{bmatrix} k_1 \\ k_2 \end{bmatrix} = \begin{bmatrix} 0 \\ 0 \end{bmatrix} \tag{4.37}
$$

Observe that the coefficient matrix of (4.37) is nothing but the operator matrix $\mathbf{W}(p)$ of (4.34) with p replaced by α. Thus the characteristic equation is obtained as

$$
\det \mathbf{W}(\alpha) = 2\alpha^2 + 5\alpha + 2 = (2\alpha + 1)(\alpha + 2) = 0 \tag{4.38}
$$

giving the characteristic roots $\alpha_1 = -0.5$ and $\alpha_2 = -2$.

From these two roots, we can construct two particular solutions

$$
\mathbf{v}_1(t) = \begin{bmatrix} \hat{v}_{n1} \\ \hat{v}_{n2} \end{bmatrix} = \begin{bmatrix} k_{11} \\ k_{21} \end{bmatrix} e^{-0.5t} = \mathbf{K}_1 e^{\alpha_1 t}, \qquad t \geq 0 \tag{4.39a}
$$

$$
\mathbf{v}_2(t) = \begin{bmatrix} \tilde{v}_{n1} \\ \tilde{v}_{n2} \end{bmatrix} = \begin{bmatrix} k_{12} \\ k_{22} \end{bmatrix} e^{-2t} = \mathbf{K}_2 e^{\alpha_2 t}, \qquad t \geq 0 \tag{4.39b}
$$

The constants k_{11}, k_{12}, k_{21}, and k_{22} in (4.39) are, however, not all independent, and relations will always exist among them serving to reduce their number to the order of the matrix differential equation (4.34), which is 2. To do this we note that these constants must satisfy the equation (4.37) for the corresponding characteristic root α_j. For $\alpha_1 = -0.5$ we obtain

$$
\begin{bmatrix} 2 & -1 \\ -1 & 0.5 \end{bmatrix} \begin{bmatrix} k_{11} \\ k_{21} \end{bmatrix} = \begin{bmatrix} 0 \\ 0 \end{bmatrix} \tag{4.40}
$$

We know, of course, that the determinant of the coefficient matrix is zero. This means that the two equations are not linearly independent, and one of them is sufficient. From the first equation of (4.40) we obtain a relation between k_{11} and k_{21} as

$$
k_{21} = 2k_{11} \tag{4.41}
$$

Similarly, for $\alpha_2 = -2$ we have from (4.37)

$$
\begin{bmatrix} -1 & 2 \\ 2 & -4 \end{bmatrix} \begin{bmatrix} k_{12} \\ k_{22} \end{bmatrix} = \begin{bmatrix} 0 \\ 0 \end{bmatrix} \tag{4.42}
$$

giving

$$k_{12} = 2k_{22} \tag{4.43}$$

Substituting (4.41) and (4.43) in (4.39), we obtain two particular solutions of (4.34):

$$\mathbf{v}_1(t) = c_1 \begin{bmatrix} 1 \\ 2 \end{bmatrix} e^{-0.5t}, \quad \mathbf{v}_2(t) = c_2 \begin{bmatrix} 2 \\ 1 \end{bmatrix} e^{-2t} \tag{4.44}$$

where $c_1 = k_{11}$ and $c_2 = k_{22}$. According to Properties 4.1 and 4.3, we can combine these two particular solutions into the complete solution of (4.34), as stated in (4.18): For $t \geq 0$,

$$\mathbf{v}_h(t) = \begin{bmatrix} v_{n1} \\ v_{n2} \end{bmatrix} = \mathbf{v}_1(t) + \mathbf{v}_2(t) = \begin{bmatrix} c_1 e^{-0.5t} + 2c_2 e^{-2t} \\ 2c_1 e^{-0.5t} + c_2 e^{-2t} \end{bmatrix} \tag{4.45}$$

4-2.1 The Zero-Input Response

In system analysis we are almost always interested in the behavior of a particular set of system variables called the *response* or sometimes the *output*. In the case of networks, the variables can be branch voltages, branch currents, cutset voltages, nodal voltages, loop currents, or a linear combination of these voltages or currents. The response is usually due to either the independent sources that are considered as inputs, to the initial conditions, or to both. The matrix differential equation (4.34) is, in fact, the nodal equation of the network of Figure 4.2(a) with the independent current source removed. Therefore we shall call such a solution of (4.34) the *zero-input response* because it is the response with no applied input. This zero-input response depends on the initial conditions and the characteristics of the system.

As can be seen from the complete solution (4.45), the homogeneous equation (4.34) has an infinite number of solutions, each of which corresponds to a choice of the arbitrary constants c_1 and c_2. The initial conditions for the network are the capacitor voltages at time $t = 0$: $v_1(0+) = 2$ V and $v_2(0+) = 0$. In terms of the initial nodal voltages, we have from Figure 4.2(a)

$$v_{n1}(0+) = v_1(0+) + v_2(0+) = 2 \text{ V}, \quad v_{n2}(0+) = v_1(0+) = 2 \text{ V} \tag{4.46}$$

The general solution (4.45) can be rewritten as

$$\begin{bmatrix} v_{n1} \\ v_{n2} \end{bmatrix} = \begin{bmatrix} e^{-0.5t} & 2e^{-2t} \\ 2e^{-0.5t} & e^{-2t} \end{bmatrix} \begin{bmatrix} c_1 \\ c_2 \end{bmatrix}, \quad t \geq 0 \tag{4.47}$$

At $t = 0$ we have

$$\begin{bmatrix} 2 \\ 2 \end{bmatrix} = \begin{bmatrix} 1 & 2 \\ 2 & 1 \end{bmatrix} \begin{bmatrix} c_1 \\ c_2 \end{bmatrix} \tag{4.48}$$

giving

$$c_1 = c_2 = \frac{2}{3} \tag{4.49}$$

Using (4.49) in (4.47), we obtain the zero-input response of the network of Figure 4.2(a):

$$\mathbf{v}_i(t) = \begin{bmatrix} v_{n1i} \\ v_{n2i} \end{bmatrix} = \begin{bmatrix} \dfrac{2}{3} e^{-0.5t} + \dfrac{4}{3} e^{-2t} \\ \dfrac{4}{3} e^{-0.5t} + \dfrac{2}{3} e^{-2t} \end{bmatrix}, \qquad t \geqq 0 \tag{4.50}$$

The initial conditions are specified by the capacitor voltages $v_1(0+)$ and $v_2(0+)$. They are also referred to as the *initial state* of the network. Therefore the zero-input response depends on the initial state since no input is applied. We demonstrate that the zero-input response is a linear function of the initial state.

Consider the situation where the characteristic roots α_j $(j = 1, 2, \ldots, n)$ are all distinct. Then from the solution form (4.33) the complete solution of the homogeneous equation (4.26) can be written as

$$\mathbf{x}_h(t) = [\mathbf{K}_1 e^{\alpha_1 t} \ \mathbf{K}_2 e^{\alpha_2 t} \ \cdots \ \mathbf{K}_n e^{\alpha_n t}] \begin{bmatrix} c_1 \\ c_2 \\ \vdots \\ c_n \end{bmatrix} \tag{4.51}$$

or more compactly as[†]

$$\mathbf{x}_h(t) = \mathbf{M}(t)\mathbf{c} \tag{4.52}$$

where the coefficient matrix is of order n. At the initial time $t = t_0$, $\mathbf{x}_h(t_0)$ is the *initial state* dictated by the initial conditions. Assume that we have two initial states $\mathbf{x}_{h1}(t_0)$ and $\mathbf{x}_{h2}(t_0)$ such that

$$\mathbf{x}_h(t_0) = \mathbf{x}_{h1}(t_0) + \mathbf{x}_{h2}(t_0) \tag{4.53}$$

The corresponding $\mathbf{c}$'s are found from (4.52) to be

$$\mathbf{c}_{h1} = \mathbf{M}^{-1}(t_0)\mathbf{x}_{h1}(t_0) \tag{4.54a}$$

$$\mathbf{c}_{h2} = \mathbf{M}^{-1}(t_0)\mathbf{x}_{h2}(t_0) \tag{4.54b}$$

since $\mathbf{M}(t)$ is nonsingular for all finite $t \geqq t_0$. The zero-input responses corresponding to the initial states $\mathbf{x}_{h1}(t_0)$ and $\mathbf{x}_{h2}(t_0)$ are obtained from (4.52) for $t \geqq t_0$

$$\mathbf{x}_{h1}(t) = \mathbf{M}(t)\mathbf{M}^{-1}(t_0)\mathbf{x}_{h1}(t_0) \tag{4.55a}$$

$$\mathbf{x}_{h2}(t) = \mathbf{M}(t)\mathbf{M}^{-1}(t_0)\mathbf{x}_{h2}(t_0) \tag{4.55b}$$

[†] For multiple characteristic roots, (4.52) remains valid, with the columns of $\mathbf{M}(t)$ now corresponding to those given in (4.69).

The zero-input response corresponding to the initial state $\mathbf{x}_h(t_0)$ is

$$\mathbf{x}_h(t) = \mathbf{M}(t)\mathbf{M}^{-1}(t_0)\mathbf{x}_h(t_0) \tag{4.56}$$

where the corresponding $\mathbf{c}$ is $\mathbf{c}_h = \mathbf{M}^{-1}(t_0)\mathbf{x}_h(t_0)$. It is straightforward to verify that

$$\mathbf{x}_h(t) = \mathbf{x}_{h1}(t) + \mathbf{x}_{h2}(t) \tag{4.57}$$

This, together with the fact that if the initial state is multiplied by a constant β, the corresponding zero-input response (4.55) or (4.56) is also multiplied by β, shows that the dependence of the zero-input response on the initial state is a linear function.

> **Property 4.4** *The zero-input response of a linear time-invariant system is a linear function of the initial state.*

We illustrate this by the following example.

EXAMPLE 4.3
Consider again the network of Figure 4.2(a). To compute the zero-input response, we remove the 1-A independent current source. The resulting network is shown in Figure 4.3, which is described by the nodal differential equation

$$\begin{bmatrix} 3 + 2p & -2 - 2p \\ -2 - 2p & 2 + 3p \end{bmatrix} \begin{bmatrix} v_{n1} \\ v_{n2} \end{bmatrix} = \begin{bmatrix} 0 \\ 0 \end{bmatrix}, \qquad t \geq 0 \tag{4.58}$$

The complete solution of (4.58), as obtained in (4.47), is repeated below:

$$\mathbf{x}_h(t) = \begin{bmatrix} v_{n1} \\ v_{n2} \end{bmatrix} = \begin{bmatrix} e^{-0.5t} & 2e^{-2t} \\ 2e^{-0.5t} & e^{-2t} \end{bmatrix} \begin{bmatrix} c_1 \\ c_2 \end{bmatrix}, \qquad t \geq 0 \tag{4.59}$$

Assume that at time $t = 0$ the network is at the initial state, with $v_1(0) = 3$ V and $v_2(0) = -2$ V, or equivalently

$$\mathbf{v}_{h1}(0) = \begin{bmatrix} \hat{v}_{n1}(0) \\ \hat{v}_{n2}(0) \end{bmatrix} = \begin{bmatrix} 1 \\ 3 \end{bmatrix} \tag{4.60}$$

Figure 4.3
Network resulting from the removal of the 1-A independent current source from the network of Figure 4.2(a).

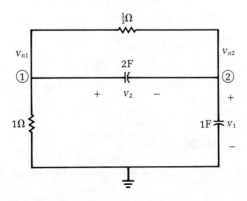

yielding from (4.59)

$$\begin{bmatrix} 1 \\ 3 \end{bmatrix} = \begin{bmatrix} 1 & 2 \\ 2 & 1 \end{bmatrix} \begin{bmatrix} c_1 \\ c_2 \end{bmatrix} \tag{4.61}$$

which can be solved to give

$$c_1 = \frac{5}{3} \quad c_2 = -\frac{1}{3} \tag{4.62}$$

The zero-input response due to the initial state $\mathbf{v}_{h1}(0)$ is found from (4.59) to be

$$\mathbf{v}_{h1}(t) = \begin{bmatrix} \hat{v}_{n1} \\ \hat{v}_{n2} \end{bmatrix} = \begin{bmatrix} \dfrac{5}{3} e^{-0.5t} - \dfrac{2}{3} e^{-2t} \\[3mm] \dfrac{10}{3} e^{-0.5t} - \dfrac{1}{3} e^{-2t} \end{bmatrix}, \quad t \geq 0 \tag{4.63}$$

Likewise if the network is at the initial state

$$\mathbf{v}_{h2}(0) = \begin{bmatrix} \tilde{v}_{n1}(0) \\ \tilde{v}_{n2}(0) \end{bmatrix} = \begin{bmatrix} 1 \\ -1 \end{bmatrix} \tag{4.64}$$

the zero-input response due to this initial state is given by

$$\mathbf{v}_{h2}(t) = \begin{bmatrix} \tilde{v}_{n1} \\ \tilde{v}_{n2} \end{bmatrix} = \begin{bmatrix} -e^{-0.5t} + 2e^{-2t} \\ -2e^{-0.5t} + e^{-2t} \end{bmatrix}, \quad t \geq 0 \tag{4.65}$$

Finally, if the network is at the initial state

$$\mathbf{v}_h(0) = \begin{bmatrix} v_{n1}(0) \\ v_{n2}(0) \end{bmatrix} = \begin{bmatrix} 2 \\ 2 \end{bmatrix} \tag{4.66}$$

which is equal to the sum of $\mathbf{v}_{h1}(0)$ and $\mathbf{v}_{h2}(0)$, the corresponding zero-input response is, according to Property 4.4, equal to the sum of the zero-input responses $\mathbf{v}_{h1}(t)$ and $\mathbf{v}_{h2}(t)$: For $t \geq 0$,

$$\mathbf{v}_h(t) = \begin{bmatrix} v_{n1} \\ v_{n2} \end{bmatrix} = \mathbf{v}_{h1}(t) + \mathbf{v}_{h2}(t) = \begin{bmatrix} \dfrac{2}{3} e^{-0.5t} + \dfrac{4}{3} e^{-2t} \\[3mm] \dfrac{4}{3} e^{-0.5t} + \dfrac{2}{3} e^{-2t} \end{bmatrix} \tag{4.67}$$

confirming (4.50).

4-2.2 Multiple Characteristic Roots

In the preceding section we indicated that if the characteristic roots α_j are all distinct we can construct n linearly independent solutions (4.32), so that the complete solution of the homogeneous equation (4.26) can be formed simply by adding these solutions as in (4.33). When the characteristic equation has multiple roots, say, $\alpha_1 = \alpha_2$, the first two solutions

$$c_1 \mathbf{K}_1 e^{\alpha_1 t} \quad \text{and} \quad c_2 \mathbf{K}_2 e^{\alpha_2 t}$$

of (4.32) become identical to within an arbitrary scalar factor. This implies that the solutions of (4.32) are not linearly independent and we do not have an adequate basis for constructing the complete solution. In this section we demonstrate how this difficulty can be circumvented.

Assume that the characteristic equation (4.31) has q distinct roots $\alpha_1, \alpha_2, \ldots, \alpha_q$, where $q < n$, having the multiplicities $n_1, n_2, \ldots, n_q$, respectively, where a root α_i is said to be of *multiplicity* n_i if there are n_i repeating roots at α_i. Then we have

$$n_1 + n_2 + \cdots + n_q = n \tag{4.68}$$

For the characteristic root α_j, the appropriate solutions are given by

$$\mathbf{K}_{j11} e^{\alpha_j t}$$
$$\mathbf{K}_{j21} t e^{\alpha_j t} + \mathbf{K}_{j22} e^{\alpha_j t}$$
$$\vdots$$
$$\mathbf{K}_{jn_j 1} t^{n_j - 1} e^{\alpha_j t} + \mathbf{K}_{jn_j 2} t^{n_j - 2} e^{\alpha_j t} + \cdots + \mathbf{K}_{jn_j n_j} e^{\alpha_j t} \tag{4.69}$$

where $j = 1, 2, \ldots, q$ and $\mathbf{K}$'s are arbitrary constant vectors. It can be shown that after (4.69) is substituted into the homogeneous equation (4.26), the constant vectors $\mathbf{K}_{j11}, \mathbf{K}_{j21}, \ldots, \mathbf{K}_{jn_j 1}$ become identical to within scalar factors. Consider, for example, $n_1 = 4$. The appropriate solutions for the fourfold root α_1 are

$$\mathbf{K}_{111} e^{\alpha_1 t}$$
$$\mathbf{K}_{121} t e^{\alpha_1 t} + \mathbf{K}_{122} e^{\alpha_1 t}$$
$$\mathbf{K}_{131} t^2 e^{\alpha_1 t} + \mathbf{K}_{132} t e^{\alpha_1 t} + \mathbf{K}_{133} e^{\alpha_1 t}$$
$$\mathbf{K}_{141} t^3 e^{\alpha_1 t} + \mathbf{K}_{142} t^2 e^{\alpha_1 t} + \mathbf{K}_{143} t e^{\alpha_1 t} + \mathbf{K}_{144} e^{\alpha_1 t} \tag{4.70}$$

where $\mathbf{K}_{111}$, $\mathbf{K}_{121}$, $\mathbf{K}_{131}$, and $\mathbf{K}_{141}$ are identical to within arbitrary scalar factors. The n solutions constructed in this way are linearly independent. The constants appearing initially in the solutions are, however, not all independent. The necessary relations among the constants can always be found by substituting the solutions into the original equation (4.26). Using these relations, the number of constants can be reduced to n, as required by Property 4.3. The complete solution is then formed simply by adding the resulting solutions. The following example will make this procedure clear.

EXAMPLE 4.4

In the network of Figure 4.4(a), switch S is open for $t < 0$. At $t = 0$, the switch is closed. Our objective is to determine the nodal voltages for all $t \geq 0$.

Figure 4.4
(a) A network with switch S closed at $t = 0$. (b) The plots of nodal voltages as functions of t.

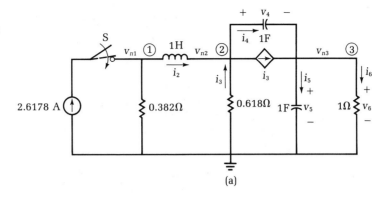

(a)

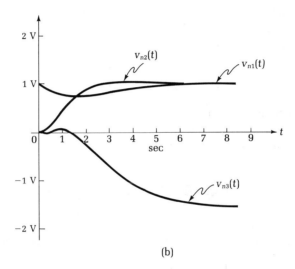

(b)

Applying the procedure outlined in Section 3-2.1, the nodal integro-differential equation is found to be, for $t \geqq 0$,

$$
\begin{bmatrix}
2.61780 + \dfrac{1}{p} & -\dfrac{1}{p} & 0 \\[2ex]
-\dfrac{1}{p} & p + \dfrac{1}{p} & -p \\[2ex]
0 & 1.61812 - p & 1 + 2p
\end{bmatrix}
\begin{bmatrix}
v_{n1} \\ v_{n2} \\ v_{n3}
\end{bmatrix}
=
\begin{bmatrix}
2.61780 \\ 0 \\ 0
\end{bmatrix}
\qquad \textbf{(4.71)}
$$

Differentiating the first two equations in (4.71) yields a system of nodal differential equations: For $t \geqq 0$,

$$
\begin{bmatrix}
2.61780p + 1 & -1 & 0 \\
-1 & p^2 + 1 & -p^2 \\
0 & 1.61812 - p & 2p + 1
\end{bmatrix}
\begin{bmatrix}
v_{n1} \\ v_{n2} \\ v_{n3}
\end{bmatrix}
=
\begin{bmatrix}
0 \\ 0 \\ 0
\end{bmatrix}
\qquad \textbf{(4.72)}
$$

the characteristic equation of which is given by

$$\det \mathbf{W}(\alpha) = 2.61780\alpha(\alpha^3 + 3\alpha^2 + 3\alpha + 1)$$

$$= 2.61780\alpha(\alpha + 1)^3 = 0 \qquad \textbf{(4.73)}$$

Thus $\alpha_1 = 0$ is a characteristic root of multiplicity $n_1 = 1$ and $\alpha_2 = -1$ is a characteristic root of multiplicity $n_2 = 3$. Note that the nodal differential equation (4.72) is of order $n = 4$.

For the characteristic root α_1, a solution can be written as

$$\mathbf{v}_{h1} = \mathbf{K}_{111}e^{0t} = \mathbf{K}_{111} = \begin{bmatrix} a_1 \\ a_2 \\ a_3 \end{bmatrix}, \qquad t \geq 0 \qquad \textbf{(4.74a)}$$

Substituting (4.74a) in (4.72) and solving for a_1, a_2, and a_3 in the resulting equations, we obtain

$$\mathbf{v}_{h1} = c_1 \begin{bmatrix} 1 \\ 1 \\ -1.61812 \end{bmatrix}, \qquad t \geq 0 \qquad \textbf{(4.74b)}$$

where c_1 is an arbitrary constant.

For the characteristic root $\alpha_2 = -1$, three additional solutions can be constructed as follows: For $t \geq 0$,

$$\mathbf{v}_{h2}(t) = \mathbf{K}_{211}e^{-t} \qquad \textbf{(4.75a)}$$

$$\mathbf{v}_{h3}(5) = \mathbf{K}_{221}te^{-t} + \mathbf{K}_{222}e^{-t} \qquad \textbf{(4.75b)}$$

$$\mathbf{v}_{h4}(t) = \mathbf{K}_{231}t^2 e^{-t} + \mathbf{K}_{232}te^{-t} + \mathbf{K}_{233}e^{-t} \qquad \textbf{(4.75c)}$$

We now demonstrate how to determine the necessary relations among the constants in (4.75).

For $\mathbf{v}_{h2}$ we write (4.75a) explicitly as

$$\mathbf{v}_{h2}(t) = \mathbf{K}_{211}e^{-t} = \begin{bmatrix} b_1 \\ b_2 \\ b_3 \end{bmatrix} e^{-t} \qquad \textbf{(4.76)}$$

As before, we substitute this in (4.72) and solve for the b's using the fact that the coefficient matrix of the resulting algebraic equations is identically singular, giving

$$\mathbf{v}_{h2}(t) = \mathbf{K}_{211}e^{-t} = c_2 \begin{bmatrix} -0.23609 \\ 0.38195 \\ 0.99999 \end{bmatrix} e^{-t} \qquad \textbf{(4.77)}$$

where c_2 is an arbitrary constant. For $\mathbf{v}_{h3}$ we set

$$\mathbf{v}_{h3}(t) = \mathbf{K}_{221}te^{-t} + \mathbf{K}_{222}e^{-t} = \begin{bmatrix} d_{11} \\ d_{21} \\ d_{31} \end{bmatrix} te^{-t} + \begin{bmatrix} d_{12} \\ d_{22} \\ d_{32} \end{bmatrix} e^{-t} \qquad \textbf{(4.78)}$$

Again, we substitute (4.78) in the origin system (4.72) and rearrange terms. The results can be written as

$$
t\begin{bmatrix} -1.61780 & -1 & 0 \\ -1 & 2 & -1 \\ 0 & 2.61812 & -1 \end{bmatrix}\begin{bmatrix} d_{11} \\ d_{21} \\ d_{31} \end{bmatrix} + \begin{bmatrix} 2.61780 & 0 & 0 \\ 0 & -2 & 2 \\ 0 & -1 & 2 \end{bmatrix}\begin{bmatrix} d_{11} \\ d_{21} \\ d_{31} \end{bmatrix}
$$

$$
+ \begin{bmatrix} -1.61780 & -1 & 0 \\ -1 & 2 & -1 \\ 0 & 2.61812 & -1 \end{bmatrix}\begin{bmatrix} d_{12} \\ d_{22} \\ d_{32} \end{bmatrix} = \begin{bmatrix} 0 \\ 0 \\ 0 \end{bmatrix} \tag{4.79}
$$

Observe that the coefficient matrix of the first term on the left-hand side is the same as that of the third term, being identically singular. To proceed, we set the first term to zero and obtain

$$
\mathbf{K}_{221} = \begin{bmatrix} d_{11} \\ d_{21} \\ d_{31} \end{bmatrix} = c_3 \begin{bmatrix} -0.23609 \\ 0.38195 \\ 0.99999 \end{bmatrix} \tag{4.80}
$$

where c_3 is an arbitrary constant. Substituting this back in (4.79) yields

$$
\begin{bmatrix} -1.61780 & -1 & 0 \\ -1 & 2 & -1 \\ 0 & 2.61812 & -1 \end{bmatrix}\begin{bmatrix} d_{12} \\ d_{22} \\ d_{32} \end{bmatrix} = c_3 \begin{bmatrix} 0.61804 \\ -1.23608 \\ -1.61803 \end{bmatrix} \tag{4.81}
$$

noting that the first term on the left-hand side of (4.79) is identically zero. Equation (4.81) can be solved to give

$$
\mathbf{K}_{222} = \begin{bmatrix} d_{12} \\ d_{22} \\ d_{32} \end{bmatrix} = c_3 \begin{bmatrix} 0 \\ -0.61804 \\ 0 \end{bmatrix} + d_{32}\begin{bmatrix} -0.23609 \\ 0.38195 \\ 0.99999 \end{bmatrix} \tag{4.82}
$$

The second term on the right-hand side is a column matrix proportional to the coefficient matrix $\mathbf{K}_{211}$ in (4.77). When the complete solution is constructed by adding the solutions $\mathbf{v}_{hk}(t)$ ($k = 1, 2, 3, 4$), these two matrices, as one appears in $\mathbf{v}_{h2}(t)$ of (4.77) and the other in $\mathbf{v}_{h3}(t)$ of (4.78), can be combined to yield the same matrix with arbitrary constant $c_2 + d_{32}$. Thus without loss of generality, we may let $d_{32} = 0$ and obtain

$$
\mathbf{K}_{222} = \begin{bmatrix} d_{12} \\ d_{22} \\ d_{32} \end{bmatrix} = c_3 \begin{bmatrix} 0 \\ -0.61804 \\ 0 \end{bmatrix} \tag{4.83}
$$

Combining (4.80) and (4.83) with (4.78), we obtain

$$
\mathbf{v}_{h3}(t) = \mathbf{K}_{221} t e^{-t} + \mathbf{K}_{222} e^{-t}
$$

$$
= c_3 \begin{bmatrix} -0.23609 \\ 0.38195 \\ 0.99999 \end{bmatrix} t e^{-t} + c_3 \begin{bmatrix} 0 \\ -0.61804 \\ 0 \end{bmatrix} e^{-t} \tag{4.84}
$$

Finally, for $\mathbf{v}_{h4}$ the process can be repeated by letting

$$\mathbf{v}_{h4}(t) = \mathbf{K}_{231}t^2e^{-t} + \mathbf{K}_{232}te^{-t} + \mathbf{K}_{233}e^{-t}$$

$$= \begin{bmatrix} g_{11} \\ g_{21} \\ g_{31} \end{bmatrix} t^2e^{-t} + \begin{bmatrix} g_{12} \\ g_{22} \\ g_{32} \end{bmatrix} te^{-t} + \begin{bmatrix} g_{13} \\ g_{23} \\ g_{33} \end{bmatrix} e^{-t} \qquad (4.85)$$

Substituting this in (4.72) and rearranging terms,

$$(t^2\mathbf{M}_1 + t\mathbf{M}_2 + \mathbf{M}_3)\mathbf{K}_{231} + (t\mathbf{M}_4 + \mathbf{M}_5)\mathbf{K}_{232} + \mathbf{M}_6\mathbf{K}_{233} = \mathbf{0} \qquad (4.86)$$

where

$$\mathbf{M}_1 = \mathbf{M}_4 = \mathbf{M}_6 = \begin{bmatrix} -1.61780 & -1 & 0 \\ -1 & 2 & -1 \\ 0 & 2.61812 & -1 \end{bmatrix} \qquad (4.87a)$$

$$\mathbf{M}_2 = \begin{bmatrix} 5.23560 & 0 & 0 \\ 0 & -4 & 4 \\ 0 & -2 & 4 \end{bmatrix} \qquad (4.87b)$$

$$\mathbf{M}_3 = \begin{bmatrix} 0 & 0 & 0 \\ 0 & 2 & -2 \\ 0 & 0 & 0 \end{bmatrix} \qquad (4.87c)$$

$$\mathbf{M}_5 = \begin{bmatrix} 2.61780 & 0 & 0 \\ 0 & -2 & 2 \\ 0 & -1 & 2 \end{bmatrix} \qquad (4.87d)$$

To determine $\mathbf{K}_{231}$, we set

$$\mathbf{M}_1\mathbf{K}_{231} = \mathbf{0} \qquad (4.88)$$

and obtain

$$\mathbf{K}_{231} = \begin{bmatrix} g_{11} \\ g_{21} \\ g_{31} \end{bmatrix} = c_4 \begin{bmatrix} -0.23609 \\ 0.38195 \\ 0.99999 \end{bmatrix} \qquad (4.89)$$

where c_4 is an arbitrary constant. For $\mathbf{K}_{232}$ we set

$$\mathbf{M}_2\mathbf{K}_{231} + \mathbf{M}_4\mathbf{K}_{232} = \mathbf{0} \qquad (4.90)$$

using $\mathbf{K}_{231}$ obtained in (4.89). This results in

$$\mathbf{K}_{232} = \begin{bmatrix} g_{12} \\ g_{22} \\ g_{32} \end{bmatrix} = c_4 \begin{bmatrix} 0 \\ -1.23608 \\ 0 \end{bmatrix} + g_{32} \begin{bmatrix} -0.23609 \\ 0.38195 \\ 0.99999 \end{bmatrix} \qquad (4.91)$$

As indicated before, since the second matrix on the right-hand side is proportional to the coefficient matrix of te^{-t} in (4.84), these two terms can be combined when the complete solution is constructed. Thus without loss of generality we may let $g_{32} = 0$ and obtain

$$\mathbf{K}_{232} = \begin{bmatrix} g_{12} \\ g_{22} \\ g_{32} \end{bmatrix} = c_4 \begin{bmatrix} 0 \\ -1.23608 \\ 0 \end{bmatrix} \tag{4.92}$$

For $\mathbf{K}_{233}$ we set

$$\mathbf{M}_3 \mathbf{K}_{231} + \mathbf{M}_5 \mathbf{K}_{232} + \mathbf{M}_6 \mathbf{K}_{233} = \mathbf{0} \tag{4.93}$$

using (4.89) and (4.92). The solution is found to be

$$\mathbf{K}_{233} = \begin{bmatrix} g_{13} \\ g_{23} \\ g_{33} \end{bmatrix} = c_4 \begin{bmatrix} 0.29183 \\ -0.47212 \\ 0 \end{bmatrix} + g_{33} \begin{bmatrix} -0.23609 \\ 0.38195 \\ 0.99999 \end{bmatrix} \tag{4.94}$$

Since the second matrix on the right-hand side is proportional to the coefficient matrix in (4.77), these two can be combined when the complete solution is constructed. Thus for simplicity we let $g_{33} = 0$, giving

$$\mathbf{K}_{233} = \begin{bmatrix} g_{13} \\ g_{23} \\ g_{33} \end{bmatrix} = c_4 \begin{bmatrix} 0.29183 \\ -0.47212 \\ 0 \end{bmatrix} \tag{4.95}$$

Substituting (4.89), (4.92), and (4.95) in (4.75c) results in

$$\mathbf{v}_{h4}(t) = \mathbf{K}_{231} t^2 e^{-t} + \mathbf{K}_{232} t e^{-t} + \mathbf{K}_{233} e^{-t}$$

$$= c_4 \begin{bmatrix} -0.23609 \\ 0.38195 \\ 0.99999 \end{bmatrix} t^2 e^{-t} + c_4 \begin{bmatrix} 0 \\ -1.23608 \\ 0 \end{bmatrix} t e^{-t} + c_4 \begin{bmatrix} 0.29183 \\ -0.47212 \\ 0 \end{bmatrix} e^{-t} \tag{4.96}$$

Observe that the coefficient matrices $\mathbf{K}_{211}$, $\mathbf{K}_{221}$, and $\mathbf{K}_{231}$ are identical to within arbitrary scalar factors.

The complete solution of the network now becomes

$$\mathbf{v}_h(t) = \mathbf{v}_{h1}(t) + \mathbf{v}_{h2}(t) + \mathbf{v}_{h3}(t) + \mathbf{v}_{h4}(t)$$
$$= \mathbf{K}_{231} t^2 e^{-t} + (\mathbf{K}_{221} + \mathbf{K}_{232}) t e^{-t} + (\mathbf{K}_{211} + \mathbf{K}_{222} + \mathbf{K}_{233}) e^{-t} \tag{4.97}$$

or explicitly for $t \geq 0$

$$v_{n1}(t) = c_1 - (0.23609 c_2 - 0.29183 c_4) e^{-t}$$
$$- 0.23609 (c_3 + c_4 t) t e^{-t} \tag{4.98a}$$

$$v_{n2}(t) = c_1' + (0.38195 c_2 - 0.61804 c_3 - 0.47212 c_4) e^{-t} + (0.38195 c_3$$
$$- 1.23608 c_4) t e^{-t} + 0.38195 c_4 t^2 e^{-t} \tag{4.98b}$$

$$v_{n3}(t) = -1.61812 c_1 + c_2 e^{-t} + c_3 t e^{-t} + c_4 t^2 e^{-t} \tag{4.98c}$$

The initial conditions are inductor current i_2 and capacitor voltages v_4 and v_5. The network is assumed to be initially relaxed. When the switch S is closed at $t = 0$, we have

$$i_2(0+) = 0 \qquad v_4(0+) = 0 \qquad v_5(0+) = 0 \tag{4.99}$$

since current cannot change instantaneously in the inductor and voltages cannot change instantaneously in the capacitors. The initial nodal voltages become

$$v_{n1}(0+) = 1, \qquad v_{n2}(0+) = 0, \qquad v_{n3}(0+) = 0 \tag{4.100}$$

resulting in $i_3(0+) = 0$ and $i_6(0+) = 0$. Applying KCL at nodes 2 and 3, we obtain the initial currents in the capacitors

$$i_4(0+) = i_5(0+) = 0 \tag{4.101}$$

or

$$i_5(0+) = pv_5(t)\big|_{t=0} = pv_{n3}(t)\big|_{t=0} = -c_2 + c_3 = 0 \tag{4.102}$$

Three additional relations among the c's are obtained by setting $t = 0$ in (4.98) and using (4.100). These relations together with (4.102) can be written in matrix form, as follows:

$$\begin{bmatrix} 1 & -0.23609 & 0 & 0.29183 \\ 1 & 0.38195 & -0.61804 & -0.47212 \\ -1.61812 & 1 & 0 & 0 \\ 0 & 1 & -1 & 0 \end{bmatrix} \begin{bmatrix} c_1 \\ c_2 \\ c_3 \\ c_4 \end{bmatrix} = \begin{bmatrix} 1 \\ 0 \\ 0 \\ 0 \end{bmatrix} \tag{4.103}$$

which can be solved to give

$$c_1 = 1, \; c_2 = c_3 = 1.61812, \; c_4 = 1.30906 \tag{4.104}$$

Finally, substituting (4.104) in (4.98) gives the desired nodal voltages for $t \geq 0$; their plots as functions of t are shown in Figure 4.4(b):

$$v_{n1}(t) = 1 - 0.38202te^{-t} - 0.30906t^2e^{-t} \tag{4.105a}$$

$$v_{n2}(t) = 1 - e^{-t} - te^{-t} + 0.5t^2e^{-t} \tag{4.105b}$$

$$v_{n3}(t) = -1.61812 + 1.61812e^{-t} + 1.61812te^{-t} + 1.30906t^2e^{-t} \tag{4.105c}$$

The zero-input response is the response when the constant current source is removed from the network of Figure 4.4(a). The general nodal voltages are still those given in (4.98) but the constants c's are different. They are again determined from the initial conditions as stated in (4.100) and (4.101) except that now we set $v_{n1}(0+) = 0$. This is equivalent to setting the 1 on the right-hand side of (4.103) to zero, and we obtain the only trivial solution $c_1 = c_2 = c_3 = c_4 = 0$. Therefore the zero-input response is identically zero for all three nodal voltages, as expected.

4-2.3   Complex Characteristic Roots

In the case in which the set of characteristic roots includes one or more pairs of conjugate complex roots, the corresponding solutions involve complex exponentials having complex coefficients. It is useful to devise a real-valued general solution. In this section we show how this can be accomplished for a pair of nonrepeating conjugate complex roots. The extension to multiple complex roots is straightforward.

Let us suppose that

$$\alpha_1 = a + jb \quad \text{and} \quad \alpha_2 = a - jb \tag{4.106}$$

are a pair of conjugate complex nonrepeating roots to the characteristic equation of a network, so that the solutions corresponding to this pair of roots in the complete solution (4.33) can be written as

$$c_1 \mathbf{K}_1 e^{(a+jb)t} + c_2 \mathbf{K}_2 e^{(a-jb)t} \tag{4.107}$$

For $\alpha = \alpha_1$ the constant vector $\mathbf{K}_1$ is a solution of (4.30) or

$$\mathbf{W}(\alpha_1)\mathbf{K}_1 = \mathbf{W}(a+jb)\mathbf{K}_1 = \mathbf{0} \tag{4.108}$$

Since all coefficients in $\mathbf{W}(\alpha)$ are real, by taking conjugates throughout we obtain

$$\overline{\mathbf{W}}(\alpha_1)\overline{\mathbf{K}}_1 = \mathbf{W}(\overline{\alpha}_1)\overline{\mathbf{K}}_1 = \mathbf{W}(\alpha_2)\overline{\mathbf{K}}_1 = \mathbf{0} \tag{4.109}$$

where the bar denotes the complex conjugate. The complex conjugate of a matrix is interpreted as the complex conjugate of each of its elements. Equation (4.109) indicates that $\overline{\mathbf{K}}_1$ is also a solution of (4.30) for $\alpha = \alpha_2$. This means that $\mathbf{K}_2$ is proportional to $\overline{\mathbf{K}}_1$ and by absorbing this proportionality constant in c_2, (4.107) can be expressed as

$$
\begin{aligned}
c_1 \mathbf{K}_1 e^{(a+jb)t} + c_2 \overline{\mathbf{K}}_1 e^{(a-jb)t} &= (\hat{c}_1 - j\hat{c}_2)\mathbf{K}_1 e^{(a+jb)t} + (\hat{c}_1 + j\hat{c}_2)\overline{\mathbf{K}}_1 e^{(a-jb)t} \\
&= \hat{c}_1[\mathbf{K}_1 e^{(a+jb)t} + \overline{\mathbf{K}}_1 e^{(a-jb)t}] \\
&\quad - j\hat{c}_2[\mathbf{K}_1 e^{(a+jb)t} - \overline{\mathbf{K}}_1 e^{(a-jb)t}] \\
&= e^{at}[\hat{c}_1(\mathbf{K}_1 e^{jbt} + \overline{\mathbf{K}}_1 e^{-jbt}) \\
&\quad - j\hat{c}_2(\mathbf{K}_1 e^{jbt} - \overline{\mathbf{K}}_1 e^{-jbt})]
\end{aligned}
\tag{4.110}
$$

where

$$\hat{c}_1 = \tfrac{1}{2}(c_1 + c_2) \quad \text{and} \quad \hat{c}_2 = j\tfrac{1}{2}(c_1 - c_2) \tag{4.111}$$

Now the expressions in the parentheses of (4.110) can be simplified by using the well-known Euler's formula:

$$e^{\pm jbt} = \cos bt \pm j \sin bt \tag{4.112}$$

The result of these simplifications is

$$\hat{c}_1 e^{at}(\mathbf{D}_1 \cos bt - \mathbf{D}_2 \sin bt) + \hat{c}_2 e^{at}(\mathbf{D}_2 \cos bt + \mathbf{D}_1 \sin bt) \tag{4.113}$$

where $\mathbf{D}_1$ and $\mathbf{D}_2$ are real vectors defined by

$$\mathbf{D}_1 = \mathbf{K}_1 + \bar{\mathbf{K}}_1 \quad \text{and} \quad \mathbf{D}_2 = -j(\mathbf{K}_1 - \bar{\mathbf{K}}_1) \qquad (4.114)$$

In this way the complete solution of the homogeneous equation can be put in the purely real form. The following example will illustrate this procedure.

EXAMPLE 4.5

The switch S in the terminated low-pass filter of Figure 4.5 is open for $t < 0$. At $t = 0$, switch S is flipped to closed position. We wish to determine the output voltage v_3 for $t \geq 0$.

For a change, we shall use loop analysis. For $t \geq 0$, the loop integro-differential equation can be obtained by inspection:

$$\begin{bmatrix} 1 + p + \dfrac{1}{p} & -\dfrac{1}{p} \\[2ex] -\dfrac{1}{p} & 1 + \dfrac{1}{p} \end{bmatrix} \begin{bmatrix} i_{m1} \\ i_{m2} \end{bmatrix} = \begin{bmatrix} 1 \\ 0 \end{bmatrix} \qquad (4.115)$$

Upon differentiation we obtain the loop differential equation

$$\begin{bmatrix} p^2 + p + 1 & -1 \\ -1 & p + 1 \end{bmatrix} \begin{bmatrix} i_{m1} \\ i_{m2} \end{bmatrix} = \begin{bmatrix} 0 \\ 0 \end{bmatrix}, \qquad t \geq 0 \qquad (4.116)$$

whose characteristic equation is

$$\det \mathbf{W}(\alpha) = \alpha(\alpha^2 + 2\alpha + 2) = \alpha(\alpha + 1 + j)(\alpha + 1 - j) = 0 \qquad (4.117)$$

giving a pair of complex conjugate roots plus a root at the origin

$$\alpha_1 = -1 + j, \quad \alpha_2 = -1 - j, \quad \alpha_3 = 0 \qquad (4.118)$$

For the root $\alpha_1 = -1 + j$, the corresponding solution can be written as

$$\mathbf{i}_{h1}'(t) = \mathbf{K}e^{-(1-j)t} = \begin{bmatrix} k_1 \\ k_2 \end{bmatrix} e^{-(1-j)t} \qquad (4.119)$$

and (4.108) becomes

$$\begin{bmatrix} -j & -1 \\ -1 & j \end{bmatrix} \begin{bmatrix} k_1 \\ k_2 \end{bmatrix} = \begin{bmatrix} 0 \\ 0 \end{bmatrix} \qquad (4.120)$$

Figure 4.5
A terminated low-pass filter with switch S closed at $t = 0$.

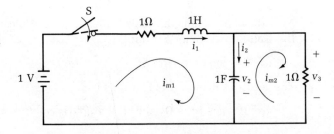

giving

$$k_2 = -jk_1 \tag{4.121}$$

Using this in (4.119) with $c_1 = k_1$, we obtain

$$\mathbf{i}_{h1}(t) = \mathbf{K}e^{-(1-j)t} = c_1 \begin{bmatrix} 1 \\ -j \end{bmatrix} e^{-(1-j)t} \tag{4.122}$$

According to the complex conjugate (4.109), the other solution is readily given by

$$\mathbf{i}_{h2}(t) = c_2 \begin{bmatrix} 1 \\ j \end{bmatrix} e^{-(1+j)t} \tag{4.123}$$

As an alternative, the two solutions $\mathbf{i}_{h1}$ and $\mathbf{i}_{h2}$ can be combined in accordance with (4.113) to give

$$\mathbf{i}_{h1}(t) + \mathbf{i}_{h2}(t) = c_1 \left(\begin{bmatrix} 2 \\ 0 \end{bmatrix} e^{-t} \cos t + \begin{bmatrix} 0 \\ 2 \end{bmatrix} e^{-t} \sin t \right)$$
$$+ c_2 \left(\begin{bmatrix} 0 \\ -2 \end{bmatrix} e^{-t} \cos t + \begin{bmatrix} 2 \\ 0 \end{bmatrix} e^{-t} \sin t \right), \qquad t \geq 0 \tag{4.124}$$

where

$$\mathbf{D}_1 = \begin{bmatrix} 2 \\ 0 \end{bmatrix} \quad \text{and} \quad \mathbf{D}_2 = \begin{bmatrix} 0 \\ -2 \end{bmatrix} \tag{4.125}$$

For the characteristic root $\alpha_3 = 0$, the corresponding solution is obtained as

$$\mathbf{i}_{h3} = c_3 \begin{bmatrix} 1 \\ 1 \end{bmatrix} e^{0t} = c_3 \begin{bmatrix} 1 \\ 1 \end{bmatrix}, \qquad t \geq 0 \tag{4.126}$$

The complete solution of the loop differential equation (4.116) is therefore

$$\mathbf{i}_h(t) = \mathbf{i}_{h1}(t) + \mathbf{i}_{h2}(t) + \mathbf{i}_{h3}(t) \tag{4.127}$$

or explicitly for $t \geq 0$,

$$\begin{bmatrix} i_{m1} \\ i_{m2} \end{bmatrix} = \begin{bmatrix} 2c_1 e^{-t} \cos t + 2c_2 e^{-t} \sin t + c_3 \\ 2c_1 e^{-t} \sin t - 2c_2 e^{-t} \cos t + c_3 \end{bmatrix} \tag{4.128}$$

To evaluate the constants c_1, c_2, and c_3, we need the initial conditions. For $t < 0$, all the passive branch voltages and currents are zero. When the switch S is closed at $t = 0$, we have $i_1(0+) = 0$ and $v_2(0+) = 0$ since neither inductor current nor capacitor voltage can change instantaneously. Therefore from Figure 4.5,

$$i_{m1}(0+) = 0, \quad i_{m2}(0+) = 0 \tag{4.129}$$

Also, applying KVL around the outer circuit gives

$$i_{m1}(t) + pi_{m1}(t) + i_{m2}(t) = 1 \tag{4.130}$$

At $t = 0$,

$$\frac{di_{m1}(t)}{dt}\bigg|_{t=0} = 1 \tag{4.131}$$

which, when i_{m1} is substituted from (4.128), yields

$$-2c_1 + 2c_2 = 1 \tag{4.132}$$

The other two constraints are obtained from (4.129), which, when combined with (4.132), can be written in matrix notation as

$$\begin{bmatrix} 2 & 0 & 1 \\ 0 & -2 & 1 \\ -2 & 2 & 0 \end{bmatrix} \begin{bmatrix} c_1 \\ c_2 \\ c_3 \end{bmatrix} = \begin{bmatrix} 0 \\ 0 \\ 1 \end{bmatrix} \tag{4.133}$$

Solving (4.133) for the c's, we obtain

$$2c_2 = c_3 = -2c_1 = \tfrac{1}{2} \tag{4.134}$$

Using these in (4.128) results in the desired loop currents for $t \geq 0$:

$$i_{m1}(t) = 0.5 - 0.5e^{-t}(\cos t - \sin t)$$

$$= 0.5 + 0.707e^{-t} \cos\left(t + \frac{5\pi}{4}\right) \tag{4.135a}$$

$$i_{m2}(t) = 0.5 - 0.5e^{-t}(\sin t + \cos t)$$

$$= 0.5 + 0.707e^{-t} \cos\left(t - \frac{5\pi}{4}\right) \tag{4.135b}$$

Once the loop currents are known, all other branch voltages and currents are thereby determined. In particular the output voltage is given by

$$v_3(t) = 1i_{m2}(t) = 0.5 + 0.707e^{-t} \cos\left(t - \frac{5\pi}{4}\right), \qquad t \geq 0 \tag{4.136}$$

A plot of v_3 as a function of t is presented in Figure 4.6. Observe that the voltage reaches its final value of 0.5 V in an oscillatory manner.

Figure 4.6
Plot of voltage v_3 as a function of t.

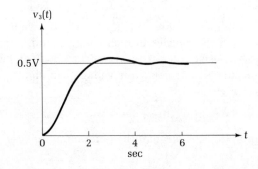

4-3 NONHOMOGENEOUS LINEAR DIFFERENTIAL EQUATIONS

As demonstrated in Section 4-1, the presence of the input exciting sources in a linear time-invariant system results in a non-homogeneous linear differential equation of the general form

$$\mathbf{W}(p)\mathbf{x}(t) = \mathbf{f}(t) \tag{4.137}$$

According to Property 4.2 the complete solution of the above equation can be written as the sum of the complementary function $\mathbf{x}_h(t)$ and a particular integral $\mathbf{x}_p(t)$:

$$\mathbf{x}(t) = \mathbf{x}_h(t) + \mathbf{x}_p(t) \tag{4.138}$$

The complementary function $\mathbf{x}_h(t)$ is the complete solution of the associated homogeneous equation

$$\mathbf{X}(p)\mathbf{x}(t) = \mathbf{0} \tag{4.139}$$

obtained from the nonhomogeneous equation (4.137) by deleting the forcing function $\mathbf{f}(t)$ corresponding to the sources and initial conditions. Techniques for computing the complementary function $\mathbf{x}_h$ were discussed in the preceding section. In this section we present a method for determining a particular or special solution to the nonhomogeneous equation (4.137) called a particular integral of (4.137) for a very important class of forcing functions. The procedure is known as the *method of undetermined coefficients*.

Remember that in applying (4.138) the particular integral $\mathbf{x}_p$ is *any* solution of (4.137). Any solution obtained from outright guessing to the most sophisticated theoretical techniques is acceptable.

To illustrate the method, suppose that we wish to find a particular integral of the nonhomogeneous equation

$$\begin{bmatrix} 3 + 2p & -2 - 2p \\ -2 - 2p & 2 + 3p \end{bmatrix} \begin{bmatrix} v_{n1} \\ v_{n2} \end{bmatrix} = \begin{bmatrix} 1 \\ 0 \end{bmatrix} e^{-t}, \qquad t \geq 0 \tag{4.140}$$

This is the nodal differential equation for the network of Figure 4.2(a), with the 1-A current source being replaced by an exponential current source e^{-t}. Since differentiating an exponential of the form e^{at} merely reproduces the function with at most a change of its coefficient, it is natural to guess a particular integral as

$$\mathbf{v}_p(t) = \begin{bmatrix} k_1 \\ k_2 \end{bmatrix} e^{-t}, \qquad t \geq 0 \tag{4.141}$$

To verify, we substitute this in (4.140) and obtain

$$\begin{bmatrix} 1 & 0 \\ 0 & -1 \end{bmatrix} \begin{bmatrix} k_1 \\ k_2 \end{bmatrix} = \begin{bmatrix} 1 \\ 0 \end{bmatrix} \tag{4.142}$$

giving $k_1 = 1$ and $k_2 = 0$. Thus the required particular integral is

$$\mathbf{v}_p(t) = \begin{bmatrix} 1 \\ 0 \end{bmatrix} e^{-t}, \qquad t \geq 0 \tag{4.143}$$

Now suppose that the exponential e^{-t} in (4.140) is replaced by $\sin t$. Guided by our previous success, we might be led to try (4.141) with e^{-t} being replaced by $\sin t$ as a particular integral. Substituting this back to the original equation results in

$$\cot t \begin{bmatrix} 2 & -2 \\ -2 & 3 \end{bmatrix} \begin{bmatrix} k_1 \\ k_2 \end{bmatrix} + \begin{bmatrix} 3 & -2 \\ -2 & 2 \end{bmatrix} \begin{bmatrix} k_1 \\ k_2 \end{bmatrix} = \begin{bmatrix} 1 \\ 0 \end{bmatrix} \tag{4.144}$$

Setting the first term on the left-hand side of (4.144) to zero gives $k_1 = k_2 = 0$, but these values cannot satisfy (4.144). The source of difficulty is not hard to identify. The differentiation of $\sin t$ introduces a new function $\cos t$, which must also be eliminated identically from the equations resulting from the substitution. To avoid this difficulty, we use

$$\mathbf{v}_p(t) = \begin{bmatrix} k_{11} \\ k_{21} \end{bmatrix} \sin t + \begin{bmatrix} k_{12} \\ k_{22} \end{bmatrix} \cos t$$

$$= \mathbf{K}_1 \sin t + \mathbf{K}_2 \cos t, \qquad t \geq 0 \tag{4.145}$$

The actual determination of $\mathbf{K}_1$ and $\mathbf{K}_2$ is a simple matter, for substituion into the given differential equation yields

$$(\mathbf{M}_1\mathbf{K}_1 + \mathbf{M}_2\mathbf{K}_2) \sin t - (\mathbf{M}_2\mathbf{K}_1 - \mathbf{M}_1\mathbf{K}_2) \cos t = \mathbf{M}_3 \sin t \tag{4.146}$$

where

$$\mathbf{M}_1 = \begin{bmatrix} 3 & -2 \\ -2 & 2 \end{bmatrix}, \qquad \mathbf{M}_2 = \begin{bmatrix} -2 & 2 \\ 2 & -3 \end{bmatrix} \tag{4.147a}$$

$$\mathbf{M}_3 = \begin{bmatrix} 1 \\ 0 \end{bmatrix} \tag{4.147b}$$

For (4.146) to hold for all t, we first set

$$\mathbf{M}_2\mathbf{K}_1 - \mathbf{M}_1\mathbf{K}_2 = \mathbf{0} \tag{4.148}$$

giving

$$\mathbf{K}_1 = \begin{bmatrix} k_{11} \\ k_{21} \end{bmatrix} = \begin{bmatrix} -\frac{5}{2} & 1 \\ -1 & 0 \end{bmatrix} \begin{bmatrix} k_{12} \\ k_{22} \end{bmatrix} \tag{4.149}$$

Substituting this $\mathbf{K}_1$ in (4.146) results in

$$\mathbf{M}_1\mathbf{K}_1 + \mathbf{M}_2\mathbf{K}_2 = \mathbf{M}_3 \tag{4.150}$$

which can be solved for $\mathbf{K}_2$,

$$\mathbf{K}_2 = \begin{bmatrix} k_{12} \\ k_{22} \end{bmatrix} = \frac{2}{5} \begin{bmatrix} -1 \\ -1 \end{bmatrix} \tag{4.151}$$

Using this value of $\mathbf{K}_2$, we obtain

$$\mathbf{K}_1 = \begin{bmatrix} k_{11} \\ k_{21} \end{bmatrix} = \frac{1}{5} \begin{bmatrix} 3 \\ 2 \end{bmatrix} \tag{4.152}$$

and finally from (4.145),

$$\mathbf{v}_p(t) = \begin{bmatrix} v_{n1} \\ v_{n2} \end{bmatrix} = \frac{1}{5} \begin{bmatrix} 3 \sin t - 2 \cos t \\ 2 \sin t - 2 \cos t \end{bmatrix}$$

$$= -\frac{1}{5} \begin{bmatrix} \sqrt{13} \cos(t + 0.98279) \\ 2\sqrt{2} \cos\left(t + \frac{\pi}{4}\right) \end{bmatrix}, \qquad t \geqq 0 \tag{4.153}$$

The reason for the success of (4.145) is that it contains a linear combination of *all* the independent terms that can be obtained from $\sin t$ by repeated differentiation. The general principle on the use of the method of undetermined coefficients for finding a particular integral is stated as follows.

Property 4.5 *If the forcing vector $\mathbf{f}(t)$ of a system possesses a finite number of independent derivatives after repeated differentiation, then in general a particular integral $\mathbf{x}_p(t)$ of the system can be found by assuming $\mathbf{x}_p(t)$ to be an arbitrary linear combination of $\mathbf{f}(t)$ and all its independent derivatives. The unknown coefficients are determined by substituting this expression into the original system.*

The class of functions which possess a finite number of linearly independent derivatives includes the following most common and useful simple functions:

constant

t^n (n a positive integer)

e^{at}

$\cos bt$

$\sin bt$

and any others obtainable from these by a finite number of additions, subtractions, and multiplications, such as

$c_1 t^n + c_2 e^{at} + c_3 \cos bt$

$c_1 e^{at} \cos bt + c_2 t^n \sin bt$

$c_1 \sin bt \cos at$

For our purposes there is no need to study rules for selecting trial functions $\mathbf{x}_p(t)$. The required form of the particular integral is given in Table 4.1.

Simultaneous Linear Differential Equations

Table 4.1

Factor in $\mathbf{f}(t)$	Necessary choice for the particular integral $\mathbf{x}_p(t)$
1. k (constant)	$\mathbf{K}$
2. at^n (n a positive integer)	$\mathbf{K}_n t^n + \mathbf{K}_{n-1} t^{n-1} + \cdots + \mathbf{K}_1 t + \mathbf{K}_0$
3. ae^{bt}	$\mathbf{K}e^{bt}$
4. $a \cos bt$	$\mathbf{K}_1 \cos bt + \mathbf{K}_2 \sin bt$
5. $a \sin bt$	
6. $at^n e^{bt} \cos ct$	$(\mathbf{K}_n t^n + \mathbf{K}_{n-1} t^{n-1} + \cdots + \mathbf{K}_1 t + \mathbf{K}_0)e^{bt} \cos ct$
7. $at^n e^{bt} \sin ct$	$\qquad + (\mathbf{H}_n t^n + \mathbf{H}_{n-1} t^{n-1} + \cdots + \mathbf{H}_1 t + \mathbf{H}_0)e^{bt} \sin ct$

When $\mathbf{f}(t)$ is composed of many terms, the appropriate choice for the particular integral $\mathbf{x}_p$ is the sum of the $\mathbf{x}_p$ expressions corresponding to these terms individually. The procedure works well with one exception which we will investigate next.

Suppose that in the nonhomogeneous network equation (4.140) we replace the exponential e^{-t} by e^{-2t}, everything else being the same. The equation becomes

$$\begin{bmatrix} 3 + 2p & -2 - 2p \\ -2 - 2p & 2 + 3p \end{bmatrix} \begin{bmatrix} v_{n1} \\ v_{n2} \end{bmatrix} = \begin{bmatrix} 1 \\ 0 \end{bmatrix} e^{-2t}, \qquad t \geq 0 \tag{4.154}$$

Proceeding as in (4.141), we let the particular integral $\mathbf{v}_p$ take the general form

$$\mathbf{v}_p(t) = \mathbf{K}e^{-2t} = \begin{bmatrix} k_1 \\ k_2 \end{bmatrix} e^{-2t}, \qquad t \geq 0 \tag{4.155}$$

Substituting this in (4.154) yields

$$-k_1 + 2k_2 = 1 \tag{4.156a}$$

$$-k_1 + 2k_2 = 0 \tag{4.156b}$$

This is impossible to solve because the two equations are inconsistent. The source of this difficulty is not hard to identify. The characteristic equation of (4.154) is found to be

$$2\alpha^2 + 5\alpha + 2 = 0 \tag{4.157}$$

with roots $\alpha_1 = -0.5$ and $\alpha_2 = -2$. As shown in Example 4.2, the complementary function $\mathbf{v}_h$ of (4.154) is given by

$$\mathbf{v}_h(t) = c_1 \begin{bmatrix} 1 \\ 2 \end{bmatrix} e^{-0.5t} + c_2 \begin{bmatrix} 2 \\ 1 \end{bmatrix} e^{-2t}, \qquad t \geq 0 \tag{4.158}$$

The solution given by (4.155) is a part of the complementary function. Therefore when it is substituted back into the differential equation (4.154), we can only find solutions for $\mathbf{K}$ that will make the left-hand side identically zero, since it is a solution of the associated homogeneous equation.

To avoid this difficulty, we follow a procedure similar to that outlined in Section 4-2.2 for multiple characteristic roots by multiplying the exponential e^{-2t} by t. In general we can state that if $\mathbf{f}(t)$ contains a term that duplicates a term in the complementary function, then a particular integral can always be chosen by multiplying the usual choice for a particular integral by the lowest power in t that will eliminate the duplication in the complementary function plus the products of the usual choice and all lower power of t. The procedure is similar to those given in (4.69) with the exponential being replaced by the function in question.

As an example suppose that the current source in the network of Figure 4.4 is replaced by an exponential current source $2.61780e^{-t}$, everything else being the same. From Example 4.4 a system of nodal differential equations can be written as follows. For $t \geq 0$,

$$
\begin{bmatrix}
2.61780p + 1 & -1 & 0 \\
-1 & p^2 + 1 & -p^2 \\
0 & 1.61812 - p & 2p + 1
\end{bmatrix}
\begin{bmatrix}
v_{n1} \\
v_{n2} \\
v_{n3}
\end{bmatrix}
=
\begin{bmatrix}
-2.61780 \\
0 \\
0
\end{bmatrix}
e^{-t} \quad \textbf{(4.159)}
$$

Since -1 is a characteristic root of multiplicity 3, the complementary function $\mathbf{v}_h$ is the sum of three solutions listed in (4.75). Thus duplication occurs between the source vector on the right-hand side of (4.159) and a vector in the complementary function. According to Table 4.1, the usual choice for a particular integral would be $\mathbf{K}_1 e^{-t}$, but this already is contained in the complementary function. The lowest power n of t that will eliminate all duplication between $t^n \mathbf{K}_1 e^{-t}$ and the complementary function is 3. Therefore we choose $t^3 \mathbf{K}_1 e^{-t}$ plus the products of $\mathbf{K}_1 e^{-t}$ and all lower powers of t—namely, t^2, t and 1. An appropriate choice for a particular integral $\mathbf{v}_p$ can be expressed as

$$
\mathbf{v}_p(t) = \mathbf{K}_1 t^3 e^{-t} + \mathbf{K}_2 t^2 e^{-t} + \mathbf{K}_3 t e^{-t} + \mathbf{K}_4 e^{-t}, \quad t \geq 0 \quad \textbf{(4.160)}
$$

Substituting this expression in (4.159) and determining the constant vectors $\mathbf{K}$'s yields the desired particular integral. The determination of these constant vectors is illustrated in the following example.

EXAMPLE 4.6

We derive a particular integral for the nodal differential equation

$$
\begin{bmatrix}
3 + 2p & -2 - 2p \\
-2 - 2p & 2 + 3p
\end{bmatrix}
\begin{bmatrix}
v_{n1} \\
v_{n2}
\end{bmatrix}
=
\begin{bmatrix}
1 \\
0
\end{bmatrix}
e^{-2t}, \quad t \geq 0 \quad \textbf{(4.161)}
$$

having characteristic roots of -0.5 and -2. As shown in Example 4.2, its complementary function can be expressed as

$$
\mathbf{v}_h(t) = c_1 \begin{bmatrix} 1 \\ 2 \end{bmatrix} e^{-0.5t} + c_2 \begin{bmatrix} 2 \\ 1 \end{bmatrix} e^{-2t}, \quad t \geq 0 \quad \textbf{(4.162)}
$$

According to Table 4.1, the usual choice for a particular integral is $\mathbf{K}_1 e^{-2t}$, but this term is already in (4.162). Thus we must multiply it by t before including it. An appropriate choice for a particular integral assumes the general form, for $t \geq 0$,

$$\mathbf{v}_p(t) = \mathbf{K}_1 t e^{-2t} + \mathbf{K}_2 e^{-2t} = \begin{bmatrix} k_{11} \\ k_{21} \end{bmatrix} t e^{-2t} + \begin{bmatrix} k_{12} \\ k_{22} \end{bmatrix} e^{-2t} \tag{4.163}$$

Substituting this expression in (4.161) gives

$$t \mathbf{M}_1 \mathbf{K}_1 + \mathbf{M}_2 \mathbf{K}_1 + \mathbf{M}_1 \mathbf{K}_2 = \mathbf{M}_3 \tag{4.164}$$

where

$$\mathbf{M}_1 = \begin{bmatrix} -1 & 2 \\ 2 & -4 \end{bmatrix}, \qquad \mathbf{M}_2 = \begin{bmatrix} 2 & -2 \\ -2 & 3 \end{bmatrix}, \qquad \mathbf{M}_3 = \begin{bmatrix} 1 \\ 0 \end{bmatrix} \tag{4.165}$$

Setting $\mathbf{M}_1 \mathbf{K}_1 = \mathbf{0}$ yields

$$k_{11} = 2k_{21} \tag{4.166}$$

Using this in (4.164),

$$\begin{bmatrix} -1 & 2 \\ 2 & -4 \end{bmatrix} \begin{bmatrix} k_{12} \\ k_{22} \end{bmatrix} = \begin{bmatrix} 1 - 2k_{21} \\ k_{21} \end{bmatrix} \tag{4.167}$$

For the two equations in (4.167) to be consistent, we require that

$$-2(1 - 2k_{21}) = k_{21} \tag{4.168}$$

giving $k_{21} = \frac{2}{3}$. From (4.167) we have

$$\mathbf{K}_2 = \begin{bmatrix} k_{12} \\ k_{22} \end{bmatrix} = \frac{1}{3} \begin{bmatrix} 1 \\ 0 \end{bmatrix} + \begin{bmatrix} 2 \\ 1 \end{bmatrix} k_{22} \tag{4.169}$$

Since the second term on the right-hand side is proportional to the coefficient matrix of e^{-2t} in (4.162), these two vectors can be combined with the arbitrary constant $c_2 + k_{22}$ when the complete solution of (4.161) is constructed. Thus without loss of generality we may let $k_{22} = 0$, obtaining

$$\mathbf{K}_1 = \frac{2}{3} \begin{bmatrix} 2 \\ 1 \end{bmatrix}, \qquad \mathbf{K}_2 = \frac{1}{3} \begin{bmatrix} 1 \\ 0 \end{bmatrix} \tag{4.170}$$

The particular integral is therefore, from (4.163),

$$\mathbf{v}_p(t) = \begin{bmatrix} v_{n1} \\ v_{n2} \end{bmatrix} = \frac{1}{3} \begin{bmatrix} 4t e^{-2t} + e^{-2t} \\ 2t e^{-2t} \end{bmatrix} \tag{4.171}$$

which can easily be verified by substituting it back in (4.161).

4-3.1 The Zero-State Response

In Section 4-2.1 we gave the name *zero-input response* as the response of a system when its input is identically zero. Therefore it is the response due to the initial state or the energy initially stored in the system. In a similar manner, the *zero-state response* is defined as the response of a system to inputs applied at some arbitrary time t_0 subject to the condition that the system be in the zero state just prior to the application of the inputs— namely, at time t_0^-. It is the response due only to the inputs. For a linear time-invariant network, if the initial voltages across all capacitors and the initial currents through all inductors are zero, the network is in the zero state. Since in calculating the zero-state response we are primarily interested in the behavior of the response for all $t \geq t_0$, we assume that all inputs and the zero-state response are identically zero for all $t < t_0$.

We now demonstrate that the zero-state response is a linear function of the input. Consider two inputs $\mathbf{f}_1$ and $\mathbf{f}_2$ that are both applied to the system described by (4.3) at t_0. Call $\mathbf{x}_{s1}$ and $\mathbf{x}_{s2}$ the corresponding zero-state responses. By definition, for all $t \geq t_0$ we have

$$\mathbf{W}(p)\mathbf{x}_{s1}(t) = \mathbf{f}_1(t) \tag{4.172}$$

with $\mathbf{x}_{s1}(t_0) = \mathbf{0}$, and

$$\mathbf{W}(p)\mathbf{x}_{s2}(t) = \mathbf{f}_2(t) \tag{4.173}$$

with $\mathbf{x}_{s2}(t_0) = \mathbf{0}$. Adding (4.172) and (4.173) and taking into account that $\mathbf{x}_{s1}(t_0) = \mathbf{0}$ and $\mathbf{x}_{s2}(t_0) = \mathbf{0}$, we see that

$$\mathbf{W}(p)[\mathbf{x}_{s1}(t) + \mathbf{x}_{s2}(t)] = \mathbf{f}_1(t) + \mathbf{f}_2(t) \tag{4.174}$$

with $\mathbf{x}_{s1}(t_0) + \mathbf{x}_{s2}(t_0) = \mathbf{0}$.

Now by definition the zero-state response to the input $\mathbf{f}_1 + \mathbf{f}_2$ applied at $t = t_0$ is the solution satisfying the differential equation

$$\mathbf{W}(p)\mathbf{x}(t) = \mathbf{f}_1(t) + \mathbf{f}_2(t) \tag{4.175}$$

with $\mathbf{x}(t_0) = \mathbf{0}$. We know from our earlier experience that (4.175) in general has an infinite number of solutions. However, if the initial conditions are specified for the solution $\mathbf{x}(t)$ at $t = t_0$, then one and only one solution results. We state this as the following property.[†]

Property 4.6 *The solution of an nth-order matrix differential equation* $\mathbf{W}(p)\mathbf{x}(t) = \mathbf{f}(t)$ *satisfying n linearly independent initial conditions is unique.*

The initial condition for the solution $\mathbf{x}(t)$ of (4.175) is $\mathbf{x}(t_0) = \mathbf{0}$, which is the same as that for the solution of (4.174). Thus according to this property,

---

[†] See, for example, L. S. Pontryagin, *Ordinary Differential Equations* (Reading, Mass.: Addison-Wesley, 1962).

(4.174) and (4.175) have the identical solution, or

$$\mathbf{x}(t) = \mathbf{x}_{s1}(t) + \mathbf{x}_{s2}(t) \tag{4.176}$$

This in conjunction with the observation that if the input signal is multiplied by an arbitrary real constant k, the corresponding zero-state response is also multiplied by k, leads to the following conclusion:

> **Property 4.7** *The zero-state response of a linear time-invariant system is a linear function of the input; that is, the dependence of the zero-state response on the input has the property of additivity and homogeneity.*

The following example will clarify the procedure.

EXAMPLE 4.7

Consider the network of Figure 4.7, as discussed in Example 4.1. The nodal differential equation is given by

$$\begin{bmatrix} 3 + 2p & -2 - 2p \\ -2 - 2p & 2 + 3p \end{bmatrix} \begin{bmatrix} v_{n1} \\ v_{n2} \end{bmatrix} = \begin{bmatrix} 1 \\ 0 \end{bmatrix}, \qquad t \geqq 0 \tag{4.177}$$

The complementary function of this equation is shown in (4.18) and is repeated below:

$$\mathbf{v}_h(t) = \begin{bmatrix} v_{n1h} \\ v_{n2h} \end{bmatrix} = \begin{bmatrix} c_1 e^{-0.5t} + 2c_2 e^{-2t} \\ 2c_1 e^{-0.5t} + c_2 e^{-2t} \end{bmatrix}, \qquad t \geqq 0 \tag{4.178}$$

To find a particular integral $\mathbf{v}_p$, we appeal to Table 4.1. An appropriate choice for $\mathbf{v}_p$ is

$$\mathbf{v}_p(t) = \begin{bmatrix} v_{n1p} \\ v_{n2p} \end{bmatrix} = \begin{bmatrix} k_1 \\ k_2 \end{bmatrix}, \qquad t \geqq 0 \tag{4.179}$$

Substituting this expression in (4.177) yields

$$k_1 = k_2 = 1 \tag{4.180}$$

Figure 4.7
A network with switch S_1 closed and switch S_2 open at $t = 0$.

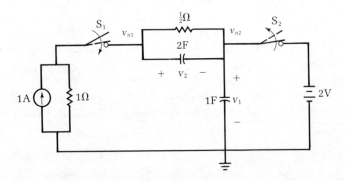

The complete solution of the nonhomogeneous equation (4.177) is obtained by combining the complementary function $\mathbf{v}_h$ and the particular integral $\mathbf{v}_p$:

$$\mathbf{v}_n(t) = \begin{bmatrix} V_{n1} \\ V_{n2} \end{bmatrix} = \mathbf{v}_h(t) + \mathbf{v}_p(t) = \begin{bmatrix} 1 + c_1 e^{-0.5t} + 2c_2 e^{-2t} \\ 1 + 2c_1 e^{-0.5t} + c_2 e^{-2t} \end{bmatrix}, \qquad t \geqq 0 \qquad \textbf{(4.181)}$$

To compute the zero-state response, we set the initial capacitor voltages v_1 and v_2 to zero at $t = 0$, giving $v_{n1}(0+) = v_{n2}(0+) = 0$. Substituting these in (4.181) after setting $t = 0$,

$$\begin{bmatrix} 1 & 2 \\ 2 & 1 \end{bmatrix} \begin{bmatrix} c_1 \\ c_2 \end{bmatrix} = \begin{bmatrix} -1 \\ -1 \end{bmatrix} \qquad\qquad \textbf{(4.182)}$$

yielding $c_1 = c_2 = -\frac{1}{3}$. The zero-state response can now be expressed as

$$\mathbf{v}_s(t) = \begin{bmatrix} V_{n1s} \\ V_{n2s} \end{bmatrix} = \frac{1}{3} \begin{bmatrix} 3 - e^{-0.5t} - 2e^{-2t} \\ 3 - 2e^{-0.5t} - e^{-2t} \end{bmatrix}, \qquad t \geqq 0 \qquad \textbf{(4.183)}$$

Suppose now that the current source in the network of Figure 4.7 is redistributed and new ones are added such as those shown in Figure 4.8. The corresponding nodal differential equation is the same as that given in (4.177) except for the forcing vectors. For the networks of Figure 4.8(a)

Figure 4.8
Networks having the same topology but different current excitations.

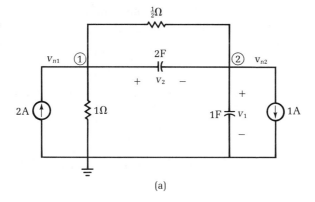

(a)

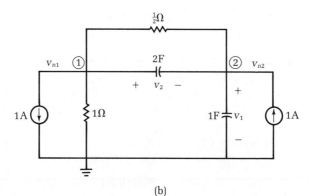

(b)

and (b), the forcing vectors are

$$\mathbf{f}_1 = \begin{bmatrix} 2 \\ -1 \end{bmatrix} \quad \text{and} \quad \mathbf{f}_2 = \begin{bmatrix} -1 \\ 1 \end{bmatrix} \tag{4.184}$$

respectively. For the network of Figure 4.8(a), the zero-state response is found to be

$$\mathbf{v}_{s1}(t) = \begin{bmatrix} V_{n1s} \\ V_{n2s} \end{bmatrix} = \frac{1}{2} \begin{bmatrix} 2 - 2e^{-2t} \\ 1 - e^{-2t} \end{bmatrix}, \quad t \geq 0 \tag{4.185}$$

For the network of Figure 4.8(b), the zero-state response is

$$\mathbf{v}_{s2}(t) = \begin{bmatrix} V_{n1s} \\ V_{n2s} \end{bmatrix} = \frac{1}{6} \begin{bmatrix} -2e^{-0.5t} + 2e^{-2t} \\ 3 - 4e^{-0.5t} + e^{-2t} \end{bmatrix}, \quad t \geq 0 \tag{4.186}$$

If the current sources of Figure 4.8(b) are applied simultaneously to the corresponding points in the network of Figure 4.8(a), the resulting forcing vector becomes

$$\mathbf{f} = \mathbf{f}_1 + \mathbf{f}_2 = \begin{bmatrix} 2 \\ -1 \end{bmatrix} + \begin{bmatrix} -1 \\ 1 \end{bmatrix} = \begin{bmatrix} 1 \\ 0 \end{bmatrix} \tag{4.187}$$

According to Property 4.7 the zero-state response due to $\mathbf{f}$ is equal to the sum of the zero-state responses due to $\mathbf{f}_1$ and $\mathbf{f}_2$ separately, yielding from (4.185) and (4.186)

$$\mathbf{v}_n(t) = \mathbf{v}_{s1}(t) + \mathbf{v}_{s2}(t) = \begin{bmatrix} V_{n1} \\ V_{n2} \end{bmatrix} = \frac{1}{3} \begin{bmatrix} 3 - e^{-0.5t} - 2e^{-2t} \\ 3 - 2e^{-0.5t} - e^{-2t} \end{bmatrix}, \quad t \geq 0 \tag{4.188}$$

confirming (4.183), since $\mathbf{f}$ is the corresponding forcing vector for the network of Figure 4.7.

4-3.2 The Complete Response: Transient and Steady-State

The response of a system to both the input and the initial conditions is called the *complete response*. If the system is described by the nonhomogeneous matrix differential equation

$$\mathbf{W}(p)\mathbf{x}(t) = \mathbf{f}(t) \tag{4.189}$$

with specified initial conditions, then the unique solution obtained from its complete solution by imposing the initial conditions is the complete response. Therefore the zero-input response and the zero-state response are special cases of the complete response. We first demonstrate that

Property 4.8 *The complete response is equal to the sum of the zero-input response and the zero-state response.*

Consider a system described by the differential equation (4.189) with initial condition $\mathbf{x}(t_0)$. The zero-input response is the response of the system when the input is identically zero. Accordingly it is the solution of the associated homogeneous differential equation

$$\mathbf{W}(p)\mathbf{x}_i(t) = \mathbf{0}, \qquad t \geq t_0 \tag{4.190}$$

with $\mathbf{x}_i(t_0) = \mathbf{x}(t_0)$. By definition the zero-state response is the solution of the nonhomogeneous differential equation

$$\mathbf{W}(p)\mathbf{x}_s(t) = \mathbf{f}(t), \qquad t \geq t_0 \tag{4.191}$$

with $\mathbf{x}_s(t_0) = \mathbf{0}$. Adding (4.190) and (4.191) and taking into account the initial conditions $\mathbf{x}_i(t_0) = \mathbf{x}(t_0)$ and $\mathbf{x}_s(t_0) = \mathbf{0}$,

$$\mathbf{W}(p)[\mathbf{x}_i(t) + \mathbf{x}_s(t)] = \mathbf{f}(t), \qquad t \geq t_0 \tag{4.192}$$

with $\mathbf{x}_i(t_0) + \mathbf{x}_s(t_0) = \mathbf{x}(t_0)$. Since $\mathbf{x}_i + \mathbf{x}_s$ satisfies both the required differential equation (4.189) and the initial condition $\mathbf{x}(t_0)$, according to Property 4.6 there is one and only one solution. It follows that the complete response can be expressed as

$$\mathbf{x}(t) = \mathbf{x}_i(t) + \mathbf{x}_s(t), \qquad t \geq t_0 \tag{4.193}$$

We illustrate this by the following example.

EXAMPLE 4.8
Consider again the network of Figure 4.7. As given by (4.50) and (4.183), the zero-input response is

$$\mathbf{v}_i(t) = \begin{bmatrix} v_{n1i} \\ v_{n2i} \end{bmatrix} = \frac{1}{3} \begin{bmatrix} 2e^{-0.5t} + 4e^{-2t} \\ 4e^{-0.5t} + 2e^{-2t} \end{bmatrix}, \qquad t \geq 0 \tag{4.194}$$

and the zero-state response is

$$\mathbf{v}_s(t) = \begin{bmatrix} v_{n1s} \\ v_{n2s} \end{bmatrix} = \frac{1}{3} \begin{bmatrix} 3 - e^{-0.5t} - 2e^{-2t} \\ 3 - 2e^{-0.5t} - e^{-2t} \end{bmatrix}, \qquad t \geq 0 \tag{4.195}$$

The complete response of the network is therefore

$$\mathbf{v}_n(t) = \mathbf{v}_i(t) + \mathbf{v}_s(t) = \begin{bmatrix} v_{n1} \\ v_{n2} \end{bmatrix} = \frac{1}{3} \begin{bmatrix} 3 + e^{-0.5t} + 2e^{-2t} \\ 3 + 2e^{-0.5t} + e^{-2t} \end{bmatrix}, \qquad t \geq 0 \tag{4.196}$$

Previously we indicated that the zero-input response is a linear function of the initial state and that the zero-state response is a linear function of the input. We might be tempted to conclude that the complete response is also a linear function of the input. This is *not* a valid assumption unless the system starts from the zero state. This can be seen from the following argument. Write $\mathbf{f} = \mathbf{f}_1 + \mathbf{f}_2$ in the nonhomogeneous equation (4.189). Call $\mathbf{x}_{s1}$ and $\mathbf{x}_{s2}$ the zero-state responses corresponding to the inputs $\mathbf{f}_1$ and $\mathbf{f}_2$, respectively. Then according to Property 4.7, the zero-state response $\mathbf{x}_s$ corresponding to $\mathbf{f}$ is $\mathbf{x}_s = \mathbf{x}_{s1} + \mathbf{x}_{s2}$. On the other hand, the zero-input

Figure 4.9
Plots of the transient, steady-state, and complete responses of the nodal voltages
(a) v_{n1} and (b) v_{n2}.

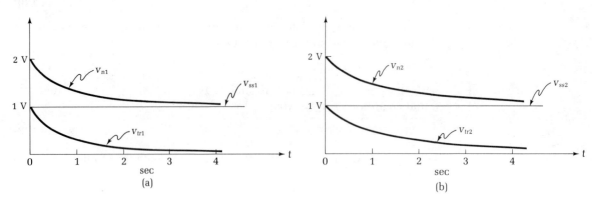

response $\mathbf{x}_i$ is the same for all three inputs $\mathbf{f}_1$, $\mathbf{f}_2$, and $\mathbf{f}$. Call $\mathbf{x}_1$, $\mathbf{x}_2$, and $\mathbf{x}$ the complete responses corresponding to the inputs $\mathbf{f}_1$, $\mathbf{f}_2$, and $\mathbf{f}$, respectively. Then $\mathbf{x}_1 = \mathbf{x}_i + \mathbf{x}_{s1}$, $\mathbf{x}_2 = \mathbf{x}_i + \mathbf{x}_{s2}$, and $\mathbf{x} = \mathbf{x}_i + \mathbf{x}_s$. Clearly $\mathbf{x} \neq \mathbf{x}_1 + \mathbf{x}_2$ unless $\mathbf{x}_i = \mathbf{0}$. This can easily be verified by combining the results obtained in Examples 4.3 and 4.7; the details are left as an exercise (Problem 4.5).

The complete response (4.196) can also be partitioned in a different way, as follows:

$$\mathbf{v}_n(t) = \begin{bmatrix} v_{n1} \\ v_{n2} \end{bmatrix} = \frac{1}{3} \begin{bmatrix} e^{-0.5t} + 2e^{-2t} \\ 2e^{-0.5t} + e^{-2t} \end{bmatrix} + \begin{bmatrix} 1 \\ 1 \end{bmatrix}, \qquad t \geq 0 \tag{4.197}$$

The first term on the right-hand side contains the decaying exponentials as represented by the curves v_{tr1} and v_{tr2} in Figure 4.9. For a sufficiently large t, this right-hand term is negligible compared with the second. For this reason we call the first term the *transient response*. The second term is known as the *steady-state response*. The steady-state response is represented by the straight lines v_{ss1} and v_{ss2} in Figure 4.9. Physically the transient part is due to the initial conditions and the sudden application of the input. If the system is well behaved in time, the transient eventually dies out. On the other hand, the steady-state response is contributed only by the input and has a waveform closely related to that of the input. For example, if the input is a constant such as that given in the above example, the steady-state response is also a constant. If the input is a sinusoid, the steady-state response is also a sinusoid of the same angular frequency, as will be demonstrated in the example given below.

In general, the complete response can be expressed as the sum of the transient response and the steady-state response:

$$\mathbf{x}(t) = \mathbf{x}_{tr}(t) + \mathbf{x}_{ss}(t) \tag{4.198}$$

The term transient response is meaningful only if the response will die out eventually as time goes on. Likewise the steady-state response is significant only if it will go on forever. For a response that does not behave in this manner, it is more convenient to represent it as the sum of the *natural response* and the *forced response*. The natural response corresponds to the complementary function and the forced response to the particular integral of the nonhomogeneous equation describing the system. In fact, if the input is not a constant or periodic, the concept of steady-state loses its significance in that it may also die out with time. For example, if we apply a decaying exponential e^{-t} as an input, the forced or "steady-state" response, which will resemble the input, will die out as time goes on. Likewise if a system possesses a pair of purely imaginary characteristic roots, the natural or "transient" response will contain terms similar to that given in (4.113) with $a = 0$, which will go on forever. It is not really ephemeral or transient. Under these conditions it would be better to call them forced response and natural response. At any rate the complete response is the sum of the zero-input response and the zero-state response.

We illustrate these concepts by the following example.

EXAMPLE 4.9

The network of Figure 4.10 is the same as that of Figure 4.1 except that the 1-V voltage source is replaced by a sinusoidal current source. The nodal differential equation becomes for $t \geq 0$

$$\begin{bmatrix} 3 + 2p & -2 - 2p \\ -2 - 2p & 2 + 3p \end{bmatrix} \begin{bmatrix} v_{n1} \\ v_{n2} \end{bmatrix} = \begin{bmatrix} 1 \\ 0 \end{bmatrix} \cos 2t \qquad \textbf{(4.199)}$$

with initial conditions $v_{n1}(0+) = v_{n2}(0+) = 2$ V. As computed in (4.45), the complementary function is given by

$$\mathbf{v}_h(t) = \begin{bmatrix} v_{n1h} \\ v_{n2h} \end{bmatrix} = \begin{bmatrix} c_1 e^{-0.5t} + 2c_2 e^{-2t} \\ 2c_1 e^{-0.5t} + c_2 e^{-2t} \end{bmatrix}, \qquad t \geq 0 \qquad \textbf{(4.200)}$$

To find a particular integral, we appeal to Table 4.1. An appropriate choice for this is

$$\mathbf{v}_p(t) = \mathbf{K}_1 \cos 2t + \mathbf{K}_2 \sin 2t = \begin{bmatrix} k_{11} \\ k_{21} \end{bmatrix} \cos 2t + \begin{bmatrix} k_{12} \\ k_{22} \end{bmatrix} \sin 2t \qquad \textbf{(4.201)}$$

Substituting this expression in (4.199) yields

$$(\mathbf{M}_1\mathbf{K}_1 - \mathbf{M}_2\mathbf{K}_2) \cos 2t + (\mathbf{M}_1\mathbf{K}_2 + \mathbf{M}_2\mathbf{K}_1) \sin 2t = \mathbf{M}_3 \cos 2t \qquad \textbf{(4.202)}$$

where

$$\mathbf{M}_1 = \begin{bmatrix} 3 & -2 \\ -2 & 2 \end{bmatrix}, \qquad \mathbf{M}_2 = \begin{bmatrix} -4 & 4 \\ 4 & -6 \end{bmatrix}, \qquad \mathbf{M}_3 = \begin{bmatrix} 1 \\ 0 \end{bmatrix} \qquad \textbf{(4.203)}$$

Figure 4.10
A network with switch S_1
closed and switch S_2 open
at $t = 0$.

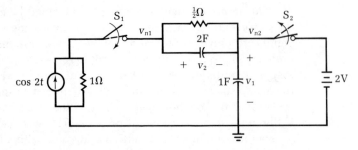

Setting $\mathbf{M}_1\mathbf{K}_2 + \mathbf{M}_2\mathbf{K}_1 = \mathbf{0}$

$$\mathbf{K}_1 = \begin{bmatrix} k_{11} \\ k_{21} \end{bmatrix} = \begin{bmatrix} 1.25 & -0.5 \\ 0.5 & 0 \end{bmatrix}\begin{bmatrix} k_{12} \\ k_{22} \end{bmatrix} \qquad (4.204)$$

Putting this back into (4.202) and solving for $\mathbf{K}_2$, we obtain

$$k_{12} = \frac{7}{17} \quad \text{and} \quad k_{22} = \frac{11}{34} \qquad (4.205)$$

and from (4.204)

$$k_{11} = \frac{6}{17} \quad \text{and} \quad k_{21} = \frac{7}{34} \qquad (4.206)$$

The particular integral can now be written from (4.201) as

$$\mathbf{v}_p(t) = \begin{bmatrix} v_{n1p} \\ v_{n2p} \end{bmatrix} = \frac{1}{34}\begin{bmatrix} 12\cos 2t + 14\sin 2t \\ 7\cos 2t + 11\sin 2t \end{bmatrix}, \qquad t \geq 0 \qquad (4.207)$$

The complete solution is the sum of the complementary function (4.200)
and the particular integral (4.207): For $t \geq t_0$,

$$\begin{bmatrix} v_{n1} \\ v_{n2} \end{bmatrix} = \begin{bmatrix} c_1 e^{-0.5t} + 2c_2 e^{-2t} + \dfrac{6}{17}\cos 2t + \dfrac{7}{17}\sin 2t \\[2ex] 2c_1 e^{-0.5t} + c_2 e^{-2t} + \dfrac{7}{34}\cos 2t + \dfrac{11}{34}\sin 2t \end{bmatrix} \qquad (4.208)$$

To determine the constants c_1 and c_2, we apply the initial conditions
$v_{n1}(0) = v_{n2}(0) = 2$ V,

$$\begin{bmatrix} 1 & 2 \\ 2 & 1 \end{bmatrix}\begin{bmatrix} c_1 \\ c_2 \end{bmatrix} = \frac{1}{34}\begin{bmatrix} 56 \\ 61 \end{bmatrix} \qquad (4.209)$$

yielding

$$c_1 = \frac{11}{17} \quad \text{and} \quad c_2 = \frac{1}{2} \qquad (4.210)$$

Substituting these back into (4.208), we obtain the complete response of the network for $t \geq 0$

$$\begin{bmatrix} v_{n1} \\ v_{n2} \end{bmatrix} = \frac{1}{34} \begin{bmatrix} 22e^{-0.5t} + 34e^{-2t} + 12 \cos 2t + 14 \sin 2t \\ 44e^{-0.5t} + 17e^{-2t} + 7 \cos 2t + 11 \sin 2t \end{bmatrix} \tag{4.211}$$

or, writing it differently,

$$\underbrace{\begin{bmatrix} v_{n1} \\ v_{n2} \end{bmatrix}}_{\substack{\text{Complete} \\ \text{response}}} = \underbrace{\frac{1}{34} \begin{bmatrix} 22e^{-0.5t} + 34e^{-2t} \\ 44e^{-0.5t} + 17e^{-2t} \end{bmatrix}}_{\text{transient response}} + \underbrace{\frac{1}{34} \begin{bmatrix} \sqrt{340} \cos(2t - 0.86217) \\ \sqrt{170} \cos(2t - 1.00407) \end{bmatrix}}_{\text{steady-state response}}$$

$$\text{Complete response} = \text{transient response} + \text{steady-state response}$$

$$\tag{4.212}$$

Observe that the steady-state response closely resembles the input with the same angular frequency, which is 2. The transient response dies out as time goes on. On the other hand, the zero-input response is the response of the network when the sinusoidal source current $\cos 2t$ is set to zero, yielding as computed in (4.67)

$$\mathbf{v}_i(t) = \begin{bmatrix} v_{n1i} \\ v_{n2i} \end{bmatrix} = \frac{1}{3} \begin{bmatrix} 2e^{-0.5t} + 4e^{-2t} \\ 4e^{-0.5t} + 2e^{-2t} \end{bmatrix}, \qquad t \geq 0 \tag{4.213}$$

The zero-state response is the response of the network when the initial conditions are set to zero. From Figure 4.10, this corresponds to setting $v_1(0) = v_2(0) = 0$ or $v_{n1}(0) = v_{n2}(0) = 0$. The complete solution of the differential equation (4.199) is given by (4.208). To compute the zero-state response, this solution is required to satisfy the zero initial condition. This results in

$$\begin{bmatrix} 1 & 2 \\ 2 & 1 \end{bmatrix} \begin{bmatrix} c_1 \\ c_2 \end{bmatrix} = \frac{1}{34} \begin{bmatrix} -12 \\ -7 \end{bmatrix} \tag{4.214}$$

Therefore

$$c_1 = -\frac{1}{51} \quad \text{and} \quad c_2 = -\frac{1}{6} \tag{4.215}$$

Finally, substituting these in (4.208) gives the zero-state response of the network for $t \geq 0$:

$$\mathbf{v}_s(t) = \begin{bmatrix} v_{n1s} \\ v_{n2s} \end{bmatrix} = \frac{1}{102} \begin{bmatrix} -2e^{-0.5t} - 34e^{-2t} + 36 \cos 2t + 42 \sin 2t \\ -4e^{-0.5t} - 17e^{-2t} + 21 \cos 2t + 33 \sin 2t \end{bmatrix} \tag{4.216}$$

The complete response is the sum of the zero-input response (4.213) and the zero-state response (4.216). For $t \geq 0$,

$$\mathbf{v}_i(t) + \mathbf{v}_s(t) = \frac{1}{34} \begin{bmatrix} 22e^{-0.5t} + 34e^{-2t} + 12 \cos 2t + 14 \sin 2t \\ 44e^{-0.5t} + 17e^{-2t} + 7 \cos 2t + 11 \sin 2t \end{bmatrix} \tag{4.217}$$

confirming (4.211). Comparing this with (4.212), we see that the transient part is due not only to the initial conditions but also to the sudden application of the current source. It is composed of the contributions from the zero-input response and a part of the zero-state response. The steady-state response is actually the particular integral of (4.199), as given in (4.207).

EXAMPLE 4.10

In the network of Figure 4.11, switch S_1 is open and switch S_2 is closed for $t < 0$. At $t = 0$, switch S_2 is flipped to open position and switch S_1 is closed. We wish to study the behavior of the inductor current i_1.

It is convenient to use the loop analysis because it involves only one loop integrodifferential equation, which is given by

$$\left(p + \frac{1}{p}\right)i_m(t) = e^{-t}\cos t, \qquad t \geq 0 \tag{4.218}$$

Upon differentiation we obtain

$$(p^2 + 1)i_m(t) = -e^{-t}(\sin t + \cos t), \qquad t \geq 0 \tag{4.219}$$

The characteristic equation is

$$\alpha^2 + 1 = 0 \tag{4.220}$$

giving a pair of purely imaginary characteristic roots $\pm j$. From (4.113) the complementary function can be expressed as

$$i_{mh}(t) = c_1 \cos t + c_2 \sin t, \qquad t \geq 0 \tag{4.221}$$

where c_1 and c_2 are arbitrary constants. To find a particular integral, we appeal to Table 4.1 and select

$$i_{mp}(t) = e^{-t}(c_3 \sin t + c_4 \cos t), \qquad t \geq 0 \tag{4.222}$$

with the coefficients c_3 and c_4 to be determined. Substituting this expression in (4.219),

$$c_3 = \frac{1}{5} \quad \text{and} \quad c_4 = -\frac{3}{5} \tag{4.223}$$

Figure 4.11
A network with switch S_1 closed and switch S_2 open at $t = 0$.

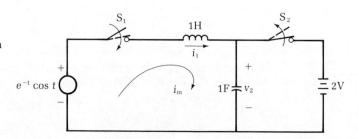

giving

$$i_{mp}(t) = \frac{1}{5} e^{-t}(\sin t - 3 \cos t), \qquad t \geq 0 \tag{4.224}$$

To determine the constants c_1 and c_2, we apply the initial conditions $i_1(0-) = i_1(0+) = 0$ and $v_2(0-) = v_2(0+) = 2$ V to the complete solution i_m obtained by combining (4.221) and (4.224):

$$i_m(t) = c_1 \cos t + c_2 \sin t + \frac{1}{5} e^{-t}(\sin t - 3 \cos t), \qquad t \geq 0 \tag{4.225}$$

The second term on the left-hand side of (4.218) is the voltage v_2 across the capacitor. At $t = 0$, (4.218) becomes

$$pi_m(t)\big|_{t=0+} + 2 = 1 \tag{4.226}$$

Using this in conjunction with the fact that $i_m(0+) = i_1(0+) = 0$, the coefficients c_1 and c_2 in (4.225) are determined as

$$c_1 = \frac{3}{5} \quad \text{and} \quad c_2 = -\frac{9}{5} \tag{4.227}$$

which when put back in (4.225) yields the complete response of the network:

$$i_m(t) = \frac{1}{5} (3 \cos t - 9 \sin t) + \frac{1}{5} e^{-t}(\sin t - 3 \cos t), \qquad t \geq 0 \tag{4.228}$$

The first term on the right-hand side will continue forever and is really not "transient." Therefore the term natural response is preferred to the term transient response. Likewise the second term decreases with time and is not really "steady-state," and we prefer the term forced response to the term steady-state response. At any rate the natural response corresponds to the complementary function of the nonhomogeneous differential equation (4.219) and the forced response to the particular integral. Figure 4.12(a) is a plot of the natural response, whereas the forced response is plotted in Figure 4.12(b). The complete response is presented in Figure 4.12(c).

Alternatively the complete response can be decomposed as the sum of the zero-input response and the zero-state response. The zero-input response is the response when the current source $e^{-t} \cos t$ is set to zero. Using the same initial conditions, we obtain from (4.221) the zero-input response:

$$i_{mi}(t) = -2 \sin t, \qquad t \geq 0 \tag{4.229}$$

For the zero-state response i_{ms}, we apply the zero initial conditions $i_1(0+)=0$ and $v_2(0+) = 0$ or $i_{ms}(0+) = 0$ and $pi_{ms}(t)\big|_{t=0+} = 1$ V to equation (4.225),

$$i_{ms}(t) = \frac{1}{5} (3 \cos t + \sin t) + \frac{1}{5} e^{-t}(\sin t - 3 \cos t), \qquad t \geq 0 \tag{4.230}$$

Figure 4.12
Plots of the natural
response (a), the forced
response (b), and the
complete response (c) of
the inductor current in the
network of Figure 4.11.

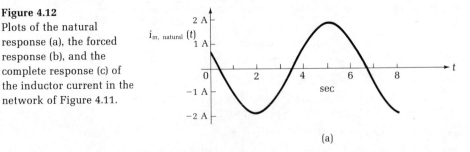

(a)

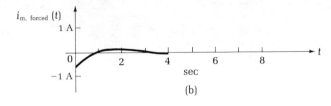

(b)

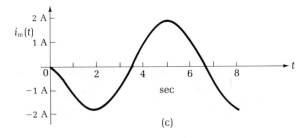

(c)

The complete response can now be expressed in two different ways:

$$i_m(t) \quad = \underbrace{\frac{\sqrt{90}}{5} \cos(t + 1.24905)} + \underbrace{\frac{\sqrt{10}}{5} e^{-t} \cos(t + 3.46334)} \quad \textbf{(4.231a)}$$

$$\begin{array}{c}\text{Complete}\\\text{response}\end{array} = \quad \text{natural response} \quad + \quad \text{forced response}$$

$$= \underbrace{-2 \sin t} + \underbrace{\frac{\sqrt{10}}{5} \cos(t - 0.32175) + \frac{\sqrt{10}}{5} e^{-t} \cos(t + 3.46334)}, \quad t \geqq 0$$

$$\begin{array}{c}\text{zero-}\\\text{input} \quad +\\\text{response}\end{array} \qquad\qquad \text{zero-state response}$$

$$\textbf{(4.231b)}$$

4-4   SUMMARY

In this chapter we presented techniques for solving a system of simultaneous
linear differential equations. We indicated that the *complete solution* of
such a system written in matrix form can be expressed as the sum of a

complementary function and a *particular integral*. The complementary function, which corresponds to the *natural response* or in particular the *transient response*, is the complete solution of the associated homogeneous equation obtained for the original system by deleting the forcing vector. If the characteristic equation has q distinct roots α_j ($j = 1, 2, \ldots, q$) having the multiplicities n_j, the complementary function is given by

$$\mathbf{x}_h(t) = \sum_{j=1}^{q} \sum_{k=1}^{n_j} \sum_{m=1}^{k} \mathbf{K}_{jkm} t^{k-m} e^{\alpha_j t} \tag{4.232}$$

The constant vectors $\mathbf{K}_{jkm}$ appearing initially in (4.232), however, may not all be independent. The necessary relations among them can always be found by substituting the solution into the associated homogeneous equation, which serves to reduce the number of arbitrary constants to the order of the original differential equation.

The particular integral, which corresponds to the *forced response* or in particular the *steady-state response* if the forcing vector is periodic or is a constant, is any solution of the original nonhomogeneous equation. Any solution obtained from outright guessing to the most sophisticated theoretical techniques is acceptable. To this end we presented a procedure, known as the method of undetermined coefficients, for determining a particular integral for a class of very useful functions contained in the forcing vectors as the input signals.

The complete response of a system can be decomposed as the sum of the *zero-input response* and the *zero-state response*. The zero-input response is the response when the input signals to the system are removed and therefore is the response due to the initial state or the energy initially stored in the system. The zero-state response is the response of the system to input signals applied at some arbitrary time t_0 subject to the condition that the system be in the zero state just prior to the application of the signals—that is, at time t_0^-. Thus the zero-state response is due only to the inputs. We proved that the zero-input response is a linear function of the initial state and that the zero-state response is a linear function of the input. However, the complete response is *not* a linear function of the input provided, of course, the system does not start from the zero state. In fact, it can be shown that the linear property for the zero-input response and the zero-state response also holds for linear time-varying systems.

Alternatively the complete response can be represented as the sum of the *natural response* or, in particular for nonsustained oscillation, the *transient response* and the *forced response* or the *steady-state response*. The natural or transient response corresponds to the solution of the associated homogeneous equation and the forced or the steady-state response to the particular solution of the original system. The transient response is made up of two parts. One part is contributed by the zero-input response corresponding to the energy initially stored in the system and the other by that certain part of the zero-state response corresponding to the suddenness of the application of the input signals.

Finally, we mention that the complete solution of a system of differential equations contains an infinite number of solutions. However, if the solution is required to satisfy all the necessary initial conditions, there can be one and only one solution. For a linear time-varying system, the coefficients in the differential equations become time dependent. It can be shown that Properties 4.1–4.4 and 4.6–4.7 remain valid. Therefore the techniques discussed in this chapter can be extended to time-varying systems.

The characteristic roots are often called the *natural frequencies*. A formal definition of the natural frequency will be presented in Chapter 12. The locations of the characteristic roots in the complex plane are important because of their close relation to system stability. The root locations together with the determination of the initial conditions will be discussed in the following chapters when the operational method of solution using the Laplace transformation is introduced.

REFERENCES AND SUGGESTED READING

Balabanian, N., and T. A. Bickart. *Electrical Network Theory.* New York: Wiley, 1969, Chapter 2.

Desoer, C. A., and E. S. Kuh. *Basic Circuit Theory.* New York: McGraw-Hill, 1969, Chapter 14.

Lago, G. V., and L. M. Benningfield. *Circuit and System Theory.* New York: Wiley, 1979, Chapter 3.

Pontryagin, L. S. *Ordinary Differential Equations.* Reading, Mass.: Addison-Wesley, 1962, Chapters 2 and 4.

Wylie, Jr., C. R. *Advanced Engineering Mathematics.* New York: McGraw-Hill, 1960, Chapters 2–4.

PROBLEMS

4.1 Given a series RLC network with $R = 4\ \Omega$, $L = 1$ H, and $C = \frac{1}{2}$ F, write the differential equation and determine the zero-input response for the voltage v_C across the capacitor for $t \geq 0$. The initial conditions are $v_C(0) = 2$ V and inductor current $i_L(0) = 3$ A.

4.2 For the series RLC network of Problem 4.1, let the input be a 1-V battery connected in series through a switch. For $t < 0$, the switch is open. At $t = 0$ the switch is flipped to closed position. Determine the zero-state response for the voltage v_C across the capacitor for $t \geq 0$. Also compute the complete response for v_C for $t \geq 0$, using the initial conditions of Problem 4.1.

4.3 Figure 3.2(a) is a small-signal equivalent network of the simple transistor amplifier of Figure 3.1. Let

$$R_1 = 50\ \Omega, \qquad R_4 = R_6 = 200\ \Omega$$
$$C_2 = 100\ \text{pF}, \qquad C_3 = 5\ \text{pF}$$
$$g_m = 0.2\ \text{mho}, \qquad L_5 = 0 \tag{4.233}$$

Write the differential equation and determine the voltage across the resistor R_6 for $t \geq 0$. The network is initially relaxed and the input voltage $v_g = 10 \cos 10^8 t$ is applied at $t = 0$.

4.4 Given a parallel RLC network with $R = 1\ \Omega$, $L = 1$ H, and $C = 2$ F, write the differential equation and determine the zero-input response for the voltage v_C across the capacitor for $t \geq 0$. The initial conditions are $v_C(0) = 2$ V and inductor current $i_L(0) = 3$ A.

Figure 4.13

A network with switch S_1 closed and switch S_2 open at $t = 0$.

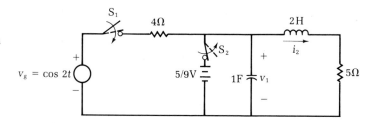

Figure 4.14

A network with switch S closed at $t = 0$.

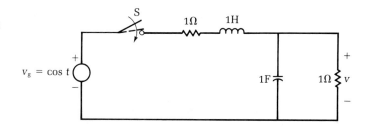

4.5 Using the results in Examples 4.3 and 4.7, verify that the complete response is not a linear function of the input.

4.6 In the network of Figure 4.13, switch S_1 is open and switch S_2 is closed for $t < 0$. At $t = 0$ switch S_1 is flipped to closed position and switch S_2 is open. The output is the voltage v_1 across the capacitor and current i_2 in the inductor. For $t \geq 0$, determine the following for $t \geq 0$:

(a) The loop differential equations.

(b) The zero-input responses for v_1 and i_2.

(c) The zero-state responses for v_1 and i_2.

(d) The transient responses, the steady-state responses, and the complete responses for v_1 and i_2.

4.7 Repeat Problem 4.6 if the voltage source $v_g = \cos 2t$ is replaced by an exponential voltage source $v_g = e^{-0.5t}$.

4.8 Repeat Problem 4.6 if the voltage source $v_g = \cos 2t$ is replaced by a decaying sinusoidal voltage source $v_g = e^{-t} \cos 2t$.

4.9 The switch S in the network of Figure 4.14 is open for $t < 0$. At $t = 0$, switch S is flipped to closed position. Compute and sketch the voltage v for $t \geq 0$.

4.10 Repeat Problem 4.9 if the voltage source $v_g = \cos t$ is replaced by a decaying sinusoidal voltage source $v_g = e^{-t} \cos t$.

4.11 Figure 3.33(a) is a practical version of a field-effect transistor (FET) Colpitts oscillator. After the biasing circuitry and the radio frequency (RF) choke have been removed, an equivalent network is shown in Figure 4.15. Let

$$C_1 = 750 \text{ pF}, \qquad C_2 = 2500 \text{ pF}, \qquad L_3 = 40 \text{ } \mu\text{H}$$
$$R = 50 \text{ k}\Omega, \qquad g_m = 6 \text{ } \mu\text{mho} \tag{4.234}$$

Figure 4.15

A small-signal equivalent network of the FET Colpitts oscillator of Figure 3.33(a).

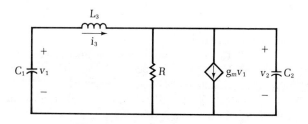

Figure 4.16
A network with switch S_1 closed and switch S_2 flipped from the left position to the right position as shown at $t = 0$.

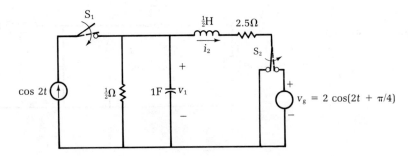

Figure 4.17
A network with switch S_1 closed and switch S_2 flipped to the lower position as shown at $t = 0$.

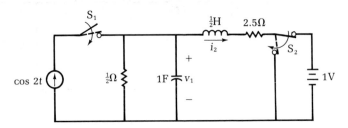

The initial conditions are $v_1(0) = 1$ V, $v_2(0) = 2$ V, and $i_3(0) = 0$. Write the nodal differential equation for $t \geq 0$. Compute and sketch the inductor current i_3 for $t \geq 0$.

4.12 In the network of Fig. 4.16, the switch S_1 is open and switch S_2 is in the left position as shown for $t < 0$. At $t = 0$, switch S_1 is flipped to closed position and switch S_2 to the right position. Verify that the complete responses for the capacitor voltage v_1 and inductor current i_2 for $t \geq 0$ are given by

$$v_1(t) = -1.548e^{-3t} + 1.048e^{-4t} + 0.57\cos(2t - 0.5) \tag{4.235a}$$

$$i_2(t) = -1.548e^{-3t} + 2.096e^{-4t} + 0.713\cos(2t - 2.448) \tag{4.235b}$$

4.13 In the network of Fig. 4.17, the switch S_1 is open and switch S_2 is in the upper position as shown for $t < 0$. At $t = 0$, switch S_1 is flipped to closed position and switch S_2 to lower position. Verify that the complete responses for the capacitor voltage v_1 and inductor current i_2 for $t \geq 0$ are given by

$$v_1(t) = -\frac{3}{10}e^{-4t} + \frac{8}{39}e^{-3t} + \frac{\sqrt{1885}}{130}\cos(2t - 0.671) \tag{4.236a}$$

$$i_2(t) = -\frac{3}{5}e^{-4t} + \frac{8}{39}e^{-3t} + \frac{1}{\sqrt{65}}\cos(2t - 1.052) \tag{4.236b}$$

Also determine the zero-input responses and the zero-state responses for v_1 and i_2 for $t \geq 0$.

4.14 In the network of Figure 4.11, the switch S_1 is open and switch S_2 is closed for $t < 0$. At $t = 0$ switch S_2 is open and switch S_1 is closed for a time interval of 2π sec. At $t = 2\pi$ sec switch S_1 is opened and remains open thereafter, as does switch S_2. Determine the capacitor voltage v_2 and inductor current i_1 for $t \geq 2\pi$. Sketch v_2 and i_1 for $t \geq 2\pi$.

4.15 Repeat Example 4.4 if the constant current source is replaced by a sinusoidal current source $\cos 2t$.

4.16 The network of Figure 4.18 is a third-order Butterworth prototype filter with equal generator and load resistance having normalized value of one ohm. The switch S is open for $t < 0$ and is closed at $t = 0$. Compute and sketch the output voltage v for $t \geq 0$.

4.17 The switch S in the network of Figure 4.19 is closed at $t = 0$. The initial capacitor voltages are zero for $t < 0$. Compute and sketch the output voltage v for $t \geq 0$.

Figure 4.18
A third-order Butterworth
prototype filter with equal
generator and load
resistors having normal-
ized values of one ohm.

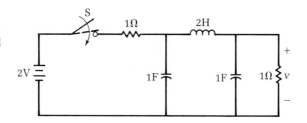

Figure 4.19
A network with switch S
closed at $t = 0$.

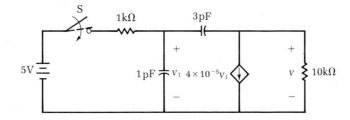

4.18 Repeat Problem 4.17 if the 5-V battery is replaced by a sinusoidal voltage source $\cos 10^8 t$.

4.19 Repeat Problem 4.16 if the 2-V battery is replaced by a sinusoidal voltage source $\cos t$.

4.20 Figure 3.29 is an equivalent network of a transistor feedback amplifier in which all initial conditions are zero. Let

$$g_m = 0.4 \text{ mho}, \qquad r_\pi = 100 \ \Omega, \qquad r_x = 50 \ \Omega$$
$$C_\pi = 100 \text{ pF}, \qquad C_\mu = 10 \text{ pF}, \qquad R_1 = 40 \text{ k}\Omega$$
$$R_2 = 4 \text{ k}\Omega, \qquad v_g = \cos 10^7 t, t \geq 0 \tag{4.237}$$

Determine the capacitor voltages for all $t \geq 0$.

4.21 Figure 4.20 is an active filter in which all initial conditions are zero. At $t = 0$ the switch S is closed. Determine and sketch the output voltage v_o for $t \geq 0$. The operational amplifier is assumed to be ideal.

4.22 The switch S in the network of Figure 4.21 is open for $t < 0$. At $t = 0$ the switch is flipped to closed position. Determine and sketch the inductor current i_1 and capacitor voltage v_2 for $t \geq 0$.

4.23 Repeat Problem 4.22 if the switch S is connected across the inductor instead of the 4 kΩ resistor, everything else being the same.

Figure 4.20
An active filter with a
voltage source applied to
its input through the
closing of the switch S at
$t = 0$.

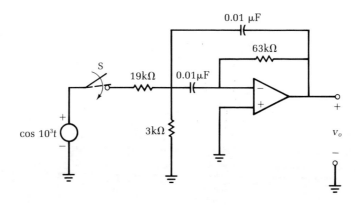

Figure 4.21
A network with switch S
closed at $t = 0$.

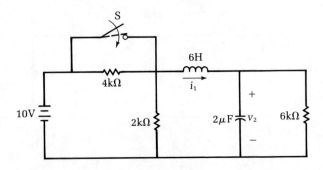

Figure 4.22
A small-signal equivalent
network of a transistor
amplifier.

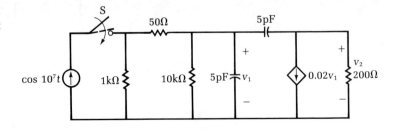

4.24 Figure 4.22 is a small-signal equivalent network of a transistor amplifier. The switch S is open for $t < 0$. At $t = 0$ the switch is flipped to the closed position. Compute and sketch the output voltage v_2 for $t \geq 0$.

CHAPTER FIVE

THE LAPLACE TRANSFORMATION

The classical method of solution of a secondary system of linear simultaneous integrodifferential equations was presented in the foregoing chapters. The technique involves the determination of the complementary function and a particular integral and provides an insight into the interpretation of differential equations and the requirements of a solution. In the present chapter we introduce the operational method of solution using Laplace transformations. As we will see, aside from conceptual advantages, the operational method is better suited to our purpose. Not only does it give the complete response in one operation, but also it specifies the initial conditions automatically. In effect the manipulation of network equations is changed from calculus to algebra.

Operational calculus was invented by the brilliant electrical engineer Oliver Heaviside (1850–1925) in his quest for the solutions of practical electrical network problems. However, his heuristic approach lacked mathematical rigor and therefore drew bitter criticism from the leading mathematicians of his time. Over the years his work has been replaced piece by piece by the method of Laplace transformation, which provides a rigorous basis for his operational method. No important errors have been found in Heaviside's results.

5-1 THE LAPLACE TRANSFORMATION

The use of transformations in solving problems is not new. The *logarithm* is a simple example of a transformation that we used previously. Suppose that we are required to multiply two large numbers $a = 472{,}819$ and $b = 942{,}675$. We might find it convenient to use logarithms by considering

$$\log ab = \log a + \log b \tag{5.1}$$

Thus we transform these numbers by taking their logarithms, which are added to give the transform of the product—that is, the logarithm of the product ab. This transform itself has little meaning. However, if we perform an *inverse transformation* by finding the antilogarithm, we have the desired result. The virtue of the method is that adding numbers is far easier than multiplying them. If this simple example of transformation is not convincing, then evaluate $(94267)^{0.472819}$ without transformations (logarithms). The Laplace transformation is an integral transformation that reduces a system of linear simultaneous *integrodifferential* equations to a system of linear simultaneous *algebraic* equations. Instead of relating a positive number a to another real number log a, the Laplace transform associates a time-domain function with another function, which is defined in the "complex-frequency plane." The transform is manipulated algebraically after the initial conditions are inserted, and then an inverse Laplace transformation is performed to yield the desired solution. In the last step we can use a table of transform pairs just as we use the table of logarithms to obtain the antilogarithm.

The Laplace transformation is a very important and effective tool for studying linear time-invariant networks, but it is of limited value for time-varying and for nonlinear networks. Its importance arises from the fact that it introduces the notion of system function and that it exhibits the close relation between the time-domain behavior of a system and its sinusoidal steady-state behavior.

Given a function of time $f(t)$ defined for all $t \geq 0$, a Laplace transform of $f(t)$ is constructed by first multiplying $f(t)$ by the factor e^{-st}, where s is a complex number called the *complex frequency*

$$s = \sigma + j\omega \tag{5.2}$$

The resulting function $f(t)e^{-st}$ is then integrated with respect to time t from $0-$ to infinity, giving

$$F(s) = \int_{0-}^{\infty} f(t)e^{-st}\,dt \tag{5.3}$$

The function on the left-hand side, denoted by the uppercase f, does not depend on t because the integral has fixed limits and is called the *Laplace transform*[†] $F(s)$ of $f(t)$. To distinguish F from f, we refer to f as the time function and to F as its Laplace transform, and write (5.3) in the form[‡]

$$F(s) = \mathscr{L}[f(t)] \tag{5.4}$$

where the script letter $\mathscr{L}$ is read "the Laplace transform of." The time function $f(t)$ and its transform $F(s)$ are called a *transform pair*. Throughout this text we shall use uppercase to denote the Laplace transforms of the lower-case time functions. Thus $V(s)$ and $I(s)$ denote the Laplace transforms of $v(t)$ and $i(t)$, respectively. The Laplace transform of a vector or a matrix is the transform of each of its elements.

[†] Named for the French mathematician Pierre Simon Laplace (1749–1827), who was professor of mathematics, Military School, Paris.
[‡] More formally, (5.3) is the *unilateral* or *one-sided Laplace transform* of $f(t)$. The *bilateral Laplace transform* of $f(t)$ will be introduced in Section 7-6.1 when the bilateral convolution is presented.

We remark that we choose $0-$ as the lower limit in (5.3) rather than just 0 or $0+$ in order to include cases where the function $f(t)$ may have a jump discontinuity at $t = 0$. The integral (5.3) does not depend on the values of $f(t)$ prior to $t = 0-$. This should not be deemed a restriction on the application of the Laplace transform, since in the usual transient studies the time origin can always be taken at the instance $t = 0$ or some finite time $t > 0$. In evaluating the integral (5.3), we perform the following limiting operations:

$$F(s) = \int_{0-}^{\infty} f(t)e^{-st}\,dt = \lim_{\substack{\epsilon \to 0 \\ \epsilon > 0}} \lim_{T \to \infty} \int_{-\epsilon}^{T} f(t)e^{-st}\,dt \qquad (5.5)$$

A sufficient condition for the integral to exist is that

$$\int_{0-}^{\infty} |f(t)|e^{-\sigma_0 t}\,dt < \infty \qquad (5.6)$$

for some finite σ_0. Clearly, if the integral converges for σ_0, it converges for all $\sigma > \sigma_0$ because the larger the positive real part σ, the faster the integral tends to zero. The region of the s-plane (the complex-frequency plane) in which the Laplace transform of a function exists is referred to as the *region of convergence*. The greatest lower bound of the values σ in the region of convergence is called the *abscissa of convergence* σ_c. In other words, the Laplace integral (5.5) converges for all s with $\sigma > \sigma_c$ and diverges for all s with $\sigma < \sigma_c$.

EXAMPLE 5.1

The *unit step function*, denoted by the symbol $u(t)$, is a function described by the equation

$$u(t) = 0, \qquad t < 0$$
$$ = 1, \qquad t > 0 \qquad (5.7)$$

and its value at $t = 0$ may be taken to be 0, $\frac{1}{2}$, or 1. However, when using Laplace transform, $u(0) = \frac{1}{2}$ is preferable as will be elaborated shortly in (5.16). A plot of $u(t)$ is shown in Figure 5.1, where a discontinuity at $t = 0$ occurs. Such a function is convenient for simulating a sudden closing or opening of a switch or gate. Observe that from (5.7) we have

$$\lim_{t \to 0-} u(t) = 0 \quad \text{and} \quad \lim_{t \to 0+} u(t) = 1 \qquad (5.8)$$

Figure 5.1
The unit-step function.

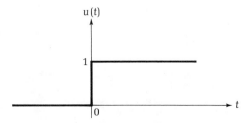

Figure 5.2
The region of convergence
of the Laplace transform
of the unit-step function.

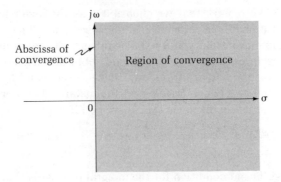

In other words, the limit of the function $u(t)$ as time approaches zero from the left is zero, whereas the limit as time approaches zero from the right is unity.

To compute the Laplace transform of the unit step function, we evaluate the integral

$$\mathscr{L}[u(t)] = \int_{0-}^{\infty} u(t)e^{-st}\,dt = \int_{0+}^{\infty} e^{-st}\,dt = -\frac{1}{s}e^{-st}\Big|_{0+}^{\infty} \tag{5.9}$$

For $t > 0$ and $\sigma > 0$, the upper limit, $t = \infty$, will make $e^{-st} = 0$, giving

$$\mathscr{L}[u(t)] = \frac{1}{s} \tag{5.10}$$

Since the upper limit, hence the integral, exists only if σ is positive, the region of convergence of $\mathscr{L}[u(t)]$ is the right half of the s-plane, as depicted in Figure 5.2, with the imaginary axis being the abscissa of convergence. Therefore the Laplace transform of $u(t)$ is defined only for Re $s = \sigma > 0$. However, by appealing to the theory of analytic continuation of a function of a complex variable,[†] which we will not treat in depth here, we can enlarge the domain of the function to all values of s except $s = 0$. It is also intuitively reasonable to consider the Laplace transform $1/s$ as defined for all values of s except $s = 0$. The process of extension is known as the *technique of analytic continuation* and will be applied to other transforms.

EXAMPLE 5.2
We compute the Laplace transform of an exponential function $f(t) = e^{at}$, where a is a real or complex number. By definition the Laplace transform of e^{at} is given by

$$\mathscr{L}[e^{at}] = \int_{0-}^{\infty} e^{at}e^{-st}\,dt = \int_{0-}^{\infty} e^{-(s-a)t}\,dt = -\frac{e^{-(s-a)t}}{s-a}\Big|_{0-}^{\infty}$$

$$= \frac{1}{s-a} \qquad \text{for Re } (s-a) > 0 \tag{5.11}$$

[†] See, for example, R. V. Churchill, *Introduction to Complex Variables and Applications*, 2d ed., (New York: McGraw-Hill, 1960), Chapter 12.

Figure 5.3
The regions of convergence of the Laplace transform of an exponential function
e^{at} with (a) Re $a \leq 0$ and (b) Re $a > 0$.

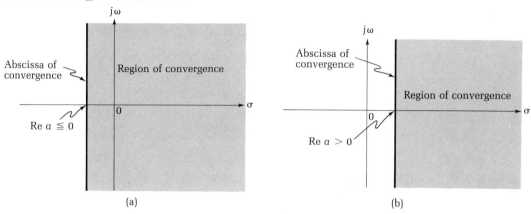

(a)   (b)

The integral exists only if Re$(s - a) > 0$. Therefore the abscissa of convergence for $\mathscr{L}[e^{at}]$ is a straight line described by Re s = Re a. Depending on the sign of Re a, the abscissa of convergence is sketched in Figure 5.3. The region of convergence is to the right of the abscissa of convergence. The Laplace transform of e^{at} is defined only for Re$(s - a) > 0$. Again, by the technique of analytic continuation, the transform $1/(s - a)$ can be enlarged to all values of s except $s = a$. The functions e^{at} and $1/(s - a)$ constitute a transform pair.

EXAMPLE 5.3
Consider the time function $f(t) = e^{at^2}$. By definition its Laplace transform is given by

$$\mathscr{L}[e^{at^2}] = \int_{0-}^{\infty} e^{at^2} e^{-st}\, dt = \int_{0-}^{\infty} e^{(at-s)t}\, dt \tag{5.12}$$

For any fixed value of s—however large, when Re $at >$ Re s, the integral increases monotonically without bound as t approaches infinity. Thus there is no value of s for which the integral exists. We say that the function e^{at^2} has no Laplace transform or it is not Laplace transformable. However, if a generator that produces this signal is for a limited range of values of t, however large the extent of this range, the function then has a Laplace transform. To see this, let us consider as the input signal the function

$$f(t) = e^{at^2}, \qquad 0 \leq t \leq t_1$$
$$= 0, \qquad t > t_1 \tag{5.13}$$

for any time t_1 as large as we wish. Then we have

$$\mathscr{L}[f(t)] = \int_{0-}^{\infty} f(t) e^{-st}\, dt = \int_{0-}^{t_1} e^{(at-s)t}\, dt \tag{5.14}$$

This integral is now well defined, and the input signal (5.13) therefore possesses a Laplace transform.

EXAMPLE 5.4

Compute the Laplace transform of the sinusoidal signal $f(t) = \sin \omega_0 t$. By definition, its transform is

$$\mathcal{L}[f(t)] = \int_{0-}^{\infty} \sin \omega_0 t \, e^{-st} \, dt = \frac{1}{2j} \int_{0-}^{\infty} (e^{j\omega_0 t} - e^{-j\omega_0 t}) e^{-st} \, dt$$

$$= \frac{1}{2j} \left[-\frac{e^{-(s-j\omega_0)t}}{s - j\omega_0} + \frac{e^{-(s+j\omega_0)t}}{s + j\omega_0} \right]_{0-}^{\infty}$$

$$= \frac{1}{2j} \left[\frac{1}{s - j\omega_0} - \frac{1}{s + j\omega_0} \right] = \frac{\omega_0}{s^2 + \omega_0^2} \qquad \text{for Re } s = \sigma > 0$$

$$(5.15)$$

The abscissa of convergence is the $j\omega$-axis and the region of convergence is the entire open right-half of the s-plane—namely, the right half of the s-plane, excluding the $j\omega$-axis. As before, by using the technique of analytic continuation, the Laplace transform $\omega_0/(s^2 + \omega_0^2)$ can be extended to all values of s except $s = \pm j\omega_0$, while the integral (5.15) is defined only for all s with $\sigma > 0$. The functions $\sin \omega_0 t$ and $\omega_0/(s^2 + \omega_0^2)$ constitute a transform pair.

Several points are important to mention at this stage. First, the Laplace integral is defined only between $0-$ and $+\infty$. The behavior of the function $f(t)$ for $t < 0$ never enters the integral and therefore has no effect on its transform. For example, the functions $f(t) = 1$ and $u(t)$ have the same transform $1/s$. The exponential functions e^{-t} and $e^{-t}u(t)$, as depicted in Figure 5.4, possess the same transform—namely, $1/(s + 1)$—even though they behave differently for $t < 0$. Second, the Laplace integral converges only in a region of the s-plane. Using the technique of analytic continuation, the domain of the function can be extended to all values of s except the singular points. This formality of extension will be assumed implicitly for all the

Figure 5.4
Plots of the exponential functions (a) e^{-t} and (b) $e^{-t}u(t)$.

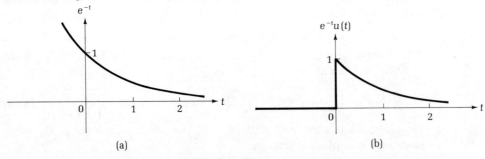

Table 5.1   Laplace Transforms of Elementary Functions

Time function $f(t)$	Laplace transform $F(s)$
$\delta(t)$	1
$u(t)$	$\dfrac{1}{s}$
$\dfrac{1}{n!}\, t^n\ (n = 1, 2, \ldots)$	$\dfrac{1}{s^{n+1}}$
e^{-at}, a real or complex	$\dfrac{1}{s + a}$
$\dfrac{1}{n!}\, t^n e^{-at}$, a real or complex	$\dfrac{1}{(s + a)^{n+1}}\quad (n = 1, 2, \ldots)$
$\cos(\omega_0 t + \theta)$	$\dfrac{s \cos\theta - \omega_0 \sin\theta}{s^2 + \omega_0^2}$
$\sin(\omega_0 t + \theta)$	$\dfrac{s \sin\theta + \omega_0 \cos\theta}{s^2 + \omega_0^2}$
$e^{-\alpha t}\cos\omega_0 t$	$\dfrac{s + \alpha}{(s + \alpha)^2 + \omega_0^2}$
$e^{-\alpha t}\sin\omega_0 t$	$\dfrac{\omega_0}{(s + \alpha)^2 + \omega_0^2}$
$\dfrac{1}{2\omega_0}\, t \sin\omega_0 t$	$\dfrac{s}{(s^2 + \omega_0^2)^2}$
$\cosh\alpha t$	$\dfrac{s}{s^2 - \alpha^2}$
$\sinh\alpha t$	$\dfrac{\alpha}{s^2 - \alpha^2}$

transforms. Third, in computing the Laplace transform, the direct-integration approach may not always be the best. It is sometimes convenient to treat complicated functions as algebraic combinations of several simple functions having Laplace transforms already known and often tabulated, as in accompanying Table 5.1. The last consideration is of the uniqueness of a Laplace transform. Can two different functions defined for all $t \geq 0-$ possess the same Laplace transform? Fortunately it can be shown[†] that if a function is defined as

$$f(t) = \begin{cases} 0, & t < 0 \\ \tfrac{1}{2}f(0+), & t = 0 \\ \tfrac{1}{2}[f(t+) + f(t-)], & t > 0 \end{cases} \tag{5.16}$$

[†] See footnote on p. 180.

Figure 5.5
A staircase function used
to illustrate conditions
under which there is one
and only one Laplace
transform.

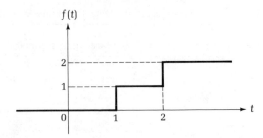

then there is one and only one Laplace transform. No two different functions can have the same Laplace transform. As an example, we take the unit step function $u(t)$. If we define

$$u(t) = \begin{cases} 0, & t < 0 \\ \frac{1}{2}, & t = 0 \\ 1, & t > 0 \end{cases} \tag{5.17}$$

then there can be one and only one Laplace transform of this function. The essential requirement is that at the points of discontinuities for $t > 0$, the values of the function are usually not defined. As an approach we choose the average value of $f(t+)$ and $f(t-)$. In Figure 5.5, if the values of the function at the points of discontinuities are defined as

$$f(1) = \tfrac{1}{2}[f(1+) + f(1-)] = \tfrac{1}{2}(1 + 0) = \tfrac{1}{2} \tag{5.18a}$$

$$f(2) = \tfrac{1}{2}[f(2+) + f(2-)] = \tfrac{1}{2}(2 + 1) = 1.5 \tag{5.18b}$$

then it possesses a unique Laplace transform.

On the other hand, if the Laplace transform $F(s)$ of a function $f(t)$ is known, then there is a unique time function defined by the integral[†]

$$f(t) = \frac{1}{2\pi j} \int_{\sigma_1 - j\infty}^{\sigma_1 + j\infty} F(s)e^{st}\,ds = \begin{cases} 0, & t < 0 \\ \tfrac{1}{2}f(0+), & t = 0 \\ \tfrac{1}{2}[f(t+) + f(t-)], & t > 0 \end{cases} \tag{5.19}$$

where $\sigma_1 \geqq 0$ and $\sigma_1 > \sigma_c$, and σ_c is the abscissa of convergence. The contour integral (5.19) that gives the inverse Laplace transformation is referred to as the *complex inversion integral*. The path of integration is along the vertical line $s = \sigma_1$ from the point $-j\infty$ to $j\infty$, as depicted in Figure 5.6. To simplify we symbolize (5.19) by

$$f(t) = \mathscr{L}^{-1}[F(s)] \tag{5.20}$$

where the script letter $\mathscr{L}^{-1}$ is read "the inverse Laplace transform of."

[†] See, for example, D. V. Widder, *Laplace Transform* (Princeton, N.J.: Princeton University Press, 1941), Chapter II.

Figure 5.6
The path of integration
used in the complex
inversion integral (5.19).

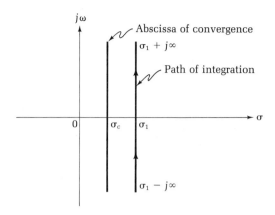

The integral (5.19) appears to be formidable to evaluate. However, in most cases we may use the table of transform pairs[†] such as the one given in Table 5.1 to find $f(t)$. This is possible because of the uniqueness of the transform and its inverse transform, as defined above. The evaluation of (5.19) and the algebraic manipulation of the transform $F(s)$ into the sum of several simple functions the inverse Laplace transforms of which are already known or tabulated will be presented later.

5-2   BASIC PROPERTIES OF THE LAPLACE TRANSFORM

In this section we discuss a number of fundamental properties of the Laplace transform that are useful in solving a system of simultaneous linear integrodifferential equations.

Linearity

Let $F_1(s)$ and $F_2(s)$ be the Laplace transforms of the time functions $f_1(t)$ and $f_2(t)$, respectively. Then for arbitrary constants c_1 and c_2 we have

$$\mathscr{L}[c_1f_1(t) + c_2f_2(t)] = c_1F_1(s) + c_2F_2(s) \tag{5.21}$$

In words this states that the Laplace transformation is a linear function. To verify this we apply definition (5.5) to the left-hand side of (5.21) and obtain

$$\mathscr{L}[c_1f_1(t) + c_2f_2(t)] = \int_{0-}^{\infty} [c_1f_1(t) + c_2f_2(t)]e^{-st}\,dt$$

$$= c_1\int_{0-}^{\infty} f_1(t)e^{-st}\,dt + c_2\int_{0-}^{\infty} f_2(t)e^{-st}\,dt$$

$$= c_1F_1(s) + c_2F_2(s) \tag{5.22}$$

[†] For more extensive tables of Laplace transforms, see M. F. Gardner and J. L. Barnes, *Transients in Linear Systems* (New York: Wiley, 1942); and G. E. Roberts and H. Kaufman, *Table of Laplace Transforms* (Philadelphia: W. B. Saunders, 1966).

EXAMPLE 5.5

We use (5.22) to find the Laplace transform of $\cos \omega_0 t$, which by Euler's formula (4.112) can be expressed as

$$\cos \omega_0 t = \tfrac{1}{2}[e^{j\omega_0 t} + e^{-j\omega_0 t}] \tag{5.23}$$

Since the transform of the exponential is known, as given in (5.11), we get from (5.22)

$$
\begin{aligned}
\mathscr{L}[\cos \omega_0 t] &= \mathscr{L}[\tfrac{1}{2}e^{j\omega_0 t} + \tfrac{1}{2}e^{-j\omega_0 t}] \\
&= \tfrac{1}{2}\mathscr{L}[e^{j\omega_0 t}] + \tfrac{1}{2}\mathscr{L}[e^{-j\omega_0 t}] \\
&= \frac{1}{2}\frac{1}{s - j\omega_0} + \frac{1}{2}\frac{1}{s + j\omega_0} = \frac{s}{s^2 + \omega_0^2}
\end{aligned} \tag{5.24}
$$

Real Differentiation

Let $F(s)$ be the Laplace transform of $f(t)$. Then

$$\mathscr{L}\left[\frac{df(t)}{dt}\right] = \mathscr{L}[pf(t)] = sF(s) - f(0-) \tag{5.25}$$

To verify this, we apply definition (5.5) to $pf(t)$ and obtain

$$\mathscr{L}\left[\frac{df(t)}{dt}\right] = \int_{0-}^{\infty} \frac{df(t)}{dt} e^{-st}\, dt \tag{5.26}$$

which can be integrated by parts by letting

$$u = e^{-st} \quad \text{and} \quad dv = df(t) \tag{5.27}$$

in the equation

$$\int_a^b u\, dv = uv \Big|_a^b - \int_a^b v\, du \tag{5.28}$$

This gives

$$
\begin{aligned}
\mathscr{L}\left[\frac{df(t)}{dt}\right] &= e^{-st}f(t)\Big|_{0-}^{\infty} + s\int_{0-}^{\infty} f(t)e^{-st}\, dt \\
&= sF(s) - f(0-)
\end{aligned} \tag{5.29}
$$

provided that we take Re s sufficiently large so that, as $t \to \infty$, $f(t)e^{-st} \to 0$.

To find the second derivative, we apply (5.25) twice, as follows:

$$
\begin{aligned}
\mathscr{L}\left[\frac{d^2 f(t)}{dt^2}\right] &= \mathscr{L}[p^2 f(t)] = \mathscr{L}[ppf(t)] = s\mathscr{L}[pf(t)] - pf(t)\big|_{t=0-} \\
&= s[sF(s) - f(0-)] - pf(t)\big|_{t=0-} \\
&= s^2 F(s) - sf(0-) - pf(t)\big|_{t=0-}
\end{aligned} \tag{5.30}
$$

Likewise we can show that for the nth derivative

$$\mathscr{L}\left[\frac{d^n f(t)}{dt^n}\right] = \mathscr{L}[p^n f(t)]$$

$$= s^n F(s) - s^{n-1} f(0-) - s^{n-2} pf(t)\big|_{t=0-} - \cdots - p^{n-1} f(t)\big|_{t=0-}$$

(5.31)

EXAMPLE 5.6

We compute the Laplace transform of the derivative of $\cos \omega_0 t$. Let $f(t) = \cos \omega_0 t$. From (5.25) in conjunction with (5.24)

$$\mathscr{L}[p(\cos \omega_0 t)] = s\frac{s}{s^2 + \omega_0^2} - 1 = -\frac{\omega_0^2}{s^2 + \omega_0^2}$$

(5.32)

Since $p(\cos \omega_0 t) = -\omega_0 \sin \omega_0 t$, (5.32) becomes

$$\mathscr{L}[\sin \omega_0 t] = \frac{\omega_0}{s^2 + \omega_0^2}$$

(5.33)

confirming (5.15).

EXAMPLE 5.7

A rectangular pulse function is defined by

$$\Delta(t) = \begin{cases} 0, & t < 0 \\ 1/\epsilon, & 0 < t < \epsilon \\ 0, & t > 0 \end{cases}$$

(5.34)

as depicted in Figure 5.7. Observe that the area under $\Delta(t)$ is 1. As $\epsilon \to 0$, the pulse function (5.34) becomes the *unit impulse* or the *Dirac delta function* after physicist P. A. M. Dirac (1902–   ), who first used it in his writings on quantum mechanics:[†]

$$\delta(t) = \begin{cases} 0 & \text{for } t \neq 0 \\ \text{singular} & \text{at } t = 0 \end{cases}$$

(5.35)

[†] P. A. M. Dirac, *The Principles of Quantum Mechanics* (Oxford: Oxford University Press, 1930).

Figure 5.7
A rectangular pulse
function.

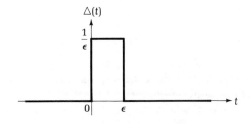

Figure 5.8
The unit-impulse or the
Dirac delta function.

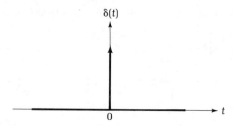

with

$$\int_{-a}^{a} \delta(t)\, dt = 1 \tag{5.36}$$

for any real $a > 0$. Thus from the definition of $\delta(t)$ and $u(t)$ we get

$$u(t) = \int_{-\infty}^{t} \delta(\tau)\, d\tau \tag{5.37}$$

and

$$\delta(t) = \frac{du(t)}{dt} = pu(t) \tag{5.38}$$

The unit impulse $\delta(t)$ is represented graphically as in Figure 5.8.

A frequently used property of the unit impulse is the *sifting property*, expressed by

$$\int_{-a}^{a} f(t)\, \delta(t)\, dt = f(0) \tag{5.39}$$

for any continuous function $f(t)$. To see this, we make use of the pulse function (5.34) to approximate the unit impulse as follows:

$$\int_{-a}^{a} f(t)\, \delta(t)\, dt = \lim_{\epsilon \to 0} \int_{-a}^{a} f(t)\, \Delta(t)\, dt = \lim_{\epsilon \to 0} \int_{0}^{\epsilon} f(t)\, \frac{1}{\epsilon}\, dt$$

$$= \lim_{\epsilon \to 0} \frac{1}{\epsilon} f(\epsilon)\epsilon = f(0) \tag{5.40}$$

To find the Laplace transform of the unit impulse, we appeal to (5.25) in conjunction with (5.10) and (5.38)

$$\mathscr{L}[\delta(t)] = \mathscr{L}[pu(t)] = s\mathscr{L}[u(t)] - u(0-) = s\,\frac{1}{s} - 0 = 1 \tag{5.41}$$

To demonstrate how to handle the Laplace transform of the derivative of a function containing discontinuities, we consider the time function $f(t) = (\cos \omega_0 t)u(t)$, which has a discontinuity at $t = 0$. The derivative of this function is found to be

$$\frac{d}{dt}[(\cos \omega_0 t)u(t)] = \cos \omega_0 t \frac{du(t)}{dt} - \omega_0(\sin \omega_0 t)u(t)$$

$$= \delta(t) - \omega_0(\sin \omega_0 t)u(t)$$

This yields

$$\mathscr{L}\left[\frac{df(t)}{dt}\right] = 1 - \frac{\omega_0^2}{s^2 + \omega_0^2} = s\frac{s}{s^2 + \omega_0^2} = s\mathscr{L}[f(t)]$$

The last equality follows from the fact that the functions $\cos \omega_0 t$ and $(\cos \omega_0 t)u(t)$ have the same Laplace transform because the behavior of a function for $t < 0$ has no effect on its transform.

Let us modify the function $f(t)$ by setting its values to -1 for $t < 0$. We thus obtain a new function $g(t)$ defined by

$$g(t) = \begin{cases} -1 & t < 0 \\ \cos \omega_0 t & t \geq 0 \end{cases}$$

Clearly the Laplace transform of $g(t)$ is

$$\mathscr{L}[g(t)] = \mathscr{L}[\cos \omega_0 t] = \frac{s}{s^2 + \omega_0^2}$$

However, the derivative of $g(t)$ is given by

$$\frac{dg(t)}{dt} = 2\delta(t) - \omega_0(\sin \omega_0 t)u(t)$$

the Laplace transform of which is

$$\mathscr{L}\left[\frac{dg(t)}{dt}\right] = 2 - \frac{\omega_0^2}{s^2 + \omega_0^2} = \frac{2s^2 + \omega_0^2}{s^2 + \omega_0^2} = 1 + s\mathscr{L}[g(t)]$$

Observe that even though the functions $f(t)$ and $g(t)$ have the same Laplace transform, the transforms of their derivatives $pf(t)$ and $pg(t)$ differ because the magnitude of the jump in $f(t)$ at $t = 0$ is 1, whereas that of $g(t)$ is 2.

Real Integration

Let $F(s)$ be the Laplace transform of a time function $f(t)$. Then

$$\mathscr{L}\left[\int_{0-}^{t} f(\tau)\, d\tau\right] = \frac{1}{s}F(s) \tag{5.42}$$

To prove (5.42) we apply definition (5.5) to yield

$$\mathscr{L}\left[\int_{0-}^{t} f(\tau)\, d\tau\right] = \int_{0-}^{\infty}\left[\int_{0-}^{t} f(\tau)\, d\tau\right]e^{-st}\, dt \tag{5.43}$$

Again, we integrate (5.43) by parts by letting

$$u = \int_{0-}^{t} f(\tau)\, d\tau \quad \text{and} \quad dv = e^{-st}\, dt \qquad \qquad \textbf{(5.44)}$$

in (5.28), getting

$$\mathscr{L}\left[\int_{0-}^{t} f(\tau)\, d\tau\right] = \frac{e^{-st}}{-s} \int_{0-}^{t} f(\tau)\, d\tau \bigg|_{0-}^{\infty} + \frac{1}{s}\int_{0-}^{\infty} f(t)e^{-st}\, dt \qquad \qquad \textbf{(5.45)}$$

If we take sufficiently large Re s so that as $t \to \infty$ the exponential e^{-st} approaches zero, the upper limit for the first term on the right-hand side is zero. At the lower limit $t = 0-$, the integral is zero. Hence the first term is zero. Equation (5.45) reduces to (5.42).

By repeated application of (5.42) we can show that the Laplace transform of the nth-order integration is given by

$$\mathscr{L}\left[\underbrace{\int_{0-}^{t}\int_{0-}^{t_2} \cdots \int_{0-}^{t_n} f(\tau)\, d\tau\, dt_n\, dt_{n-1} \cdots dt_2}_{n \text{ integrals}}\right] = \frac{1}{s^n}F(s) \qquad \qquad \textbf{(5.46)}$$

EXAMPLE 5.8

We compute the Laplace transforms of the functions obtained from the unit step function by successive integration.

$$r(t) = \int_{0-}^{t} u(\tau)\, d\tau = tu(t) \qquad \qquad \textbf{(5.47)}$$

The function $r(t)$ is known as the *unit ramp*, as depicted in Figure 5.9. Appealing to (5.42) in conjunction with (5.10) gives the Laplace transform of the unit ramp:

$$\mathscr{L}[r(t)] = \mathscr{L}[tu(t)] = \frac{1}{s}\mathscr{L}[u(t)] = \frac{1}{s} \cdot \frac{1}{s} = \frac{1}{s^2} \qquad \qquad \textbf{(5.48)}$$

The integration of $r(t)$ from $0-$ to t is $t^2 u(t)/2!$, the Laplace transform of which, as obtained from (5.42) and (5.48), is

$$\mathscr{L}\left[\frac{1}{2!}t^2 u(t)\right] = \frac{1}{s}\mathscr{L}[r(t)] = \frac{1}{s} \cdot \frac{1}{s^2} = \frac{1}{s^3} \qquad \qquad \textbf{(5.49)}$$

Continuing this way, we can show that

$$\mathscr{L}\left[\frac{1}{n!}t^n u(t)\right] = \frac{1}{s^{n+1}} \qquad \qquad \textbf{(5.50)}$$

as indicated in Table 5.1.

Figure 5.9
The unit ramp.

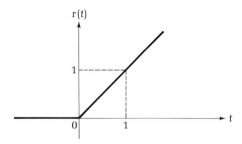

Figure 5.10
A delayed unit-step
function.

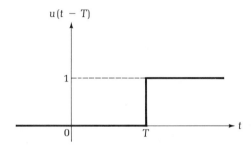

Real Translation (Shifting Theorem)

Let $F(s)$ be the Laplace transform of $f(t)u(t)$. We are interested in finding the Laplace transform of a function $f_d(t)$ which is identical to $f(t)u(t)$ except that it is *delayed* or *shifted* by T sec, $T > 0$. We remark that we use $f(t)u(t)$ instead of $f(t)$ in order to suppress the behavior of $f(t)$ prior to $t = 0-$. This is necessary in applying the shifting theorem to be presented below because the Laplace transform suppresses the nature of the time function prior to $t = 0-$. Figure 5.10 is the unit step function delayed by T sec and can be expressed mathematically as $u_d(t) = u(t - T)$. In general, the delayed function $f_d(t)$ can be written as

$$f_d(t) = f(t - T)u(t - T) \tag{5.51}$$

for all t, the Laplace transform of which is from (5.5)

$$\int_{0-}^{\infty} f_d(t)e^{-st}\, dt = \int_{0-}^{\infty} f(t - T)u(t - T)e^{-st}\, dt$$

$$= \int_{-T}^{\infty} f(x)u(x)e^{-s(x+T)}\, dx = e^{-sT} \int_{0-}^{\infty} f(x)u(x)e^{-sx}\, dx$$

$$= e^{-sT} \int_{0-}^{\infty} f(t)e^{-st}\, dt = e^{-sT}F(s) \tag{5.52}$$

Thus we have the shifting theorem with $T > 0$

$$\mathscr{L}[f_d(t)] = \int_{0-}^{\infty} f(t - T)u(t - T)e^{-st}\, dt = e^{-sT}F(s) \tag{5.53}$$

Figure 5.11
A rectangular pulse
function (a), which can be
considered as the sum of
two unit-step functions
(b).

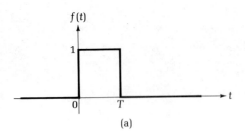

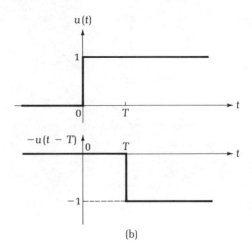

(a)   (b)

EXAMPLE 5.9

We compute the Laplace transform of the pulse function $f(t)$ as shown in
Figure 5.11(a). This pulse signal can be considered as the sum of the two
unit step functions $u(t)$ and $-u(t - T)$ as illustrated in Figure 5.11(b). By
(5.21) and (5.53), we have

$$\mathscr{L}[f(t)] = \mathscr{L}[u(t) - u(t - T)] = \mathscr{L}[u(t)] - e^{-sT}\mathscr{L}[u(t)]$$

$$= \frac{1}{s} - e^{-sT}\frac{1}{s} = \frac{1}{s}(1 - e^{-sT}) \tag{5.54}$$

Alternatively we can use (5.38) and (5.53) and obtain

$$\mathscr{L}[pf(t)] = \mathscr{L}[\delta(t) - \delta(t - T)] = \mathscr{L}[\delta(t)] - e^{-sT}\mathscr{L}[\delta(t)]$$

$$= 1 - e^{-sT} \tag{5.55}$$

From (5.25), the left-hand side is $s\mathscr{L}[f(t)]$ since $f(0-) = 0$. Substituting this
in (5.55) and solving for $\mathscr{L}[f(t)]$ yield (5.54).

5-3   SOLUTION OF SECONDARY SYSTEMS OF
NETWORK EQUATIONS

In this section we demonstrate how the Laplace transformation is used to
solve the secondary systems of network equations of the form

$$\mathbf{W}(p)\mathbf{x}(t) = \mathbf{f}(t) \tag{5.56}$$

as given in (4.3), where $\mathbf{W}(p)$ is the coefficient matrix of the differential
operator p and the integration operator $1/p$, $\mathbf{x}(t)$ is the unknown vector, and
$\mathbf{f}(t)$ is the forcing vector of the sources. Equation (5.56) denotes either the

nodal equations, the cutset equations, or the loop equations and, in general, is a set of linear simultaneous integrodifferential equations. Consider, for example, the nodal system of equations for the network of Figure 3.16(a), as given by (3.56) and repeated below with $t_0 = 0$:

$$
\begin{bmatrix}
G_1 + (C_2 + C_3)p & -C_3p & 0 \\
g_m - C_3p & G_4 + C_3p + \dfrac{1}{L_5p} & -\dfrac{1}{L_5p} \\
0 & -\dfrac{1}{L_5p} & G_6 + \dfrac{1}{L_5p}
\end{bmatrix}
\begin{bmatrix}
v_{n1}(t) \\
v_{n2}(t) \\
v_{n3}(t)
\end{bmatrix}
=
\begin{bmatrix}
G_1 v_g(t) \\
-i_5(0) \\
i_5(0)
\end{bmatrix}
$$

$$(5.57)$$

Now we take the Laplace transform of (5.57) by applying formulas (5.21), (5.25), and (5.42). The result is given by

$$
\begin{bmatrix}
G_1 + (C_2 + C_3)s & -C_3s & 0 \\
g_m - C_3s & G_4 + C_3s + \dfrac{1}{L_5s} & -\dfrac{1}{L_5s} \\
0 & -\dfrac{1}{L_5s} & G_6 + \dfrac{1}{L_5s}
\end{bmatrix}
\begin{bmatrix}
V_{n1}(s) \\
V_{n2}(s) \\
V_{n3}(s)
\end{bmatrix}
$$

$$
=
\begin{bmatrix}
G_1 V_g(s) \\
0 \\
0
\end{bmatrix}
+
\begin{bmatrix}
C_2 v_{n1}(0-) + C_3[v_{n1}(0-) - v_{n2}(0-)] \\
-i_5(0)/s + C_3[v_{n2}(0-) - v_{n1}(0-)] \\
i_5(0)/s
\end{bmatrix}
\qquad (5.58)
$$

This can be written more compactly in matrix notation as

$$
\mathbf{Y}_n(s)\mathbf{V}_n(s) = \mathbf{J}_{n1}(s) + \mathbf{J}_{n2}(s) = \mathbf{J}_n(s) \qquad (5.59)
$$

where $V_{nk}(s)$ ($k = 1, 2, 3$) and $V_g(s)$ are the Laplace transforms of $v_{nk}(t)$ and $v_g(t)$, respectively. It is significant to observe that (5.58) is a system of linear simultaneous *algebraic* equations, which is much easier to handle than the original integrodifferential equations. The coefficient matrix $\mathbf{Y}_n(s)$, which is identical in form to the node-admittance matrix operator as defined in (5.57) except that p is replaced by s, is simply referred to as the *node-admittance matrix* by dropping the word operator. The r-vector $\mathbf{V}_n(s)$, denoting the Laplace transforms of the node-to-datum voltages, is called the *node-to-datum voltage vector*. Strictly speaking, it should be called the Laplace transform of the node-to-datum voltage vector, but the words "the Laplace transform of" are usually omitted in common practice. This is similarly valid for other terms to be defined. The right-hand side of (5.58) is composed of two terms. The first term $\mathbf{J}_{n1}(s)$ is due to the contribution of the independent current source, and the second $\mathbf{J}_{n2}(s)$ to the initial conditions. Together they constitute the *nodal current-source vector* $\mathbf{J}_n(s)$. Note that the vector $\mathbf{J}_n(s)$ is not merely the Laplace transform of $\mathbf{j}_n(t)$ in (3.50); it also includes the contributions due to initial capacitor voltages.

In general, we take the Laplace transform on both sides of (5.56),

$$\mathbf{W}(s)\mathbf{X}(s) = \mathbf{F}(s) + \mathbf{h} \tag{5.60}$$

where $\mathbf{X}(s)$ and $\mathbf{F}(s)$ are the Laplace transforms of $\mathbf{x}(t)$ and $\mathbf{f}(t)$, respectively, and $\mathbf{h}$ is a vector that includes the contributions due to initial conditions. The coefficient matrix $\mathbf{W}(s)$ in the complex frequency variable s is obtained from $\mathbf{W}(p)$ with s replacing p. This follows directly from the differentiation rule (5.25) and the integration rule (5.42). As in nodal analysis, (5.60) denotes the loop equation, rewritten as

$$\mathbf{Z}_m(s)\mathbf{I}_m(s) = \mathbf{E}_m(s) \tag{5.61}$$

The coefficient matrix $\mathbf{Z}_m(s)$ is called the *loop-impedance matrix*, and the m-vector $\mathbf{I}_m(s)$ of the Laplace transforms of the loop currents is called the *loop-current vector*. The m-vector $\mathbf{E}_m(s)$, which includes the contribution due to initial conditions, is the *loop voltage-source vector*. Finally, (5.60) also represents the cutset system of equations, rewritten as

$$\mathbf{Y}_c(s)\mathbf{V}_c(s) = \mathbf{J}_c(s) \tag{5.62}$$

We call $\mathbf{Y}_c(s)$ the *cutset-admittance matrix*, $\mathbf{V}_c(s)$ the *cutset-voltage vector*, and $\mathbf{J}_c(s)$ the *cutset current-source vector* containing the contributions due to initial conditions.

Our conclusion is that after the Laplace transformation is taken, the secondary system of integrodifferential equations becomes a system of simultaneous linear algebraic equations, as given in (5.60). The unknown transform vector $\mathbf{X}(s)$ can be obtained immediately by inverting the matrix $\mathbf{W}(s)$:

$$\mathbf{X}(s) = \mathbf{W}^{-1}(s)[\mathbf{F}(s) + \mathbf{h}] \tag{5.63}$$

provided that $\mathbf{W}(s)$ is not identically singular. Therefore the complete solution of $\mathbf{X}$ involves no more than the computation of the inverse of a given matrix. The remaining problem is to obtain the time function $\mathbf{x}(t)$ from $\mathbf{X}(s)$ or the inverse Laplace transform of $\mathbf{X}(s)$. This will be the subject to be studied in Section 5-4.

EXAMPLE 5.10

The network of Figure 4.2(a) is redrawn in Figure 5.12. The nodal differential equation as given by (4.14) is

$$\begin{bmatrix} 3 + 2p & -2 - 2p \\ -2 - 2p & 2 + 3p \end{bmatrix} \begin{bmatrix} v_{n1}(t) \\ v_{n2}(t) \end{bmatrix} = \begin{bmatrix} 1 \\ 0 \end{bmatrix}, \qquad t \geq 0 \tag{5.64}$$

with initial capacitor voltages $v_1(0-) = 2$ V and $v_2(0-) = 0$. Taking the Laplace transform of both sides yields

$$\begin{bmatrix} 3 + 2s & -2 - 2s \\ -2 - 2s & 2 + 3s \end{bmatrix} \begin{bmatrix} V_{n1}(s) \\ V_{n2}(s) \end{bmatrix} = \begin{bmatrix} 1/s + 2v_{n1}(0-) - 2v_{n2}(0-) \\ -2v_{n1}(0-) + 3v_{n2}(0-) \end{bmatrix} \tag{5.65}$$

Figure 5.12
A network used to write the nodal differential equation (5.64).

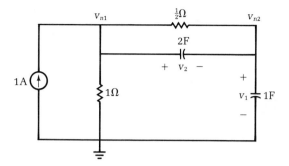

Since from Figure 5.12, $v_{n1}(0-) = v_1(0-) + v_2(0-) = 2$ V and $v_{n2}(0-) = v_1(0-) = 2$ V, (5.65) is simplified to

$$\begin{bmatrix} 3 + 2s & -2 - 2s \\ -2 - 2s & 2 + 3s \end{bmatrix} \begin{bmatrix} V_{n1}(s) \\ V_{n2}(s) \end{bmatrix} = \begin{bmatrix} 1/s \\ 2 \end{bmatrix} \tag{5.66}$$

Equation (5.66) can now be solved for the nodal voltages V_{n1} and V_{n2} by inverting the node-admittance matrix:

$$\begin{bmatrix} V_{n1}(s) \\ V_{n2}(s) \end{bmatrix} = \frac{1}{2s^2 + 5s + 2} \begin{bmatrix} 2 + 3s & 2 + 2s \\ 2 + 2s & 3 + 2s \end{bmatrix} \begin{bmatrix} 1/s \\ 2 \end{bmatrix} \tag{5.67}$$

giving the Laplace transforms of the nodal voltages

$$V_{n1}(s) = \frac{4s^2 + 7s + 2}{s(2s^2 + 5s + 2)} \tag{5.68a}$$

$$V_{n2}(s) = \frac{4s^2 + 8s + 2}{s(2s^2 + 5s + 2)} \tag{5.68b}$$

To find the time-domain nodal voltages $v_{n1}(t)$ and $v_{n2}(t)$, we must perform the inverse Laplace transformation. This will be elaborated in the following section.

EXAMPLE 5.11

The loop integrodifferential equation for the network of Figure 4.5, which is redrawn in Figure 5.13, is given by

$$\begin{bmatrix} 1 + p + \dfrac{1}{p} & -\dfrac{1}{p} \\ -\dfrac{1}{p} & 1 + \dfrac{1}{p} \end{bmatrix} \begin{bmatrix} i_{m1}(t) \\ i_{m2}(t) \end{bmatrix} = \begin{bmatrix} 1 \\ 0 \end{bmatrix}, \qquad t \geq 0 \tag{5.69}$$

Figure 5.13
A network used to write
the loop integrodifferential
equation (5.69).

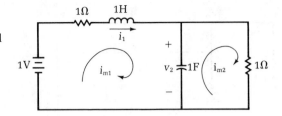

with initial conditions $i_1(0-) = 0$ and $v_2(0-) = 0$. Taking the Laplace transform of both sides of (5.69) yields a system of linear algebraic equations in the frequency domain as

$$\begin{bmatrix} 1 + s + 1/s & -1/s \\ -1/s & 1 + 1/s \end{bmatrix} \begin{bmatrix} I_{m1}(s) \\ I_{m2}(s) \end{bmatrix} = \begin{bmatrix} 1/s + i_{m1}(0-) \\ 0 \end{bmatrix} = \begin{bmatrix} 1/s \\ 0 \end{bmatrix} \qquad \textbf{(5.70)}$$

since $i_{m1}(0-) = i_{m2}(0-) = 0$. Equation (5.70) can now be solved for the Laplace transforms of the loop currents by inverting the loop-impedance matrix. The result is given by

$$\begin{bmatrix} I_{m1}(s) \\ I_{m2}(s) \end{bmatrix} = \frac{1}{s^2 + 2s + 2} \begin{bmatrix} s + 1 & 1 \\ 1 & s^2 + s + 1 \end{bmatrix} \begin{bmatrix} 1/s \\ 0 \end{bmatrix} \qquad \textbf{(5.71)}$$

or

$$I_{m1}(s) = \frac{s + 1}{s(s^2 + 2s + 2)} \qquad \textbf{(5.72a)}$$

$$I_{m2}(s) = \frac{1}{s(s^2 + 2s + 2)} \qquad \textbf{(5.72b)}$$

Performing the inverse Laplace transformation yields the time-domain loop currents $i_{m1}(t)$ and $i_{m2}(t)$. This will be discussed later.

EXAMPLE 5.12
Consider the same network of Figure 5.12 except that the 1-A current source is replaced by a sinusoidal current source $\cos 2t$, everything else being the same. The nodal differential equation becomes

$$\begin{bmatrix} 3 + 2p & -2 - 2p \\ -2 - 2p & 2 + 3p \end{bmatrix} \begin{bmatrix} v_{n1}(t) \\ v_{n2}(t) \end{bmatrix} = \begin{bmatrix} \cos 2t \\ 0 \end{bmatrix}, \qquad t \geqq 0 \qquad \textbf{(5.73)}$$

with initial conditions $v_{n1}(0-) = v_{n2}(0-) = 2$ V. Using (5.24), the Laplace transform of (5.73) is obtained as

$$\begin{bmatrix} 3 + 2s & -2 - 2s \\ -2 - 2s & 2 + 3s \end{bmatrix} \begin{bmatrix} V_{n1}(s) \\ V_{n2}(s) \end{bmatrix} = \begin{bmatrix} s/(s^2 + 4) \\ 2 \end{bmatrix} \qquad \textbf{(5.74)}$$

giving the Laplace transforms of the nodal voltages

$$V_{n1}(s) = \frac{4s^3 + 7s^2 + 18s + 16}{(s^2 + 4)(2s^2 + 5s + 2)}$$ (5.75a)

$$V_{n2}(s) = \frac{4s^3 + 8s^2 + 18s + 24}{(s^2 + 4)(2s^2 + 5s + 2)}$$ (5.75b)

5-4 PARTIAL-FRACTION EXPANSION

The examples in the preceding section demonstrate the ease with which the transforms of the nodal voltages or loop currents are computed. They are ratios of two polynomials in s known as the *rational functions* in s. The crux of the problem is then to find the corresponding time functions. One way is to apply the complex inversion integral (5.19), which is generally complicated and difficult to evaluate. A simpler way is to manipulate algebraically the transform first into a sum of several simple functions with inverse Laplace transforms that are either known or tabulated. The time function is then obtained by adding these inverse Laplace transforms. The procedure is perfectly legitimate because the uniqueness property guarantees us that if we find a time function $f(t)$ by using any method whatsoever, the Laplace transform $F(s)$ of which is the desired one, then $f(t)$ is *the* only time function that corresponds to $F(s)$. In this section we present a general method for breaking up any rational function into simple components. The method is known as the *partial-fraction expansion*.

Consider a rational function with numerator and denominator polynomials designated by $P(s)$ and $Q(s)$, respectively:

$$F(s) = \frac{P(s)}{Q(s)} = \frac{b_m s^m + b_{m-1} s^{m-1} + \cdots + b_1 s + b_0}{a_n s^n + a_{n-1} s^{n-1} + \cdots + a_1 s + a_0}$$ (5.76)

where a's and b's are real numbers. The polynomials $P(s)$ and $Q(s)$ can also be expressed in the factored form in terms of their zeros, as follows:

$$F(s) = K \frac{\prod_{h=1}^{M} (s + d_h)^{m_h}}{\prod_{k=1}^{N} (s + s_k)^{n_k}}$$ (5.77)

where $-d_h$ $(h = 1, 2, \ldots, M)$ are M distinct zeros of $P(s)$, and $-s_k$ $(k = 1, 2, \ldots, N)$ are N distinct zeros of $Q(s)$. The $-d_h$'s are called the *zeros* of the rational function $F(s)$, and the $-s_k$'s the *poles* of $F(s)$. Here we implicitly assume that the numerator and denominator polynomials $P(s)$ and $Q(s)$ have no common factors. If they do, these common factors must be removed first.

Then we have

$$m_1 + m_2 + \cdots + m_M = m \tag{5.78a}$$

$$n_1 + n_2 + \cdots + n_N = n \tag{5.78b}$$

If $m_h = 1$, $-d_h$ is said to be a *simple zero* of $F(s)$. Otherwise $-d_h$ is a *multiple zero of order* m_h. Likewise if $n_k = 1$, $-s_k$ is a *simple pole* of $F(s)$; otherwise it is a *multiple pole of order* n_k.

As the first step in the expansion of a rational function $F(s)$, we check to see that the degree of the numerator polynomial $P(s)$ is less than that of the denominator polynomial $Q(s)$. If this condition is not satisfied, we divide $P(s)$ by $Q(s)$ to obtain an expression in the form

$$F(s) = \frac{P(s)}{Q(s)} = c_{m-n}s^{m-n} + \cdots + c_1 s + c_0 + \frac{P_\alpha(s)}{Q(s)} \tag{5.79}$$

The first $m - n + 1$ terms on the right-hand side are the quotient and $P_\alpha(s)$ is the remainder, the degree of which is less than that of $Q(s)$. A rational function is said to be *proper* if the degree of the numerator polynomial is less than the degree of the denominator polynomial. The new rational function $P_\alpha(s)/Q(s)$ is proper and has now been prepared for further expansion. In the remainder of this section we assume that all rational functions are proper.

The next step is to factor the denominator $Q(s)$, as was shown in (5.77). With this, $P_\alpha(s)/Q(s)$ can be expanded in the general form

$$F_\alpha(s) = \frac{P_\alpha(s)}{Q(s)} = \frac{P_\alpha(s)}{\displaystyle\prod_{k=1}^{N}(s + s_k)^{n_k}}$$

$$= \frac{K_{11}}{(s + s_1)} + \frac{K_{12}}{(s + s_1)^2} + \cdots + \frac{K_{1n_1}}{(s + s_1)^{n_1}}$$

$$+ \frac{K_{21}}{(s + s_2)} + \frac{K_{22}}{(s + s_2)^2} + \cdots + \frac{K_{2n_2}}{(s + s_2)^{n_2}}$$

$$+ \cdots + \frac{K_{N1}}{(s + s_N)} + \frac{K_{N2}}{(s + s_N)^2} + \cdots + \frac{K_{Nn_N}}{(s + s_N)^{n_N}} \tag{5.80}$$

or, more compactly, using the summation notation

$$F_\alpha(s) = \frac{P_\alpha(s)}{Q(s)} = \frac{P_\alpha(s)}{\displaystyle\prod_{k=1}^{N}(s + s_k)^{n_k}} = \sum_{i=1}^{N} \sum_{j=1}^{n_i} \frac{K_{ij}}{(s + s_i)^j} \tag{5.81}$$

Using this in (5.79), the resulting expression is known as the *partial-fraction expansion* of $F(s)$.

As an example, consider the proper rational function

$$F(s) = \frac{(s + 3)^4}{(s + 1)^2(s + 2)^3} \tag{5.82}$$

The partial-fraction expansion of this function has the form

$$F(s) = \frac{K_{11}}{(s+1)} + \frac{K_{12}}{(s+1)^2} + \frac{K_{21}}{(s+2)} + \frac{K_{22}}{(s+2)^2} + \frac{K_{23}}{(s+2)^3} \tag{5.83}$$

In (5.80), the coefficients $K_{11}, K_{21}, \ldots, K_{N1}$ on top of the factors $(s+s_1)$, $(s+s_2), \ldots, (s+s_N)$ are called the *residues* of $F(s)$ at the poles $-s_1$, $-s_2, \ldots, -s_N$, respectively. The others do not have this particular name. In the expansion (5.83) the residues at the poles -1 and -2 are K_{11} and K_{21}, respectively. To complete the expansion we must ascertain the coefficients K_{ij}. Once the coefficients K_{ij} are known, the inverse Laplace transform of $F_\alpha(s)$ can be written immediately by appealing to (5.21) and Table 5.1, using the transform pair

$$\frac{1}{n!} t^n e^{-at} = \mathcal{L}^{-1}\left[\frac{1}{(s+a)^{n+1}}\right] \tag{5.84}$$

Thus from (5.81) we obtain the desired time function

$$f_\alpha(t) = \mathcal{L}^{-1}[F_\alpha(s)] = \sum_{i=1}^{N} \sum_{j=1}^{n_i} \frac{K_{ij}}{(j-1)!} t^{j-1} e^{-s_i t} \tag{5.85}$$

Therefore the problem is to find the coefficients K_{ij}. We now proceed to derive formulas for computing these coefficients. To this end two cases are distinguished, each being presented in a section below.

5-4.1 Simple Poles

Assume that all the poles of $F_\alpha(s)$ are simple. The partial-fraction expansion (5.81) is simplified to

$$F_\alpha(s) = \frac{P_\alpha(s)}{Q(s)} = \frac{P_\alpha(s)}{(s+s_1)(s+s_2)\cdots(s+s_n)}$$

$$= \frac{K_1}{s+s_1} + \frac{K_2}{s+s_2} + \cdots + \frac{K_n}{s+s_n} \tag{5.86}$$

Suppose that we wish to compute K_1. We multiply both sides by the factor $s+s_1$ and then substitute $s = -s_1$. The resulting equation becomes

$$\frac{P_\alpha(-s_1)}{(-s_1+s_2)(-s_1+s_3)\cdots(-s_1+s_n)} = K_1 + 0 + \cdots + 0 \tag{5.87}$$

or, more compactly, as

$$K_1 = (s+s_1)F_\alpha(s)\big|_{s=-s_1} \tag{5.88}$$

Likewise the formula for computing the residue K_j at the pole $-s_j$ is

$$K_j = (s+s_j)F_\alpha(s)\big|_{s=-s_j} \tag{5.89}$$

for $j = 1, 2, \ldots, n$.

We illustrate this procedure by the following example.

EXAMPLE 5.13

We determine the partial-fraction expansions of the transform nodal voltages $V_{n1}(s)$ and $V_{n2}(s)$, as given in (5.68), together with their inverse Laplace transforms.

From (5.68a) the proper rational function $V_{n1}(s)$ can be expanded in partial fractions in the form

$$V_{n1}(s) = \frac{4s^2 + 7s + 2}{s(2s^2 + 5s + 2)} = \frac{1}{2}\frac{4s^2 + 7s + 2}{s(s + 2)(s + 0.5)}$$

$$= \frac{K_1}{s} + \frac{K_2}{s + 2} + \frac{K_3}{s + 0.5} \qquad (5.90)$$

To compute K_1 we multiply both sides by s and then set $s = 0$, getting

$$K_1 = sV_{n1}(s)\Big|_{s=0} = \frac{1}{2}\frac{4s^2 + 7s + 2}{(s + 2)(s + 0.5)}\Big|_{s=0} = 1 \qquad (5.91)$$

Likewise to compute K_2 we multiply both sides by the factor $s + 2$ and then set $s = -2$. The result is found to be

$$K_2 = (s + 2)V_{n1}(s)\Big|_{s=-2} = \frac{1}{2}\frac{4s^2 + 7s + 2}{s(s + 0.5)}\Big|_{s=-2} = \frac{2}{3} \qquad (5.92)$$

Finally, for K_3 we have

$$K_3 = (s + 0.5)V_{n1}(s)\Big|_{s=-0.5} = \frac{1}{2}\frac{4s^2 + 7s + 2}{s(s + 2)}\Big|_{s=-0.5} = \frac{1}{3} \qquad (5.93)$$

Substituting these in (5.90) yields the partial-fraction expansion

$$V_{n1}(s) = \frac{1}{s} + \frac{2}{3(s + 2)} + \frac{1}{3(s + 0.5)} \qquad (5.94)$$

From Table 5.1 the inverse Laplace transform of $V_{n1}(s)$ is obtained as

$$v_{n1}(t) = \mathscr{L}^{-1}[V_{n1}(s)] = 1 + \frac{2}{3}e^{-2t} + \frac{1}{3}e^{-0.5t}, \qquad t \geq 0 \qquad (5.95)$$

confirming (4.20) with $c_1 = c_2 = \frac{1}{3}$.

In a similar manner, from (5.68b) the partial-fraction expansion of $V_{n2}(s)$ is given by

$$V_{n2}(s) = \frac{4s^2 + 8s + 2}{s(2s^2 + 5s + 2)} = \frac{1}{s} + \frac{1}{3(s + 2)} + \frac{2}{3(s + 0.5)} \qquad (5.96)$$

obtaining

$$v_{n2}(t) = \mathscr{L}^{-1}[V_{n2}(s)] = 1 + \frac{1}{3}e^{-2t} + \frac{2}{3}e^{-0.5t}, \qquad t \geq 0 \qquad (5.97)$$

An important special situation in the expansion (5.86) is that $F_\alpha(s)$ possesses a pair of complex conjugate poles, say, at $-\sigma_1 \pm j\omega_1$. For this case the partial-fraction expansion (5.86) would contain terms of the form

$$F_\alpha(s) = \frac{P_\alpha(s)}{Q(s)} = \frac{K_1}{s + \sigma_1 + j\omega_1} + \frac{K_2}{s + \sigma_1 - j\omega_1} + \cdots \qquad (5.98)$$

Using the formula (5.89) for simple poles, we have

$$K_1 = (s + \sigma_1 + j\omega_1) F_\alpha(s)\big|_{s = -\sigma_1 - j\omega_1} \qquad (5.99a)$$

$$K_2 = (s + \sigma_1 - j\omega_1) F_\alpha(s)\big|_{s = -\sigma_1 + j\omega_1} \qquad (5.99b)$$

Since $F_\alpha(s)$ is a rational function of s with real coefficients, we have $\bar{F}_\alpha(s) = F_\alpha(\bar{s})$ with the bar denoting the complex conjugate. Thus from (5.99) we obtain

$$K_2 = \bar{K}_1 \qquad (5.100)$$

In words it states that when the poles are conjugates, so are the residues. We now express K_1 and K_2 in polar form

$$K_1 = |K_1| e^{j\theta}, \qquad \theta = \arg K_1 \qquad (5.101a)$$

$$K_2 = \bar{K}_1 = |K_1| e^{-j\theta} \qquad (5.101b)$$

The inverse Laplace transforms of the first two terms on the right-hand side of (5.98) can now be combined to give

$$K_1 e^{-(\sigma_1 + j\omega_1)t} + K_2 e^{-(\sigma_1 - j\omega_1)t} = |K_1| e^{-\sigma_1 t} [e^{-j(\omega_1 t - \theta)} + e^{j(\omega_1 t - \theta)}]$$

$$= 2|K_1| e^{-\sigma_1 t} \cos(\omega_1 t - \theta) \qquad (5.102)$$

This formula is very useful in computing the inverse Laplace transform because it gives the corresponding time function for a pair of complex conjugate poles.

EXAMPLE 5.14

We compute the inverse Laplace transforms for the loop currents given in (5.72). From (5.72a) we have

$$I_{m1}(s) = \frac{s + 1}{s(s^2 + 2s + 2)} = \frac{s + 1}{s(s + 1 + j)(s + 1 - j)}$$

$$= \frac{K_1}{s} + \frac{K_2}{s + 1 + j} + \frac{\bar{K}_2}{s + 1 - j} \qquad (5.103)$$

To compute K_1 we multiply both sides by s and then set $s = 0$, yielding

$$K_1 = sI_{m1}(s)\big|_{s=0} = \frac{s + 1}{s^2 + 2s + 2}\bigg|_{s=0} = \frac{1}{2} \qquad (5.104)$$

Likewise, to compute K_2 we multiply both sides of (5.103) by the factor $s + 1 + j$ and then set $s = -1 - j$. The result is given by

$$K_2 = (s + 1 + j)I_{m1}(s)\big|_{s=-1-j} = \frac{s+1}{s(s+1-j)}\bigg|_{s=-1-j}$$

$$= -\tfrac{1}{4}(1-j) = \tfrac{1}{4}\sqrt{2}e^{j3\pi/4} \qquad\qquad (5.105)$$

To verify that the residue at the pole $-1 + j$ is $\bar{K}_2$, we compute

$$\bar{K}_2 = (s + 1 - j)I_{m1}(s)\big|_{s=-1+j} = \frac{s+1}{s(s+1+j)}\bigg|_{s=-1+j}$$

$$= -\tfrac{1}{4}(1+j) = \tfrac{1}{4}\sqrt{2}e^{-j3\pi/4} \qquad\qquad (5.106)$$

Using (5.102), the inverse Laplace transform of $I_{m1}(s)$ is obtained from (5.103) in conjunction with (5.104) and (5.105):

$$i_{m1}(t) = \mathscr{L}^{-1}[I_{m1}(s)] = \tfrac{1}{2} + \tfrac{1}{4}\sqrt{2}e^{-(1+j)t+j3\pi/4}$$

$$+ \tfrac{1}{4}\sqrt{2}e^{-(1-j)t-j3\pi/4}$$

$$= 0.5 + 0.707e^{-t}\cos\left(t - \frac{3\pi}{4}\right), \qquad t \geq 0 \qquad (5.107)$$

In a similar fashion, from (5.72b) the transform $I_{m2}(s)$ can be expanded in partial fractions

$$I_{m2}(s) = \frac{1}{s(s^2 + 2s + 2)} = \frac{1}{s(s+1+j)(s+1-j)}$$

$$= \frac{\tfrac{1}{2}}{s} - \frac{\tfrac{1}{4}(1+j)}{s+1+j} - \frac{\tfrac{1}{4}(1-j)}{s+1-j} \qquad\qquad (5.108)$$

giving

$$i_{m2}(t) = \mathscr{L}^{-1}[I_{m2}(s)] = 0.5 + 0.707e^{-t}\cos\left(t + \frac{3\pi}{4}\right), \qquad t \geq 0 \qquad (5.109)$$

5-4.2 Multiple Poles

In the case of multiple poles, the process is similar but more complicated. To illustrate the general procedure, we consider first the partial-fraction expansion of the rational function

$$F_\alpha(s) = \frac{P_\alpha(s)}{(s+s_1)^2(s+s_2)^3} = \frac{K_{11}}{s+s_1} + \frac{K_{12}}{(s+s_1)^2}$$

$$+ \frac{K_{21}}{s+s_2} + \frac{K_{22}}{(s+s_2)^2} + \frac{K_{23}}{(s+s_2)^3} \qquad\qquad (5.110)$$

To compute K_{23} we multiply both sides by $(s + s_2)^3$ and then set $s = -s_2$. This would make all the terms on the right-hand side zero except the last one in the resulting equation, which is K_{23}, giving

$$K_{23} = \frac{P_\alpha(s)}{(s + s_1)^2}\bigg|_{s = -s_2} = \frac{P_\alpha(-s_2)}{(-s_2 + s_1)^2} = (s + s_2)^3 F_\alpha(s)\bigg|_{s = -s_2} \qquad \textbf{(5.111)}$$

To compute K_{22} we apply the same technique by multiplying both sides of (5.110) by $(s + s_2)^3$. But before we set $s = -s_2$, we take derivative on both sides. The result is that all the terms on the right-hand side of the resulting equation will be zero except the one next to the last term, which is K_{22}. Thus we have

$$1! \, K_{22} = \frac{d}{ds}\left[\frac{P_\alpha(s)}{(s + s_1)^2}\right]\bigg|_{s = -s_2} = \frac{d}{ds}[(s + s_2)^3 F_\alpha(s)]\bigg|_{s = -s_2} \qquad \textbf{(5.112)}$$

Following this pattern we see that to obtain K_{21} we multiply both sides of (5.110) by $(s + s_2)^3$, take derivative twice, and then set $s = -s_2$. The result is given by

$$2! \, K_{21} = \frac{d^2}{ds^2}\left[\frac{P_\alpha(s)}{(s + s_1)^2}\right]\bigg|_{s = -s_2} = \frac{d^2}{ds^2}[(s + s_2)^3 F_\alpha(s)]\bigg|_{s = -s_2} \qquad \textbf{(5.113)}$$

The above process clearly can be extended if $-s_2$ is a pole of order n_2 instead of 3. In this case the corresponding partial-fraction expansion would contain n_2 terms of the form $K_{2j}/(s + s_2)^j$, $j = 1, 2, \ldots, n_2$. The coefficients K_{2j} are determined by the formula

$$K_{2j} = \frac{1}{(n_2 - j)!} \cdot \frac{d^{n_2 - j}}{ds^{n_2 - j}}[(s + s_2)^{n_2} F_\alpha(s)]\bigg|_{s = -s_2} \qquad \textbf{(5.114)}$$

To consider the general expansion of (5.81), the coefficients K_{ij} are given by the formula

$$K_{ij} = \frac{1}{(n_i - j)!} \cdot \frac{d^{n_i - j}}{ds^{n_i - j}}[(s + s_i)^{n_i} F_\alpha(s)]\bigg|_{s = -s_i} \qquad \textbf{(5.115)}$$

where for $j = n_i$, d^0/ds^0 means that no derivative is taken.

EXAMPLE 5.15

The nodal integrodifferential equation for the initially relaxed network of Figure 4.4(a) is given by (4.71). The Laplace transform of (4.71) is found to be

$$\begin{bmatrix} 2.61780 + 1/s & -1/s & 0 \\ -1/s & s + 1/s & -s \\ 0 & 1.61812 - s & 1 + 2s \end{bmatrix}\begin{bmatrix} V_{n1}(s) \\ V_{n2}(s) \\ V_{n3}(s) \end{bmatrix} = \begin{bmatrix} 2.61780/s \\ 0 \\ 0 \end{bmatrix} \qquad \textbf{(5.116)}$$

which can be solved to give

$$\begin{bmatrix} V_{n1}(s) \\ V_{n2}(s) \\ V_{n3}(s) \end{bmatrix} = \frac{1}{s(s + 1)^3}\begin{bmatrix} s^3 + 2.61812s^2 + 2s + 1 \\ 2s + 1 \\ s - 1.61812 \end{bmatrix} \qquad \textbf{(5.117)}$$

Thus the transforms of the nodal voltages have a simple pole at $s = 0$ and a pole of order 3 at $s = -1$. They are proper rational functions and can be expanded in partial fractions in the form of (5.81). For $V_{n1}(s)$, we have

$$V_{n1}(s) = \frac{s^3 + 2.61812s^2 + 2s + 1}{s(s+1)^3} = \frac{K_{11}}{s} + \frac{K_{21}}{s+1}$$

$$+ \frac{K_{22}}{(s+1)^2} + \frac{K_{23}}{(s+1)^3} \tag{5.118}$$

The coefficients K_{11} and K_{21} are the residues at the poles 0 and -1, respectively. The residue K_{11} is computed in the same manner as in the preceding section and is obtained as $K_{11} = 1$. To compute K_{23} we multiply both sides of (5.118) by $(s+1)^3$ and then set $s = -1$, getting

$$K_{23} = (s+1)^3 V_{n1}(s)\big|_{s=-1} = (s^3 + 2.61812s^2 + 2s + 1)/s\big|_{s=-1}$$

$$= -0.61812 \tag{5.119}$$

For K_{22} we again multiply both sides of (5.118) by $(s+1)^3$. But before we set $s = -1$, we take derivative on both sides. The result is

$$K_{22} = \frac{d}{ds}[(s+1)^3 V_{n1}(s)]\bigg|_{s=-1} = 2s + 2.61812 - 1/s^2\big|_{s=-1}$$

$$= -0.38188 \tag{5.120}$$

Finally, for K_{21} we again multiply both sides by $(s+1)^3$, take derivative twice, and then set $s = -1$, yielding

$$K_{21} = \frac{1}{2!}\frac{d^2}{dt^2}[(s+1)^3 V_{n1}(s)]\bigg|_{s=-1} = 1 + 1/s^3\big|_{s=-1} = 0 \tag{5.121}$$

Substituting the K's in (5.118) and applying (5.84), the inverse Laplace transform of $V_{n1}(s)$ is found to be

$$v_{n1}(t) = \mathscr{L}^{-1}[V_{n1}(s)] = K_{11} + K_{21}e^{-t} + K_{22}te^{-t} + \tfrac{1}{2}K_{23}t^2e^{-t}$$

$$= 1 - 0.38188te^{-t} - 0.30906t^2e^{-t}, \quad t \geq 0 \tag{5.122}$$

Likewise the partial-fraction expansions of $V_{n2}(s)$ and $V_{n3}(s)$ and their inverse Laplace transforms are obtained as follows:

$$v_{n2}(t) = \mathscr{L}^{-1}[V_{n2}(s)] = \mathscr{L}^{-1}\left[\frac{1}{s} - \frac{1}{s+1} - \frac{1}{(s+1)^2} + \frac{1}{(s+1)^3}\right]$$

$$= 1 - e^{-t} - te^{-t} + \tfrac{1}{2}t^2e^{-t}, \quad t \geq 0 \tag{5.123}$$

$$v_{n3}(t) = \mathscr{L}^{-1}[V_{n3}(s)]$$

$$= \mathscr{L}^{-1}\left[-\frac{1.61812}{s} + \frac{1.61812}{s+1} + \frac{1.61812}{(s+1)^2} + \frac{2.61812}{(s+1)^3}\right]$$

$$= -1.61812 + 1.61812e^{-t}$$

$$+ 1.61812te^{-t} + 1.30906t^2e^{-t}, \quad t \geq 0 \tag{5.124}$$

These results confirm (4.105) obtained by the classical method to within computational accuracy.

We remark that instead of using the above argument, the coefficients K_{ij} can be obtained directly from formula (5.115). For example, in applying (5.115) to (5.118) we have $i = 2$, $n_i = 3$, $s_i = 1$, and $j = 1, 2, 3$, giving (5.119) to (5.122) directly.

To avoid differentiation, we make use of the partial-fraction expansion via simple poles. This technique is best illustrated by taking the previous example.

EXAMPLE 5.16

We wish to obtain (5.118) via the partial-fraction expansions of the simple poles. To this end let

$$F_1(s) = \frac{s^3 + 2.61812s^2 + 2s + 1}{s(s+1)}, \qquad F_2(s) = \frac{F_1(s)}{s+1} \qquad (5.125)$$

The partial-fraction expansion of $F_1(s)$ is obtained as

$$F_1(s) = s + 1.61812 + \frac{1}{s} - \frac{0.61812}{s+1} \qquad (5.126)$$

Substituting this in F_2 corresponds to multiplying each term on the right-hand side of (5.126) by $1/(s+1)$. The first and third terms in the resulting equation can again be expanded via simple poles, as follows:

$$
\begin{aligned}
F_2(s) &= \frac{s}{s+1} + \frac{1.61812}{s+1} + \frac{1}{s(s+1)} - \frac{0.61812}{(s+1)^2} \\
&= \left[1 - \frac{1}{s+1}\right] + \frac{1.61812}{s+1} + \left[\frac{1}{s} - \frac{1}{s+1}\right] - \frac{0.61812}{(s+1)^2} \\
&= 1 - \frac{0.38188}{s+1} + \frac{1}{s} - \frac{0.61812}{(s+1)^2} \qquad (5.127)
\end{aligned}
$$

The above process can now be repeated by observing that

$$
\begin{aligned}
V_{n1}(s) &= \frac{F_2(s)}{s+1} = \frac{1}{s+1} - \frac{0.38188}{(s+1)^2} + \frac{1}{s(s+1)} - \frac{0.61812}{(s+1)^3} \\
&= \frac{1}{s+1} - \frac{0.38188}{(s+1)^2} + \left[\frac{1}{s} - \frac{1}{s+1}\right] - \frac{0.61812}{(s+1)^3} \\
&= \frac{1}{s} - \frac{0.38188}{(s+1)^2} - \frac{0.61812}{(s+1)^3} \qquad (5.128)
\end{aligned}
$$

confirming (5.119)–(5.121). We have thus obtained the partial-fraction expansion of a multiple pole via the repeated expansions of the simple poles. The technique is valid in general.

5-4.3   A Noniterative Method for the Partial-Fraction Expansion

A proper rational function $F_\alpha(s)$, having both simple and multiple poles, always admits the partial-fraction expansion (5.81), the coefficients K_{ij} of which are determined by the formula (5.115). However, in using formula (5.115) for a high-order pole it is necessary to work out the derivatives of the functions up to an order that is one unit lower than that of the suppressed pole. As can easily be verified by working out a few examples, the operations become more and more tedious as the order of poles is increased. Moreover, because of the interdependence, computational errors propagate. In this section we shall present a technique that avoids the necessity of passing through the lower-order derivatives. This technique can easily be programmed for a digital computer, as implemented in Section 5-5.1 as subroutine PFRAC.

Consider a proper rational function $F(s)$ and its partial-fraction expansion:

$$F(s) = K \frac{\displaystyle\prod_{h=1}^{M} (s + d_h)^{m_h}}{\displaystyle\prod_{k=1}^{N} (s + s_k)^{n_k}} = \sum_{k=1}^{N} \sum_{j=1}^{n_k} \frac{K_{kj}}{(s + s_k)^j} \tag{5.129}$$

where from (5.115) the coefficients K_{kj} are determined by

$$K_{kj} = \frac{1}{(n_k - j)!} \cdot \frac{d^{n_k - j}}{ds^{n_k - j}} \left[(s + s_k)^{n_k} F(s) \right] \Big|_{s = -s_k} \tag{5.130}$$

To simplify the notation, let

$$F_k(s) = (s + s_k)^{n_k} F(s), \qquad k = 1, 2, \ldots, N \tag{5.131}$$

$$\frac{d^k F(s)}{ds^k} = F^{(k)}(s) \tag{5.132}$$

$$C_{ki} = \frac{1}{i!} F_k^{(i)}(-s_k) \tag{5.133}$$

giving

$$K_{kj} = C_{k(n_k - j)} \tag{5.134}$$

for $k = 1, 2, \ldots, N$ and $j = 1, 2, \ldots, n_k$. Our objective here is to determine K_{kj} via (5.133) and (5.134).

To this end we compute the derivative of $F_k(s)$. Consider first the simpler situation where $N = M = 2$. Then we have

$$F_1^{(1)}(s) = F_1(s)\left[\frac{m_1}{s+d_1} + \frac{m_2}{s+d_2} - \frac{n_2}{s+s_2}\right] \tag{5.135a}$$

$$F_2^{(1)}(s) = F_2(s)\left[\frac{m_1}{s+d_1} + \frac{m_2}{s+d_2} - \frac{n_1}{s+s_1}\right] \tag{5.135b}$$

which can be combined to give

$$F_k^{(1)}(s) = F_k(s)\left[\sum_{h=1}^{2}\frac{m_h}{s+d_h} - \sum_{\substack{i=1\\i\neq k}}^{2}\frac{n_i}{s+s_i}\right], \qquad k=1,2 \tag{5.136}$$

It is not difficult to see that in the general situation of (5.129) the first derivative of $F(s)$ can be written as

$$F_k^{(1)}(s) = F_k(s)\left[\sum_{h=1}^{M}\frac{m_h}{s+d_h} - \sum_{\substack{i=1\\i\neq k}}^{N}\frac{n_i}{s+s_i}\right] \tag{5.137}$$

for $k = 1, 2, \ldots, N$. Stating it differently, (5.137) shows that the first derivative of $F_k(s)$ is equal to the product of $F_k(s)$ and a proper rational function defined by

$$G_k(s) = \sum_{h=1}^{M}\frac{m_h}{s+d_h} - \sum_{\substack{i=1\\i\neq k}}^{N}\frac{n_i}{s+s_i} \tag{5.138}$$

or

$$F_k^{(1)}(s) = F_k(s)G_k(s) \tag{5.139}$$

Since

$$\frac{d^r}{ds^r}\left(\frac{1}{s+a}\right) = (-1)^r\frac{r!}{(s+a)^{r+1}} \tag{5.140}$$

the rth derivative of $G_k(s)$ is found to be

$$G_k^{(r)}(s) = (-1)^r r!\left[\sum_{h=1}^{M}\frac{m_h}{(s+d_h)^{r+1}} - \sum_{\substack{i=1\\i\neq k}}^{N}\frac{n_i}{(s+s_i)^{r+1}}\right] \tag{5.141}$$

On the other hand, by using Leibniz rule,[†] the higher-order derivatives of $F_k(s)$ can be expressed in terms of the lower-order derivatives, as follows:

$$F_k^{(r+1)}(s) = \sum_{j=0}^{r}\binom{r}{j}F_k^{(j)}G_k^{(r-j)} \tag{5.142}$$

[†] W. Kaplan, *Advanced Calculus*, 2d ed. (Reading, Mass.: Addison-Wesley, 1973), p. 26. The rule is named after Gottfried Wilhelm von Leibniz (1646–1716).

where

$$\binom{r}{j} = \frac{r!}{(r-j)!j!} \tag{5.143}$$

By starting from (5.139) and by assigning successively to r the values $1, 2, \ldots, i-1$, we obtain a system of i linear algebraic equations:

$$-F_k^{(1)} = -F_k G_k$$

$$G_k F_k^{(1)} - F_k^{(2)} = -F_k G_k^{(1)}$$

$$2G_k^{(1)} F_k^{(1)} + G_k F_k^{(2)} - F_k^{(3)} = -F_k G_k^{(2)}$$

$$\vdots$$

$$(i-1)G_k^{(i-2)}F_k^{(1)} + \binom{i-1}{2}G_k^{(i-3)}F_k^{(2)} + \binom{i-1}{3}G_k^{(i-4)}F_k^{(3)} + \cdots$$

$$+ (i-1)G_k^{(1)}F_k^{(i-2)} + G_k F_k^{(i-1)} - F_k^{(i)} = -F_k G_k^{(i-1)} \tag{5.144}$$

If $F_k^{(1)}, F_k^{(2)}, \ldots, F_k^{(i)}$ are considered as the unknowns and the others as the known quantities, (5.144) can be put in matrix notation as

$$\begin{bmatrix} -1 & 0 & \cdots & 0 & 0 \\ G_k & -1 & \cdots & 0 & 0 \\ 2G_k^{(1)} & G_k & \cdots & 0 & 0 \\ \vdots & \vdots & \cdots & \vdots & \vdots \\ (i-1)G_k^{(i-2)} & \binom{i-1}{2}G_k^{(i-3)} & \cdots & G_k & -1 \end{bmatrix} \begin{bmatrix} F_k^{(1)} \\ F_k^{(2)} \\ F_k^{(3)} \\ \vdots \\ F_k^{(i)} \end{bmatrix} = -F_k \begin{bmatrix} G_k \\ G_k^{(1)} \\ G_k^{(2)} \\ \vdots \\ G_k^{(i-1)} \end{bmatrix} \tag{5.145}$$

We can now solve for $F_k^{(i)}$ by applying Cramer's rule[†] and noting that the determinant of the coefficient matrix is equal to $(-1)^i$: For $i > 1$,

$$F_k^{(i)} = (-1)^{i-1} F_k \det \begin{bmatrix} -1 & 0 & \cdots & 0 & G_k \\ G_k & -1 & \cdots & 0 & G_k^{(1)} \\ 2G_k^{(1)} & G_k & \cdots & 0 & G_k^{(2)} \\ \vdots & \vdots & \cdots & \vdots & \vdots \\ (i-1)G_k^{(i-2)} & \binom{i-1}{2}G_k^{(i-3)} & \cdots & G_k & G_k^{(i-1)} \end{bmatrix} \tag{5.146}$$

which expresses the ith derivative of F_k in terms of F_k, G_k, and the derivatives of G_k, where $G_k^{(0)} = G_k$. For $i = 1$, $F_k^{(1)} = F_k G_k$.

[†] C. R. Wylie, Jr., *Advanced Engineering Mathematics*, 3d ed. (New York: McGraw-Hill, 1966), p. 453.

We note that (5.146) in general is a function of s. To compute K_{kj} through (5.133) and (5.134), we substitute $s = -s_k$ in (5.146) and obtain

$$K_{k(n_k - i)} = C_{ki} = \frac{1}{i!} F_k^{(i)}(-s_k), \qquad i = 0, 1, \ldots, n_k - 1 \qquad \text{(5.147)}$$

To summarize we present the following steps for computing the coefficients K_{kj}:

Step 1. For each k, compute $F_k(s)$, $G_k(s)$, and $G_k^{(r)}(s)$ for $r = 1, 2, \ldots, n_k - 2$, $n_k > 2$, using (5.131), (5.138), and (5.141). For $n_k \leq 2$, only $F_k(s)$ and $G_k(s)$ are required.

Step 2. Compute the values of the functions obtained in Step 1 at the pole $-s_k$.

Step 3. Using the values obtained in Step 2, compute $F_k^{(i)}$ by (5.146) for $i = 0, 1, \ldots, n_k - 1$.

Step 4. Compute K_{kj} by (5.147) for $j = 1, 2, \ldots, n_k$.

We illustrate the above procedure by the following examples.

EXAMPLE 5.17

We wish to expand the proper rational function

$$F(s) = \frac{(s + 3)^4}{(s + 1)^2(s + 2)^3} \qquad \text{(5.148)}$$

into partial fractions, as indicated in (5.83). From (5.148) we have

$$d_1 = 3, s_1 = 1, s_2 = 2, m_1 = 4, n_1 = 2$$
$$n_2 = 3, M = 1, N = 2$$

Step 1. From (5.131), we obtain

$$F_1(s) = \frac{(s + 3)^4}{(s + 2)^3}, \qquad F_2(s) = \frac{(s + 3)^4}{(s + 1)^2} \qquad \text{(5.149)}$$

and from (5.138)

$$G_1(s) = \frac{4}{s + 3} - \frac{3}{s + 2} \qquad \text{(5.150a)}$$

$$G_2(s) = \frac{4}{s + 3} - \frac{2}{s + 1} \qquad \text{(5.150b)}$$

Hence from (5.141)

$$G_2^{(1)}(s) = -\frac{4}{(s + 3)^2} + \frac{2}{(s + 1)^2} \qquad \text{(5.151)}$$

Step 2. The values of the functions obtained in Step 1 at the poles $-s_k$ are computed, as follows:

$$F_1(-s_1) = F_1(-1) = 16 \tag{5.152a}$$

$$F_2(-s_2) = F_2(-2) = 1 \tag{5.152b}$$

$$G_1(-s_1) = G_1(-1) = -1 \tag{5.152c}$$

$$G_2(-s_2) = G_2(-2) = 6 \tag{5.152d}$$

$$G_2^{(1)}(-s_2) = G_2^{(1)}(-2) = -2 \tag{5.152e}$$

Step 3. Using (5.139) and (5.146), we compute $F_k^{(i)}(-s_k)$ as follows:

$$F_1^{(1)}(-s_1) = F_1(-s_1)G_1(-s_1) = 16 \times (-1) = -16 \tag{5.153a}$$

$$F_2^{(1)}(-s_2) = F_2(-s_2)G_2(-s_2) = 1 \times 6 = 6 \tag{5.153b}$$

$$F_2^{(2)}(-s_2) = -F_2(-s_2) \det \begin{bmatrix} -1 & G_2(-s_2) \\ G_2(-s_2) & G_2^{(1)}(-s_2) \end{bmatrix}$$

$$= -1 \cdot \det \begin{bmatrix} -1 & 6 \\ 6 & -2 \end{bmatrix} = 34 \tag{5.153c}$$

Step 4. From (5.147) we compute the coefficients K_{kj}:

$$K_{11} = C_{11} = F_1^{(1)}(-s_1) = -16 \tag{5.154a}$$

$$K_{12} = C_{10} = F_1(-s_1) = 16 \tag{5.154b}$$

$$K_{21} = C_{22} = \tfrac{1}{2}F_2^{(2)}(-s_2) = 17 \tag{5.154c}$$

$$K_{22} = C_{21} = F_2^{(1)}(-s_2) = 6 \tag{5.154d}$$

$$K_{23} = C_{20} = F_2(-s_2) = 1 \tag{5.154e}$$

Substituting these in (5.83) gives the desired partial-fraction expansion of $F(s)$.

The procedure appears to be "more" complicated in relation to formula (5.130), but in practical computation it is simple and easy to use because all the algebraic work reduces to finding the expressions for F_k, G_k and its derivatives; the rest involves only elementary arithmetic. Moreover, the procedure is readily amenable to digital computation. A computer program based on this will be presented in the next section. On the other hand, the operations of formula (5.130) become more and more tedious as the order of poles increases and also the cumulative errors may be considerable. If $F(s)$ is given as a ratio of two polynomials, another root-finding program is required to factor the polynomials. This represents a disadvantage with respect to formula (5.130), which requires that only the denominator polynomial be in factored form.

5-5 COMPUTER SOLUTIONS AND PROGRAMS

In this section we present computer programs in FORTRAN WATFIV[†] that implement the above procedure and compute the inverse Laplace transform of a rational function. The results may either be printed or plotted as desired.

5-5.1 Subroutine PFRAC

1. Purpose. The program, as presented in Figure 5.14, is used for the evaluation of the coefficients K_{kj} of a proper real rational function $F(s)$ having both simple and multiple poles,

$$F(s) = K \frac{\displaystyle\prod_{h=1}^{M} (s + d_h)^{m_h}}{\displaystyle\prod_{k=1}^{N} (s + s_k)^{n_k}} = \sum_{k=1}^{N} \sum_{j=1}^{n_k} \frac{K_{kj}}{(s + s_k)^j} \qquad (5.155)$$

2. Method. A noniterative method, as described in Section 5-4.3, is used for computing the coefficients K_{kj}. The method avoids the necessity of expressing the higher-order derivatives of $F(s)$ in terms of the lower-order derivatives.

3. Usage. The program consists of a subroutine, PFRAC, called by

CALL PFRAC (AK, M, N, NKM, ME, NE, D, S, CK, C, G, GG)

AK	is the constant K
M	is the number of distinct zeros
N	is the number of distinct poles
NKM	is the maximum order of the poles—that is, NKM = $\max\limits_k n_k$
ME	is the name of the orders of the zeros $-d_h$ with ME(MH) = m_h
NE	is the name of the orders of the poles $-s_k$ with NE(KP) = n_k
D	is the name of the zeros $-d_h$ with D(MH) = d_h
S	is the name of the poles $-s_k$ with S(KP) = s_k
CK	is the name of the coefficients K_{kj} with CK(K, J) = K_{kj}
C	is the name of a two-dimensional temporary storage array of dimension (20, NKM)
G	is the name of a one-dimensional temporary storage array of dimension NKM
GG	is the name of a one-dimensional temporary storage array of dimension NKM

[†] The WATFIV compiler was developed at the University of Waterloo as a very fast form of FORTRAN compiler. The name WATFIV was chosen to stand for *WAT*erloo Fortran *IV*. This new compiler incorporates language features described in IBM's two SRL manuals, C28-6515 and C28-6817.

Figure 5.14
Subroutine PFRAC used
for obtaining the partial-
fraction expansion of a
proper real rational
function having both
simple and multiple poles.

```
C
C
C      SUBROUTINE PFRAC
C
C      PURPOSE
C          THE PROGRAM IS FOR THE EVALUATION OF THE COEFFICIENTS CK(K,J) OF A
C          PROPER REAL RATIONAL FUNCTION F(S), HAVING BOTH SIMPLE AND MULTIPLE
C          POLES.
C
C      METHOD
C          A NONITERATIVE METHOD FOR OBTAINING THE PARTIAL FRACTION EXPANSION
C          OF F(S).
C
C      USAGE
C          CALL PFRAC (AK,M,N,NKM,ME,NE,D,S,CK,C,G,GG)
C
C          AK    -THE CONSTANT K
C          M     -THE NUMBER OF DISTINCT ZEROS
C          N     -THE NUMBER OF DISTINCT POLES
C          NKM   -THE MAXIMUM ORDER OF THE POLES
C          ME    -THE ORDERS OF THE ZEROS
C          NE    -THE ORDERS OF THE POLES
C          D     -THE ZEROS
C          S     -THE POLES
C          CK    -THE COEFFICIENTS
C          C     -A TWO-DIMENSIONAL TEMPORARY STORAGE ARRAY OF DIMENSION 20 BY
C                  NKM
C          G     -A ONE-DIMENSIONAL TEMPORARY STORAGE ARRAY OF DIMENSION NKM
C          GG    -A ONE-DIMENSIONAL TEMPORARY STORAGE ARRAY OF DIMENSION NKM
C
C      SUBROUTINES AND FUNCTION SUBPROGRAMS REQUIRED
C          NONE
C
C      REMARKS
C          THE NUMBER OF DISTINCT POLES CANNOT EXCEED TWENTY.  OTHERWISE,
C             NEW DIMENSIONAL STATEMENTS ARE REQUIRED FOR CK AND C.
C          THE CALLING PROGRAM MUST CONTAIN THE DIMENSIONAL STATEMENTS FOR C,
C             G AND GG.
C          THE INPUT DATA ARE AK, M, N, NKM, ME(I) OF ME, NE(J) OF NE, D(I) OF D
C             AND S(J) OF S.  THE OUTPUT IS CK(K,J) OF CK.
       SUBROUTINE PFRAC (AK,M,N,NKM,ME,NE,D,S,CK,C,G,GG)
       DIMENSION ME(M),NE(N),D(M),S(N),CK(20,NKM),C(20,NKM),G(NKM),
      C  GG(NKM)
       COMPLEX*16 U,E,GM,GN,D,S,CK,C,G,GG
       DOUBLE PRECISION RP,PI,PII,AK
C
C      COMPUTE C(K,1)
   1   DO 22 K=1,N
       U=(1.0,0.0)
       IF (M)4,4,2
   2   DO 3 MH=1,M
   3   U=U*(-S(K)+D(MH))**ME(MH)
   4   E=(1.0,0.0)
       DO 6 KP=1,N
       IF (KP-K) 5,6,5
   5   E=E*(-S(K)+S(KP))**NE(KP)
   6   CONTINUE
       C(K,1)=AK*U/E
C
C      COMPUTE C(K,I)
       N1=NE(K)-1
       IF (N1) 22,22,7
   7   DO 15 NR1=1,N1
       NR=NR1-1
       GM=(0.0,0.0)
       IF (M) 10,10,8
   8   DO 9 MH=1,M
   9   GM=GM+ME(MH)/(-S(K)+D(MH))**NR1
  10   GN=(0.0,0.0)
       DO 12 KP=1,N
       IF (KP-K) 11,12,11
  11   GN=GN+NE(KP)/(-S(K)+S(KP))**NR1
  12   CONTINUE
       RP=1.0
       IF(NR) 15,15,13
  13   DO 14 LL=1,NR
  14   RP=RP*LL
  15   G(NR1)=(-1.0)**NR*RP*(GM-GN)
       DO 21 II=1,N1
```

Figure 5.14 (continued)

```
         GG(II)=G(II)
         IIM1=II-1
         IF(IIM1) 19,19,16
   16    DO 18 I=1,IIM1
         PI=1.0
         DO 17 J=1,I
   17    PI=PI*(II-J)/J
         L=II-I
   18    GG(II)=GG(II)+PI*G(L)*GG(I)
   19    PII=1.0
         DO 20 IP=1,II
   20    PII=PII*IP
         IIP1=II+1
   21    C(K,IIP1)=1.0/PII*C(K,1)*GG(II)
   22    CONTINUE
   C
   C     REARRANGE THE SUBSCRIPTS OF C AND STORE IN CK
         DO 23 K=1,N
         JF=NE(K)
         DO 23 J=1,JF
   23    CK(K,J)=C(K,NE(K)-J+1)
         RETURN
         END
```

4. Remarks. The number of distinct poles cannot exceed 20. Otherwise, new dimensional statements are required for CK(20, NKM) and C(20, NKM). The calling program must contain the dimensional statements for C, G, and GG. The input data are AK, M, N, NKM, ME(I) of ME, NE(J) of NE, D(I) of D, and S(J) of S. The output is CK(K, J) of CK. No user-supplied subprogram is required. If $F(s)$ is devoid of zeros, set $M = 1$, $d_h = 1$ and $m_h = 0$ in (5.155).

EXAMPLE 5.18

We shall use the subroutine PFRAC to obtain the partial-fraction expansion of the proper real rational function

$$F(s) = \frac{(s + 5)^3(s + 6 + j4)^4(s + 6 - j4)^4}{(s + 4 + j5)^5(s + 4 - j5)^5(s + 1)^4(s + 6)} \tag{5.156}$$

The main program and the computer output are presented, respectively, in Figures 5.15 and 5.16. The computer used was an IBM 370 Model 158 with total execution time 0.08 sec.

Figure 5.15

The main program for computing the partial-fraction expansion of a proper real rational function, using subroutine PFRAC.

```
         DIMENSION ME(10),NE(10),D(10),S(10),CK(20,10),C(20,10),G(10),
   C     GG(10)
         COMPLEX*16 D,S,CK,C,G,GG
         DOUBLE PRECISION AK
         READ,AK,M,N,NKM,(ME(I),I=1,M),(NE(J),J=1,N),(D(I),I=1,M),
   C     (S(J),J=1,N)
         CALL PFRAC (AK,M,N,NKM,ME,NE,D,S,CK,C,G,GG)
         PRINT 1
   1     FORMAT ('-',18X,'REAL PART',9X,'IMAGINARY PART'/)
         DO 2 K=1,N
         JF=NE(K)
         DO 2 J=1,JF
   2     PRINT 3,K,J,CK(K,J)
   3     FORMAT (' CK(',I2,',',I2,')=',2D20.8)
         STOP
         END
```

Figure 5.16
The computer output
showing the coefficients
$K_{ij} = CK(I, J)$.

	REAL PART	IMAGINARY PART
CK(1, 1)=	0.14359690D-01	-0.25132699D-02
CK(1, 2)=	0.38708333D-01	0.56902122D-02
CK(1, 3)=	0.73730263D-01	0.15621514D-01
CK(1, 4)=	0.82928637D-01	0.62313794D-02
CK(1, 5)=	0.37862980D-01	-0.67864037D-02
CK(2, 1)=	0.14359690D-01	0.25132699D-02
CK(2, 2)=	0.38708333D-01	-0.56902122D-02
CK(2, 3)=	0.73730263D-01	-0.15621514D-01
CK(2, 4)=	0.82928637D-01	-0.62313794D-02
CK(2, 5)=	0.37862980D-01	0.67864037D-02
CK(3, 1)=	-0.28714268D-01	0.00000000D 00
CK(3, 2)=	0.33848614D-01	0.00000000D 00
CK(3, 3)=	0.51207693D 00	0.00000000D 00
CK(3, 4)=	0.79606918D 00	0.00000000D 00
CK(4, 1)=	-0.51122246D-05	0.00000000D 00

(a)

5-5.2   Subroutine INVLAP

1. Purpose.   The program, as presented in Figure 5.17, is used to compute the inverse Laplace transform $f(t)$ of a proper real rational function $F(s)$ expanded in partial fractions:

$$f(t) = \mathcal{L}^{-1}\left[\sum_{k=1}^{N}\sum_{j=1}^{n_k} \frac{K_{kj}}{(s+s_k)^j}\right] = \sum_{k=1}^{N}\sum_{j=1}^{n_k} \frac{K_{kj}}{(j-1)!}\, t^{j-1} e^{-s_k t} \qquad \textbf{(5.157)}$$

2. Usage.   The program consists of a subroutine, INVLAP, called by

CALL INVLAP (N, NKM, NE, CK, S, FT, T)

N   is the number of distinct poles

NKM   is the maximum order of the poles—that is, NKM = $\max\limits_{k} n_k$

NE   is the name of the orders of the poles $-s_k$ with NE(K) = n_k

CK   is the name of the coefficients K_{kj} with CK(K, J) = K_{kj}

S   is the name of the poles $-s_k$ with S(K) = s_k

FT   is the value of the function $f(t)$ at $t = T$

T   is time $t = T$

3. Remarks.   The number of distinct poles cannot exceed 20. Otherwise a new dimensional statement is required for CK(20, NKM). The input data are N, NKM, NE(J) of NE, CK(K, J) of CK, S(J) of S and T. The output is FT. No user-supplied subprogram is required.

EXAMPLE 5.19

We wish to compute the inverse Laplace transform of the proper real rational function $F(s)$ of (5.156). We use the subroutine PFRAC to expand $F(s)$ in partial fractions; its coefficients K_{kj} = CK(K, J) are shown in Figure 5.16 as computed for $F(s)$ of (5.156). Once the K_{kj}'s are determined, the inverse Laplace transform is obtained by using the subroutine INVLAP.

Figure 5.17
Subroutine INVLAP used to compute the inverse Laplace transform of a real proper rational function expanded in partial fractions.

```
C
C
C      SUBROUTINE INVLAP
C
C      PURPOSE
C          THE PROGRAM IS USED TO COMPUTE THE INVERSE LAPLACE TRANSFORM F(T) OF
C          A PROPER REAL RATIONAL FUNCTION F(S) EXPANDED IN PARTIAL FRACTIONS
C
C      USAGE
C          CALL INVLAP (N,NKM,NE,CK,S,FT,T)
C
C          N      -THE NUMBER OF DISTINCT POLES
C          NKM    -THE MAXIMUM ORDER OF THE POLES
C          NE     -THE ORDERS OF THE POLES
C          CK     -THE COEFFICIENTS
C          S      -THE POLES
C          FT     -THE VALUE OF THE FUNCTION F(T)
C          T      -THE TIME
C
C      SUBROUTINES AND FUNCTION SUBPROGRAMS REQUIRED
C          FUNCTION CDABS(X)     -ABSOLUTE VALUE OF A COMPLEX X
C          FUNCTION DATAN2(X,Y)  -ARCTANGENT X/Y
C          FUNCTION DCOS(X)      -COSINE X
C          FUNCTION DEXP(X)      -EXPONENTIAL X
C
C      REMARKS
C          THE NUMBER OF DISTINCT POLES CANNOT EXCEED TWENTY.  OTHERWISE, A NEW
C              DIMENSIONAL STATEMENT IS REQUIRED FOR CK.
C          THE INPUT DATA ARE N, NKM, NE(J) OF NE, CK(K,J) OF CK, S(J) OF S AND T.
C              THE OUTPUT IS FT.
       SUBROUTINE INVLAP (N,NKM,NE,CK,S,FT,T)
       DIMENSION NE(N),CK(20,NKM),S(N)
       COMPLEX*16 CK,S
       DOUBLE PRECISION F,FT,ST,DREAL,DIMAG,DEXP,DCOS,DATAN2,CDABS,DIMSK
       FT=0.0
       DO 9 K=1,N
       DIMSK=DIMAG(S(K))
       IF (DIMSK) 9,1,4
    1  NK=NE(K)
       F=DREAL(CK(K,1))
       IF (NK.EQ.1) GO TO 7
       DO 3 J=2,NK
       JP=1
       JM1=J-1
       DO 2 L=1,JM1
    2  JP=JP*L
    3  F=F+DREAL(CK(K,J))*T**JM1/FLOAT(JP)
       GO TO 7
    4  NK=NE(K)
       F=2.0*CDABS(CK(K,1))*DCOS(DIMSK*T-DATAN2(DIMAG(CK(K,1)),
      C    DREAL(CK(K,1))))
       IF (NK.EQ.1) GO TO 7
       DO 6 J=2,NK
       JP=1
       JM1=J-1
       DO 5 L=1,JM1
    5  JP=JP*L
    6  F=F+2.0*CDABS(CK(K,J))*DCOS(DIMSK*T-DATAN2(DIMAG(CK(K,J)),
      C    DREAL(CK(K,J))))*T**JM1/FLOAT(JP)
    7  ST=DREAL(S(K))*T
       IF (ST-140.0) 8,8,9
    8  F=F*DEXP(-ST)
       FT=FT+F
    9  CONTINUE
       RETURN
       END
       FUNCTION DREAL(X)
C
C      COMPUTE THE REAL PART OF A COMPLEX X
       COMPLEX*16 X
       DOUBLE PRECISION DREAL
       DREAL=X
       RETURN
       END
       FUNCTION DIMAG(X)
C
C      COMPUTE THE IMAGINARY PART OF A COMPLEX X
       COMPLEX*16 X
       DOUBLE PRECISION DIMAG
       DIMAG=X*(0.0,-1.0)
       RETURN
       END
```

Figure 5.18

The main program used to compute the inverse Laplace transform of a proper real rational function, using sub-routines PFRAC and INVLAP.

```
      DIMENSION ME(10),NE(10),D(10),S(10),CK(20,10),C(20,10),G(10),
     C  GG(10),XY(200,2),JXY(2)
      COMPLEX*16 D,S,CK,C,G,GG
      DOUBLE PRECISION AK,FT
      READ,AK,M,N,NKM,(ME(I),I=1,M),(NE(J),J=1,N),(D(I),I=1,M),
     C  (S(J),J=1,N),TO,DT,TF
      CALL PFRAC (AK,M,N,NKM,ME,NE,D,S,CK,C,G,GG)
      PRINT 1
1     FORMAT ('-',18X,'REAL PART',9X,'IMAGINARY PART'/)
      DO 2 K=1,N
      JF=NE(K)
      DO 2 J=1,JF
2     PRINT 3,K,J,CK(K,J)
3     FORMAT (' CK(',I2,',',I2,')=',2D20.8)
4     T=TO
      I=1
5     CALL INVLAP (N,NKM,NE,CK,S,FT,T)
      XY(I,1)=T
      XY(I,2)=FT
      T=T+DT
      I=I+1
      IF (T-TF) 5,5,6
6     N=I-1
      JXY(1)=1
      JXY(2)=2
      CALL PLOT (XY,JXY,N,200,1)
      STOP
      END
```

The main program that combines the two subroutines PFRAC and INVLAP for computing the inverse Laplace transform $f(t)$ is presented in Figure 5.18. The coefficients K_{kj} in the partial-fraction expansion of $F(s)$ are displayed in Figure 5.19(a) as CK(K, J) $= K_{kj}$. The values of $f(t)$ for $t = 0$,

Figure 5.19

The computer output showing the coefficients $K_{ij} =$ CK(I, J) in (a), and a plot of $f(t)$ as a function of t in (c) the values of which are shown in (b).

	REAL PART	IMAGINARY PART
CK(1, 1)=	0.14359690D-01	-0.25132699D-02
CK(1, 2)=	0.38708333D-01	0.56902122D-02
CK(1, 3)=	0.73730263D-01	0.15621514D-01
CK(1, 4)=	0.82928637D-01	0.62313794D-02
CK(1, 5)=	0.37862980D-01	-0.67864037D-02
CK(2, 1)=	0.14359690D-01	0.25132699D-02
CK(2, 2)=	0.38708333D-01	-0.56902122D-02
CK(2, 3)=	0.73730263D-01	-0.15621514D-01
CK(2, 4)=	0.82928637D-01	-0.62313794D-02
CK(2, 5)=	0.37862980D-01	0.67864037D-02
CK(3, 1)=	-0.28714268D-01	0.00000000D 00
CK(3, 2)=	0.33848614D-01	0.00000000D 00
CK(3, 3)=	0.51207693D 00	0.00000000D 00
CK(3, 4)=	0.79606918D 00	0.00000000D 00
CK(4, 1)=	-0.51122246D-05	0.00000000D 00

(a)

T=	0.00000000E 00	FT=	-0.55839165D-16
T=	0.10000000E 01	FT=	0.14557166D 00
T=	0.20000000E 01	FT=	0.28730109D 00
T=	0.30000000E 01	FT=	0.29669722D 00
T=	0.40000000E 01	FT=	0.23251176D 00
T=	0.50000000E 01	FT=	0.15582356D 00
T=	0.60000000E 01	FT=	0.94317147D-01
T=	0.70000000E 01	FT=	0.53128735D-01
T=	0.80000000E 01	FT=	0.28366644D-01
T=	0.90000000E 01	FT=	0.14529955D-01
T=	0.10000000E 02	FT=	0.72000573D-02

(b)

† The program does not print out $f(t) =$ FT.

Figure 5.19   (continued)

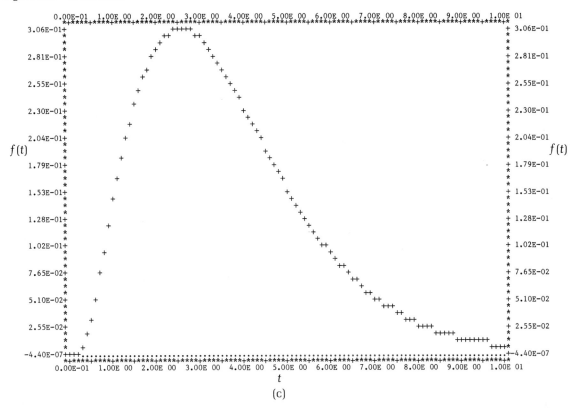

(c)

1, 2, ..., 10 are shown in Figure 5.19(b).[†] The total execution time for an IBM 370 Model 158 computer was 0.23 sec. Figure 5.19(c) is a computer plot of $f(t)$, using subroutine PLOT.[†]

EXAMPLE 5.20

We shall use the subroutines PFRAC and INVLAP to compute the inverse Laplace transform of the real proper rational function

$$F(s) = \frac{4(s+2+j8)^8(s+2-j8)^8(s+1+j2)^3(s+1-j2)^3}{(s+11+j12)^6(s+11-j12)^6(s+j4)^5(s-j4)^5(s+15)} \quad \textbf{(5.158)}$$

The main program that uses both subroutines PFRAC and INVLAP is the same as in Figure 5.18. For illustrative purposes the coefficients K_{kj} in the partial-fraction expansion of $F(s)$ are displayed in Figure 5.20(a) as $CK(K, J) = K_{kj}$. The computer output[‡] is the inverse Laplace transform $f(t)$

[†] The subroutine PLOT is described in Appendix A.
[‡] The program does not print out $f(t) = FT$.

Figure 5.20

The computer output of the partial-fraction expansion coefficients (a) for the real proper rational function (5.158), the inverse Laplace transform of which is shown in (b).

	REAL PART	IMAGINARY PART					
CK(1, 1)=	-0.81904151D 01	-0.32778469D 02					
CK(1, 2)=	0.23809127D 03	0.37401206D 03					
CK(1, 3)=	-0.22248578D 04	-0.32324009D 04	T=	0.00000000E 00	FT=	0.40000000D 01	
CK(1, 4)=	0.99275678D 04	0.19058995D 05	T=	0.10000000E 01	FT=	-0.62191465D-02	
CK(1, 5)=	-0.17410968D 05	-0.63826622D 05	T=	0.20000000E 01	FT=	0.31738315D-02	
CK(1, 6)=	-0.36171219D 03	0.87921681D 05	T=	0.30000000E 01	FT=	0.79079523D-01	
CK(2, 1)=	-0.81904151D 01	0.32778469D 02	T=	0.40000000E 01	FT=	-0.26397593D 00	
CK(2, 2)=	0.23809127D 03	-0.37401206D 03	T=	0.50000000E 01	FT=	0.33060967D 00	
CK(2, 3)=	-0.22248578D 04	0.32324009D 04	T=	0.60000000E 01	FT=	0.17432950D 00	
CK(2, 4)=	0.99275678D 04	-0.19058995D 05	T=	0.70000000E 01	FT=	-0.13904937D 01	
CK(2, 5)=	-0.17410968D 05	0.63826622D 05	T=	0.80000000E 01	FT=	0.24142627D 01	
CK(2, 6)=	-0.36171219D 03	-0.87921681D 05	T=	0.90000000E 01	FT=	-0.14339600D 01	
CK(3, 1)=	0.69533417D-03	-0.25155397D-04	T=	0.10000000E 02	FT=	-0.25902562D 01	
CK(3, 2)=	0.16871540D-02	0.20747413D-03	T=	0.11000000E 02	FT=	0.77107956D 01	
CK(3, 3)=	0.35845181D-02	0.77544250D-03	T=	0.12000000E 02	FT=	-0.88041282D 01	
CK(3, 4)=	0.42425242D-02	0.14294931D-02	T=	0.13000000E 02	FT=	0.12232110D 01	
CK(3, 5)=	0.43626669D-02	0.10053026D-02	T=	0.14000000E 02	FT=	0.13425507D 02	
CK(4, 1)=	0.69533417D-03	0.25155397D-04	T=	0.15000000E 02	FT=	-0.24730536D 02	
CK(4, 2)=	0.16871540D-02	-0.20747413D-03	T=	0.16000000E 02	FT=	0.19023041D 02	
CK(4, 3)=	0.35845181D-02	-0.77544250D-03	T=	0.17000000E 02	FT=	0.83122064D 01	
CK(4, 4)=	0.42425242D-02	-0.14294931D-02	T=	0.18000000E 02	FT=	-0.43072318D 02	
CK(4, 5)=	0.43626669D-02	-0.10053026D-02	T=	0.19000000E 02	FT=	0.56366185D 02	
CK(5, 1)=	0.20379440D 02	0.00000000D 00	T=	0.20000000E 02	FT=	-0.25263119D 02	

(a)   (b)

Figure 5.21

The computer plot of the inverse Laplace transform of the real proper rational function (5.158).

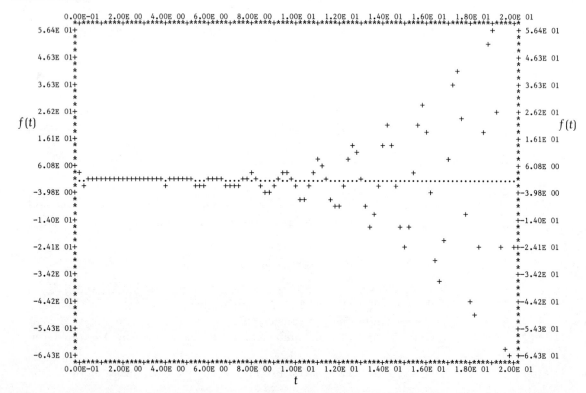

that is obtained for $t = 0, 1, 2, \ldots, 20$ and is shown in Figure 5.20(b). The total execution time for an IBM 370 Model 158 computer was 0.51 sec, including the time required for the computation of the coefficients K_{kj}. Figure 5.21 is a computer plot of $f(t)$ as a function of t, using subroutine PLOT.[†]

5-6 THE COMPLEX INVERSION INTEGRAL

In this section we show how to evaluate the general complex inversion integral (5.19), repeated below as

$$f(t) = \frac{1}{2\pi j} \int_{\sigma_1 - j\infty}^{\sigma_1 + j\infty} F(s)e^{st}\,ds = \begin{cases} 0, & t < 0 \\ \frac{1}{2}f(0+), & t = 0 \\ \frac{1}{2}[f(t+) + f(t-)], & t > 0 \end{cases} \qquad \textbf{(5.159)}$$

where $\sigma_1 \geqq 0$ and $\sigma_1 > \sigma_c$, and σ_c is the abscissa of convergence. Unlike the techniques discussed in the preceding section for the rational functions, the inversion integral can be used to determine the inverse Laplace transform $f(t)$ of a general function $F(s)$, which may or may not be rational. The path of integration, as previously depicted in Figure 5.6, is known as the *Bromwich path*.[‡]

When the function $F(s)$ alone is given, we do not generally know the abscissa of convergence σ_c. However, we do know that $F(s)$ is analytic for $\sigma > \sigma_c$. Thus we can take the path of integration to be a vertical line to the right of all the singular points of $F(s)$, as shown in Figure 5.6. In order to evaluate this integral, we need a result from theory of complex variables known as the residue theorem.[§]

> **Residue Theorem** *If a function $F(s)$ is analytic within and on a closed contour C except at a finite number of singular points in the interior of C, then*
>
> $$\int_C F(s)\,ds = 2\pi j(r_1 + r_2 + \cdots + r_k) \qquad \textbf{(5.160)}$$
>
> *where $r_1, r_2, \ldots, r_k$ are the residues of $F(s)$ at its singular points within the contour C.*

In order to apply this theorem, we have to close the contour of integration of (5.159) to the left, as depicted in Figure 5.22. The integral on the closed contour, according to the residue theorem, is $2\pi j$ times the sum of the residues

---

[†] The subroutine PLOT is described in Appendix A.
[‡] Named after T. J. I'A. Bromwich, who made many significant contributions to the theory of Laplace transformation.
[§] C. R. Wylie, Jr., *Advanced Engineering Mathematics*, 3d ed. (New York: McGraw-Hill, 1966), Chapter 16.

Figure 5.22
The closed contour used
to evaluate the general
complex inversion integral
(5.159).

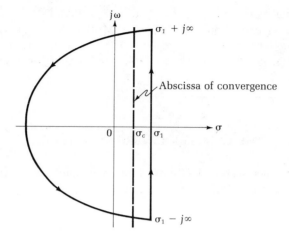

of $F(s)e^{st}$ at the enclosed singular points. If the integrand $F(s)e^{st}$ satisfies *Jordan's lemma*,[†] which states that the infinite semicircular arc to the left will not contribute anything to the integral for $t > 0$ provided that $F(s)$ vanishes uniformly as the radius of the semicircle approaches infinity, the inversion integral (5.159) becomes

$$f(t) = \sum \text{residues of } F(s)e^{st} \text{ at the finite singularities of } F(s) \qquad \textbf{(5.161)}$$

This is a most useful result. For simple functions like proper rational functions, Jordan's lemma is always satisfied. Thus the above formula is always applicable for proper rational functions.

We next consider the computation of residues. Let $F(s)$ be a function that has a pole of order n at $-s_0$. Then the residue of this pole is determined by the formula[†]

$$r_0 = \frac{1}{(n-1)!} \cdot \frac{d^{n-1}}{ds^{n-1}} [(s+s_0)^n F(s)] \Big|_{s=-s_0} \qquad \textbf{(5.162)}$$

which is very similar to (5.130) except that (5.162) is not restricted to rational functions. We illustrate the above results by the following example.

EXAMPLE 5.21
We use the inversion integral to determine a time-domain nodal voltage $v_{n1}(t)$ from its Laplace transform

$$V_{n1}(s) = \frac{s^3 + 2.61812s^2 + 2s + 1}{s(s+1)^3} \qquad \textbf{(5.163)}$$

as given by (5.118). We first compute the residues of $V_{n1}(s)e^{st}$ at the poles of $V_{n1}(s)$, which are at 0 and -1. For the simple pole at 0, the residue is

---

[†] See, for example, K. Knopp, *Theory of Functions*, vol. I (New York: Dover, 1945). The lemma is named after Camille Jordan (1838–1922).

obtained from (5.162) as

$$r_1 = sV_{n1}(s)e^{st}\big|_{s=0} = 1, \qquad t \geqq 0 \tag{5.164}$$

For the third-order pole at -1, the residue is computed by

$$
\begin{aligned}
r_2 &= \frac{1}{2!} \cdot \frac{d^2}{ds^2} \left[(s+1)^3 V_{n1}(s)e^{st}\right]\Bigg|_{s=-1} \\
&= [(1 + 1/s^3)e^{st} + (2s + 2.61812 - 1/s^2)te^{st} \\
&\quad + (\tfrac{1}{2}s^2 + 1.30906s + 1 + 1/2s)t^2 e^{st}]\big|_{s=-1} \\
&= -0.38188te^{-t} - 0.30906t^2 e^{-t}, \qquad t \geqq 0
\end{aligned}
\tag{5.165}
$$

Appealing to (5.161), the inverse Laplace transform of (5.163) is found to be

$$v_{n1}(t) = r_1 + r_2 = 1 - 0.38188te^{-t} - 0.30906t^2 e^{-t}, \qquad t \geqq 0 \tag{5.166}$$

confirming (5.122).

5-7 SUMMARY

In this chapter we introduced the operational method of solution of the secondary systems of network equations using the Laplace transformation. Not only does it give the complete response in one operation, but also it specifies the initial conditions automatically.

The *Laplace transformation* is an integral transform which relates a time-domain function with another function which is defined in the complex-frequency plane. The transform is manipulated algebraically after the initial conditions are inserted, and then an inverse Laplace transformation by way of the complex inversion integral—by the use of a table of transforms or by any other means—is performed to yield the time-domain solution. For this we discussed a number of fundamental properties of the Laplace transform that are useful in solving a system of simultaneous linear integrodifferential equations. Specifically, they are linearity, real differentiation, real integration, and real translation. Moreover, the Laplace transformation is unique in that two essentially different functions $f(t)$ cannot lead to the same transform $F(s)$.

We demonstrated how the Laplace transform is used to solve the network equations. The main conclusion is that after taking the Laplace transform, a system of linear simultaneous integrodifferential equations becomes a system of linear algebraic equations, which can be solved to obtain the desired transform variables. In order to recover the time-domain functions, one simple way is to manipulate algebraically these transforms first into sums of several simple functions the inverse Laplace transforms of which are either known or tabulated. The time function is then obtained by adding these inverse Laplace transforms. This procedure is perfectly legitimate because of the uniqueness property of the Laplace transformation. It states that if a time function is found using any method whatsoever and its Laplace transform is the desired one, then the time function is *the* only one.

To help manipulate the transforms, we presented a general technique for breaking up a rational function into simple components called the *partial-fraction expansion*. In the case of simple poles, the procedure is exceedingly simple. For multiple poles, we derived a formula for computing the coefficients in the expansion, which involves the derivatives of the function after the suppression of the pole under consideration. However, in using this formula for a high-order pole, it is necessary to work out the derivatives of the functions up to an order that is one unit lower than that of the suppressed pole. As a result the operations become more and more tedious as the order of poles is increased. Furthermore, because of interdependence, computational errors propagate and may be considerable. To avoid this difficulty, we introduced a noniterative method that circumvents the necessity of passing through the lower-order derivatives. This method can easily be implemented on a digital computer. The associated computer program together with the program for computing the inverse Laplace transform was discussed. The partial-fraction expansion program requires that the numerator and denominator polynomials be given in factored form. If not, a root-finding program is required.

Finally, we showed how to evaluate the complex inversion integral by applying the residue theorem. Network functions, theorems, and other properties of transform networks will be presented in the next chapter.

REFERENCES AND SUGGESTED READING

Bellman, R. E., R. E. Kalaba and J. A. Lockett. *Numerical Inversion of the Laplace Transform.* New York: American Elsevier, 1966, Chapters 1 and 2.

Brugia, O. "A Noniterative Method for the Partial Fraction Expansion of a Rational Function with High-Order Poles," *SIAM Rev.* 7: (1965) 381–87.

Churchill, R. V. *Complex Variables and Applications.* 2d ed. New York: McGraw-Hill, 1960, Chapters 7 and 12.

Churchill, R. V. *Operational Mathematics.* 2d ed. New York: McGraw-Hill, 1958, Chapters 1, 2, and 6.

Desoer, C. A., and E. S. Kuh. *Basic Circuit Theory.* New York: McGraw-Hill, 1969, Chapter 13.

Karni, S. *Intermediate Network Analysis.* Boston, Mass.: Allyn & Bacon, 1971, Chapter 2 and Appendix B.

Knopp, K. *Theory of Functions.* New York: Dover, 1945, Part I.

Kuo, F. F. *Network Analysis and Synthesis.* 2d ed. New York: Wiley, 1966, Chapter 6.

Papoulis, A. *Circuits and Systems: A Modern Approach.* New York: Holt, Rinehart & Winston, 1980, Chapter 2.

Spiegel, M. R. *Laplace Transforms.* New York: McGraw-Hill, 1967.

Van Valkenburg, M. E. *Network Analysis.* 3d ed. Englewood Cliffs, N.J.: Prentice-Hall, 1974, Chapter 7.

PROBLEMS

5.1 Verify that the Laplace transform of $\cosh \alpha t$ is

$$\mathcal{L}[\cosh \alpha t] = \frac{s}{s^2 - \alpha^2} \tag{5.167}$$

5.2 Verify that the Laplace transform of $e^{-\alpha t} \sin \omega_0 t$ is

$$\mathcal{L}[e^{-\alpha t} \sin \omega_0 t] = \frac{\omega_0}{(s + \alpha)^2 + \omega_0^2} \tag{5.168}$$

5.3 Verify (5.31) for $n = 3$ and $n = 4$.

5.4 Determine the Laplace transforms of the following time functions $f(t)$, where $f(t) = 0$ for $t < 0$:

(a) $\sin^2 \omega_0 t$ **(b)** $(1/t) \sin \omega_0 t$ **(c)** $\sin \alpha t \sin \beta t$ **(d)** $e^{-\alpha t^2}$ **(e)** $t \sinh \alpha t$

5.5 Verify (5.46) for $n = 3$ and $n = 4$.

5.6 Using the Laplace transformation method, solve the integrodifferential equation (4.140) with $v_{n1}(0-) = v_{n2}(0-) = 2$ V.

5.7 Using the Laplace transformation method, solve the integrodifferential equation (4.154) with $v_{n1}(0-) = v_{n2}(0-) = 2$ V.

5.8 Compute the inverse Laplace transforms of $V_{n1}(s)$ and $V_{n2}(s)$ of (5.75).

5.9 Work Problem 4.3, using the Laplace transformation method.

5.10 Work Problem 4.11, using the Laplace transformation method.

5.11 Work Problem 4.12, using the Laplace transformation method.

5.12 Work Problem 4.13, using the Laplace transformation method.

5.13 Work Problem 4.14, using the Laplace transformation method.

5.14 Work Problem 4.20, using the Laplace transformation method.

5.15 Work Problem 4.22, using the Laplace transformation method.

5.16 Work Problem 4.24, using the Laplace transformation method.

5.17 Find the inverse Laplace transforms of the following rational functions by means of the partial-fraction expansions:

(a) $\dfrac{3s^2 + 2s + 9}{2s^2 + 4s + 5}$ **(b)** $\dfrac{s^2}{(s^2 + 9)(s + 4)}$

(c) $\dfrac{s}{s^3 + 3s^2 + 6s + 4}$ **(d)** $\dfrac{(s^2 + 2)^2(s + 3)}{s^2(s + 1)^3(s + 5)}$

(e) $\dfrac{(s + 4)^2(s + 5)(s^2 + 1)}{(s + 1 + j2)^2(s + 1 - j2)^2}$

5.18 Repeat Problem 5.17 using formula (161).

5.19 Find the inverse Laplace transforms for the following functions:

(a) $\dfrac{1 + e^{-s}}{s(s + 3)}$ **(b)** $\dfrac{e^{-2s} - se^{-s}}{s^2 + 6s + 5}$

5.20 Find the Laplace transforms for the waveforms shown in Figure 5.23.

5.21 If $F(s)$ is the Laplace transform of $f(t)$, prove the following identities with a being a real and positive constant:

(a) $\mathcal{L}[f(at)] = \dfrac{1}{a} F(s/a)$ (5.169a)

(b) $\mathcal{L}[e^{at} f(t)] = F(s - a)$ (5.169b)

(c) $\mathcal{L}[t f(t)] = -dF(s)/ds$ (5.169c)

Figure 5.23
Signal waveforms the Laplace transforms of which are to be ascertained.

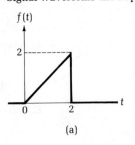

(a)

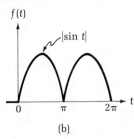

(b)

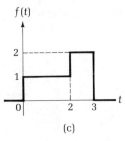

(c)

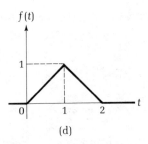

(d)

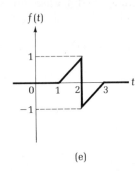

(e)

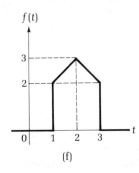

(f)

5.22 Using the subroutine PFRAC, obtain the partial-fraction expansions of the following rational functions:

(a) $$\frac{(s + 3 + j2)^3(s + 3 - j2)^3(s^2 + 1)^2}{(s + 4)^3(s + 1 + j2)^4(s + 1 - j2)^4(s + 3)^2} \tag{5.170a}$$

(b) $$\frac{(s + 3)^5(s + 2 + j5)^3(s + 2 - j5)^3(s + 9)^2(s + 4)}{(s + 7 + j8)^5(s + 7 - j8)^5(s + 2 + j4)^3(s + 2 - j4)^3} \tag{5.170b}$$

5.23 Using the subroutines PFRAC and INVLAP, compute the inverse Laplace transforms of the rational functions of Problem 5.22(a) for $t = 0, 1, 2, \ldots, 20$, and of Problem 5.22(b) for $t = 0, 0.25, 0.50, \ldots, 5$.

5.24 Show that

$$\mathscr{L}[(1 - e^{-t})^5] = \frac{5!}{s(s + 1)(s + 2)(s + 3)(s + 4)(s + 5)} \tag{5.171}$$

which can be generalized to yield

$$\mathscr{L}[(1 - e^{-t})^n] = \frac{n!}{s(s + 1)(s + 2) \cdots (s + n)} \tag{5.172}$$

5.25 Let $F(s)$ be the Laplace transform of $f(t)$. Prove that

$$\mathscr{L}\left[\frac{f(t)}{t}\right] = \int_s^\infty F(s)\, ds \tag{5.173}$$

5.26 Determine the regions of convergence of the Laplace transforms of the following functions:

(a) $e^{-\alpha t}u(-t)$ (b) $e^{-\alpha|t|}$ (c) $e^{-\alpha t^2}u(t)$

where α is real and positive and $u(t)$ is the unit-step function. Also find the abscissas of convergence.

5.27 Prove that for real and positive α

$$\mathscr{L}\left[e^{\alpha t}\frac{d^n}{dt^n}\left(\frac{t^n}{n!}e^{-2\alpha t}\right)\right] = \frac{(s-\alpha)^n}{(s+\alpha)^{n+1}}$$

(5.174)

[Hint: Appeal to (5.31) and (5.169b).]

5.28 Using (5.173) find the Laplace transforms of the following functions:
(a) $t^{-1}\cos\omega_0 t$ **(b)** $t^{-1}(1-e^{-\alpha t})$ **(c)** $t^{-1}(\sinh\alpha t + \cosh\alpha t)$

5.29 Prove that for any positive integer n

$$\mathscr{L}[t^n f(t)] = (-1)^n\frac{d^n F(s)}{ds^n}$$

(5.175)

where $F(s)$ is the Laplace transform of $f(t)$.

5.30 Show that for any positive integer n

$$\mathscr{L}\left[\frac{d^n\delta(t)}{dt^n}\right] = s^n$$

(5.176)

where $\delta(t)$ the unit impulse.

CHAPTER SIX

NETWORK ANALYSIS

In the previous chapter, we saw how the Laplace transform is used to solve secondary systems of network equations. However, the mere computation of the solution of a secondary system is one of the many important applications for this elegant tool of network analysis. Our purpose here is to associate analytic functions with a network and hence to make use of the extensive theory of analytic functions developed by mathematicians over the last century. More specifically, in this chapter we shall define the network function and study the different ways of its representation. We shall study several very general and useful network theorems and the characterizations and representations of a two-port network. The frequency response of a network will also be discussed.

6-1 TRANSFORM IMPEDANCE AND TRANSFORM NETWORKS

In Section 5-3 we demonstrated that upon taking the Laplace transform on both sides of the secondary system of integrodifferential equations

$$\mathbf{W}(p)\mathbf{x}(t) = \mathbf{f}(t) \tag{6.1}$$

as given in (4.3) we obtain a system of linear algebraic equations

$$\mathbf{W}(s)\mathbf{X}(s) = \mathbf{F}(s) + \mathbf{h}(s) \tag{6.2}$$

where $\mathbf{X}(s)$ and $\mathbf{F}(s)$ denote the Laplace transforms of $\mathbf{x}(t)$ and $\mathbf{f}(t)$, respectively, and $\mathbf{h}(s)$ is a vector that includes the contributions due to initial conditions. The coefficient matrix $\mathbf{W}(s)$ in the complex frequency variable s is obtained from $\mathbf{W}(p)$, with s replacing p. In the following we show how to formulate (6.2) directly from an equivalent network of transformed variables, thus bypassing the necessity of writing the integrodifferential equations. Such an analysis is often referred to as analysis in the *frequency domain*, in contrast to the analysis with integrodifferential equations (6.1), which is called analysis in the *time domain*. We next determine network representations

Figure 6.1
Transform network
representations of a
resistor.

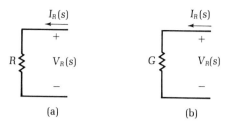

(a)

(b)

for each of the network elements in terms of transform impedance or admittance and sources due to initial conditions.

The ideal voltage source $v_g(t)$ and current source $i_g(t)$ can be represented by their transforms $V_g(s) = \mathscr{L}[v_g(t)]$ and $I_g(s) = \mathscr{L}[i_g(t)]$. The time-domain expression relating voltage and current for the resistor is given by Ohm's law in the form

$$v_R(t) = Ri_R(t) \quad \text{or} \quad i_R(t) = Gv_R(t), \qquad G = \frac{1}{R} \tag{6.3}$$

Transforming both equations, the corresponding transform equations are

$$V_R(s) = RI_R(s) \quad \text{or} \quad I_R(s) = GV_R(s) \tag{6.4}$$

The transform network representations of (6.4) are shown in Figure 6.1. For an inductor, the time-domain v-i relationships are described by

$$v_L(t) = L\frac{di_L(t)}{dt} \tag{6.5a}$$

$$i_L(t) = \frac{1}{L}\int_{0-}^{t} v_L(\tau)\,d\tau + i_L(0-) \tag{6.5b}$$

The equivalent frequency-domain expressions are

$$V_L(s) = LsI_L(s) - Li_L(0-) \tag{6.6a}$$

$$I_L(s) = \frac{1}{Ls}V_L(s) + \frac{i_L(0-)}{s} \tag{6.6b}$$

where, as before, the uppercase $V_L(s)$ and $I_L(s)$ denote the Laplace transforms of the corresponding lowercase time-domain variables $v_L(t)$ and $i_L(t)$. This convention will be used throughout the remainder of the book unless stated specifically otherwise. The transform network representations for an inductor are depicted in Figure 6.2. For a capacitor, the v-i relationships are given by

$$v_C(t) = \frac{1}{C}\int_{0-}^{t} i_C(\tau)\,d\tau + v_C(0-) \tag{6.7a}$$

$$i_C(t) = C\frac{dv_C(t)}{dt} \tag{6.7b}$$

Figure 6.2
Transform network
representations of an
inductor.

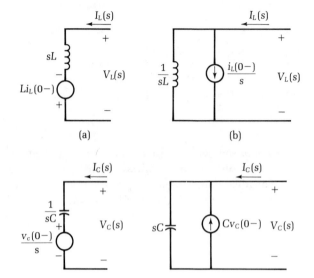

(a)

(b)

Figure 6.3
Transform network
representations of a
capacitor.

(a)

(b)

The frequency-domain expressions are

$$V_C(s) = \frac{1}{Cs} I_C(s) + \frac{v_C(0-)}{s}$$

(6.8a)

$$I_C(s) = CsV_C(s) - Cv_C(0-)$$

(6.8b)

The transform network representations for a capacitor are shown in Figure 6.3. The entities R, Ls, and $1/Cs$ are referred to as the *impedances* and G, $1/Ls$, and Cs the *admittances* for the resistor, the inductor, and the capacitor, respectively. The term *immittance* refers to either the impedance or admittance.

The process of representing the three passive network elements by their transform networks may be extended to more complicated networks. For example, a voltage-controlled current source described by the equation

$$i_2(t) = g_m v_1(t)$$

(6.9)

as depicted in Figure 6.4(a), has the frequency-domain counterpart

$$I_2(s) = g_m V_1(s)$$

(6.10)

which can be represented by the transform network of Figure 6.4(b).

As another example, consider the transformer of Figure 6.5(a) where terminal voltage and current relations are described by the equations

$$v_1(t) = L_1 \frac{di_1(t)}{dt} + M \frac{di_2(t)}{dt}$$

(6.11a)

$$v_2(t) = M \frac{di_1(t)}{dt} + L_2 \frac{di_2(t)}{dt}$$

(6.11b)

Figure 6.4
A voltage-controlled current source (a) and its frequency-domain representation.

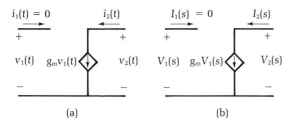

(a)   (b)

Figure 6.5
A transformer (a) and its frequency-domain representation (b).

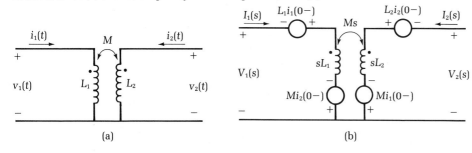

(a)   (b)

The corresponding transform equations are

$$V_1(s) = L_1 s I_1(s) - L_1 i_1(0-) + M s I_2(s) - M i_2(0-) \qquad \textbf{(6.12a)}$$

$$V_2(s) = M s I_1(s) - M i_1(0-) + L_2 s I_2(s) - L_2 i_2(0-) \qquad \textbf{(6.12b)}$$

Equations (6.12) can be represented by the transform network of Figure 6.5(b) in which the initial conditions are described by the four transform voltage sources.

As a result we see that working from a transform network diagram we can write the loop, nodal, or cutset equations (6.2) in the frequency domain directly, thus avoiding the necessity of first writing the integrodifferential equations. In fact, with minor modifications the rules for writing the loop, nodal, and cutset equations, as outlined in Sections 3-1.1, 3-2.1, and 3-2.2, remain valid. We illustrate this by the following examples.

EXAMPLE 6.1

Consider the transistor amplifier of Figure 6.6(a) with its equivalent network shown in Figure 6.6(b) after removing the biasing circuit. Using the complex-frequency representations of Figures 6.1 to 6.4, the transform network of Figure 6.6(b) is presented in Figure 6.7. We shall write the frequency-domain nodal equations directly from Figure 6.7, applying the rules of Section 3-2.1.

The transform nodal voltages V_{nk} ($k = 1, 2, 3$) are as indicated in Figure 6.7. First treat the controlled source $g_m V_2$ as if it were independent.

Figure 6.6
A transistor amplifier (a) and its equivalent network (b) after the removal of the biasing circuitry.

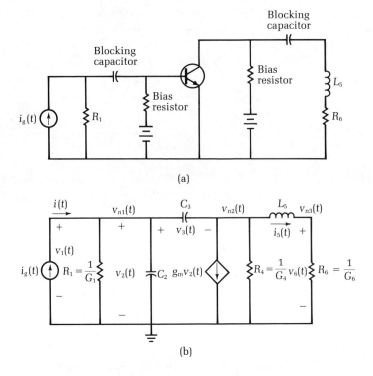

(a)

(b)

Figure 6.7
The transform network of the network of Figure 6.6(b).

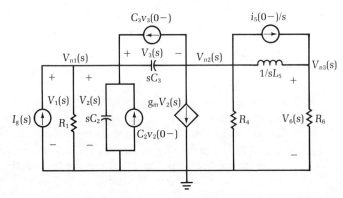

The resulting nodal equations can be written directly from Figure 6.7 by inspection, as follows. Let $G_i = 1/R_i$, $i = 1, 4, 6$.

$$
\begin{bmatrix}
G_1 + (C_2 + C_3)s & -C_3 s & 0 \\
-C_3 s & G_4 + C_3 s + 1/L_5 s & -1/L_5 s \\
0 & -1/L_5 s & G_6 + 1/L_5 s
\end{bmatrix}
\begin{bmatrix}
V_{n1}(s) \\
V_{n2}(s) \\
V_{n3}(s)
\end{bmatrix}
$$

$$
=
\begin{bmatrix}
I_g(s) + C_2 v_2(0-) + C_3 v_3(0-) \\
-C_3 v_3(0-) - g_m V_2(s) - i_5(0-)/s \\
i_5(0-)/s
\end{bmatrix}
\tag{6.13}
$$

Since $V_2 = V_{n1}$, the controlled current source $g_m V_2$ on the right-hand side of (6.13) can be rewritten as $g_m V_{n1}$. Using this in (6.13) and regrouping the terms, we have the desired transform nodal equation

$$
\begin{bmatrix}
G_1 + (C_2 + C_3)s & -C_3 s & 0 \\
g_m - C_3 s & G_4 + C_3 s + 1/L_5 s & -1/L_5 s \\
0 & -1/L_5 s & G_6 + 1/L_5 s
\end{bmatrix}
\begin{bmatrix}
V_{n1} \\
V_{n2} \\
V_{n3}
\end{bmatrix}
$$

$$
= \begin{bmatrix} I_g \\ 0 \\ 0 \end{bmatrix} + \begin{bmatrix} C_2 v_2(0-) + C_3 v_3(0-) \\ -C_3 v_3(0-) - i_5(0-)/s \\ i_5(0-)/s \end{bmatrix} \tag{6.14}
$$

which is in the standard form of (6.2). Equation (6.14) can also be deduced from (3.56) by taking the Laplace transform on both sides with $V_g = R_1 I_g$ and $t_0 = 0-$.

EXAMPLE 6.2

We shall write the loop equation for the network of Figure 6.6. An equivalent network of Figure 6.6(b) for the loop equation is shown in Figure 6.8(a). Using the frequency-domain representations of Figures 6.1 to 6.4, the transform network of Figure 6.8(a) is presented in Figure 6.8(b). Choosing the three transform loop currents I_{mk} ($k = 1, 2, 3$) as indicated in Figure 6.8(b), the transform loop equation can be written directly by inspection, as follows. First treat the controlled source $g_m R_4 V_2$ as if it were an independent

Figure 6.8
An equivalent network (a) of Figure 6.6(b) and its frequency-domain representation (b).

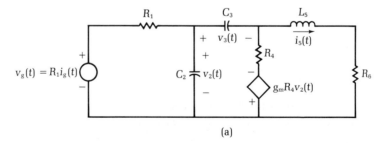

source. Then applying the rules outlined in Section 3-1.1, we have

$$\begin{bmatrix} R_1 + 1/C_2 s & -1/C_2 s & 0 \\ -1/C_2 s & R_4 + 1/C_2 s + 1/C_3 s & -R_4 \\ 0 & -R_4 & R_4 + R_6 + L_5 s \end{bmatrix} \begin{bmatrix} I_{m1}(s) \\ I_{m2}(s) \\ I_{m3}(s) \end{bmatrix}$$

$$= \begin{bmatrix} V_g(s) - v_2(0-)/s \\ g_m R_4 V_2(s) + v_2(0-)/s - v_3(0-)/s \\ -g_m R_4 V_2(s) + L_5 i_5(0-) \end{bmatrix} \tag{6.15}$$

Since the controlled voltage source can be expressed as

$$g_m R_4 V_2(s) = g_m R_4 \left[\frac{1}{C_2 s} I_2(s) + \frac{v_2(0-)}{s} \right]$$

$$= \frac{g_m R_4}{C_2 s} (I_{m1} - I_{m2}) + \frac{g_m R_4 v_2(0-)}{s} \tag{6.16}$$

the term $g_m R_4 V_2$ on the right-hand side of (6.15) can be eliminated. After substituting (6.16) in (6.15) and regrouping, we obtain

$$\begin{bmatrix} R_1 + 1/C_2 s & -1/C_2 s & 0 \\ -(1 + g_m R_4)/C_2 s & R_4 + (1 + g_m R_4)/C_2 s + 1/C_3 s & -R_4 \\ g_m R_4/C_2 s & -R_4 - g_m R_4/C_2 s & R_4 + R_6 + L_5 s \end{bmatrix} \begin{bmatrix} I_{m1} \\ I_{m2} \\ I_{m3} \end{bmatrix}$$

$$= \begin{bmatrix} V_g \\ 0 \\ 0 \end{bmatrix} + \begin{bmatrix} -v_2(0-)/s \\ (1 + g_m R_4)v_2(0-)/s - v_3(0-)/s \\ -g_m R_4 v_2(0-)/s + L_5 i_5(0-) \end{bmatrix} \tag{6.17}$$

which again is in the standard form of (6.2). The corresponding time-domain loop equation was computed previously and is given by (3.22).

6-2   NETWORK FUNCTIONS

In the foregoing we demonstrated the process of writing the transformed secondary system of network equations

$$\mathbf{W}(s)\mathbf{X}(s) = \mathbf{F}(s) + \mathbf{h}(s) \tag{6.18}$$

directly from the transform network by inspection. Once this is accomplished, the unknown transform vector $\mathbf{X}(s)$ can be obtained immediately by inverting the matrix $\mathbf{W}(s)$:

$$\mathbf{X}(s) = \mathbf{W}^{-1}(s) [\mathbf{F}(s) + \mathbf{h}(s)] \tag{6.19}$$

provided that det $\mathbf{W}(s)$ is not identically zero.

Consider a linear time-invariant network that contains a single independent voltage or current source as the input with arbitrary waveform.

Assume that all initial conditions in the network have been set to zero. Let the zero-state response be either a voltage across any two nodes of the network or a current in any branch of the network. Then the *network function H(s)* is defined by

$$H(s) = \frac{\text{the Laplace transform of the zero-state response}}{\begin{array}{c}\text{the Laplace transform of the input}\\ \text{or another zero-state response}\end{array}} \tag{6.20}$$

Network functions generally fall into two classes depending on whether the terminals to which the response relates are the same or different from the input terminals. For the same pair of terminals, it is referred to as the *driving-point* or *input function*; and for different pairs of terminals, the *transfer function*. Since the input and the response may either be a current or a voltage, the network function may be a *driving-point impedance*, a *driving-point admittance*, a *transfer impedance*, a *transfer admittance*, a *transfer voltage ratio*, or a *transfer current ratio*. Our primary objective here is to obtain some general and broad properties of network functions, recognizing that each of the network functions mentioned above has its own distinct characteristics.

We shall illustrate the ways of determining the network functions by the following example.

EXAMPLE 6.3
Consider the network of Figure 6.6(b), with its nodal equation obtained in Example 6.1. Setting all initial conditions to zero, reduces (6.14) to

$$\begin{bmatrix} G_1 + (C_2 + C_3)s & -C_3 s & 0 \\ g_m - C_3 s & G_4 + C_3 s + 1/L_5 s & -1/L_5 s \\ 0 & -1/L_5 s & G_6 + 1/L_5 s \end{bmatrix} \begin{bmatrix} V_{n1} \\ V_{n2} \\ V_{n3} \end{bmatrix} = \begin{bmatrix} I_g \\ 0 \\ 0 \end{bmatrix} \tag{6.21}$$

The determinant of the coefficient matrix—the nodal determinant—is found to be

$$\det \mathbf{Y}_n(s) = C_2 C_3 G_6 s^2 + \left[\left(G_4 G_6 + \frac{C_3}{L_5}\right)C_2 + G_1 C_3 G_6 + (G_4 + g_m)C_3 G_6\right]s$$

$$+ G_1 \left(G_4 G_6 + \frac{C_3}{L_5}\right) + \frac{C_2(G_4 + G_6)}{L_5} + \frac{C_3(G_4 + G_6 + g_m)}{L_5}$$

$$+ \frac{G_1(G_4 + G_6)}{L_5 s} \tag{6.22}$$

By Cramer's rule, the nodal voltage V_{n1} is obtained as

$$V_{n1}(s) = \frac{C_3 G_6 s + G_4 G_6 + C_3/L_5 + (G_4 + G_6)/L_5 s}{\det \mathbf{Y}_n(s)} I_g(s) \tag{6.23}$$

After setting the initial conditions to zero, the driving-point impedance facing the current source I_g is defined as the ratio of the transform voltage V_1 across I_g to the transform of the input current, yielding from (6.23)

$$Z_{in}(s) = \frac{V_1(s)}{I_g(s)} = \frac{C_3 G_6 s + G_4 G_6 + C_3/L_5 + (G_4 + G_6)/L_5 s}{\det \mathbf{Y}_n(s)} \tag{6.24}$$

where $V_1 = V_{n1}$.

Likewise the transfer impedance between the resistor R_6 and the source is defined as the ratio of the transform voltage V_6 across the resistor R_6 to the transform of the input current source I_g. Appealing once more to the Cramer's rule for V_{n3}

$$Z_{g6}(s) = \frac{V_6(s)}{I_g(s)} = \frac{V_{n3}(s)}{I_g(s)} = \frac{C_3/L_5 - g_m/L_5 s}{\det \mathbf{Y}_n(s)} \tag{6.25}$$

Finally, from (6.24) and (6.25) we obtain the transfer voltage-ratio function

$$G_{16}(s) = \frac{V_6(s)}{V_1(s)} = \frac{C_3/L_5 - g_m/L_5 s}{C_3 G_6 s + G_4 G_6 + C_3/L_5 + (G_4 + G_6)/L_5 s} \tag{6.26}$$

An important observation from the above example is that by Cramer's rule a network function can be expressed as the ratio of a cofactor, the sum of a number of weighted cofactors and the coefficient determinant, or the reciprocal of this ratio or the ratio of any two of these combinations. As a result the network functions are rational functions of the complex frequency variable s with *real* coefficients. Thus we can write in general

$$H(s) = \frac{P(s)}{Q(s)} = \frac{b_m s^m + b_{m-1} s^{m-1} + \cdots + b_1 s + b_0}{a_n s^n + a_{n-1} s^{n-1} + \cdots + a_1 s + a_0} \tag{6.27}$$

as in (5.76). The coefficients a's and b's are real because each one is the sum of products of element values of resistors, inductors, capacitors, and so on, and these element values are all real numbers. In factored form we have

$$H(s) = K \frac{\displaystyle\prod_{h=1}^{M} (s + d_h)^{m_h}}{\displaystyle\prod_{k=1}^{N} (s + s_k)^{n_k}} \tag{6.28}$$

where K is a real scale factor. As in (5.77), $-d_h$ ($h = 1, 2, \ldots, M$) are the M distinct zeros of $H(s)$ and $-s_k$ ($k = 1, 2, \ldots, N$) are the N distinct poles of $H(s)$. Since any polynomial $P(s)$ with real coefficients has the property that $\overline{P(s)} = P(\bar{s})$ for all s, we have

$$\overline{H(s)} = \overline{H(\sigma + j\omega)} = H(\bar{s}) = H(\sigma - j\omega) \tag{6.29}$$

Thus the poles and zeros of a network function must either be real or occur in complex conjugate pairs. More precisely if $s_k = \sigma_k + j\omega_k$ is a pole, then $\bar{s}_k = \sigma_k - j\omega_k$ is also a pole. Likewise if $d_h = \sigma_h + j\omega_h$ is a zero, then $\bar{d}_h = \sigma_h - j\omega_h$ is also a zero. Equation (6.29) states that a network function assumes

Figure 6.9
The pole-zero pattern of a
network function.

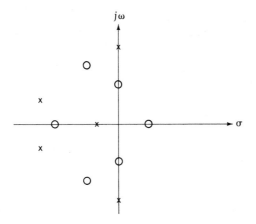

conjugate values at conjugate points in the complex frequency plane. This
is known as the *reflection property.*

Apart from the real constant K, a network function $H(s)$ is completely
determined by its poles and zeros. Therefore we generally represent $H(s)$
in the s-plane as in Figure 6.9. The small circles refer to the zeros of $H(s)$
and the small crosses to the poles of $H(s)$. Such a diagram is called the
pole-zero pattern.

6-3 NETWORK THEOREMS

In this section we restate and prove several fundamental and general theorems for linear time-invariant networks in the frequency domain.

Consider an arbitrary linear time-invariant network with a single input.
We can always write the loop, nodal, or cutset equations so that the input
appears only in the first equation:

$$\mathbf{W}(s)\mathbf{X}(s) = \mathbf{F}(s) + \mathbf{h}(s) \tag{6.30}$$

where

$$\mathbf{F}(s) = [F_1, 0, \ldots, 0]' \tag{6.31}$$

$$\mathbf{h}(s) = [h_1, h_2, \ldots, h_n]' \tag{6.32}$$

The single input F_1 is the transform of an input voltage or current. Suppose
that the kth row variable X_k of $\mathbf{X}(s)$ is the desired response. By appealing
to Cramer's rule, we obtain from (6.30)

$$X_k(s) = \frac{\Delta_{1k}(s)}{\Delta(s)} F_1(s) + \frac{\sum\limits_{j=1}^{n} \Delta_{jk}(s)h_j}{\Delta(s)} \tag{6.33}$$

where $\Delta(s) = \det \mathbf{W}(s)$ and $\Delta_{jk}(s)$ is the cofactor of (j, k) element of $\mathbf{W}(s)$.
The first term on the right-hand side of (6.33) is the Laplace transform of
the zero-state response, and the second term is the Laplace transform of the

zero-input response. The reason is that the coefficients h_j are due to initial conditions. When the initial conditions are set to zero, all h_j are zero and the second term in (6.33) drops out. On the other hand, when the input F_1 is set to zero for all s, the first term drops out. In fact, the ratio Δ_{1k}/Δ is the network function relating the response X_k to the input F_1. Thus we have the following.

Property 6.1 *For a linear time-invariant network, the Laplace transform of the complete response is the sum of the Laplace transform of the zero-state response and the Laplace transform of the zero-input response. The network function, when multiplied by the transform of the input, gives the transform of the zero-state response.*

By the linearity of Laplace transforms, the above result also implies that the complete response in the time-domain is the sum of the zero-state response and the zero-input response, a fact that was demonstrated in Chapter 4.

EXAMPLE 6.4

The nodal equation for the transistor amplifier of Figure 6.6(a) is given by (6.14). For simplicity let $G_1 = G_4 = G_6 = 1$ mho, $C_2 = C_3 = 1$ F, $L_5 = 1$ H, $v_2(0-) = v_3(0-) = 1$ V, and $i_5(0-) = 1$ A. Using these in (6.14) yields

$$
\begin{bmatrix}
1 + 2s & -s & 0 \\
g_m - s & 1 + s + 1/s & -1/s \\
0 & -1/s & 1 + 1/s
\end{bmatrix}
\begin{bmatrix}
V_{n1} \\
V_{n2} \\
V_{n3}
\end{bmatrix}
=
\begin{bmatrix}
I_g \\
0 \\
0
\end{bmatrix}
+
\begin{bmatrix}
2 \\
-1 - 1/s \\
1/s
\end{bmatrix}
\qquad \textbf{(6.34)}
$$

Assume that the nodal voltage V_{n3} is the desired response. Applying Cramer's rule

$$
V_{n3}(s) = \frac{s - g_m}{s\Delta(s)} I_g(s) + \frac{s^2 + (3 + g_m)s - 2g_m}{s\Delta(s)}
\qquad \textbf{(6.35)}
$$

where the nodal determinant is found to be

$$
\Delta(s) = s^2 + (4 + g_m)s + 6 + g_m + 2/s
\qquad \textbf{(6.36)}
$$

The final result becomes

$$
\underset{\substack{\text{Laplace transform of the} \\ \text{complete response}}}{V_{n3}(s)} \quad = \quad \underset{\substack{\text{Laplace transform of the zero-state response}}}{\frac{s - g_m}{s^3 + (4 + g_m)s^2 + (6 + g_m)s + 2} I_g(s)}
$$

$$
+ \underset{\substack{\text{Laplace transform of the} \\ \text{zero-input response}}}{\frac{s^2 + (3 + g_m)s - 2g_m}{s^3 + (4 + g_m)s^2 + (6 + g_m)s + 2}}
\qquad \textbf{(6.37)}
$$

The coefficient of I_g of the first term on the right-hand side of (6.37) is the transfer impedance relating the response V_{n3} to input I_g.

6-3.1   The Principle of Superposition

The principle of superposition is intimately tied up with the concept of linearity. We cannot really discuss one without discussing the other. The principle is applicable to any linear network, whether it is time invariant or time varying. It is fundamental in characterizing network behavior and is extremely useful in solving linear network problems. For our purposes we shall restrict ourselves to the class of linear time-invariant networks.

The validity of the superposition principle is readily established by considering the general expression for a response in an arbitrary network characterized by its secondary system of network equations (6.30).

We first generalize the result (6.33) for a single input to n input excitations F_i ($i = 1, 2, \ldots, n$). Then from (6.30) we have

$$X_k(s) = \frac{\displaystyle\sum_{i=1}^{n} \Delta_{ik}(s)F_i(s)}{\Delta(s)} + \frac{\displaystyle\sum_{j=1}^{n} \Delta_{jk}(s)h_j}{\Delta(s)} \tag{6.38}$$

Observe that to compute the complete response transform X_k, we may consider each of the transform sources F_i one at a time and then add the partial responses so determined to obtain X_k. This is in essence the *superposition principle*.

> **Property 6.2**   *For a linear system, the zero-state response due to all the independent sources acting simultaneously is equal to the sum of the zero-state responses due to each independent source acting one at a time. If, in addition, the system is time invariant, the same holds in the frequency domain.*

Two aspects of superposition are important to emphasize. The first is the *additivity* property described in Property 6.2. The other is the *homogeneity* property, which states that if all sources are multiplied by a constant, the response is also multiplied by the same constant.

Different versions of the superposition principle were established earlier in Chapter 4. Specifically, Properties 4.4, 4.7, and 4.8 state that in a linear time-invariant system the zero-input response is a linear function of the initial state, the zero-state response is a linear function of the input, and the complete response is the sum of the zero-input response and the zero-state response. Thus the complete response of a linear network[†] to a number of excitations applied simultaneously is the sum of the responses of the network when each of the excitations is applied individually. This statement remains valid even if we consider the initial capacitor voltage and inductor currents themselves to be separate excitations. Of course, the controlled sources cannot be considered as separate excitations. In the case of linear time-invariant networks, the same holds in the frequency domain or in the transform network.

[†] The presence of any initial condition can render a network nonlinear (see p. 26).

EXAMPLE 6.5

In the network of Figure 6.10, let

$$
\begin{aligned}
i_g(t) &= 0, && t < 0 \\
&= \cos 2t, && t \geq 0
\end{aligned}
\tag{6.39a}
$$

$$
\begin{aligned}
v_g(t) &= 0, && t < 0 \\
&= 2 \cos\left(2t + \frac{\pi}{4}\right), && t \geq 0
\end{aligned}
\tag{6.39b}
$$

We shall apply the superposition principle to compute the voltage $v_1(t)$ across the capacitor.

The transform network of Figure 6.10 is presented in Figure 6.11(a). The transform response V_{1a} due to I_g alone is shown in Figure 6.11(b) with the voltage source being short circuited. Likewise the transform response V_{1b} due to V_g alone is presented in Figure 6.11(c) with the current source being open circuited. By the principle of superposition, the transform of the complete response is simply the sum of V_{1a} and V_{1b}:

$$
V_1(s) = V_{1a}(s) + V_{1b}(s)
\tag{6.40}
$$

where

$$
V_{1a}(s) = \frac{s+5}{s^2 + 7s + 12} I_g(s) = \frac{s(s+5)}{(s+3)(s+4)(s^2+4)}
\tag{6.41}
$$

$$
V_{1b}(s) = \frac{2}{s^2 + 7s + 12} V_g(s) = \frac{2\sqrt{2}(s-2)}{(s+3)(s+4)(s^2+4)}
\tag{6.42}
$$

were computed in the usual way from Figure 6.11(b) and (c). Note that the coefficient of I_g in (6.41) is the transfer impedance relating V_{1a} to I_g, and that of V_g in (6.42) is the transfer voltage ratio between V_{1b} and V_g. They are of course all network functions. Combining (6.40) with (6.41) and (6.42) yields

$$
V_1(s) = \frac{s^2 + (5 + 2\sqrt{2})s - 4\sqrt{2}}{(s+3)(s+4)(s^2+4)} = -\frac{6 + 10\sqrt{2}}{13(s+3)} + \frac{1 + 3\sqrt{2}}{5(s+4)}
$$

$$
+ \frac{(34 + 22\sqrt{2})s + 27 + 6\sqrt{2}}{130(s^2+4)}
\tag{6.43}
$$

Figure 6.10
A network used to illustrate the super-position principle.

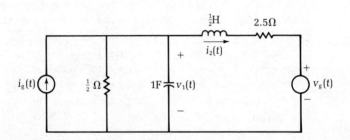

Figure 6.11
(a) The transform network of Figure 6.10. (b) The transform network used to compute the response due to the current source alone with the voltage source being short circuited. (c) The transform network used to compute the response due to the voltage source alone with the current source being open circuited.

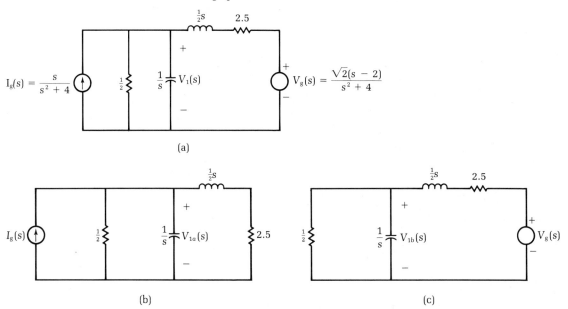

with its inverse Laplace transform found from Table 5.1 as

$$v_1(t) = -1.54940e^{-3t} + 1.04853e^{-4t} + 0.57042\cos(2t - 0.49898) \qquad \textbf{(6.44)}$$

6-3.2 Thévenin's Theorem and Norton's Theorem

In network analysis we are frequently faced with the problem of determining the current or voltage of a small part of the network instead of all currents and voltages. In such cases it would be convenient and useful to replace the part of the network that is not of interest by a simpler equivalent. In this section we shall present such equivalents known as Thévenin's and Norton's theorems.[†]

Consider Figure 6.12(a), which shows two parts N_1 and N_2 of a network. The two networks are connected only at the points shown and there is no magnetic or controlled source coupling between them. Otherwise they are completely general with any kind linear elements, sources, or initial conditions. For instance, the network of Figure 6.8(a) can be partitioned along

---

[†] The original theorem was first proposed in 1883 by the French telegraph engineer Léon Charles Thévenin (1857–1926) in the journal *Annales Télégraphiques*. The dual of the theorem is due to Edward Lawry Norton (1898–) of the Bell Laboratories.

Figure 6.12
Networks used to justify
the Thévenin's theorem.

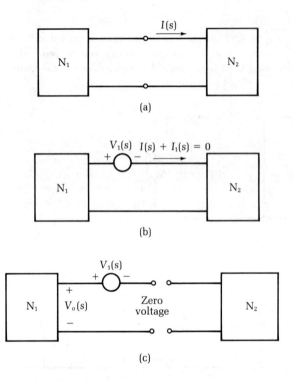

the two terminals immediately to the right of the capacitor C_3. The theorem
is not applicable in this case because there is a controlled source coupling
between $g_m R_4 v_2$ and the voltage across C_2. We will initially assume that
there are no independent sources and initial conditions in N_2. The current
from N_1 to N_2 has the transform $I(s)$, as indicated in Figure 6.12(a). In Figure
6.12(b) a voltage source and its transform $V_1(s)$ is added. By the principle of
superposition, the total current will be the sum of the original current I and
the current I_1 due to V_1 alone with all other independent sources and initial
conditions set to zero. We now adjust the voltage V_1 so that the total current
$I + I_1$ is zero, as shown in Figure 6.12(b). Under this condition there is no
current into N_2 and the voltage at its terminals will be zero. With zero
current the two networks N_1 and N_2 can be broken apart as depicted in
Figure 6.12(c) without affecting the conditions in N_1, and the voltage transform at the broken terminals will remain zero. If V_o is the voltage transform
at the terminals of N_1, as indicated in Figure 6.12(c) when the broken terminals are open circuited, then we must have $V_o - V_1 = 0$, giving us the
value of V_1 as

$$V_1(s) = V_o(s) \tag{6.45}$$

In other words, V_1 is the value of the voltage transform at the open-circuit
terminals of N_1 for the zero current condition of Figure 6.12(b) to apply.

The above argument demonstrates that if we place a voltage source with
transform $V_1 = V_o$ in the position and with the reference polarity as shown

Figure 6.13

Reversing the polarity of
the voltage source V_1 in
(a), resulting in a network
(b) that is equivalent to
that of Figure 6.12(a)
insofar as the right-hand
subnetwork N_2 is
concerned.

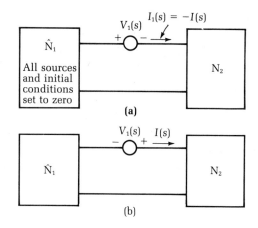

(a)

(b)

in Figure 6.13(a) and set all independent sources and initial conditions to
zero, we will get a current transform $I_1 = -I$. If we reverse the polarity of
V_1, the current transform becomes $I_1 = I$ as shown in Figure 6.13(b). This
is exactly the same current transform in the original network of Figure
6.12(a). Therefore insofar as the right-hand network N_2 is concerned, the
networks of Figure 6.12(a) and Figure 6.13(b) are completely equivalent.
Our conclusion is that the network N_1 can be replaced by an equivalent
network consisting of the series combination of a voltage source the voltage
transform of which is the open-circuit voltage transform of N_1 and a network
$\hat{N}_1$ derived from the original network N_1 by setting all independent sources
and all initial conditions to zero. The sourceless network $\hat{N}_1$ can be represented by its driving-point impedance Z_{eq}. This leads to Figure 6.14 as the
final equivalent representation of N_1 and is called the *Thévenin's equivalent
network*, where $V_{eq} = V_1$.

In the case where the right-hand network N_2 contains independent
sources and initial conditions, the same conclusion can be reached but the
length and complexity of the proof do not warrant its inclusion here. For
the purpose of future reference, we state the above result as

Thévenin's Theorem *A two-terminal linear network that has no
magnetic or controlled source coupling to an external network can be
represented equivalently by a simple series network consisting of a*

Figure 6.14

The Thévenin's equivalent
network.

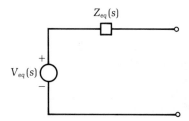

Figure 6.15
The short-circuit current transform (a) and the Norton's equivalent network (b).

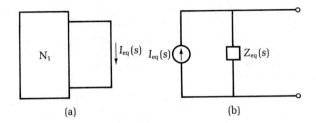

(a)   (b)

voltage source V_{eq} and an impedance Z_{eq} (Figure 6.14), where V_{eq} is the voltage transform at the two terminals when they are left open and Z_{eq} is the driving-point impedance at the two terminals when all independent sources and all initial conditions are set to zero.

We emphasize that the equivalence is valid only at the terminals and does not extend to any of the voltages or currents inside the original two-terminal network.

The dual theorem to Thévenin's theorem was given by Norton[†] and can be stated as follows:

Norton's Theorem   *A two-terminal linear network that has no magnetic or controlled source coupling to an external network can be represented equivalently by a simple network consisting of a current source I_{eq} and a parallel impedance Z_{eq} [Figure 6.15(b)], where I_{eq} is the current transform between the terminals when they are short circuited, as depicted in Figure 6.15(a), and Z_{eq} is the driving-point impedance at the two terminals when all independent sources and initial conditions are set to zero.*

To prove Norton's theorem, we need merely to find the Thévenin's equivalent network of Figure 6.15(b), and the result follows immediately.

EXAMPLE 6.6
We apply Thévenin's theorem to compute the voltage transform V_1 in the network of Figure 6.11(a), which is redrawn in Figure 6.16. The equivalent voltage transform V_{eq} is computed by the network of Figure 6.17(a). To compute V_{eq}, it is convenient to convert the parallel combination of the current source and the $\frac{1}{2}\Omega$ resistor into its Thévenin's equivalent. The resulting network is shown in Figure 6.17(b), from which we immediately obtain the open-circuit voltage V_{eq}:

$$V_{eq}(s) = \frac{s^2 + (5 + 2\sqrt{2})s - 4\sqrt{2}}{2(s + 6)(s^2 + 4)} \qquad (6.46)$$

[†] See footnote, page 235.

Figure 6.16
A network used to
illustrate Thévenin's
theorem.

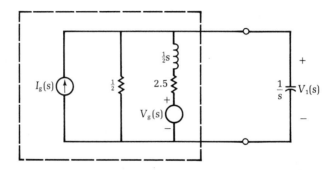

Figure 6.17
Networks used to
compute Thévenin's
equivalent voltage
transform for the two-
terminal network shown
in Figure 6.16.

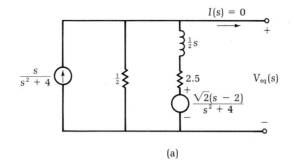

(a)

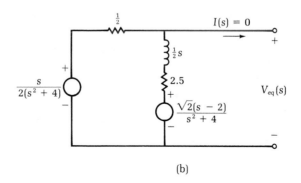

(b)

Figure 6.18
Network used to compute
Thévenin's equivalent
impedance for the two-
terminal network shown
in Figure 6.16.

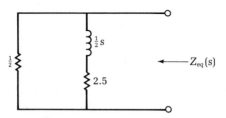

The equivalent impedance Z_{eq} is the driving-point impedance of the network
of Figure 6.18, which yields

$$Z_{eq}(s) = \frac{s + 5}{2(s + 6)} \qquad\qquad (6.47)$$

The final simplified equivalent network of Figure 6.11(a) is presented in Figure 6.19, from which we obtain the desired voltage transform

$$V_1(s) = \frac{V_{eq}(s)}{Z_{eq}(s) + 1/s} \cdot \frac{1}{s} = \frac{s^2 + (5 + 2\sqrt{2})s - 4\sqrt{2}}{(s+3)(s+4)(s^2+4)} \qquad (6.48)$$

confirming (6.43).

Figure 6.19

The Thévenin's equivalent network for the two-terminal network shown in Figure 6.16.

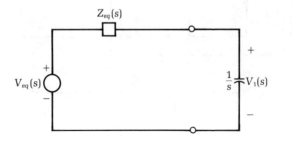

6-4 TWO-PORT NETWORKS

In the preceding section, we showed that a two-terminal pair network can be represented by its Thévenin's or Norton's equivalent network. The equivalence is maintained only at the terminal pair, not inside the original network. The two associated terminals are frequently called a *port*, suggesting a port of entry into the network. Thus a two-terminal pair network is referred to as a *one-port network* or simply a *one-port*, as represented in Figure 6.20(a). If a network is accessible through two ports such as shown in Figure 6.20(b), the network is called a *two-port network* or simply a *two-port*. This nomenclature can be extended to networks having *n* accessible ports called the *n-port networks* or *n-ports*, as depicted in Figure 6.20(c). Our emphasis in this section will be two-ports because they are the most common and the most important of this class of networks.

Fundamental to the concept of a *port* is the assumption that the instantaneous current entering one terminal of the port is always equal to the

Figure 6.20

Symbolic representations of a one-port network (a), a two-port network (b), and an n-port network (c).

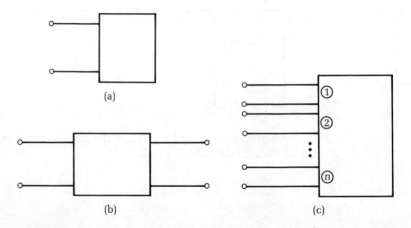

Figure 6.21
A general representation of a two-port network with port voltages and currents shown explicitly.

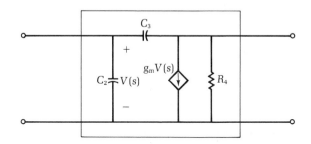

Figure 6.22
The representation of an amplifier as a two-port network.

instantaneous current leaving the other terminal of the port. This assumption is crucial in subsequent derivations and the resulting conclusions. If it is violated, the terminal pair do not constitute a port. Figure 6.21 is a general representation of a two-port that is electrically and magnetically isolated except at the two ports. By focusing attention on the two ports, we are interested in the behavior of the network only at the two ports. Our discussion will be entirely in terms of the transform network, under the assumption that the two-port is devoid of independent sources inside and has zero initial conditions. Therefore we establish the sign convention for the references of port voltages and currents as in Figure 6.21.

An example of the use of the two-port concepts is given by Figure 6.6(a). The amplifier can be represented in the frequency domain by the two-port network of Figure 6.22. The output port is terminated in the series connection of L_5s and R_6, and the input port is connected to the Norton's equivalent, which is the parallel combination of I_g and R_1.

In Figure 6.21, there are four variables V_1, V_2, I_1, and I_2 associated with a two-port. We can specify any two of the four variables as being independent; the other two become dependent. The dependence of two of the four variables on the other two is described in a number of ways depending on the choice of the independent variables. Some common choices are to be described below. As we shall see, the names are chosen to reflect their dimensions or principal applications.

6-4.1 Short-Circuit Admittance Parameters

Suppose that we choose V_1 and V_2 as the independent variables. Then the port currents I_1 and I_2 are related to V_1 and V_2 by the equation

$$\begin{bmatrix} I_1 \\ I_2 \end{bmatrix} \triangleq \begin{bmatrix} y_{11} & y_{12} \\ y_{21} & y_{22} \end{bmatrix} \begin{bmatrix} V_1 \\ V_2 \end{bmatrix} \tag{6.49}$$

Figure 6.23
Representation of a two-
port network in terms of
its y-parameters.

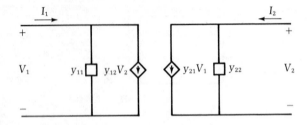

Figure 6.24
Networks used to compute
the short-circuit
admittance parameters
of a two-port network.

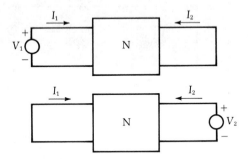

Equation (6.49) can be represented equivalently by the network of Figure 6.23. The four admittance parameters y_{ij} (i, j = 1, 2) are known as the *short-circuit admittance parameters* or simply the *y-parameters*. The coefficient matrix is called the *short-circuit admittance matrix*. Observe that if either V_1 or V_2 is zero, the four parameters y_{ij} may be defined as

$$y_{11} = \frac{I_1}{V_1}\bigg|_{V_2=0}, \qquad y_{12} = \frac{I_1}{V_2}\bigg|_{V_1=0}$$

$$y_{21} = \frac{I_2}{V_1}\bigg|_{V_2=0}, \qquad y_{22} = \frac{I_2}{V_2}\bigg|_{V_1=0} \tag{6.50}$$

The name "short circuit" becomes obvious. In computing y_{11} and y_{21}, the port V_2 is short circuited, whereas for y_{12} and y_{22} the port V_1 is short circuited, as depicted in Figure 6.24.

EXAMPLE 6.7
We compute the y-parameters for the two-port network of Figure 6.22. Following the pattern of connections shown in Figure 6.24, the y-parameters are obtained as follows:

$$y_{11} = (C_2 + C_3)s, \qquad y_{12} = -C_3 s$$

$$y_{21} = g_m - C_3 s, \qquad y_{22} = \frac{1}{R_4} + C_3 s \tag{6.51}$$

yielding the short-circuit admittance matrix

$$\mathbf{Y}_{sc}(s) = \begin{bmatrix} (C_2 + C_3)s & -C_3 s \\ g_m - C_3 s & 1/R_4 + C_3 s \end{bmatrix} \tag{6.52}$$

Figure 6.25

Representation of a two-port network in terms of its z-parameters.

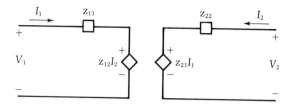

6-4.2 Open-Circuit Impedance Parameters

To express V_1 and V_2 in terms of I_1 and I_2, we may begin by computing the inverse of the short-circuit admittance matrix in (6.49). For this we let $\Delta_y = y_{11}y_{22} - y_{12}y_{21}$ and obtain

$$\begin{bmatrix} V_1 \\ V_2 \end{bmatrix} = \frac{1}{\Delta_y} \begin{bmatrix} y_{22} & -y_{12} \\ -y_{21} & y_{11} \end{bmatrix} \begin{bmatrix} I_1 \\ I_2 \end{bmatrix}$$

$$\triangleq \begin{bmatrix} z_{11} & z_{12} \\ z_{21} & z_{22} \end{bmatrix} \begin{bmatrix} I_1 \\ I_2 \end{bmatrix} \tag{6.53}$$

The last step defines a new matrix called the *open-circuit impedance matrix*, the elements of which are the *open-circuit impedance parameters* or simply the *z-parameters*. The reason for the name again becomes apparent by noting from (6.53) that

$$z_{11} = \frac{V_1}{I_1}\bigg|_{I_2 = 0}, \qquad z_{12} = \frac{V_1}{I_2}\bigg|_{I_1 = 0}$$

$$z_{21} = \frac{V_2}{I_1}\bigg|_{I_2 = 0}, \qquad z_{22} = \frac{V_2}{I_2}\bigg|_{I_1 = 0} \tag{6.54}$$

The condition $I_2 = 0$ or $I_1 = 0$ is met by having the corresponding pair of terminals open. As before, (6.53) can be represented by the equivalent network of Figure 6.25. This can easily be verified by writing the two loop equations of the network.

EXAMPLE 6.8

We compute the z-parameters for the two-port network of Figure 6.22 by means of (6.54). From (6.54), we have

$$z_{11} = \frac{V_1}{I_1}\bigg|_{I_2 = 0} = \frac{1/R_4 + C_3 s}{C_2 C_3 s^2 + (C_2/R_4 + C_3/R_4 + g_m C_3)s} \tag{6.55a}$$

$$z_{21} = \frac{V_2}{I_1}\bigg|_{I_2 = 0} = \frac{C_3 s - g_m}{C_2 C_3 s^2 + (C_2/R_4 + C_3/R_4 + g_m C_3)s} \tag{6.55b}$$

which are obtained from the network of Figure 6.26. In a similar manner we can compute z_{12} and z_{22}. The open-circuit impedance matrix $\mathbf{Z}_{oc}$, which is

Figure 6.26
Network used to illustrate
the computations of the
open-circuit impedance
parameters z_{11} and z_{21}.

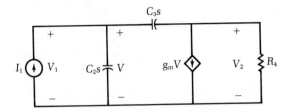

Figure 6.26
Network used to illustrate
the computations of the
open-circuit impedance
parameters z_{11} and z_{21}.

the inverse of $\mathbf{Y}_{sc}$ of (6.52), is found to be

$$\mathbf{Z}_{oc}(s) = \frac{1}{q(s)} \begin{bmatrix} C_3s + 1/R_4 & C_3s \\ C_3s - g_m & (C_2 + C_3)s \end{bmatrix} \tag{6.56a}$$

where

$$q(s) = C_2 C_3 s^2 + \left(\frac{C_2}{R_4} + \frac{C_3}{R_4} + g_m C_3 \right) s \tag{6.56b}$$

6-4.3   The Hybrid Parameters

The hybrid parameters to be studied in this section are widely used in
electronic circuits, especially in modeling transistor networks. In this case
we choose I_1 and V_2 as the independent variables, resulting in the defining
equation

$$\begin{bmatrix} V_1 \\ I_2 \end{bmatrix} \triangleq \begin{bmatrix} h_{11} & h_{12} \\ h_{21} & h_{22} \end{bmatrix} \begin{bmatrix} I_1 \\ V_2 \end{bmatrix} \tag{6.57}$$

The coefficient matrix is called the *hybrid matrix*; its elements are the *hybrid
parameters* or simply *h-parameters*. Equation (6.57) can be represented
equivalently by the network of Figure 6.27. The interpretation of the *h*-
parameters can easily be determined from (6.57), as follows:

$$h_{11} = \left. \frac{V_1}{I_1} \right|_{V_2=0}, \qquad h_{12} = \left. \frac{V_1}{V_2} \right|_{I_1=0}$$

$$h_{21} = \left. \frac{I_2}{I_1} \right|_{V_2=0}, \qquad h_{22} = \left. \frac{I_2}{V_2} \right|_{I_1=0} \tag{6.58}$$

Figure 6.27
Representation of a two-
port network in terms of
its *h*-parameters.

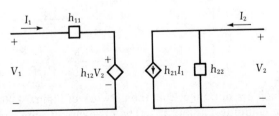

Figure 6.28
Representation of a two-port network in terms of its g-parameters.

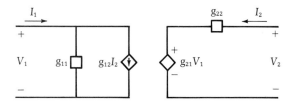

Thus h_{11} is the short-circuit input impedance, h_{21} is the short-circuit forward current ratio, h_{12} is the open-circuit reverse voltage ratio, and h_{22} is the open-circuit output admittance. These parameters are not only dimensionally mixed but also under a mixed set of terminal conditions. For this reason they are called "hybrid" parameters.

The *inverse hybrid parameters* or *g-parameters* are defined by the matrix equation

$$\begin{bmatrix} I_1 \\ V_2 \end{bmatrix} \triangleq \begin{bmatrix} g_{11} & g_{12} \\ g_{21} & g_{22} \end{bmatrix} \begin{bmatrix} V_1 \\ I_2 \end{bmatrix} \tag{6.59}$$

The coefficient matrix, being the inverse of the hybrid matrix, is the *inverse hybrid matrix*, the elements of which can be interpreted as follows:

$$g_{11} = \left. \frac{I_1}{V_1} \right|_{I_2=0}, \qquad g_{12} = \left. \frac{I_1}{I_2} \right|_{V_1=0}$$

$$g_{21} = \left. \frac{V_2}{V_1} \right|_{I_2=0}, \qquad g_{22} = \left. \frac{V_2}{I_2} \right|_{V_1=0} \tag{6.60}$$

A two-port characterizing the g-parameters is presented in Figure 6.28.

EXAMPLE 6.9
We compute the h-parameters for the two-port network of Figure 6.22 by means of (6.58). The results for h_{11} and h_{21} are found from Figure 6.29:

$$h_{11} = \left. \frac{V_1}{I_1} \right|_{V_2=0} = \frac{1}{(C_2 + C_3)s} \tag{6.61a}$$

$$h_{21} = \left. \frac{I_2}{I_1} \right|_{V_2=0} = \frac{g_m - C_3 s}{(C_2 + C_3)s} \tag{6.61b}$$

Figure 6.29
Network used to illustrate the computations of the hybrid parameters h_{11} and h_{21}.

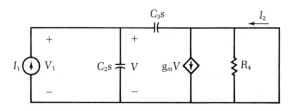

Likewise we can compute h_{12} and h_{22}. The resulting hybrid matrix is

$$\mathbf{H}_h(s) = \frac{1}{(C_2 + C_3)s} \begin{bmatrix} 1 & C_3 s \\ g_m - C_3 s & q(s) \end{bmatrix} \tag{6.62}$$

where $q(s)$ is defined in (6.56b).

6-4.4 Transmission Parameters

Another useful set of parameters, which express the input entities in terms of the output ones, is defined by the relation

$$\begin{bmatrix} V_1 \\ I_1 \end{bmatrix} \triangleq \begin{bmatrix} A & B \\ C & D \end{bmatrix} \begin{bmatrix} V_2 \\ -I_2 \end{bmatrix} \tag{6.63}$$

The coefficient matrix is the *transmission matrix*, the elements of which are known by a variety of names including the *transmission*, the *ABCD*, and the *chain parameters*. For our purposes we shall use the name transmission parameters. Note that there is a negative sign associated with I_2, which is a consequence of our choice of reference for I_2 in Figure 6.21. The elements of the inverse of the transmission matrix are called the *inverse transmission parameters*.

From (6.63) we can interpret the transmission parameters as follows:

$$A = \frac{V_1}{V_2}\bigg|_{I_2=0}, \qquad -B = \frac{V_1}{I_2}\bigg|_{V_2=0}$$

$$C = \frac{I_1}{V_2}\bigg|_{I_2=0}, \qquad -D = \frac{I_1}{I_2}\bigg|_{V_2=0} \tag{6.64}$$

EXAMPLE 6.10

We again consider the two-port network of Figure 6.22 by computing its transmission parameters through (6.64). For the parameters A and C, we open-circuit the output port of Figure 6.22 and obtain

$$A = \frac{V_1}{V_2}\bigg|_{I_2=0} = \frac{C_3 s + 1/R_4}{C_3 s - g_m} \tag{6.65a}$$

$$C = \frac{I_1}{V_2}\bigg|_{I_2=0} = \frac{q(s)}{C_3 s - g_m} \tag{6.65b}$$

where $q(s)$ is given by (6.56b). Likewise we can compute B and D by short-circuiting the output port. The resulting transmission matrix is found to be

$$\mathbf{T}(s) = \frac{1}{C_3 s - g_m} \begin{bmatrix} C_3 s + 1/R_4 & 1 \\ q(s) & (C_2 + C_3)s \end{bmatrix} \tag{6.66}$$

6-4.5 Interrelations Among the Parameter Sets

In the preceding sections we presented various ways of representing the external behavior of a two-port network. Each of these representations finds useful applications. For future reference we will tabulate the interrelationships among the various parameter sets. The results are presented in Table 6.1 and are valid for linear time-invariant two-ports.

Table 6.1 Conversion Chart for Two-Port Parameters, $\Delta_x = x_{11}x_{22} - x_{12}x_{21}$.

To		From: Z		Y		T		T^{-1}		H		H^{-1}	
z-parameters	Z	z_{11}	z_{12}	$\dfrac{y_{22}}{\Delta_y}$	$-\dfrac{y_{12}}{\Delta_y}$	$\dfrac{A}{C}$	$\dfrac{\Delta_T}{C}$	$\dfrac{D'}{C'}$	$\dfrac{1}{C'}$	$\dfrac{\Delta_h}{h_{22}}$	$\dfrac{h_{12}}{h_{22}}$	$\dfrac{1}{g_{11}}$	$-\dfrac{g_{12}}{g_{11}}$
		z_{21}	z_{22}	$-\dfrac{y_{21}}{\Delta_y}$	$\dfrac{y_{11}}{\Delta_y}$	$\dfrac{1}{C}$	$\dfrac{D}{C}$	$\dfrac{\Delta_{T'}}{C'}$	$\dfrac{A'}{C'}$	$\dfrac{h_{21}}{h_{22}}$	$\dfrac{1}{h_{22}}$	$\dfrac{g_{21}}{g_{11}}$	$\dfrac{\Delta_g}{g_{11}}$
y-parameters	Y	$\dfrac{z_{22}}{\Delta_z}$	$-\dfrac{z_{12}}{\Delta_z}$	y_{11}	y_{12}	$\dfrac{D}{B}$	$-\dfrac{\Delta_T}{B}$	$\dfrac{A'}{B'}$	$\dfrac{1}{B'}$	$\dfrac{1}{h_{11}}$	$-\dfrac{h_{12}}{h_{11}}$	$\dfrac{\Delta_g}{g_{22}}$	$\dfrac{g_{12}}{g_{22}}$
		$-\dfrac{z_{21}}{\Delta_z}$	$\dfrac{z_{11}}{\Delta_z}$	y_{21}	y_{22}	$-\dfrac{1}{B}$	$\dfrac{A}{B}$	$-\dfrac{\Delta_{T'}}{B'}$	$\dfrac{D'}{B'}$	$\dfrac{h_{21}}{h_{11}}$	$\dfrac{\Delta_h}{h_{11}}$	$-\dfrac{g_{21}}{g_{22}}$	$\dfrac{1}{g_{22}}$
Transmission parameters	T	$\dfrac{z_{11}}{z_{21}}$	$\dfrac{\Delta_z}{z_{21}}$	$-\dfrac{y_{22}}{y_{21}}$	$-\dfrac{1}{y_{21}}$	A	B	$\dfrac{D'}{\Delta_{T'}}$	$\dfrac{B'}{\Delta_{T'}}$	$-\dfrac{\Delta_h}{h_{21}}$	$-\dfrac{h_{11}}{h_{21}}$	$\dfrac{1}{g_{21}}$	$\dfrac{g_{22}}{g_{21}}$
		$\dfrac{1}{z_{21}}$	$\dfrac{z_{22}}{z_{21}}$	$-\dfrac{\Delta_y}{y_{21}}$	$-\dfrac{y_{11}}{y_{21}}$	C	D	$\dfrac{C'}{\Delta_{T'}}$	$\dfrac{A'}{\Delta_{T'}}$	$-\dfrac{h_{22}}{h_{21}}$	$-\dfrac{1}{h_{21}}$	$\dfrac{g_{11}}{g_{21}}$	$\dfrac{\Delta_g}{g_{21}}$
Inverse transmission parameters	T^{-1}	$\dfrac{z_{22}}{z_{12}}$	$\dfrac{\Delta_z}{z_{12}}$	$-\dfrac{y_{11}}{y_{12}}$	$-\dfrac{1}{y_{12}}$	$\dfrac{D}{\Delta_T}$	$\dfrac{B}{\Delta_T}$	A'	B'	$\dfrac{1}{h_{12}}$	$\dfrac{h_{11}}{h_{12}}$	$-\dfrac{\Delta_g}{g_{12}}$	$-\dfrac{g_{22}}{g_{12}}$
		$\dfrac{1}{z_{12}}$	$\dfrac{z_{11}}{z_{12}}$	$-\dfrac{\Delta_y}{y_{12}}$	$-\dfrac{y_{22}}{y_{12}}$	$\dfrac{C}{\Delta_T}$	$\dfrac{A}{\Delta_T}$	C'	D'	$\dfrac{h_{22}}{h_{12}}$	$\dfrac{\Delta_h}{h_{12}}$	$-\dfrac{g_{11}}{g_{12}}$	$-\dfrac{1}{g_{12}}$
h-parameters	H	$\dfrac{\Delta_z}{z_{22}}$	$\dfrac{z_{12}}{z_{22}}$	$\dfrac{1}{y_{11}}$	$-\dfrac{y_{12}}{y_{11}}$	$\dfrac{B}{D}$	$\dfrac{\Delta_T}{D}$	$\dfrac{B'}{A'}$	$\dfrac{1}{A'}$	h_{11}	h_{12}	$\dfrac{g_{22}}{\Delta_g}$	$-\dfrac{g_{12}}{\Delta_g}$
		$-\dfrac{z_{21}}{z_{22}}$	$\dfrac{1}{z_{22}}$	$\dfrac{y_{21}}{y_{11}}$	$\dfrac{\Delta_y}{y_{11}}$	$-\dfrac{1}{D}$	$\dfrac{C}{D}$	$-\dfrac{\Delta_{T'}}{A'}$	$\dfrac{C'}{A'}$	h_{21}	h_{22}	$-\dfrac{g_{21}}{\Delta_g}$	$\dfrac{g_{11}}{\Delta_g}$
g-parameters	H^{-1}	$\dfrac{1}{z_{11}}$	$-\dfrac{z_{12}}{z_{11}}$	$\dfrac{\Delta_y}{y_{22}}$	$\dfrac{y_{12}}{y_{22}}$	$\dfrac{C}{A}$	$-\dfrac{\Delta_T}{A}$	$\dfrac{C'}{D'}$	$\dfrac{1}{D'}$	$\dfrac{h_{22}}{\Delta_h}$	$-\dfrac{h_{12}}{\Delta_h}$	g_{11}	g_{12}
		$\dfrac{z_{21}}{z_{11}}$	$\dfrac{\Delta_z}{z_{11}}$	$-\dfrac{y_{21}}{y_{22}}$	$\dfrac{1}{y_{22}}$	$\dfrac{1}{A}$	$\dfrac{B}{A}$	$\dfrac{\Delta_{T'}}{D'}$	$\dfrac{B'}{D'}$	$-\dfrac{h_{21}}{\Delta_h}$	$\dfrac{h_{11}}{\Delta_h}$	g_{21}	g_{22}

EXAMPLE 6.11

Consider a typical n-channel planar silicon field-effect transistor (FET) 2N3819. For small signal operation, the transistor can be represented by a two-port as shown in Figure 6.22. The common-source specifications for the 2N3819 as given by the manufacturer are listed below:

1. Forward transfer admittance at 1 kHz — min 2×10^{-3}, max 6.5×10^{-3} mhos
2. Output admittance at 1 kHz — max 5×10^{-5} mhos
3. Input capacitance (output shorted) — max 8 pF
4. Reverse transfer capacitance (input shorted) — max 4 pF
5. Forward transfer admittance at 100 MHz — min 1.6×10^{-3} mhos

These specifications were obtained under the steady-state condition, which corresponds to the $j\omega$-axis of the complex-frequency s-plane. Thus by replacing s by $j\omega$ in (6.51) we have sufficient $j\omega$-axis information to determine the y-parameters for the common-source FET. The first and the second specifications state that at low frequencies, $0.002 < g_m < 0.0065$ and $1/R_4 < 5 \times 10^{-5}$ mhos. Let us choose $g_m = 0.004$ and $R_4 = 25$ kΩ. The third and the fourth specifications give $C_2 + C_3 < 8$ pF and $C_3 < 4$ pF. Let us use $C_2 = 4$ pF and $C_3 = 3$ pF. Substituting these in (6.52) yields

$$\mathbf{Y}_{sc}(j\omega) = \begin{bmatrix} j(C_2 + C_3)\omega & -jC_3\omega \\ g_m - jC_3\omega & 1/R_4 + jC_3\omega \end{bmatrix}$$

$$= 10^{-12} \begin{bmatrix} j7\omega & -j3\omega \\ 4 \times 10^9 - j3\omega & 4 \times 10^7 + j3\omega \end{bmatrix} \quad \textbf{(6.67)}$$

To determine the h-parameters we use Table 6.1, giving

$$\mathbf{H}_h(j\omega) = \frac{1}{y_{11}} \begin{bmatrix} 1 & -y_{12} \\ y_{21} & y_{11}y_{22} - y_{12}y_{21} \end{bmatrix}$$

$$= \frac{1}{7} \begin{bmatrix} -j10^{12}/\omega & 3 \\ -3 - j4 \times 10^9/\omega & 12.28 \times 10^{-5} + j12\omega 10^{-12} \end{bmatrix} \quad \textbf{(6.68)}$$

Alternatively we can use (6.62) to compute the h-parameters directly.

6-4.6 Interconnection of Two-Port Networks

Simple two-ports are interconnected to yield more complicated and practical two-port networks. In this section we study three useful and common connections of two-port networks.

Two two-ports are said to be connected in *cascade* or *tandem* if the output terminals of one two-port are the input terminals of the other, as depicted in Figure 6.30. The cascade connection is most conveniently

Figure 6.30

Symbolic representation
of two two-ports
connected in cascade.

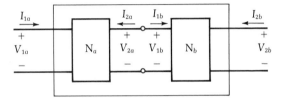

described by the transmission parameters. From Figure 6.30 we have for N_b

$$\begin{bmatrix} V_{2a} \\ -I_{2a} \end{bmatrix} = \begin{bmatrix} V_{1b} \\ I_{1b} \end{bmatrix} = \begin{bmatrix} A_b & B_b \\ C_b & D_b \end{bmatrix} \begin{bmatrix} V_{2b} \\ -I_{2b} \end{bmatrix} \qquad \text{(6.69)}$$

and for N_a

$$\begin{bmatrix} V_{1a} \\ I_{1a} \end{bmatrix} = \begin{bmatrix} A_a & B_a \\ C_a & D_a \end{bmatrix} \begin{bmatrix} V_{2a} \\ -I_{2a} \end{bmatrix} \qquad \text{(6.70)}$$

where the subscripts a and b are used to distinguish the transmission
parameters of N_a and N_b. Substituting (6.69) in (6.70) gives

$$\begin{bmatrix} V_{1a} \\ I_{1a} \end{bmatrix} = \begin{bmatrix} A_a & B_a \\ C_a & D_a \end{bmatrix} \begin{bmatrix} A_b & B_b \\ C_b & D_b \end{bmatrix} \begin{bmatrix} V_{2b} \\ -I_{2b} \end{bmatrix} \qquad \text{(6.71)}$$

showing that the coefficient matrix, being the product of two matrices, is
the transmission matrix of the composite two-port network. Thus we have

$$\begin{bmatrix} A & B \\ C & D \end{bmatrix} = \begin{bmatrix} A_a & B_a \\ C_a & D_a \end{bmatrix} \begin{bmatrix} A_b & B_b \\ C_b & D_b \end{bmatrix} \qquad \text{(6.72)}$$

In words it states that the transmission matrix of two two-ports connected
in cascade is equal to the product of the transmission matrices of the individual two-ports.

Another useful connection, called a *parallel connection*, is depicted
in Figure 6.31. This connection forces the equality of the terminal voltages
of the two-ports. From Figure 6.31 we have

$$\begin{bmatrix} I_1 \\ I_2 \end{bmatrix} = \begin{bmatrix} I_{1a} \\ I_{2a} \end{bmatrix} + \begin{bmatrix} I_{1b} \\ I_{2b} \end{bmatrix} = \begin{bmatrix} y_{11a} & y_{12a} \\ y_{21a} & y_{22a} \end{bmatrix} \begin{bmatrix} V_1 \\ V_2 \end{bmatrix} + \begin{bmatrix} y_{11b} & y_{12b} \\ y_{21b} & y_{22b} \end{bmatrix} \begin{bmatrix} V_1 \\ V_2 \end{bmatrix}$$

$$= \begin{bmatrix} y_{11a} + y_{11b} & y_{12a} + y_{12b} \\ y_{21a} + y_{21b} & y_{22a} + y_{22b} \end{bmatrix} \begin{bmatrix} V_1 \\ V_2 \end{bmatrix} \qquad \text{(6.73)}$$

Figure 6.31

Symbolic representation
of two two-ports
connected in parallel.

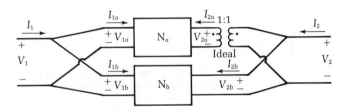

Figure 6.32
Brune's test for parallel connection of two two-ports.

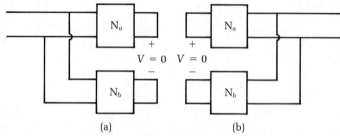

(a)   (b)

Figure 6.33
A bridged-T network (a) and the equivalent representation as the parallel connection of two two-ports (b).

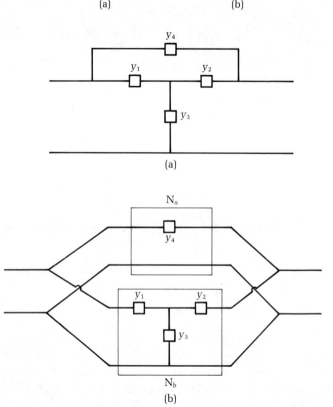

showing that the short-circuit admittance matrix of the composite two-port network is the sum of the short-circuit admittance matrices of the individual two-ports. An ideal transformer is used in the connection to make sure that the nature of the ports are not altered after the connection, recalling the requirement of a port as stated at the beginning of Section 6.4.

In many situations the ideal transformer can be removed if the *Brune's test* as shown in Figure 6.32 is satisfied[†]: the voltage marked V is zero. The bridged-T network of Figure 6.33(a) can be viewed as the parallel

---

[†] See, for example, L. Weinberg, *Network Analysis and Synthesis* (New York: McGraw-Hill, 1962), Chapter 1, pp. 15–22. The test is named after Otto Brune, who is now Principal Research Officer of the National Research Laboratories, Pretoria, South Africa.

Figure 6.34

Symbolic representation of two two-ports connected in series.

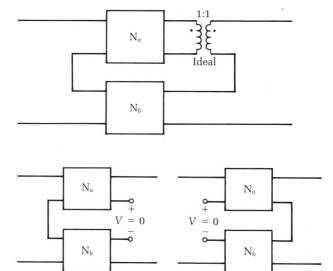

Figure 6.35

Brune's test for series connection of two two-ports.

connection of two two-ports as shown in Figure 6.33(b). It is easy to verify that the Brune's test is satisfied and the short-circuit admittance matrices can be added to give the overall admittance matrix. On the other hand, if the bottom conductor in the component two-port N_a is replaced by a nonzero impedance, Brune's condition will not be satisfied in the resulting two-port and an ideal transformer is required.

Finally, let us consider the series connection. Two two-ports are said to be connected in *series* if they are connected as shown in Figure 6.34. As before, we can show that the open-circuit impedance matrix of the composite two-port is simply the sum of the open-circuit impedance matrices of the individual two-ports. The ideal transformer in Figure 6.34 can be removed if Brune's test for series connection as shown in Figure 6.35 is satisfied: the voltage marked V is zero.

Variations of the parallel and series connections are possible such as *series-parallel* and *parallel-series connections*. As one might expect, the *h*- and *g*-parameters of the component two-ports are added to give the overall *h*- and *g*-parameters. However, we shall not pursue this topic further here.

EXAMPLE 6.12

Figure 6.36(a) shows a three-terminal device N with a series feedback provided by the admittances y_a and y_b. We wish to compute the open-circuit impedance matrix of the overall two-port in terms of the y-parameters y_{ij} ($i, j = 1, 2$) of N and y_a and y_b.

The two-port network of Figure 6.36(a) can first be viewed as the parallel combination of two-ports $\tilde{N}_a$ and N, as depicted in Figure 6.36(b), and the

Figure 6.36
A three-terminal device (a)
and its equivalent
representation as the
parallel and series
combinations of three
two-ports (b).

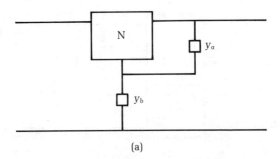

(a)

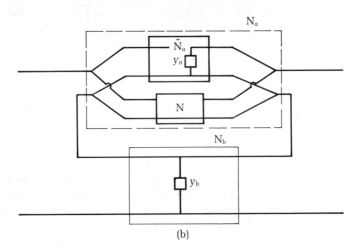

(b)

resulting two-port N_a is then in series with two-port N_b. It is easy to verify
that Brune's tests for parallel and series connections, as shown in Figures
6.32 and 6.35, are all satisfied. No ideal transformers are required. The
open-cirucit impedance matrix of the composite two-port is obtained as

$$\mathbf{Z}(s) = \left\{ \begin{bmatrix} y_{11} & y_{12} \\ y_{21} & y_{22} \end{bmatrix} + \begin{bmatrix} 0 & 0 \\ 0 & y_a \end{bmatrix} \right\}^{-1} + \frac{1}{y_b} \begin{bmatrix} 1 & 1 \\ 1 & 1 \end{bmatrix} \qquad \textbf{(6.74)}$$

The terms inside the curly brackets on the right-hand side are the short-
circuit admittance matrices of N and $\tilde{N}_a$. The inverse gives the open-circuit
impedance matrix of the composite two-port N_a formed by N and $\tilde{N}_a$. The
last term on the right-hand side is the impedance matrix of N_b. After simplification the desired impedance matrix is found to be

$$\mathbf{Z}(s) = \frac{1}{y_{11}(y_{22} + y_a) - y_{12}y_{21}} \begin{bmatrix} y_{22} + y_a & -y_{12} \\ -y_{21} & y_{11} \end{bmatrix} + \frac{1}{y_b} \begin{bmatrix} 1 & 1 \\ 1 & 1 \end{bmatrix} \qquad \textbf{(6.75)}$$

6-5 FREQUENCY RESPONSE

For a general value of the complex frequency s, a network function will take on the complex value. However, we are not usually interested in the behavior of the network function for all complex values of s. Rather the network behavior on the $j\omega$-axis as ω varies from 0 to ∞ is of particular significance because it describes the network in the sinusoidal steady state. In this way we will be able to exhibit the sinusoidal steady-state properties from very low frequencies to very high frequencies. The quantities that might be plotted as functions of ω are the real part, the imaginary part, the magnitude, and the phase of a network function. We refer to any of these as a *frequency response function*. In engineering practice little use is made of the real or imaginary part of the response function, and for this reason we will concentrate on the magnitude and phase response functions.

For the complex frequency $s = \sigma + j\omega$, ω is known by a variety of names including *real frequency*, *radian frequency*, or simply *frequency*. For a fixed ω, a network function $H(j\omega)$ is usually a complex number, which may be written in polar form as

$$H(j\omega) = |H(j\omega)|e^{j\phi(\omega)} = |H(j\omega)| \, \underline{/\phi(\omega)} \tag{6.76}$$

where $|H(j\omega)|$ is called the *magnitude* or *amplitude* and $\phi(\omega)$ the *phase* or *angle* of the network function at the frequency ω.

In general, a network function is a rational function with real coefficients as shown in (6.27). To obtain the frequency response, we first find its poles and zeros so that the function can be factored as

$$H(j\omega) = K \frac{\displaystyle\prod_{h=1}^{M} (j\omega + d_h)^{m_h}}{\displaystyle\prod_{k=1}^{N} (j\omega + s_k)^{n_k}} \tag{6.77}$$

To compute the magnitude of $H(j\omega)$ at any value of ω, we find the magnitude of each of its zero and pole factors and compute

$$|H(j\omega)| = |K| \frac{\displaystyle\prod_{h=1}^{M} |j\omega + d_h|^{m_h}}{\displaystyle\prod_{k=1}^{N} |j\omega + s_k|^{n_k}} \tag{6.78}$$

Likewise the phase of $H(j\omega)$ at any value of ω is found from the phases of its zero and pole factors, as follows:

$$\phi(\omega) = \arg H(j\omega) = \sum_{h=1}^{M} m_h \arg(j\omega + d_h) - \sum_{k=1}^{N} n_k \arg(j\omega + s_k) \tag{6.79}$$

where K in (6.77) is assumed to be positive. In the case K is negative, π has to be added to (6.79). Thus from (6.78) and (6.79) we have $|H(-j\omega)| =$

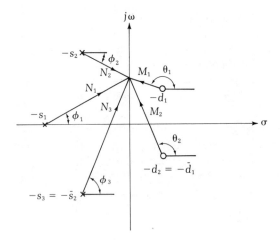

$|H(j\omega)|$ and $\phi(-j\omega) = -\phi(j\omega)$, meaning that the magnitude is an even function whereas the phase is an odd function.

The above procedure can also be performed graphically. Consider a network function $H(s)$ that when evaluated on the $j\omega$-axis is given by

$$H(j\omega) = \frac{(j\omega + d_1)(j\omega + \bar{d}_1)}{(j\omega + s_1)(j\omega + s_2)^2(j\omega + \bar{s}_2)^2} \tag{6.80}$$

Its pole-zero pattern is presented in Figure 6.37. For a fixed value of ω, the complex number $(j\omega + d_1)$ can be represented by a directed line from the point $-d_1$ to $j\omega$. The magnitude of this factor is simply the length M_1 of the directed line and its phase is the one labeled θ_1 in the diagram. In a similar way the other factors can be represented as shown in Figure 6.37. The magnitude and phase of $H(j\omega)$ become

$$H(j\omega) = |H(j\omega)|e^{j\phi(\omega)} = \frac{M_1 M_2}{N_1 N_2^2 N_3^2} \underline{/\theta_1 + \theta_2 - \phi_1 - 2\phi_2 - 2\phi_3} \tag{6.81}$$

In other words, to compute the magnitude of a rational function at any point on the $j\omega$-axis, we first measure the distances from the given point to each of the zeros and poles. The desired magnitude will be the product of the line lengths to the zeros—each being raised to the power equal to the order of the zero, and divided by the product of the line lengths to the poles—each being raised to the power equal to the order of the pole. Likewise the phase of the function at any point on the $j\omega$-axis is found by adding the phases of all the zero factors—each being multiplied by a factor equal to the order of the zero, and subtracting from this the sum of the phases of all the pole factors—each being multiplied by a factor equal to the order of the pole. After computing the magnitude and phase for a number of points on the $j\omega$-axis, we will be able to sketch the frequency response. We illustrate this by the following example.

EXAMPLE 6.13

Consider the tuned circuit shown in Figure 6.38(a), the input admittance of which is found to be

$$Y(s) = \frac{RCs + 1}{RLCs^2 + Ls + R} = \frac{1}{L} \cdot \frac{s + 2\beta}{s^2 + 2\beta s + \omega_0^2} \qquad (6.82)$$

where

$$\beta = \frac{1}{2RC}, \qquad \omega_0^2 = \frac{1}{LC} \qquad (6.83)$$

Assuming that $\omega_0 > \beta$, $Y(s)$ has a pair of complex conjugate poles at

$$s_1, \bar{s}_1 = -\beta \pm j\sqrt{\omega_0^2 - \beta^2} \qquad (6.84)$$

The pole-zero pattern of $Y(s)$ is shown in Figure 6.38(b). Using the geometrical construction procedure outlined above, the magnitude and phase of $Y(j\omega)$ can be sketched as shown in Figure 6.39. The maximum magnitude is attained

Figure 6.38
A tuned circuit (a) and its pole-zero pattern (b).

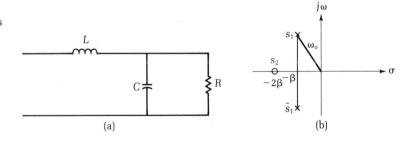

Figure 6.39
The magnitude and phase plots of the input admittance of the tuned circuit of Figure 6.38(a) as functions of ω, $\omega \geqq 0$.

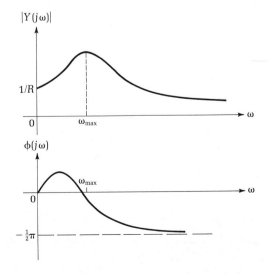

Figure 6.40

The complete magnitude and phase plots of the input admittance of the tuned circuit of Figure 6.38(a) as functions of ω.

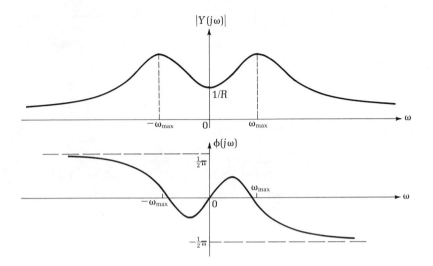

in the neighborhood of s_1 because the length from s_1 to a point in this neighborhood, being in the denominator of the magnitude function, is near its minimum. For very large ω, a zero factor and a pole factor will cancel out, leaving the function approximately proportional to $1/j\omega$. Thus the magnitude will eventually drop to zero and the phase will approach $-\pi/2$. Note that these sketches are drawn only for the positive values of ω because the magnitude is an even function of ω and the phase an odd function. Once the plots for the positive values of ω are known, the complete response can easily be attained by duplication such as shown in Figure 6.40. Hence it is necessary only to show the behavior for positive ω.

In most situations, especially where $H(j\omega)$ represents a transfer function, it is convenient to consider the logarithm of $H(j\omega)$ instead of $H(j\omega)$. Taking the logarithm on both sides of (6.76) yields

$$\ln H(j\omega) = \alpha(\omega) + j\phi(\omega) \tag{6.85}$$

and

$$\alpha(\omega) = \ln |H(j\omega)| \quad \text{nepers} \tag{6.86}$$

The function $\alpha(\omega)$ is called the *logarithmic gain* or simply the *gain*, and is measured in units called *nepers*. Alternatively if $\alpha(\omega)$ is measured in decibels,[†] we write

$$\alpha'(\omega) = 20 \log |H(j\omega)| \quad \text{dB} \tag{6.87}$$

[†] A *bel* is the natural logarithm of a power ratio of 10 and equals 10 *decibel*, abbreviated as dB.

giving

$$\alpha'(\omega) = \frac{20}{\ln 10}\,\alpha(\omega) \approx 8.686\,\alpha(\omega) \tag{6.88}$$

Thus a neper is approximately 8.686 dB. From (6.77) we have

$$20\,\log\,H(j\omega) = 20\,\log\,K + \sum_{h=1}^{M} 20m_h\,\log(j\omega + d_h)$$

$$-\sum_{k=1}^{N} 20n_k\,\log(j\omega + s_k) \tag{6.89}$$

giving

$$\alpha'(\omega) = 20\,\log\,|K| + \sum_{h=1}^{M} 20m_h\,\log\,|j\omega + d_h|$$

$$-\sum_{k=1}^{N} 20n_k\,\log\,|j\omega + s_k| \tag{6.90}$$

$$\phi(\omega) = \sum_{h=1}^{M} m_h\,\arg(j\omega + d_h) - \sum_{k=1}^{N} n_k\,\arg(j\omega + s_k) \tag{6.91}$$

where π has to be added in (6.91) in the case where K is negative.

Here we see the advantage of using logarithms. Not only has multiplication been changed to addition, but each pole factor and each zero factor also appear separately. We shall next study the contribution to the frequency response by the various factors. Once this is accomplished, the overall behavior of a network function on the $j\omega$-axis is then obtained by merely adding the responses of the components.

(a) First-Order Factor.   The function under consideration is $20\,\log(j\omega + a)$, where a is real. It may either be a zero factor or a pole factor. The corresponding gain and phase are given by

$$\alpha_i'(\omega) = 10\,\log(\omega^2 + a^2) \tag{6.92}$$

$$\phi_i'(\omega) = \tan^{-1}\frac{\omega}{a} \tag{6.93}$$

At $\omega = 0$, $\alpha_i' = 20\,\log\,|a|$ and $\phi_i' = 0$. For small values of ω for which $\omega \ll |a|$, $\alpha_i' \approx 20\,\log\,|a|$ and $\phi_i' \approx \omega/a$. For large values of ω such that $\omega \gg |a|$, we have

$$\alpha_i' \approx 20\,\log\,\omega \quad \text{dB} \tag{6.94a}$$

$$\phi_i' \approx \pm 90° \tag{6.94b}$$

Thus for large values of ω, α_i' becomes a linear function of $\log\,\omega$ with a slope of 20 dB. This means that α_i' rises at the rate of 20 dB per decade or

Figure 6.41
The magnitude response
(a) and phase response (b)
given by (6.92) and (6.93)
showing the actual curve
and the low-frequency
and high-frequency
asymptotes.

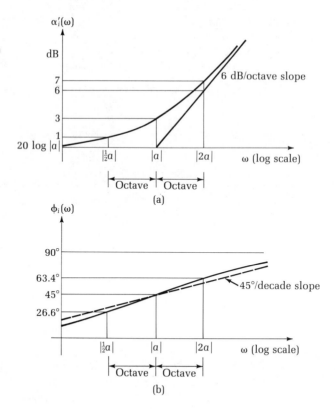

equivalently 6 dB per octave.[†] Figure 6.41(a) shows the asymptotic behavior for small and large values of ω. If the asymptotic values of α_i' apply for all values of log ω, the point of intersection of the asymptotic values occurs when (6.94a) is equal to 20 log$|a|$, the small ω asymptote. This gives the point of intersection of the asymptotes at log ω = log $|a|$. We refer to this point as the *break frequency* or *corner frequency*. On a linear scale, this point corresponds to $\omega = |a|$. The greatest inaccuracy in using the asymptotic values instead of the actual values occurs at the break frequency. The value of α_i' at the break frequency is from (6.92)

$$\alpha_i'(|a|) = 10 \log 2a^2 = 20 \log |a| + 10 \log 2 \tag{6.95}$$

Since 10 log 2 is approximately 3 dB, a 3 dB error will be made at the break frequency if the asymptotic values are used. Likewise at $\omega = \frac{1}{2}|a|$ and $\omega = 2|a|$, the errors are approximately 1 dB, as depicted in Figure 6.41(a). Our conclusion is that knowing a we can ascertain the break frequency, from which two asymptotic lines are drawn—one with zero slope and the

---

[†] A frequency change by a factor of 10 is a *decade*, and a frequency change by a factor of 2 is an *octave*.

other with 6 dB/octave slope. At the break frequency, the true response is displaced by 3 dB from the intersection of asymptotes. An octave below and above the break frequency, the true value is displaced by 1 dB from the asymptotic lines.

As to the phase ϕ_i, the procedure is similar. At the break frequency $\omega = a > 0$, $\phi_i'(a) = 45°$. At one octave below and above the break frequency, $\phi_i'(\frac{1}{2}a) = 26.6°$ and $\phi_i'(2a) = 63.4°$. The result is sketched in Figure 6.41(b). For $a < 0$, multiply all angles described above by -1.

(b) Second-Order Factor. In the case of complex conjugate poles or zeros, we shall consider $s^2 + as + b$, where a and b are real and, in addition, b is positive. Its contributions to the gain and phase are

$$\alpha_i'(\omega) = 10 \log [(b - \omega^2)^2 + a^2\omega^2] \tag{6.96a}$$

$$\phi_i'(\omega) = \tan^{-1} \frac{a\omega}{b - \omega^2} \tag{6.96b}$$

The low-frequency asymptote for gain is again $20 \log b$, but the high-frequency asymptote is now changed to

$$\alpha_i'(\omega) \approx 10 \log \omega^4 = 40 \log \omega \tag{6.97}$$

showing that the slope of the straight line is now 40 dB per decade instead of 20 or, equivalently, 12 dB per octave. The break frequency is found by setting (6.97) to $20 \log b$, giving $\omega = \sqrt{b}$. It is straightforward to confirm that $\sqrt{b}$ is the distance from the origin to the pole or zero. At the break frequency $\omega = \sqrt{b}$, the magnitude becomes

$$\alpha_i'(\sqrt{b}) = 10 \log a^2 b \tag{6.98}$$

As a result the actual value of α_i' depends very much on the values of a. Figure 6.42(a) shows the plots of α_i' as a function of ω with $\gamma = \frac{1}{2}a/\sqrt{b}$ as the parameter. Observe that since $-a$ is twice the real part of the pole or zero, the asymptotic values form a fairly good approximation to the actual value if the zero or pole is not very close to the $j\omega$-axis.

As to the phase (6.96b), we see that for $a > 0$ it varies from 0 at low frequencies to 180° at high frequencies, with the value 90° at the break frequency for all values of a. Equation (6.96b) is plotted in Figure 6.42(b) with $\gamma = \frac{1}{2}a/\sqrt{b}$ as the parameter. Notice that the phase characteristics change most significantly near the break frequency and the smaller the value of γ or a, the more abrupt the change. For $a < 0$ multiply all angles described above by -1.

So far our discussion is centered on the behaviors of simple poles or zeros. For a higher-order pole or zero, the corresponding gain and phase are to be multiplied by the order of the pole or zero. Also, all the equations obtained above are for zero factors. For pole factors we simply multiply all

Figure 6.42
The magnitude response
(a) and phase response (b)
given by (6.96) for several
values of γ.

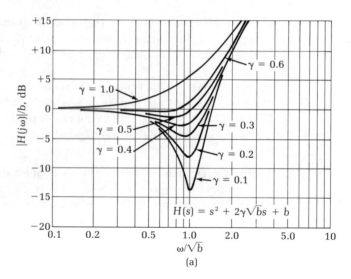

(a)

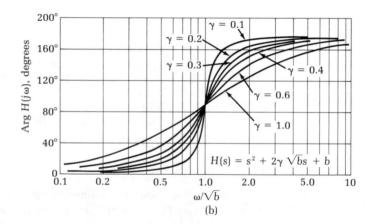

(b)

equations by -1, everything else being the same. For example, for a third-order real pole the high-frequency asymptote for gain will be a straight line of slope -18 dB/octave, and the phase at high frequencies will approach $-270°$. The contributions of the various factors to the overall gain and phase characteristics (6.90) and (6.91) can now be added to yield an approximate representation of the overall gain and phase. This representation is known as the *Bode plot* or *Bode diagram* after the inventor.[†] Note that the right half of the s-plane zeros or poles correspond to $a < 0$.

[†] H. W. Bode, *Network Analysis and Feedback Amplifier Design* (Princeton, N.J.: Van Nostrand, 1945), Chapter XV. The plot is named after Hendrik Wade Bode (1905–) of Bell Laboratories, who was also a professor at Harvard University.

Figure 6.43
A voltage-shunt feedback amplifier (a) and its transform network (b) used to compute the open-loop current ratio $-I_2/I_g$.

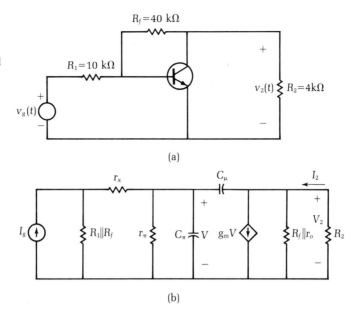

(a)

(b)

Example 6.14

Consider the voltage-shunt feedback amplifier of Figure 6.43(a). Assume that the transistor is represented by its hybrid-pi model with

$$g_m = 0.4 \text{ mho}, \qquad r_x = 50 \ \Omega, \qquad r_\pi = 250 \ \Omega$$
$$C_\pi = 195 \text{ pF}, \qquad C_\mu = 5 \text{ pF}, \qquad r_o = 50 \text{ k}\Omega \qquad \text{(6.99)}$$

It can be shown that the open-loop current ratio $-I_2/I_g$ can be computed by the transform network of Figure 6.43(b). Using nodal analysis, the open-loop current ratio is found to be

$$H(s) = -\frac{I_2}{I_g} = -\frac{5 \times 10^7 (4 \times 10^9 - 0.05s)}{1.962s^2 + 418.524 \times 10^7 \, s + 24.485 \times 10^{14}}$$

$$= 12.74 \times 10^5 \, \frac{s - 8 \times 10^{10}}{(s + 5.85 \times 10^5)(s + 21.33 \times 10^8)} \qquad \text{(6.100)}$$

The function has two real poles on the negative σ-axis and a real zero on the positive σ-axis. The zero frequency gain is $20 \log 81.68 = 38.24$ dB, and the first break frequency occurs at $\omega = 5.85 \times 10^5$. The break is downward with a slope of -6 dB/octave. At $\omega = 21.33 \times 10^8$ there is another break downward with a slope of -6 dB/octave. Finally, at $\omega = 8 \times 10^{10}$ there is a break upward with a slope of 6 dB/octave. The asymptotic plots at these breaks are shown in Figure 6.44(a), which when summed gives the asymptotic plot of gain of $H(j\omega)$ as depicted in Figure 6.44(b). The actual gain responses are also presented in Figure 6.44 by the dashed curves. It is clear that the

Figure 6.44
(a) The individual
magnitude responses for
the four factors in (6.100).
(b) The addition of the
four individual responses
(a) to give the magnitude
response of (6.100).

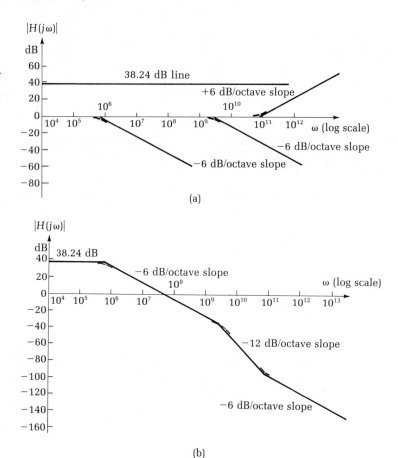

(a)

(b)

Figure 6.45
The individual phase
responses (a) for the four
factors in (6.100) and
their sum (b) to give the
phase response of (6.100).

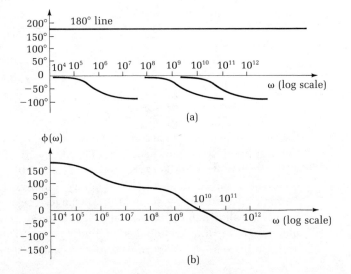

(a)

(b)

asymptotic behavior approximates the actual behavior very closely. A similar summation of the phase characteristics gives the phase response as shown in Figure 6.45. Observe the nicety in using a logarithmic scale for the frequency ω and the decibels for the gain.

Example 6.15

The normalized transfer function of a feedback amplifier is given by

$$H(s) = \frac{3(s + 2)}{(s + 1)(s + 3)(s^2 + 2s + 2)} \tag{6.101}$$

The function has a pair of complex conjugate poles at $-1 \pm j1$. The zero frequency gain is $20 \log 1 = 0$ dB and the break frequency for the zero occurs at $\omega = 2$. The break frequencies for the poles occur at $\omega = 1$, 3, and $\sqrt{2}$. Therefore there is a break upward at $\omega = 2$ with a slope of 6 dB/octave and two breaks downward at $\omega = 1$ and 3 with slope of -6 dB/octave. Finally, at $\omega = \sqrt{2}$ there is another break downward with a slope of -12 dB/octave. The results are presented in Figure 6.46(a) with the dashed curves denoting the actual gains of the factors, and these are summed to give the total gain characteristics of Figure 6.46(b). A similar summation of the phase characteristics yields the phase response also shown in Figure 6.46(b) by the dashed line.

Figure 6.46

The individual magnitude responses (a) for the four factors in (6.101), and the magnitude and phase responses (b) of (6.101).

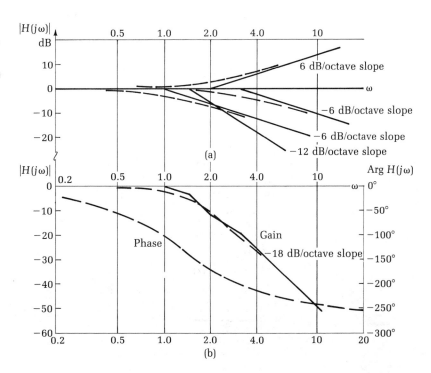

Figure 6.47
Subroutine FRESP used
for the evaluation of the
magnitude and phase
responses of a real
rational function at a
fixed frequency ω.

```
C
C
C      SBUROUTINE FRESP
C
C      PURPOSE
C          THE PROGRAM IS FOR THE EVALUATION OF THE MAGNITUDE AND PHASE
C          OF A REAL RATIONAL FUNCTION AT A FIXED RADIAN FREQUENCY OMEGA.
C
C      USAGE
C          CALL FRESP (M,N,B,A,W,FMAG,FPHASE)
C
C          M      -THE DEGREE OF THE NUMERATOR POLYNOMIAL PLUS ONE
C          N      -THE DEGREE OF THE DENOMINATOR POLYNOMIAL PLUS ONE
C          B      -THE NUMERATOR COEFFICIENTS
C          A      -THE DENOMINATOR COEFFICIENTS
C          W      -FREQUENCY OMEGA IN RADIANS PER SECOND
C          FMAG   -MAGNITUDE IN DB
C          FPHASE -PHASE IN DEGREES
C
C      SUBROUTINES AND FUNCTION SUBPROGRAMS REQUIRED
C          NONE
C
C      REMARKS
C          THE INPUT DATA ARE M, N, B(I) OF B, A(I) OF A, AND W.  THE
C              OUTPUTS ARE FMAG AND FPHASE.
       SUBROUTINE FRESP (M,N,B,A,W,FMAG,FPHASE)
       DOUBLE PRECISION A(N),B(M),W,FMAG,FPHASE,DREAL,DIMAG,CDABS,
C      DLOG10,DATAN2
       COMPLEX*16 S,SB,SA,POLY
       S=(0.0,1.0)*W
       SB=POLY(M,B,S)
       SA=POLY(N,A,S)
       FMAG=CDABS(SB)/CDABS(SA)
       FMAG=20.0*DLOG10(FMAG)
       FPHASE=DATAN2(DIMAG(SB),DREAL(SB))-DATAN2(DIMAG(SA),DREAL(SA))
       FPHASE=FPHASE*360.0/6.2831853071796
       RETURN
       END


       FUNCTION POLY(N,A,S)
C
C      COMPUTE THE REAL AND IMAGINARY PARTS OF A POLYNOMIAL OF DEGREE N-1
C      WITH COEFFICIENTS A FOR A FIXED VALUE OF THE COMPLEX VARIABLE S.
C
C      A      -THE COEFFICIENTS
C      N      -THE DEGREE OF THE POLYNOMIAL PLUS ONE
C
       DOUBLE PRECISION A(N)
       COMPLEX*16 S,POLY
       POLY=0.0
       IF(N-1) 400,100,100
  100  POLY=A(1)
       IF(N-1) 400,400,200
  200  DO 300 I=2,N
  300  POLY=POLY+A(I)*S**(I-1)
  400  RETURN
       END
       FUNCTION DREAL(X)
C
C      COMPUTE THE REAL PART OF A COMPLEX X
       COMPLEX*16 X
       DOUBLE PRECISION DREAL
       DREAL=X
       RETURN
       END
       FUNCTION DIMAG(X)
C
C      COMPUTE THE IMAGINARY PART OF A COMPLEX X
       COMPLEX*16 X
       DOUBLE PRECISION DIMAG
       DIMAG=X*(0.0,-1.0)
       RETURN
       END
```

6-6 COMPUTER SOLUTIONS AND PROGRAMS

In this section we shall present computer programs in WATFIV that evaluate the magnitude and phase of a real rational function at a fixed frequency ω. Programs for the Bode plot will also be described.

6-6.1 Subroutine FRESP

1. Purpose. The program is for the evaluation of the magnitude and phase of a real rational function at a fixed frequency ω:[†]

$$H(j\omega) = |H(j\omega)| \underline{/\phi(\omega)} = \frac{b_m s^{m-1} + b_{m-1} s^{m-2} + \cdots + b_2 s + b_1}{a_n s^{n-1} + a_{n-1} s^{n-2} + \cdots + a_2 s + a_1} \quad (6.102)$$

The program is listed in Figure 6.47.

2. Usage. The program consists of a subroutine, FRESP, called by

CALL FRESP (M, N, B, A, W, FMAG, FPHASE)

M	is the degree of the numerator polynomial plus one
N	is the degree of the denominator polynomial plus one
B	is the name of the numerator coefficients b_i
A	is the name of the denominator coefficients a_i
W	is the frequency ω
FMAG	is the magnitude $|H(j\omega)|$ in dB
FPHASE	is the phase $\phi(\omega)$ in degrees

3. Remarks. Since arctangent is a multiple-valued function, its representation is not unique. The input data are M, N, B(I) of B, A(I) of A, and W. The outputs are FMAG and FPHASE.

4. Subroutines and Function Subprograms Required. None

EXAMPLE 6.16
Consider the transfer current ratio $H(s)$ of (6.100) associated with the voltage-shunt feedback amplifier of Figure 6.43(a). We use the subroutine FRESP to compute its frequency response.

The main program and the computer output are presented, respectively, in Figures 6.48 and 6.49. A portion of the tabulated magnitude and phase is shown in Figure 6.49(a). Using the subroutine PLOT described in Appendix A, the magnitude and phase plots are presented in Figure 6.49 (b) and (c). The results are consistent with those shown in Figures 6.44(b) and 6.45(b). The total execution time for an IBM 370 Model 158 computer was 2.67 sec.

[†] In relation to (6.27), the degrees of the polynomials are represented in a slightly different form for ease of programming.

Figure 6.48
The main program used to compute and plot the magnitude and phase responses of a real rational function, using subroutines FRESP and PLOT.

```
      DIMENSION B(20),A(20),XY(200,3),JXY(2)
      DOUBLE PRECISION B,A,W,WO,WF,X,DX,FMAG,FPHASE,DLOG10
      READ,M,N,(B(I),I=1,M),(A(I),I=1,N),WO,WF,NW
      X=DLOG10(WO)
      DX=(DLOG10(WF)-X)/DFLOAT(NW-1)
      I=1
1     W=10.0**X
      CALL FRESP(M,N,B,A,W,FMAG,FPHASE)
      PRINT 2,W,FMAG,FPHASE
2     FORMAT (8H0OMEGA =,D20.8,8H  RADS/S/6X,11HMAGNITUDE =,D20.8,
     C   4H DB,10X,7HPHASE =,D20.8,9H DEGREES)
      XY(I,1)=X
      XY(I,2)=FMAG
      XY(I,3)=FPHASE
      X=X+DX
      I=I+1
      IF (I-NW) 1,1,3
3     JXY(1)=1
      JXY(2)=2
      CALL PLOT (XY,JXY,NW,200,1)
      PRINT 4
4     FORMAT (1H0,52X,17HOMEGA (LOG SCALE)///51X,
     C   22HMAGNITUDE VS FREQUENCY)
      JXY(1)=1
      JXY(2)=3
      CALL PLOT (XY,JXY,NW,200,1)
      PRINT 5
5     FORMAT (1H0,52X,17HOMEGA (LOG SCALE)///52X,
     C   18HPHASE VS FREQUENCY)
      STOP
      END
```

Figure 6.49
(a) Computer printout of a portion of the tabulated magnitude and phase responses of the current ratio (6.100). (b) Computer plot of the magnitude response of the current ratio (6.100). (c) Computer plot of the phase response of the current ratio (6.100).

```
OMEGA =     0.10000000D 05  RADS/S
    MAGNITUDE =      0.38312549D 02  DB        PHASE =      0.17901266D 03  DEGREES
OMEGA =     0.30199517D 05  RADS/S
    MAGNITUDE =      0.38302096D 02  DB        PHASE =      0.17702068D 03  DEGREES
OMEGA =     0.91201084D 05  RADS/S
    MAGNITUDE =      0.38207911D 02  DB        PHASE =      0.17106751D 03  DEGREES
OMEGA =     0.27542287D 06  RADS/S
    MAGNITUDE =      0.37431842D 02  DB        PHASE =      0.15460664D 03  DEGREES
OMEGA =     0.83176377D 06  RADS/S
    MAGNITUDE =      0.33465633D 02  DB        PHASE =      0.12488470D 03  DEGREES
OMEGA =     0.25118864D 07  RADS/S
    MAGNITUDE =      0.25362652D 02  DB        PHASE =      0.10294147D 03  DEGREES
OMEGA =     0.75857758D 07  RADS/S
    MAGNITUDE =      0.15963151D 02  DB        PHASE =      0.94166115D 02  DEGREES
OMEGA =     0.22908677D 08  RADS/S
    MAGNITUDE =      0.63852686D 01  DB        PHASE =      0.90819457D 02  DEGREES
OMEGA =     0.69183097D 08  RADS/S
    MAGNITUDE =     -0.32163144D 01  DB        PHASE =      0.88573031D 02  DEGREES
OMEGA =     0.20892961D 09  RADS/S
    MAGNITUDE =     -0.12852934D 02  DB        PHASE =      0.84414075D 02  DEGREES
OMEGA =     0.63095734D 09  RADS/S
    MAGNITUDE =     -0.22775619D 02  DB        PHASE =      0.73119068D 02  DEGREES
OMEGA =     0.19054607D 10  RADS/S
    MAGNITUDE =     -0.34557725D 02  DB        PHASE =      0.46872089D 02  DEGREES
OMEGA =     0.57543994D 10  RADS/S
    MAGNITUDE =     -0.50769882D 02  DB        PHASE =      0.16226190D 02  DEGREES
OMEGA =     0.17378008D 11  RADS/S
    MAGNITUDE =     -0.69298053D 02  DB        PHASE =     -0.52576051D 01  DEGREES
OMEGA =     0.52480746D 11  RADS/S
    MAGNITUDE =     -0.87086128D 02  DB        PHASE =     -0.30937665D 02  DEGREES
OMEGA =     0.15848932D 12  RADS/S
    MAGNITUDE =     -0.10091026D 03  DB        PHASE =     -0.62445801D 02  DEGREES
OMEGA =     0.47863009D 12  RADS/S
    MAGNITUDE =     -0.11137560D 03  DB        PHASE =     -0.80255734D 02  DEGREES
OMEGA =     0.10000000D 13  RADS/S
    MAGNITUDE =     -0.11786749D 03  DB        PHASE =     -0.85303858D 02  DEGREES
```

(a)

Figure 6.49 (continued)

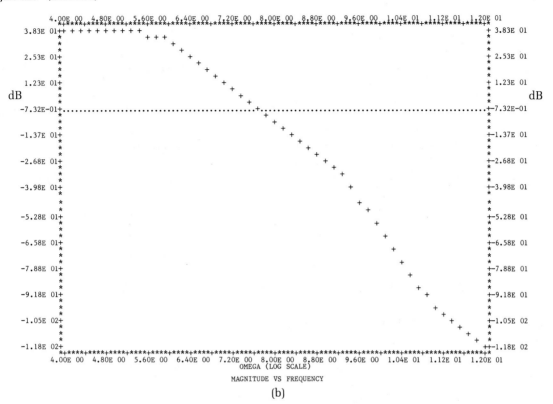

(b)

(c)

EXAMPLE 6.17

We shall use the subroutines FRESP and PLOT to compute and plot the frequency response of the normalized transfer function of (6.101) considered earlier in Example 6.15.

The main program that uses both subroutines FRESP and PLOT is the same as in Figure 6.48. The magnitude and phase values and their plots are presented in Figure 6.50. The discontinuity in the phase plot of Figure 6.50(c) is due to an alternative representation of the phase function. At $\omega = 2$, for example, the phase is $191.3°$ or equivalently $-168.7°$. The results are consistent with those shown in Figure 6.46(b). The total execution time was 2.99 sec.

6-6.2　Subroutine FBODE

1. Purpose.　The program, as listed in Figure 6.51, is for the evaluation of the magnitude and phase of a real rational function $H(j\omega)$, in factored form as in (6.77), at a fixed frequency ω.

2. Method.　The program is based on the Bode plot.

Figure 6.50

(a) Computer printout of a portion of the tabulated magnitude and phase responses of the normalized transfer function (6.101). (b) Computer plot of the magnitude response of the normalized transfer function (6.101), using subroutines FRESP and PLOT. (c) Computer plot of the phase response of the normalized transfer function (6.101), using subroutines FRESP and PLOT.

```
OMEGA =      0.20000000D 00  RADS/S
     MAGNITUDE =      -0.14811570D 00  DB          PHASE =      -0.20948035D 02  DEGREES
OMEGA =      0.26365135D 00  RADS/S
     MAGNITUDE =      -0.25569044D 00  DB          PHASE =      -0.27560095D 02  DEGREES
OMEGA =      0.34756017D 00  RADS/S
     MAGNITUDE =      -0.43977805D 00  DB          PHASE =      -0.36214981D 02  DEGREES
OMEGA =      0.45817353D 00  RADS/S
     MAGNITUDE =      -0.75315478D 00  DB          PHASE =      -0.47504386D 02  DEGREES
OMEGA =      0.60399034D 00  RADS/S
     MAGNITUDE =      -0.12863585D 01  DB          PHASE =      -0.62165380D 02  DEGREES
OMEGA =      0.79621434D 00  RADS/S
     MAGNITUDE =      -0.22049123D 01  DB          PHASE =      -0.81059163D 02  DEGREES
OMEGA =      0.10496149D 01  RADS/S
     MAGNITUDE =      -0.38215441D 01  DB          PHASE =      -0.10481216D 03  DEGREES
OMEGA =      0.13836619D 01  RADS/S
     MAGNITUDE =      -0.66094387D 01  DB          PHASE =      -0.13245781D 03  DEGREES
OMEGA =      0.18240217D 01  RADS/S
     MAGNITUDE =      -0.10859727D 02  DB          PHASE =       0.19980857D 03  DEGREES
OMEGA =      0.24045289D 01  RADS/S
     MAGNITUDE =      -0.16296142D 02  DB          PHASE =       0.17593568D 03  DEGREES
OMEGA =      0.31697864D 01  RADS/S
     MAGNITUDE =      -0.22422575D 02  DB          PHASE =       0.15691266D 03  DEGREES
OMEGA =      0.41785923D 01  RADS/S
     MAGNITUDE =      -0.28927427D 02  DB          PHASE =       0.14195098D 03  DEGREES
OMEGA =      0.55084574D 01  RADS/S
     MAGNITUDE =      -0.35669312D 02  DB          PHASE =       0.13014900D 03  DEGREES
OMEGA =      0.72615611D 01  RADS/S
     MAGNITUDE =      -0.42574041D 02  DB          PHASE =       0.12086481D 03  DEGREES
OMEGA =      0.95726018D 01  RADS/S
     MAGNITUDE =      -0.49589903D 02  DB          PHASE =       0.11362015D 03  DEGREES
OMEGA =      0.12619147D 02  RADS/S
     MAGNITUDE =      -0.56678271D 02  DB          PHASE =       0.10801646D 03  DEGREES
OMEGA =      0.16635275D 02  RADS/S
     MAGNITUDE =      -0.63811936D 02  DB          PHASE =       0.10371234D 03  DEGREES
```

(a)

Figure 6.50 (*continued*)

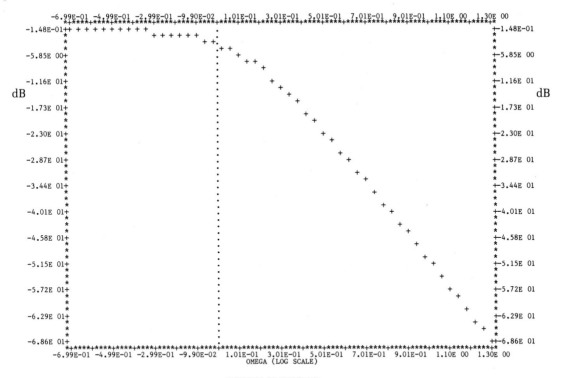

(b)

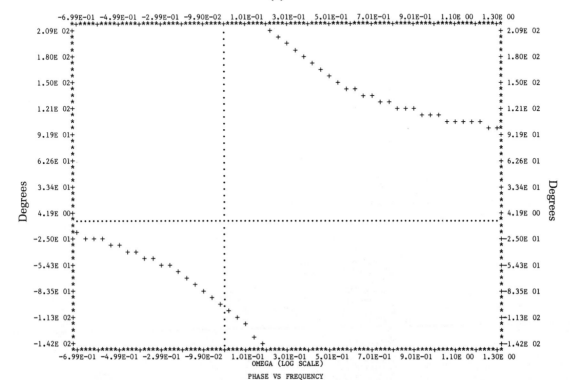

(c)

Figure 6.51

Subroutine FBODE used for the evaluation of the magnitude and phase responses of a real rational function in factored form at a fixed frequency ω, using Bode plot.

```
C
C
C     SUBROUTINE FBODE
C
C     PURPOSE
C         THE PROGRAM IS FOR THE EVALUATION OF THE MAGNITUDE AND PHASE OF A
C         REAL RATIONAL FUNCTION IN FACTORED FORM AT A FIXED FREQUENCY.
C
C     METHOD
C         THE PROGRAM IS BASED ON THE BODE PLOT.
C
C     USAGE
C         CALL   FBODE (AK,M,N,ME,NE,D,S,W,FMAG,FPHASE)
C
C         AK      -THE SCALE FACTOR K
C         M       -THE NUMBER OF DISTINCT ZEROS
C         N       -THE NUMBER OF DISTINCT POLES
C         ME      -THE ORDERS OF THE ZEROS
C         NE      -THE ORDERS OF THE POLES
C         D       -THE ZEROS
C         S       -THE POLES
C         W       -THE FREQUENCY IN RADIANS PER SECOND
C         FMAG    -THE MAGNITUDE IN DB
C         FPHASE  -THE PHASE IN DEGREES
C
C     SUBROUTINES AND FUNCTION SUBPROGRAMS REQUIRED
C         NONE
C
C     REMARKS
C         THE REAL RATIONAL FUNCTION MUST BE GIVEN IN FACTORED FORM.  OTHERWISE,
C         USE A ROOT FINDING SUBROUTINE TO FACTOR OUT THE POLYNOMIALS
C         FIRST.
C         THE INPUT DATA ARE AK, M, N, ME(I) OF ME, NE(J) OF NE, D(I) OF D, S(J)
C         OF S AND W.  THE OUTPUTS ARE FMAG AND FPHASE.
C
      SUBROUTINE FBODE (AK,M,N,ME,NE,D,S,W,FMAG,FPHASE)
      DIMENSION ME(M),NE(N),D(M),S(N)
      DOUBLE PRECISION AK,W,FMAG,FPHASE,SMAG,SPHASE,DR,DI,A,B,
     C DREAL,DIMAG,DFLOAT,DABS
      COMPLEX*16 D,S
      FMAG=0.0
      FPHASE=0.0
      DO 3 I=1,M
      DR=DREAL(D(I))
      DI=DIMAG(D(I))
      IF(DI) 3,1,2
1     FMAG=FMAG+DFLOAT(ME(I))*10.0*DLOG10(W**2+DR**2)
      FPHASE=FPHASE+DFLOAT(ME(I))*DATAN2(W,DR)
      GO TO 3
2     A=2.0*DR
      B=DR**2+DI**2
      A=A*W
      B=B-W**2
      FMAG=FMAG+DFLOAT(ME(I))*10.0*DLOG10(B**2+A**2)
      FPHASE=FPHASE+DFLOAT(ME(I))*DATAN2(A,B)
3     CONTINUE
      SMAG=0.0
      SPHASE=0.0
      DO 6 J=1,N
      DR=DREAL(S(J))
      DI=DIMAG(S(J))
      IF(DI) 6,4,5
4     SMAG=SMAG+DFLOAT(NE(J))*10.0*DLOG10(W**2+DR**2)
      SPHASE=SPHASE+DFLOAT(NE(J))*DATAN2(W,DR)
      GO TO 6
5     A=2.0*DR
      B=DR**2+DI**2
      A=A*W
      B=B-W**2
      SMAG=SMAG+DFLOAT(NE(J))*10.0*DLOG10(B**2+A**2)
      SPHASE=SPHASE+DFLOAT(NE(J))*DATAN2(A,B)
6     CONTINUE
      FMAG=20.0*DLOG10(DABS(AK))+FMAG-SMAG
      FPHASE=(FPHASE-SPHASE)*360.0/6.2831853071796
      IF(AK.LT.0.0) FPHASE=FPHASE+180.0
      RETURN
      END
      FUNCTION DREAL(X)
C
C     COMPUTE THE REAL PART OF A COMPLEX X
      COMPLEX*16 X
      DOUBLE PRECISION DREAL
```

```
       DREAL=X
       RETURN
       END
       FUNCTION DIMAG(X)
C
C      COMPUTE THE IMAGINARY PART OF A COMPLEX X
       COMPLEX*16 X
       DOUBLE PRECISION DIMAG
       DIMAG=X*(0.0,-1.0)
       RETURN
       END
```

3. Usage.   The program consists of a subroutine, FBODE, called by

CALL FBODE (AK, M, N, ME, NE, D, S, W, FMAG, FPHASE)

AK	is the scale factor K
M	is the number of distinct zeros $-d_h$
N	is the number of distinct poles $-s_k$
ME	is the name of the orders of the zeros $-d_h$ with ME(MH) $= m_h$
NE	is the name of the orders of the poles $-s_k$ with NE(KP) $= n_k$
D	is the name of the zeros $-d_h$ with D(MH) $= d_h$
S	is the name of the poles $-s_k$ with S(KP) $= s_k$
W	is the frequency ω
FMAG	is the magnitude $|H(j\omega)|$ in dB
FPHASE	is the phase $\phi(\omega)$ in degrees

4. Remarks.   The real rational function $H(s)$ must be given in factored form as in (6.77). Otherwise use a root-finding subroutine to factor out the polynomials first. The imput data are AK, M, N, ME(I) of ME, NE(J) of NE, D(I) of D, S(J) of S, and W. The outputs are FMAG and FPHASE. If the function $H(s)$ is devoid of zeros, set $M = 1$, $d_h = 1$ and $m_h = 0$ in (6.77).

5. Subroutines and Function Subprograms Required.   None except a root-finding subroutine for a polynomial may be required.

EXAMPLE 6.18

We use the subroutines FBODE and PLOT to compute and plot the frequency response of the transfer current ratio $H(s)$ of (6.100).

The main program and the computer output are presented in Figures 6.52 and 6.53. A portion of the tabulated magnitude and phase is shown in Figure 6.53. The magnitude and phase plots are identical to those given in Figures 6.49(b) and 6.49(c) and are omitted here. The total execution time for an IBM Model 158 computer was 2.68 sec.

EXAMPLE 6.19

We repeat Example 6.17 by using subroutine FBODE. The main program and the magnitude plot are the same as those shown in Figures 6.52 and 6.50(b). The magnitude and phase values and the phase plot are shown in Figure 6.54. Observe that the phase is expressed in terms of the negative angles, thus avoiding a discontinuity occurring in the phase plot of Figure 6.50(c).

Figure 6.52

The main program used to compute and plot the magnitude and phase responses of a real rational function, using subroutines FBODE and PLOT.

```
C
      DIMENSION ME(10),NE(10),D(10),S(10),XY(200,3),JXY(2)
      DOUBLE PRECISION AK,W,WO,WF,X,DX,FMAG,FPHASE,DLOG10
      COMPLEX*16 D,S
      READ,AK,M,N,(ME(I),I=1,M),(NE(J),J=1,N),(D(I),I=1,M),(S(J),J=1,N),
  C   WO,WF,NW
      X=DLOG10(WO)
      DX=(DLOG10(WF)-X)/DFLOAT(NW-1)
      I=1
  1   W=10.0**X
      CALL FBODE (AK,M,N,ME,NE,D,S,W,FMAG,FPHASE)
      PRINT 2,W,FMAG,FPHASE
  2   FORMAT (8H0OMEGA =,D20.8,8H  RADS/S/6X,11HMAGNITUDE =,D20.8,
  C   4H DB,10X,7HPHASE =,D20.8,9H  DEGREES)
      XY(I,1)=X
      XY(I,2)=FMAG
      XY(I,3)=FPHASE
      X=X+DX
      I=I+1
      IF (I-NW) 1,1,3
  3   JXY(1)=1
      JXY(2)=2
      CALL PLOT (XY,JXY,NW,200,1)
      PRINT 4
  4   FORMAT (1H0,52X,17HOMEGA (LOG SCALE)///51X,
  C   22HMAGNITUDE VS FREQUENCY)
      JXY(1)=1
      JXY(2)=3
      CALL PLOT (XY,JXY,NW,200,1)
      PRINT 5
  5   FORMAT (1H0,52X,17HOMEGA (LOG SCALE)///52X,
  C   18HPHASE VS FREQUENCY)
      STOP
      END
```

Figure 6.53

Computer printout of a portion of the tabulated magnitude and phase responses of the transfer current ratio (6.100).

```
OMEGA =      0.10000000D 05  RADS/S
    MAGNITUDE =      0.38240985D 02  DB          PHASE =      0.17902040D 03  DEGREES
OMEGA =      0.30199517D 05  RADS/S
    MAGNITUDE =      0.38230696D 02  DB          PHASE =      0.17704400D 03  DEGREES
OMEGA =      0.91201084D 05  RADS/S
    MAGNITUDE =      0.38137963D 02  DB          PHASE =      0.17113644D 03  DEGREES
OMEGA =      0.27542287D 06  RADS/S
    MAGNITUDE =      0.37372749D 02  DB          PHASE =      0.15478096D 03  DEGREES
OMEGA =      0.83176377D 06  RADS/S
    MAGNITUDE =      0.33439930D 02  DB          PHASE =      0.12509673D 03  DEGREES
OMEGA =      0.25118864D 07  RADS/S
    MAGNITUDE =      0.25355974D 02  DB          PHASE =      0.10304080D 03  DEGREES
OMEGA =      0.75857758D 07  RADS/S
    MAGNITUDE =      0.15959564D 02  DB          PHASE =      0.94200611D 02  DEGREES
OMEGA =      0.22908677D 08  RADS/S
    MAGNITUDE =      0.63820395D 01  DB          PHASE =      0.90831050D 02  DEGREES
OMEGA =      0.69183097D 08  RADS/S
    MAGNITUDE =     -0.32195025D 01  DB          PHASE =      0.88577206D 02  DEGREES
OMEGA =      0.20892961D 09  RADS/S
    MAGNITUDE =     -0.12856103D 02  DB          PHASE =      0.84416456D 02  DEGREES
OMEGA =      0.63095734D 09  RADS/S
    MAGNITUDE =     -0.22778663D 02  DB          PHASE =      0.73122631D 02  DEGREES
OMEGA =      0.19054607D 10  RADS/S
    MAGNITUDE =     -0.34560131D 02  DB          PHASE =      0.46877973D 02  DEGREES
OMEGA =      0.57543994D 10  RADS/S
    MAGNITUDE =     -0.50771526D 02  DB          PHASE =      0.16230004D 02  DEGREES
OMEGA =      0.17378008D 11  RADS/S
    MAGNITUDE =     -0.69299511D 02  DB          PHASE =     -0.52561915D 01  DEGREES
OMEGA =      0.52480746D 11  RADS/S
    MAGNITUDE =     -0.87087562D 02  DB          PHASE =     -0.30937191D 02  DEGREES
OMEGA =      0.15848932D 12  RADS/S
    MAGNITUDE =     -0.10091169D 03  DB          PHASE =     -0.62445644D 02  DEGREES
OMEGA =      0.47863009D 12  RADS/S
    MAGNITUDE =     -0.11137703D 03  DB          PHASE =     -0.80255682D 02  DEGREES
OMEGA =      0.10000000D 13  RADS/S
    MAGNITUDE =     -0.11786892D 03  DB          PHASE =     -0.85303834D 02  DEGREES
```

Figure 6.54

(a) Computer printout of a portion of the tabulated magnitude and phase responses of the normalized transfer function (6.101), using subroutine FBODE. (b) Computer plot of the phase response of the normalized transfer function (6.101), using subroutines FBODE and PLOT.

```
OMEGA =      0.20000000D 00  RADS/S
    MAGNITUDE =      -0.14811570D 00  DB        PHASE =     -0.20948035D 02  DEGREES
OMEGA =      0.26365135D 00  RADS/S
    MAGNITUDE =      -0.25569044D 00  DB        PHASE =     -0.27560095D 02  DEGREES
OMEGA =      0.34756017D 00  RADS/S
    MAGNITUDE =      -0.43977805D 00  DB        PHASE =     -0.36214981D 02  DEGREES
OMEGA =      0.45817353D 00  RADS/S
    MAGNITUDE =      -0.75315478D 00  DB        PHASE =     -0.47504386D 02  DEGREES
OMEGA =      0.60399034D 00  RADS/S
    MAGNITUDE =      -0.12863585D 01  DB        PHASE =     -0.62165380D 02  DEGREES
OMEGA =      0.79621434D 00  RADS/S
    MAGNITUDE =      -0.22049123D 01  DB        PHASE =     -0.81059163D 02  DEGREES
OMEGA =      0.10496149D 01  RADS/S
    MAGNITUDE =      -0.38215441D 01  DB        PHASE =     -0.10481216D 03  DEGREES
OMEGA =      0.13836619D 01  RADS/S
    MAGNITUDE =      -0.66094387D 01  DB        PHASE =     -0.13245781D 03  DEGREES
OMEGA =      0.18240217D 01  RADS/S
    MAGNITUDE =      -0.10859727D 02  DB        PHASE =     -0.16019143D 03  DEGREES
OMEGA =      0.24045289D 01  RADS/S
    MAGNITUDE =      -0.16296142D 02  DB        PHASE =     -0.18406432D 03  DEGREES
OMEGA =      0.31697864D 01  RADS/S
    MAGNITUDE =      -0.22422575D 02  DB        PHASE =     -0.20308734D 03  DEGREES
OMEGA =      0.41785923D 01  RADS/S
    MAGNITUDE =      -0.28927427D 02  DB        PHASE =     -0.21804902D 03  DEGREES
OMEGA =      0.55084574D 01  RADS/S
    MAGNITUDE =      -0.35669312D 02  DB        PHASE =     -0.22985100D 03  DEGREES
OMEGA =      0.72615611D 01  RADS/S
    MAGNITUDE =      -0.42574041D 02  DB        PHASE =     -0.23913519D 03  DEGREES
OMEGA =      0.95726018D 01  RADS/S
    MAGNITUDE =      -0.49589903D 02  DB        PHASE =     -0.24637985D 03  DEGREES
OMEGA =      0.12619147D 02  RADS/S
    MAGNITUDE =      -0.56678271D 02  DB        PHASE =     -0.25198354D 03  DEGREES
OMEGA =      0.16635275D 02  RADS/S
    MAGNITUDE =      -0.63811936D 02  DB        PHASE =     -0.25628766D 03  DEGREES
OMEGA =      0.20000000D 02  RADS/S
    MAGNITUDE =      -0.68583746D 02  DB        PHASE =     -0.25857832D 03  DEGREES
```

(a)

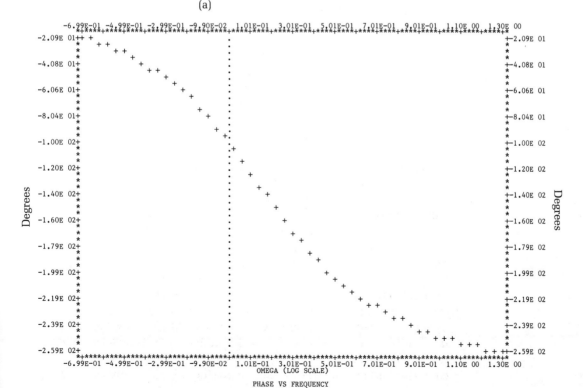

(b)

6-7   SUMMARY

We began this chapter by considering the network representations of various network elements in terms of their *transform immittances* and *transform sources* due to initial conditions. A *transform network* is an interconnection of these network representations. As a result we can write the secondary system of network equations in the frequency domain directly from the transform network, thus bypassing the necessity of first writing down the integrodifferential equations. However, the determination and solution of a secondary system is only one of the many important applications of the Laplace transform. In fact, we can associate analytic functions with a network, thus allowing us to make use of the extensive theory of analytic functions developed by mathematicians over the last century. For this we introduced the notion of the *network function*, which is defined as the ratio of the Laplace transform of the zero-state response to the Laplace transform of the input, or the ratio of the Laplace transforms of any two zero-state responses. Depending on whether the terminals to which the response relates are the same or different from the input terminals, network functions generally fall into two general classes: *driving-point functions* and *transfer functions*. One of the important properties of the network functions is that they are rational functions in the complex frequency variable s with real coefficients. As a result, a network function assumes conjugate values at conjugate points in the s-plane. Apart from a scale factor, a network function is completely determined by its *poles* and *zeros*, which can be represented by the *pole-zero pattern*.

We restated and proved several important theorems in the frequency domain. Specifically we showed that the *zero-state response* of a linear network to a number of excitations applied simultaneously is the sum of the zero-state responses when each of the excitations is applied individually. In the case of linear time-invariant networks, the same holds in the frequency domain. This is known as the *principle of superposition* and is intimately tied up with the concept of linearity. We presented Thévenin's and Norton's theorems, which state that a two-terminal network having no magnetic or controlled source coupling with an external network can be represented equivalently by a simple network consisting of either the series or parallel combination of an equivalent voltage or current source and an impedance.

In the situations where we are interested in the behavior of a network only at the two accessible ports, we use the *two-port* concepts. There are a number of ways of characterizing a two-port, depending on the choice of the independent variables. Some common choices were discussed: the *y-parameters*, the *z-parameters*, the *h-parameters*, the *g-parameters*, and the *transmission parameters*. It was found that when two two-ports are connected in cascade, it is most convenient to use the transmission parameters; the resulting transmission matrix is the product of the transmission matrices of the individual two-ports. In the case of a parallel connection, the short-circuit admittance matrix of the composite two-port network is the sum of the short-circuit admittance matrices of the component two-ports. Likewise in a series connection, the z-parameters of the component two-ports are added to give the over-all z-parameters. Thus these parameters are all useful in various situations. For example, the h-parameters are widely used in the modeling of transistors.

In engineering practice it is of paramount importance to know the behavior of a network function on the $j\omega$-axis as ω varies from 0 to ∞, because it not only describes the network in the sinusoidal steady state, but also, as will be shown in the following chapter, it contains all the information for calculating the zero-state response due to an *arbitrary* input. For this we presented the frequency response functions and showed how to plot their magnitudes and phases as functions of ω. These plots can be measured with great precision in the laboratory, and therefore are of great practical significance.

In the case of a transfer function, it is convenient to consider the logarithm of the transfer function instead of the transfer function itself. The advantage is that not only multiplication has been changed to addition, but each pole factor and each zero factor also appear separately. To compute the complete frequency response, one needs only to plot the contributions of the various factors. The sum of these contributions will give the complete response. The process is called the *Bode plot*, which involves only two typical factors: the first-order factor and the second-order factor. Techniques for rapid estimate of the contributions of these two factors were presented in terms of the break frequencies and asymptotes with the aid of four charts. The result approximates the actual behavior quite closely.

Finally, we presented computer programs for the determination of the frequency response which plot the magnitude and the phase of a network function as functions of the frequency variable.

REFERENCES AND SUGGESTED READING

Balabanian, N., and T. A. Bickart. *Electrical Network Theory*. New York: Wiley, 1969, Chapter 3.

Bode, H. W. *Network Analysis and Feedback Amplifier Design*. Princeton, N.J.: Van Nostrand, 1945, Chapter XV.

Chen, W. K. *Active Network and Feedback Amplifier Theory*. New York: McGraw-Hill, 1980, Chapters 1 and 6.

Desoer, C. A., and E. S. Kuh. *Basic Circuit Theory*. New York: McGraw-Hill, 1969, Chapters 15–17.

Jensen, R. W., and B. O. Watkins. *Network Analysis: Theory and Computer Methods*. Englewood Cliffs, N.J.: Prentice-Hall, 1974, Chapter 9.

Kuo, F. F. *Network Analysis and Synthesis*. 2d ed. New York: Wiley, 1966, Chapters 7–9.

Papoulis, A. *Circuits and Systems: A Modern Approach*. New York: Holt, Rinehart & Winston, 1980, Chapter 5.

Van Valkenburg, M. E. *Network Analysis*. 3d ed. Englewood Cliffs, N.J.: Prentice-Hall, 1974, Chapters 9–13.

PROBLEMS

6.1 Write the nodal system of equations for the transform network of Figure 6.43(b), where the element values are given by (6.99) and Figure 6.43(a). Verify that the transfer current ratio $-I_2/I_g$ is given by (6.100).

6.2 Compute the open-loop voltage ratio V_2/V_g of the feedback amplifier of Figure 6.43(a) using the transform network of Figure 6.43(b), where V_2 and V_g are the Laplace transforms of v_2 and v_g, respectively.

6.3 The transfer impedance relating V_{1a} to I_g in the network of Figure 6.11(b) was obtained in Example 6.5 and is given by

$$Z_{1g}(s) = \frac{s+5}{s^2 + 7s + 12} \qquad\qquad\qquad (6.103)$$

    (a) Plot the magnitude $|Z_{1g}(j\omega)|$ as a function of ω.
    (b) Plot the phase of $Z_{1g}(j\omega)$ as a function of ω.

6.4 Write down the frequency-domain nodal equations of Figure 4.2(a). Compute the input impedance facing the 1-A current source and also the transfer impedance relating the transform of v_1 to the transform of the 1-A current source. Plot the frequency responses of these network functions.

6.5 Using transform network technique, write the nodal equations for the network of Figure 4.4(a) for $t \geq 0$. Compute the driving-point impedance facing the current source and the transfer current ratio relating the transform of i_6 to the transform of the current source. Plot the frequency responses (magnitude and phase plots) of these functions.

6.6 Figure 4.18 is a third-order Butterworth prototype filter with equal generator and load resistance having normalized value of one ohm. Compute the driving-point impedance facing the 2-V battery and the transfer voltage ratio relating the transform of v to the transform of the voltage source. Plot the frequency responses (magnitude and phase plots) of these network functions.

6.7 Using transform network technique, write the loop system of equations for the network of Figure 4.19 for $t \geq 0$. Compute the driving-point impedance facing the voltage source and the transfer voltage ratio relating the transform of v to the transform of the voltage source. Plot the magnitudes and phases of these functions versus ω.

6.8 Using transform network technique, write the nodal system of equations for the network of Figure 4.22 for $t \geq 0$. Compute the driving-point impedance facing the current source and the transfer impedance relating the transform of v_2 to the transform of the current source. Plot the magnitudes and phases of these functions versus ω.

6.9 Determine the Thévenin's equivalent network looking into the two terminals of the 1-F capacitor in the network of Figure 4.2(a). Use this to compute the capacitor voltage $v_1(t)$, $t \geq 0$. What is the Norton's equivalent network?

6.10 Applying Thévenin's and Norton's theorems, compute the output voltage $v_6(t)$ in the network of Figure 4.4(a) for $t \geq 0$.

6.11 Determine the Thévenin's equivalent network looking into the two terminals of the 10 kΩ resistor in the network of Figure 4.19 for $t \geq 0$. Use this to compute the resistor voltage $v(t)$ as indicated for $t \geq 0$.

6.12 Determine the Norton's equivalent network looking into the two terminals of the 200 Ω resistor in the network of Figure 4.22 for $t \geq 0$. Use this to compute the resistor voltage $v_2(t)$ as indicated for $t \geq 0$.

6.13 A capacitor inside a linear time-invariant one-port N is charged and then discharged. This is repeated twice. We first short-circuit the terminals of N and measure the current as shown in Figure 6.55(a). The short-circuit current is found to be

$$i(t) = 2e^{-2t} \cos(2t + \tfrac{1}{4}\pi) \qquad\qquad\qquad (6.104a)$$

The second time we open circuit the terminals and measure the voltage as depicted in Figure 6.55(b). The open-circuit voltage is determined to be

$$v(t) = 3e^{-t} \cos(3t + \tfrac{1}{2}\pi) + e^{-2t} \qquad\qquad\qquad (6.104b)$$

Determine the driving-point impedance at the terminals of the one-port.

Figure 6.55
A one-port network with
known short-circuit
terminal current and
open-circuit terminal
voltage.

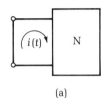

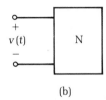

(a)   (b)

6.14 The operational amplifier given in Figure 6.56 is a multiple-input and a single-output network. Assume that the operational amplifier is ideal. Use the superposition principle to find the transform V_o of v_o in terms of the transforms V_1, V_2, V_3, and V_4 of the inputs v_1, v_2, v_3, and v_4 and the known resistances in Figure 6.56.

6.15 Figure 6.57 is a low-frequency, small-signal model of a common-emitter amplifier. Find the Thévenin's equivalent network from the two terminals of the resistor R. What is the Norton's equivalent?

6.16 A typical set of element values for the transistor model N of Figure 6.57 is listed below:

$$r_b = 50 \ \Omega, \qquad r_e = 5 \ \Omega$$
$$r_c = 1 \ M\Omega, \qquad \alpha = 0.98$$

(6.105)

Using these values compute the h-parameters and the y-parameters of the associated two-port network N of Figure 6.57.

6.17 A one-port network N is composed only of linear time-invariant elements and sinusoidal sources at a frequency of 60 Hz. When the terminals of N are short circuited, the short-circuit current is found to be 15 $\underline{/30°}$ A. When

Figure 6.56
A multiple-input and a
single output network
constructed from an
operational amplifier.

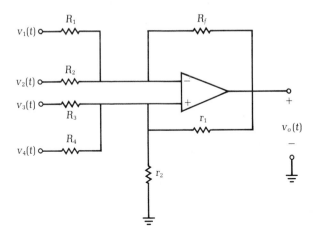

Figure 6.57
An equivalent network of
a transistor amplifier.

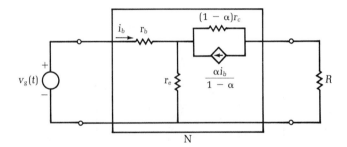

N is terminated by an impedance of $40 + j15\ \Omega$ at 60 Hz, the terminal voltage is measured as $60\ \underline{/-50°}$ V. Determine the Thévenin's and Norton's equivalent networks.

6.18 Compute the z-parameters and the transmission parameters of the two-port network N shown in Figure 6.57.

6.19 Determine one set of two-port parameters that is easiest to calculate for the transformer of Figure 6.5(a), and compute these parameters.

6.20 A simple RC twin-Tee used in the design of equalizers is shown in Figure 6.58. Write a set of two-port parameters that is easiest to calculate for this two-port network.

6.21 Show that a two-port network characterized by its z-parameters z_{ij} or y-parameters y_{ij} can be represented by the Tee equivalent or the Pi equivalent of Figure 6.59, provided that $z_{12} = z_{21}$ or $y_{12} = y_{21}$.

6.22 A typical interstage RC coupling network of an audio amplifier is shown in Figure 6.60. Compute the transmission matrix of this two-port network. Also determine the z-parameters of the coupling network.

6.23 Refer to the two-port network of Figure 6.30. Show that

$$y_{12} = \frac{-y_{12a}y_{12b}}{y_{22a} + y_{11b}}, \qquad z_{12} = \frac{z_{12a}z_{12b}}{z_{22a} + z_{11b}} \tag{6.106}$$

where subscripts a and b refer to the component two-ports N_a and N_b.

6.24 It is claimed that the network of Figure 6.61 is equivalent to that of Figure 6.6(b). Is this claim correct? If so, what are the advantages in characterizing the network between the source i_g and the load L_5 and R_6. Give your reasons and justifications.

Figure 6.58
A simple RC twin-Tee used in the design of equalizers.

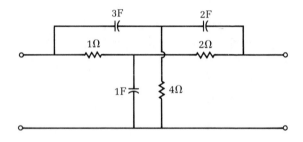

Figure 6.59
Representations of a reciprocal two-port network in terms of its z-parameters and y-parameters.

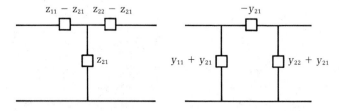

Figure 6.60
A typical interstage RC coupling network of an audio amplifier.

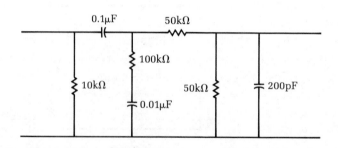

Figure 6.61
The representation of the
network of Figure 6.6(b) as
the parallel combination
of two two-ports
terminated in the same
current source and load.

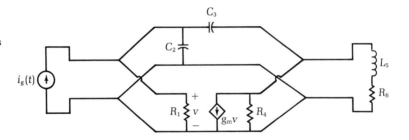

Figure 6.62
Symbolic representation
of a two-port network
terminated in a voltage
source and a load
impedance.

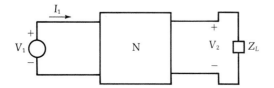

6.25 The two-port network N of Figure 6.62 is terminated in an impedance $Z_L = 1/Y_L$. Show that

$$\frac{V_2}{V_1} = -\frac{y_{21}}{y_{22} + Y_L}, \qquad \frac{V_2}{I_1} = \frac{z_{21}Z_L}{z_{22} + Z_L} \tag{6.107}$$

6.26 Use the subroutine FRESP to plot the magnitude and phase of the transfer voltage ratio

$$G_{12}(s) = \frac{0.09476}{s^6 + 1.15918s^5 + 2.17185s^4 + 1.58976s^3 + 1.17186s^2 + 0.43237s + 0.09476} \tag{6.108}$$

associated with the sixth-order Chebyshev response as functions of frequency ω for the range $0 \leqq \omega \leqq 4$.

6.27 Repeat Problem 6.26 for the transfer voltage ratio

$$G_{12}(s) = \frac{1}{s^6 + 3.86370s^5 + 7.46410s^4 + 9.14162s^3 + 7.46410s^2 + 3.86370s + 1} \tag{6.109}$$

associated with the sixth-order Butterworth response.

6.28 Repeat Problem 6.26 for the network function

$$S(s) = \frac{s^4 + 0.26633s^3 + 1.12458s^2 + 0.19996s + 0.18935}{s^4 + 1.22218s^3 + 1.84256s^2 + 1.17835s + 0.56789} \tag{6.110}$$

associated with the fourth-order elliptic response.

6.29 The transfer voltage ratio of a double-tuned network is given below. Determine its frequency response (magnitude and phase plots) by using both the Bode plots and the subroutine FRESP.

$$H(s) = \frac{0.5s^3}{(s + 4 + j200)(s + 4 - j200)(s + 4 + j215)(s + 4 - j215)} \tag{6.111}$$

6.30 Show that the minimum magnitude of the curves shown in Figure 6.42(a) occurs at the frequency

$$\omega_m/\sqrt{b} = \sqrt{1 - 2\gamma^2}, \qquad \gamma \leqq 1/\sqrt{2} \tag{6.112}$$

having the minimum magnitude

$$20 \log (2\gamma b\sqrt{1 - \gamma^2}) \quad \text{dB} \tag{6.113}$$

6.31 Applying the principle of superposition, verify that the inductor current i_2 in the network of Figure 6.10 is given by

$$i_2(t) = -1.548e^{-3t} + 2.096e^{-4t} + 0.713 \cos(2t - 2.448) \qquad \text{(6.114)}$$

for $i_g(t) = (\cos 2t)u(t)$ and $v_g(t) = 2[\cos(2t + \pi/4)] u(t)$

6.32 Repeat Problem 6.26 by using the subroutine FBODE.

6.33 Using the subroutine FBODE, repeat Problem 6.27.

6.34 Using the subroutine FBODE, repeat Problem 6.28.

6.35 Use subroutine FBODE to compute the frequency response of the transfer function $H(s)$ of (6.111).

CHAPTER SEVEN

INTEGRAL SOLUTION: CONVOLUTION

The principle of superposition was introduced in the preceding chapter and is intimately tied up with the concept of linearity. It states that for a linear system the zero-state response due to all the independent sources acting simultaneously is equal to the sum of the zero-state responses due to each independent source acting one at a time (Property 6.2). If, in addition, the system is time-invariant, the same holds in the frequency domain. Other versions of the superposition principle were established in Chapter 4 as Properties 4.4, 4.7, and 4.8, which state that in a linear time-invariant system the zero-input response is a linear function of the initial state, the zero-state response is a linear function of the input, and the complete response is the sum of the zero-input response and the zero-state response. For a given initial state, the Laplace transform of the zero-input response is a rational function. Hence its time-domain solution can be found routinely with partial-fraction expansions. When the system is initially relaxed, its behavior can be characterized by an appropriate transfer function. As a result, it is not even necessary that the system be given to compute the zero-state response, as long as its transfer function is known.

In this chapter we shall be concerned with the problem of determining the zero-state response of a system to an arbitrary driving source when the system is not given, but when its response to some standard input function is known.

7-1 THE CONVOLUTION THEOREM

In this section we shall derive the convolution theorem, leaving its applications to network problems to the following sections.

The Convolution Theorem   Let $F_1(s)$ and $F_2(s)$ be the Laplace transforms of the functions $f_1(t)$ and $f_2(t)$, which are zero for $t < 0$, respectively. The convolution $f(t)$ of $f_1(t)$ and $f_2(t)$, written as

$$f(t) = f_1(t) * f_2(t) \tag{7.1}$$

is defined by the convolution integral

$$f(t) \triangleq \int_{0-}^{t+} f_1(\tau) f_2(t - \tau) \, d\tau, \qquad t \geq 0 \tag{7.2}$$

Then the Laplace transform $F(s)$ of $f(t)$ is given by

$$F(s) = F_1(s) F_2(s) \tag{7.3}$$

Proof.   By definition, we have

$$F(s) = \int_{0-}^{\infty} f(t) e^{-st} \, dt = \int_{0-}^{\infty} \left[\int_{0-}^{t+} f_1(\tau) f_2(t - \tau) \, d\tau \right] e^{-st} \, dt \tag{7.4}$$

Recall that $f_1(t)$ and $f_2(t)$ are zero when their arguments are negative. Thus for a fixed value of t, $f_2(t - \tau)$ is identically zero for $\tau > t$, and we may extend the upper limit of the inner integral to infinity. Also our usual assumptions about the functions we transform are sufficient to permit the order of integration in (7.4) to be interchanged:

$$F(s) = \int_{0-}^{\infty} \left[\int_{0-}^{\infty} f_1(\tau) f_2(t - \tau) \, d\tau \right] e^{-st} \, dt$$

$$= \int_{0-}^{\infty} \left[\int_{0-}^{\infty} f_1(\tau) f_2(t - \tau) e^{-st} \, dt \right] d\tau \tag{7.5}$$

If we let $x = t - \tau$ so that

$$e^{-st} = e^{-s(x+\tau)} \tag{7.6}$$

the integrals (7.5) can be split as

$$F(s) = \int_{0-}^{\infty} f_1(\tau) e^{-s\tau} \left[\int_{0-}^{\infty} f_2(x) e^{-sx} dx \right] d\tau$$

$$= \left[\int_{0-}^{\infty} f_1(\tau) e^{-s\tau} \, d\tau \right] \left[\int_{0-}^{\infty} f_2(x) e^{-sx} \, dx \right] = F_1(s) F_2(s) \tag{7.7}$$

Equation (7.7) states that the Laplace transform of the convolution of two functions equals the product of their Laplace transforms. Observe that the limits of integration in (7.2) is from $0-$ to $t+$. This is deliberate for the following reasons. If $f_1(t)$ has an impulse at the origin, it must be included in the computation of the convolution integral (7.2). Likewise if $f_2(t)$ has an impulse at the origin, $f_2(t - \tau)$ as a function of τ has an impulse at $\tau = t$ and this impulse must be included in the computation of the convolution integral.

An alternative form of the convolution integral (7.2) can be obtained by a simple change of variable. As before, let $x = t - \tau$. The integral (7.2) can be rewritten as

$$f(t) = \int_{0-}^{t+} f_1(\tau) f_2(t - \tau) \, d\tau = -\int_{t+}^{0-} f_1(t - x) f_2(x) \, dx$$

$$= \int_{0-}^{t+} f_1(t - x) f_2(x) \, dx, \qquad t \geq 0 \tag{7.8}$$

Since x is the integration dummy variable, we can replace it by τ again and obtain the following two alternative forms:

$$f(t) = \int_{0-}^{t+} f_1(\tau)f_2(t-\tau)\,d\tau = \int_{0-}^{t+} f_1(t-\tau)f_2(\tau)\,d\tau \tag{7.9}$$

for $t \geq 0$.

EXAMPLE 7.1

We wish to compute the inverse Laplace transform of the proper rational function

$$F(s) = \frac{s}{(s^2+1)^2} \tag{7.10}$$

Let

$$F_1(s) = \frac{1}{s^2+1}, \qquad F_2(s) = \frac{s}{s^2+1} \tag{7.11}$$

Then $F(s) = F_1(s)F_2(s)$. From Table 5.1 the inverse Laplace transforms of $F_1(s)$ and $F_2(s)$ are found to be

$$f_1(t) = \mathscr{L}^{-1}[F_1(s)] = \sin t \tag{7.12a}$$

$$f_2(t) = \mathscr{L}^{-1}[F_2(s)] = \cos t \tag{7.12b}$$

By appealing to the convolution theorem, we obtain the inverse Laplace transform $f(t)$ of $F(s)$:

$$f(t) = \int_{0-}^{t+} f_1(\tau)f_2(t-\tau)\,d\tau = \int_0^t \sin\tau\,\cos(t-\tau)\,d\tau$$

$$= \int_0^t (\cos t \sin\tau \cos\tau + \sin t \sin^2\tau)\,d\tau$$

$$= \tfrac{1}{2}t \sin t \tag{7.13}$$

7-1.1   Properties of the Convolution Integral

Some properties of the convolution integral are derived in this section.

Property 7.1   *The operation of convolution is commutative, distributive, and associative:*

$$f_1(t) * f_2(t) = f_2(t) * f_1(t) \tag{7.14a}$$

$$f(t) * [f_1(t) + f_2(t) + \cdots + f_k(t)] = f(t) * f_1(t) + f(t) * f_2(t)$$
$$+ \cdots + f(t) * f_k(t) \tag{7.14b}$$

$$f_1(t) * [f_2(t) * f_3(t)] = [f_1(t) * f_2(t)] * f_3(t) \tag{7.14c}$$

We shall verify only (7.14c), leaving the other two as exercises. Let $G_1(s)$ and $G_2(s)$ be the Laplace transforms of the functions $g_1(t) = f_2(t) * f_3(t)$ and $g_2(t) = f_1(t) * f_2(t)$, respectively. By the convolution theorem, we have

$$G_1(s) = F_2(s)F_3(s), \qquad G_2(s) = F_1(s)F_2(s) \tag{7.15}$$

where $F_i(s)$ ($i = 1, 2, 3$) denote the Laplace transforms of $f_i(t)$. This gives

$$\begin{aligned}
\mathscr{L}\{f_1(t) * [f_2(t) * f_3(t)]\} &= \mathscr{L}[f_1(t) * g_1(t)] = F_1(s)G_1(s) \\
&= F_1(s)F_2(s)F_3(s) = G_2(s)F_3(s) = \mathscr{L}[g_2(t) * f_3(t)] \\
&= \mathscr{L}\{[f_1(t) * f_2(t)] * f_3(t)\}
\end{aligned} \tag{7.16}$$

Taking the inverse Laplace transform on both sides yields the desired identity (7.14c).

Property 7.2[†]   *If the functions $f_1(t)$ and $f_2(t)$ are differentiable for $t > 0$ and continuous for $t = 0$, then their convolution is differentiable for $t > 0$:*

$$\dot{f}(t) = \frac{df(t)}{dt} = \int_0^t f_1(\tau)\dot{f}_2(t - \tau)\, d\tau + f_1(t)f_2(0) \tag{7.17a}$$

$$= \int_0^t \dot{f}_1(t - \tau)f_2(\tau)\, d\tau + f_1(0)f_2(t), \qquad t > 0 \tag{7.17b}$$

where the dot indicates differentiation with respect to t.

To prove, we apply Leibnitz's rule for differentiation under an integral,[‡] which states that if

$$h(t) = \int_a^b g(t, \tau)\, d\tau \tag{7.18}$$

where a and b are differentiable functions of t and where $g(t, \tau)$ and $\partial g(t, \tau)/\partial t$ are continuous in t and τ, then

$$\dot{h}(t) = \int_a^b \frac{\partial g(t, \tau)}{\partial t}\, d\tau + g(t, b)\dot{b}(t) - g(t, a)\dot{a}(t) \tag{7.19}$$

Applying this to (7.9) with $h(t) = f(t)$, $g(t, \tau) = f_1(\tau)f_2(t - \tau)$, or $f_1(t - \tau)f_2(\tau)$, $a = 0-$ and $b = t+$, (7.17) follows.

Observe that from (7.17) we do not really need the hypothesis that both $f_1(t)$ and $f_2(t)$ are differentiable. In fact, if either function is differentiable and the other continuous, then their convolution is differentiable. In terms

---

[†] This property is stated in terms of the ordinary point functions and can be extended to include functions like the unit impulse known as a *distribution*. We will use this extension freely in the text and omit its justification.

[‡] See, for example, C. R. Wylie, Jr., *Advanced Engineering Mathematics*, 3d ed. (New York: McGraw-Hill, 1966), p. 274. The rule is named after Gottfried Wilhelm von Leibnitz (1646–1716).

of the convolution operation, (7.17) can be rewritten as

$$\dot{f}(t) = f_1(t) * \dot{f}_2(t) + f_1(t)f_2(0) = \dot{f}_1(t) * f_2(t) + f_1(0)f_2(t) \tag{7.20}$$

Property 7.3   Let $f(t) = f_1(t) * f_2(t)$ and write

$$g_1(t) = f_1(t - T_1)u(t - T_1), \qquad T_1 \geq 0 \tag{7.21a}$$

$$g_2(t) = f_2(t - T_2)u(t - T_2), \qquad T_2 \geq 0 \tag{7.21b}$$

$$g(t) = f(t - T_1 - T_2)u(t - T_1 - T_2) \tag{7.21c}$$

where $u(t)$ denotes the unit step function. Then

$$g(t) = g_1(t) * g_2(t) \tag{7.22}$$

We use convolution theorem to prove this property. As before, we use the capital letters to denote the Laplace transforms of the corresponding lowercase time functions. Using the shift theorem (5.53), we get

$$G_1(s) = e^{-sT_1}F_1(s), \qquad G_2(s) = e^{-sT_2}F_2(s) \tag{7.23}$$

and from the convolution theorem

$$\begin{aligned}
\mathscr{L}[g_1(t) * g_2(t)] &= G_1(s)G_2(s) = e^{-s(T_1 + T_2)}F_1(s)F_2(s) \\
&= e^{-s(T_1+T_2)}\mathscr{L}[f_1(t) * f_2(t)] = e^{-s(T_1+T_2)}\mathscr{L}[f(t)] \\
&= e^{-s(T_1+T_2)}F(s) = \mathscr{L}[f(t - T_1 - T_2)u(t - T_1 - T_2)]
\end{aligned} \tag{7.24}$$

The last equality follows again from the shift theorem (5.53). Taking the inverse Laplace transform on both sides yields (7.22).

Property 7.3 states that if the functions $f_1(t)$ and $f_2(t)$ are delayed by T_1 and T_2 sec, respectively. Then the convolution of the two delayed functions equals the convolution of the original functions, delayed by $T_1 + T_2$ sec.

We illustrate the above properties by the following examples.

EXAMPLE 7.2

The signal $f_1(t)$ of Figure 7.1(a) is to be convolved with the signal $f_2(t)$ of Figure 7.1(b). We do this by applying the above properties. The signal $f_1(t)$ can be expressed in terms of the unit-step functions as

$$f_1(t) = 2u(t - 2) - 2u(t - 6) \tag{7.25}$$

From Properties 7.1 and 7.3, we have

$$\begin{aligned}
f_1(t) * f_2(t) &= 2u(t - 2) * f_2(t) - 2u(t - 6) * f_2(t) \\
&= 2g(t - 2)u(t - 2) - 2g(t - 6)u(t - 6)
\end{aligned} \tag{7.26}$$

Figure 7.1

Two given rectangular pulse signals (a, b) and the convolution (c) of the unit step and the pulse signal shown in (b).

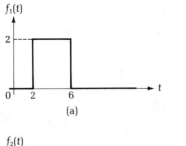

(a)

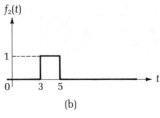

(b)

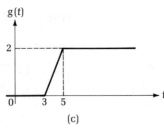

(c)

Figure 7.2

Signals (a, b) obtained from Figure 7.1(c) by scaling and delaying operations and their sum (c) to give the convolution of the rectangular pulse signals of Figure 7.1(a) and (b).

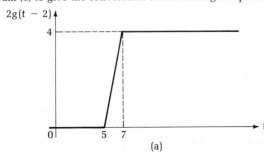

(a)

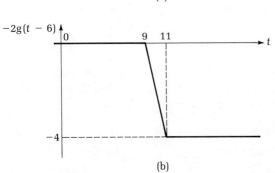

(b)

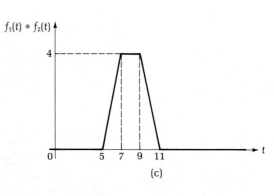

(c)

where

$$g(t) = u(t) * f_2(t) = \int_{0-}^{t+} u(t - \tau)f_2(\tau)\, d\tau = \int_{0-}^{t+} f_2(\tau)\, d\tau$$

$$= 0, \qquad t < 3$$
$$= t - 3, \qquad 3 \leq t \leq 5$$
$$= 2, \qquad t > 5 \tag{7.27}$$

A plot of $g(t)$ is shown in Figure 7.1(c). The functions $2g(t - 2)u(t - 2)$ and $-2g(t - 6)u(t - 6)$ are presented in Figure 7.2(a) and (b). According to (7.26), their sum, as shown in Figure 7.2(c), is the desired convolution of $f_1(t)$ and $f_2(t)$.

EXAMPLE 7.3

Consider again the signals $f_1(t)$ and $f_2(t)$ of Figure 7.1(a) and (b). Suppose that we take derivative of (7.25). From (5.38) we get

$$\dot{f}_1(t) = 2\delta(t - 2) - 2\delta(t - 6) \tag{7.28}$$

where $\delta(t)$ denotes the unit impulse function. We wish to compute the convolution of $\dot{f}_1(t)$ and $f_2(t)$.

First we show that the convolution of the unit impulse and any function $g(t)$ results in the same $g(t)$.

$$\delta(t) * g(t) = g(t) \tag{7.29}$$

Since the Laplace transform of $\delta(t)$ is 1, by convolution theorem we have

$$\delta(t) * g(t) = \mathscr{L}^{-1}[1G(s)] = \mathscr{L}^{-1}[G(s)] = g(t) \tag{7.30}$$

as asserted, where $G(s) = \mathscr{L}[g(t)]$. Using this and Property 7.3 gives $\delta(t) * f_2(t)$, $2\delta(t - 2) * f_2(t)$, and $-2\delta(t - 6) * f_2(t)$ as shown in Figure 7.3(a)–(c), respectively. Finally, applying Property 7.1 we obtain the desired convolution $\dot{f}_1(t) * f_2(t)$ as indicated in Figure 7.3(d), which is the sum of Figure 7.3(b) and (c). As a check, we apply (7.20)[†] and obtain

$$\dot{f}(t) = \dot{f}_1(t) * f_2(t) \tag{7.31}$$

where $f(t) = f_1(t) * f_2(t)$ was computed in Example 7.2 and is shown in Figure 7.2(c). It is straightforward to verify that $\dot{f}(t)$, being the slope of $f(t)$, is that shown in Figure 7.3(d).

7-1.2 Graphical Interpretation of the Convolution Integral

While the operation of the convolution is clear, it is important that we have a feeling for the steps described in the convolution integral. In this section

[†] See the footnote of Property 7.2.

Figure 7.3

The convolutions (a, b, c) of various impulse functions and the rectangular pulse signal $f_2(t)$ of Figure 7.1(b), and the addition of the two signals (b) and (c) to give the convolution (d) of the derivative of the rectangular pulse signal of Figure 7.1(a) and $f_2(t)$.

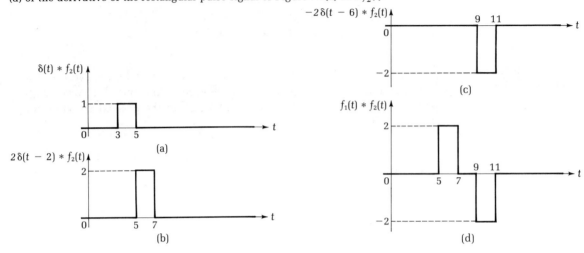

we show by examples that the convolution operation may be interpreted in terms of four steps: (1) folding, (2) translating, (3) multiplying, and (4) integrating. The process is equivalent to the German word *Faltung*, and therefore the convolution integral is also known as the *Faltung integral*.

Consider the signals $f_1(t)$ and $f_2(t)$ of Figure 7.4(a). We wish to determine $f(t) = f_1(t) * f_2(t)$ by its graphical interpretation. In Figure 7.4(b) we have $f_1(\tau)$ and $f_2(\tau)$. They are "folded" in (c) about the line $\tau = 0$ and "translated" in (d) for some typical but fixed value of t, say $t = 4$ s. In (e) of Figure 7.4, we perform the indicated multiplication within the integrals in (7.9). Finally, the integration of the cross-hatched area in (e) gives a point on the curve $f(t)$ for the value of t selected in step (d), which in the present situation is 4. At $t = 4$ s, the shaded area is 1, giving $f(4) = 1$ as indicated in (f). By repeating the above process for various values of t, we obtain the response $f(t)$ as shown in Figure 7.4(f).

As another example consider the signals $f_1(t)$ and $f_2(t)$ of Figure 7.5(a). With a change of variable, $f_1(\tau)$ and $f_2(\tau)$ are shown in (b), "folded" in (c) about the line $\tau = 0$, and "translated" in (d) for some fixed value of t, say $t = 2$ s. In (e) we perform the indicated multiplication within the integrals in (7.9). Finally, the integration of the shaded area in (e) gives a point on the curve of $f(t)$ at $t = 2$ s—that is, $f(2) = 0.8$, as indicated in Figure 7.5(f). If we carry out the above process for several values of t—say $t = 1, 2, 3, 4$, as depicted in Figure 7.6—we obtain the response $f(t)$ in (e). For illustrative

Figure 7.4
Graphical interpretation of the convolution of the unit step and a rectangular pulse signal.

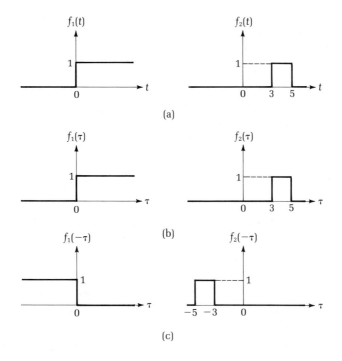

(a)

(b)

(c)

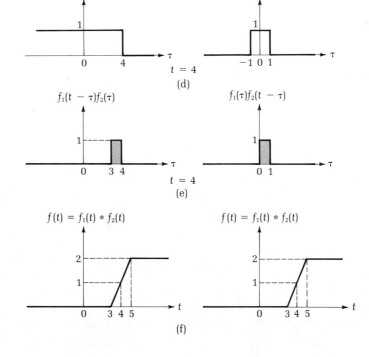

$f_1(t - \tau)$ $f_2(t - \tau)$

0   4 -1 0 1

$t = 4$

(d)

$f_1(t - \tau)f_2(\tau)$ $f_1(\tau)f_2(t - \tau)$

0   3 4   0 1

$t = 4$

(e)

$f(t) = f_1(t) * f_2(t)$ $f(t) = f_1(t) * f_2(t)$

0   3 4 5   0   3 4 5

(f)

Figure 7.5
Graphical interpretation
of the convolution of the
unit step and a ramp
function at $t = 2$ s.

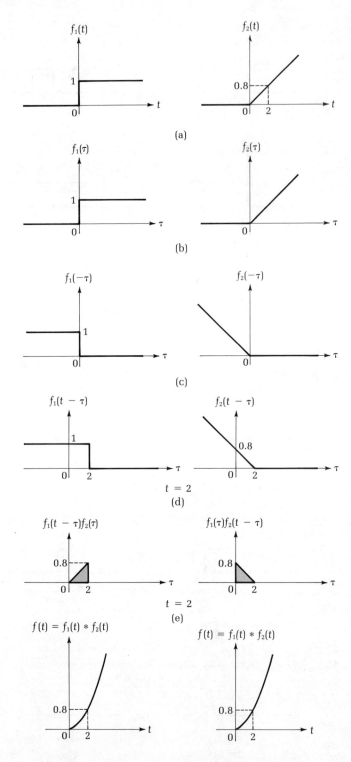

Figure 7.6

Graphical interpretation of the convolution of the unit step and a ramp function at (a) $t = 1$ s, (b) $t = 2$ s, (c) $t = 3$ s, and (d) $t = 4$ s, which are plotted in (e).

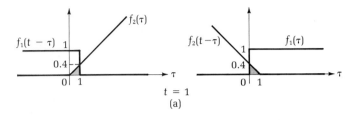

$t = 1$
(a)

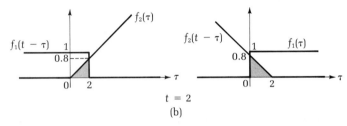

$t = 2$
(b)

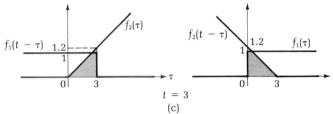

$t = 3$
(c)

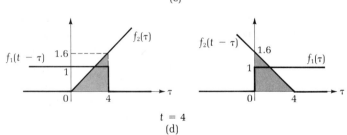

$t = 4$
(d)

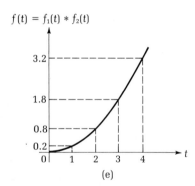

(e)

purposes we compute $f(t) = f_1(t) * f_2(t)$ as follows:

$$f(t) = 0.4 \int_0^t u(\tau)(t - \tau)u(t - \tau) \, d\tau = 0.4 \int_0^t u(t - \tau)\tau u(\tau) \, d\tau$$

$$= 0.2t^2 u(t) \tag{7.32}$$

confirming Figure 7.5(f) or 7.6(e).

Before we apply the convolution theorem to network problems, we interpret graphically the convolution of the unit impulse and any function $g(t)$. Figure 7.7(a) shows the unit impulse $\delta(t)$ to be convolved with a function

Figure 7.7
The convolution (c) of the unit impulse (a) and an arbitrary function (b) showing the sifting property of the unit impulse.

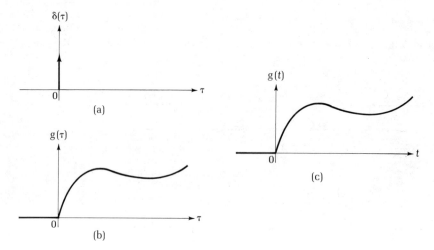

$g(t)$ in (b). As the impulse is folded and translated across $g(\tau)$, the product of the two curves is zero everywhere except at the point t, where the impulse is located. Due to the sifting property (5.39) of the impulse, the value of the integral of the product at the location of the impulse is $g(t)$. Therefore $\delta(t) * g(t)$ is the same as $g(t)$, as shown in (c).

7-2   THE ZERO-STATE RESPONSE

We now apply the convolution theorem to determine the zero-state response of a linear time-invariant network to an arbitrary driving source when its response to some standard function is known. The standard functions we shall consider will be the unit impulse, the unit step, and the sinusoidal functions.

An initially relaxed network N is driven by an excitation $e(t)$, which may either be a voltage source or a current source. Assume that the network is characterized by the pertinent transfer function $H(s)$. Our objective is to compute a particular response $r(t)$. As before, let $E(s)$ and $R(s)$ denote the Laplace transforms of $e(t)$ and $r(t)$. Then

$$R(s) = H(s)E(s) \qquad\qquad\qquad (7.33)$$

In the following we consider the responses due to several standard functions.

7-2.1   The Impulse Response

Let us consider the situation in which the excitation $e(t)$ is a unit impulse $\delta(t)$. The response of the network due to this excitation, written as $r_\delta(t)$, is called the *impulse response*. Since $\mathscr{L}[\delta(t)] = 1$, the Laplace transform of

$r_\delta(t)$ becomes

$$R_\delta(s) = H(s) \tag{7.34}$$

giving

$$r_\delta(t) = \mathscr{L}^{-1}[R_\delta(s)] = \mathscr{L}^{-1}[H(s)] \tag{7.35}$$

Equation (7.35) states that the inverse Laplace transform of the network transfer function equals the impulse response of the network.

Now consider the response $r(t)$ of the network to any excitation $e(t)$. Substituting (7.34) in (7.33), we obtain

$$R(s) = H(s)E(s) = R_\delta(s)E(s) \tag{7.36}$$

the inverse Laplace transform of which is the convolution of $r_\delta(t)$ and $e(t)$:

$$r(t) = \int_{0-}^{t+} r_\delta(\tau)e(t - \tau)\, d\tau = \int_{0-}^{t+} r_\delta(t - \tau)e(\tau)\, d\tau \tag{7.37}$$

This is a very important result. It states that once the impulse response is known, the zero-state response to any other driving source $e(t)$ is determined.

EXAMPLE 7.4

In the voltage-shunt feedback amplifier of Figure 7.8(a), assume that the transistor is represented by its hybrid-pi model with

$$
\begin{array}{lll}
g_m = 0.4 \text{ mho}, & r_x = 50 \ \Omega, & r_\pi = 250 \ \Omega \\
C_\pi = 195 \text{ pF}, & C_\mu = 5 \text{ pF}, & r_o = 50 \text{ k}\Omega
\end{array} \tag{7.38}
$$

Figure 7.8
A voltage-shunt feedback amplifier (a) and the network (b) used to compute the open-loop current ratio $-I_2(s)/I_g(s)$.

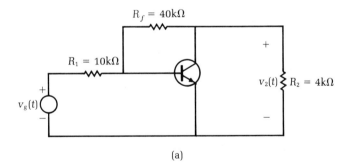

(a)

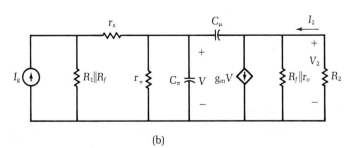

(b)

It can be shown that the open-loop current ratio $-I_2(s)/I_g(s)$ can be computed by the transform network of Figure 7.8(b). Using nodal analysis the open-loop current ratio is found to be

$$H(s) = -\frac{I_2(s)}{I_g(s)} = 12.74 \times 10^5 \frac{s - 8 \times 10^{10}}{(s + 5.85 \times 10^5)(s + 21.33 \times 10^8)} \qquad (7.39)$$

Because the impulse response equals the inverse Laplace transform of the transfer function $H(s)$, by partial-fraction expansion of $H(s)$ the impulse response of the network is found to be

$$r_\delta(t) = \mathscr{L}^{-1}\left[\left(\frac{-47.795}{s + 5.85 \times 10^5} + \frac{49.069}{s + 21.33 \times 10^8}\right)10^6\right]$$

$$= -47.795 \times 10^6 e^{-5.85 \times 10^5 t} + 49.069 \times 10^6 e^{-21.33 \times 10^8 t} \qquad (7.40)$$

Suppose that we wish to determine the response of the network due to a unit-step current. For this we let $e(t) = u(t)$, and from (7.37) we get

$$r(t) = \int_0^t r_\delta(\tau)u(t - \tau)\,d\tau = \int_0^t r_\delta(t - \tau)u(\tau)\,d\tau$$

$$= \int_0^t r_\delta(\tau)\,d\tau$$

$$= 81.702e^{-5.85 \times 10^5 t} - 0.023e^{-21.33 \times 10^8 t} - 81.679, \qquad t \geq 0 \quad (7.41)$$

7-2.2   The Step Response

In the preceding section, we demonstrated that the zero-state response of a network to an arbitrary excitation can be determined from a knowledge of the response of the same network to a unit impulse. In the present section, we show that the same conclusion can be reached from a knowledge of the response of the network to a unit-step excitation.

Let a unit-step excitation be applied to an initially relaxed network, and denote the response due to this unit step by $r_u(t)$, with transform $R_u(s)$. The response $r_u(t)$ is known as the *step response*. Since $\mathscr{L}[u(t)] = 1/s$, we have from (7.33)

$$R_u(s) = \frac{H(s)}{s} \qquad (7.42)$$

giving

$$r_u(t) = \mathscr{L}^{-1}\left[\frac{H(s)}{s}\right] = \mathscr{L}^{-1}\left[\frac{R_\delta(s)}{s}\right] \qquad (7.43)$$

or

$$r_u(t) = \int_0^t r_\delta(\tau)\,d\tau \qquad (7.44)$$

Thus the step response is the integral of the impulse response. On the other hand, if we write (7.42) as

$$R_\delta(s) = H(s) = sR_u(s) \tag{7.45}$$

we obtain

$$r_\delta(t) = \dot{r}_u(t) + r_u(0-)\,\delta(t) \tag{7.46}$$

To compute the zero-state response due to an arbitrary excitation, we rewrite (7.33) as

$$R(s) = s\left[\frac{H(s)}{s}\right]E(s) = [sR_u(s)]E(s) \tag{7.47}$$

Since

$$
\begin{aligned}
\mathscr{L}^{-1}[sR_u(s)] &= \mathscr{L}^{-1}\{[sR_u(s) - r_u(0-)] + r_u(0-)\} \\
&= \dot{r}_u(t) + r_u(0-)\,\delta(t)
\end{aligned} \tag{7.48}
$$

we can now use the convolution theorem to obtain the inverse Laplace transform of (7.47), yielding

$$
\begin{aligned}
r(t) &= \int_{0-}^{t+} [\dot{r}_u(\tau) + r_u(0-)\,\delta(\tau)]e(t-\tau)\,d\tau \\
&= \int_{0-}^{t+} \dot{r}_u(\tau)e(t-\tau)\,d\tau + r_u(0-)e(t) \tag{7.49a} \\
&= \int_{0-}^{t+} \dot{r}_u(t-\tau)e(\tau)\,d\tau + r_u(0-)e(t) \tag{7.49b}
\end{aligned}
$$

Likewise, if we write $R(s) = R_u(s)[sE(s)]$, we can show that

$$
\begin{aligned}
r(t) &= \int_{0-}^{t+} r_u(t-\tau)\dot{e}(\tau)\,d\tau + e(0-)r_u(t) \tag{7.50a} \\
&= \int_{0-}^{t+} r_u(\tau)\dot{e}(t-\tau)\,d\tau + e(0-)r_u(t) \tag{7.50b}
\end{aligned}
$$

The above expressions (7.49) and (7.50) were originally used in dynamics by DuHamel in 1833 and are known as the *DuHamel Integrals*[†] or the *Superposition Integrals*. They state in effect that the zero-state response of a network to an arbitrary excitation can be found simply from a knowledge of the step response.

EXAMPLE 7.5

In the network of Figure 7.9, the excitation is the current source $i_g(t)$ and the responses are the nodal voltages $v_{n1}(t)$ and $v_{n2}(t)$. As shown in Example 4.7, the zero-state nodal voltages to a unit-step current source $i_g(t) = u(t)$

[†] Named after Jean Marie Constant DuHamel (1797–1872).

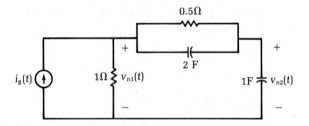

were found to be

$$v_{u1}(t) = v_{n1}(t)\big|_{i_g(t)=u(t)} = 1 - \tfrac{1}{3}e^{-0.5t} - \tfrac{2}{3}e^{-2t}, \qquad t \geq 0 \tag{7.51a}$$

$$v_{u2}(t) = v_{n2}(t)\big|_{i_g(t)=u(t)} = 1 - \tfrac{2}{3}e^{-0.5t} - \tfrac{1}{3}e^{-2t}, \qquad t \geq 0 \tag{7.51b}$$

We wish to compute the nodal voltages to the excitation

$$i_g(t) = \cos 2t, \qquad t \geq 0$$
$$= 0, \qquad t < 0 \tag{7.52}$$

To compute the nodal voltage $v_{n1}(t)$, we appeal to (7.50a), which results in

$$
\begin{aligned}
v_{n1}(t) &= \int_{0-}^{t+} v_{u1}(t-\tau)i_g(\tau)\,d\tau + i_g(0-)v_{u1}(t) \\
&= \int_0^t [1 - \tfrac{1}{3}e^{-0.5(t-\tau)} - \tfrac{2}{3}e^{-2(t-\tau)}](-2\sin 2\tau)\,d\tau \\
&\quad + 1[1 - \tfrac{1}{3}e^{-0.5t} - \tfrac{2}{3}e^{-2t}]u(t) \\
&= \tfrac{1}{102}[36\cos 2t + 42\sin 2t - 2e^{-0.5t} - 34e^{-2t}]u(t) \tag{7.53}
\end{aligned}
$$

Likewise we can compute the nodal voltage $v_{n2}(t)$ by (7.50a). For a change we shall use (7.49b). First we compute the derivative of $v_{u2}(t)$ with respect to t, obtaining from (7.51b)

$$\dot{v}_{u2}(t) = \tfrac{1}{3}e^{-0.5t} + \tfrac{2}{3}e^{-2t}, \qquad t \geq 0 \tag{7.54}$$

Using this in (7.49b), we obtain

$$
\begin{aligned}
v_{n2}(t) &= \int_{0-}^{t+} \dot{v}_{u2}(t-\tau)i_g(\tau)\,d\tau + v_{u2}(0-)i_g(t) \\
&= \int_0^t [\tfrac{1}{3}e^{-0.5(t-\tau)} + \tfrac{2}{3}e^{-2(t-\tau)}](\cos 2\tau)\,d\tau + 0\cos 2t \\
&= \tfrac{1}{102}[21\cos 2t + 33\sin 2t - 17e^{-2t} - 4e^{-0.5t}]u(t) \tag{7.55}
\end{aligned}
$$

The above results are the same as those obtained in (4.216).

7-2.3 Representation of the Zero-State Response

In the preceding two sections, we have demonstrated that the zero-state response of a network to an arbitrary excitation can be expressed in terms of the impulse or step response and the input. This *representation* of the zero-state solution is given in the form of a convolution. The restriction to impulse and step functions was really unnecessary. In fact, we can represent the zero-state response in terms of the response of the network to any driver and the input. To see this let us drive the network by our old familiar sine function with $e(t) = \sin \omega t$. Let $r_s(t)$ denote the corresponding response, with Laplace transform $R_s(s)$. We emphasize that this zero-state response includes both the steady state and transient, not merely the steady state. Since $\mathscr{L}[\sin \omega t] = \omega/(s^2 + \omega^2)$, (7.36) becomes

$$R_s(s) = \frac{\omega}{s^2 + \omega^2} R_\delta(s) \tag{7.56}$$

Solving for $R_\delta(s)$ yields

$$R_\delta(s) = \frac{s^2 + \omega^2}{\omega} R_s(s) \tag{7.57}$$

giving

$$r_\delta(t) = \mathscr{L}^{-1}\left[\frac{s^2 + \omega^2}{\omega} R_s(s)\right] \tag{7.58}$$

Thus knowing the response of the network to a sine function we know the impulse response $r_\delta(t)$. Once we know the impulse response, we can determine the response to any arbitrary excitation. Going one step further, we do not have to use a sine function; any Laplace-transformable function would do. Hence the following uniqueness property can now be stated:

> **Property 7.4** *If two linear time-invariant systems have the same zero-state response to a Laplace-transformable input function, then their zero-state responses to any arbitrary Laplace-transformable input excitation will be the same.*

We illustrate the above result by the following example.

EXAMPLE 7.6

The response of a network to the excitation

$$\begin{aligned} e(t) &= 0, & t &< 0 \\ &= \cos t, & t &\geq 0 \end{aligned} \tag{7.59}$$

is found to be

$$r(t) = \tfrac{1}{82}(9 \cos t - \sin t + \tfrac{1}{9}e^{-t/9}), \qquad t \geq 0 \tag{7.60}$$

We wish to determine the unit-step response of the network.

The Laplace transforms of $e(t)$ and $r(t)$ are given by

$$E(s) = \frac{s}{s^2 + 1} \tag{7.61}$$

$$R(s) = \frac{1}{82}\left(\frac{9s}{s^2 + 1} - \frac{1}{s^2 + 1} + \frac{1}{9s + 1}\right)$$

$$= \frac{s^2}{(s^2 + 1)(9s + 1)} \tag{7.62}$$

From (7.36) we obtain the impulse response of the network as follows:

$$r_\delta(t) = \mathscr{L}^{-1}[R_\delta(s)] = \mathscr{L}^{-1}\left[\frac{R(s)}{E(s)}\right]$$

$$= \mathscr{L}^{-1}\left[\frac{s^2}{(s^2 + 1)(9s + 1)} \cdot \frac{s^2 + 1}{s}\right] = \mathscr{L}^{-1}\left(\frac{s}{9s + 1}\right)$$

$$= \tfrac{1}{9}\delta(t) - \tfrac{1}{81}e^{-t/9}, \qquad t \geq 0 \tag{7.63}$$

Knowing the impulse response, we can obtain the step response by a simple integration:

$$r_u(t) = \int_{0-}^{t+} r_\delta(\tau)\,d\tau = \int_{0-}^{t}\left[\frac{\delta(\tau)}{9} - \frac{e^{-\tau/9}}{81}\right]d\tau$$

$$= \tfrac{1}{9}e^{-t/9}u(t) \tag{7.64}$$

7-3   THE PRINCIPLE OF SUPERPOSITION

In the foregoing we have related the zero-state response of a network to the impulse response or the step response through a convolution integral. In the present section, we show that these integrals can be interpreted as the superposition of the impulse or step responses of various strengths, suitably displaced.

Consider the excitation $e(t)$ as shown in Figure 7.10. We approximate this excitation by a series of step functions, as illustrated in the figure. The staircase function can be represented by a series of step functions, suitably displaced and up to a temporarily fixed time $t = n\Delta\tau$:

$$e(\hat{t}) \cong e(0+)u(\hat{t}) + \Delta e_1 u(\hat{t} - \Delta\tau) + \Delta e_2 u(\hat{t} - 2\Delta\tau) + \cdots$$

$$+ \Delta e_n u(\hat{t} - n\Delta\tau), \qquad 0 \leq \hat{t} \leq t \tag{7.65}$$

where Δe_k is the height of the step increment at $t = k\Delta\tau$. Since the network is linear and time invariant, the response due to a series of step inputs is equal to the sum of responses due to each step input. If $r_u(t)$ is the response to the unit step $u(t)$, then the response to a step $\Delta e_k u(t - k\Delta\tau)$ will be

Figure 7.10

The approximation of an excitation by a series of step functions.

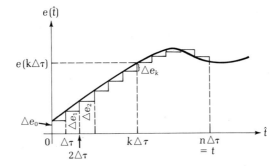

$\Delta e_k r_u(t - k\Delta\tau)$. Therefore we can write the response to the input approximated by the staircase function $e(\hat{t})$ at time $t = n\Delta\tau$ as

$$r(t) \cong e(0+)r_u(t) + \Delta e_1 r_u(t - \Delta\tau) + \Delta e_2 r_u(t - 2\Delta\tau)$$
$$+ \cdots + \Delta e_n r_u(t - n\Delta\tau) \tag{7.66}$$

$$= e(0+)r_u(t) + \frac{\Delta e_1}{\Delta\tau}\, r_u(t - \Delta\tau)\, \Delta\tau + \frac{\Delta e_2}{\Delta\tau}\, r_u(t - 2\Delta\tau)\, \Delta\tau$$

$$+ \cdots + \frac{\Delta e_n}{\Delta\tau}\, r_u(t - n\Delta\tau)\, \Delta\tau \tag{7.67}$$

If $\Delta\tau$ is small, $\Delta e_k/\Delta\tau$ can be approximated by the slope of $e(t)$ at $k\Delta\tau$,

$$\frac{\Delta e_k}{\Delta\tau} \cong \left.\frac{de(t)}{dt}\right|_{t=k\Delta\tau} = \dot{e}(k\Delta\tau) \tag{7.68}$$

Using this in (7.67) yields

$$r(t) \cong e(0+)r_u(t) + \dot{e}(\Delta\tau)r_u(t - \Delta\tau)\,\Delta\tau + \dot{e}(2\Delta\tau)r_u(t - 2\Delta\tau)\,\Delta\tau$$
$$+ \cdots + \dot{e}(n\Delta\tau)r_u(t - n\Delta\tau)\,\Delta\tau$$

$$= e(0+)r_u(t) + \sum_{k=1}^{n} \dot{e}(k\Delta\tau)r_u(t - k\Delta\tau)\,\Delta\tau \tag{7.69}$$

Let us now concentrate on a particular point $\tau = k\Delta\tau$. If we let $\Delta\tau$ get smaller, we will have to increase k proportionately so that the product $k\Delta\tau$ will stay the same. In the limit as $\Delta\tau$ approaches zero, (7.69) becomes

$$r(t) = e(0+)r_u(t) + \lim_{\substack{\Delta\tau\to 0 \\ n\to\infty}} \sum_{k=1}^{n} \dot{e}(k\Delta\tau)r_u(t - k\Delta\tau)\,\Delta\tau$$

$$= e(0+)r_u(t) + \int_{0+}^{t} \dot{e}(\tau)r_u(t - \tau)\, d\tau \tag{7.70}$$

In the development we have assumed that the excitation $e(t)$ is continuous for $t > 0$. In the case where it has discontinuities of values E_j occurring at times T_j, respectively, the excitation can then be written as

$$e(t) = \hat{e}(t) + \sum_j E_j u(t - T_j) \tag{7.71}$$

where $\hat{e}(t)$ is the continuous part of the excitation for $t > 0$. Adding the response due to the discontinuities to the response already found for the continuous part, the total response becomes

$$r(t) = \sum_j E_j r_u(t - T_j) + \int_{0+}^t \dot{\hat{e}}(\tau) r_u(t - \tau)\, d\tau \tag{7.72}$$

where $j = 0, 1, 2, \ldots$, with $E_0 = e(0+)$ and $T_0 = 0$. In particular if $e(t)$ has no discontinuities except at $t = 0$, (7.72) reduces to (7.70), which under the stipulated condition is identical to (7.50a). Thus the zero-state response to an arbitrary excitation $e(t)$ can be regarded as the superposition of the responses to a series of step functions that approximate the excitation.

A similar development can now be carried out by representing the excitation $e(t)$ as a sum of impulse functions of various strengths. To this end we divide the time axis into equal intervals of length $\Delta\tau$. Express the excitation $e(t)$ as a sum

$$e(t) = e_0(t) + e_1(t) + \cdots + e_k(t) + \cdots \tag{7.73}$$

where each component $e_k(t)$ equals $e(t)$ in the interval $k\Delta\tau \leq t < (k + 1)\Delta\tau$ and equals zero otherwise, as depicted in Figure 7.11.

Figure 7.11
The representation (a) of an excitation as a sum of pulses of various shapes and strengths with kth such pulse shown in (b).

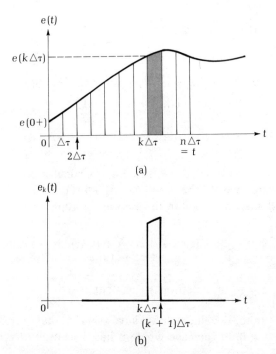

Figure 7.12

The representation of an excitation by a sequence of impulses of various strengths.

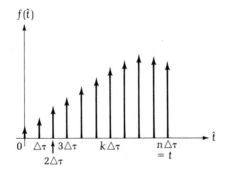

Now consider a sequence of impulses as shown in Figure 7.12 at the points $0, \Delta\tau, \ldots, n\Delta\tau$. The strength of the impulse at the point $k\Delta\tau$ is $\Delta\tau e(k\Delta\tau)$, being equal to the area of a rectangle formed by the base $\Delta\tau$ and the height $e(k\Delta\tau)$ of the curve $e(t)$ at the point $k\Delta\tau$. If $\Delta\tau$ is sufficiently small, $\Delta\tau e(k\Delta\tau)$ approximates the area under $e_k(t)$, as shown by the cross-hatched area in Figure 7.11(a). The heights of the arrows in Figure 7.12 were drawn proportional to the strengths of the impulses, even though all the impulses are of infinite height. The sequence of impulses of Figure 7.12 can be represented by

$$f(\hat{t}) = [e(0+)\,\Delta\tau]\,\delta(\hat{t}) + [e(\Delta\tau)\,\Delta\tau]\,\delta(\hat{t} - \Delta\tau) + \cdots$$
$$+ [e(n\Delta\tau)\,\Delta\tau]\,\delta(\hat{t} - n\Delta\tau), \qquad 0 \le \hat{t} \le t \tag{7.74}$$

If $r_\delta(t)$ is the response to the unit impulse $\delta(t)$, then the response to an impulse $[e(k\Delta\tau)\,\Delta\tau]\,\delta(t - k\Delta\tau)$ will be $[e(k\Delta\tau)\,\Delta\tau]r_\delta(t - k\Delta\tau)$. Therefore we can write the response to the sequence $f(\hat{t})$ at time $t = n\Delta\tau$ as

$$r(t) = [e(0+)\,\Delta\tau]r_\delta(t) + [e(\Delta\tau)\,\Delta\tau]r_\delta(t - \Delta\tau) + \cdots$$
$$+ [e(n\Delta\tau)\,\Delta\tau]r_\delta(t - n\Delta\tau) \tag{7.75}$$

As before, let us concentrate on a particular point $\tau = k\Delta\tau$ by diminishing $\Delta\tau$ and at the same time increasing k proportionately so that the product $k\Delta\tau$ remains the same. In the limit as $\Delta\tau$ approaches zero, (7.74) becomes

$$f(\hat{t}) = \lim_{\substack{\Delta\tau \to 0 \\ n \to \infty}} \sum_{k=0}^{n} e(k\Delta\tau)\,\delta(\hat{t} - k\Delta\tau)\,\Delta\tau$$

$$= \int_{0+}^{t} e(\tau)\,\delta(\hat{t} - \tau)\,d\tau = \int_{0+}^{t} e(\hat{t} - \tau)\,\delta(\tau)\,d\tau$$

$$= e(\hat{t}) \tag{7.76}$$

and (7.75) becomes the zero-state response to the excitation $e(t)$:

$$r(t) = \lim_{\substack{\Delta\tau \to 0 \\ n \to \infty}} \sum_{k=0}^{n} e(k\Delta\tau)r_\delta(t - k\Delta\tau)\,\Delta\tau$$

$$= \int_{0+}^{t} e(\tau)r_\delta(t - \tau)\,d\tau \tag{7.77}$$

Our conclusion is that we can interpret the convolution integrals in (7.37) as expressing the zero-state response of a network to an excitation as the superposition of the zero-state responses to a sequence of impulses representing the input.

7-4 NUMERICAL CONVOLUTION

One of the important applications of the preceding interpretations of the convolution integrals is its use in the numerical computation of network response.

Given a linear time-invariant network, suppose that we wish to compute the zero-state response of the network to an input excitation that cannot be represented as a sum of simple functions. Or suppose its analytic representation may be very complicated. For instance, the input may be given as a curve or a set of points. In such cases the Laplace transform of the input and the resulting inverse Laplace transform of the transform response will be very difficult to find. It is more meaningful to approximate the input by a sequence of impulse or step functions.

Let $r_\delta(t)$ be the impulse response of the network. This impulse response may have been found experimentally by using a short pulse as a good approximation to the unit impulse. We now select a suitable interval T so that the variation of the input $e(t)$ and $r_\delta(t)$ over the interval is sufficiently small to be negligible. Then we can approximate the input by the expression

$$\tilde{e}(t) = \sum_k e(kT)T\,\delta(t - kT) \tag{7.78}$$

The function $\tilde{e}(t)$ is referred to as a *time series*. Now, appealing to the convolution theorem, the zero-state response of the network to the time series (7.78) can be written as

$$r(t) = \sum_{k=0}^{n} e(kT)r_\delta(t - kT)T, \qquad nT \leq t < (n + 1)T \tag{7.79}$$

In particular the response at the point nT is given by

$$r(nT) = \sum_{k=0}^{n} e(kT)r_\delta[(n - k)T]T \tag{7.80}$$

Finally, the zero-state response can then be approximated by a time series

$$\tilde{r}(t) = \sum_{n}\sum_{k=0}^{n} e(kT)r_\delta[(n - k)T]T\,\delta(t - nT) \tag{7.81}$$

We illustrate the above procedure by the following examples.

Figure 7.13

The input (a) to a network having the unit impulse response approximated by (b).

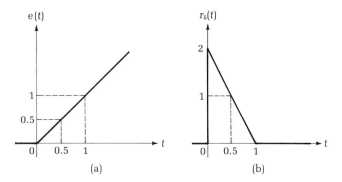

(a)   (b)

EXAMPLE 7.7

We use (7.80) to compute the approximate response of a network the input of which is shown in Figure 7.13(a) and the unit impulse response is approximated by Figure 7.13(b).

To illustrate the above procedure, let us first compute the approximate response at $t = 0.5$ s. For our purposes we choose the interval $T = 0.05$, which is too large for any accuracy but suffices to illustrate the point.

First we tabulate the values of $e(kT)$ and $r_\delta(kT)$ from Figure 7.13 and arrange them in rows as shown in Table 7.1. Rows 3 and 4 are obtained from Figure 7.13. Row 5 corresponds to folding and translating of $r_\delta(kT)$ and is found simply by copying row 4 backwards. The last row is formed by multiplying the corresponding entries in rows 3 and 5.

Finally, according to (7.80) the approximate zero-state response at $t = 10T = 0.5$ s is simply the sum of entries in the last row times the interval width T:

$$r(0.5) = 4.70 \times 0.05 = 0.235 \tag{7.82}$$

For other values of t, the whole process has to be repeated. However, there is a simple scheme for carrying out numerical convolution for many values of t. Suppose that we wish to compute the response for $t = 0, 0.05,$

Table 7.1

k	0	1	2	3	4	5	6	7	8	9	10
kT	0.00	0.05	0.10	0.15	0.20	0.25	0.30	0.35	0.40	0.45	0.50
$e(kT)$	0.00	0.05	0.10	0.15	0.20	0.25	0.30	0.35	0.40	0.45	0.50
$r_\delta(kT)$	2.00	1.90	1.80	1.70	1.60	1.50	1.40	1.30	1.20	1.10	1.00
$r_\delta[(10 - k)T]$	1.00	1.10	1.20	1.30	1.40	1.50	1.60	1.70	1.80	1.90	2.00
$e(kT)r_\delta[(10 - k)T]$	0.00	0.06	0.12	0.20	0.28	0.38	0.48	0.60	0.72	0.86	1.00

Table 7.2

k	0	1	2	3	4	5	6	7	8	9	10
kT	0.00	0.05	0.10	0.15	0.20	0.25	0.30	0.35	0.40	0.45	0.50
$e(kT)$	0.00	0.05	0.10	0.15	0.20	0.25	0.30	0.35	0.40	0.45	0.50
$r_\delta(kT)$	2.00	1.90	1.80	1.70	1.60	1.50	1.40	1.30	1.20	1.10	1.00
	0.00	0.10	0.20	0.30	0.40	0.50	0.60	0.70	0.80	0.90	1.00
		0.00	0.095	0.19	0.285	0.38	0.475	0.57	0.665	0.76	0.855
			0.00	0.09	0.18	0.27	0.36	0.45	0.54	0.63	0.72
				0.00	0.085	0.17	0.255	0.34	0.425	0.51	0.595
					0.00	0.08	0.16	0.24	0.32	0.40	0.48
						0.00	0.075	0.15	0.225	0.30	0.375
							0.00	0.07	0.14	0.21	0.28
								0.00	0.065	0.13	0.195
									0.00	0.06	0.12
										0.00	0.055
											0.00
$r(kT)$	0.00	0.05	0.015	0.029	0.048	0.07	0.096	0.126	0.159	0.195	0.235
exact	0.00	0.003	0.01	0.021	0.037	0.057	0.081	0.108	0.139	0.172	0.208

0.10, ..., 0.5. As before, we tabulate the values of $e(kT)$ and $r_\delta(kT)$ from Figure 7.13 with $T = 0.05$ and arrange them in rows, as shown in Table 7.2. Row 5 to row 15 are obtained by multiplying the entries of row 3 by the entries of row 4, one at a time. The multiplication starts at the first element of row 4 and proceeds to the right, and the beginning of each row of products is immediately below the fourth-row element that is multiplying the third row. For instance, row 5 is obtained by multiplying row 3 by the first element of row 4, which is 2. Row 6 is found by multiplying row 3 by the second element of row 4, which is 1.90, and the first element of the products is

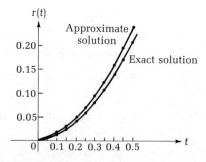

Figure 7.14
The exact and approximate solutions for the network that has the input excitation and impulse response shown in Figure 7.13.

located immediately below the entry 1.90. The zero-state response is obtained by summing the columns from row 5 to row 15 and multiplying by the interval width $T = 0.05$. The result is shown in the row next to the last row. The exact values of the response are tabulated in the last row. Both the exact solution and the approximate solution are shown in Figure 7.14. For a more accurate response, a smaller interval width is required.

For illustrative purposes we compute the analytic convolution of $e(t)$ and $r_\delta(t)$. The input $e(t)$ and impulse response $r_\delta(t)$ can be written analytically as

$$e(t) = tu(t) \tag{7.83}$$

$$r_\delta(t) = 2(1 - t)[u(t) - u(t - 1)] \tag{7.84}$$

yielding

$$r(t) = \int_0^t (t - \tau)u(t - \tau)2(1 - \tau)[u(\tau) - u(\tau - 1)]\, d\tau$$

$$= 2 \int_0^t (t - \tau)(1 - \tau)\, d\tau - 2u(t - 1) \int_1^t (t - \tau)(1 - \tau)\, d\tau$$

$$= t^2\left(1 - \frac{t}{3}\right)[u(t) - u(t - 1)] + (t - \tfrac{1}{3})u(t - 1) \tag{7.85}$$

EXAMPLE 7.8

The impulse response of the open-loop equivalent network of Figure 7.8(b) is shown in Figure 7.15, where the zero-state response is the output current $i_2(t) = \mathcal{L}^{-1}[I_2(s)]$ as indicated in Figure 7.8(b). Suppose that we wish to determine the response to a step-input current source $i_g(t) = \mathcal{L}^{-1}[I_g(s)] = 0.01u(t)$ at $t = 1\ \mu s$.

As in the previous example, we tabulate the values of $r_\delta(t)$ and $i_g(t)$ at the points $t = 0, T, 2T, \ldots, 10T$, with T chosen as 10^{-7}, and arrange them in rows as shown in Table 7.3. The zero-state response to the input $i_g(t) = 0.01u(t)$ at $t = 1\ \mu s$ is simply the sum of entries in the last row times $10^4\, T$:

$$i_2(10^{-6}) = -0.35\ \text{A} \tag{7.86}$$

As a check we compute $i_2(10^{-6})$ from (7.41), obtaining

$$i_2(10^{-6}) = -0.36\ \text{A} \tag{7.87}$$

Figure 7.15
The impulse response of the open-loop equivalent network of Figure 7.8(b).

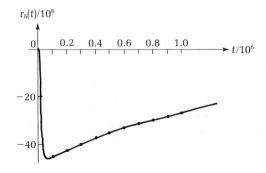

Table 7.3

k	0	1	2	3	4	5	6	7	8	9	10
$kT \times 10^6$	0.0	0.1	0.2	0.3	0.4	0.5	0.6	0.7	0.8	0.9	1.0
$r_\delta(kT)10^{-6}$	1.27	−45.1	−42.5	−40.1	−37.8	−35.7	−33.6	−31.7	−29.9	−28.2	−26.6
$i_g(kT)$	0.01	0.01	0.01	0.01	0.01	0.01	0.01	0.01	0.01	0.01	0.01
$i_g[(10-k)T]$	0.01	0.01	0.01	0.01	0.01	0.01	0.01	0.01	0.01	0.01	0.01
$r_\delta(kT) \times 10^{-4}$ $i_g[(10-k)T]$	1.27	−45.1	−42.5	−40.1	−37.8	−35.7	−33.6	−31.7	−29.9	−28.2	−26.6

Figure 7.16
The input to the network of Figure 7.8(b)

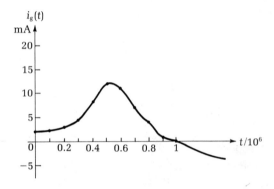

Table 7.4

k	0	1	2	3	4	5	6	7	8	9	10
$kT \times 10^6$	0	0.1	0.2	0.3	0.4	0.5	0.6	0.7	0.8	0.9	1.0
$r_\delta(kT)10^{-6}$	1.3	−45.1	−42.5	−40.1	−37.8	−35.7	−33.6	−31.7	−29.9	−28.2	−26.6
$i_g(kT)10^3$	2	2.2	2.8	4.5	8.2	12.1	11.6	7.0	4.5	0.5	0.0
$i_g[(10-k)T]10^3$	0	0.5	4.5	7.0	11.6	12.1	8.2	4.5	2.8	2.2	2.0
$i_g[(10-k)T]$ $r_\delta(kT)10^{-5}$	0	−0.23	−1.91	−2.81	−4.38	−4.32	−2.76	−1.43	−0.84	−0.62	−0.53

Now suppose that the network is driven by the current excitation of it has to be multiplied by a factor of 0.01.

Now suppose that the network is driven by the current excitation of Figure 7.16. We wish to determine the response $i_2(t)$ at $t = 1\ \mu s$. For our purposes we choose the time interval $T = 10^{-7}$ and tabulate the values of $r_\delta(t)$ and $i_g(t)$ at $t = 0, T, 2T, \ldots, 10T$, as shown in Table 7.4. Note that rows 4 and 5 are given in milliampere. The zero-state response to the input current of Figure 7.16 is the sum of entries in the last row times $10^5 T$, giving

$$i_2(10^{-6}) = -0.198 \text{ A} \tag{7.88}$$

7-5 COMPUTER SOLUTIONS AND PROGRAMS

In this section we shall present computer programs in WATFIV that perform convolution of two functions.

7-5.1 Subroutine CONV

1. Purpose. The program is designed to perform convolution $f(t)$ of two given functions $f_1(t)$ and $f_2(t)$ and is listed in Figure 7.17.

2. Method. The program is based on the numerical convolution discussed in Section 7-4.

Figure 7.17
Subroutine CONV designed to perform convolution of two given functions.

```
C
C
C      SUBROUTINE CONV
C
C      PURPOSE
C          THE PROGRAM IS DESIGNED TO PERFORM CONVOLUTION OF TWO FUNCTIONS.
C
C      METHOD
C          THE PROGRAM IS BASED ON THE CONVOLUTION INTEGRAL.
C
C      USAGE
C          CALL CONV (T,H,FT)
C
C          T      -THE TIME
C          H      -THE STEPSIZE
C          FT     -THE VALUE OF THE CONVOLUTION INTEGRAL AT TIME T
C
C      SUBROUTINES AND FUNCTION SUBPROGRAMS REQUIRED
C          FUNCTION FUNCT1(X), PROVIDED BY THE USER.
C          FUNCTION FUNCT2(X), PROVIDED BY THE USER.
C
C      REMARKS
C          DOUBLE PRECISION IS USED IN ALL THE COMPUTATION.
C          THE INPUT DATA ARE T AND H.  THE OUTPUT IS FT.
C
       SUBROUTINE CONV (T,H,FT)
       DOUBLE PRECISION T,H,FT,F1(2),F2(2),DLAMDA,FUNCT1,FUNCT2
       FT=0.0
       DLAMDA=0.0
       F1(1)=FUNCT1(DLAMDA)
       F2(1)=FUNCT2(T-DLAMDA)
    1  DLAMDA=DLAMDA+H
       IF (DLAMDA-T) 2,2,3
    2  F1(2)=FUNCT1(DLAMDA)
       F2(2)=FUNCT2(T-DLAMDA)
       FT=FT+0.5*(F1(2)*F2(2)+F1(1)*F2(1))*H
       F1(1)=F1(2)
       F2(1)=F2(2)
       GO TO 1
    3  CONTINUE
       RETURN
       END
```

3. Usage. The program consists of a subroutine, CONV, called by

$$\text{CALL CONV (T, H, FT)}$$

T   is the time t

H   is the stepsize or interval width used in the integration

FT   is the value of the convolution integral at time $\text{T} = t$

4. Remarks. Double precision is used in all the computations. The input data are T and H. The output is FT.

5. Subroutines and Function Subprograms Required. FUNCTION FUNCT1 (X), provided by the user, is the given $f_1(t)$. FUNCTION FUNCT2 (X), provided by the user, is the given $f_2(t)$.

Figure 7.18

The main program used to perform convolution of two given functions using subroutines CONV and PLOT.

```
      DOUBLE PRECISION T,TF,DT,H,FT
      DIMENSION XY(200,2),JXY(2)
      READ,TF,DT,H
      I=1
      T=0.0
1     CALL CONV (T,H,FT)
      PRINT 2,T,FT
2     FORMAT (3H T=,D15.8,10X,5HF(T)=,D20.12)
      XY(I,1)=T
      XY(I,2)=FT
      I=I+1
      T=T+DT
      IF (T-TF) 1,1,3
3     JXY(1)=1
      JXY(2)=2
      N=I-1
      CALL PLOT (XY,JXY,N,200,1)
      STOP
      END
```

EXAMPLE 7.9

We use the subroutine CONV to obtain the convolution of the functions

$$f_1(t) = \tfrac{1}{2}(\sinh 2t)u(t) \tag{7.89a}$$

$$f_2(t) = (\cosh 3t)u(t) \tag{7.89b}$$

Figure 7.19

Function subprograms for functions (7.89).

```
FUNCTION FUNCT1 (X)
DOUBLE PRECISION X,FUNCT1
FUNCT1=0.5*DSINH(2.0*X)
RETURN
END

FUNCTION FUNCT2 (X)
DOUBLE PRECISION X,FUNCT2
FUNCT2=DCOSH(3.0*X)
RETURN
END
```

for t ranged from 0 to 2. For our purposes we choose the stepsize H = 0.005 and $t = 0, 0.05, 0.1, \ldots, 2$.

The main program and the required function subprograms are presented in Figures 7.18 and 7.19, respectively. The computer output is shown in Figures 7.20 and 7.21. Figure 7.20 is the printout of $f(t) = f_1(t) * f_2(t)$ for various values of t. The same function is plotted in Figure 7.21 as a function of t. The total execution time for an IBM 370 Model 158 computer was 9.74 sec.

Suppose now that we increase the stepsize from 0.005 to 0.01, everything else being the same. The computer time was decreased from 9.74 sec to

Figure 7.20
Computer printout of the convolution of the two functions (7.89).

T= 0.00000000D 00	F(T)= 0.000000000000D 00		
T= 0.50000000D-01	F(T)= 0.125337524821D-02		
T= 0.10000000D 00	F(T)= 0.505429905314D-02		
T= 0.15000000D 00	F(T)= 0.115262108210D-01		
T= 0.20000000D 00	F(T)= 0.208783517985D-01		
T= 0.25000000D 00	F(T)= 0.334111158625D-01		
T= 0.30000000D 00	F(T)= 0.495237175705D-01		
T= 0.35000000D 00	F(T)= 0.697243592976D-01		
T= 0.40000000D 00	F(T)= 0.946431383403D-01		
T= 0.45000000D 00	F(T)= 0.125047999464D 00		
T= 0.50000000D 00	F(T)= 0.161864110005D 00		
T= 0.55000000D 00	F(T)= 0.206197114934D 00		
T= 0.60000000D 00	F(T)= 0.259360820130D 00		
T= 0.65000000D 00	F(T)= 0.322909955515D 00		
T= 0.70000000D 00	F(T)= 0.398678788110D 00		
T= 0.75000000D 00	F(T)= 0.488826491114D 00		
T= 0.80000000D 00	F(T)= 0.595890331979D 00		
T= 0.85000000D 00	F(T)= 0.722847923641D 00		
T= 0.90000000D 00	F(T)= 0.873189992791D 00		
T= 0.95000000D 00	F(T)= 0.105100536205D 01		
T= 0.10000000D 01	F(T)= 0.126108012472D 01		
T= 0.10500000D 01	F(T)= 0.150901331785D 01		
T= 0.11000000D 01	F(T)= 0.180135177899D 01		
T= 0.11500000D 01	F(T)= 0.214574731318D 01		
T= 0.12000000D 01	F(T)= 0.255113980891D 01		
T= 0.12500000D 01	F(T)= 0.302797053738D 01		
T= 0.13000000D 01	F(T)= 0.358843056098D 01		
T= 0.13500000D 01	F(T)= 0.424674998138D 01		
T= 0.14000000D 01	F(T)= 0.501953469171D 01		
T= 0.14500000D 01	F(T)= 0.592615838417D 01		
T= 0.15000000D 01	F(T)= 0.698921882637D 01		
T= 0.15500000D 01	F(T)= 0.823506888777D 01		
T= 0.16000000D 01	F(T)= 0.969443450303D 01		
T= 0.16500000D 01	F(T)= 0.114031337419D 02		
T= 0.17000000D 01	F(T)= 0.134029134596D 02		
T= 0.17500000D 01	F(T)= 0.157424226803D 02		
T= 0.18000000D 01	F(T)= 0.184783449796D 02		
T= 0.18500000D 01	F(T)= 0.216767157491D 02		
T= 0.19000000D 01	F(T)= 0.254144544335D 02		
T= 0.19500000D 01	F(T)= 0.297811467166D 02		
T= 0.20000000D 01	F(T)= 0.348811173132D 02		

Figure 7.21
Computer plot of the convolution of the two functions (7.89).

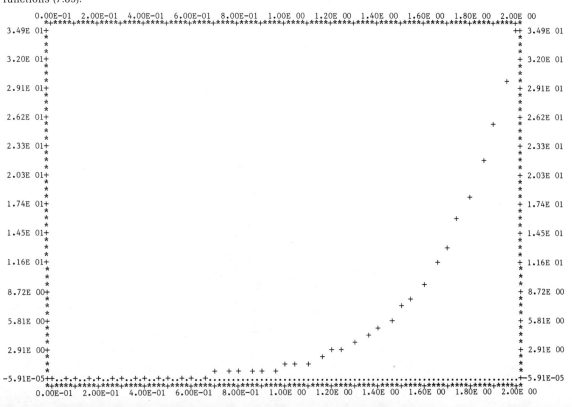

Table 7.5

		$f(t)$	
t	H = 0.01	H = 0.005	Exact
0.5	0.161859052	0.161864110	0.161865796
1.0	1.261040718	1.261080125	1.261093261
1.5	6.989000425	6.989218826	6.989291631
2.0	34.88002734	34.88111731	34.88148066

5.68 sec. The resulting $f(t)$ for several values of t are listed in Table 7-5. The exact solution was computed from (7.9) and is given by

$$f(t) = \int_0^t \tfrac{1}{2} \sinh 2\tau \cosh 3(t - \tau)\, d\tau$$

$$= \tfrac{1}{5}(\cosh 3t - \cosh 2t)u(t) \tag{7.90}$$

Figure 7.22
Computer printout of the convolution of the two functions (7.91).

```
T= 0.00000000D 00      F(T)=  0.000000000000D 00
T= 0.10000000D 00      F(T)=  0.499167083234D-02
T= 0.20000000D 00      F(T)=  0.198669330795D-01
T= 0.30000000D 00      F(T)=  0.443280309992D-01
T= 0.40000000D 00      F(T)=  0.778836684617D-01
T= 0.50000000D 00      F(T)=  0.119856384651D 00
T= 0.60000000D 00      F(T)=  0.169392742019D 00
T= 0.70000000D 00      F(T)=  0.225476190533D 00
T= 0.80000000D 00      F(T)=  0.286942436360D 00
T= 0.90000000D 00      F(T)=  0.352497109332D 00
T= 0.10000000D 01      F(T)=  0.420735492404D 00
T= 0.11000000D 01      F(T)=  0.490164048034D 00
T= 0.12000000D 01      F(T)=  0.559223451580D 00
T= 0.13000000D 01      F(T)=  0.626312820521D 00
T= 0.14000000D 01      F(T)=  0.689814810992D 00
T= 0.15000000D 01      F(T)=  0.748121239953D 00
T= 0.16000000D 01      F(T)=  0.799658882433D 00
T= 0.17000000D 01      F(T)=  0.842915088885D 00
T= 0.18000000D 01      F(T)=  0.876462867790D 00
T= 0.19000000D 01      F(T)=  0.898985083303D 00
T= 0.20000000D 01      F(T)=  0.909297426826D 00
T= 0.21000000D 01      F(T)=  0.906369834981D 00
T= 0.22000000D 01      F(T)=  0.889346044202D 00
T= 0.23000000D 01      F(T)=  0.857560994003D 00
T= 0.24000000D 01      F(T)=  0.810555816661D 00
T= 0.25000000D 01      F(T)=  0.748090180130D 00
T= 0.26000000D 01      F(T)=  0.670151783368D 00
T= 0.27000000D 01      F(T)=  0.576962838316D 00
T= 0.28000000D 01      F(T)=  0.468983410218D 00
T= 0.29000000D 01      F(T)=  0.346911527360D 00
T= 0.30000000D 01      F(T)=  0.211680012090D 00
T= 0.31000000D 01      F(T)=  0.644500267716D-01
T= 0.32000000D 01      F(T)= -0.933986294841D-01
T= 0.33000000D 01      F(T)= -0.260280395336D 00
T= 0.34000000D 01      F(T)= -0.434419873446D 00
T= 0.35000000D 01      F(T)= -0.613870648457D 00
T= 0.36000000D 01      F(T)= -0.796536797931D 00
T= 0.37000000D 01      F(T)= -0.980196860681D 00
T= 0.38000000D 01      F(T)= -0.116252999279D 01
T= 0.39000000D 01      F(T)= -0.134114401041D 01
T= 0.40000000D 01      F(T)= -0.151360499062D 01
```

EXAMPLE 7.10

As shown in Example 7.1, the inverse Laplace transform of the rational function $s/(s^2 + 1)^2$ is the convolution of the functions

$$f_1(t) = \sin t \qquad\qquad\qquad\qquad \textbf{(7.91a)}$$

$$f_2(t) = \cos t \qquad\qquad\qquad\qquad \textbf{(7.91b)}$$

We use the subroutine CONV to find $f(t) = f_1(t) * f_2(t)$ for $t = 0, 0.1, \ldots, 4$, choosing the stepsize H = 0.01. The computer output is presented in Figures 7.22 and 7.23. The total execution time for an IBM 370 Model 158 computer was 8.14 sec. The resulting approximation and the exact solution are listed in Table 7.6 for several values of t. This shows that the approximate solution is accurate to at least 10 significant figures.

Figure 7.23

Computer plot of the convolution of the two functions (7.91).

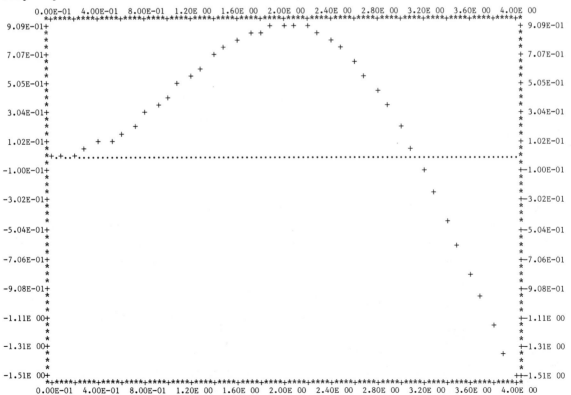

Table 7.6

	$f(t)$	
t	H = 0.01	Exact
1	0.4207354924	0.4207354924
2	0.9092974268	0.9092974268
3	0.2116800121	0.2116800121
4	−1.5136049906	−1.5136049906

7-6   BILATERAL CONVOLUTION

In applying the convolution theorem, we assume that the functions to be convolved are zero for $t < 0$. In many instances it is convenient to consider functions that extend into negative time. Such signals or functions can be studied and analyzed by means of the bilateral convolution. The *bilateral convolution* of two functions $f_1(t)$ and $f_2(t)$ is defined by the integral[†]

$$f(t) = f_1(t) * f_2(t) = \int_{-\infty}^{\infty} f_1(\tau)f_2(t - \tau)\,d\tau \tag{7.92}$$

for every t positive or negative. Following (7.8) we can show that an equivalent expression for (7.92) is

$$f(t) = \int_{-\infty}^{\infty} f_1(t - \tau)f_2(\tau)\,d\tau \tag{7.93}$$

indicating that the bilateral convolution is again commutative. In fact, it is also distributive and associative, as in the unilateral case.

In the special situation, if $f_2(t) = 0$ for $t < 0$, then

$$f(t) = \int_{-\infty}^{t} f_1(\tau)f_2(t - \tau)\,d\tau = -\int_{\infty}^{0} f_1(t - x)f_2(x)\,dx \tag{7.94}$$

since $f_2(t - \tau) = 0$ for $\tau > t$, where the second integral is obtained by a simple change of variable $\tau = t - x$. Now interchanging the limits and replacing the dummy variable x by τ yield

$$f(t) = \int_{0}^{\infty} f_1(t - \tau)f_2(\tau)\,d\tau \tag{7.95}$$

Finally, if $f_1(t) = 0$ and $f_2(t) = 0$ for $t < 0$, (7.95) reduces to the conventional or the unilateral convolution integral (7.9).

[†] Unless stated to the contrary, the term *convolution* means the unilateral convolution.

EXAMPLE 7.11

We compute the bilateral convolution of the two rectangular pulses as shown in Figure 7.24. The rectangular pulses are described by the equation

$$f_1(t) = \tfrac{1}{2}f_2(t) = u(t + T) - u(t - T) \tag{7.96}$$

giving

$$f(t) = 2 \int_{-\infty}^{\infty} [u(\tau + T) - u(\tau - T)][u(t - \tau + T) - u(t - \tau - T)]\, d\tau \tag{7.97}$$

The integrand is a sum of four products. The limits on the integrals of the products are determined by the regions of the τ-axis over which the step functions are nonzero. They can be seen by sketching out the four functions involved, as depicted in Figure 7.25. For $2T > t > 0$ we have

$$f(t) = 2 \int_{-T}^{t+T} u(\tau + T)u(t - \tau + T)\, d\tau - 2 \int_{T}^{t+T} u(\tau - T)u(t - \tau + T)\, d\tau$$

$$- 2 \int_{-T}^{t-T} u(\tau + T)u(t - \tau - T)\, d\tau + 0$$

$$= 2(t + 2T) - 2t - 2t = 4T - 2t \tag{7.98}$$

Figure 7.24
Two given rectangular pulses.

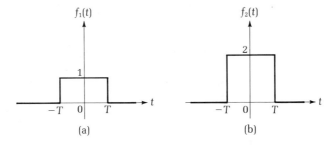

(a)   (b)

Figure 7.25
Advanced, delayed, and transposed unit-step functions.

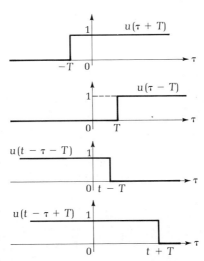

Figure 7.26
The bilateral convolution
of the two rectangular
pulses of Figure 7.24.

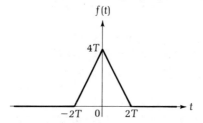

For $t > 2T$ the first three integrals are the same as in (7.98). The last one, instead of zero, becomes

$$2 \int_{T}^{t-T} u(\tau - T)u(t - \tau - T)\,d\tau = 2(t - 2T) \tag{7.99}$$

giving

$$f(t) = (4T - 2t) + (2t - 4T) = 0 \tag{7.100}$$

Likewise for $-2T < t < 0$ we have

$$f(t) = 2 \int_{-T}^{t+T} u(\tau + T)u(t - \tau + T)\,d\tau - 0 - 0 + 0$$

$$= 2t + 4T \tag{7.101}$$

and for $t < -2T$

$$f(t) = 0 \tag{7.102}$$

The above results can be combined and rewritten as

$$f(t) = (4T - 2t)[u(t) - u(t - 2T)] + (2t + 4T)[u(t + 2T) - u(t)] \tag{7.103}$$

A plot of $f(t)$ as a function of t is presented in Figure 7.26. The same result is readily obtained from the suitably modified graphical interpretation of the convolution integral outlined in Section 7-1.2.

7-6.1 The Bilateral Laplace Transform and Convolution Theorem

The Laplace transform, as defined in (5.3), involves the values of the function $f(t)$ for every t from $0-$ to ∞. The behavior of the function $f(t)$ for $t < 0$ never enters the Laplace integral and therefore has no effect on its transform. In many applications it is useful to consider the values of the function $f(t)$ for all t, not merely for nonnegative values of t. This leads to the *bilateral Laplace transform* of $f(t)$, defined by the integral

$$F(s) = \int_{-\infty}^{\infty} f(t)\,e^{-st}\,dt \tag{7.104}$$

If $f(t) = 0$ for $t < 0$, (7.104) becomes the conventional or unilateral Laplace integral (5.3).

Following the same reasoning as in the proof of (7.3), we can show that if $F_1(s)$ and $F_2(s)$ are the bilateral Laplace transforms of the functions $f_1(t)$ and $f_2(t)$, respectively, then the bilateral Laplace transform $F(s)$ of their bilateral convolution $f(t)$ equals $F_1(s)F_2(s)$:

$$F(s) = F_1(s)F_2(s) \tag{7.105}$$

where

$$f(t) = f_1(t) * f_2(t) = \int_{-\infty}^{\infty} f_1(\tau)f_2(t-\tau)\,d\tau \tag{7.106}$$

The result is known as the *bilateral convolution theorem*.

EXAMPLE 7.12

Consider the simple RC network of Figure 7.27. The output is the voltage $v(t)$ across the capacitor as indicated in the figure. The impulse response is found to be

$$v_\delta(t) = \frac{1}{RC} e^{-t/RC} u(t) \tag{7.107}$$

We shall determine the response of the network to a rectangular pulse voltage as shown in Figure 7.24(a):

$$v_g(t) = u(t+T) - u(t-T) \tag{7.108}$$

The bilateral Laplace transforms $V_\delta(s)$ and $V_g(s)$ of $v_\delta(t)$ and $v_g(t)$ are obtained as follows:

$$V_\delta(s) = \int_{-\infty}^{\infty} \left(\frac{1}{RC}\right) e^{-t/RC} u(t) e^{-st}\,dt = \frac{1}{RC}\int_0^{\infty} e^{-t/RC} e^{-st}\,dt$$

$$= \frac{1}{1+RCs} \tag{7.109a}$$

$$V_g(s) = \int_{-\infty}^{\infty} [u(t+T) - u(t-T)]e^{-st}\,dt$$

$$= \int_{-T}^{\infty} e^{-st}\,dt - \int_{T}^{\infty} e^{-st}\,dt = \frac{1}{s}e^{Ts} - \frac{1}{s}e^{-Ts} \tag{7.109b}$$

Figure 7.27
A simple RC network excited by a voltage source.

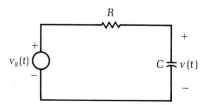

The transfer function of the network, defined as the ratio of $V(s)$ to $V_g(s)$, is given by

$$\frac{V(s)}{V_g(s)} = \frac{1}{1 + RCs} \tag{7.110}$$

The response transform to a rectangular pulse voltage becomes

$$V(s) = \frac{1}{s(1 + RCs)} (e^{Ts} - e^{-Ts})$$

$$= \frac{1}{s} (e^{Ts} - e^{-Ts}) - \frac{RC}{1 + RCs} (e^{Ts} - e^{-Ts}) \tag{7.111}$$

Appealing to the *shifting theorem*, which states that if a function $f(t)$ is delayed by T sec, the bilateral Laplace transform of the delayed function $f(t - T)$ is related to that of $f(t)$ by (Problem 7.10):

$$\mathcal{L}[f(t - T)] = \int_{-\infty}^{\infty} f(t - T)e^{-st}\,dt = e^{-Ts}F(s) \tag{7.112}$$

The inverse bilateral Laplace transform of (7.111) is found to be

$$v(t) = u(t + T) - u(t - T) - e^{-(t + T)/RC}u(t + T)$$
$$+ e^{-(t - T)/RC}u(t - T) \tag{7.113}$$

where we have used the transforms

$$\int_{-\infty}^{\infty} u(t)e^{-st}\,dt = \frac{1}{s} \tag{7.114}$$

$$\int_{-\infty}^{\infty} e^{at}u(t)e^{-st}\,dt = \frac{1}{s + a} \tag{7.115}$$

A plot of the output voltage $v(t)$ as a function of t is presented in Figure 7.28 together with the input rectangular pulse. We remark that the proof of the shifting theorem (7.112) is identical to that of (5.53) except that the lower limit is changed from $0-$ to $-\infty$. Also the constant T in (7.112) may either be positive or negative, whereas in (5.53) T cannot be negative.

As in the unilateral case, the impulse response $v_\delta(t)$ equals the inverse bilateral Laplace transform of the transfer function $V(s)/V_g(s)$ of (7.110), or

$$\frac{V(s)}{V_g(s)} = V_\delta(s) \tag{7.116}$$

as given in (7.109a) and (7.110).

Figure 7.28
A plot of the capacitor voltage as a function of t due to a rectangular pulse input to the network of Figure 7.27.

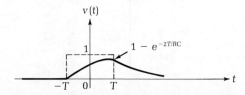

7-6.2   The Zero-State Response

In Section 7-2.1 we demonstrated that if the impulse response $r_\delta(t)$ to a system is known, then the zero-state response to any other driving source $e(t)$ is determined by (7.37). More specifically it states that if an input excitation $e(t)$ is applied to a system at $t = 0$, the resulting response $r(t)$ is given by the integral

$$r(t) = \int_{0-}^{t} e(\tau) r_\delta(t - \tau)\, d\tau = \int_{0-}^{t} e(t - \tau) r_\delta(\tau)\, d\tau \qquad (7.117)$$

Suppose that the input is applied not at $t = 0$ but at $t = -\infty$. The bilateral Laplace transform response $R(s)$ can be expressed in terms of the transforms $R_\delta(s)$ and $E(s)$ of the impulse response $r_\delta(t)$ and input $e(t)$:

$$R(s) = R_\delta(s) E(s) \qquad (7.118)$$

where $R_\delta(s)$ equals the transfer function. Applying the bilateral convolution theorem (7.106) gives

$$r(t) = \int_{-\infty}^{\infty} e(\tau) r_\delta(t - \tau)\, d\tau = \int_{-\infty}^{\infty} e(t - \tau) r_\delta(\tau)\, d\tau$$

$$= \int_{-\infty}^{t} e(\tau) r_\delta(t - \tau)\, d\tau = \int_{0}^{\infty} e(t - \tau) r_\delta(\tau)\, d\tau \qquad (7.119)$$

because $r_\delta(t) = 0$ for $t < 0$. This is the bilateral convolution of the input with the impulse response of the system. It again shows that the zero-state response of a system to any driving source is determined once the impulse response is known.

EXAMPLE 7.13

We use (7.119) to compute the capacitor voltage $v(t)$ of Figure 7.27 to a rectangular pulse voltage $v_g(t)$ of (7.108). From (7.107) and (7.108), we obtain the output voltage from (7.119), as follows:

$$v(t) = \int_{-\infty}^{t} [u(\tau + T) - u(\tau - T)]\left(\frac{1}{RC}\right) e^{-(t-\tau)/RC} u(t - \tau)\, d\tau \qquad (7.120)$$

Equation (7.120) can be written as a sum of two integrals. The limits on these integrals are determined by regions of the τ-axis over which the step functions are nonzero and can be seen by sketching out the three functions involved, as depicted in Figure 7.29. This leads to

$$v(t) = \frac{1}{RC} u(t + T) \int_{-T}^{t} e^{-(t-\tau)/RC}\, d\tau$$

$$- \frac{1}{RC} u(t - T) \int_{T}^{t} e^{-(t-\tau)/RC}\, d\tau$$

$$= [1 - e^{-(t+T)/RC}] u(t + T) - [1 - e^{-(t-T)/RC}] u(t - T) \qquad (7.121)$$

confirming (7.113).

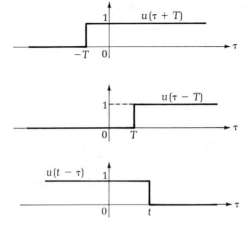

7-6.3   An Additional Property[†]

There is an additional property of the bilateral convolution that is frequently
used to check the validity of a particular calculation. Consider the areas of
the functions $f_1(t)$ and $f_2(t)$ over the interval $-\infty < t < \infty$, which are
assumed to be bounded:

$$A_1 = \int_{-\infty}^{\infty} f_1(t)\, dt \tag{7.122}$$

$$A_2 = \int_{-\infty}^{\infty} f_2(t)\, dt \tag{7.123}$$

The area of the convolution resultant $f(t) = f_1(t) * f_2(t)$ over the same interval
is determined by the integral (Problem 7.18):

$$A = \int_{-\infty}^{\infty} f(t)\, dt = \int_{-\infty}^{\infty} \int_{-\infty}^{\infty} f_1(\tau) f_2(t - \tau)\, d\tau\, dt$$

$$= \int_{-\infty}^{\infty} f_1(\tau) \left[\int_{-\infty}^{\infty} f_2(t - \tau)\, dt \right] d\tau = \int_{-\infty}^{\infty} f_1(\tau) \left[\int_{-\infty}^{\infty} f_2(x)\, dx \right] d\tau$$

$$= \int_{-\infty}^{\infty} f_1(\tau) A_2\, d\tau = A_2 \int_{-\infty}^{\infty} f_1(\tau)\, d\tau = A_1 A_2 \tag{7.124}$$

In words it states that the area of a convolution product is equal to the
product of the areas of the components.

EXAMPLE 7.14
Consider the two rectangular pulses of Figure 7.24. The area are found to be

$$A_1 = 2T, \qquad A_2 = 4T \tag{7.125}$$

[†] As indicated in Problem 7.18, this property remains valid in the case of unilateral convolution.

The convolution product of $f_1(t)$ and $f_2(t)$ is given by (7.103), the area of which can be determined directly from Figure 7.26, getting

$$A = 8T^2 = A_1 A_2 \qquad (7.126)$$

EXAMPLE 7.15
We use the problem of Example 7.13 to verify (7.124). The impulse response of the network of Figure 7.27 is found to be

$$v_\delta(t) = \frac{1}{RC}\, e^{-t/RC}\, u(t) \qquad (7.127)$$

The response of the network to a rectangular pulse voltage of Figure 7.24(a),

$$v_g(t) = u(t + T) - u(t - T) \qquad (7.128)$$

was computed in Example 7.13 and is given by

$$v(t) = [1 - e^{-(t+T)/RC}]u(t+T) - [1 - e^{-(t-T)/RC}]u(t-T) \qquad (7.129)$$

The areas of $v_\delta(t)$, $v_g(t)$, and $v(t)$ are computed by integrating over the interval $-\infty < t < \infty$, obtaining

$$A_1 = \int_{-\infty}^{\infty} \left(\frac{1}{RC}\right) e^{-t/RC} u(t)\, dt = \frac{1}{RC} \int_0^{\infty} e^{-t/RC}\, dt = 1 \qquad (7.130)$$

$$A_2 = \int_{-\infty}^{\infty} [u(t+T) - u(t-T)]\, dt = 2T \qquad (7.131)$$

$$A = \int_{-\infty}^{\infty} v(t)\, dt = \lim_{\lambda \to \infty} \left\{ \int_{-T}^{\lambda} [1 - e^{-(t+T)/RC}]\, dt \right.$$
$$\left. - \int_T^{\lambda} [1 - e^{-(t-T)/RC}]\, dt \right\}$$
$$= \lim_{\lambda \to \infty} [RCe^{-(T+\lambda)/RC} - RCe^{(T-\lambda)/RC} + 2T] = 2T \qquad (7.132)$$

Thus we have

$$A = A_1 A_2 \qquad (7.133)$$

7-7 SUMMARY

We began this chapter by introducing the *convolution integral* and showed that the Laplace transform of the convolution product of two functions equals the product of their Laplace transforms. The result is known as the *convolution theorem*. Some properties of the convolution integral were discussed. We demonstrated by examples that the convolution operation may be interpreted graphically in terms of folding, translating, multiplying, and integrating. The process permits a graphical computation of the convolution integral.

The convolution theorem was applied to determine the zero-state response of a linear time-invariant network to an arbitrary driving source when its response to some standard function is known. The standard functions considered were the unit impulse, the unit step, and the sinusoidal functions. Specifically we showed that the *zero-state response* of a network to an arbitrary excitation can be expressed in terms of the impulse or step response and the input. In fact, the zero-state response equals the convolution of the excitation with the impulse response of the system. This representation of the zero-state solution is important in that once the impulse response is known, the response to any arbitrary excitation can be determined. In other words, if two linear time-invariant systems have the same zero-state response to a Laplace-transformable input, then their zero-state responses to any arbitrary Laplace-transformable input will be the same. This establishes the uniqueness property for linear time-invariant systems.

The representation of the zero-state response of a system in terms of the input and the impulse or the step response can also be interpreted as the superposition of the responses to impulses or steps of various strengths, suitably displaced. Specifically, we showed that the zero-state response to an arbitrary excitation can be regarded as the superposition of the responses to a sequence of impulses or steps that approximates the excitation. One of the important applications of these interpretations is its use in the numerical computation of system response. In many situations the excitations are either very complicated or cannot be represented analytically. For instance, the input may be given as a curve or a set of points. In such cases the Laplace transform of the input and the resulting inverse Laplace transform of the transform response will be very difficult, if not impossible, to find. The above interpretation provides an elegant solution where the input is approximated by a sequence of impulse or step functions and is represented by a time series. Based on this a computer program was written in WATFIV that performs convolution of two functions. The program can readily be modified to accommodate the convolution of two sequences of numbers.

Finally, we introduced the *bilateral convolution* and the *bilateral Laplace transformation*, which take into account functions that extend into negative time. We stated the *bilateral convolution theorem*, which essentially is the same as the unilateral convolution theorem except by adding the word bilateral when we refer to convolution or Laplace transformation. As a result the zero-state response of a system to any driving source, applied at any time, is determined once the impulse response is known.

REFERENCES AND SUGGESTED READING

Balabanian, N., and T. A. Bickart. *Electrical Network Theory.* New York: Wiley, 1969, Chapter 5.

Bracewell, R. N. *The Fourier Transform and Its Applications.* 2d ed. New York: McGraw-Hill, 1978, Chapter 3.

Guillemin, E. A. *Theory of Linear Physical Systems*. New York: Wiley, 1963, Chapter 13.

Kuo, F. F. *Network Analysis and Synthesis*. 2d ed. New York: Wiley, 1966, Chapter 7.

Lago, G. V., and L. M. Benningfield. *Circuit and System Theory*. New York: Wiley, 1979, Chapters 6 and 9.

Liu, C. L., and J. W. S. Liu. *Linear Systems Analysis*. New York: McGraw-Hill, 1975, Chapters 4 and 9.

Mason, S. J., and H. J. Zimmermann. *Electronic Circuits, Signals, and Systems*. New York: Wiley, 1960, Chapter 7.

McGillem, C. D., and G. R. Cooper. *Continuous and Discrete Signal and System Analysis*. New York: Holt, Rinehart & Winston, 1974, Chapters 4 and 6.

Papoulis, A. *Circuits and Systems: A Modern Approach*. New York: Holt, Rinehart & Winston, 1980, Chapter 4.

Van Valkenburg, M. E. *Network Analysis*. 3d ed. Englewood Cliffs, N.J.: Prentice-Hall, 1974, Chapter 8.

PROBLEMS

7.1 The impulse response of a network is found to be

$$r_\delta(t) = e^{-2t} \tag{7.134}$$

Determine the zero-state response of the network to the input excitation $e(t) = (\cos t)u(t)$.

7.2 Using convolution theorem, find the inverse Laplace transforms for the following functions:

(a) $F(s) = \dfrac{1}{(s+a)(s^2+1)}$ **(b)** $F(s) = \dfrac{K}{(s+a)^2(s+b)}$ **(c)** $F(s) = \dfrac{a_1 s + a_0}{(s+a)^2}$

7.3 A network having an impulse response $r_\delta(t)$ approximated by the straight lines as shown in Figure 7.30(a) is driven by a source $e(t)$ of Figure 7.30(b). Determine the zero-state response of the network for (a) $T = 1$ s, (b) $T = 4$ s, and (c) $T = 5$ s. For each case sketch the response.

7.4 Repeat Problem 7.3 if the input excitation is a rectangular pulse of Figure 7.1(b).

7.5 Repeat Problem 7.3 for the input having the waveform of Figure 7.31.

Figure 7.30
The input (b) to a network the impulse response of which can be approximated by the straight lines as shown in (a).

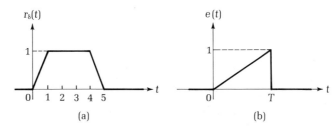

Figure 7.31
A given excitation waveform.

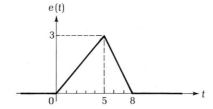

Figure 7.32
The approximate impulse
response of a network.

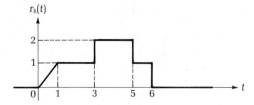

7.6 Consider a network having an approximate impulse response as shown in Figure 7.32. Determine the zero-state response to the input signal of Figure 7.31. Sketch the response.

7.7 Repeat Problem 7.6 for the input signal of Figure 7.30(b).

7.8 Repeat Example 7.12 for the input signal $v_g(t)$ having the waveform of Figure 7.31.

7.9 Repeat of Example 7.13 for the input voltage $v_g(t)$ having the waveform of Figure 7.30(b).

7.10 Using the bilateral convolution integral, prove the shifting theorem of (7.112).

7.11 Applying the convolution theorem, prove the real integration theorem of (5.42).

7.12 Using graphical technique, determine the convolution of $e(t)$ of Figure 7.31 with itself.

7.13 Rework Example 7.7 using a time interval $T = 0.02$ s.

7.14 Example 7.8 shows the steps in the determination of the response of an amplifier at a particular time t by making use of the approximate convolution. Carry out the process for enough values of t so that the response $i_2(t)$ can be plotted.

7.15 Using subroutine CONV and subroutine PLOT, compute and plot the convolution of the input signal $e(t)$ and the impulse response $r_\delta(t)$ of Figure 7.13. Compare the results with those obtained in Example 7.7.

7.16 The step response of a network is found to be

$$r_u(t) = 0.02e^{-2t} + 0.05 \tag{7.135}$$

Determine the zero-state response of the network to the rectangular pulse input of Figure 7.1(b).

7.17 Repeat Problem 7.16 for the input having the waveform of Figure 7.31.

7.18 Show that if A, A_1, and A_2 are the areas of the functions $f(t)$, $f_1(t)$, and $f_2(t)$ in the interval $0 \leq t < \infty$, respectively, and if $f(t)$ is the unilateral convolution of $f_1(t)$ and $f_2(t)$, then

$$A = A_1 A_2 \tag{7.136}$$

showing that (7.124) remains valid if the time interval $-\infty < t < \infty$ is replaced by $0 \leq t < \infty$ and the bilateral convolution is replaced by the unilateral convolution.

7.19 Let

$$f_1(t) = \cos \omega t \tag{7.137}$$

$$f_2(t) = e^{-\alpha t} \tag{7.138}$$

Show that

$$f_1(t) * f_2(t) = \frac{1}{\alpha^2 + \omega^2} (\alpha \cos \omega t + \omega \sin \omega t - \alpha e^{-\alpha t}) \tag{7.139}$$

7.20 Let

$$f_1(t) = \sin \omega t \tag{7.140}$$

$$f_2(t) = e^{-\alpha t} \tag{7.141}$$

Show that

$$f_1(t) * f_2(t) = \frac{1}{\alpha^2 + \omega^2} (\alpha \sin \omega t - \omega \cos \omega t + \omega e^{-\alpha t}) \tag{7.142}$$

7.21 In the network of Figure 7.27, let $R = 1\,k\Omega$ and $C = 0.1\,\mu F$. Determine the zero-state response $v(t)$ of the network to the input voltage $v_g(t) = 5\sin 10^4\,t$ by making use of the convolution integral. Sketch the response $v(t)$ as a function of t.

7.22 Repeat Problem 7.21 by making use of the subroutine CONV and subroutine PLOT.

7.23 Determine the bilateral convolution of the signal $f(t)$ shown in Figure 7.26 with itself. Sketch $f(t) * f(t)$ as a function of t.

7.24 Verify the following:

(a) $[e^{-t}u(t)] * [e^{-3t}u(t)] = \frac{1}{2}(e^{-t} - e^{-3t})u(t)$ (7.143)

(b) $[t^3u(t)] * [t^5u(t)] = \dfrac{3!5!}{9!}\,t^9$ (7.144)

(c) $[tu(t)] * [e^{-2t}u(t)] = (\frac{1}{2}t - \frac{1}{4} + \frac{1}{4}e^{-2t})u(t)$ (7.145)

7.25 The excitation $e(t)$ of a linear time-invariant system is shown in Figure 7.33(b). The system's impulse response $r_\delta(t)$ can be approximated by the straight lines of Figure 7.33(a). Sketch the zero-state response of the system to $e(t)$.

7.26 Repeat Problem 7.25 for the input excitation $e(t)$ having the waveform of Figure 7.13(a).

7.27 Consider a network having the impulse response approximated by the straight lines of Figure 7.32. Determine and sketch the zero-state response of the network to the input having the waveform shown in Figure 7.33(b).

7.28 Repeat Problem 7.27 using subroutine CONV and subroutine PLOT.

7.29 Repeat Problem 7.27 for the input having the waveform of Figure 7.13(a).

7.30 A system having the impulse response as shown in Figure 7.33(a) is known as a *finite time integrator*. Using the convolution integral, express the output of this system to an arbitrary excitation. Verify the expression for the input having the waveform of Figure 7.33(b).

7.31 Compute the bilateral convolution of the two functions shown in Figure 7.34.

7.32 Verify the following results:

(a) $[t^2u(t)] * [e^tu(t)] = (2e^t - t^2 - 2t - 2)u(t)$ (7.146)

(b) $[e^tu(t)] * u(t) = (e^t - 1)u(t)$ (7.147)

Figure 7.33
The input (b) to a network the impulse response of which can be approximated by the straight lines as shown in (a).

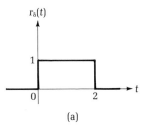

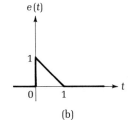

Figure 7.34
Two given functions the bilateral convolution of which is to be ascertained.

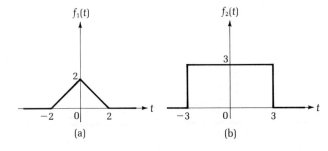

(c) $[be^{-bt}] * e^{bt} = \sinh bt, \; b$ real   **(7.148)**

(d) $[\omega e^{-j\omega t}] * e^{j\omega t} = \sin \omega t$ **(7.149)**

7.33 In Problem 7.19, let $\alpha = 1$ and $\omega = 2$. Using the subroutines CONV and PLOT, compute and plot the convolution $f_1(t) * f_2(t)$. Verify the results using (7.139).

7.34 In Problem 7.20, let $\alpha = 2$ and $\omega = 1$. Using the subroutines CONV and PLOT, compute and plot $f_1(t) * f_2(t)$ as a function of t. Verify the results using (7.142).

7.35 Prove the identities (7.14a) and (7.14b).

7.36 Let

$$f^{(n)}(t) = \frac{d^n f(t)}{dt^n}$$ **(7.150)**

Show that if the functions $f_1(t)$ and $f_2(t)$ have n derivatives, then their convolution $f(t) = f_1(t) * f_2(t)$ will have $n + 1$ derivatives. The $(n + 1)$th derivative is given by

$$f^{(n+1)}(t) = \int_0^t f_1(\tau) f_2^{(n+1)}(t - \tau) \, d\tau + f_1^{(n)}(t) f_2(0)$$
$$+ f_1^{(n-1)}(t) f_2^{(1)}(0) + \cdots + f_1(t) f_2^{(n)}(0)$$ **(7.151)**

where $f_2^{(k)}(0) = f_2^{(k)}(t)|_{t=0}$.

CHAPTER EIGHT

FOURIER SERIES AND SIGNAL SPECTRA

In Chapter 4 we studied the time-domain solution of network equations. In Chapter 6 the same problem was reformulated in the frequency domain, using the Laplace transformation as a tool. In the preceding chapter we showed that the zero-state response of a system to an arbitrary excitation is determined once its impulse response is known. In the present chapter we shall study the periodic signal in terms of its *frequency content*. The notion of frequency content of a periodic signal is extremely important in engineering problems and is the basis for much of the jargon used by electrical engineers to communicate among themselves. The steady-state response of a system to a periodic signal input that ideally started at the beginning of time and will last forever is considered in the following chapter.

8-1 PERIODIC SIGNALS

A signal $f(t)$ is said to be *periodic* if

$$f(t + T) = f(t) \tag{8.1}$$

for every t. The number T is sometimes referred as a *period*. From this definition we see that if T is a period, then all of its multiples are also periods of $f(t)$ because

$$f(t + nT) = f(t) \tag{8.2}$$

for every integer n. For our purposes the term *period* is defined as the smallest positive number T for which (8.1) holds. Examples of signals that fulfill the requirement (8.1) are presented in Figure 8.1 with each having period T as indicated. A train of rectangular pulses is illustrated in (a), (b) shows a train of triangular pulses, and (c) depicts a full-wave rectified waveform.

The most important periodic signals, insofar as we are concerned, are the familiar sine and cosine functions:

$$\sin \omega_0 t \quad \text{and} \quad \cos \omega_0 t$$

326

Fourier Series and Signal Spectra

Figure 8.1

Illustration of three periodic signals: (a) rectangular pulse train, (b) triangular pulse train, and (c) full-wave rectified waveform. Notice that the waveforms (a) and (c) are functions with even symmetry.

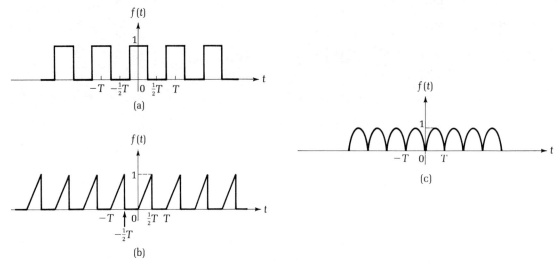

where ω_0 is the angular or radian frequency. The period T of each is given by

$$T = \frac{2\pi}{\omega_0} \tag{8.3}$$

Consider two periodic signals $f_1(t)$ and $f_2(t)$ with respective periods T_1 and T_2. A *common period* of $f_1(t)$ and $f_2(t)$, if it exists, is the smallest positive number T satisfying

$$f_1(t + T) = f_1(t) \tag{8.4a}$$

$$f_2(t + T) = f_2(t) \tag{8.4b}$$

We shall examine the conditions which the numbers T_1 and T_2 must satisfy in order that the functions $f_1(t)$ and $f_2(t)$ have a common period. From (8.2), for $f_1(t)$ and $f_2(t)$ to have a common period T, it is necessary and sufficient that

$$n_1 T_1 = n_2 T_2 \tag{8.5}$$

where n_1 and n_2 are integers, requiring that the ratio

$$\frac{T_1}{T_2} = \frac{n_2}{n_1} \tag{8.6}$$

be a rational number. The common period is the least common multiple of T_1 and T_2. If n_1 and n_2 are relatively prime, the common period T is given by

$$T = n_1 T_1 = n_2 T_2 \qquad (8.7)$$

If, on the other hand, the ratio (8.6) is an irrational number, then $f_1(t)$ and $f_2(t)$ do not have a common period.

Let $h(t)$ be an arbitrary function. If $f(t)$ is a periodic signal with a period T, then the function defined by the equation

$$g(t) = h[f(t)] \qquad (8.8)$$

is periodic with period not greater than T, because

$$g(t + T) = h[f(t + T)] = h[f(t)] = g(t) \qquad (8.9)$$

The period of $g(t)$, being the smallest number T satisfying (8.9), is not necessarily the smallest number T satisfying (8.1), the period of $f(t)$. The period of $g(t)$ might be a fraction of the period of $f(t)$.

We illustrate the above results by the following examples.

EXAMPLE 8.1

Consider the sinusoidal signals

$$f_1(t) = \cos \omega_0 t \qquad (8.10)$$

$$f_2(t) = \sin 2\omega_0 t \qquad (8.11)$$

Their periods are found to be $T_1 = 2\pi/\omega_0$ and $T_2 = \pi/\omega_0$, yielding

$$\frac{T_1}{T_2} = \frac{2}{1} = 2 \qquad (8.12)$$

Because the ratio of their periods is a rational number, the two signals have a common period equal to

$$T = 1T_1 = 2T_2 = \frac{2\pi}{\omega_0} \qquad (8.13)$$

On the other hand, if we consider the signals $f_1(t)$ and

$$f_3(t) = \sin\left(\frac{8\sqrt{2}\omega_0 t}{5}\right) \qquad (8.14)$$

then $f_1(t)$ and $f_3(t)$ do not have a common period because the ratio of their periods $T_1 = 2\pi/\omega_0$ and $T_3 = 5\pi/(4\sqrt{2}\omega_0)$ is given by

$$\frac{T_1}{T_3} = \frac{8\sqrt{2}}{5} \qquad (8.15)$$

an irrational number. Plots of the functions $f_1(t)$, $f_2(t)$, and $f_3(t)$ are presented in Figure 8.2.

Figure 8.2

Plots of three periodic sinusoidal signals: (a) and (b) have a common period $T = 2\pi/\omega_0$; however, (a) and (c) do not have a common period.

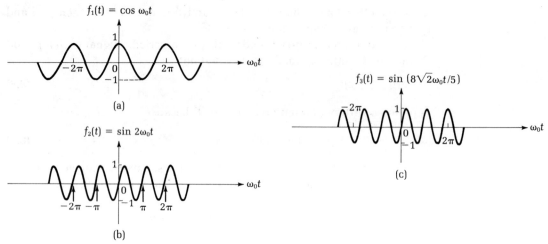

EXAMPLE 8.2

Consider the periodic signal

$$f(t) = \sin t \tag{8.16}$$

with period $T_1 = 2\pi$. Let $h(t) = t^2$. Then the function defined by

$$g(t) = h[f(t)] = [f(t)]^2 = \sin^2 t \tag{8.17}$$

is also periodic with period $T_2 = \pi$. The signals $f(t)$ and $g(t)$ are shown in Figure 8.3. The period T_2 of $g(t)$ is only half that of $f(t)$.

Now consider $h(t) = t^3$. The function $g(t)$ in this case is

$$g(t) = h[f(t)] = \sin^3 t \equiv \frac{3}{4}\sin t - \frac{1}{4}\sin 3t \tag{8.18}$$

The right-hand side of (8.18) is obtained by using a trigonometric identity. In (8.18) let $f_1(t) = \frac{3}{4}\sin t$ and $f_2(t) = -\frac{1}{4}\sin 3t$ with respective periods $T_1 = 2\pi$ and $T_2 = 2\pi/3$. Because the ratio of their periods

$$\frac{T_1}{T_2} = 3 \tag{8.19}$$

is a rational number, the functions $f_1(t)$ and $f_2(t)$ have a common period

$$T = 1T_1 = 3T_2 = 2\pi \tag{8.20}$$

The functions $f_1(t)$ and $f_2(t)$ together with $g(t)$ of (8.18) are plotted in Figure 8.4 with periods as indicated.

Figure 8.3
Representation of the periodic signals (a) $f(t) = \sin t$ and (b) $g(t) = h[f(t)]$ where $h(t) = t^2$.

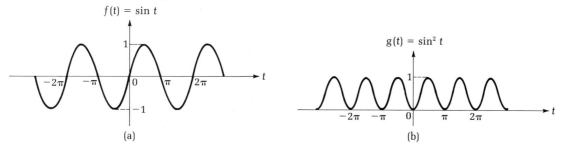

(a)

(b)

Figure 8.4
Illustration of the periodic signals (a) $f_1(t) = \frac{3}{4} \sin t$, (b) $f_2(t) = -\frac{1}{4} \sin 3t$, and (c) $g(t) = f_1(t) + f_2(t)$ with a common period $T = 2\pi$.

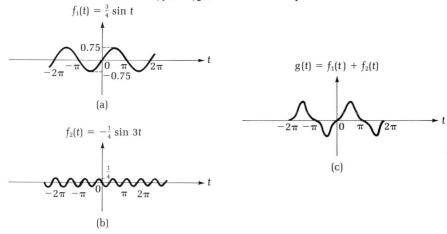

(a)

(b)

(c)

We now proceed to demonstrate how to construct a periodic function from an arbitrary signal. Given an arbitrary function $h(t)$ and its displacements located at the multiples of a preassigned period T,

$$\ldots, h(t + 2T), h(t + T), h(t), h(t - T), h(t - 2T), \ldots \qquad \textbf{(8.21)}$$

the sum

$$g(t) = \sum_{n = -\infty}^{\infty} h(t + nT) \qquad \textbf{(8.22)}$$

is also a periodic function because

$$g(t + T) = \sum_{n = -\infty}^{\infty} h(t + T + nT) = \sum_{n = -\infty}^{\infty} h(t + nT) = g(t) \qquad \textbf{(8.23)}$$

Notice that $h(t)$ does not always equal $g(t)$ in the interval $-\frac{1}{2}T < t < \frac{1}{2}T$ because $h(t)$ may not be zero for $|t| > \frac{1}{2}T$. We illustrate this by the following example.

EXAMPLE 8.3

Consider the signal of Figure 8.5(a):

$$h(t) = e^{-|t|} \tag{8.24}$$

having displacements $h(t + T)$, $h(t)$, and $h(t - T)$ as depicted in Figure 8.5(b) with dashed lines. The sum $g(t)$ of (8.22) is a periodic signal as also sketched in the figure with a solid line. It is evident that $h(t)$, in this case, does not equal $g(t)$ in the interval $-\frac{1}{2}T < t < \frac{1}{2}T$.

On the other hand, if $h(t)$ is a rectangular pulse as shown in Figure 8.6(a), the sum of its displacements $g(t)$ of (8.22) is a periodic signal of period T [Figure 8.6(b)]. In this case $h(t)$ equals $g(t)$ in the interval $-\frac{1}{2}T < t < \frac{1}{2}T$.

Figure 8.5
Illustration of the case where the periodic signal (b) $g(t) = \sum_{n=-\infty}^{\infty} h(t + nT)$ does not equal $h(t)$ given in (a) as $h(t) = e^{-|t|}$ in the interval from $t = -\frac{1}{2}T$ to $t = \frac{1}{2}T$.

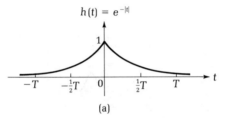

(a)

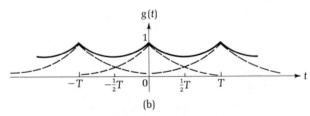

(b)

Figure 8.6
Illustration of the case where $h(t)$ is a rectangular pulse (a) giving a periodic signal (b) $g(t) = \sum_{n=-\infty}^{\infty} h(t + nT)$ that is equal to $h(t)$ over the interval from $t = -\frac{1}{2}T$ to $t = \frac{1}{2}T$.

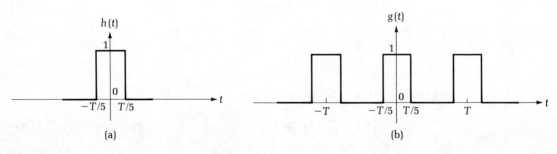

(a)   (b)

8-2 FOURIER SERIES

In this section we indicate that a periodic signal $f(t)$ with period T satisfying the *Dirichlet conditions*, to be discussed in Section 8-7, can be expressed as a series of sine and cosine functions as follows:

$$f(t) = a_0 + a_1 \cos \omega_0 t + a_2 \cos 2\omega_0 t + \cdots + a_n \cos n\omega_0 t + \cdots$$
$$+ b_1 \sin \omega_0 t + b_2 \sin 2\omega_0 t + \cdots + b_n \sin n\omega_0 t + \cdots$$

$$= a_0 + \sum_{n=1}^{\infty} (a_n \cos n\omega_0 t + b_n \sin n\omega_0 t) \qquad (8.25)$$

Such a sum is called *Fourier series* named after the French mathematician, Jean Baptiste Joseph Fourier (1768–1830), who first undertook the systematic study of such expansions in his work on heat transfer. The various terms in (8.25) are referred to as the *harmonics*. The nth harmonic is the term

$$a_n \cos n\omega_0 t + b_n \sin n\omega_0 t = c_n \cos (n\omega_0 t + \theta_n) \qquad (8.26)$$

in polar form, where, from Figure 8.7,

$$c_n = \sqrt{a_n^2 + b_n^2} \qquad (8.27)$$

$$\theta_n = -\tan^{-1} \frac{b_n}{a_n} \qquad (8.28)$$

for $n \geq 1$. The first harmonic

$$a_1 \cos \omega_0 t + b_1 \sin \omega_0 t = c_1 \cos (\omega_0 t + \theta_1) \qquad (8.29)$$

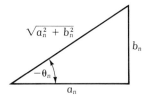

Figure 8.7
Representation of the relationships between the rectangular and the polar forms of the Fourier series expansion coefficients.

is also called the *fundamental*. Its *frequency* ω_0 is the *fundamental frequency*. Observe that all other frequencies are multiples of the fundamental frequency. The constant term a_0 is the 0th harmonic, corresponding to the dc component of the signal.

Using (8.26) in (8.25) yields the equivalent form for the Fourier series

$$f(t) = a_0 + c_1 \cos (\omega_0 t + \theta_1) + c_2 \cos (2\omega_0 t + \theta_2)$$
$$+ \cdots + c_n \cos (n\omega_0 t + \theta_n) + \cdots$$

$$= c_0 + \sum_{n=1}^{\infty} c_n \cos (n\omega_0 t + \theta_n) \qquad (8.30)$$

with $c_0 = a_0$.

We remark that if a Fourier series is known, its fundamental might not appear explicitly in (8.30). To ascertain the fundamental frequency ω_0, we need only to find the greatest common divisor of all other frequencies. As an example consider the Fourier series

$$32 \cos^6 20\pi t = 10 + 15 \cos 40\pi t + 6 \cos 80\pi t + \cos 120\pi t \qquad (8.31)$$

which is a special form of (8.30). Since 40π is the greatest common divisor of the frequencies 40π, 80π, and 120π, the fundamental frequency ω_0 of the

Fourier series representation of $32 \cos^6 20\pi t$ is 40π. The term $15 \cos 40\pi t$ is the fundamental, $6 \cos 80\pi t$ the second harmonic, and $\cos 120\pi t$ the third harmonic. The dc value is 10. All other harmonics are zero.

Observe that to represent a periodic function by a Fourier series (8.30), the set of numbers c_0, c_n, and θ_n contains all the necessary information. The coefficient c_n is the *amplitude* and θ_n the *phase* of the nth harmonic. The plot of c_n as a function of n or $n\omega_0$ is called the *amplitude spectrum*, and the plot of θ_n as a function of n or $n\omega_0$ is the *phase spectrum*. Together they are called the *frequency spectrum*. The Fourier analysis amounts to determining the frequency spectrum and finding the number of terms to be included in a truncated series so that the partial sum will represent the signal within an allowable error. If the series (8.30) converges rapidly, only a few terms will suffice.

EXAMPLE 8.4

The waveform shown in Figure 8.8 consists of a train of voltage pulses. For this periodic voltage $v_g(t)$ of period $T = 4$, the Fourier series is found to be (see Example 8.5)

$$v_g(t) = \frac{3}{4} - \frac{1}{\pi} \cos \frac{\pi t}{2} + \frac{1}{3\pi} \cos \frac{3\pi t}{2} + \cdots$$

$$+ \frac{3}{\pi} \sin \frac{\pi t}{2} - \frac{1}{\pi} \sin \pi t + \frac{1}{\pi} \sin \frac{3\pi t}{2} + \cdots \qquad (8.32)$$

Figure 8.8
A periodic signal which consists of a train of voltage pulses.

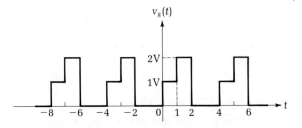

Figure 8.9
The amplitude (a) and phase (b) spectra of the signal given in Figure 8.8.

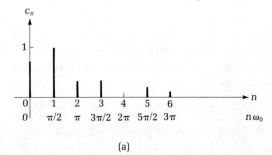

(a)

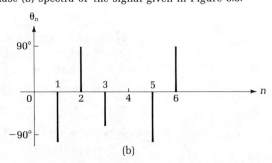

(b)

where the fundamental frequency $\omega_0 = \pi/2$. Appealing to (8.26), we obtain

$$v_g(t) = \frac{3}{4} + \frac{\sqrt{10}}{\pi} \cos\left(\frac{\pi t}{2} - 108.4°\right) + \frac{1}{\pi}\cos(\pi t + 90°)$$

$$+ \frac{\sqrt{10}}{3\pi}\cos\left(\frac{3\pi t}{2} - 71.6°\right) + \cdots \tag{8.33}$$

The amplitude and phase spectra of $v_g(t)$ are presented in Figure 8.9.

8-3 EVALUATION OF FOURIER COEFFICIENTS

In the preceding section, we have indicated that a periodic signal may be represented as an infinite series of sinusoids of harmonically related frequencies. The a and b coefficients in the Fourier series (8.25) are called the *Fourier coefficients*, the evaluation of which is the subject of this section.

To evaluate the Fourier coefficients, we use the *orthogonality property* of the functions $\cos n\omega_0 t$ and $\sin m\omega_0 t$ with integer values for n and m. These functions are orthogonal over the interval of one period T, say, from t_0 to $t_0 + T$. For our purposes we shall use the value $t_0 = 0$ or $t_0 = -\frac{1}{2}T$ but with the understanding that any period may be used simply by replacing the limits 0 to T in the integrals to follow by t_0 to $t_0 + T$. The following definite integrals are valid for all integers m and n satisfying the given restrictions:

$$\int_0^T \sin m\omega_0 t\, dt = 0 \tag{8.34}$$

$$\int_0^T \cos n\omega_0 t\, dt = 0, \qquad n \neq 0 \tag{8.35}$$

$$\int_0^T \sin m\omega_0 t \cos n\omega_0 t\, dt = 0 \tag{8.36}$$

$$\int_0^T \sin m\omega_0 t \sin n\omega_0 t\, dt = 0, \qquad m \neq n \tag{8.37a}$$

$$= \tfrac{1}{2}T, \qquad m = n \tag{8.37b}$$

$$\int_0^T \cos m\omega_0 t \cos n\omega_0 t\, dt = 0, \qquad m \neq n \tag{8.38a}$$

$$= \tfrac{1}{2}T, \qquad m = n \tag{8.38b}$$

With these integrals the determination of the Fourier coefficients a_n and b_n will now proceed as follows. First multiply both sides of (8.25) by a suitable factor, integrate the resulting expression term by term over the interval of time from 0 to T, then make use of the integrals (8.34)–(8.38), which will render nearly all of the terms of the infinite series to the value zero, to obtain the desired formulas.

To evaluate a_0 we simply integrate both sides of (8.25) from 0 to T, the multiplying factor being 1. This gives

$$\int_0^T f(t)\,dt = \int_0^T a_0\,dt + \sum_{n=1}^{\infty} \int_0^T (a_n \cos n\omega_0 t + b_n \sin n\omega_0 t)\,dt$$

$$= a_0 T + 0 = a_0 T \tag{8.39}$$

or

$$a_0 = \frac{1}{T}\int_0^T f(t)\,dt = \frac{1}{T}\int_{t_0}^{t_0+T} f(t)\,dt \tag{8.40}$$

This indicates that a_0 is simply the mean or average value of $f(t)$ over a period. The mean value of a signal over a period is referred to as the *dc value* of the signal.

To compute a_n we multiply both sides of (8.25) by the factor $\cos m\omega_0 t$ and then integrate from 0 to T, yielding

$$\int_0^T f(t)\cos m\omega_0 t\,dt = \int_0^T a_0 \cos m\omega_0 t\,dt$$

$$+ \sum_{n=1}^{\infty}\int_0^T (a_n \cos n\omega_0 t + b_n \sin n\omega_0 t)\cos m\omega_0 t\,dt \tag{8.41}$$

According to (8.34) through (8.38), all the terms on the right-hand side of (8.41) are zero, except the one of the form of (8.38b) corresponding to $m=n$. The value of this nonzero term is $\frac{1}{2}T$. Thus we obtain

$$a_m = \frac{2}{T}\int_0^T f(t)\cos m\omega_0 t\,dt = \frac{2}{T}\int_{t_0}^{t_0+T} f(t)\cos m\omega_0 t\,dt \tag{8.42}$$

for $m = 1, 2, \ldots$.

Finally, to calculate the coefficients b_n we follow the same pattern by multiplying both sides of (8.25) by the factor $\sin m\omega_0 t$ and integrate from 0 to T to give

$$\int_0^T f(t)\sin m\omega_0 t\,dt = \int_0^T a_0 \sin m\omega_0 t\,dt$$

$$+ \sum_{n=1}^{\infty}\int_0^T (a_n \cos n\omega_0 t + b_n \sin n\omega_0 t)\sin m\omega_0 t\,dt \tag{8.43}$$

The only nonzero integral on the right-hand side of this equation is the one of the form (8.37b) corresponding to $m=n$, which has the value $\frac{1}{2}T$. Thus the value of b_m is given by the equation

$$b_m = \frac{2}{T}\int_0^T f(t)\sin m\omega_0 t\,dt = \frac{2}{T}\int_{t_0}^{t_0+T} f(t)\sin m\omega_0 t\,dt \tag{8.44}$$

for $m = 1, 2, \ldots$. We remark that in (8.42) and (8.44) the value of t_0 is arbitrary; any real value can be used.

Our conclusion is that if a periodic signal $f(t)$ is represented as an infinite series in (8.25), then the a and b coefficients can be evaluated by the formulas (8.40), (8.42), and (8.44). This, however, does not prove that any particular periodic signal can be expanded into an infinite series as in (8.25). To prove that a particular periodic signal has a Fourier series expansion, we must show that if the a and b coefficients as computed above are inserted into (8.25), then the infinite series converges to $f(t)$. It can be shown that if a periodic signal satisfies the Dirichlet conditions to be discussed in Section 8-7, this is always the case. However, the proof will not be given here.

We illustrate the computation of the Fourier coefficients by the following examples.

EXAMPLE 8.5

We will now calculate the coefficients for the Fourier series expansion of the train of voltage pulses of Figure 8.8. The voltage waveform can be written as

$$
\begin{aligned}
v_g(t) &= 1, && 0 < t < 1 \\
&= 2, && 1 < t < 2 \\
&= 0, && 2 < t < 4
\end{aligned}
\tag{8.45}
$$

The period of the signal $v_g(t)$ is $T = 4$, giving the fundamental frequency $\omega_0 = 2\pi/T = \pi/2$. To compute a_0, we use (8.40) and obtain

$$
a_0 = \frac{1}{4}\int_0^1 1\, dt + \frac{1}{4}\int_1^2 2\, dt = \frac{3}{4}
\tag{8.46}
$$

which is the average value over one period. Alternatively a_0 can be obtained directly from Figure 8.8 by inspection without using (8.40)

The values of a_m are computed from (8.42), as follows:

$$
\begin{aligned}
a_m &= \frac{2}{4}\left[\int_0^1 \cos \tfrac{1}{2}m\pi t\, dt + \int_1^2 2\cos \tfrac{1}{2}m\pi t\, dt\right] \\
&= \frac{1}{m\pi}\left[\sin \tfrac{1}{2}m\pi + 2\sin m\pi - 2\sin \tfrac{1}{2}m\pi\right] \\
&= -\frac{1}{m\pi}\sin \tfrac{1}{2}m\pi, \qquad m = 1, 2, \dots
\end{aligned}
\tag{8.47}
$$

Likewise the coefficients b_m are calculated from (8.44):

$$
\begin{aligned}
b_m &= \frac{2}{4}\left[\int_0^1 \sin \tfrac{1}{2}m\pi t\, dt + \int_1^2 2\sin \tfrac{1}{2}m\pi t\, dt\right] \\
&= -\frac{1}{m\pi}\left[\cos \tfrac{1}{2}m\pi + 2\cos m\pi - 2\cos \tfrac{1}{2}m\pi - 1\right] \\
&= \begin{cases} \dfrac{3}{m\pi}, & m \text{ odd} \\[2mm] \dfrac{(-1)^{m/2} - 1}{m\pi}, & m \text{ even} \end{cases}
\end{aligned}
\tag{8.48}
$$

Substituting these into (8.25) gives the desired Fourier series representation (8.32) of the voltage pulses of Figure 8.8:

$$v_g(t) = \frac{3}{4} + \frac{1}{\pi} \sum_{k=1}^{\infty} \left\{ \left[\frac{(-1)^k - 1}{2k} \right] \sin k\pi t \right.$$

$$\left. + \frac{1}{2k-1} \left[(-1)^k \cos \frac{(2k-1)\pi t}{2} + 3 \sin \frac{(2k-1)\pi t}{2} \right] \right\} \quad \textbf{(8.49)}$$

EXAMPLE 8.6

We wish to represent the rectangular-wave voltage signal of Figure 8.10 by a Fourier series. The waveform can be written as

$$v(t) = 1, \qquad 0 < t < \frac{T}{2}$$

$$= -1, \qquad \frac{T}{2} < t < T \qquad \textbf{(8.50)}$$

The Fourier coefficients are computed from (8.40), (8.42), and (8.44) with $\omega_0 = 2\pi/T$, as follows:

$$a_0 = \frac{1}{T} \left[\int_0^{T/2} dt - \int_{T/2}^T dt \right] = 0 \qquad \textbf{(8.51)}$$

$$a_m = \frac{2}{T} \left[\int_0^{T/2} \cos m\omega_0 t \, dt - \int_{T/2}^T \cos m\omega_0 t \, dt \right] = 0 \qquad \textbf{(8.52)}$$

$$b_m = \frac{2}{T} \left[\int_0^{T/2} \sin m\omega_0 t \, dt - \int_{T/2}^T \sin m\omega_0 t \, dt \right]$$

$$= \frac{1}{m\pi} [-2 \cos m\pi + \cos 2m\pi + 1]$$

$$= \begin{cases} 0, & m \text{ even} \\ \dfrac{4}{m\pi}, & m \text{ odd} \end{cases} \qquad \textbf{(8.53)}$$

Figure 8.10
Representation of a rectangular-wave voltage signal in the time domain. Notice that this waveform has odd symmetry.

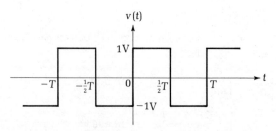

Figure 8.11
Amplitude and phase
spectra of the cosine
Fourier series repre-
sentation of the time
domain rectangular-wave
signal of Figure 8.10.

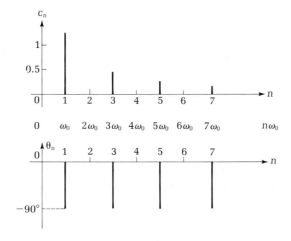

Substituting these in (8.25), the desired Fourier series is

$$v(t) = \frac{4}{\pi} \left[\sin \omega_0 t + \frac{1}{3} \sin 3\omega_0 t + \frac{1}{5} \sin 5\omega_0 t + \cdots \right]$$

$$= \frac{4}{\pi} \sum_{n=1}^{\infty} \frac{1}{2n - 1} \sin [(2n - 1)\omega_0 t] \tag{8.54}$$

with the amplitude and phase spectra presented in polar form in Figure
8.11 as a function of n. The horizontal axis may be frequency scaled so that
the plots appear as a function of $n\omega_0$. We remark that in plotting the amplitude and phase spectra, we convert (8.54) to the form of (8.30). In doing so
we used the relation $\sin n\omega_0 t \equiv \cos (n\omega_0 t - \pi/2)$.

Observe that in (8.54) all the cosine terms are missing. We next show
that this is a direct consequence of certain symmetries in the signal. We will
make use of this information in simplifying the computation of the Fourier
coefficients, thus avoiding the needless work of evaluating integrals that
result in zero values.

8-4   SYMMETRY PROPERTIES OF FOURIER SERIES

In this section we show that if a signal possesses certain identifiable waveform properties, these properties may be used to simplify the evaluation of
the Fourier coefficients.

A function $f(t)$ is said to be *even* if it satisfies the condition

$$f(-t) = f(t) \tag{8.55}$$

Examples of the even functions are the familiar functions $\cos n\omega_0 t$ and the constant a_0. Other examples are $\sin^2 t$, t^{2n}, $|\sin t|$, and $e^{-|t|}$. Geometrically a function is even if its graph is symmetric in the vertical axis. For instance, the waveforms of Figures 8.1(a) and (c) are even.

On the other hand, a function $f(t)$ is said to be *odd* if it satisfies the condition

$$f(-t) = -f(t) \tag{8.56}$$

Examples of the odd functions are $\sin n\omega_0 t$, t, and $\tan^{-1} t$. Geometrically, a function is odd if its graph is symmetric in the origin. The rectangular-wave voltage of Figure 8.10 is odd because it is symmetric with respect to the origin.

In the following we shall investigate in detail what effect the symmetry of $f(t)$ has on the evaluation of the Fourier coefficients. Before we do this, however, we recognize the following relationships:

Even function × odd function = odd function

Even function × even function = even function

Odd function × odd function = even function

For any even function $f_e(t)$,

$$\int_{-t_0}^{t_0} f_e(t)\, dt = 2 \int_0^{t_0} f_e(t)\, dt \tag{8.57}$$

For any odd function $f_o(t)$,

$$\int_{-t_0}^{t_0} f_o(t)\, dt = 0 \tag{8.58}$$

Equations (8.57) and (8.58) hold for any t_0.

EVEN FUNCTIONS

If $f(t)$ is an even function—namely $f(t) = f(-t)$—we show that the Fourier series coefficient formulas (8.40), (8.42), and (8.44) reduce to

$$a_0 = \frac{2}{T} \int_0^{T/2} f(t)\, dt \tag{8.59}$$

$$a_m = \frac{4}{T} \int_0^{T/2} f(t) \cos m\omega_0 t\, dt \tag{8.60}$$

$$b_m = 0 \tag{8.61}$$

We shall only verify (8.60) and (8.61), since the time average (8.59) is readily apparent from the symmetry. To prove (8.60), we choose $t_0 = -T/2$ in

(8.42), so that

$$a_m = \frac{2}{T} \int_{-T/2}^{T/2} f(t) \cos m\omega_0 t \, dt$$

$$= \frac{2}{T} \int_{-T/2}^{0} f(t) \cos m\omega_0 t \, dt + \frac{2}{T} \int_{0}^{T/2} f(t) \cos m\omega_0 t \, dt$$

$$= \frac{2}{T} \int_{0}^{T/2} f(-t) \cos(-m\omega_0 t) \, dt + \frac{2}{T} \int_{0}^{T/2} f(t) \cos m\omega_0 t \, dt$$

$$= \frac{2}{T} \int_{0}^{T/2} f(t) \cos m\omega_0 t \, dt + \frac{2}{T} \int_{0}^{T/2} f(t) \cos m\omega_0 t \, dt$$

$$= \frac{4}{T} \int_{0}^{T/2} f(t) \cos m\omega_0 t \, dt \tag{8.62}$$

Likewise from (8.44) we have

$$b_m = \frac{2}{T} \int_{-T/2}^{T/2} f(t) \sin m\omega_0 t \, dt$$

$$= \frac{2}{T} \int_{0}^{T/2} f(-t) \sin(-m\omega_0 t) \, dt + \frac{2}{T} \int_{0}^{T/2} f(t) \sin m\omega_0 t \, dt$$

$$= -\frac{2}{T} \int_{0}^{T/2} f(t) \sin m\omega_0 t \, dt + \frac{2}{T} \int_{0}^{T/2} f(t) \sin m\omega_0 t \, dt$$

$$= 0 \tag{8.63}$$

In words it states that the Fourier series representation of an even periodic signal contains only cosine terms and a constant.

EXAMPLE 8.7
Figure 8.12 is a full-wave rectifier circuit, the output voltage waveform of which is shown in Figure 8.13. We wish to determine its Fourier series and the corresponding amplitude and phase spectra.

The voltage waveform is an even periodic signal with period π. The fundamental frequency is found to be $\omega_0 = 2\pi/T = 2$. Thus the symmetry (8.55) requires that $b_m = 0$ for all m. To find a_0 and a_m, we use (8.59) and

Figure 8.12
Circuit of a typical
full-wave rectifier.

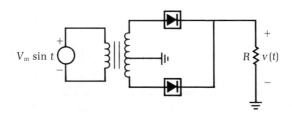

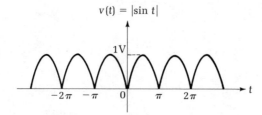

Figure 8.13
Output voltage waveform
of the rectifier circuit of
Figure 8.12. Note that this
waveform has even
symmetry.

(8.60) to obtain

$$a_0 = \frac{2}{\pi} \int_0^{\pi/2} \sin t \, dt = \frac{2}{\pi} \tag{8.64}$$

$$a_m = \frac{4}{\pi} \int_0^{\pi/2} \sin t \cos m\omega_0 t \, dt$$

$$= \frac{2}{\pi}\left[\frac{1}{1-2m} + \frac{1}{1+2m}\right] = \frac{4}{(1-4m^2)\pi} \tag{8.65}$$

The desired Fourier series is

$$v(t) = \frac{2}{\pi} - \frac{4}{\pi}\left[\frac{1}{3}\cos 2t + \frac{1}{15}\cos 4t + \frac{1}{35}\cos 6t + \cdots\right]$$

$$= \frac{2}{\pi} + \sum_{m=1}^{\infty} \frac{4}{(1-4m^2)\pi}\cos 2mt \tag{8.66}$$

The amplitude and phase spectra of the full-wave rectified sine waveform
are presented in Figure 8.14 as a function of n and $n\omega_0$.

Figure 8.14
Amplitude (a) and phase (b) spectra of the full-wave rectified waveform of Figure 8.13.

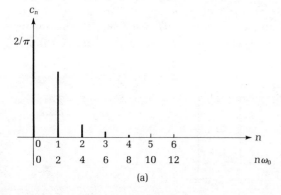

(a)

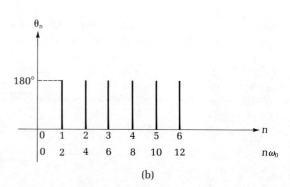

(b)

Odd Functions

By a similar analysis, it is straightforward to show that if $f(t)$ is odd—namely, $f(-t) = -f(t)$—the Fourier series coefficient formulas reduce to

$$a_0 = 0 \tag{8.67}$$

$$a_m = 0 \tag{8.68}$$

$$b_m = \frac{4}{T} \int_0^{T/2} f(t) \sin m\omega_0 t \, dt \tag{8.69}$$

The justification is left as an exercise (Problem 8.11). Thus the Fourier series expansion of an odd periodic signal contains only sine terms.

EXAMPLE 8.8

The rectangular-wave voltage signal $v(t)$ of Figure 8.10 is odd. Therefore $a_0 = 0$ and $a_m = 0$ for all m. To compute b_m, we use (8.69) and obtain

$$b_m = \frac{4}{T} \int_0^{T/2} 1 \sin m\omega_0 t \, dt = \frac{2[1 - (-1)^m]}{m\pi}$$

$$= \begin{cases} 0, & m \text{ even} \\ \dfrac{4}{m\pi}, & m \text{ odd} \end{cases} \tag{8.70}$$

confirming (8.53).

The advantages of the symmetry formulas for the evaluation of the Fourier series coefficients are that, if a signal waveform possesses a certain symmetry, not only do we know in advance what coefficients will be zero but also the values of the other coefficients can be obtained by integrating over only half of the period.

Half-Wave Symmetry

A periodic signal $f(t)$ of period T is said to possess *odd half-wave symmetry*[†] if it satisfies the condition

$$f(t) = -f\left(t + \frac{T}{2}\right) \tag{8.71}$$

Geometrically, this means that the signal waveform with t increasing from $-T/2$ to 0 is the negative of the waveform with t increasing from 0 to $T/2$.

[†] For a reason that we will make clear shortly, we consider the odd rather than even half-wave symmetry initially.

Figure 8.15
Representation of two signal waveforms that exhibit odd half-wave symmetry.

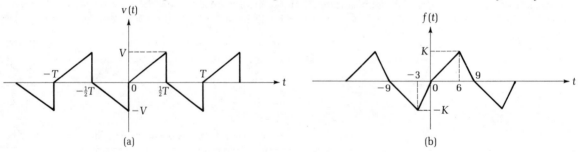

(a)   (b)

Two such signals are shown in Figure 8.15. In this case, the Fourier series expansion contains only odd harmonics:

$$a_0 = 0, \qquad a_m = b_m = 0, \qquad\qquad m \text{ even} \tag{8.72}$$

$$a_m = \frac{4}{T} \int_0^{T/2} f(t) \cos m\omega_0 t\, dt, \qquad m \text{ odd} \tag{8.73}$$

$$b_m = \frac{4}{T} \int_0^{T/2} f(t) \sin m\omega_0 t\, dt, \qquad m \text{ odd} \tag{8.74}$$

We shall verify the above only for a_m, leaving the others as an exercise (Problem 8.15). From (8.42) we have

$$a_m = \frac{2}{T} \int_0^{T/2} f(t) \cos m\omega_0 t\, dt + \frac{2}{T} \int_{T/2}^{T} f(t) \cos m\omega_0 t\, dt$$

$$= \frac{2}{T} \int_0^{T/2} f(t) \cos m\omega_0 t\, dt + \frac{2}{T} \int_0^{T/2} f\left(x + \frac{T}{2}\right) \cos\left[m\omega_0 \left(x + \frac{T}{2}\right)\right] dx$$

$$= \frac{2}{T} \int_0^{T/2} f(t) \cos m\omega_0 t\, dt - \frac{2}{T} \int_0^{T/2} f(x) \cos m\omega_0 x \cos m\pi\, dx$$

$$= \frac{2}{T} \int_0^{T/2} f(t) \cos m\omega_0 t\, dt - (-1)^m \frac{2}{T} \int_0^{T/2} f(t) \cos m\omega_0 t\, dt$$

$$= \begin{cases} 0, & m \text{ even} \\ \dfrac{4}{T} \displaystyle\int_0^{T/2} f(t) \cos m\omega_0 t\, dt, & m \text{ odd} \end{cases} \tag{8.75}$$

Likewise $f(t)$ possesses the *even half-wave* symmetry if

$$f(t) = f\left(t + \frac{T}{2}\right) \tag{8.76}$$

Figure 8.16
Examples of sawtooth signal waveforms that possess even half-wave symmetry.

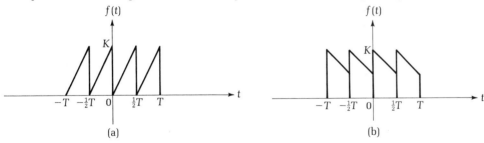

(a)   (b)

In this case the signal waveform with t increasing from $-T/2$ to 0 is the same as the waveform with t increasing from 0 to $T/2$. Examples of waveforms having even half-wave symmetry are illustrated in Figure 8.16. Notice that the period of $f(t)$ is actually $T/2$ instead of T. As a result the formulas for the Fourier series coefficients are obtained simply by replacing m by $2m$ in (8.42) and (8.44) and applying (8.57). This gives

$$a_m = b_m = 0, \qquad\qquad\qquad m \text{ odd} \qquad\qquad\qquad\qquad \textbf{(8.77)}$$

$$a_0 = \frac{2}{T} \int_0^{T/2} f(t)\, dt \qquad\qquad\qquad\qquad\qquad\qquad \textbf{(8.78a)}$$

$$a_m = \frac{4}{T} \int_0^{T/2} f(t) \cos m\omega_0 t\, dt, \qquad m \text{ even} \qquad\qquad \textbf{(8.78b)}$$

$$b_m = \frac{4}{T} \int_0^{T/2} f(t) \sin m\omega_0 t\, dt, \qquad m \text{ even} \qquad\qquad \textbf{(8.79)}$$

Thus only even harmonics exist. We emphasize the difference between an even or odd harmonic and an even or odd function. For instance, $a_3 \cos 3\omega_0 t$ is an odd harmonic but an even function, and $a_4 \sin 4\omega_0 t$ is an even harmonic but an odd function.

EXAMPLE 8.9

We compute the Fourier series of the signal waveform of Figure 8.15(a). The waveform possesses the odd half-wave symmetry and therefore (8.72) through (8.74) apply, yielding

$$a_0 = 0, \qquad a_m = b_m = 0, \qquad m \text{ even} \qquad\qquad\qquad \textbf{(8.80)}$$

$$a_m = \frac{4}{T} \int_0^{T/2} \left(\frac{2Vt}{T}\right) \cos m\omega_0 t\, dt = -\frac{4V}{m^2\pi^2}, \qquad m \text{ odd} \qquad \textbf{(8.81)}$$

$$b_m = \frac{4}{T} \int_0^{T/2} \left(\frac{2Vt}{T}\right) \sin m\omega_0 t\, dt = \frac{2V}{m\pi}, \qquad m \text{ odd} \qquad \textbf{(8.82)}$$

8-5   EXPONENTIAL FOURIER SERIES

The trigonometric Fourier series discussed in the foregoing can be expressed in an equivalent form in terms of the exponentials. The exponential Fourier series will be considered in detail in this section.

From Euler's formula (4.112), the cosine and sine functions may be expressed as

$$\cos n\omega_0 t = \frac{1}{2}(e^{jn\omega_0 t} + e^{-jn\omega_0 t}) \tag{8.83}$$

$$\sin n\omega_0 t = -j\frac{1}{2}(e^{jn\omega_0 t} - e^{-jn\omega_0 t}) \tag{8.84}$$

Substituting these into the trigonometric Fourier series (8.25) results in

$$f(t) = a_0 + \sum_{n=1}^{\infty} \frac{1}{2}(a_n - jb_n)e^{jn\omega_0 t} + \frac{1}{2}(a_n + jb_n)e^{-jn\omega_0 t} \Bigg]$$

$$= \tilde{c}_0 + \sum_{n=1}^{\infty} [\tilde{c}_n e^{jn\omega_0 t} + \tilde{c}_{-n} e^{-jn\omega_0 t}] \tag{8.85}$$

where

$$\tilde{c}_0 = a_0, \qquad \tilde{c}_n = \frac{1}{2}(a_n - jb_n), \qquad \tilde{c}_{-n} = \frac{1}{2}(a_n + jb_n) \tag{8.86}$$

To simplify the expression (8.85), we let n take both positive and negative values including zero. In this way (8.85) can be simplified by rewriting

$$f(t) = \sum_{n=-\infty}^{\infty} \tilde{c}_n e^{jn\omega_0 t} \tag{8.87}$$

which is known as the *exponential form* of the Fourier series. The coefficients $\tilde{c}_n$ can be computed from (8.42) and (8.44) using (8.86) or more directly as follows:

$$\tilde{c}_n = \frac{1}{T}\int_0^T f(t)\cos n\omega_0 t\, dt - \frac{j}{T}\int_0^T f(t)\sin n\omega_0 t\, dt$$

$$= \frac{1}{T}\int_0^T f(t)(\cos n\omega_0 t - j\sin n\omega_0 t)\, dt$$

$$= \frac{1}{T}\int_0^T f(t)e^{-jn\omega_0 t}\, dt \tag{8.88}$$

More generally, for an arbitrary t_0,

$$\tilde{c}_n = \frac{1}{T}\int_{t_0}^{t_0+T} f(t)e^{-jn\omega_0 t}\, dt \tag{8.89}$$

This equation holds for n positive, negative, or zero. The coefficients $\tilde{c}_n$ are generally complex and are to be distinguished from the real c_n used in

Section 8-2 by the tilde. If we express $\tilde{c}_n$ in exponential form

$$\tilde{c}_n = |\tilde{c}_n| e^{j\phi_n} \tag{8.90}$$

then

$$\tilde{c}_{-n} = |\tilde{c}_n| e^{-j\phi_n} \tag{8.91}$$

and from (8.86) the amplitude is

$$|\tilde{c}_n| = \frac{1}{2}\sqrt{a_n^2 + b_n^2} = \frac{c_n}{2} \tag{8.92}$$

and the phase angle is

$$\phi_n = \tan^{-1}\frac{-b_n}{a_n} \tag{8.93}$$

for all $n > 0$ with $\tilde{c}_0 = a_0$.

If the exponential Fourier series is to be constructed in the form of (8.87), then the set of numbers $|\tilde{c}_n|$ and ϕ_n contains all the information needed. This information can be displayed by plotting $|\tilde{c}_n|$ and ϕ_n as a function of n or $n\omega_0$ known, as before, as the *amplitude* and *phase spectra*. These plots differ from the previous ones in that spectra are displayed with both positive and negative n, while the plots for the Fourier form (8.30) show only the nonnegative n. Notice, however, that the amplitudes $|\tilde{c}_n|$ of the coefficients of the exponential Fourier series are one-half the amplitudes c_n of the polar form of the cosine expansion of (8.30).

EXAMPLE 8.10

We compute the exponential Fourier series of the rectangular-wave voltage of Figure 8.10. The waveform over the period can be written as

$$v(t) = 1, \qquad 0 < t < \frac{T}{2}$$

$$= -1, \qquad \frac{T}{2} < t < T \tag{8.94}$$

From (8.88) we calculate the complex coefficients

$$\tilde{c}_n = \frac{1}{T}\int_0^{T/2} e^{-jn\omega_0 t}\,dt - \frac{1}{T}\int_{T/2}^{T} e^{-jn\omega_0 t}\,dt$$

$$= \frac{j}{n\pi}[(-1)^n - 1]$$

$$= \begin{cases} 0, & n \text{ even} \\[2mm] -\dfrac{j2}{n\pi}, & n \text{ odd} \end{cases} \tag{8.95}$$

Alternatively a_n and b_n could have been obtained from (8.52) and (8.53), and then the coefficients $\tilde{c}_n$ could have been calculated using (8.86). The result, of course, would have been the same. The resulting exponential Fourier series of the rectangular-wave voltage is

$$
\begin{aligned}
v(t) = j\frac{4}{\pi} \Bigg[\cdots &+ \frac{1}{10} e^{-j5\omega_0 t} + \frac{1}{6} e^{-j3\omega_0 t} + \frac{1}{2} e^{-j\omega_0 t} \\
&- \frac{1}{2} e^{j\omega_0 t} - \frac{1}{6} e^{j3\omega_0 t} - \frac{1}{10} e^{j5\omega_0 t} - \cdots \Bigg]
\end{aligned}
$$

$$
= -\frac{j2}{\pi} \sum_{n=-\infty}^{\infty} \frac{1}{2n-1} e^{j(2n-1)\omega_0 t} \tag{8.96}
$$

The amplitude and phase spectra for the rectangular-wave signal are presented in Figure 8.17 with $T = 2$.

When a signal waveform possesses odd half-wave symmetry (8.71), the coefficient formula (8.89) reduces to

$$
\tilde{c}_n = \frac{2}{T} \int_0^{T/2} f(t) e^{-jn\omega_0 t}\, dt, \qquad n \text{ odd} \tag{8.97a}
$$

$$
= 0, \qquad n \text{ even} \tag{8.97b}
$$

Similarly if a waveform has even half-wave symmetry (8.76), the exponential Fourier coefficients become

$$
\tilde{c}_n = \frac{2}{T} \int_0^{T/2} f(t) e^{-jn\omega_0 t}\, dt, \qquad n \text{ even} \tag{8.98a}
$$

$$
= 0, \qquad n \text{ odd} \tag{8.98b}
$$

EXAMPLE 8.11

Consider the waveform of Figure 8.15(a), which possesses odd half-wave symmetry. Applying (8.97) yields

$$
\tilde{c}_n = 0, \qquad n \text{ even} \tag{8.99a}
$$

$$
\tilde{c}_n = \frac{2}{T} \int_0^{T/2} \left(\frac{2Vt}{T} \right) e^{-jn\omega_0 t}\, dt
$$

$$
= -\frac{(2 + jn\pi)V}{n^2 \pi^2}, \qquad n \text{ odd} \tag{8.99b}
$$

The exponential Fourier series is

$$
v(t) = -\frac{V}{\pi^2} \sum_{n=-\infty}^{\infty} \frac{2 + j(2n-1)\pi}{(2n-1)^2} e^{j(2n-1)\omega_0 t} \tag{8.100}
$$

Figure 8.17

Amplitude (a) and (a′) and phase (b) and (b′) spectra of the exponential Fourier series representation of the time-domain rectangular-wave signal of Figure 8.10. Compare these spectra with the cosine form of Figure 8.11.

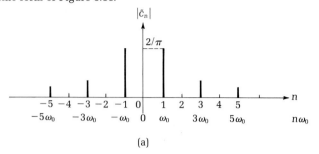

(a)

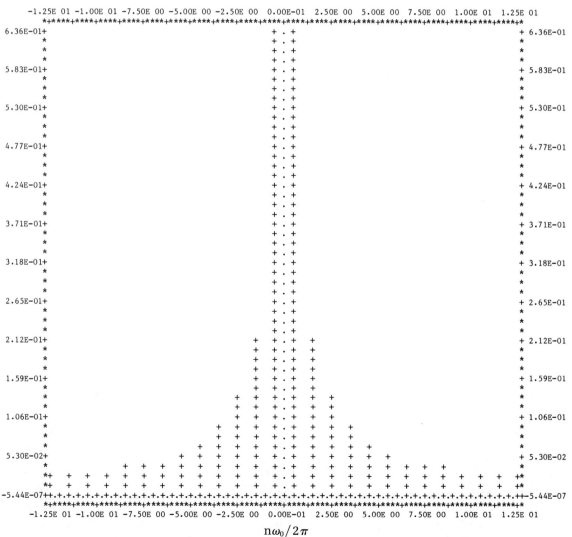

$$n\omega_0/2\pi$$

(a′)

Figure 8.17 (continued)

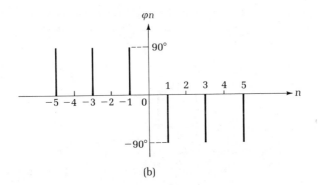

(b)

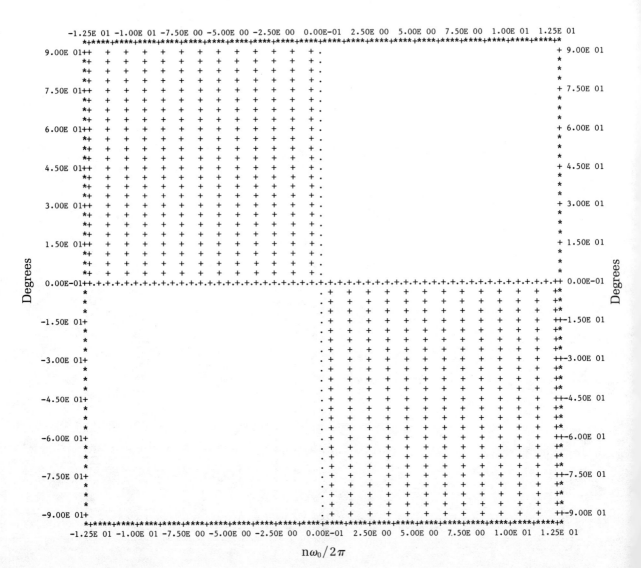

$n\omega_0/2\pi$

(b')

Figure 8.18

Two periodic waveforms representing half-wave rectifier outputs of differing phase.

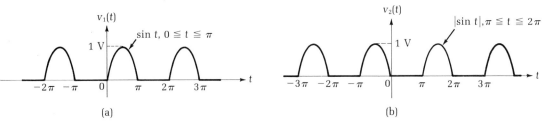

(a)   (b)

8-6   MANIPULATION OF SIGNALS

When a periodic signal $f(t)$ is translated, transposed, differentiated, or integrated, the Fourier coefficients of the resulting signal can be obtained from those of $f(t)$. In this section we shall derive such relations in terms of the exponential Fourier coefficients.

Consider the sum of two periodic signals, each of which has a period T. Their sum $f(t) = f_1(t) + f_2(t)$ is also periodic; the period of the resulting waveform is T or T/k for some positive integer k. To demonstrate this, consider the two half-wave rectified waveforms of Figure 8.18, each of which has a period 2π. Their sum is the full-wave rectified sine wave of Figure 8.13 with period π. It is clear that the nth Fourier coefficient of $f(t)$, expressed as

$$f(t) = \sum_{n=-\infty}^{\infty} \tilde{c}_{n1} e^{jn\omega_0 t} + \sum_{n=-\infty}^{\infty} \tilde{c}_{n2} e^{jn\omega_0 t} = \sum_{n=-\infty}^{\infty} (\tilde{c}_{n1} + \tilde{c}_{n2}) e^{jn\omega_0 t} \quad \textbf{(8.101)}$$

equals the sum of the nth Fourier coefficients of $f_1(t)$ and $f_2(t)$.

We next investigate the effect on the frequency spectrum of a periodic signal caused by a time displacement of the signal. Let $f(t)$ be a periodic signal of period T, which has a Fourier series given by

$$f(t) = \sum_{n=-\infty}^{\infty} \tilde{c}_n e^{jn\omega_0 t} \quad \textbf{(8.102)}$$

We consider the Fourier series of the time-delayed periodic signal $f(t - t_0)$ obtained from $f(t)$ by a time displacement of t_0. From (8.102) we have

$$f(t - t_0) = \sum_{n=-\infty}^{\infty} \tilde{c}_n e^{jn\omega_0(t - t_0)}$$

$$= \sum_{n=-\infty}^{\infty} (\tilde{c}_n e^{-jn\omega_0 t_0}) e^{jn\omega_0 t}$$

$$= \sum_{n=-\infty}^{\infty} \tilde{c}_n' e^{jn\omega_0 t} \quad \textbf{(8.103)}$$

where

$$\tilde{c}_n' = \tilde{c}_n e^{-jn\omega_0 t_0} \tag{8.104}$$

This gives the following property.

Property 8.1 *The nth Fourier coefficient of $f(t - t_0)$ equals the nth Fourier coefficient of $f(t)$ multiplied by $\exp(-jn\omega_0 t_0)$.*

The result shows that the amplitude spectrum is unchanged due to a time displacement. The phases, however, are different. The nth harmonic of the displaced signal lags behind by an amount of $n\omega_0 t_0$ radians. Physically, this simply means that if a periodic signal is shifted in time, the waveform is unchanged except that it is delayed by t_0 seconds. For a sinusoidal signal of frequency $n\omega_0$, a delay of t_0 seconds is equivalent to a phase shift of $-n\omega_0 t_0$ radians:

$$\sin n\omega_0(t - t_0) = \sin(n\omega_0 t - n\omega_0 t_0) \tag{8.105}$$

If a periodic signal $f(t)$ of period T is contracted in time by a factor κ, the resultant signal, $f(\kappa t)$, is also periodic but has a contracted period T/κ. The fundamental frequency of $f(\kappa t)$ is therefore $\kappa\omega_0$, where ω_0 is the fundamental frequency of $f(t)$. For values of the scaling factor κ less than unity, the signal is extended in time, the period is extended, and the harmonics are scaled down. The Fourier coefficients $\tilde{c}_n''$ of $f(\kappa t)$ can be computed by

$$\tilde{c}_n'' = \frac{\kappa}{T} \int_0^{T/\kappa} f(\kappa t)e^{-jn\kappa\omega_0 t}\, dt = \frac{1}{T} \int_0^T f(\tau)e^{-jn\omega_0 \tau}\, d\tau = \tilde{c}_n \tag{8.106}$$

where $\tau = \kappa t$. We conclude the following.

Property 8.2 *The Fourier coefficients of a periodic signal do not change when the time scale is changed. However, the frequencies of the harmonics are changed by a factor equal to the scaling factor.*

If we differentiate a periodic signal $f(t)$ of period T, the resultant signal is also periodic with period T. For $n \neq 0$ the Fourier coefficient of $\dot{f}(t) = df(t)/dt$ can be computed by

$$\frac{1}{T} \int_0^T \dot{f}(t)e^{-jn\omega_0 t}\, dt = \frac{1}{T} [f(t)e^{-jn\omega_0 t}]\Big|_0^T$$

$$+ \frac{jn\omega_0}{T} \int_0^T f(t)e^{-jn\omega_0 t}\, dt$$

$$= \frac{1}{T} [f(T)e^{-j2n\pi} - f(0)] + jn\omega_0\tilde{c}_n = jn\omega_0\tilde{c}_n \tag{8.107}$$

where $\tilde{c}_n$ is the nth Fourier coefficient of $f(t)$. The integral was evaluated by parts by letting $u = \exp(-jn\omega_0 t)$ and $dv = df$. Therefore we see that upon differentiation of a periodic signal, both its amplitude and its phase spectra change in the frequency domain. This gives the following.

Property 8.3   *For* $n \neq 0$ *the nth Fourier coefficient of* $\dot{f}(t)$ *equals the nth Fourier coefficient of* $f(t)$ *multiplied by* $jn\omega_0$.

Similarly for a periodic signal of period T defined over the interval $0 \leq t \leq T$ by the equation

$$g(t) = \int_0^t f(\tau)\, d\tau \qquad\qquad (8.108)$$

we can show that (Problem 8.18) the following obtains.

Property 8.4   *For* $n \neq 0$ *the nth Fourier coefficient of* $g(t)$ *equals the nth Fourier coefficient of* $f(t)$ *divided by* $jn\omega_0$.

Thus the harmonics in $\dot{f}(t)$ are more prominent and those in $\int f(t)\, dt$ are less significant than the corresponding harmonics in $f(t)$. The differentiation of a signal accentuates its fast variations, whereas integration of the signal reduces the effect of the fast variations. Notice also in the frequency domain the 90° phase lead in the phase spectrum due to differentiation and the 90° phase lag due to integration.

EXAMPLE 8.12
Consider the periodic rectangular-wave voltage $v(t)$ of Figure 8.10. Suppose that we shift the horizontal axis upward by 1 volt; the resulting negatively biased periodic voltage signal $v_1(t)$ is shown in Figure 8.19(a), giving

Figure 8.19
(a) Periodic rectangular-wave signal of Figure 8.10 biased negatively and possessing odd symmetry. (b) Periodic triangular-wave signal with even symmetry. (c) Periodic wave form that is the sum of rectangular (a) and triangular (b) waveforms and possesses odd half-wave symmetry.

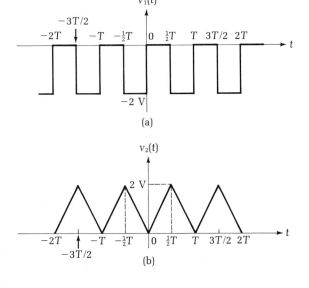

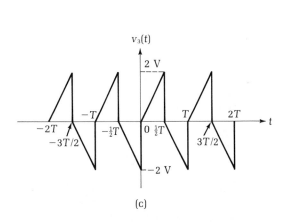

$v_1(t) = v(t) - 1$. Since in the Fourier series for $v_1(t)$, -1 combines only with a_0 or c_0 of $v(t)$, the shifting of the horizontal axis affects only the dc value. Thus from (8.54), the Fourier series for $v_1(t)$ is

$$v_1(t) = -1 + \frac{4}{\pi} \sum_{n=1}^{\infty} \frac{1}{2n-1} \sin[(2n-1)\omega_0 t] \qquad (8.109)$$

A triangular or sawtooth voltage waveform $v_2(t)$ is shown in Figure 8.19(b). This is an even function, so the symmetry condition requires that $b_n = 0$ for all n. From (8.59) and (8.60), we obtain $a_0 = 1$ and

$$a_n = \frac{4}{T} \int_0^{T/2} \left(\frac{4t}{T}\right) \cos n\omega_0 t \, dt$$

$$= \frac{4}{n^2\pi^2}[(-1)^n - 1] = \begin{cases} 0, & n \text{ even} \\ -\dfrac{8}{n^2\pi^2}, & n \text{ odd} \end{cases} \qquad (8.110)$$

giving the Fourier series of $v_2(t)$ as

$$v_2(t) = 1 - \frac{8}{\pi^2} \sum_{n=1}^{\infty} \frac{1}{(2n-1)^2} \cos[(2n-1)\omega_0 t] \qquad (8.111)$$

Observe that the voltage waveform $v_3(t)$ of Figure 8.19(c) is the sum of the voltage waveforms of (a) and (b). The Fourier coefficients of $v_3(t)$ are simply the sums of the corresponding Fourier coefficients of $v_1(t)$ and $v_2(t)$, yielding

$$v_3(t) = \frac{4}{\pi} \sum_{n=1}^{\infty} \left\{ \frac{-2}{(2n-1)^2\pi} \cos[(2n-1)\omega_0 t] \right.$$

$$\left. + \frac{1}{2n-1} \sin[(2n-1)\omega_0 t] \right\} \qquad (8.112)$$

To express (8.112) in exponential form, we apply (8.86) and obtain

$$\tilde{c}_{2n-1} = \frac{1}{2}(a_{2n-1} - jb_{2n-1}) = -\frac{4}{(2n-1)^2\pi^2} - j\frac{2}{(2n-1)\pi}$$

$$= -\frac{4 + j2\pi(2n-1)}{(2n-1)^2\pi^2} \qquad (8.113)$$

The resulting exponential Fourier series is

$$v_3(t) = -\frac{2}{\pi^2} \sum_{n=-\infty}^{\infty} \frac{2 + j(2n-1)\pi}{(2n-1)^2} e^{j(2n-1)\omega_0 t} \qquad (8.114)$$

confirming (8.100) for $V = 2$.

Figure 8.20
Periodic triangular waveform (a) the derivative of which is the rectangular waveform (b).

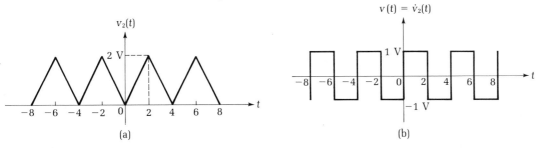

(a)

(b)

EXAMPLE 8.13

In the periodic triangular waveform of Figure 8.19(b), let $T = 4$. This is redrawn in Figure 8.20(a). with the slope shown in Figure 8.20(b). Since the voltage signal $v(t)$ of (b) is the derivative of that of (a), according to Property 8.3 the nth $(n \neq 0)$ exponential Fourier coefficient $\tilde{c}_n$ of $v(t)$ equals the nth exponential Fourier coefficient of $v_2(t)$, which from (8.110) is given by

$$\frac{1}{2}(a_n - jb_n) = -\frac{4}{n^2\pi^2}, \qquad n \text{ odd} \tag{8.115}$$

multiplied by $jn\omega_0$, where $\omega_0 = 2\pi/T = \pi/2$. Hence we have

$$\tilde{c}_n = -\frac{j2}{n\pi}, \qquad n \text{ odd} \tag{8.116}$$

confirming (8.95).

EXAMPLE 8.14

We wish to compute the exponential Fourier series for the sawtooth voltage waveform $v_a(t)$ of Figure 8.21(a). For this we first consider the waveform $v_2(t)$ of Figure 8.20(a). The period of this waveform $v_2(t)$ is contracted by a factor 4 and the resulting waveform, as shown in Figure 8.21(b), can be written as

$$v_b(t) = v_2(4t) \tag{8.117}$$

By Property 8.2 the exponential Fourier coefficients of $v_b(t)$ are the same as those of $v_2(t)$, but the frequencies of the harmonics are increased from the original $2\pi/4 = \pi/2$ to $4(\pi/2) = 2\pi$. Thus from (8.115), the Fourier series of $v_b(t)$ is given by

$$v_b(t) = 1 - \frac{4}{\pi^2} \sum_{n=-\infty}^{\infty} \frac{1}{(2n-1)^2} e^{j2(2n-1)\pi t} \tag{8.118}$$

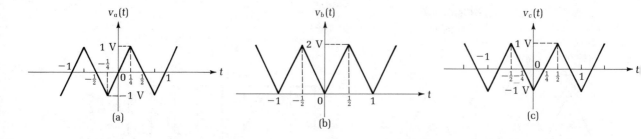

(a)　(b)　(c)

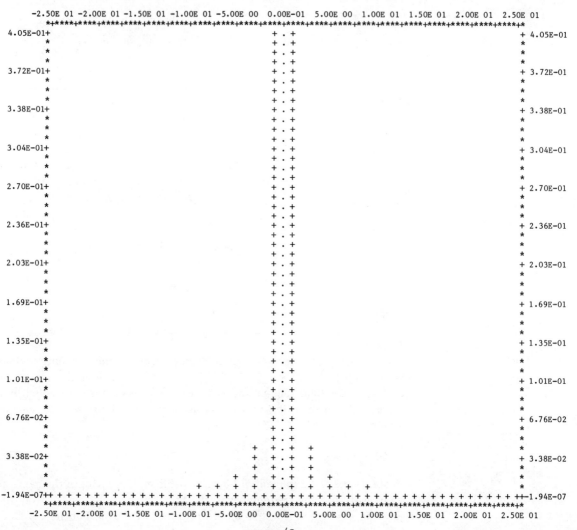

$$n\omega_0/2\pi$$

(d)

Figure 8.21

The periodic triangular-wave voltage of (a) can be achieved by the following steps: (1) contract $v_2(t)$ of Figure 8.20(a) by a factor of 4 to obtain $v_b(t)$ shown in (b), (2) negatively bias $v_b(t)$ by 1 V to obtain $v_c(t)$ shown in (c), and (3) delay $v_c(t)$ by $\frac{1}{4}$ sec to obtain $v_a(t)$ shown in (a). (d) Amplitude spectrum of the periodic triangular waveform of (a). (e) Phase spectrum of the periodic triangular waveform of (a).

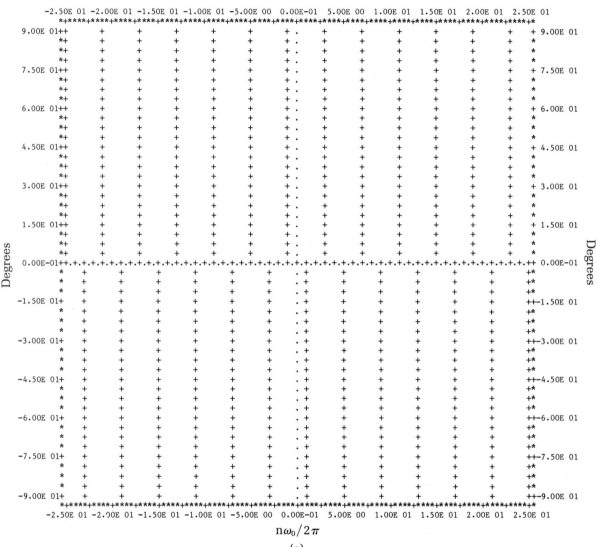

$$n\omega_0/2\pi$$

(e)

We next shift the horizontal axis of Figure 8.21(b) upward by one volt, and obtain the waveform of (c). This is equivalent to adding -1 to (8.118) to yield the Fourier series of $v_c(t)$:

$$v_c(t) = -\frac{4}{\pi^2} \sum_{n=-\infty}^{\infty} \frac{1}{(2n-1)^2} e^{j2(2n-1)\pi t} \tag{8.119}$$

Finally, to produce the waveform of Figure 8.21(a), we move the vertical axis of (c) to the right by $\frac{1}{4}$; this is equivalent to shifting the waveform to the left by $\frac{1}{4}$. The resulting waveform is shown in Figure 8.21(a) with $v_a(t) = v_c(t + \frac{1}{4})$. By Property 8.1 the Fourier coefficients of $v_a(t)$ and those of $v_c(t)$ are related only by $\exp(jn\pi/2)$. Consequently

$$v_a(t) = -\frac{4}{\pi^2} \sum_{n=-\infty}^{\infty} \frac{1}{(2n-1)^2} e^{j(2n-1)\pi/2} e^{j2(2n-1)\pi t}$$

$$= \frac{j4}{\pi^2} \sum_{n=-\infty}^{\infty} \frac{(-1)^n}{(2n-1)^2} e^{j(4n-2)\pi t} \tag{8.120}$$

The corresponding trigonometric Fourier series becomes

$$v_a(t) = \frac{8}{\pi^2} \sum_{n=1}^{\infty} \frac{(-1)^{n+1}}{(2n-1)^2} \sin[(4n-2)\pi t] \tag{8.121}$$

The amplitude and phase spectra of $v_a(t)$ are presented in Figures 8.21(d) and (e).

It is important to note that the symmetry property of a waveform is not intrinsic. It depends upon its relation to the axes of the coordinate system. For instance, in Figure 8.21(a) to (c) all three waveforms are the same periodic triangular wave. If the axes are chosen as in Figure 8.21(a), the waveform represents an odd function; if chosen as in (b) and (c), the waveforms are even functions. On the other hand, if the vertical axis of Figure 8.21(c) is shifted to the right by $\frac{1}{5}$, the resulting waveform represents a function which is neither odd nor even. The chief value in choosing an appropriate coordinate system is that it allows considerable simplification in the calculation of the Fourier series coefficients. Once the coefficients are obtained, the compensation for the translation of the axes can be achieved by reversing the process, as illustrated above.

8-7 CONVERGENCE OF THE FOURIER SERIES

At this stage we must be careful not to delude ourselves with the belief that we have proved that any periodic signal can be represented by a Fourier series. What we have accomplished so far is merely that *if* a periodic signal has an expansion of the form (8.25) or (8.87) for which term-by-term integration is valid, then the coefficients in the series can be computed by the formulas derived. Questions concerning the existence of the Fourier series and the conditions under which it will validly represent the signal that was used to generate the series are many, difficult, and by no means completely answered. However, for almost any conceivable practical application, these questions are answered by the famous *theorem of Dirichlet,*[†] named after the French mathematician Peter Gustave Lejeune Dirichlet[‡] (1805–1859).

Dirichlet Conditions *Let $f(t)$ be a bounded periodic function. If in any one period $f(t)$ has at most a finite number of local maxima and minima and a finite number of points of discontinuity, then the Fourier series of $f(t)$ converges to $f(t)$ at all points where $f(t)$ is continuous, and the series converges to*

$$f(t_i) = \tfrac{1}{2}[f(t_i+) + f(t_i-)] \qquad (8.122)$$

at the points t_i of discontinuity. The function $f(t)$ is said to be bounded if

$$\int_0^T |f(t)| \, dt < \infty \qquad (8.123)$$

over the period T.

Thus a function need not be continuous in order to possess a valid Fourier expansion. This means that a waveform may consist of a number of disjointed arcs of different curves such as shown in Figure 8.22. The points of discontinuity occur at $t_1 = \tfrac{1}{2}T$ and $t_2 = 3T/4$. At each of these points, the series assumes a value midway between the two values of the function, regardless of the definition or the lack of definition of the function at these points. In this case the result is

$$f(\tfrac{1}{2}T) = \tfrac{1}{2}(K_1 + K_2) \qquad (8.124a)$$

$$f\left(\frac{3T}{4}\right) = \frac{1}{2} K_1 \qquad (8.124b)$$

as indicated by the dots in Figure 8.22.

Examples of functions that do not satisfy the Dirichlet conditions are $\sin(1/t)$ and $t^2 \sin(1/t)$.

[†] For a proof of this theorem, see, for example, H. S. Carslaw, *Fourier Series* (New York: Dover, 1930), pp. 225–32.
[‡] He was born of a French family in Duren, Germany, educated at Cologne and Paris (1822–1827), and became a friend of J. B. J. Fourier. He was appointed a professor at the University of Berlin in 1839 and was invited as successor to Karl F. Gauss to the University of Göttingen in 1855, where he spent his last four years as a professor.

Figure 8.22
Example of a signal with discontinuities at $t = \frac{1}{2}T$ and $t = 3T/4$.

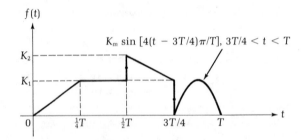

In practical applications the infinite Fourier series is truncated after a finite number of terms—say, the nth term. Let $s_n(t)$ be this partial sum. The rate at which this partial sum approaches the value of the function is important in that the more rapid the convergence of the partial sum, the fewer the number of terms required to attain the desired accuracy. The rate of convergence is directly related to the rate of decrease in the magnitude of the Fourier coefficients. The law governing the manner in which these coefficients decrease with increasing n is stated below without proof:[†]

> **Property 8.5** *The Fourier coefficients of a periodic function satisfying the Dirichlet conditions always approach zero at least as fast as M/n—that is,*
>
> $$|a_n| < \frac{M}{n} \quad \text{and} \quad |b_n| < \frac{M}{n} \tag{8.125}$$
>
> *where M is a constant independent of n. If the function has a jump discontinuity, then either $|a_n|$ or $|b_n|$ or, in general, both can decrease no faster than this. In general, if a function and its $k - 1$ derivatives satisfy the Dirichlet conditions and are everywhere continuous, then the coefficients $|a_n|$ and $|b_n|$ tend to zero at least as fast as M/n^{k+1}; that is,*
>
> $$|a_n| < \frac{M}{n^{k+1}} \quad \text{and} \quad |b_n| < \frac{M}{n^{k+1}} \tag{8.126}$$

For instance, all the Fourier series coefficients obtained so far tend to zero at least as fast as M/n. The periodic rectangular waveform of Figure 8.20(b) contains points of discontinuity, and thus the magnitude of its Fourier coefficients can diminish no faster than M/n for some constant M, as can be verified from (8.116).

More intuitively, Property 8.5 asserts that the smoother the signal, the faster its Fourier series converges. For most practical waveforms, it can be

[†] Carslaw, *Fourier series*, pp. 269–71.

shown[†] that if k is the number of times a function must be differentiated to produce a jump discontinuity, then the inequalities (8.126) hold for the magnitudes of its Fourier coefficients. For instance, a periodic triangular waveform such as the one shown in Figure 8.20(a) has Fourier coefficients that fall off as $1/n^2$, whereas a periodic rectangular waveform has coefficients that fall off as $1/n$, as previously indicated.

It is often possible to obtain a new Fourier series representation by means of term-by-term differentiation or integration of a known series. If a periodic function satisfies the Dirichlet conditions, it is permissible to integrate a Fourier series term by term to yield a series that is the Fourier series for the integral of the original time function. However, for differentiation the requirements are much more restrictive and are stated as:

Property 8.6 *If a periodic function* $f(t)$ *and its derivative* $\dot{f}(t)$ *satisfy the Dirichlet conditions and if* $f(t)$ *is everywhere continuous, then wherever it exists* $\dot{f}(t)$ *can be found by term-by-term differentiation of the Fourier series of* $f(t)$.

As an example, consider the periodic triangular waveform $v_a(t)$ of Figure 8.21(a); its Fourier series was found to be

$$v_a(t) = -\frac{8}{\pi^2} \sum_{n=1}^{\infty} \frac{(-1)^n}{(2n-1)^2} \sin\,[2(2n-1)\pi t] \qquad\qquad (8.127)$$

Integrating this series term-by-term yields

$$v_p(t) = \int_0^t v_a(\tau)\,d\tau$$

$$= -\frac{8}{\pi^2} \sum_{n=1}^{\infty} \frac{(-1)^n}{(2n-1)^2} \int_0^t \sin\,[2(2n-1)\pi\tau]\,d\tau$$

$$= \frac{4}{\pi^3} \sum_{n=1}^{\infty} \frac{(-1)^n}{(2n-1)^3} \{\cos\,[2(2n-1)\pi t] - 1\}$$

$$= \frac{1}{8} + \frac{4}{\pi^3} \sum_{n=1}^{\infty} \frac{(-1)^n}{(2n-1)^3} \cos\,[2(2n-1)\pi t] \qquad\qquad (8.128)$$

In arriving at (8.128), we made use of the relationship

$$\sum_{n=1}^{\infty} \frac{(-1)^{n+1}}{(2n-1)^3} = \frac{\pi^3}{32} \qquad\qquad (8.129)$$

The integral $\int_0^t v_a(\tau)\,d\tau$ in (8.128) can be evaluated over the time intervals

---

[†] I. S. Sokolnikoff and R. M. Redheffer, *Mathematics of Physics and Modern Engineering* (New York: McGraw-Hill, 1958), p. 211.

Figure 8.23
Four waveforms that are related through differentiation in going from parabolic waveform (a), triangular waveform (b), and rectangular waveform (c) to the impulse train (d); or through integration in going from (d) to (a).

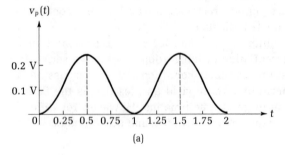

(a)

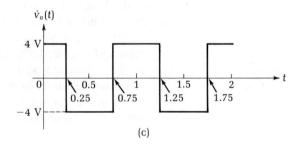

(c)

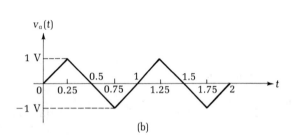

(b)

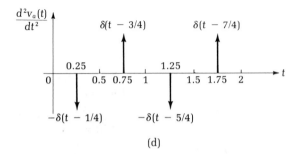

(d)

(0, 0.25), (0.25, 0.75), and (0.75, 1); the results are given by

$$v_p(t) = 2t^2, \qquad 0 \leq t < 0.25$$
$$= -2t^2 + 2t - 0.25, \qquad 0.25 \leq t < 0.75 \qquad \textbf{(8.130)}$$
$$= 2t^2 - 4t + 2, \qquad 0.75 \leq t < 1$$

The voltage $v_p(t)$ has the parabolic waveform of Figure 8.23(a). As a check we compute the area under $v_p(t)$ over a period and obtain the constant term a_0 of (8.128), as follows:

$$a_0 = \int_0^{0.25} 2t^2 \, dt + \int_{0.25}^{0.75} (-2t^2 + 2t - 0.25) \, dt$$
$$+ \int_{0.75}^1 (2t^2 - 4t + 2) \, dt$$
$$= \frac{1}{8} \qquad \textbf{(8.131)}$$

On the other hand, the derivative of the periodic triangular waveform $v_a(t)$ of Figure 8.23(b) leads to the Fourier series of the periodic rectangular

Table 8.1

Waveforms	k	Upper bound of falling of $\|a_n\|$ and $\|b_n\|$ with increasing n
Parabolic wave	2	$1/n^3$
Triangular wave	1	$1/n^2$
Square wave	0	$1/n$

waveform of Figure 8.23(c):

$$\dot{v}_a(t) = -\frac{16}{\pi} \sum_{n=1}^{\infty} \frac{(-1)^n}{2n-1} \cos\left[2(2n-1)\pi t\right] \tag{8.132}$$

The operation is permissible because $v_a(t)$ and $\dot{v}_a(t)$ satisfy the requirements of Property 8.6. Note that $\dot{v}_a(t)$ is not a continuous function of time. If we take derivative on both sides of (8.132), we obtain

$$\frac{d^2 v_a(t)}{dt^2} = 32 \sum_{n=1}^{\infty} (-1)^n \sin\left[2(2n-1)\pi t\right] \tag{8.133}$$

This series is seen to diverge for certain values of t. For instance, if $t = \frac{1}{4}$, (8.133) becomes

$$\frac{d^2 v_a(t)}{dt^2} = -32 \sum_{n=1}^{\infty} 1 = -\infty \tag{8.134}$$

Thus (8.133) is an incorrect representation of a train of impulses denoted by $d^2 v_a(t)/dt^2$ of Figure 8.23(d).

The application of (8.126) to the waveforms of Figure 8.23 is summarized in Table 8.1, using k as the number of times a function must be differentiated to produce a jump discontinuity.

8-8 GIBBS PHENOMENA

In representing a periodic signal by a truncated Fourier series $s_n(t)$, the resulting *truncation error* is the difference between the signal $f(t)$ and the partial sum $s_n(t)$:

$$\epsilon_n = f(t) - s_n(t) \tag{8.135}$$

Figure 8.24
Periodic square-wave
signal with even
symmetry.

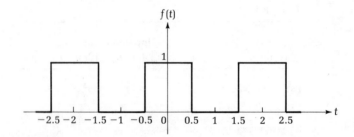

A useful measure of this error is the *mean-square error* defined by

$$E_n = \frac{1}{T} \int_0^T [\epsilon_n(t)]^2 \, dt \tag{8.136}$$

It can be shown that the truncated Fourier series $s_n(t)$ minimizes the value of E_n in the sense that no smaller E_n can be found for another trigonometric polynomial with the same number of terms.

If $f(t)$ is discontinuous and possesses jumps at the points t_i, the value of its Fourier series at t_i converges to

$$f(t_i) = \tfrac{1}{2}[f(t_i+) + f(t_i-)] \tag{8.137}$$

as indicated in (8.122). If the partial sum $s_n(t)$ is used for $f(t)$, then near the discontinuity points t_i the truncation error ϵ_n is appreciable no matter how many terms are chosen.

Consider, for example, the square wave of Figure 8.24, which has a Fourier series representation:

$$f(t) = \frac{1}{2} - \frac{2}{\pi} \sum_{n=1}^{\infty} \frac{(-1)^n}{2n-1} \cos[(2n-1)\pi t] \tag{8.138a}$$

$$= \frac{1}{2} - \frac{1}{\pi} \sum_{n=-\infty}^{\infty} \frac{(-1)^n}{2n-1} e^{j(2n-1)\pi t} \tag{8.138b}$$

The partial sums are given by

$$s_M(t) = \frac{1}{2} - \frac{2}{\pi} \sum_{n=1}^{M} \frac{(-1)^n}{2n-1} \cos[(2n-1)\pi t] \tag{8.139a}$$

$$= \frac{1}{2} - \frac{1}{\pi} \sum_{n=1-M}^{M} \frac{(-1)^n}{2n-1} e^{j(2n-1)\pi t} \tag{8.139b}$$

for $M = 1, 2, \ldots$. For a fixed M, there are $M + 1$ terms in (8.139a) and $2M + 1$ terms in (8.139b). The Fourier coefficients a_n, b_n, and $\tilde{c}_n$ and the frequency spectrum are given in Figure 8.25. For $M = 5, 9, 13, 17, 21,$ and 25,

Figure 8.25
(a) Trigonometric Fourier coefficients of the periodic square waveform of Figure 8.24.
(b) Exponential Fourier coefficients of the periodic square waveform of Figure 8.24.
(c) Amplitude spectrum of the periodic square waveform of Figure 8.24. (d) Phase
angles of the exponential Fourier coefficients listed in (b). (e) Phase spectrum of the
periodic square waveform of Figure 8.24.

(a)	(b)	
A(0)= 0.500000000000D 00	C(-25)= 0.126275030293D-01	-0.176580972067D-15
A(1)= 0.636611394765D 00	C(-24)= 0.929173404884D-15	-0.308364445090D-16
A(2)= -0.183725257230D-14	C(-23)= -0.137430836923D-01	0.165284452791D-15
A(3)= -0.212181457453D 00	C(-22)= -0.910022057710D-15	0.675015598972D-16
A(4)= 0.185540471875D-14	C(-21)= 0.150695467664D-01	-0.139832589952D-15
A(5)= 0.127282063815D 00	C(-20)= 0.877881101147D-15	-0.791033905045D-17
A(6)= -0.179600778694D-14	C(-19)= -0.166734891710D-01	0.136335387424D-15
A(7)= -0.908870311398D-01	C(-18)= -0.868471961013D-15	0.299205105136D-16
A(8)= 0.186606285979D-14	C(-17)= 0.186528472964D-01	-0.105693231944D-15
A(9)= 0.706601159608D-01	C(-16)= 0.899835761459D-15	-0.185962356625D-16
A(10)= -0.167726943445D-14	C(-15)= -0.211577899868D-01	0.134864341916D-15
A(11)= -0.577823420157D-01	C(-14)= -0.885569395592D-15	0.324462678947D-16
A(12)= 0.171118674785D-14	C(-13)= 0.244308973515D-01	-0.154182222545D-15
A(13)= 0.488617947030D-01	C(-12)= 0.855593373927D-15	-0.280608869474D-16
A(14)= -0.177113879118D-14	C(-11)= -0.288911710079D-01	0.119459997450D-15
A(15)= -0.423155799736D-01	C(-10)= -0.838634717226D-15	0.508204589522D-16
A(16)= 0.179967152292D-14	C(-9)= 0.353300579804D-01	-0.139138700561D-15
A(17)= 0.373056945929D-01	C(-8)= 0.933031429895D-15	-0.238697950294D-15
A(18)= -0.173694392203D-14	C(-7)= -0.454435155699D-01	0.142746925391D-15
A(19)= -0.333469783420D-01	C(-6)= -0.898003893468D-15	0.353300722011D-15
A(20)= 0.175576220229D-14	C(-5)= 0.636410319075D-01	-0.178463351208D-15
A(21)= 0.301390935329D-01	C(-4)= 0.927702359377D-15	-0.489643048329D-15
A(22)= -0.182004411542D-14	C(-3)= -0.106090728726D 00	-0.153765888911D-15
A(23)= -0.274861673845D-01	C(-2)= -0.918626286150D-15	0.605064609527D-15
A(24)= 0.185834680977D-14	C(-1)= 0.318305697383D 00	0.559635671138D-15
A(25)= 0.252550060587D-01	C(0)= 0.500000000000D 00	0.000000000000D 00
B(1)= 0.111927134228D-14	C(1)= 0.318305697383D 00	-0.559635671138D-15
B(2)= 0.121012921905D-14	C(2)= -0.918626286150D-15	-0.605064609527D-15
B(3)= -0.307531777821D-16	C(3)= -0.106090728726D 00	0.153765888911D-16
B(4)= -0.979286096658D-15	C(4)= 0.927702359377D-15	0.489643048329D-15
B(5)= -0.356936702417D-15	C(5)= 0.636410319075D-01	0.178463351208D-15
B(6)= 0.706601444023D-15	C(6)= -0.898003893468D-15	-0.353300722011D-15
B(7)= 0.285493850782D-15	C(7)= -0.454435155699D-01	-0.142746925391D-15
B(8)= -0.477395900589D-15	C(8)= 0.933031429895D-15	0.238697950294D-15
B(9)= -0.278277401122D-15	C(9)= 0.353300579804D-01	0.139138700561D-15
B(10)= 0.101640917904D-15	C(10)= -0.838634717226D-15	-0.508204589522D-16
B(11)= 0.238919994899D-15	C(11)= -0.288911710079D-01	-0.119459997450D-15
B(12)= -0.561217738948D-16	C(12)= 0.855593373927D-15	0.280608869474D-16
B(13)= -0.308364445090D-15	C(13)= 0.244308973515D-01	0.154182222545D-15
B(14)= 0.648925357893D-16	C(14)= -0.885569395592D-15	-0.324462678947D-16
B(15)= 0.269728683833D-15	C(15)= -0.211577899868D-01	-0.134864341916D-15
B(16)= -0.371924713249D-16	C(16)= 0.899835761459D-15	0.185962356625D-16
B(17)= -0.211386463889D-15	C(17)= 0.186528472964D-01	0.105693231944D-15
B(18)= 0.598410210273D-16	C(18)= -0.868471961013D-15	-0.299205105136D-16
B(19)= 0.272670774848D-15	C(19)= -0.166734891710D-01	-0.136335387424D-15
B(20)= -0.158206781009D-16	C(20)= 0.877881101147D-15	0.791033905045D-17
B(21)= -0.279665179903D-15	C(21)= 0.150695467664D-01	0.139832589952D-15
B(22)= 0.135003119794D-15	C(22)= -0.910022057710D-15	-0.675015598972D-16
B(23)= 0.330568905582D-15	C(23)= -0.137430836923D-01	-0.165284452791D-15
B(24)= -0.616728890179D-16	C(24)= 0.929173404884D-15	0.308364445090D-16
B(25)= -0.353161944133D-15	C(25)= 0.126275030293D-01	0.176580972067D-15

(a)   (b)

Figure 8.25 *(continued)*

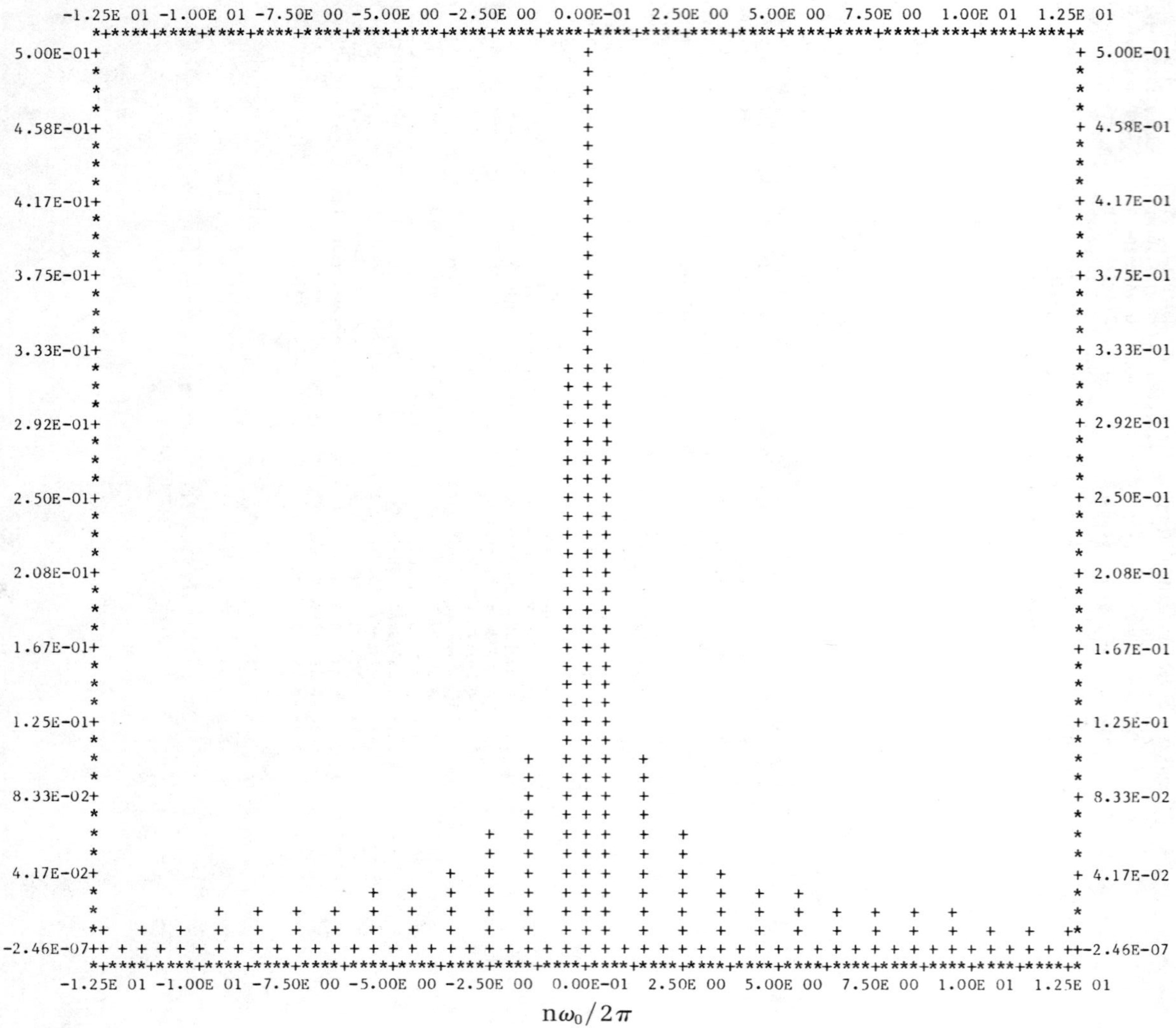
-1.25E 01 -1.00E 01 -7.50E 00 -5.00E 00 -2.50E 00 0.00E-01 2.50E 00 5.00E 00 7.50E 00 1.00E 01 1.25E 01
5.00E-01
4.58E-01
4.17E-01
3.75E-01
3.33E-01
2.92E-01
2.50E-01
2.08E-01
1.67E-01
1.25E-01
8.33E-02
4.17E-02
-2.46E-07
$n\omega_0/2\pi$

(c)

PHI(-25)=	0.00000000E 00
PHI(-24)=	0.00000000E 00
PHI(-23)=	0.18000000E 03
PHI(-22)=	0.00000000E 00
PHI(-21)=	0.00000000E 00
PHI(-20)=	0.00000000E 00
PHI(-19)=	0.18000000E 03
PHI(-18)=	0.00000000E 00
PHI(-17)=	0.00000000E 00
PHI(-16)=	0.00000000E 00
PHI(-15)=	0.18000000E 03
PHI(-14)=	0.00000000E 00
PHI(-13)=	0.00000000E 00
PHI(-12)=	0.00000000E 00
PHI(-11)=	0.18000000E 03
PHI(-10)=	0.00000000E 00
PHI(-9)=	0.00000000E 00
PHI(-8)=	0.00000000E 00
PHI(-7)=	0.18000000E 03
PHI(-6)=	0.00000000E 00
PHI(-5)=	0.00000000E 00
PHI(-4)=	0.00000000E 00
PHI(-3)=	-0.18000000E 03
PHI(-2)=	0.00000000E 00
PHI(-1)=	0.00000000E 00
PHI(0)=	0.00000000E 00
PHI(1)=	0.00000000E 00
PHI(2)=	0.00000000E 00
PHI(3)=	0.18000000E 03
PHI(4)=	0.00000000E 00
PHI(5)=	0.00000000E 00
PHI(6)=	0.00000000E 00
PHI(7)=	-0.18000000E 03
PHI(8)=	0.00000000E 00
PHI(9)=	0.00000000E 00
PHI(10)=	0.00000000E 00
PHI(11)=	-0.18000000E 03
PHI(12)=	0.00000000E 00
PHI(13)=	0.00000000E 00
PHI(14)=	0.00000000E 00
PHI(15)=	-0.18000000E 03
PHI(16)=	0.00000000E 00
PHI(17)=	0.00000000E 00
PHI(18)=	0.00000000E 00
PHI(19)=	-0.18000000E 03
PHI(20)=	0.00000000E 00
PHI(21)=	0.00000000E 00
PHI(22)=	0.00000000E 00
PHI(23)=	-0.18000000E 03
PHI(24)=	0.00000000E 00
PHI(25)=	0.00000000E 00

(d)

Figure 8.25 (continued)

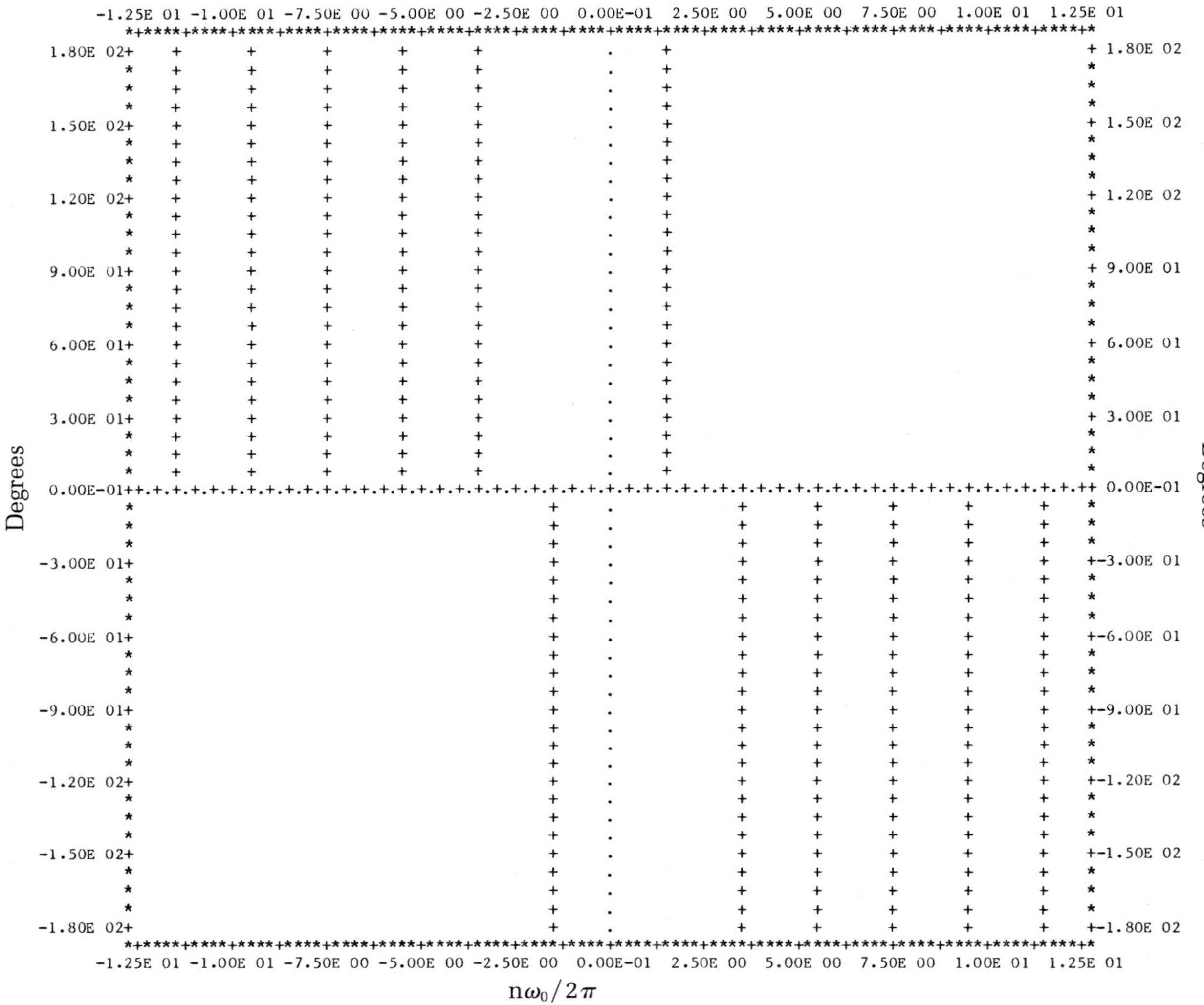

(e)

365

Figure 8.26

(a) Square-wave approximation resulting from truncated Fourier series of 11 terms in Eq. (8.138). (b) Square-wave approximation resulting from truncated Fourier series of 19 terms in Eq. (8.138). (c) Square-wave approximation resulting from truncated Fourier series of 27 terms in Eq. (8.138). (d) Square-wave approximation resulting from truncated Fourier series of 35 terms in Eq. (8.138). (e) Square-wave approximation resulting from truncated Fourier series of 43 terms in Eq. (8.139). (f) Square-wave approximation resulting from truncated Fourier series of 51 terms in Eq. (8.138).

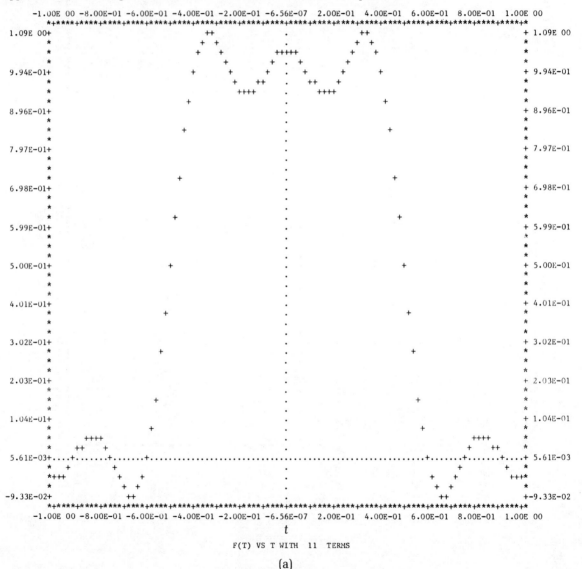

F(T) VS T WITH 11 TERMS

(a)

Figure 8.26 (continued)

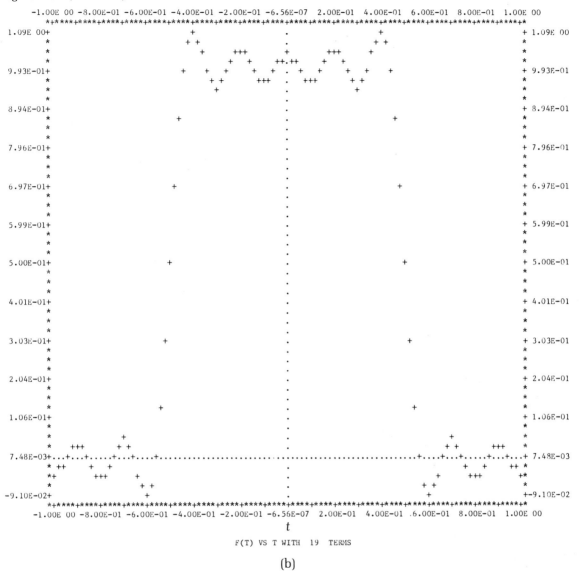

F(T) VS T WITH 19 TERMS

(b)

the partial sums are plotted in Figures 8.26(a) through (f) for 11, 19, 27, 35, 43, and 51 terms of (8.139b), respectively.

Near the points of jumps at $t = \pm \frac{1}{2}$, the partial sums $s_M(t)$ exhibit overshoot and oscillations that deviate considerably from $f(t)$. It is seen from the figures that as M increases, the approximation improves everywhere except near the vicinity of the discontinuities, where the deviation from the true waveform becomes narrower but not any smaller in amplitude. Even in

Figure 8.26 (*continued*)

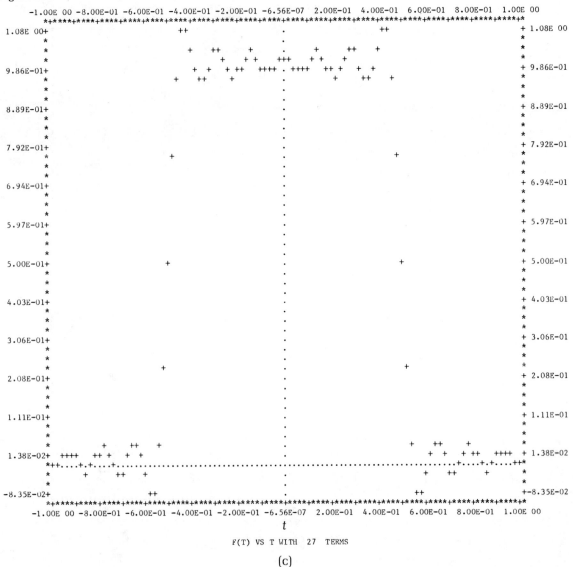

F(T) VS T WITH 27 TERMS

(c)

the limit as $M \to \infty$ there remains a discrepancy at the points of discontinuity. The maximum value of this difference $f(t) - s_M(t)$ is about 9 percent of the discontinuity jump $f(t_i +) - f(t_i -)$ of $f(t)$ at each t_i. This behavior of $s_M(t)$ is known as the *Gibbs phenomenon*, after Sir Josiah Willard Gibbs (1839–1903), who first investigated it. For illustrative purposes we present the partial sums

Figure 8.26   (*continued*)

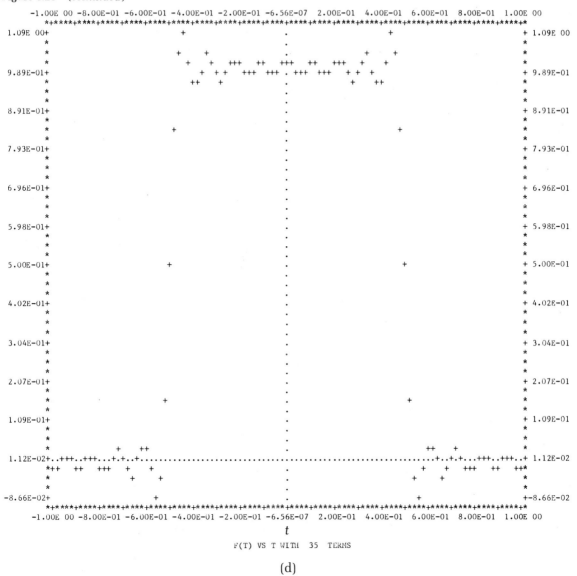

F(T) VS T WITH 35 TERMS

(d)

$s_{21}(t)$ and $s_{25}(t)$ in Figure 8.27 for t from -1 to 1 with time increment of 0.02. It is seen that the maximum deviation occurs at $t = \pm 0.46$ with $s_{21}(\pm 0.46) = 1.0813$ and $s_{25}(\pm 0.46) = 1.0881$, showing a maximum deviation of about 8.8 percent of the discontinuity jump, which is 1. This is a consequence of the Gibbs phenomenon.

Figure 8.26 (*continued*)

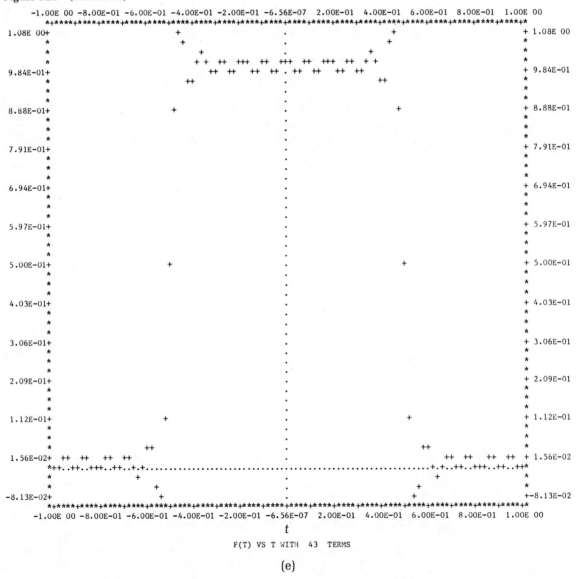

F(T) VS T WITH 43 TERMS

(e)

As another example consider the sawtooth waveform of Figure 8.28, having a Fourier series found to be

$$f(t) = \frac{1}{\pi} \sum_{n=1}^{\infty} \frac{(-1)^{n+1}}{n} \sin 2n\pi t \qquad (8.140a)$$

$$= \frac{j}{2\pi} \sum_{\substack{n=-\infty \\ n \neq 0}}^{\infty} \frac{(-1)^n}{n} e^{j2n\pi t} \qquad (8.140b)$$

Figure 8.26   (*continued*)

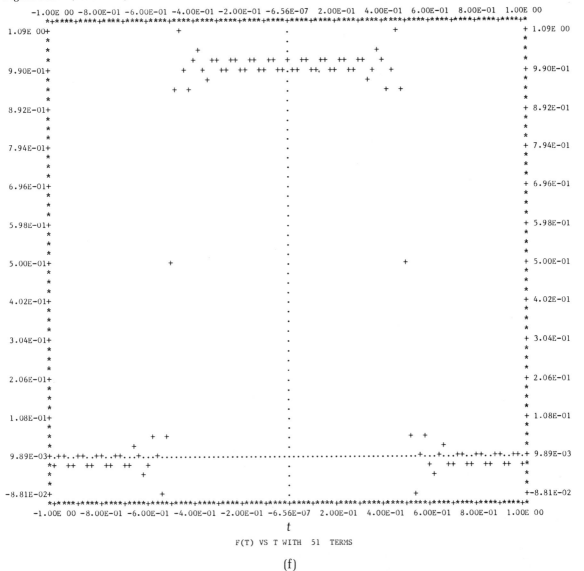

F(T) VS T WITH   51   TERMS

(f)

The partial sums are given by

$$s_M(t) = \frac{1}{\pi} \sum_{n=1}^{M} \frac{(-1)^{n+1}}{n} \sin 2n\pi t \qquad \text{(8.141a)}$$

$$= \frac{j}{2\pi} \sum_{\substack{n=-M \\ n \neq 0}}^{M} \frac{(-1)^n}{n} e^{j2n\pi t} \qquad \text{(8.141b)}$$

Figure 8.27

(a) Testing of partial sums resulting from truncated Fourier series representation of the square waveform of Figure 8.24 with 43 terms. (b) Testing of partial sums resulting from truncated Fourier series representation of the square waveform of Figure 8.24 with 51 terms.

```
F(T) WITH  43 TERMS
T= -0.10000000D 01     F(T)= -0.143467684460D-01     T=  0.80000000D-01     F(T)=  0.101091349734D 01
T= -0.98000000D 00     F(T)= -0.265298032798D-02     T=  0.10000000D 00     F(T)=  0.101206746687D 01
T= -0.96000000D 00     F(T)=  0.134751606249D-01     T=  0.12000000D 00     F(T)=  0.993182238300D 00
T= -0.94000000D 00     F(T)=  0.771742702370D-02     T=  0.14000000D 00     F(T)=  0.984740562273D 00
T= -0.92000000D 00     F(T)= -0.109135236433D-01     T=  0.16000000D 00     F(T)=  0.100143555796D 01
T= -0.90000000D 00     F(T)= -0.120674438606D-01     T=  0.18000000D 00     F(T)=  0.101689212699D 01
T= -0.88000000D 00     F(T)=  0.681779792092D-02     T=  0.20000000D 00     F(T)=  0.100491216161D 01
T= -0.86000000D 00     F(T)=  0.152594274845D-01     T=  0.22000000D 00     F(T)=  0.983391690604D 00
T= -0.84000000D 00     F(T)= -0.143560035234D-02     T=  0.24000000D 00     F(T)=  0.988142722673D 00
T= -0.82000000D 00     F(T)= -0.168921324952D-01     T=  0.26000000D 00     F(T)=  0.101407461429D 01
T= -0.80000000D 00     F(T)= -0.491211784757D-02     T=  0.28000000D 00     F(T)=  0.101902141018D 01
T= -0.78000000D 00     F(T)=  0.166083326501D-01     T=  0.30000000D 00     F(T)=  0.991084253208D 00
T= -0.76000000D 00     F(T)=  0.118572379630D-01     T=  0.32000000D 00     F(T)=  0.973924860039D 00
T= -0.74000000D 00     F(T)= -0.140746561668D-01     T=  0.34000000D 00     F(T)=  0.100053644647D 01
T= -0.72000000D 00     F(T)= -0.190213820058D-01     T=  0.36000000D 00     F(T)=  0.103286622050D 01
T= -0.70000000D 00     F(T)=  0.891580700899D-02     T=  0.38000000D 00     F(T)=  0.101239849945D 01
T= -0.68000000D 00     F(T)=  0.260751311896D-01     T=  0.40000000D 00     F(T)=  0.960155006500D 00
T= -0.66000000D 00     F(T)= -0.536523596722D-03     T=  0.42000000D 00     F(T)=  0.966132101168D 00
T= -0.64000000D 00     F(T)= -0.328662421358D-01     T=  0.44000000D 00     F(T)=  0.105043507245D 01
T= -0.62000000D 00     F(T)= -0.123984078736D-01     T=  0.46000000D 00     F(T)=  0.108129136459D 01
T= -0.60000000D 00     F(T)=  0.398450641471D-01     T=  0.48000000D 00     F(T)=  0.895197275466D 00
T= -0.58000000D 00     F(T)=  0.338677960788D-01     T=  0.50000000D 00     F(T)=  0.500000411213D 00
T= -0.56000000D 00     F(T)= -0.504352399786D-01     T=  0.52000000D 00     F(T)=  0.104803309734D 00
T= -0.54000000D 00     F(T)= -0.812912537997D-01     T=  0.54000000D 00     F(T)= -0.812912537998D-01
T= -0.52000000D 00     F(T)=  0.104803309734D 00     T=  0.56000000D 00     F(T)= -0.504352399786D-01
T= -0.50000000D 00     F(T)=  0.500000411213D 00     T=  0.58000000D 00     F(T)=  0.338677960788D-01
T= -0.48000000D 00     F(T)=  0.895197275466D 00     T=  0.60000000D 00     F(T)=  0.398450641471D-01
T= -0.46000000D 00     F(T)=  0.108129136459D 01     T=  0.62000000D 00     F(T)= -0.123984078736D-01
T= -0.44000000D 00     F(T)=  0.105043507245D 01     T=  0.64000000D 00     F(T)= -0.328662421358D-01
T= -0.42000000D 00     F(T)=  0.966132101168D 00     T=  0.66000000D 00     F(T)= -0.536523596723D-03
T= -0.40000000D 00     F(T)=  0.960155006500D 00     T=  0.68000000D 00     F(T)=  0.260751311896D-01
T= -0.38000000D 00     F(T)=  0.101239849945D 01     T=  0.70000000D 00     F(T)=  0.891580700899D-02
T= -0.36000000D 00     F(T)=  0.103286622050D 01     T=  0.72000000D 00     F(T)= -0.190213820058D-01
T= -0.34000000D 00     F(T)=  0.100053644647D 01     T=  0.74000000D 00     F(T)= -0.140746561668D-01
T= -0.32000000D 00     F(T)=  0.973924860039D 00     T=  0.76000000D 00     F(T)=  0.118572379630D-01
T= -0.30000000D 00     F(T)=  0.991084253208D 00     T=  0.78000000D 00     F(T)=  0.166083326501D-01
T= -0.28000000D 00     F(T)=  0.101902141018D 01     T=  0.80000000D 00     F(T)= -0.491211784757D-02
T= -0.26000000D 00     F(T)=  0.101407461429D 01     T=  0.82000000D 00     F(T)= -0.168921324952D-01
T= -0.24000000D 00     F(T)=  0.988142722673D 00     T=  0.84000000D 00     F(T)= -0.143560035234D-02
T= -0.22000000D 00     F(T)=  0.983391690604D 00     T=  0.86000000D 00     F(T)=  0.152594274845D-01
T= -0.20000000D 00     F(T)=  0.100491216161D 01     T=  0.88000000D 00     F(T)=  0.681779792092D-02
T= -0.18000000D 00     F(T)=  0.101689212699D 01     T=  0.90000000D 00     F(T)= -0.120674438606D-01
T= -0.16000000D 00     F(T)=  0.100143555796D 01     T=  0.92000000D 00     F(T)= -0.109135236433D-01
T= -0.14000000D 00     F(T)=  0.984740562273D 00     T=  0.94000000D 00     F(T)=  0.771742702370D-02
T= -0.12000000D 00     F(T)=  0.993182238300D 00     T=  0.96000000D 00     F(T)=  0.134751606249D-01
T= -0.10000000D 00     F(T)=  0.101206746687D 01     T=  0.98000000D 00     F(T)= -0.265298032798D-02
T= -0.80000000D-01     F(T)=  0.101091349734D 01     T=  0.10000000D 01     F(T)= -0.143467684460D-01
T= -0.60000000D-01     F(T)=  0.992282540978D 00
T= -0.40000000D-01     F(T)=  0.986524853184D 00
T= -0.20000000D-01     F(T)=  0.100265301697D 01
T=  0.15612511D-15     F(T)=  0.101434676845D 01
T=  0.20000000D-01     F(T)=  0.100265301697D 01
T=  0.40000000D-01     F(T)=  0.986524853184D 00
T=  0.60000000D-01     F(T)=  0.992282540978D 00
```

(a)

Figure 8.27 (continued)

```
F(T) WITH 51 TERMS
T= -0.10000000D 01      F(T)= -0.121156071202D-01
T= -0.98000000D 00      F(T)=  0.791950390514D-03
T= -0.96000000D 00      F(T)=  0.121075100279D-01
T= -0.94000000D 00      F(T)= -0.240091087180D-02
T= -0.92000000D 00      F(T)= -0.120821846379D-01
T= -0.90000000D 00      F(T)=  0.408853267321D-02
T= -0.88000000D 00      F(T)=  0.120362054480D-01
T= -0.86000000D 00      F(T)= -0.591905159662D-02
T= -0.84000000D 00      F(T)= -0.119627286661D-01
T= -0.82000000D 00      F(T)=  0.797813027832D-02
T= -0.80000000D 00      F(T)=  0.118491387701D-01
T= -0.78000000D 00      F(T)= -0.103910262285D-01
T= -0.76000000D 00      F(T)= -0.116718409313D-01
T= -0.74000000D 00      F(T)=  0.133573319898D-01
T= -0.72000000D 00      F(T)=  0.113839656093D-01
T= -0.70000000D 00      F(T)= -0.172251596996D-01
T= -0.68000000D 00      F(T)= -0.108828699672D-01
T= -0.66000000D 00      F(T)=  0.226708904463D-01
T= -0.64000000D 00      F(T)=  0.990906970995D-02
T= -0.62000000D 00      F(T)= -0.312140639075D-01
T= -0.60000000D 00      F(T)= -0.764669219624D-02
T= -0.58000000D 00      F(T)=  0.471094391174D-01
T= -0.56000000D 00      F(T)=  0.375742863293D-03
T= -0.54000000D 00      F(T)= -0.881268647065D-01
T= -0.52000000D 00      F(T)=  0.522788777797D-01
T= -0.50000000D 00      F(T)=  0.500000485569D 00
T= -0.48000000D 00      F(T)=  0.947721716746D 00
T= -0.46000000D 00      F(T)=  0.108812682912D 01
T= -0.44000000D 00      F(T)=  0.999624062221D 00
T= -0.42000000D 00      F(T)=  0.952890597637D 00
T= -0.40000000D 00      F(T)=  0.100764680658D 01
T= -0.38000000D 00      F(T)=  0.103121402694D 01
T= -0.36000000D 00      F(T)=  0.990090851322D 00
T= -0.34000000D 00      F(T)=  0.977329146601D 00
T= -0.32000000D 00      F(T)=  0.101088292852D 01
T= -0.30000000D 00      F(T)=  0.101722512262D 01
T= -0.28000000D 00      F(T)=  0.988615989476D 00
T= -0.26000000D 00      F(T)=  0.986642705111D 00
T= -0.24000000D 00      F(T)=  0.101167187582D 01
T= -0.22000000D 00      F(T)=  0.101039098912D 01
T= -0.20000000D 00      F(T)=  0.988150834235D 00
T= -0.18000000D 00      F(T)=  0.992021906841D 00
T= -0.16000000D 00      F(T)=  0.101196274909D 01
T= -0.14000000D 00      F(T)=  0.100591901447D 01
T= -0.12000000D 00      F(T)=  0.987963779841D 00
T= -0.10000000D 00      F(T)=  0.995911504453D 00
T= -0.80000000D-01      F(T)=  0.101208219418D 01
T= -0.60000000D-01      F(T)=  0.100240087374D 01
T= -0.40000000D-01      F(T)=  0.987892485278D 00
T= -0.20000000D-01      F(T)=  0.999208086738D 00
T=  0.15612511D-15      F(T)=  0.101211560712D 01
T=  0.20000000D-01      F(T)=  0.999208086738D 00
T=  0.40000000D-01      F(T)=  0.987892485278D 00
T=  0.60000000D-01      F(T)=  0.100240087374D 01
```

```
T=  0.80000000D-01      F(T)=  0.101208219418D 01
T=  0.10000000D 00      F(T)=  0.995911504453D 00
T=  0.12000000D 00      F(T)=  0.987963779841D 00
T=  0.14000000D 00      F(T)=  0.100591901447D 01
T=  0.16000000D 00      F(T)=  0.101196274909D 01
T=  0.18000000D 00      F(T)=  0.992021906841D 00
T=  0.20000000D 00      F(T)=  0.988150834235D 00
T=  0.22000000D 00      F(T)=  0.101039098912D 01
T=  0.24000000D 00      F(T)=  0.101167187582D 01
T=  0.26000000D 00      F(T)=  0.986642705111D 00
T=  0.28000000D 00      F(T)=  0.988615989476D 00
T=  0.30000000D 00      F(T)=  0.101722512262D 01
T=  0.32000000D 00      F(T)=  0.101088292852D 01
T=  0.34000000D 00      F(T)=  0.977329146601D 00
T=  0.36000000D 00      F(T)=  0.990090851322D 00
T=  0.38000000D 00      F(T)=  0.103121402694D 01
T=  0.40000000D 00      F(T)=  0.100764680658D 01
T=  0.42000000D 00      F(T)=  0.952890597637D 00
T=  0.44000000D 00      F(T)=  0.999624062221D 00
T=  0.46000000D 00      F(T)=  0.108812682912D 01
T=  0.48000000D 00      F(T)=  0.947721716746D 00
T=  0.50000000D 00      F(T)=  0.500000485568D 00
T=  0.52000000D 00      F(T)=  0.522788777797D-01
T=  0.54000000D 00      F(T)= -0.881268647065D-01
T=  0.56000000D 00      F(T)=  0.375742863293D-03
T=  0.58000000D 00      F(T)=  0.471094391174D-01
T=  0.60000000D 00      F(T)= -0.764669219624D-02
T=  0.62000000D 00      F(T)= -0.312140639075D-01
T=  0.64000000D 00      F(T)=  0.990906970995D-02
T=  0.66000000D 00      F(T)=  0.226708904463D-01
T=  0.68000000D 00      F(T)= -0.108828699672D-01
T=  0.70000000D 00      F(T)= -0.172251596996D-01
T=  0.72000000D 00      F(T)=  0.113839656093D-01
T=  0.74000000D 00      F(T)=  0.133573319898D-01
T=  0.76000000D 00      F(T)= -0.116718409313D-01
T=  0.78000000D 00      F(T)= -0.103910262285D-01
T=  0.80000000D 00      F(T)=  0.118491387701D-01
T=  0.82000000D 00      F(T)=  0.797813027832D-02
T=  0.84000000D 00      F(T)= -0.119627286661D-01
T=  0.86000000D 00      F(T)= -0.591905159662D-02
T=  0.88000000D 00      F(T)=  0.120362054480D-01
T=  0.90000000D 00      F(T)=  0.408853267321D-02
T=  0.92000000D 00      F(T)= -0.120821846379D-01
T=  0.94000000D 00      F(T)= -0.240091087180D-02
T=  0.96000000D 00      F(T)=  0.121075100279D-01
T=  0.98000000D 00      F(T)=  0.791950390514D-03
T=  0.10000000D 01      F(T)= -0.121156071202D-01
```

(b)

The Fourier coefficients a_n, b_n, and $\tilde{c}_n$ and the amplitude and phase spectra are shown in Figure 8.29. As before, the partial sums $s_M(t)$ are plotted in Figures 8.30(a) through (f) for 3, 11, 19, 27, 35, and 43 terms of (8.141b). As expected, they exhibit oscillations near the points of discontinuity, which are at $t = \pm\frac{1}{2}$. As M increases, the approximation improves everywhere

Figure 8.28

Sawtooth waveform in the time domain that will be represented by a truncated Fourier series having coefficients and frequency spectra shown in Figure 8.29.

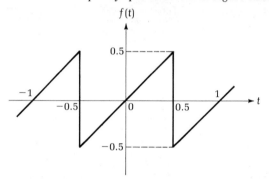

Figure 8.29

(a) Trigonometric Fourier coefficients of the sawtooth waveform of Figure 8.28. (b) Exponential Fourier coefficients of the sawtooth waveform of Figure 8.28. (c) Amplitude spectrum of the periodic sawtooth waveform of Figure 8.28. (d) Phase spectrum of the periodic sawtooth waveform of Figure 8.28.

A(0)= -0.688447120662D-16	
A(1)= -0.191768578141D-14	B(1)= 0.318511282056D 00
A(2)= 0.205299421254D-14	B(2)= -0.159558270615D 00
A(3)= -0.195221018000D-14	B(3)= 0.106709630427D 00
A(4)= 0.197867321656D-14	B(4)= -0.803884387010D-01
A(5)= -0.197309489009D-14	B(5)= 0.646797630752D-01
A(6)= 0.199537418179D-14	B(6)= -0.542790186531D-01
A(7)= -0.199850348688D-14	B(7)= 0.469131662301D-01
A(8)= 0.204187157378D-14	B(8)= -0.414460160235D-01
A(9)= -0.204653151724D-14	B(9)= 0.372466663702D-01
A(10)= 0.207738238376D-14	B(10)= -0.339368833813D-01
A(11)= -0.209333503690D-14	B(11)= 0.312762778394D-01
A(12)= 0.213510870884D-14	B(12)= -0.291048982165D-01
A(13)= -0.215914185700D-14	B(13)= 0.273123241268D-01
A(14)= 0.221876447294D-14	B(14)= -0.258200107837D-01
A(15)= -0.224779345464D-14	B(15)= 0.245707015675D-01
A(16)= 0.230099164123D-14	B(16)= -0.235218174574D-01
A(17)= -0.234762508997D-14	B(17)= 0.226411870270D-01
A(18)= 0.241636775948D-14	B(18)= -0.219042082340D-01
A(19)= -0.246191275427D-14	B(19)= 0.212919162229D-01
A(20)= 0.254409102717D-14	B(20)= -0.207896420995D-01
A(21)= -0.261075883135D-14	B(21)= 0.203860681782D-01

(a)

Figure 8.29   (continued)

C(-21)= -0.130679100438D-14 0.101930340891D-01
C(-20)= 0.126474947073D-14 -0.103948210498D-01
C(-19)= -0.123372853328D-14 0.106459581114D-01
C(-18)= 0.120046265956D-14 -0.109521041170D-01
C(-17)= -0.117365948115D-14 0.113205935135D-01
C(-16)= 0.114617601902D-14 -0.117609087287D-01
C(-15)= -0.112406679825D-14 0.122853507838D-01
C(-14)= 0.110085105349D-14 -0.129100053919D-01
C(-13)= -0.108135879914D-14 0.136561620634D-01
C(-12)= 0.106039543125D-14 -0.145524491083D-01
C(-11)= -0.104904851146D-14 0.156381389197D-01
C(-10)= 0.103301082285D-14 -0.169684416906D-01
C(-9)= -0.102244941816D-14 0.186233331851D-01
C(-8)= 0.100692194234D-14 -0.207230080118D-01
C(-7)= -0.999540864021D-15 0.234565831150D-01
C(-6)= 0.987159700388D-15 -0.271395093266D-01
C(-5)= -0.985220891797D-15 0.323398815376D-01
C(-4)= 0.971139018875D-15 -0.401942193505D-01
C(-3)= -0.976955444647D-15 0.533548152133D-01
C(-2)= 0.100360555922D-14 -0.797791353073D-01
C(-1)= -0.959455146051D-15 0.159255641028D 00
C(0)= -0.688447120662D-16 0.000000000000D 00

C(1)= -0.958842890707D-15 -0.159255641028D 00
C(2)= 0.102649710627D-14 0.797791353073D-01
C(3)= -0.976105090002D-15 -0.533548152133D-01
C(4)= 0.989336608280D-15 0.401942193505D-01
C(5)= -0.986547445044D-15 -0.323398815376D-01
C(6)= 0.997687090894D-15 0.271395093266D-01
C(7)= -0.999251743441D-15 -0.234565831150D-01
C(8)= 0.102093578689D-14 0.207230080118D-01
C(9)= -0.102326575862D-14 -0.186233331851D-01
C(10)= 0.103869119188D-14 0.169684416906D-01
C(11)= -0.104666751845D-14 -0.156381389197D-01
C(12)= 0.106755435442D-14 0.145524491083D-01
C(13)= -0.107957092850D-14 -0.136561620634D-01
C(14)= 0.110938223647D-14 0.129100053919D-01
C(15)= -0.112389672732D-14 -0.122853507838D-01
C(16)= 0.115049582062D-14 0.117609087287D-01
C(17)= -0.117381254498D-14 -0.113205935135D-01
C(18)= 0.120818387974D-14 0.109521041170D-01
C(19)= -0.123095637713D-14 -0.106459581114D-01
C(20)= 0.127204551359D-14 0.103948210498D-01
C(21)= -0.130537941567D-14 -0.101930340891D-01

(b)

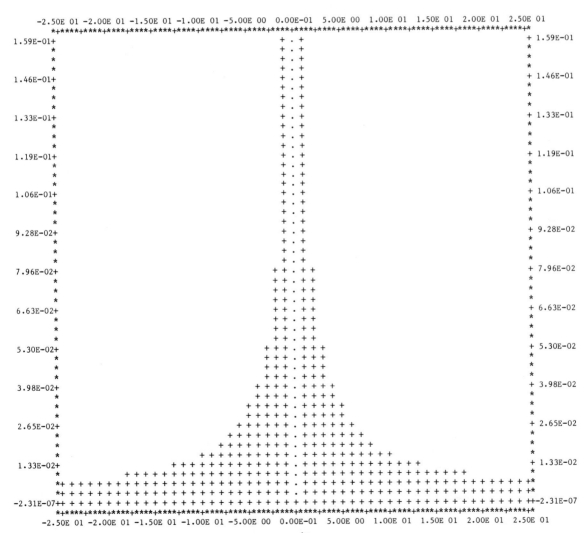

$$n\omega_0/2\pi$$

(c)

Figure 8.29 (continued)

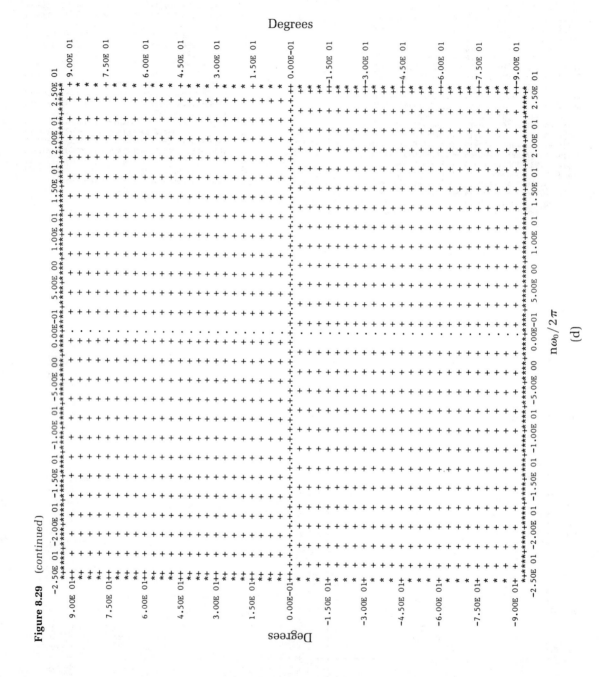

(d)

except near the vicinity of the discontinuities. For our purposes we present the partial sums $s_{21}(t) = F(T)$ in Figure 8.31 from $t = -\frac{1}{2}$ to $t = 0.49$ with time increments of 0.01. It is seen that the maximum deviation occurs at $t = \pm 0.48$ with

$$s_{21}(\pm 0.48) = \pm 0.5996 \tag{8.142}$$

or about 10 percent of the discontinuity jump, which is 1.

Figure 8.30

(a) Sawtooth approximation resulting from truncated Fourier series of 3 terms in Eq. (8.140). (b) Sawtooth approximation resulting from truncated Fourier series of 11 terms in Eq. (8.140). (c) Sawtooth approximation resulting from truncated Fourier series of 19 terms in Eq. (8.140). (d) Sawtooth approximation resulting from truncated Fourier series of 27 terms in Eq. (8.140). (e) Sawtooth approximation resulting from truncated Fourier series of 35 terms in Eq. (8.140). (f) Sawtooth approximation resulting from truncated Fourier series of 43 terms in Eq. (8.140).

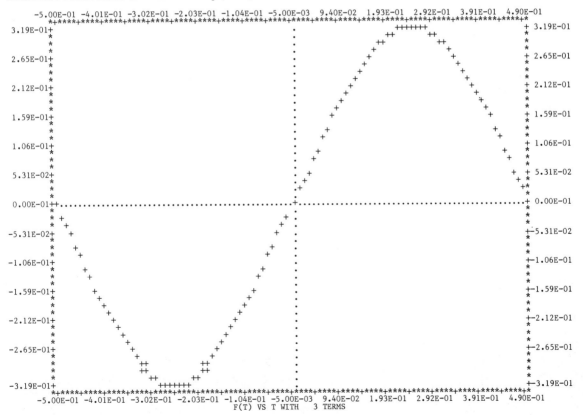

Figure 8.30 (continued)

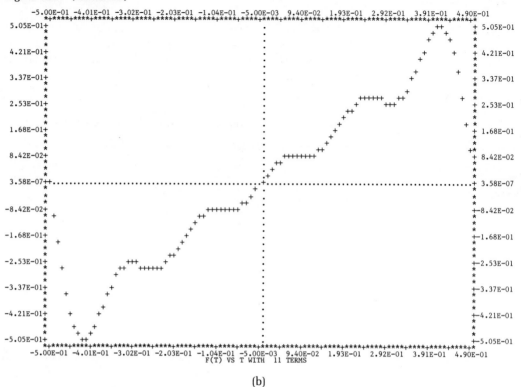

(b)

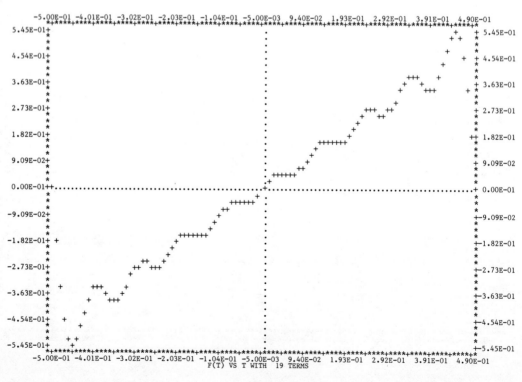

(c)

Figure 8.30 (continued)

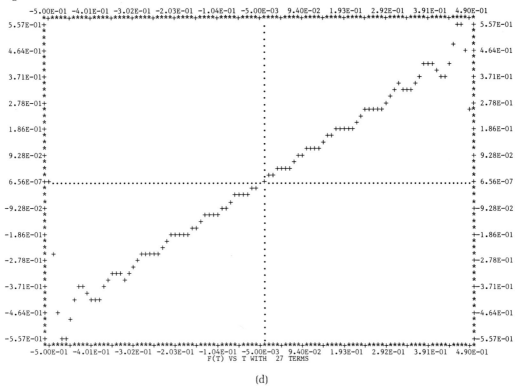

(d)

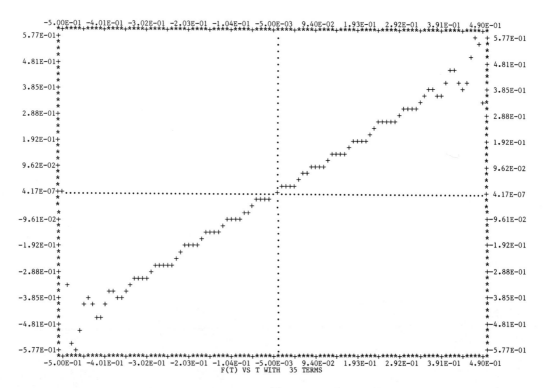

(e)

Figure 8.30   (*continued*)

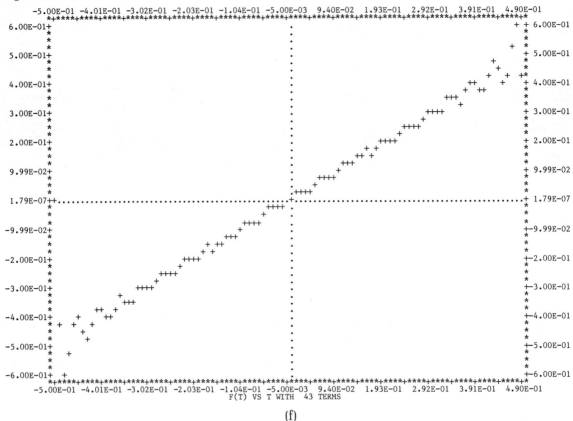

(f)

As a consequence of the Gibbs phenomenon, if a signal is applied to the input of an ideal low-pass filter with cutoff frequency ω_c, as depicted in Figure 8.32, the resulting output is a truncated Fourier series of the signal with all harmonics above ω_c being eliminated. This output exhibits oscillations near the discontinuity points no matter how large the cutoff frequency is.

Finally, as an example of no jump discontinuity, we consider the periodic triangular waveform of Figure 8.21(a); its Fourier series as given by (8.127) is repeated below:

$$v_a(t) = \frac{8}{\pi^2} \sum_{n=1}^{\infty} \frac{(-1)^{n+1}}{(2n-1)^2} \sin\left[2(2n-1)\pi t\right] \qquad \textbf{(8.143a)}$$

$$= \frac{j4}{\pi^2} \sum_{n=-\infty}^{\infty} \frac{(-1)^n}{(2n-1)^2} e^{j2(2n-1)\pi t} \qquad \textbf{(8.143b)}$$

Figure 8.31

Listing of the partial sums of the truncated Fourier series representation of the sawtooth waveform of Figure 8.28.

```
F(T) WITH  43 TERMS
```

T	F(T)	T	F(T)
-0.50000000D 00	0.952885309050D-13		
-0.49000000D 00	-0.418177244887D 00		
-0.48000000D 00	-0.599556175585D 00		
-0.47000000D 00	-0.535307731059D 00		
-0.46000000D 00	-0.414346594028D 00		
-0.45000000D 00	-0.393284153813D 00		
-0.44000000D 00	-0.451177995468D 00	0.30000000D-01	0.219763304873D-01
-0.43000000D 00	-0.475446295730D 00	0.40000000D-01	0.321594251610D-01
-0.42000000D 00	-0.428620500982D 00	0.50000000D-01	0.546552530044D-01
-0.41000000D 00	-0.377518739909D 00	0.60000000D-01	0.699587517835D-01
-0.40000000D 00	-0.380364084841D 00	0.70000000D-01	0.696569676974D-01
-0.39000000D 00	-0.409107794263D 00	0.80000000D-01	0.697500263064D-01
-0.38000000D 00	-0.404144394444D 00	0.90000000D-01	0.858433705288D-01
-0.37000000D 00	-0.363366012875D 00	0.10000000D 00	0.108609547427D 00
-0.36000000D 00	-0.336403490035D 00	0.11000000D 00	0.118034411164D 00
-0.35000000D 00	-0.346205596008D 00	0.12000000D 00	0.114736586573D 00
-0.34000000D 00	-0.359326331744D 00	0.13000000D 00	0.119440159392D 00
-0.33000000D 00	-0.341334594869D 00	0.14000000D 00	0.140741423131D 00
-0.32000000D 00	-0.307290501638D 00	0.15000000D 00	0.161204101818D 00
-0.31000000D 00	-0.294459869549D 00	0.16000000D 00	0.164197718530D 00
-0.30000000D 00	-0.305129290347D 00	0.17000000D 00	0.160264809626D 00
-0.29000000D 00	-0.306398591620D 00	0.18000000D 00	0.171320377976D 00
-0.28000000D 00	-0.282129265331D 00	0.19000000D 00	0.196270584647D 00
-0.27000000D 00	-0.255689260125D 00	0.20000000D 00	0.211849580488D 00
-0.26000000D 00	-0.251998854539D 00	0.21000000D 00	0.208721401827D 00
-0.25000000D 00	-0.260015598547D 00	0.22000000D 00	0.206989203805D 00
-0.24000000D 00	-0.251745483711D 00	0.23000000D 00	0.225526910344D 00
-0.23000000D 00	-0.225526910344D 00	0.24000000D 00	0.251745483711D 00
-0.22000000D 00	-0.206989203805D 00	0.25000000D 00	0.260015598547D 00
-0.21000000D 00	-0.208721401827D 00	0.26000000D 00	0.251998854539D 00
-0.20000000D 00	-0.211849580488D 00	0.27000000D 00	0.255689260125D 00
-0.19000000D 00	-0.196270584647D 00	0.28000000D 00	0.282129265331D 00
-0.18000000D 00	-0.171320377976D 00	0.29000000D 00	0.306398591620D 00
-0.17000000D 00	-0.160264809626D 00	0.30000000D 00	0.305129290347D 00
-0.16000000D 00	-0.164197718530D 00	0.31000000D 00	0.294459869549D 00
-0.15000000D 00	-0.161204101818D 00	0.32000000D 00	0.307290501638D 00
-0.14000000D 00	-0.140741423131D 00	0.33000000D 00	0.341334594869D 00
-0.13000000D 00	-0.119440159392D 00	0.34000000D 00	0.359326331744D 00
-0.12000000D 00	-0.114736586573D 00	0.35000000D 00	0.346205596008D 00
-0.11000000D 00	-0.118034411164D 00	0.36000000D 00	0.336403490035D 00
-0.10000000D 00	-0.108609547427D 00	0.37000000D 00	0.363366012875D 00
-0.90000000D-01	-0.858433705288D-01	0.38000000D 00	0.404144394444D 00
-0.80000000D-01	-0.697500263064D-01	0.39000000D 00	0.409107794263D 00
-0.70000000D-01	-0.696569676974D-01	0.40000000D 00	0.380364084841D 00
-0.60000000D-01	-0.699587517835D-01	0.41000000D 00	0.377518739909D 00
-0.50000000D-01	-0.546552530044D-01	0.42000000D 00	0.428620500982D 00
-0.40000000D-01	-0.321594251610D-01	0.43000000D 00	0.475446295730D 00
-0.30000000D-01	-0.219763304873D-01	0.44000000D 00	0.451177995468D 00
-0.20000000D-01	-0.243060555540D-01	0.45000000D 00	0.393284153813D 00
-0.10000000D-01	-0.198659219925D-01	0.46000000D 00	0.414346594028D 00
0.34174052D-15	-0.142186356846D-14	0.47000000D 00	0.535307731059D 00
0.10000000D-01	0.198659219925D-01	0.48000000D 00	0.599556175585D 00
0.20000000D-01	0.243060555540D-01	0.49000000D 00	0.418177244887D 00

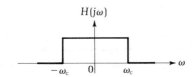

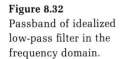

Figure 8.32

Passband of idealized low-pass filter in the frequency domain.

Figure 8.33
(a) Triangular-wave approximation resulting from truncated Fourier series of 3 terms in Eq. (8.143). (b) Triangular-wave approximation resulting from truncated Fourier series of 11 terms in Eq. (8.143). (c) Triangular-wave approximation resulting from truncated Fourier series of 19 terms in Eq. (8.143).

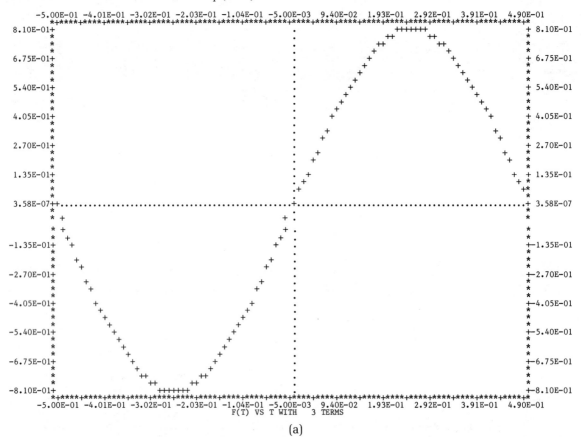

(a)

The partial sums are given by

$$s_M(t) = \frac{8}{\pi^2} \sum_{n=1}^{M} \frac{(-1)^{n+1}}{(2n-1)^2} \sin\left[2(2n-1)\pi t\right] \qquad \textbf{(8.144a)}$$

$$= \frac{j4}{\pi^2} \sum_{n=1-M}^{M} \frac{(-1)^n}{(2n-1)^2} e^{j2(2n-1)\pi t} \qquad \textbf{(8.144b)}$$

As before, the Fourier coefficients a_n, b_n, and $\tilde{c}_n$ were computed for M up to 21, and the partial sums $s_M(t)$ are plotted in Figure 8.33(a) to (c) for 3, 11, and 19 terms. The sum converges rapidly to the true waveform with only

Figure 8.33 (*continued*)

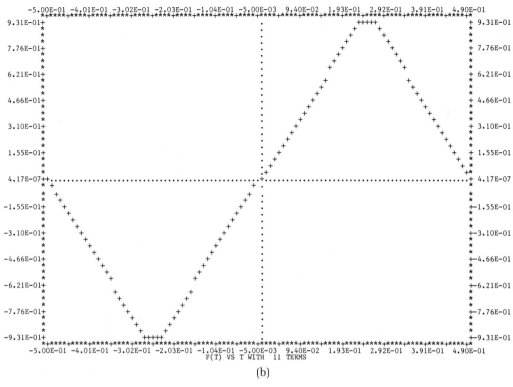

(b)

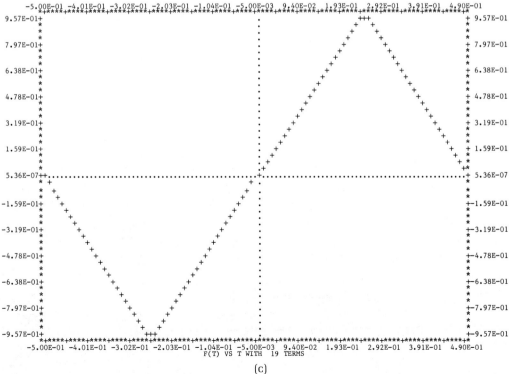

(c)

Figure 8.34
(a) Trigonometric Fourier coefficients of the periodic triangular waveform of Fig. 8.21(a).
(b) Exponential Fourier coefficients of the periodic triangular waveform of Fig. 8.21(a).

```
A(  0)=  0.888178419700D-17
A(  1)= -0.517030862568D-14
A(  2)=  0.133226762955D-15
A(  3)=  0.177080572428D-14
A(  4)=  0.210942374679D-16
A(  5)= -0.997535387626D-15
A(  6)=  0.185962356625D-16
A(  7)=  0.717620407542D-15
A(  8)=  0.103389519168D-16
A(  9)= -0.519723153403D-15
A( 10)=  0.135308431126D-16
A( 11)=  0.419976553534D-15
A( 12)=  0.181452075587D-16
A( 13)= -0.319640147683D-15
A( 14)= -0.164104840827D-16
A( 15)=  0.260763632909D-15
A( 16)=  0.308433834029D-16
A( 17)= -0.229954943975D-15
A( 18)= -0.422578638748D-16
A( 19)=  0.199111560573D-15
A( 20)=  0.133226762955D-16
A( 21)= -0.113797860024D-15
```

```
B(  1)=  0.810035398363D 00
B(  2)=  0.254934962030D-14
B(  3)= -0.895232670165D-01
B(  4)=  0.139471767469D-16
B(  5)=  0.318706980465D-01
B(  6)=  0.840230662824D-15
B(  7)= -0.159715286507D-01
B(  8)=  0.327515792264D-16
B(  9)=  0.941049085105D-02
B( 10)=  0.509592368303D-15
B( 11)= -0.606387710736D-02
B( 12)=  0.671684929898D-16
B( 13)=  0.412226247547D-02
B( 14)=  0.383582055008D-15
B( 15)= -0.287377905011D-02
B( 16)=  0.129340982369D-15
B( 17)=  0.200753335228D-02
B( 18)=  0.333483241022D-15
B( 19)= -0.136266043225D-02
B( 20)=  0.139714628755D-15
B( 21)=  0.848270325306D-03
```

(a)

```
C( -21)= -0.596744875736D-16   0.424135162653D-03
C( -20)=  0.832667268469D-18   0.663878674256D-16
C( -19)=  0.958781665172D-16  -0.681330216123D-03
C( -18)=  0.149186218934D-16   0.163792590602D-15
C( -17)= -0.120545934346D-15   0.100376667614D-02
C( -16)=  0.147451495458D-16   0.635602681598D-16
C( -15)=  0.126825633329D-15  -0.143688952505D-02
C( -14)= -0.102175212735D-16   0.186239912381D-15
C( -13)= -0.165735480895D-15   0.206113123774D-02
C( -12)=  0.692154666915D-17   0.358046925442D-16
C( -11)=  0.207629052840D-15  -0.303419355368D-02
C( -10)= -0.568989300120D-17   0.259237076250D-15
C(  -9)= -0.264042260278D-15   0.470524542553D-02
C(  -8)=  0.454497550706D-17   0.235922392733D-16
C(  -7)=  0.354664214663D-15  -0.798576432535D-02
C(  -6)=  0.667868538251D-17   0.423983764763D-15
C(  -5)= -0.490718576884D-15   0.159353490232D-01
C(  -4)=  0.108246744901D-16   0.780625564190D-17
C(  -3)=  0.865418847695D-15  -0.447616335083D-01
C(  -2)=  0.260902410787D-16   0.125786533967D-14
C(  -1)= -0.258182364377D-14   0.405017699182D 00
C(   0)=  0.888178419700D-17   0.000000000000D 00
C(   1)= -0.258515431284D-14  -0.405017699182D 00
C(   2)=  0.666133814775D-16  -0.127467481015D-14
C(   3)=  0.885402862139D-15   0.447616335083D-01
C(   4)=  0.105471187339D-16  -0.697358837343D-17
C(   5)= -0.498767693813D-15  -0.159353490232D-01
C(   6)=  0.929811783124D-17  -0.420115331412D-15
C(   7)=  0.358810203771D-15   0.798576432535D-02
C(   8)=  0.516947595841D-17  -0.163757896132D-16
C(   9)= -0.259861576701D-15  -0.470524542553D-02
C(  10)=  0.676542155631D-17  -0.254796184151D-15
C(  11)=  0.209988276767D-15   0.303419355368D-02
C(  12)=  0.907260377936D-17  -0.335842464949D-16
C(  13)= -0.159820073842D-15  -0.206113123774D-02
C(  14)= -0.820524204137D-17  -0.191791027504D-15
C(  15)=  0.130381816454D-15   0.143688952505D-02
C(  16)=  0.154216917014D-16  -0.646704911844D-16
C(  17)= -0.114977471988D-15  -0.100376667614D-02
C(  18)=  0.211289319374D-16  -0.166741620511D-15
C(  19)=  0.995557802863D-16   0.681330216123D-03
C(  20)=  0.666133814775D-17  -0.698573143776D-16
C(  21)= -0.568989300120D-16  -0.424135162653D-03
```

(b)

19 terms. The needed Fourier coefficients are shown in Figure 8.34. The amplitude and phase spectra are given in Figure 8.21(d) and (e).

8-9 COMPUTER SOLUTIONS AND PROGRAMS

In this section we shall present computer programs in WATFIV that evaluate the Fourier series coefficients for a periodic signal $f(t)$.

8-9.1 Subroutine FOURIR

1. Purpose. The program, as listed in Figure 8.35, is for the evaluation of the trigonometric and/or exponential Fourier coefficients.

Figure 8.35
FORTRAN IV/WATFIV
subroutine for computing
trigonometric or
exponential Fourier series
coefficients.

```
C
C
C      SUBROUTINE FOURIR
C
C      PURPOSE
C          THE PROGRAM IS FOR THE EVALUATION OF FOURIER SERIES COEFFICIENTS OF
C              A PERIODIC FUNCTION.
C
C      METHOD
C          THE PROGRAM IS BASED ON THE STANDARD FORMULAS FOR COMPUTING THE
C              FOURIER SERIES COEFFICIENTS.
C
C      USAGE
C          CALL FOURIR (T,NS,N,NCHOCE,A,B,C)
C
C          T      -PERIOD
C          NS     -THE HIGHEST HARMONIC REQUIRED
C          N      -NUMBER OF DIVISIONS OF T USED
C          NCHOCE -0   COMPUTE AND PRINT COEFFICIENTS C
C                  1   COMPUTE AND PRINT COEFFICIENTS C, A AND B
C          A      -TRIGONOMETRIC FOURIER SERIES COEFFICIENTS A
C          B      -TRIGONOMETRIC FOURIER SERIES COEFFICIENTS B
C          C      -EXPONENTIAL FOURIER SERIES COEFFICIENTS C
C
C      SUBROUTINES AND FUNCTION SUBPROGRAMS REQUIRED
C          FUNCTION FT(T), PROVIDED BY THE USER.
C
C      REMARKS
C          DOUBLE PRECISION IS USED IN ALL THE COMPUTATIONS.
C          THE INPUT DATA ARE T, NS, N AND NCHOCE.  THE OUTPUTS ARE C AND/OR A
C              AND B.
C
       SUBROUTINE FOURIR (T,NS,N,NCHOCE,A,B,C)
       DOUBLE PRECISION T,A(1),B(1),T1,FT,DREAL,DIMAG,DFLOAT
       COMPLEX*16 C(1),DC,W,CDEXP
       T1=T/DFLOAT(N)
       I=1
       NC=-NS
       DC=(0.0,1.0)*6.2831853071796/DFLOAT(N)
       W=CDEXP(DC)
1      C(I)=(0.0,0.0)
       DO 2 M=1,N
2      C(I)=C(I)+FT((M-1)*T1)*W**(-(M-1)*NC)
       C(I)=C(I)/DFLOAT(N)
       PRINT 3,NC,C(I)
3      FORMAT (3H C(,I4,2H)=,2D20.12)
       I=I+1
       NC=NC+1
       IF (NC-NS) 1,1,4
4      IF (NCHOCE-1) 9,5,5
5      A(NS+1)=C(NS+1)
       PRINT 6,0,A(NS+1)
6      FORMAT (3H-A(,I4,2H)=,D20.12)
       DO 7 NC=1,NS
       I=NS+NC+1
       A(I)=DREAL(2.0*C(I))
       B(I)=-DIMAG(2.0*C(I))
7      PRINT 8,NC,A(I),NC,B(I)
8      FORMAT (3H A(,I4,2H)=,D20.12,10X,2HB(,I4,2H)=,D20.12)
9      RETURN
       END
       FUNCTION FT(T)
       DOUBLE PRECISION T,FT
       FT=0.0
       IF (T.LT.0.5) FT=1.0
       IF (T.GT.1.5) FT=1.0
       RETURN
       END
       FUNCTION DREAL(X)
C
C      COMPUTE THE REAL PART OF A COMPLEX X
       COMPLEX*16 X
       DOUBLE PRECISION DREAL
       DREAL=X
       RETURN
       END
       FUNCTION DIMAG(X)
C
C      COMPUTE THE IMAGINARY PART OF A COMPLEX X
       COMPLEX*16 X
       DOUBLE PRECISION DIMAG
       DIMAG=X*(0.0,-1.0)
       RETURN
       END
```

2. Method.   The program is based on the standard formulas (8.42), (8.44), and (8.89) for computing the Fourier series coefficients.

3. Usage.   The program consists of a subroutine FOURIR called by

$$\text{CALL FOURIR (T, NS, N, NCHOCE, A, B, C)}$$

T	is the period
NS	is the highest harmonic required
N	is the number of divisions of T used
NCHOCE	$= 0$, compute and print coefficients $\tilde{c}_n$
	$= 1$, compute and print coefficients $\tilde{c}_n$, a_n, and b_n
A	is the name of the trigonometric Fourier coefficients a_n
B	is the name of the trigonometric Fourier coefficients b_n
C	is the name of the exponential Fourier coefficients $\tilde{c}_n$

4. Remarks.   Double precision is used in all the computations. If the given periodic signal has jump discontinuities, integration must be performed over the various segments where the signal is continuous. Subroutine FOURI2 is designed for this purpose. The input data are T, NS, N, and NCHOCE. The outputs are C or A, B, and C.

5. Subroutines and Function Subprograms Required.   FUNCTION FT (T), provided by the user, defines the periodic signal $f(t)$ over the first period.

Figure 8.36
Computer program for computing Fourier series coefficients of a periodic function.

```
      DOUBLE PRECISION A(200),B(200),T,TS,TI,TF,DT,WO,FR
      DIMENSION XY(200,2),JXY(2)
      COMPLEX*16 C(200),F
      READ,NS,N,NCHOCE,T,TI,TF,DT
      WO=6.2831853071796/T
      CALL FOURIR (T,NS,N,NCHOCE,A,B,C)
      DO 8 L=1,21,4
      LC=L*2+1
      PRINT 1,LC
1     FORMAT (10H1F(T) WITH,I4,6H TERMS)
      TS=TI
      J=1
2     F=(0.0,0.0)
      I=NS-L+1
      NC=-L
3     F=F+C(I)*CDEXP((0.0,1.0)*NC*WO*TS)
      I=I+1
      NC=NC+1
      IF (NC-L) 3,3,4
4     XY(J,1)=TS
      XY(J,2)=F
      FR=F
      PRINT 5,TS,FR
5     FORMAT (3H T=,D16.8,10X,5HF(T)=,D20.12)
      J=J+1
      TS=TS+DT
      IF (TS-TF) 2,2,6
6     JXY(1)=1
      JXY(2)=2
      J=J-1
      CALL PLOT (XY,JXY,J,200,1)
      PRINT 7,LC
7     FORMAT (1H-,48X,14HF(T) VS T WITH,I4,6H TERMS)
8     CONTINUE
      STOP
      END
```

Figure 8.37

Function subprogram for computing Fourier series coefficients of the periodic triangular waveform of Figure 8.21(a).

```
FUNCTION FT(T)
DOUBLE PRECISION T,FT
FT=4.0*T
IF (T.GT.0.25) FT=-4.0*T+2.0
IF (T.GT.0.75) FT=4.0*T-4.0
RETURN
END
```

EXAMPLE 8.15

We use the subroutine FOURIR to compute the Fourier coefficients a_n, b_n, and $\tilde{c}_n$ of the triangular voltage waveform $v_a(t)$ of Figure 8.21(a) for n up to 21 and plot the partial sums $s_M(t)$ for 3, 11, and 19 terms of (8.144b). For our purposes we choose the stepsize to be 0.02 and t from -0.5 to 0.49 with time increment of 0.01. Then $T = 1$, $NS = 21$, and $NCHOCE = 1$.

The main program and the required function subprogram are listed in Figures 8.36 and 8.37, respectively. The computer output was given earlier in Figures 8.33 and 8.34. Figure 8.34 is the printout of the Fourier coefficients a_n, b_n, and $\tilde{c}_n$. The partial sums $s_M(t)$ for 3, 11, and 19 terms of (8.144b) are plotted in Figure 8.33(a) to (c), respectively, as a function of t.

8-9.2 Subroutine FOURI2

1. Purpose. The program is for the evaluation of the Fourier coefficients a_n, b_n, and $\tilde{c}_n$ of a periodic function $f(t)$ with points of discontinuity and is listed in Figure 8.38.

2. Method. The program is based on the standard formulas (8.42), (8.44), and (8.89) for computing the Fourier coefficients. The integration is performed over the various segments where $f(t)$ is continuous.

3. Usage. The program consists of a subroutine, FOURI2, called by

CALL FOURI2 (M, T, NS, N, NCHOCE, A, B, C, D)

M	is the number of segments used in the integration plus one
T	is the name of a one-dimensional array used to define the $M - 1$ segments for integration
NS	is the highest harmonic required
N	is the number of divisions chosen for each segment
NCHOCE	= 0, compute and print coefficients $\tilde{c}_n$
	= 1, compute and print coefficients $\tilde{c}_n$, a_n, and b_n
A	is the name of the trigonometric Fourier coefficients a_n
B	is the name of the trigonometric Fourier coefficients b_n
C	is the name of the exponential Fourier coefficients $\tilde{c}_n$
D	is a one-dimensional storage array

4. Remarks. Double precision is used in all the computations. The input data are M, T, NS, N, and NCHOCE. The outputs are C or A, B, and C.

5. Subroutines and Function Subprograms Required. FUNCTION FT(K, T), provided by the user, defines $f(t)$ for the kth segment in the first period.

Figure 8.38
FORTRAN IV/WATFIV
subroutine for evaluation
of the Fourier series
coefficients in the
representation of a
periodic function with
points of discontinuity.

```
C
C
C      SUBROUTINE FOURI2
C
C      PURPOSE
C         THE PROGRAM IS FOR THE EVALUATION OF FOURIER SERIES
C            COEFFICIENTS OF A PERIODIC FUNCTION WITH DISCONTINUITIES.
C
C      METHOD
C         THE PROGRAM IS BASED ON THE STANDARD FORMULAS FOR COMPUTING
C            THE FOURIER SERIES COEFFICIENTS.
C
C      USAGE
C         CALL FOURI2(M,T,NS,N,NCHOCE,A,B,C,D)
C
C         M        -THE NUMBER OF SEGMENTS REQUIRED PLUS ONE
C         T        -THE NAME OF A ONE-DIMENSIONAL ARRAY USED TO DEFINE
C                    THE M-1 SEGMENTS.
C         NS       -THE HIGHEST HARMONIC REQUIRED
C         N        -THE NUMBER OF DIVISIONS CHOSEN FOR EACH SEGMENT
C         NCHOCE   -0   COMPUTE AND PRINT COEFFICIENTS C
C                   1   COMPUTE AND PRINT COEFFICIENTS C, A AND B
C         A        -TRIGONOMETRIC FOURIER SERIES COEFFICIENTS A
C         B        -TRIGONOMETRIC FOURIER SERIES COEFFICIENTS B
C         C        -EXPONENTIAL FOURIER SERIES COEFFICIENTS C
C         D        -A ONE-DIMENSIONAL STORAGE ARRAY
C
C      SUBROUTINE AND FUNCTION SUBPROGRAMS REQUIRED
C         FUNCTION FT(K,T), PROVIDED BY THE USER.
C
C      REMARKS
C         DOUBLE PRECISION IS USED IN ALL THE COMPUTATIONS.
C         THE INPUT DATA ARE M,T,NS,N AND NCHOCE.  THE OUTPUTS
C            ARE C OR A, B, AND C.
C
       SUBROUTINE FOURI2(M,T,NS,N,NCHOCE,A,B,C,D)
       DOUBLE PRECISION T(1),A(1),B(1),T1,FT
       COMPLEX*16 C(1),D(1),DC,DC1,CDEXP
       K=1
       NS2=2*NS+1
       DO 1 I=1,NS2
     1 C(I)=(0.0,0.0)
     2 T1=(T(K+1)-T(K))/DFLOAT(N)
       I=1
       NC=-NS
       DC=(0.0D00,6.2831853071796D00)/(T(M)-T(1))
     3 DC1=DC*DFLOAT(NC)
       D(I)=0.5*(FT(K,T(K))*CDEXP(-DC1*T(K))+FT(K,T(K+1))
      1*CDEXP(-DC1*T(K+1)))
       DO 4 L=2,N
     4 D(I)=D(I)+FT(K,(T(K)+(L-1)*T1))*CDEXP(DC1*(DFLOAT(1-L)*T1-T(K)))
       D(I)=D(I)*T1/(T(M)-T(1))
       I=I+1
       NC=NC+1
       IF(NC.LE.NS) GO TO 3
       DO 5 I=1,NS2
     5 C(I)=C(I)+D(I)
       K=K+1
       IF(K.LT.M) GO TO 2
       DO  6 I=1,NS2
       NC=I-NS-1
     6 PRINT  7,NC,C(I)
     7 FORMAT(1X,´C(´,I4,´)=´,2(D20.12,10X))
       IF(NCHOCE.EQ.0) GO TO 11
       A(NS+1)=C(NS+1)
       PRINT 8,A(NS+1)
     8 FORMAT(´1´,´A(    0)=´,D20.12)
       DO  9 I=1,NS
       NC=I+NS+1
       A(I)=2.0*C(NC)
       B(I)=(0.0,2.0)*C(NC)
     9 PRINT 10,I,A(I),I,B(I)
    10 FORMAT(1X,´A(´,I4,´)=´,D20.12,10X,´B(´,I4,´)=´,D20.12)
    11 RETURN
       END
```

Figure 8.39
Periodic staircase function
in the time domain as an
example of a function
with discontinuities.

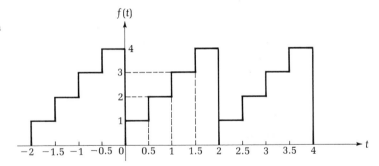

EXAMPLE 8.16

We use the subroutine FOURI2 to analyze the periodic staircase signal
$f(t)$ of Figure 8.39, which is defined by

$$
\begin{aligned}
f(t) &= 1, & 0 < t < 0.5 \\
&= 2, & 0.5 < t < 1 \\
&= 3, & 1 < t < 1.5 \\
&= 4, & 1.5 < t < 2
\end{aligned}
\tag{8.145}
$$

The waveform is divided into four segments defined by the one-dimensional array T = [0, 0.5, 1, 1.5, 2]. The stepsize for each segment is chosen to be 0.01, giving N = 50. Also let NS = 25 and NCHOCE = 1. Since there are four segments over a period, we have M = 5. The main program† and the required function subprogram are listed in Figure 8.40(a) and (b). The computer outputs are shown in Figure 8.41 through 8.43. The Fourier coefficients $\tilde{c}_n$, a_n, and b_n are listed in Figure 8.41, and the amplitude and phase spectra of the staircase waveform are displayed in Figure 8.42. Finally, the partial sums of the truncated Fourier series are exhibited in Figure 8.43 for various numbers of terms.

EXAMPLE 8.17

We use the subroutine FOURI2 to analyze the periodic signal $f(t)$ of Figure 8.44, defined by the equation

$$
\begin{aligned}
f(t) &= \cos 20\pi t, & 0 < t < 0.2 \\
&= 0, & 0.2 < t < 1
\end{aligned}
\tag{8.146}
$$

For our purposes let N = 50, NS = 50, and NCHOCE = 1. Then M = 3 and T = [0, 0.2, 1].

† The subroutine PLOT2 for plotting the frequency spectrum is given in Appendix B.

Figure 8.40

(a) Computer program for computing Fourier series coefficients of a periodic function with discontinuities. (b) Function subprogram for computing Fourier series coefficients of the periodic staircase signal of Figure 8.39.

```
      DOUBLE PRECISION A(100),B(100),T(20),TS,DT,WO,FR
      DIMENSION XY(200,2),JXY(2)
      COMPLEX*16 C(200),D(200),F
      READ,NS,N,NCHOCE,M
      READ,(T(I),I=1,M),DT
      WO=6.283185071796D00/(T(M)-T(1))
      CALL FOURI2(M,T,NS,N,NCHOCE,A,B,C,D)
      NS2=2*NS+1
      DO 11 I=1,NS2
      XY(I,1)=DFLOAT(I-NS-1)/(T(M)-T(1))
   11 XY(I,2)=CDABS(C(I))
      CALL PLOT2(XY,NS2,200)
      DO 8 L=1,21,4
      LC=L*2+1
      PRINT 1,LC
    1 FORMAT('1','F(T) WITH',I4,' TERMS')
      TS=T(1)
      J=1
    2 F=(0.0,0.0)
      I=NS-L+1
      NC=-L
    3 F=F+C(I)*CDEXP((0.0,1.0)*NC*WO*TS)
      I=I+1
      NC=NC+1
      IF(NC-L) 3,3,4
    4 XY(J,1)=TS
      XY(J,2)=F
      FR=F
      PRINT 5,TS,FR
    5 FORMAT(1X,'T=',D16.8,10X,'F(T)=',D20.12)
      J=J+1
      TS=TS+DT
      IF(TS-T(M)) 2,2,6
    6 JXY(1)=1
      JXY(2)=2
      J=J-1
      CALL PLOT(XY,JXY,J,200,1)
      PRINT 7,LC
    7 FORMAT(//50X,'F(T) VS T WITH', I4,2X,'TERMS'//)
    8 CONTINUE
      STOP
      END
```

(a)

```
      FUNCTION FT(K,T)
      DOUBLE PRECISION T,FT
      FT=DFLOAT(K)
      RETURN
      END
```

(b)

The main program[†] is the same as that given in Figure 8.40(a). The required function subprogram is listed in Figure 8.45. The computer outputs are presented in Figures 8.46–8.48. As before, Figure 8.46 lists the Fourier coefficients $\tilde{c}_n$, a_n, and b_n, whereas the amplitude and phase spectra are displayed in Figure 8.47. The partial sums of the truncated Fourier series are plotted in Figure 8.48 for 11, 19, 43, and 99 terms of the series.

[†] The subroutine PLOT2 for plotting the frequency spectrum is given in Appendix B.

Figure 8.41

(a) Trigonometric Fourier coefficients of the periodic staircase signal of Figure 8.39.

(b) Exponential Fourier coefficients of the periodic staircase signal of Figure 8.39.

```
A(   0)=   0.250000000000D 01
A(   1)=   0.184019466332D-13          B(   1)= -0.127313482326D 01
A(   2)=   0.194581850632D-13          B(   2)= -0.636410319075D 00
A(   3)=   0.172084568817D-13          B(   3)= -0.424098975794D 00
A(   4)=   0.917349529672D-14          B(   4)= -0.174082970261D-14
A(   5)=   0.175137682135D-13          B(   5)= -0.254124094723D 00
A(   6)=   0.172088732153D-13          B(   6)= -0.211577899868D 00
A(   7)=   0.177913239696D-13          B(   7)= -0.181157733725D 00
A(   8)=   0.123370758054D-13          B(   8)= -0.196620497661D-14
A(   9)=   0.161537450083D-13          B(   9)= -0.140527324581D 00
A(  10)=   0.179127546129D-13          B(  10)= -0.126275030294D 00
A(  11)=   0.168545732926D-13          B(  11)= -0.114594832934D 00
A(  12)=   0.978120362483D-14          B(  12)= -0.230510055488D-15
A(  13)=   0.170054942350D-13          B(  13)= -0.965763470439D-01
A(  14)=   0.182999448928D-13          B(  14)= -0.894748565842D-01
A(  15)=   0.188477705665D-13          B(  15)= -0.833059954018D-01
A(  16)=   0.111011200232D-13          B(  16)= -0.161870516990D-14
A(  17)=   0.157061863515D-13          B(  17)= -0.731076870930D-01
A(  18)=   0.170585767734D-13          B(  18)= -0.688404515334D-01
A(  19)=   0.160531310467D-13          B(  19)= -0.650110160260D-01
A(  20)=   0.779390441075D-14          B(  20)=  0.699024171880D-15
A(  21)=   0.168996761030D-13          B(  21)= -0.584152197860D-01
A(  22)=   0.177045877958D-13          B(  22)= -0.555521370783D-01
A(  23)=   0.165770175364D-13          B(  23)= -0.529284642057D-01
A(  24)=   0.911049014007D-14          B(  24)= -0.798527910462D-15
A(  25)=   0.150712775593D-13          B(  25)= -0.482842712475D-01
```

(a)

```
C( -25)=   0.753563877964D-14          -0.241421356237D-01
C( -24)=   0.455524507004D-14          -0.399263955231D-15
C( -23)=   0.828850876822D-14          -0.264643321029D-01
C( -22)=   0.885229389791D-14          -0.277760685391D-01
C( -21)=   0.844983805148D-14          -0.292076098930D-01
C( -20)=   0.389695220537D-14           0.349512085940D-15
C( -19)=   0.802656552334D-14          -0.325055080130D-01
C( -18)=   0.852928838668D-14          -0.344202257667D-01
C( -17)=   0.785309317575D-14          -0.365538435465D-01
C( -16)=   0.555056001161D-14          -0.809352584952D-15
C( -15)=   0.942388528324D-14          -0.416529977009D-01
C( -14)=   0.914997244639D-14          -0.447374282921D-01
C( -13)=   0.850274711750D-14          -0.482881735219D-01
C( -12)=   0.489060181241D-14          -0.115255027744D-14
C( -11)=   0.842728664630D-14          -0.572974164672D-01
C( -10)=   0.895637730647D-14          -0.631375151468D-01
C(  -9)=   0.807687250415D-14          -0.702636622904D-01
C(  -8)=   0.616853790270D-14          -0.983102488306D-15
C(  -7)=   0.889566198481D-14          -0.905788668624D-01
C(  -6)=   0.860443660766D-14          -0.105788949934D 00
C(  -5)=   0.875688410673D-14          -0.127062047362D 00
C(  -4)=   0.458674764836D-14          -0.870414851306D-15
C(  -3)=   0.860422844084D-14          -0.212049487897D 00
C(  -2)=   0.972909253161D-14          -0.318205159538D 00
C(  -1)=   0.920097331658D-14          -0.636567411629D 00
C(   0)=   0.250000000000D 01           0.000000000000D 00
C(   1)=   0.920097331658D-14           0.636567411629D 00
C(   2)=   0.972909253161D-14           0.318205159538D 00
C(   3)=   0.860422844084D-14           0.212049487897D 00
C(   4)=   0.458674764836D-14           0.870414851306D-15
C(   5)=   0.875688410673D-14           0.127062047362D 00
C(   6)=   0.860443660766D-14           0.105788949934D 00
C(   7)=   0.889566198481D-14           0.905788668624D-01
C(   8)=   0.616853790270D-14           0.983102488306D-15
C(   9)=   0.807687250415D-14           0.702636622904D-01
C(  10)=   0.895637730647D-14           0.631375151468D-01
C(  11)=   0.842728664630D-14           0.572974164672D-01
C(  12)=   0.489060181241D-14           0.115255027744D-15
C(  13)=   0.850274711750D-14           0.482881735219D-01
C(  14)=   0.914997244639D-14           0.447374282921D-01
C(  15)=   0.942388528324D-14           0.416529977009D-01
C(  16)=   0.555056001161D-14           0.809352584952D-15
C(  17)=   0.785309317575D-14           0.365538435465D-01
C(  18)=   0.852928838668D-14           0.344202257667D-01
C(  19)=   0.802656552334D-14           0.325055080130D-01
C(  20)=   0.389695220537D-14          -0.349512085940D-15
C(  21)=   0.844983805148D-14           0.292076098930D-01
C(  22)=   0.885229389791D-14           0.277760685391D-01
C(  23)=   0.828850876822D-14           0.264643321029D-01
C(  24)=   0.455524507004D-14           0.399263955231D-15
C(  25)=   0.753563877964D-14           0.241421356237D-01
```

(b)

Figure 8.42

(a) Amplitude spectrum of Fourier series representation of the periodic discontinuous staircase signal of Figure 8.39. (b) Phase spectrum of Fourier series representation of the periodic discontinuous staircase signal of Figure 8.39.

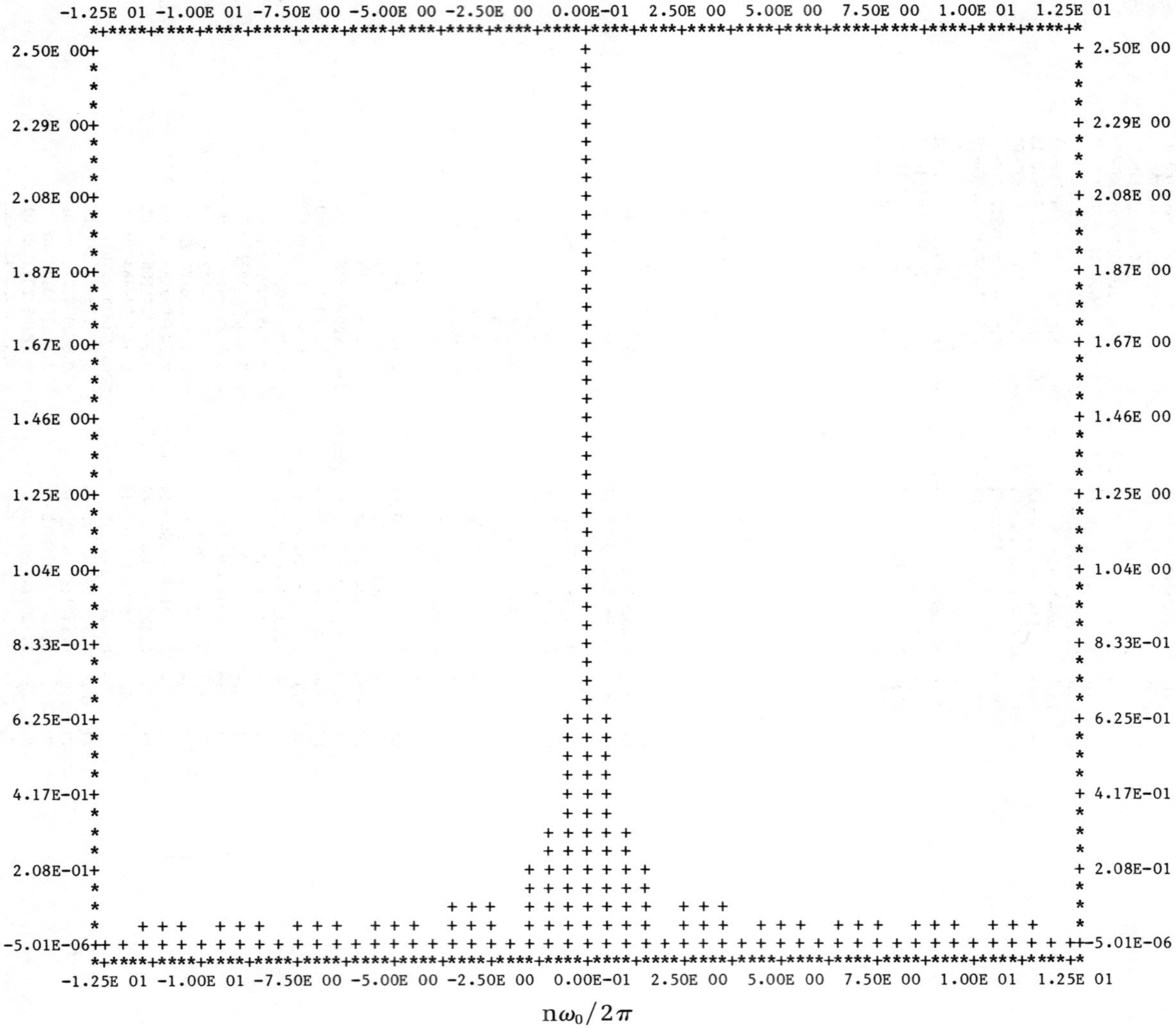

(a)

$n\omega_0/2\pi$

Figure 8.42 (*continued*)

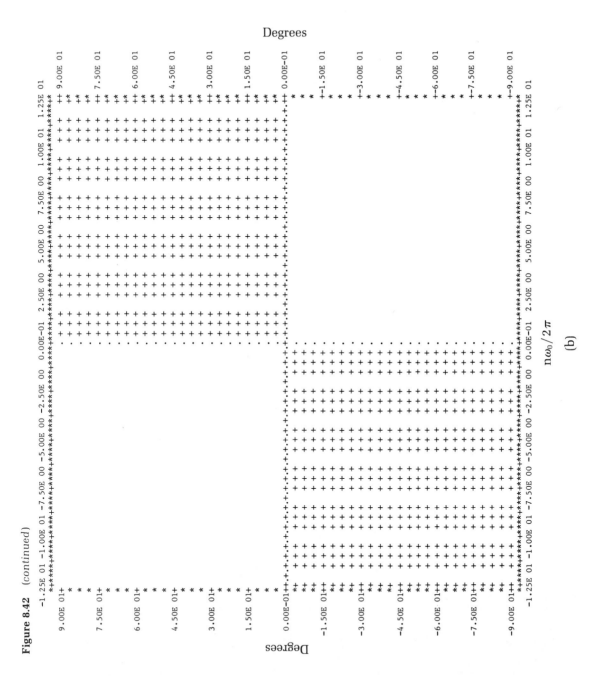

(b)

393

Figure 8.43

(a) Approximation to the staircase function of Figure 8.39 by a truncated Fourier series containing 3 terms. (b) Approximation to the staircase function of Figure 8.39 by a truncated Fourier series containing 11 terms. (c) Approximation to the staircase function of Figure 8.39 by a truncated Fourier series containing 19 terms. (d) Approximation to the staircase function of Figure 8.39 by a truncated Fourier series containing 27 terms. (e) Approximation to the staircase function of Figure 8.39 by a truncated Fourier series containing 35 terms. (f) Approximation to the staircase function of Figure 8.39 by a truncated Fourier series containing 43 terms.

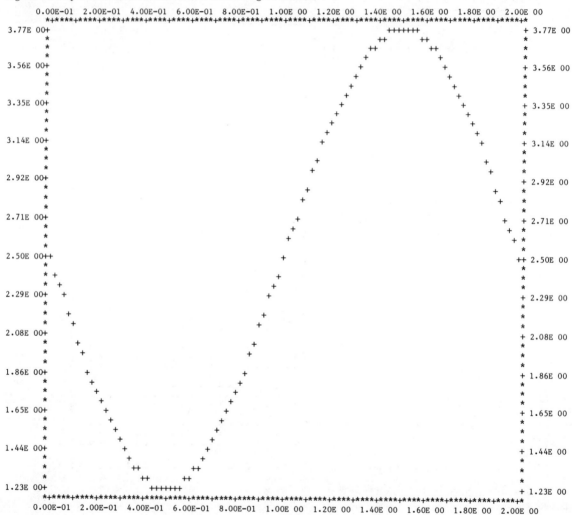

F(T) VS T WITH   3   TERMS

(a)

Figure 8.43 (continued)

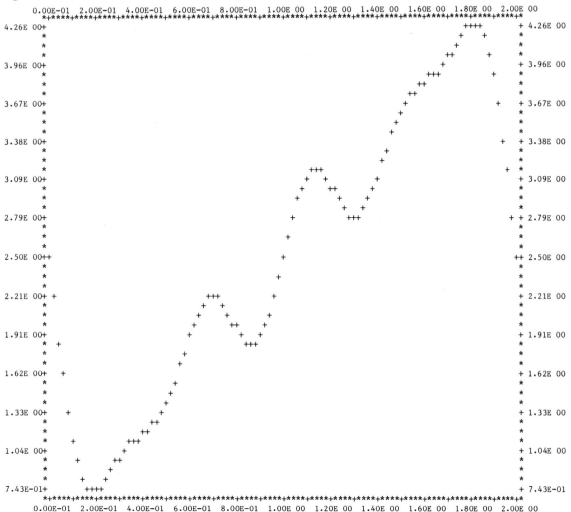

F(T) VS T WITH 11 TERMS

(b)

8-10 SUMMARY

In this chapter we demonstrated that a *periodic signal* satisfying the *Dirichlet conditions* can be expressed as an infinite series of sine and cosine functions or of exponentials called the *Fourier series*. Fourier analysis amounts to determining the *Fourier coefficients* and finding the number of terms to be included in a truncated Fourier series so that the partial sum will represent

Figure 8.43 *(continued)*

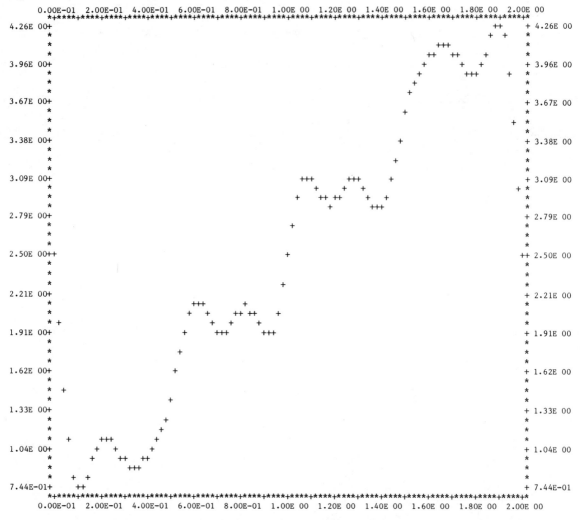

F(T) VS T WITH 19 TERMS

(c)

the signal within an allowable error. This information may be conveniently displayed by means of *frequency spectrum*—namely, the values of amplitude and phase versus frequency or series term number. Formulas for evaluating the Fourier coefficients were derived. We showed that if a signal possesses certain identifiable properties of waveforms, such as *evenness*, *oddness*, or

Figure 8.43   (*continued*)

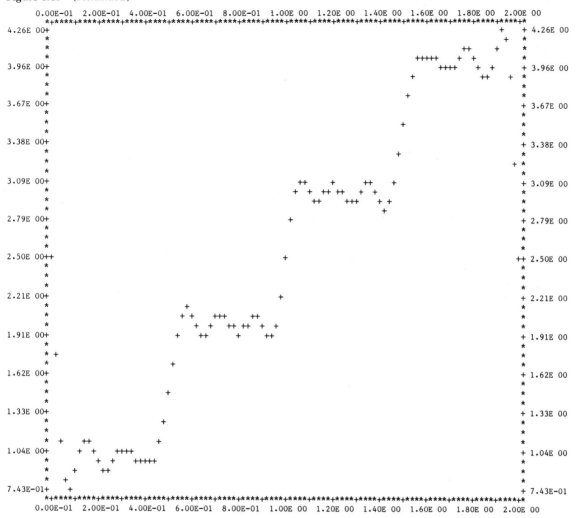

F(T) VS T WITH 27   TERMS

(d)

half-wave symmetry, these properties may be used to simplify the evaluation of its Fourier coefficients. This simplification may also be attained by signal manipulation in terms of *scaling, translation, transposition, differentiation,* or *integration*. Relationships between the Fourier coefficients of the manipulated and the original signals were presented.

Figure 8.43 *(continued)*

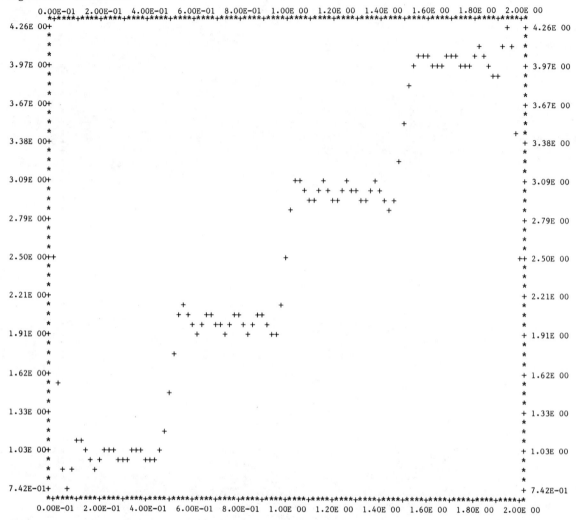

F(T) VS T WITH 35 TERMS

(e)

Questions concerning the existence of the Fourier series and the conditions under which they will represent the signals that generated them were answered by the theorem of Dirichlet. Dirichlet conditions cover almost any conceivable practical application. In order to possess a valid Fourier expansion, a signal need not be continuous; its waveform may consist of

Figure 8.43 *(continued)*

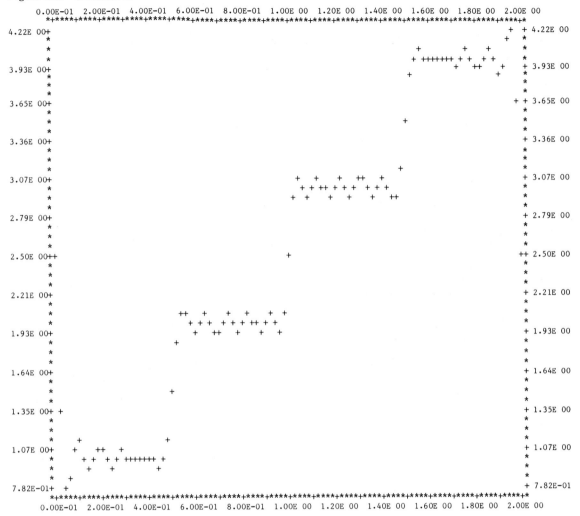

F(T) VS T WITH 43 TERMS

(f)

any finite number of disjoint arcs of different curves. At each point of discontinuity, the series assumes a value midway between the two values of the signal. Also we discussed the rate of convergence of the Fourier coefficients. We indicated that for a periodic signal that satisfies the Dirichlet conditions, its nth Fourier coefficient approaches zero at least as fast as

Figure 8.44

Periodic oscillatory pulses in the time domain.

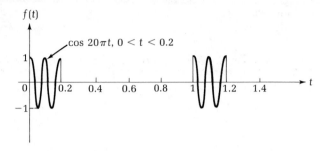

Figure 8.45

Function subprogram for computing Fourier series coefficients of the periodic oscillatory pulses of Figure 8.44.

```
FUNCTION FT(K,T)
DOUBLE PRECISION T,FT
FT=0.0
IF(K.EQ.1) FT=DCOS(62.83185071796D00*T)
RETURN
END
```

Figure 8.46

(a) Trigonometric Fourier coefficients of the periodic oscillatory pulses of Figure 8.44.

(b) Exponential Fourier coefficients of the periodic oscillatory pulses of Figure 8.44.

```
A(  0)= -0.372650603833D-08
A(  1)= -0.471392925489D-03        B(  1)= -0.153167273888D-03
A(  2)= -0.153693819286D-02        B(  2)= -0.111665642994D-02
A(  3)= -0.233472915624D-02        B(  3)= -0.321348949541D-02
A(  4)= -0.195880991746D-02        B(  4)= -0.602862093425D-02
A(  5)=  0.795250382657D-08        B(  5)= -0.853028763583D-02
A(  6)=  0.309885833789D-02        B(  6)= -0.953728007198D-02
A(  7)=  0.606534573804D-02        B(  7)= -0.834822051744D-02
A(  8)=  0.723984807312D-02        B(  8)= -0.526005106963D-02
A(  9)=  0.530084978387D-02        B(  9)= -0.172234746178D-02
A( 10)= -0.995047866043D-08        B( 10)= -0.116934240069D-14
A( 11)= -0.740380516978D-02        B( 11)= -0.240564560416D-02
A( 12)= -0.142384815069D-01        B( 12)= -0.103448708259D-01
A( 13)= -0.170887668379D-01        B( 13)= -0.235206875216D-01
A( 14)= -0.128746357549D-01        B( 14)= -0.396240996030D-01
A( 15)=  0.170861484445D-07        B( 15)= -0.546935555941D-01
A( 16)=  0.208282173276D-01        B( 16)= -0.641025976453D-01
A( 17)=  0.464321717760D-01        B( 17)= -0.639083646529D-01
A( 18)=  0.717713875566D-01        B( 18)= -0.521449367320D-01
A( 19)=  0.911972191847D-01        B( 19)= -0.296317477988D-01
A( 20)=  0.999999981665D-01        B( 20)=  0.235383604137D-07
A( 21)=  0.957786696585D-01        B( 21)=  0.311204000147D-01
A( 22)=  0.791985583016D-01        B( 22)=  0.575411467136D-01
A( 23)=  0.538831731065D-01        B( 23)=  0.741638573128D-01
A( 24)=  0.254539571034D-01        B( 24)=  0.783392772632D-01
A( 25)= -0.133596620285D-07        B( 25)=  0.705250676390D-01
A( 26)= -0.175608338862D-01        B( 26)=  0.540466558090D-01
A( 27)= -0.247360480507D-01        B( 27)=  0.340462367282D-01
A( 28)= -0.219619755126D-01        B( 28)=  0.159563035096D-01
A( 29)= -0.122316926285D-01        B( 29)=  0.397431563224D-02
A( 30)=  0.603375770558D-08        B( 30)=  0.205213623872D-14
A( 31)=  0.102619808162D-01        B( 31)=  0.333432143827D-02
A( 32)=  0.153964169880D-01        B( 32)=  0.111861552391D-01
A( 33)=  0.143671711109D-01        B( 33)=  0.197747206072D-01
A( 34)=  0.833021385268D-02        B( 34)=  0.256377743590D-01
A( 35)= -0.367259064821D-08        B( 35)=  0.267130027136D-01
A( 36)= -0.742810208712D-02        B( 36)=  0.228613370888D-01
A( 37)= -0.114109684769D-01        B( 37)=  0.157058463948D-01
A( 38)= -0.108664819164D-01        B( 38)=  0.789495912736D-02
A( 39)= -0.641248172919D-02        B( 39)=  0.208354073389D-02
A( 40)=  0.253748415108D-08        B( 40)=  0.114797060746D-14
A( 41)=  0.588657925915D-02        B( 41)=  0.191266631791D-02
A( 42)=  0.915297773763D-02        B( 42)=  0.665002920974D-02
A( 43)=  0.881102866753D-02        B( 43)=  0.121273434607D-01
A( 44)=  0.525033197507D-02        B( 44)=  0.161588664129D-01
A( 45)= -0.188442859572D-08        B( 45)=  0.172492300121D-01
A( 46)= -0.490118394858D-02        B( 46)=  0.150842876525D-01
A( 47)= -0.767636379764D-02        B( 47)=  0.105560060104D-01
A( 48)= -0.743885128955D-02        B( 48)=  0.540464064853D-02
A( 49)= -0.445979896790D-02        B( 49)=  0.144907602619D-02
A( 50)=  0.146741718071D-08        B( 50)=  0.797750754344D-15
```

(a)

Figure 8.46 (continued)

C(-50)=	0.733708590356D-09	0.398875377172D-15	C(5)= 0.397625191328D-08	0.426514381792D-02
C(-49)=	-0.222989948395D-02	0.724538013094D-03	C(6)= 0.154942916895D-02	0.476864003599D-02
C(-48)=	-0.371942564477D-02	0.270232032427D-02	C(7)= 0.303267286902D-02	0.417411025872D-02
C(-47)=	-0.383818189882D-02	0.528280300522D-02	C(8)= 0.361992403656D-02	0.263002553481D-02
C(-46)=	-0.245059197429D-02	0.754214382623D-02	C(9)= 0.265042489193D-02	0.861173730888D-03
C(-45)=	-0.942214297861D-09	0.862461500603D-02	C(10)= -0.497523933021D-08	0.584671200343D-15
C(-44)=	0.262516598754D-02	0.807943320644D-02	C(11)= -0.370190258489D-02	0.120282280208D-02
C(-43)=	0.440551433377D-02	0.606367173035D-02	C(12)= -0.711924075346D-02	0.517243541296D-02
C(-42)=	0.457648886882D-02	0.332501460487D-02	C(13)= -0.854438341894D-02	0.117603437608D-01
C(-41)=	0.294328962957D-02	0.956333158955D-03	C(14)= -0.643731787744D-02	0.198120498015D-01
C(-40)=	0.126874207554D-08	0.573985303731D-15	C(15)= 0.854307422224D-08	0.273467777971D-01
C(-39)=	-0.320624086459D-02	0.104177036694D-02	C(16)= 0.104141086638D-01	0.320512988226D-01
C(-38)=	-0.543324095819D-02	0.394747956368D-02	C(17)= 0.232160858880D-01	0.319541823264D-01
C(-37)=	-0.570548423845D-02	0.785292319738D-02	C(18)= 0.358856937783D-01	0.260724683660D-01
C(-36)=	-0.371405104356D-02	0.114306685444D-01	C(19)= 0.455986095923D-01	0.148158738994D-01
C(-35)=	-0.183629532410D-08	0.133565013568D-01	C(20)= 0.499999990832D-01	-0.117691802068D-07
C(-34)=	0.416510692634D-02	0.128188871795D-01	C(21)= 0.478893348293D-01	-0.155602000074D-01
C(-33)=	0.718358555547D-02	0.988736030359D-02	C(22)= 0.395992791508D-01	-0.287705733568D-01
C(-32)=	0.769820849400D-02	0.559307761955D-02	C(23)= 0.269415865532D-01	-0.370819286564D-01
C(-31)=	0.513099040809D-02	0.166716071913D-02	C(24)= 0.127269785517D-01	-0.391696386316D-01
C(-30)=	0.301687885279D-08	0.102606811936D-14	C(25)= -0.667983101424D-08	-0.352625338195D-01
C(-29)=	-0.611584631427D-02	0.198715781612D-02	C(26)= -0.878041694312D-02	-0.270233279045D-01
C(-28)=	-0.109809877563D-01	0.797815175480D-02	C(27)= -0.123680240253D-01	-0.170231183641D-01
C(-27)=	-0.123680240253D-01	0.170231183641D-01	C(28)= -0.109809877563D-01	-0.797815175480D-02
C(-26)=	-0.878041694312D-02	0.270233279045D-01	C(29)= -0.611584631427D-02	-0.198715781612D-02
C(-25)=	-0.667983101424D-08	0.352625338195D-01	C(30)= 0.301687885279D-08	-0.102606811936D-14
C(-24)=	0.127269785517D-01	0.391696386316D-01	C(31)= 0.513099040809D-02	-0.166716071913D-02
C(-23)=	0.269415865532D-01	0.370819286564D-01	C(32)= 0.769820849400D-02	-0.559307761955D-02
C(-22)=	0.395992791508D-01	0.287705733568D-01	C(33)= 0.718358555547D-02	-0.988736030359D-02
C(-21)=	0.478893348293D-01	0.155602000074D-01	C(34)= 0.416510692634D-02	-0.128188871795D-01
C(-20)=	0.499999990832D-01	0.117691802068D-07	C(35)= -0.183629532410D-08	-0.133565013568D-01
C(-19)=	0.455986095923D-01	-0.148158738994D-01	C(36)= -0.371405104356D-02	-0.114306685444D-01
C(-18)=	0.358856937783D-01	-0.260724683660D-01	C(37)= -0.570548423845D-02	-0.785292319738D-02
C(-17)=	0.232160858880D-01	-0.319541823264D-01	C(38)= -0.543324095819D-02	-0.394747956368D-02
C(-16)=	0.104141086638D-01	-0.320512988226D-01	C(39)= -0.320624086459D-02	-0.104177036694D-02
C(-15)=	0.854307422224D-08	-0.273467777971D-01	C(40)= 0.126874207554D-08	-0.573985303731D-15
C(-14)=	-0.643731787744D-02	-0.198120498015D-01	C(41)= 0.294328962957D-02	-0.956333158955D-03
C(-13)=	-0.854438341894D-02	-0.117603437608D-01	C(42)= 0.457648886882D-02	-0.332501460487D-02
C(-12)=	-0.711924075346D-02	-0.517243541296D-02	C(43)= 0.440551433377D-02	-0.606367173035D-02
C(-11)=	-0.370190258489D-02	-0.120282280208D-02	C(44)= 0.262516598754D-02	-0.807943320644D-02
C(-10)=	-0.497523933021D-08	-0.584671200343D-15	C(45)= -0.942214297861D-09	-0.862461500603D-02
C(-9)=	0.265042489193D-02	-0.861173730888D-03	C(46)= -0.245059197429D-02	-0.754214382623D-02
C(-8)=	0.361992403656D-02	-0.263002553481D-02	C(47)= -0.383818189882D-02	-0.528280300522D-02
C(-7)=	0.303267286902D-02	-0.417411025872D-02	C(48)= -0.371942564477D-02	-0.270232032427D-02
C(-6)=	0.154942916895D-02	-0.476864003599D-02	C(49)= -0.222989948395D-02	-0.724538013094D-03
C(-5)=	0.397625191328D-08	-0.426514381792D-02	C(50)= 0.733708590356D-09	-0.398875377172D-15
C(-4)=	-0.979404958732D-03	-0.301431046712D-02		
C(-3)=	-0.116736457812D-02	-0.160674474770D-02		
C(-2)=	-0.768469096430D-03	-0.558328214972D-03		
C(-1)=	-0.235696462745D-03	-0.765836369441D-04		
C(0)=	-0.372650603833D-08	0.000000000000D 00		
C(1)=	-0.235696462745D-03	0.765836369441D-04		
C(2)=	-0.768469096430D-03	0.558328214972D-03		
C(3)=	-0.116736457812D-02	0.160674474770D-02		
C(4)=	-0.979404958732D-03	0.301431046712D-02		

(b)

Figure 8.47

(a) Amplitude spectrum of the periodic time function of Figure 8.44, which is composed of a train of oscillatory pulses. (b) Phase spectrum of the periodic time function of Figure 8.44, which is composed of a train of oscillatory pulses. (c) Phase angles of the exponential Fourier coefficients listed in Figure 8.46(b).

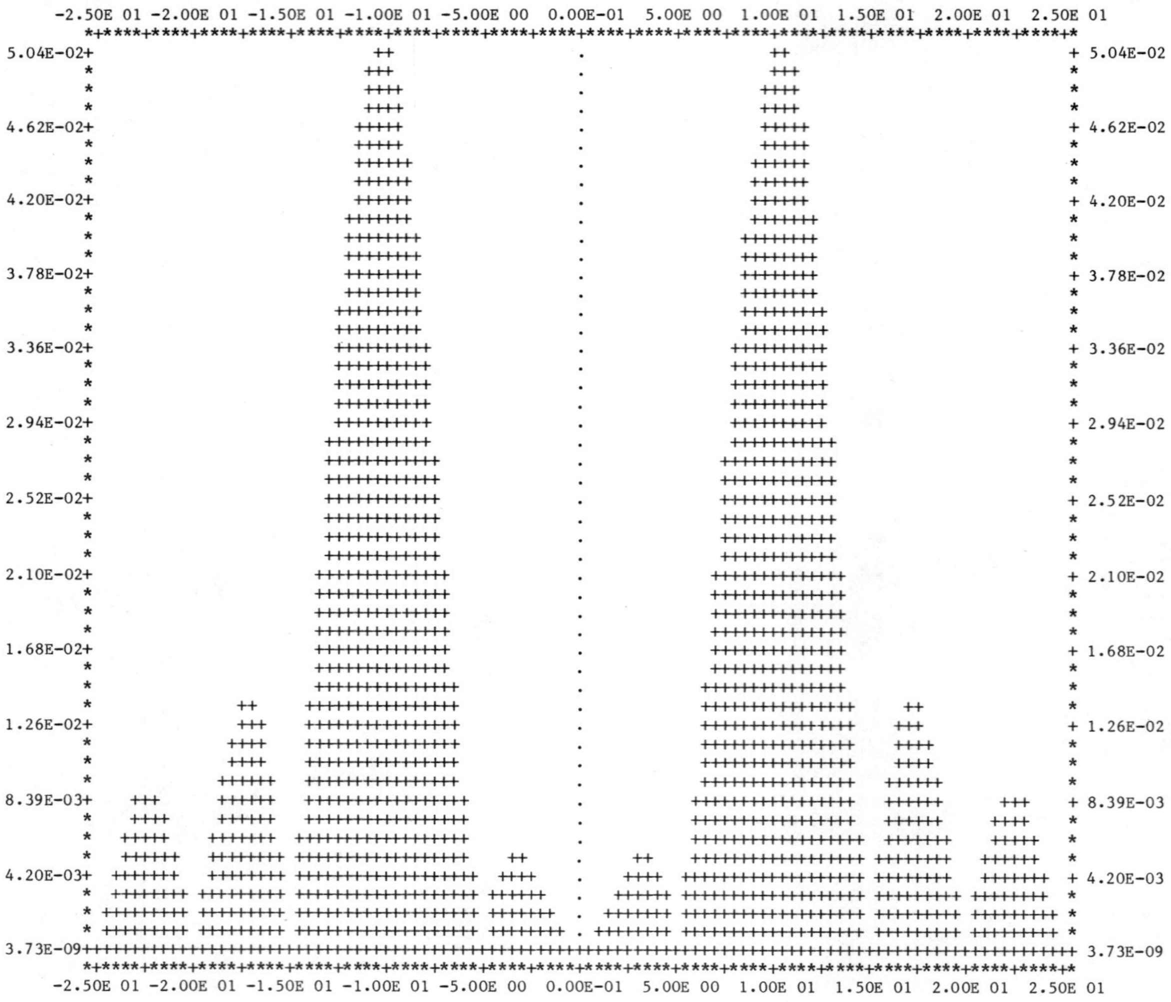

$n\omega_0/2\pi$

(a)

Figure 8.47 (continued)

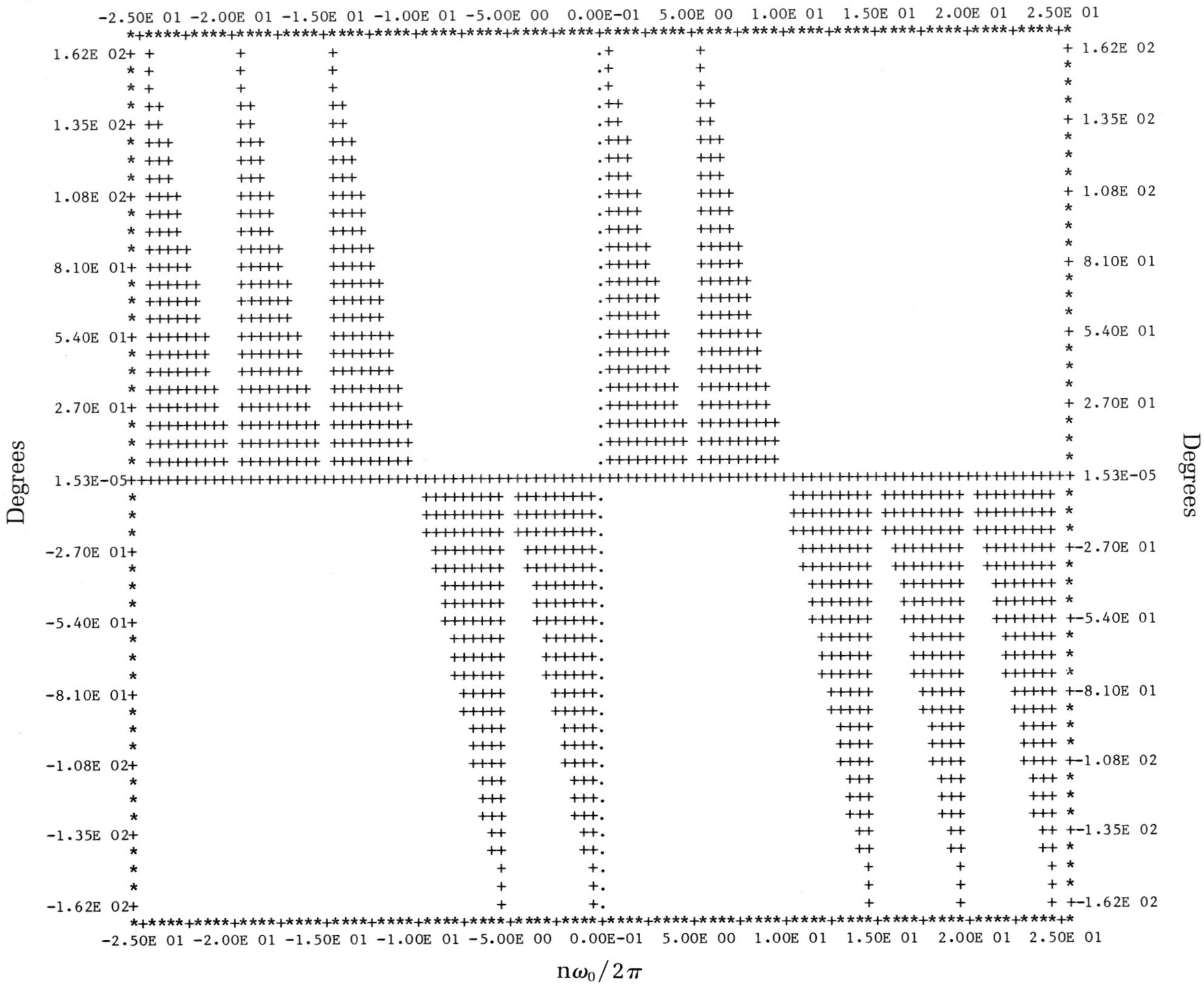

(b)

Figure 8.47   (continued)

PHI(−50)=	0.00000000E 00
PHI(−49)=	0.16200000E 03
PHI(−48)=	0.14400000E 03
PHI(−47)=	0.12600000E 03
PHI(−46)=	0.10800000E 03
PHI(−45)=	0.90000000E 02
PHI(−44)=	0.72000000E 02
PHI(−43)=	0.54000000E 02
PHI(−42)=	0.36000000E 02
PHI(−41)=	0.18000000E 02
PHI(−40)=	0.00000000E 00
PHI(−39)=	0.16200000E 03
PHI(−38)=	0.14400000E 03
PHI(−37)=	0.12600000E 03
PHI(−36)=	0.10800000E 03
PHI(−35)=	0.90000000E 02
PHI(−34)=	0.72000000E 02
PHI(−33)=	0.54000000E 02
PHI(−32)=	0.36000000E 02
PHI(−31)=	0.18000000E 02
PHI(−30)=	0.00000000E 00
PHI(−29)=	0.16200000E 03
PHI(−28)=	0.14400000E 03
PHI(−27)=	0.12600000E 03
PHI(−26)=	0.10800000E 03
PHI(−25)=	0.90000000E 02
PHI(−24)=	0.72000000E 02
PHI(−23)=	0.54000000E 02
PHI(−22)=	0.36000000E 02
PHI(−21)=	0.18000000E 02
PHI(−20)=	0.00000000E 00
PHI(−19)=	−0.17999980E 02
PHI(−18)=	−0.35999980E 02
PHI(−17)=	−0.53999980E 02
PHI(−16)=	−0.71999980E 02
PHI(−15)=	−0.89999980E 02
PHI(−14)=	−0.10799990E 03
PHI(−13)=	−0.12599990E 03
PHI(−12)=	−0.14399990E 03
PHI(−11)=	−0.16199990E 03
PHI(−10)=	0.00000000E 00
PHI(−9)=	−0.17999960E 02
PHI(−8)=	−0.35999950E 02
PHI(−7)=	−0.53999950E 02
PHI(−6)=	−0.71999950E 02
PHI(−5)=	−0.89999930E 02
PHI(−4)=	−0.10799990E 03
PHI(−3)=	−0.12599990E 03
PHI(−2)=	−0.14399980E 03
PHI(−1)=	−0.16199970E 03
PHI(0)=	0.00000000E 00
PHI(1)=	0.16199970E 03
PHI(2)=	0.14399980E 03
PHI(3)=	0.12599990E 03
PHI(4)=	0.10799990E 03

PHI(5)=	0.89999930E 02
PHI(6)=	0.71999950E 02
PHI(7)=	0.53999950E 02
PHI(8)=	0.35999950E 02
PHI(9)=	0.17999960E 02
PHI(10)=	0.00000000E 00
PHI(11)=	0.16199990E 03
PHI(12)=	0.14399990E 03
PHI(13)=	0.12599990E 03
PHI(14)=	0.10799990E 03
PHI(15)=	0.89999980E 02
PHI(16)=	0.71999980E 02
PHI(17)=	0.53999980E 02
PHI(18)=	0.35999980E 02
PHI(19)=	0.17999980E 02
PHI(20)=	0.00000000E 00
PHI(21)=	−0.18000000E 02
PHI(22)=	−0.36000000E 02
PHI(23)=	−0.54000000E 02
PHI(24)=	−0.72000000E 02
PHI(25)=	−0.90000000E 02
PHI(26)=	−0.10800000E 03
PHI(27)=	−0.12600000E 03
PHI(28)=	−0.14400000E 03
PHI(29)=	−0.16200000E 03
PHI(30)=	0.00000000E 00
PHI(31)=	−0.18000000E 02
PHI(32)=	−0.36000000E 02
PHI(33)=	−0.54000000E 02
PHI(34)=	−0.72000000E 02
PHI(35)=	−0.90000000E 02
PHI(36)=	−0.10800000E 03
PHI(37)=	−0.12600000E 03
PHI(38)=	−0.14400000E 03
PHI(39)=	−0.16200000E 03
PHI(40)=	0.00000000E 00
PHI(41)=	−0.18000000E 02
PHI(42)=	−0.36000000E 02
PHI(43)=	−0.54000000E 02
PHI(44)=	−0.72000000E 02
PHI(45)=	−0.90000000E 02
PHI(46)=	−0.10800000E 03
PHI(47)=	−0.12600000E 03
PHI(48)=	−0.14400000E 03
PHI(49)=	−0.16200000E 03
PHI(50)=	0.00000000E 00

(c)

M/n, where M is a constant independent of n. In general, the smoother the signal, the faster its Fourier series converges.

In approximating a signal by a truncated Fourier series, an anomalous behavior known as the *Gibbs phenomenon* appears. This behavior is observed near the points of jump discontinuity where the partial sum deviates considerably from the signal waveform no matter how many terms are used. The maximum value of this discrepancy is about 9 percent of the discontinuity jump.

Figure 8.48

(a) Approximation to the oscillatory pulse waveform of Figure 8.44 by a truncated Fourier series containing 11 terms. (b) Approximation to the oscillatory pulse waveform of Figure 8.44 by a truncated Fourier series containing 19 terms. (c) Approximation to the oscillatory pulse waveform of Figure 8.44 by a truncated Fourier series containing 43 terms. (d) Approximation to the oscillatory pulse waveform of Figure 8.44 by a truncated Fourier series containing 99 terms.

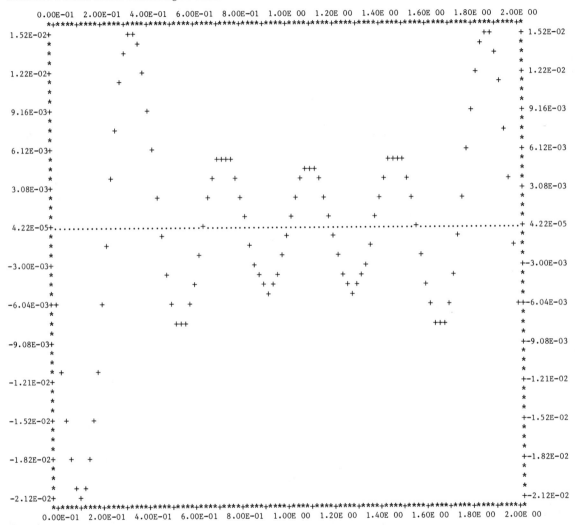

F(T) VS T WITH 11 TERMS

(a)

Figure 8.48   (*continued*)

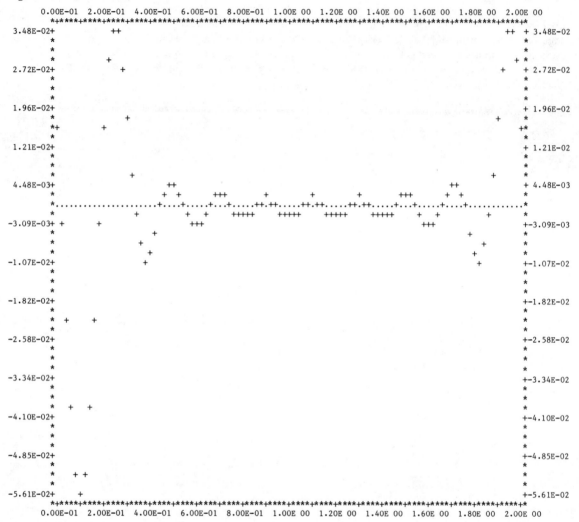

F(T) VS T WITH 19   TERMS

(b)

Finally, we presented computer programs in WATFIV for the evaluation of the Fourier coefficients. Using subroutine PLOT2, the amplitude and phase spectra can be displayed. Examples were given of the approximation to selected continuous and discontinuous functions by truncated Fourier series of various numbers of terms. The application of Fourier analysis to ascertain the steady-state response of a system to a periodic signal input and other related topics will be the subject of the following chapter.

Figure 8.48 (continued)

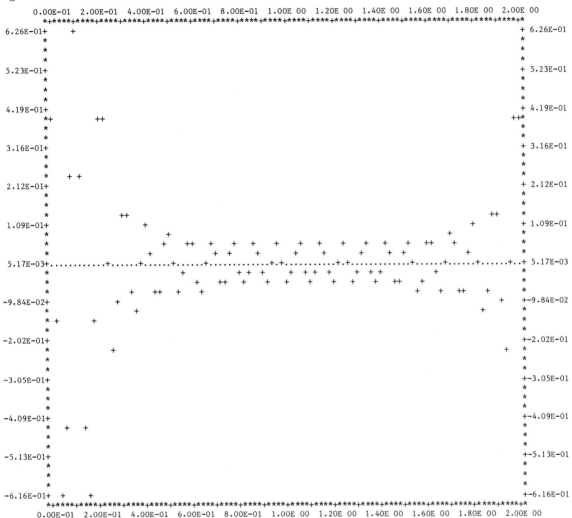

F(T) VS T WITH 43 TERMS

(C)

REFERENCES AND SUGGESTED READING

Bracewell, R. N. *The Fourier Transform and Its Applications.* 2d ed. New York: McGraw-Hill, 1978, Chapter 10.

Guillemin, E. A. *Theory of Linear Physical Systems.* New York: Wiley, 1963, Chapter 11.

Gupta, S. C., J. W. Bayless, and B. Peikari. *Circuit Analysis with Computer Applications to Problem Solving.* New York: Intext, 1972.

Figure 8.48   (continued)

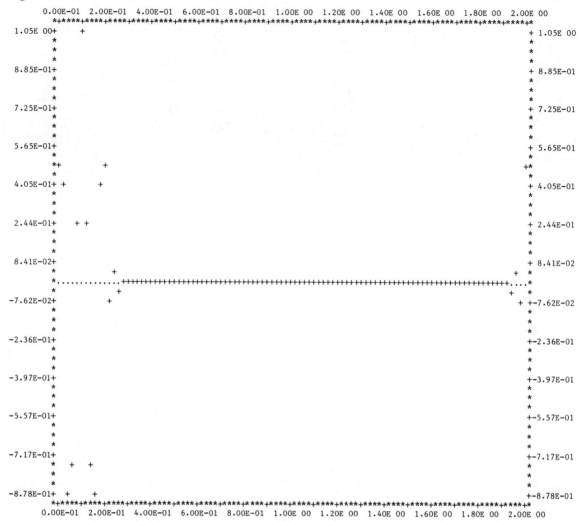

F(T) VS T WITH 99   TERMS

(d)

Kuo, F. F. *Network Analysis and Synthesis.* 2d ed. New York: Wiley, 1966, Chapter 3.

Lathi, B. P. *Signals, Systems, and Controls.* New York: Intext, 1974, Chapter 3.

Liu, C. L., and J. W. S. Liu. *Linear Systems Analysis.* New York: McGraw-Hill, 1975, Chapter 6.

Papoulis, A. *Circuits and Systems: A Modern Approach.* New York: Holt, Rinehart & Winston, 1980, Chapter 7.

Van Valkenburg, M. E. *Network Analysis.* 3d ed. Englewood Cliffs, N.J.: Prentice-Hall, 1974, Chapter 15.

Wylie, Jr., C. R. *Advanced Engineering Mathematics.* 3d ed. New York: McGraw-Hill, 1966, Chapter 6.

PROBLEMS

8.1 Calculate the trigonometric Fourier coefficients for the half-wave rectified waveform of Figure 8.18(a). Find the Fourier series and plot the amplitude and phase spectra of the signal.

8.2 Repeat Problem 8.1 for the signal waveform of Figure 8.18(b).

8.3 Determine the exponential Fourier coefficients for the half-wave rectified waveform of Figure 8.18(a). Find the Fourier series and plot the amplitude and phase spectra of the signal.

8.4 For the sawtooth waveform of Figure 8.1(b), determine the Fourier coefficients (both trigonometric and exponential) and plot the amplitude and phase spectra.

8.5 Repeat Problem 8.4 for the waveform of Figure 8.3(b).

8.6 For the periodic signal of Figure 8.15(b), determine the trigonometric and the exponential Fourier coefficients, find the Fourier series, and plot the amplitude and phase spectra.

8.7 For the staircase waveform of Figure 8.39, determine the exponential Fourier coefficients for the nth harmonic. Compare the result with those listed in Figure 8.41(b).

8.8 Repeat Problem 8.7 for a train of cosine pulses of Figure 8.44 and compare the result with those listed in Figure 8.46(b).

8.9 Given the periodic waveform

$$f(t) = e^{-t}, \qquad 0 < t < T \tag{8.147}$$

determine the exponential Fourier coefficients and plot the amplitude and phase spectra.

8.10 For the periodic signal of Figure 8.49, verify that the Fourier coefficients are as given below:

(a) $a_0 = \frac{3}{4}$ $\tag{8.148}$

(b) $a_n = 0, \quad n$ even $\tag{8.149a}$

$$= -\frac{1}{n\pi}(-1)^{(n+1)/2}, \quad n \text{ odd} \tag{8.149b}$$

(c) $b_n = \frac{3}{n\pi}, \quad n$ odd $\tag{8.150a}$

$$= \frac{1}{n\pi}[1 - (-1)^{n/2}], \quad n \text{ even} \tag{8.150b}$$

(d) $f(t) = \frac{3}{4} + \frac{1}{\pi}\left[\sqrt{10} \cos\left(\frac{\pi t}{2} - 71.56°\right) + \cos(\pi t - 90°) + \frac{\sqrt{10}}{3}\cos\left(\frac{3\pi t}{2} - 108.44°\right) + \cdots \right]$ $\tag{8.151}$

Figure 8.49
Periodic staircase
waveform for
Problem 8.10.

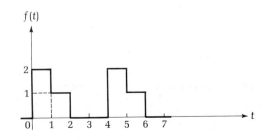

Figure 8.50
Periodic trapezoidal
waveform for Problem
8.12.

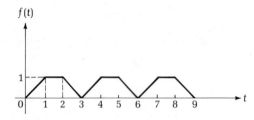

Figure 8.51
Train of triangular pulses
for Problem 8.13.

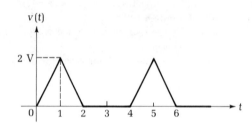

Figure 8.52
Periodic rectangular
waveform for Problem
8.14.

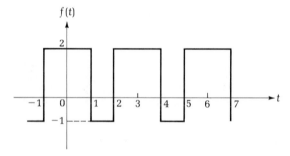

8.11 Derive the coefficient formulas (8.67) through (8.69).

8.12 Consider the trapezoidal waveform of Figure 8.50. Show that its Fourier series is given by

$$f(t) = \frac{2}{3} - \frac{9}{2\pi^2}\left[\cos\frac{2\pi t}{3} + \frac{1}{4}\cos\frac{4\pi t}{3} + \frac{1}{16}\cos\frac{8\pi t}{3} + \frac{1}{25}\cos\frac{10\pi t}{3} + \cdots\right] \tag{8.152}$$

8.13 The waveform shown in Figure 8.51 is a train of triangular pulses. For this periodic signal, verify that its Fourier coefficients are given below:

(a) $a_0 = \frac{1}{2}$ $\tag{8.153}$

(b) $a_n = \dfrac{4}{n^2\pi^2}\left[2\cos\dfrac{n\pi}{2} - (-1)^n - 1\right]$ $\tag{8.154}$

(c) $b_n = \dfrac{8}{n^2\pi^2}\sin\dfrac{n\pi}{2}$ $\tag{8.155}$

(d) $\bar{c}_n = \dfrac{2}{n^2\pi^2}\left[2\cos\dfrac{n\pi}{2} - (-1)^n - 1 - j2\sin\dfrac{n\pi}{2}\right]$ $\tag{8.156}$

8.14 Show that the Fourier series for the rectangular-wave signal of Figure 8.52 is given by

$$f(t) = 1 + \frac{6}{\pi}\sum_{n=1}^{\infty}\frac{1}{n}\sin\frac{2n\pi}{3}\cos\frac{2n\pi t}{3} \tag{8.157}$$

8.15 Derive the coefficient formulas (8.72) and (8.74).

8.16 Given the periodic waveform

$$f(t) = \left| \cos \frac{\omega_0 t}{2} \right|, \qquad 0 < t < T \tag{8.158}$$

show that its Fourier series is given by

$$f(t) = \frac{2}{\pi} + \frac{4}{\pi} \left[\frac{1}{3} \cos \omega_0 t - \frac{1}{15} \cos 2\omega_0 t + \frac{1}{35} \cos 3\omega_0 t - \cdots \right] \tag{8.159}$$

8.17 Find the trigonometric Fourier series of $\sin^4 t$ and $\cos^5 2t$.

8.18 Prove Property 8.4.

8.19 Show that

$$\left| \sin \frac{\omega_0 t}{2} \right| = \frac{2}{\pi} \sum_{n=-\infty}^{\infty} \frac{1}{1 - 4n^2} e^{jn\omega_0 t} \tag{8.160}$$

8.20 Repeat Example 8.15 for the waveform of Figure 8.18(a).

8.21 Repeat Example 8.15 for the waveform of Figure 8.3(b).

8.22 Repeat Example 8.16 for the waveform of Figure 8.15(b).

8.23 Repeat Example 8.16 for the waveform of Figure 8.49 (see Problem 8.10).

8.24 Repeat Example 8.16 for the waveform of Figure 8.50 (see Problem 8.12).

8.25 Repeat Example 8.16 for the waveform of Figure 8.51 (see Problem 8.13).

8.26 Repeat Example 8.16 for the waveform of Figure 8.52 (see Problem 8.14).

8.27 Express the exponential Fourier coefficients of $f(-t)$ in terms of the coefficients $\tilde{c}_n$ of $f(t)$.

8.28 Determine whether the following functions are periodic or nonperiodic. Find the period in the case of periodic functions:

(a) $\alpha \sin 20\pi t + \beta \sin 40\pi t$

(b) $\alpha \cos 20\pi t + \beta \sin 40\pi t$

(c) $\alpha \sin \sqrt{3} t + \beta \sin 10 t$

(d) $\beta (\alpha \sin 2t)^3$

(e) $(\alpha \sin t + \beta \sin 2t)^2$

8.29 Given the periodic signal

$$\begin{aligned} f(t) &= \cos 300\pi t, & |t| < 1 \\ &= 0, & 1 < |t| < 2 \end{aligned} \tag{8.161}$$

determine the exponential Fourier coefficients and plot the amplitude and phase spectra.

8.30 Repeat Example 8.16 for the periodic signal of (8.161).

CHAPTER NINE

―――――

SYSTEM RESPONSE AND DISCRETE FOURIER SERIES

In this chapter we continue the study on the signal representation by a Fourier series and its applications. We consider the steady-state response of a system to a periodic signal input, the discrete Fourier series, the fast Fourier transform, and other related topics.

9-1 STEADY-STATE RESPONSE TO PERIODIC EXCITATIONS

A periodic signal that ideally started at the beginning of time and will last forever is applied to a system. We shall determine the steady-state response of the system to this input, using Fourier series as a tool.

Consider the block diagram of Figure 9.1 with input $e(t)$ and response $r(t)$. Denote by $E(s)$ and $R(s)$ the Laplace transforms of $e(t)$ and $r(t)$, respectively. The system is characterized by its transfer function $H(s)$ defined by relation

$$\frac{R(s)}{E(s)} = H(s) \tag{9.1}$$

On the $j\omega$-axis, the transfer function becomes $H(j\omega)$.

Suppose that a periodic signal $e(t)$ of period T is applied to the system. The input signal can be expressed by a Fourier series as

$$e(t) = c_0 + \sum_{n=1}^{\infty} c_n \cos(n\omega_0 t + \theta_n) \tag{9.2}$$

Figure 9.1
System block diagram with input $e(t)$ and response $r(t)$.

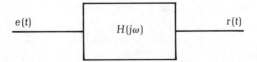

412

Figure 9.2
Block diagram showing the input-output relation with the input excitation being approximated by n sinusoidal sources having harmonically related frequencies.

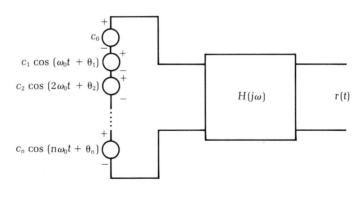

Figure 9.3
Block diagram showing the input-output phasor relation corresponding to the mth harmonic.

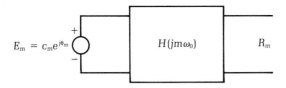

where c_n, being real, are assumed to be nonnegative with the possible minus sign to be incorporated in θ_n. The signal $e(t)$ can be approximated by the partial sum

$$e(t) \cong s_n(t) = c_0 + \sum_{m=1}^{n} c_m \cos(m\omega_0 t + \theta_m) \tag{9.3}$$

As a result, the input $e(t)$ may be replaced by n sinusoidal signal sources in series,[†] each having an amplitude c_m and phase θ_m and each at harmonically related frequencies $m\omega_0$, as depicted in Figure 9.2. For the multigenerator input, the total response may be determined by superposition. Consider the mth harmonic

$$e_m(t) = c_m \cos(m\omega_0 t + \theta_m) \tag{9.4}$$

with $e_0(t) = c_0$, the phasor representation of which is the familiar

$$E_m = c_m e^{j\theta_m} \tag{9.5}$$

Applying this input to the system of Figure 9.1, the corresponding phasor response is given by (Figure 9.3)

$$R_m = H(jm\omega_0)E_m = c_m H(jm\omega_0)e^{j\theta_m} \tag{9.6}$$

[†] Assume for the time being that $e(t)$ is a voltage source. If it is a current source, the component sources are connected in parallel.

or in time-domain

$$r_m(t) = c_m|H(jm\omega_0)| \cos[m\omega_0 t + \theta_m + \underline{/H(jm\omega_0)}] \tag{9.7}$$

By superposition principle the total response due to the dc source c_0 and the n sinusoidal signal sources is the sum of the responses due to the individual ones, and we obtain from (9.7)

$$r(t) \cong c_0 H(0) + \sum_{m=1}^{n} c_m|H(jm\omega_0)| \cos[m\omega_0 t + \theta_m + \underline{/H(jm\omega_0)}] \tag{9.8}$$

In the limit as n approaches infinity the steady-state response becomes

$$r(t) = c_0 H(0) + \sum_{m=1}^{\infty} c_m|H(jm\omega_0)| \cos[m\omega_0 t + \theta_m + \underline{/H(jm\omega_0)}] \tag{9.9}$$

In terms of exponentials we have from (8.92) $\tilde{c}_0 = c_0$, $\tilde{c}_{\pm k} = \frac{1}{2}c_k e^{\pm j\theta_k}$, $k \neq 0$, and

$$r(t) = \sum_{m=-\infty}^{\infty} \tilde{c}_m H(jm\omega_0)e^{jm\omega_0 t} \tag{9.10}$$

The $\tilde{c}_m$ are in fact the exponential Fourier coefficients of $e(t)$. This shows that if a periodic signal with Fourier coefficients $\tilde{c}_m$ is applied to a system, the resulting output is also a periodic signal the Fourier coefficients of which equal $\tilde{c}_m H(jm\omega_0)$, assuming of course that the input is applied at $t = -\infty$.

We illustrate this by the following examples.

EXAMPLE 9.1

The triangular voltage waveform $v_g(t)$ of Figure 9.4(a) is applied to the RC network as shown in (b). We wish to determine the steady-state voltage $v(t)$ across the resistor.

The period of the triangular voltage is $T = 1$ ms, giving the fundamental frequency $\omega_0 = 2\pi/T = 2\pi \times 10^3$ rad/s. The Fourier series representation of

Figure 9.4
The application of a triangular voltage waveform (a) to an RC network (b).

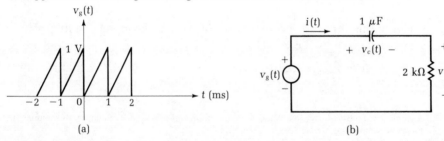

$v_g(t)$ is found to be

$$v_g(t) = \frac{1}{2} - \frac{1}{\pi} \sum_{n=1}^{\infty} \frac{1}{n} \sin n\omega_0 t$$

$$= \frac{1}{2} + \frac{1}{\pi} \sum_{n=1}^{\infty} \frac{1}{n} \cos\left(n\omega_0 t + \frac{\pi}{2}\right) \tag{9.11}$$

yielding the corresponding phasor input voltages

$$E_0 = \frac{1}{2}, \qquad E_n = \frac{1}{n\pi} e^{j\pi/2} \tag{9.12}$$

The desired network transfer function is given by

$$\frac{V(j\omega)}{V_g(j\omega)} = H(j\omega) = \frac{j2 \times 10^{-3}\,\omega}{1 + j2 \times 10^{-3}\,\omega} \tag{9.13}$$

Thus the phasor response due to the nth harmonic input is obtained as

$$R_n = H(jn\omega_0)E_n = \frac{j2 \times 10^{-3}\,n\omega_0}{1 + j2 \times 10^{-3}\,n\omega_0} \cdot \frac{e^{j\pi/2}}{n\pi}$$

$$= \frac{-4}{\sqrt{1 + 16n^2\pi^2}}\, e^{-j\tan^{-1} 4n\pi}, \qquad n \geq 0 \tag{9.14}$$

with $R_0 = 0$. The corresponding time-domain response is given by

$$r_n(t) = \frac{-4}{\sqrt{1 + 16n^2\pi^2}} \cos\left(2n\pi \times 10^3\, t - \tan^{-1} 4n\pi\right) \tag{9.15}$$

Summing up these component responses yields the steady-state response

$$v(t) = -4 \sum_{n=1}^{\infty} \frac{1}{\sqrt{1 + 16n^2\pi^2}} \cos\left(2n\pi \times 10^3\, t - \tan^{-1} 4n\pi\right) \tag{9.16}$$

which converges to

$$v(t) = 2 - \frac{e^{-500t}}{1 - e^{-0.5}}, \qquad 0 < t < 10^{-3} \tag{9.17}$$

To see this we recompute the Fourier series of the periodic signal $v(t)$ of (9.17), getting from (8.40), (8.42), and (8.44)

$$a_0 = 10^3 \int_0^{10^{-3}} \left[2 - \frac{e^{-500t}}{1 - e^{-0.5}} \right] dt = 0 \tag{9.18}$$

$$a_n = 2 \times 10^3 \int_0^{10^{-3}} \left[2 - \frac{e^{-500t}}{1 - e^{-0.5}} \right] \cos\left(2n\pi \times 10^3\, t\right) dt$$

$$= -\frac{4}{1 + 16n^2\pi^2} \tag{9.19}$$

Figure 9.5
Plots of input voltage,
capacitor voltage, and
resistor voltage of the
network of Figure 9.4(b).

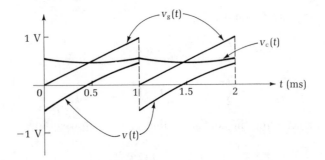

$$b_n = 2 \times 10^3 \int_0^{10^{-3}} \left[2 - \frac{e^{-500t}}{1 - e^{-0.5}} \right] \sin (2n\pi \times 10^3 \, t) \, dt$$

$$= -\frac{16n\pi}{1 + 16n^2\pi^2} \tag{9.20}$$

The Fourier series is therefore

$$v(t) = -4 \sum_{n=1}^{\infty} \frac{1}{1 + 16n^2\pi^2} [\cos 2n\pi \times 10^3 \, t + 4n\pi \sin 2n\pi \times 10^3 \, t]$$

$$= -4 \sum_{n=1}^{\infty} \frac{1}{\sqrt{1 + 16n^2\pi^2}} \cos (2n\pi \times 10^3 \, t - \tan^{-1} 4n\pi) \tag{9.21}$$

Knowing the voltage $v(t)$, the capacitor voltage $v_c(t)$ is determined as

$$v_c(t) = t - v(t) = t + \frac{e^{-500t}}{1 - e^{-0.5}} - 2, \qquad 0 < t < 10^{-3} \tag{9.22}$$

The plots of $v_c(t)$, $v(t)$, and the input signal are presented in Figure 9.5.

EXAMPLE 9.2
A typical network that produces a full-wave rectified sine wave was shown
in Figure 8.12, the output of which is shown in Figure 9.6(a) for a general
period T. This signal is fed into the input of a low-pass filter as shown in
Figure 9.6(b). The filter output is dc signal with ripples. Thus Figure 9.6(b)
is a device that converts ac to dc. We wish to determine the output voltage
$v_0(t)$ when the input to the rectifier is $\sin \omega_0 t/2$.

Refer to the waveform of Figure 9.6(a), the Fourier series of which is
found to be

$$v(t) = \left| \sin \frac{\omega_0 t}{2} \right| = \frac{2}{\pi} + \frac{4}{\pi} \sum_{n=1}^{\infty} \frac{1}{1 - 4n^2} \cos n\omega_0 t \tag{9.23}$$

The desired voltage transfer ratio is

$$\frac{V_0(j\omega)}{V(j\omega)} = H(j\omega) = \frac{R_2}{R_1 + R_2 + jR_1R_2C\omega} \tag{9.24}$$

Figure 9.6
An ac-to-dc converter (b) composed of a rectifier, the output of which is a full-wave rectified sine wave (a), and a low-pass filter, the output of which is dc signal with ripples.

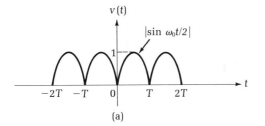

(a)

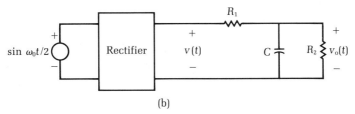

(b)

From (9.10) the steady-state output voltage can be written as

$$v_o(t) = \sum_{n=-\infty}^{\infty} \frac{2R_2}{\pi(1 - 4n^2)(R_1 + R_2 + jR_1R_2 Cn\omega_0)} \, e^{jn\omega_0 t} \tag{9.25}$$

where from (8.86) in conjunction with (9.23)

$$\tilde{c}_n = \frac{2}{(1 - 4n^2)\pi} \tag{9.26}$$

9-2   THE POWER SPECTRUM OF PERIODIC SIGNALS

In this section we determine the effective or rms value of a nonsinusoidal but periodic signal. Let $f(t)$ be a periodic signal with period T, which can be represented by a Fourier series with $\tilde{c}_n = |\tilde{c}_n| e^{j\phi_n}$:

$$f(t) = \sum_{n=-\infty}^{\infty} \tilde{c}_n e^{jn\omega_0 t} = \tilde{c}_0 + \sum_{n=1}^{\infty} 2|\tilde{c}_n| \cos(n\omega_0 t + \phi_n) \tag{9.27}$$

The *effective value* of $f(t)$ is defined by the equation

$$F_{\text{eff}} = \left[\frac{1}{T} \int_{t_0}^{t_0 + T} f^2(t)\, dt \right]^{1/2} \tag{9.28}$$

The operation involves extracting the square *root* of the *mean* (average) of the *squared* signal. Therefore F_{eff} is frequently referred to as the *root-mean-square value*, abbreviated rms value. For our purposes we choose $t_0 = 0$. Substituting (9.27) in (9.28) yields

$$F_{\text{eff}}^2 = \frac{1}{T} \int_0^T \left[\tilde{c}_0 + \sum_{n=1}^{\infty} 2|\tilde{c}_n| \cos(n\omega_0 t + \phi_n) \right]^2 dt \tag{9.29}$$

To evaluate this infinite sum of integrals, we appeal to formulas (8.34) to (8.38). As a result we see that the only nonzero terms on the right-hand side are those integrals having the forms of (8.37b) and (8.38b), which have the value $\frac{1}{2}T$. This leads to

$$F_{\text{eff}}^2 = \tilde{c}_0^2 + \sum_{n=1}^{\infty} 2|\tilde{c}_n|^2 \tag{9.30}$$

From (8.86) we see that with a bar denoting the complex conjugate

$$|\tilde{c}_n|^2 = \tilde{c}_n \overline{\tilde{c}_n} = \tilde{c}_n \tilde{c}_{-n} \tag{9.31}$$

As is well known, for a sinusoidal signal $A \cos(\omega t + \phi)$ with maximum value A, its effective value is $A/\sqrt{2}$. Thus from (9.27) the effective value of the nth harmonic is $2|\tilde{c}_n|/\sqrt{2} = \sqrt{2}|\tilde{c}_n|$ or

$$F_{n,\text{eff}} \triangleq \sqrt{2}|\tilde{c}_n| \tag{9.32}$$

Using this in (9.30), we get

$$F_{\text{eff}}^2 = F_{0,\text{eff}}^2 + \sum_{n=1}^{\infty} F_{n,\text{eff}}^2 \tag{9.33}$$

In words it states that the effective or rms value of a periodic signal equals the square root of the sum of the squares of the effective values of the harmonic components of the signal, including the dc component.

We next determine the input power to a network when the excitation signal is nonsinusoidal but periodic. Let $v(t)$ and $i(t)$ be the port voltage and current of a one-port network. Their Fourier series representations are given by

$$v(t) = V_0 + \sum_{n=1}^{\infty} V_n \cos(n\omega_0 t + \alpha_n) \tag{9.34}$$

$$i(t) = I_0 + \sum_{n=1}^{\infty} I_n \cos(n\omega_0 t + \beta_n) \tag{9.35}$$

The instantaneous power entering the one-port is

$$p(t) = v(t)i(t) \tag{9.36}$$

which, when integrated over a period, gives the average power

$$
\begin{aligned}
P_{\text{ave}} &= \frac{1}{T} \int_0^T v(t)i(t)\, dt \\[2mm]
&= \frac{1}{T} \int_0^T \left[V_0 + \sum_{n=1}^{\infty} V_n \cos(n\omega_0 t + \alpha_n) \right] \left[I_0 + \sum_{m=1}^{\infty} I_m \cos(m\omega_0 t + \beta_m) \right] dt \\[2mm]
&= V_0 I_0 + \frac{1}{2} \sum_{n=1}^{\infty} V_n I_n \cos(\beta_n - \alpha_n)
\end{aligned}
\tag{9.37}
$$

The integrals were again evaluated by appealing to formula (8.35) and the extension

$$\int_0^T \cos(n\omega_0 t + \alpha_n) \cos(m\omega_0 t + \beta_m)\, dt = 0, \qquad m \neq n \tag{9.38a}$$

$$= \tfrac{1}{2}T \cos(\beta_n - \alpha_n), \qquad m = n \tag{9.38b}$$

Because the average power of the nth harmonic is

$$P_{n,\text{ave}} = \tfrac{1}{2}V_n I_n \cos(\beta_n - \alpha_n) \tag{9.39}$$

the series (9.37) may be written as

$$P_{\text{ave}} = P_0 + \sum_{n=1}^{\infty} P_{n,\text{ave}} \tag{9.40}$$

In words it states that the total average power for a periodic signal equals the sum of the average power of its harmonic components, including the dc power P_0. It is significant to observe that there are no contributions to the average power from voltage at one frequency and current at another. As before, the average power of a periodic signal may be displayed by constructing an average power spectrum. There are two conventions for plotting the power spectra. The one-sided spectrum is constructed by drawing a line of length $P_{n,\text{ave}}$ at the discrete frequency $n\omega_0$, as illustrated in Figure 9.7(a). On the other hand, if we assign half of $P_{n,\text{ave}}$ to the positive

Figure 9.7
A one-sided power spectrum (a) and a two-sided power spectrum (b).

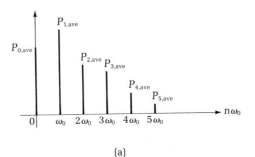

(a)

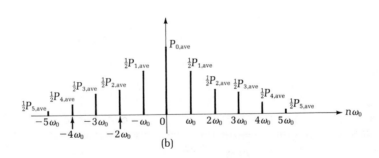

(b)

Figure 9.8
One-sided (a) and two-sided power spectra for the network of Figure 9.4(b).

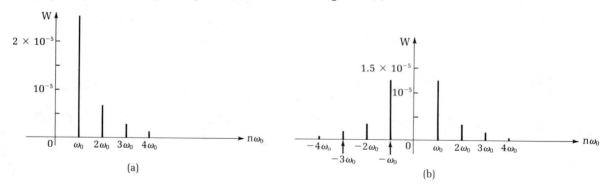

(a)

(b)

frequency $n\omega_0$ and the other half to the negative frequency $-n\omega_0$, we obtain the double-sided power spectrum as shown in Figure 9.7(b). Such spectra are commonly used in the study of random signals and other related advanced studies.

As an example we plot the power spectra of the problem in Example 9.1. Refer to the network of Figure 9.4(b). The resistor voltage and current are found to be

$$v(t) = -4 \sum_{n=1}^{\infty} \frac{1}{\sqrt{1 + 16n^2\pi^2}} \cos (2n\pi \times 10^3 t - \tan^{-1} 4n\pi) \qquad (9.41)$$

$$i(t) = -2 \times 10^{-3} \sum_{n=1}^{\infty} \frac{1}{\sqrt{1 + 16n^2\pi^2}} \cos (2n\pi \times 10^3 t - \tan^{-1} 4n\pi)$$

$$(9.42)$$

Thus from (9.39) the average power of the nth harmonic is determined as

$$P_{n,\text{ave}} = \frac{V_n I_n \cos (\beta_n - \alpha_n)}{2} = \frac{4}{1 + 16n^2\pi^2} \times 10^{-3} \qquad (9.43)$$

Two plots of the power spectra are represented in Figure 9.8. Because of the capacitor, there is no contribution from the dc component.

9-3 CIRCULAR CONVOLUTION

In Chapter 7 we studied in detail the convolution integral and its related properties. In the case of two periodic signals, their convolution will be considered in this section.

Given two periodic signals $f_1(t)$ and $f_2(t)$ with the same period T, the integral

$$f(t) = f_1(t) * f_2(t) \triangleq \int_{-T/2}^{T/2} f_1(t - \tau) f_2(\tau) \, d\tau \qquad (9.44)$$

is called the *circular convolution* of $f_1(t)$ and $f_2(t)$. The resultant signal $f(t)$ is also periodic because

$$f(t + T) = \int_{-T/2}^{T/2} f_1(t - \tau + T) f_2(\tau) \, d\tau = \int_{-T/2}^{T/2} f_1(t - \tau) f_2(\tau) \, d\tau \qquad (9.45)$$

Furthermore the operation is commutative because by a simple change of variable we can show that

$$f_1(t) * f_2(t) = f_2(t) * f_1(t) \qquad (9.46)$$

Let $f(t)$, $f_1(t)$, and $f_2(t)$ be represented by their exponential Fourier series, as follows:

$$f(t) = \sum_{n=-\infty}^{\infty} \tilde{c}_n e^{jn\omega_0 t} \qquad (9.47)$$

$$f_1(t) = \sum_{n=-\infty}^{\infty} \tilde{c}'_n e^{jn\omega_0 t} \qquad (9.48)$$

$$f_2(t) = \sum_{n=-\infty}^{\infty} \tilde{c}''_n e^{jn\omega_0 t} \qquad (9.49)$$

We shall express the Fourier coefficients $\tilde{c}_n$ in terms of $\tilde{c}'_n$ and $\tilde{c}''_n$. To this end, we compute $\tilde{c}_n$ from the coefficient formula (8.89):

$$\tilde{c}_n = \frac{1}{T} \int_{-T/2}^{T/2} f(t) e^{-jn\omega_0 t} \, dt$$

$$= \frac{1}{T} \int_{-T/2}^{T/2} \int_{-T/2}^{T/2} f_1(t - \tau) f_2(\tau) e^{-jn\omega_0 t} \, d\tau \, dt$$

$$= \frac{1}{T} \int_{-T/2}^{T/2} \int_{-T/2}^{T/2} \sum_{m=-\infty}^{\infty} \tilde{c}'_m e^{jm\omega_0(t-\tau)} f_2(\tau) e^{-jn\omega_0 t} \, d\tau \, dt$$

$$= \sum_{m=-\infty}^{\infty} \tilde{c}'_m \int_{-T/2}^{T/2} e^{j(m-n)\omega_0 t} \, dt \left[\frac{1}{T} \int_{-T/2}^{T/2} f_2(\tau) e^{-jm\omega_0 \tau} \, d\tau \right]$$

$$= \sum_{m=-\infty}^{\infty} \tilde{c}'_m \tilde{c}''_m \left[\int_{-T/2}^{T/2} e^{j(m-n)\omega_0 t} \, dt \right] = T \tilde{c}'_n \tilde{c}''_n \qquad (9.50)$$

In arriving at (9.50), use was made of the relation

$$\int_{t_0}^{t_0+T} e^{j(n-m)\omega_0 t} \, dt = 0, \qquad n \neq m$$

$$= T, \qquad n = m \qquad (9.51)$$

for any t_0, which follows directly from (8.34) and (8.35). This gives the *circular convolution theorem* as follows.

Property 9.1   *The Fourier coefficients of the circular convolution of two periodic signals with the same period T equal T times the product of the corresponding Fourier coefficients of the component signals.*

We next derive an expression for the Fourier coefficients of the product of two periodic signals $f_1(t)$ and $f_2(t)$. As before, we compute the Fourier coefficients $\tilde{c}_n$ of the resultant signal $f_1(t)f_2(t)$ by means of (8.89), as follows:

$$\tilde{c}_n = \frac{1}{T} \int_{-T/2}^{T/2} f_1(t)f_2(t)e^{-jn\omega_0 t}\,dt$$

$$= \frac{1}{T} \int_{-T/2}^{T/2} f_1(t) \left[\sum_{m=-\infty}^{\infty} \tilde{c}''_m e^{jm\omega_0 t} \right] e^{-jn\omega_0 t}\,dt$$

$$= \sum_{m=-\infty}^{\infty} \tilde{c}''_m \left[\frac{1}{T} \int_{-T/2}^{T/2} f_1(t)e^{-j(n-m)\omega_0 t}\,dt \right]$$

$$= \sum_{m=-\infty}^{\infty} \tilde{c}''_m \tilde{c}'_{n-m} \tag{9.52}$$

obtaining

Property 9.2   *The Fourier coefficients $\tilde{c}_n$ of the product $f(t)$ of two periodic signals $f_1(t)$ and $f_2(t)$ with the same period equal the discrete convolution of their Fourier coefficients $\tilde{c}'_n$ and $\tilde{c}''_n$ [also see (9.159)]:*

$$\tilde{c}_n = \sum_{m=-\infty}^{\infty} \tilde{c}'_{n-m}\tilde{c}''_m \tag{9.53}$$

In particular, for $\tilde{c}_0$ we have

$$\tilde{c}_0 = \frac{1}{T} \int_{-T/2}^{T/2} f_1(t)f_2(t)\,dt = \sum_{m=-\infty}^{\infty} \tilde{c}'_{-m}\tilde{c}''_m \tag{9.54}$$

Parseval's Formula:[†]

$$\frac{1}{T} \int_{-T/2}^{T/2} f_1(t)f_2(t)\,dt = \sum_{m=-\infty}^{\infty} \tilde{c}'_{-m}\tilde{c}''_m \tag{9.55}$$

If $f_1(t) = f_2(t)$, then $\tilde{c}'_m = \tilde{c}''_m$. Since $\tilde{c}'_{-m}$ is the conjugate of $\tilde{c}'_m$ or

$$\tilde{c}'_{-m} = \bar{\tilde{c}}'_m \tag{9.56}$$

[†] Named after Marc Antoine Parseval-Deschènes (?–1836).

from (9.55) we have

$$\frac{1}{T}\int_{-T/2}^{T/2} f_1^2(t)\,dt = \sum_{m=-\infty}^{\infty} |\tilde{c}_m'|^2 \tag{9.57}$$

Property 9.3 *If $\tilde{c}_n$ are the Fourier coefficients of a periodic signal $f(t)$ with period T, then*

$$\frac{1}{T}\int_{-T/2}^{T/2} f^2(t)\,dt = \sum_{m=-\infty}^{\infty} |\tilde{c}_m|^2 \tag{9.58}$$

Equation (9.58) is known as the *energy theorem.*

We illustrate the applications of the above formulas by the following examples. First, we derive the useful identities

$$\int_{-T/2}^{T/2} \sin m\omega_0 t \cos n\omega_0 t\,dt = 0 \tag{9.59}$$

$$\int_{-T/2}^{T/2} \sin m\omega_0 t \sin n\omega_0 t\,dt = 0, \qquad m \neq n \tag{9.60a}$$

$$= \tfrac{1}{2}T, \qquad m = n \tag{9.60b}$$

as previously given by (8.36) and (8.37). Note that there is a change of the integration limits that will not affect the outcome.

To verify (9.59) we let $f_1(t) = \sin m\omega_0 t$ and $f_2(t) = \cos n\omega_0 t$ with respective periods $T_1 = 2\pi/m\omega_0$ and $T_2 = 2\pi/n\omega_0$. Consider the ratio

$$\frac{T_1}{T_2} = \frac{n}{m} = \frac{\hat{n}}{\hat{m}} \tag{9.61}$$

where $\hat{m}$ and $\hat{n}$ are relatively prime. The common period of $f_1(t)$ and $f_2(t)$ may be written as

$$T = \hat{m}T_1 = \hat{n}T_2 \tag{9.62}$$

Using this period T, the Fourier coefficients of $f_1(t)$ and $f_2(t)$ are found to be

$$\tilde{c}_k' = \overline{\tilde{c}}_{-k}' = -j\tfrac{1}{2}, \qquad k = \hat{m} \tag{9.63}$$
$$= 0, \qquad \text{otherwise}$$

$$\tilde{c}_k'' = \tilde{c}_{-k}'' = \tfrac{1}{2}, \qquad k = \hat{n} \tag{9.64}$$
$$= 0, \qquad \text{otherwise}$$

Substituting these in Parseval's formula (9.55) yields

$$\int_{-T/2}^{T/2} \sin m\omega_0 t \cos n\omega_0 t\,dt$$

$$= T(\tilde{c}_{\hat{n}}'\tilde{c}_{-\hat{n}}'' + \tilde{c}_{-\hat{n}}'\tilde{c}_{\hat{n}}'') = 0, \qquad \hat{m} = \hat{n} \tag{9.65a}$$

$$= T(\tilde{c}_{\hat{n}}'\tilde{c}_{-\hat{n}}'' + \tilde{c}_{-\hat{n}}'\tilde{c}_{\hat{n}}'' + \tilde{c}_{\hat{m}}'\tilde{c}_{-\hat{m}}'' + \tilde{c}_{-\hat{m}}'\tilde{c}_{\hat{m}}'') = 0, \qquad \hat{m} \neq \hat{n} \tag{9.65b}$$

Similarly, to verify (9.60) we let $f_1(t) = \sin m\omega_0 t$ and $f_2(t) = \sin n\omega_0 t$ with common period T as defined in (9.62) and obtain

$$\tilde{c}'_k = \bar{\tilde{c}}'_{-k} = -j\tfrac{1}{2}, \qquad k = \hat{m}$$
$$= 0, \qquad \text{otherwise} \tag{9.66}$$

$$\tilde{c}''_k = \bar{\tilde{c}}''_{-k} = -j\tfrac{1}{2}, \qquad k = \hat{n}$$
$$= 0, \qquad \text{otherwise} \tag{9.67}$$

Again, appealing to (9.55) we get

$$\int_{-T/2}^{T/2} \sin m\omega_0 t \sin n\omega_0 t \, dt$$

$$= T(\tilde{c}'_{\hat{n}}\tilde{c}''_{-\hat{n}} + \tilde{c}'_{-\hat{n}}\tilde{c}''_{\hat{n}}) = \tfrac{1}{2}T, \qquad \hat{m} = \hat{n} \tag{9.68a}$$

$$= T(\tilde{c}'_{\hat{n}}\tilde{c}''_{-\hat{n}} + \tilde{c}'_{-\hat{n}}\tilde{c}''_{\hat{n}} + \tilde{c}'_{\hat{m}}\tilde{c}''_{-\hat{m}} + \tilde{c}'_{-\hat{m}}\tilde{c}''_{\hat{m}}) = 0, \qquad \hat{m} \neq \hat{n} \tag{9.68b}$$

proving (9.60). Likewise we can verify the validity of (8.38).

As an application of the energy theorem (9.58), we shall recompute the effective value of a periodic signal $f(t)$ with period T. From (9.58) we have

$$F_{\text{eff}}^2 = \frac{1}{T}\int_0^T f^2(t)\, dt = \sum_{m=-\infty}^{\infty} |\tilde{c}_m|^2, \tag{9.69}$$

giving

$$F_{\text{eff}}^2 = c_0^2 + 2\sum_{m=1}^{\infty} |\tilde{c}_m|^2 \tag{9.70}$$

as in (9.30).

We next apply Parseval's formula to the derivation of the average input power to a network when the excitation signal is nonsinusoidal but periodic. Let $v(t)$ and $i(t)$ be the port voltage and current of a one-port network. Their Fourier series representations are shown in (9.34) and (9.35), which can be rewritten in exponential forms as

$$v(t) = \sum_{n=-\infty}^{\infty} \tilde{V}_n e^{jn\omega_0 t} \tag{9.71}$$

$$i(t) = \sum_{n=-\infty}^{\infty} \tilde{I}_n e^{jn\omega_0 t} \tag{9.72}$$

where $\tilde{V}_0 = V_0$, $\tilde{I}_0 = I_0$, and for $n > 0$

$$\tilde{V}_n = \bar{\tilde{V}}_{-n} = \frac{1}{2}V_n e^{j\alpha_n} \tag{9.73}$$

$$\tilde{I}_n = \bar{\tilde{I}}_{-n} = \frac{1}{2}I_n e^{j\beta_n} \tag{9.74}$$

Applying the Parseval's formula (9.55) gives the average power

$$P_{ave} = \frac{1}{T} \int_0^T v(t)i(t)\,dt = \sum_{m=-\infty}^{\infty} \tilde{V}_{-m}\tilde{I}_m$$

$$= V_0 I_0 + \frac{1}{2} \sum_{n=1}^{\infty} V_n I_n \cos(\beta_n - \alpha_n) \tag{9.75}$$

confirming (9.37).

Suppose that the one-port network is characterized by its driving-point impedance $Z(s)$. Then $\tilde{V}_0 = Z(0)\tilde{I}_0$ and

$$\tilde{V}_n = Z(jn\omega_0)\tilde{I}_n \tag{9.76}$$

and (9.75) becomes

$$P_{ave} = \sum_{m=-\infty}^{\infty} \tilde{V}_{-m}\tilde{I}_m = \tilde{V}_0\tilde{I}_0 + 2\sum_{n=1}^{\infty} \operatorname{Re} \tilde{V}_{-n}\tilde{I}_n$$

$$= \tilde{V}_0\tilde{I}_0 + 2\sum_{n=1}^{\infty} \operatorname{Re} \bar{\tilde{V}}_n\tilde{I}_n = \tilde{V}_0\tilde{I}_0 + 2\sum_{n=1}^{\infty} \operatorname{Re} Z(-jn\omega_0)|\tilde{I}_n|^2$$

$$= Z(0)I_0^2 + 2\sum_{n=1}^{\infty} |\tilde{I}_n|^2 \operatorname{Re} Z(jn\omega_0)$$

$$= I_0^2 R(0) + \frac{1}{2}\sum_{n=1}^{\infty} I_n^2 \operatorname{Re} Z(jn\omega_0) \tag{9.77}$$

where Re denotes "the real part of," and $R(0) = Z(0)$. This is an extension of the familiar result in ac analysis, which states that if $i(t) = I_m \cos \omega_0 t$ is the current flowing into a network with input resistance $R(\omega_0)$, then the average power delivered to the network is $\frac{1}{2}I_m^2 R(\omega_0)$.

9-4   DISCRETE FOURIER SERIES

In the foregoing we have shown that a periodic signal $f(t)$ with period T satisfying the Dirichlet conditions can be expressed by a Fourier series

$$f(t) = \sum_{n=-\infty}^{\infty} \tilde{c}_n e^{jn\omega_0 t} \tag{9.78}$$

for which coefficients $\tilde{c}_n$ are determined by

$$\tilde{c}_n = \frac{1}{T} \int_0^T f(t)e^{-jn\omega_0 t}\,dt \tag{9.79}$$

However, in practical applications the infinite series must be approximated by a partial sum

$$f(t) \cong s_M(t) = \sum_{n=-M}^{M} \tilde{c}_n e^{jn\omega_0 t} \tag{9.80}$$

The partial sum $s_M(t)$ is a periodic function having no harmonics larger than M and is called a *trigonometric polynomial* of order M. Also in evaluating the Fourier coefficients $\tilde{c}_n$ by a digital computer, the integral in (9.79) must be approximated by a finite sum. In this section we examine the computational aspects of these problems and the underlying approximations.

As is well known, in approximating the integral (9.79) by a finite sum, we divide the period T into N equal subintervals

$$T_0 = \frac{T}{N} \tag{9.81}$$

although it is not necessary that they be equal, but the problem is easier to formulate if they are. The choice of the value of the integer N depends on the smoothness of the signal $f(t)$ and the complexity of the required computations. To simplify the notation, let

$$w = e^{j2\pi/N} \tag{9.82}$$

Then

$$w^N = 1, \qquad w^{n+N} = w^n \tag{9.83}$$

Using this to approximate the integral (9.79), we obtain

$$\tilde{c}_n \cong \frac{1}{NT_0} \sum_{m=0}^{N-1} f(mT_0)e^{-jn\omega_0 mT_0}T_0$$

$$= \frac{1}{N} \sum_{m=0}^{N-1} f(mT_0)e^{-j2mn\,\pi/N} \tag{9.84}$$

giving

$$\tilde{c}_n \cong \frac{1}{N} \sum_{m=0}^{N-1} f(mT_0)w^{-mn} \tag{9.85}$$

and from (9.80)

$$f(mT_0) \cong \sum_{n=-M}^{M} \tilde{c}_n w^{mn}, \qquad m = 0, 1, 2, \ldots, N-1 \tag{9.86}$$

Thus as t takes the values mT_0, $m = 0, 1, 2, \ldots, N-1$, the samples $f(mT_0)$ of $f(t)$ are approximated by (9.86). In fact, the computer subroutines presented in the preceding chapter and the programs for computing the partial sums are based on these approximations.

9-4.1 Aliasing Error

In approximating the Fourier coefficient $\tilde{c}_n$ by (9.85), the difference between the approximation and $\tilde{c}_n$ is called the *aliasing error*:

$$\hat{\epsilon}_n = \left[\frac{1}{N} \sum_{m=0}^{N-1} f(mT_0) w^{-mn} \right] - \tilde{c}_n \tag{9.87}$$

In this section we express this error in terms of the Fourier coefficients $\tilde{c}_n$ of $f(t)$.

Setting $t = mT_0$ in (9.78), the Fourier series of $f(t)$ becomes

$$f(mT_0) = \sum_{n=-\infty}^{\infty} \tilde{c}_n w^{mn} \tag{9.88}$$

As can be seen from (9.83),

$$w^{m(n+kN)} = w^{mn} \tag{9.89}$$

for any integer k. The term w^{mn} is a common factor of all the terms having coefficients of the form $\tilde{c}_{n+kN}$. As a result (9.88) can be expanded and factored as

$$
\begin{aligned}
f(mT_0) = {}& (\cdots + \tilde{c}_{-N} + \tilde{c}_0 + \tilde{c}_N + \cdots) \\
& + (\cdots + \tilde{c}_{-N+1} + \tilde{c}_1 + \tilde{c}_{N+1} + \cdots) w^m \\
& \cdots \\
& + (\cdots + \tilde{c}_{-N+n} + \tilde{c}_n + \tilde{c}_{N+n} + \cdots) w^{mn} \\
& \cdots \\
& + (\cdots + \tilde{c}_{-1} + \tilde{c}_{N-1} + \tilde{c}_{2N-1} + \cdots) w^{m(N-1)}
\end{aligned} \tag{9.90}
$$

To simplify the notation, let

$$\hat{c}_n = \sum_{k=-\infty}^{\infty} \tilde{c}_{n+kN} \tag{9.91}$$

The numbers $\hat{c}_n$ are known as the *aliased coefficients* of $f(t)$. This gives

$$
\begin{aligned}
f(mT_0) &= \hat{c}_0 + \hat{c}_1 w^m + \hat{c}_2 w^{2m} + \cdots + \hat{c}_n w^{nm} + \cdots + \hat{c}_{N-1} w^{(N-1)m} \\
&= \sum_{n=0}^{N-1} \hat{c}_n w^{mn}
\end{aligned} \tag{9.92}
$$

Setting $m = 0, 1, 2, \ldots, N-1$ in (9.92) yields a system of N linear equations, as follows:

$$
\begin{aligned}
\hat{c}_0 + \hat{c}_1 + \cdots + \hat{c}_n + \cdots + \hat{c}_{N-1} &= f(0) \\
\hat{c}_0 + \hat{c}_1 w + \cdots + \hat{c}_n w^n + \cdots + \hat{c}_{N-1} w^{N-1} &= f(T_0) \\
\cdots \\
\hat{c}_0 + \hat{c}_1 w^{N-1} + \cdots + \hat{c}_n w^{(N-1)n} + \cdots + \hat{c}_{N-1} w^{(N-1)^2} &= f(NT_0 - T_0)
\end{aligned} \tag{9.93}
$$

In matrix notation we have

$$\begin{bmatrix} 1 & 1 & \cdots & 1 & \cdots & 1 \\ 1 & w & \cdots & w^n & \cdots & w^{N-1} \\ \vdots & \vdots & \cdots & \vdots & \cdots & \vdots \\ 1 & w^{N-1} & \cdots & w^{(N-1)n} & \cdots & w^{(N-1)^2} \end{bmatrix} \begin{bmatrix} \hat{c}_0 \\ \hat{c}_1 \\ \vdots \\ \hat{c}_n \\ \vdots \\ \hat{c}_{N-1} \end{bmatrix} = \begin{bmatrix} f(0) \\ f(T_0) \\ \vdots \\ f(NT_0 - T_0) \end{bmatrix}$$

(9.94)

or more compactly

$$\mathbf{W}\hat{\mathbf{c}} = \mathbf{f}$$

(9.95)

Equation (9.95) relates the N aliased coefficients $\hat{c}_n$ of $f(t)$ to its samples $f(mT_0)$ and can be solved to express $\hat{c}_n$ in terms of $f(mT_0)$. For this purpose we multiply the first equation of (9.93) or (9.94) by 1, the second by w^{-n}, and the $(m+1)$th by w^{-mn}, and add all the equations together. The sum of terms containing $\hat{c}_k$ is found to be

$$[1 + w^k w^{-n} + \cdots + w^{km} w^{-mn} + \cdots + w^{k(N-1)} w^{-(N-1)n}]\hat{c}_k$$

(9.96)

for $k = 0, 1, \ldots, N - 1$. The sum of terms inside the brackets is recognized as a geometric progression with ratio w^{k-n} and equals

$$1 + w^{k-n} + \cdots + w^{(N-1)(k-n)} = \frac{1 - w^{N(k-n)}}{1 - w^{k-n}}$$

(9.97)

For $k \neq n$, $w^{k-n} \neq 1$ and from (9.83) $w^{N(k-n)} = 1$. Hence the right-hand side of (9.97) is zero for $k \neq n$. For $k = n$, the sum is N. Therefore we have for $N > 1$

$$1 + w^{k-n} + \cdots + w^{(N-1)(k-n)} = 0, \qquad k \neq n$$

(9.98a)

$$= N, \qquad k = n$$

(9.98b)

Our conclusion is that if we multiply the $(m+1)$th equation of (9.93) by w^{-mn} $(m = 0, 1, \ldots, N - 1)$ and then add all the resulting equations, the sums of terms containing $\hat{c}_k$ are all zero except for $k = n$, in which case the equation becomes

$$N\hat{c}_n = f(0) + f(T_0)w^{-n} + \cdots + f(mT_0)w^{-mn} + \cdots$$
$$+ f[(N-1)T_0]w^{-(N-1)n}$$

(9.99)

or

$$\hat{c}_n = \frac{1}{N} \sum_{m=0}^{N-1} f(mT_0)w^{-mn}$$

(9.100)

Property 9.4 *The aliasing error for the nth Fourier coefficient $\tilde{c}_n$ of a periodic signal $f(t)$ equals the difference between the nth aliased coefficient $\hat{c}_n$ of $f(t)$ and $\tilde{c}_n$.*

In other words, in approximating the Fourier coefficient $\tilde{c}_n$ by (9.85), the error equals the sum of all the coefficients of the form $\tilde{c}_{n+kN}$ for all $k \neq 0$, because

$$\hat{\epsilon}_n = \hat{c}_n - \tilde{c}_n = \left[\sum_{k=-\infty}^{\infty} \tilde{c}_{n+kN} \right] - \tilde{c}_n = \sum_{\substack{k=-\infty \\ k \neq 0}}^{\infty} \tilde{c}_{n+kN} \tag{9.101}$$

As a result, if

$$\tilde{c}_n = 0 \quad \text{for} \quad |n| \geq \frac{N}{2} \tag{9.102}$$

then

$$\hat{\epsilon}_n = \hat{c}_n - \tilde{c}_n = 0 \quad \text{for} \quad |n| < \frac{N}{2} \tag{9.103}$$

Therefore if $f(t)$ is a trigonometric polynomial of order $M < N/2$, then $\tilde{c}_n = 0$ for $|n| \geq N/2$ and $\tilde{c}_n = \hat{c}_n$ for $|n| < N/2$, the error being zero. In practice if

$$\tilde{c}_n \cong 0 \quad \text{for} \quad |n| \geq \frac{N}{2} \tag{9.104}$$

then

$$\tilde{c}_n \cong \hat{c}_n = \frac{1}{N} \sum_{m=0}^{N-1} f(mT_0) w^{-mn} \quad \text{for} \quad |n| < \frac{N}{2} \tag{9.105}$$

The aliasing error in general depends on N. It decreases with the increasing value of N.

We illustrate the above result by the following examples.

EXAMPLE 9.3

Consider a train of rectangular pulses as shown in Figure 9.9, the Fourier series of which is found to be

$$f(t) = \sum_{n=-\infty}^{\infty} \frac{\sin 0.1 n\pi}{n\pi} e^{jn\pi t} \tag{9.106}$$

Figure 9.9
A train of rectangular pulses.

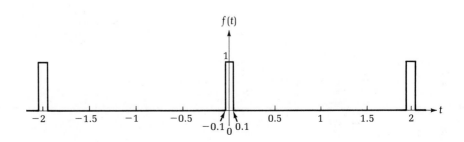

We compare the Fourier coefficients

$$\tilde{c}_n = \frac{\sin 0.1 n\pi}{n\pi} \tag{9.107}$$

of this signal with its aliased coefficients $\hat{c}_n$. For our purposes we choose $N = 500$. The aliased coefficients $\hat{c}_n$ are computed from (9.100) using the subroutine FOURI2, as discussed in the preceding chapter. The $\hat{c}_n$ are presented in Figure 9.10 for $|n| \leq 50$. Using these and (9.107), the aliasing

Figure 9.10

The aliased coefficients for a train of rectangular pulses of Figure 9.9.

C(-50)= -0.179550818658D-15	0.385802501057D-17	C(5)= 0.636410319075D-01	-0.213717932240D-17
C(-49)= 0.194358110311D-02	0.621724893790D-17	C(6)= 0.504312103204D-01	-0.269784194984D-16
C(-48)= 0.377896787621D-02	0.791033905045D-17	C(7)= 0.367645763802D-01	-0.426048085700D-16
C(-47)= 0.531890119960D-02	0.369149155688D-17	C(8)= 0.233675319033D-01	-0.416333634234D-16
C(-46)= 0.639681953813D-02	0.138777878078D-18	C(9)= 0.109175883282D-01	-0.257294185957D-16
C(-45)= 0.688404515334D-02	-0.427435864481D-17	C(10)= -0.195149452153D-15	-0.283106871279D-17
C(-44)= 0.670405713873D-02	-0.702216063075D-17	C(11)= -0.892786282882D-02	-0.324740234703D-17
C(-43)= 0.584236060288D-02	-0.438538094727D-17	C(12)= -0.155619315503D-01	-0.288657986403D-17
C(-42)= 0.435081899511D-02	-0.402455846427D-17	C(13)= -0.197650111452D-01	-0.457966997658D-17
C(-41)= 0.234579457670D-02	-0.585642645490D-17	C(14)= -0.215678191595D-01	-0.971445146547D-18
C(-40)= 0.203559391565D-15	-0.516253706451D-17	C(15)= -0.211577899868D-01	0.610622663544D-18
C(-39)= -0.247144728388D-02	-0.482947015712D-17	C(16)= -0.188568848091D-01	0.396904731303D-17
C(-38)= -0.482971015060D-02	-0.491273688397D-17	C(17)= -0.150904704563D-01	0.496824803520D-17
C(-37)= -0.683410904695D-02	-0.527355936697D-18	C(18)= -0.103499695740D-01	0.571764857682D-17
C(-36)= -0.826527835144D-02	0.185962356625D-17	C(19)= -0.515239150937D-02	0.638378239159D-17
C(-35)= -0.894748565842D-02	0.388578058619D-17	C(20)= 0.200756078428D-15	0.752176099184D-17
C(-34)= -0.876798157787D-02	0.691113832829D-17	C(21)= 0.465674604836D-02	0.697358837343D-17
C(-33)= -0.769142669182D-02	0.999200722163D-17	C(22)= 0.845021250912D-02	0.699440505514D-17
C(-32)= -0.576780737440D-02	0.954791801178D-17	C(23)= 0.111183882622D-01	0.571764857682D-17
C(-31)= -0.313277681373D-02	0.101585406753D-16	C(24)= 0.125180229238D-01	0.366373598126D-17
C(-30)= -0.205169214951D-15	0.788258347484D-17	C(25)= 0.126275030294D-01	0.194289029309D-18
C(-29)= 0.335421208793D-02	0.779931674799D-17	C(26)= 0.115397251994D-01	-0.113797860024D-17
C(-28)= 0.661298458795D-02	0.371924713249D-17	C(27)= 0.944603398037D-02	-0.255351295664D-17
C(-27)= 0.944603398037D-02	0.255351295664D-17	C(28)= 0.661298458795D-02	-0.371924713249D-17
C(-26)= 0.115397251994D-01	0.113797860024D-17	C(29)= 0.335421208793D-02	-0.779931674799D-17
C(-25)= 0.126275030294D-01	-0.194289029309D-18	C(30)= -0.205169214951D-15	-0.788258347484D-17
C(-24)= 0.125180229238D-01	-0.366373598126D-17	C(31)= -0.313277681373D-02	-0.101585406753D-16
C(-23)= 0.111183882622D-01	-0.571764857682D-17	C(32)= -0.576780737440D-02	-0.954791801178D-17
C(-22)= 0.845021250912D-02	-0.699440505514D-17	C(33)= -0.769142669182D-02	-0.999200722163D-17
C(-21)= 0.465674604836D-02	-0.697358837343D-17	C(34)= -0.876798157787D-02	-0.691113832829D-17
C(-20)= 0.200756078428D-15	-0.752176099184D-17	C(35)= -0.894748565842D-02	-0.388578058619D-17
C(-19)= -0.515239150937D-02	-0.638378239159D-17	C(36)= -0.826527835144D-02	-0.185962356625D-17
C(-18)= -0.103499695740D-01	-0.571764857682D-17	C(37)= -0.683410904695D-02	0.527355936697D-18
C(-17)= -0.150904704563D-01	-0.496824803520D-17	C(38)= -0.482971015060D-02	0.491273688397D-17
C(-16)= -0.188568848091D-01	-0.396904731303D-17	C(39)= -0.247144728388D-02	0.482947015712D-17
C(-15)= -0.211577899868D-01	-0.610622663544D-18	C(40)= 0.203559391565D-15	0.516253706451D-17
C(-14)= -0.215678191595D-01	0.971445146547D-18	C(41)= 0.234579457670D-02	0.585642645490D-17
C(-13)= -0.197650111452D-01	0.457966997658D-17	C(42)= 0.435081899511D-02	0.402455846427D-17
C(-12)= -0.155619315503D-01	0.288657986403D-17	C(43)= 0.584236060288D-02	0.438538094727D-17
C(-11)= -0.892786282882D-02	0.324740234703D-17	C(44)= 0.670405713873D-02	0.702216063075D-17
C(-10)= -0.195149452153D-15	0.283106871279D-17	C(45)= 0.688404515334D-02	0.427435864481D-17
C(-9)= 0.109175883282D-01	0.257294185957D-17	C(46)= 0.639681953813D-02	-0.138777878078D-17
C(-8)= 0.233675319033D-01	0.416333634234D-16	C(47)= 0.531890119960D-02	-0.369149155688D-17
C(-7)= 0.367645763802D-01	0.426048085700D-16	C(48)= 0.377896787621D-02	-0.791033905045D-17
C(-6)= 0.504312103204D-01	0.269784194984D-16	C(49)= 0.194358110311D-02	-0.621724893790D-17
C(-5)= 0.636410319075D-01	0.213717932240D-17	C(50)= -0.179550818658D-15	-0.385802501057D-17
C(-4)= 0.756667370881D-01	0.444089209850D-17		
C(-3)= 0.858292024852D-01	0.333066907388D-17		
C(-2)= 0.935440041088D-01	0.460742555219D-17		
C(-1)= 0.983618698976D-01	0.507927033766D-17		
C(0)= 0.100000000000D 00	0.000000000000D 00		
C(1)= 0.983618698976D-01	-0.507927033766D-17		
C(2)= 0.935440041088D-01	-0.460742555219D-17		
C(3)= 0.858292024852D-01	-0.333066907388D-17		
C(4)= 0.756667370881D-01	-0.444089209850D-17		

Table 9.1

$|n|$	$\tilde{c}_n$	$\hat{c}_n$	$\hat{\epsilon}_n = \hat{c}_n - \tilde{c}_n$
0	0.1000000000	0.1000000000	0.0000000000
1	0.0983631643	0.0983618698	-0.0000012945
2	0.0935489284	0.0935440041	-0.0000049243
3	0.0858393691	0.0858292024	-0.0000101667
4	0.0756826729	0.0756667370	-0.0000159359
5	0.0636619772	0.0636410319	-0.0000209453
6	0.0504551152	0.0504312103	-0.0000239049
7	0.0367883011	0.0367645763	-0.0000237248
8	0.0233872321	0.0233675319	-0.0000197002
9	0.0109292405	0.0109175883	-0.0000116522
10	0.0000000000	0.0000000000	0.0000000000
11	-0.0089421058	-0.0089278628	0.0000142430
12	-0.0155914881	-0.0155619315	0.0000295566
13	-0.0198090852	-0.0197650111	0.0000440741
14	-0.0216236208	-0.0215678191	0.0000558017
15	-0.0212206591	-0.0211577899	0.0000628692
16	-0.0189206682	-0.0188568848	0.0000637834
20	0.0000000000	0.0000000000	0.0000000000
26	0.0116434881	0.0115397251	-0.0001037630
32	-0.0058468080	-0.0057678073	0.0000790007
38	-0.0049236278	-0.0048297101	0.0000939177
44	0.0068802430	0.0067040571	-0.0001761859
49	0.0020074115	0.0019435811	-0.0000638304

errors are shown in Table 9.1 together with the values of $\tilde{c}_n$ and $\hat{c}_n$. Thus $\hat{c}_n$ is a good approximation of $\hat{c}_n$ for $|n| < N/2 = 250$. The amplitude and phase spectra of the pulse signal are presented in Figure 9.11.

To complete our analysis, we reconstruct the signal from its partial sum approximation (9.86). For $M = 5, 9, 21,$ and 49, the results are displayed in Figure 9.12, using subroutine PLOT. For $M = 49$ there are 99 terms in the expansion of (9.86), and their sums are displayed in Figure 9.13. Observe that the maximum deviation occurs approximately at $mT_0 = \pm 0.08$ where

Figure 9.11
(a) The amplitude spectrum for a train of rectangular pulses of Figure 9.9. (b) The
phases of the Fourier coefficients for a train of rectangular pulses of Figure 9.9. (c)
The phase spectrum for a train of rectangular pulses of Figure 9.9.

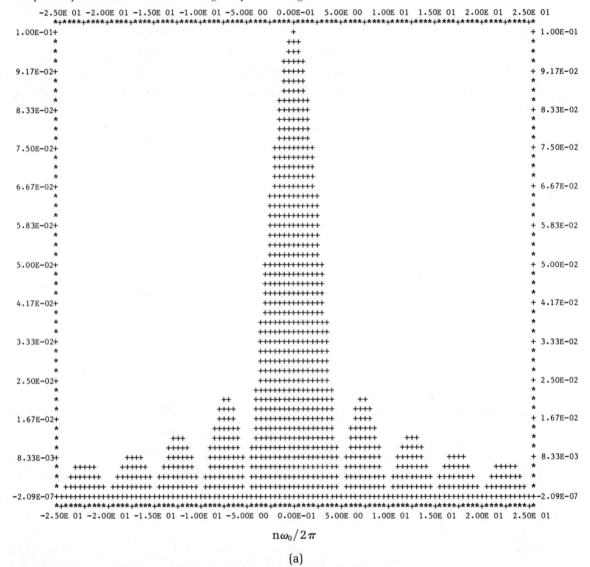

(a)

$$T_0 = 1/250 \text{ or}$$

$$f(\pm 0.08) = 1.09668 \tag{9.108}$$

showing a 9.7 percent deviation. This is due to the Gibbs phenomenon, as
previously discussed.

Figure 9.11
(continued)

```
PHI( -50)=    0.00000000E 00          PHI(    5)=    0.00000000E 00
PHI( -49)=    0.00000000E 00          PHI(    6)=    0.00000000E 00
PHI( -48)=    0.00000000E 00          PHI(    7)=    0.00000000E 00
PHI( -47)=    0.00000000E 00          PHI(    8)=    0.00000000E 00
PHI( -46)=    0.00000000E 00          PHI(    9)=    0.00000000E 00
PHI( -45)=    0.00000000E 00          PHI(   10)=    0.00000000E 00
PHI( -44)=    0.00000000E 00          PHI(   11)=  - 0.18000000E 03
PHI( -43)=    0.00000000E 00          PHI(   12)=  - 0.18000000E 03
PHI( -42)=    0.00000000E 00          PHI(   13)=  - 0.18000000E 03
PHI( -41)=    0.00000000E 00          PHI(   14)=    0.18000000E 03
PHI( -40)=    0.00000000E 00          PHI(   15)=    0.18000000E 03
PHI( -39)=  - 0.18000000E 03          PHI(   16)=    0.18000000E 03
PHI( -38)=  - 0.18000000E 03          PHI(   17)=    0.18000000E 03
PHI( -37)=  - 0.18000000E 03          PHI(   18)=    0.18000000E 03
PHI( -36)=  - 0.18000000E 03          PHI(   19)=    0.18000000E 03
PHI( -35)=    0.18000000E 03          PHI(   20)=    0.00000000E 00
PHI( -34)=    0.18000000E 03          PHI(   21)=    0.00000000E 00
PHI( -33)=    0.18000000E 03          PHI(   22)=    0.00000000E 00
PHI( -32)=    0.18000000E 03          PHI(   23)=    0.00000000E 00
PHI( -31)=    0.18000000E 03          PHI(   24)=    0.00000000E 00
PHI( -30)=    0.00000000E 00          PHI(   25)=    0.00000000E 00
PHI( -29)=    0.00000000E 00          PHI(   26)=    0.00000000E 00
PHI( -28)=    0.00000000E 00          PHI(   27)=    0.00000000E 00
PHI( -27)=    0.00000000E 00          PHI(   28)=    0.00000000E 00
PHI( -26)=    0.00000000E 00          PHI(   29)=    0.00000000E 00
PHI( -25)=    0.00000000E 00          PHI(   30)=    0.00000000E 00
PHI( -24)=    0.00000000E 00          PHI(   31)=  - 0.18000000E 03
PHI( -23)=    0.00000000E 00          PHI(   32)=  - 0.18000000E 03
PHI( -22)=    0.00000000E 00          PHI(   33)=  - 0.18000000E 03
PHI( -21)=    0.00000000E 00          PHI(   34)=  - 0.18000000E 03
PHI( -20)=    0.00000000E 00          PHI(   35)=    0.18000000E 03
PHI( -19)=  - 0.18000000E 03          PHI(   36)=    0.18000000E 03
PHI( -18)=  - 0.18000000E 03          PHI(   37)=    0.18000000E 03
PHI( -17)=  - 0.18000000E 03          PHI(   38)=    0.18000000E 03
PHI( -16)=  - 0.18000000E 03          PHI(   39)=    0.18000000E 03
PHI( -15)=  - 0.18000000E 03          PHI(   40)=    0.00000000E 00
PHI( -14)=  - 0.18000000E 03          PHI(   41)=    0.00000000E 00
PHI( -13)=    0.18000000E 03          PHI(   42)=    0.00000000E 00
PHI( -12)=    0.18000000E 03          PHI(   43)=    0.00000000E 00
PHI( -11)=    0.18000000E 03          PHI(   44)=    0.00000000E 00
PHI( -10)=    0.00000000E 00          PHI(   45)=    0.00000000E 00
PHI(  -9)=    0.00000000E 00          PHI(   46)=    0.00000000E 00
PHI(  -8)=    0.00000000E 00          PHI(   47)=    0.00000000E 00
PHI(  -7)=    0.00000000E 00          PHI(   48)=    0.00000000E 00
PHI(  -6)=    0.00000000E 00          PHI(   49)=    0.00000000E 00
PHI(  -5)=    0.00000000E 00          PHI(   50)=    0.00000000E 00
PHI(  -4)=    0.00000000E 00
PHI(  -3)=    0.00000000E 00
PHI(  -2)=    0.00000000E 00
PHI(  -1)=    0.00000000E 00
PHI(   0)=    0.00000000E 00
PHI(   1)=    0.00000000E 00
PHI(   2)=    0.00000000E 00
PHI(   3)=    0.00000000E 00
PHI(   4)=    0.00000000E 00
```

(b)

Finally, we mention that the aliased coefficients $\hat{c}_n$ are defined by the infinite series (9.91). For $n = 0$, 1, and 2, they are given by

$$\hat{c}_0 = \cdots + \tilde{c}_{-1000} + \tilde{c}_{-500} + \tilde{c}_0 + \tilde{c}_{500} + \tilde{c}_{1000} + \cdots$$

$$= \cdots + \frac{\sin(-100\pi)}{-1000\pi} + \frac{\sin(-50\pi)}{-500\pi} + 0.1$$

$$+ \frac{\sin 50\pi}{500\pi} + \frac{\sin 100\pi}{1000\pi} + \cdots$$

$$= 0.1 \qquad\qquad\qquad\qquad\qquad\qquad\qquad \textbf{(9.109 a)}$$

Figure 9.11 (continued)

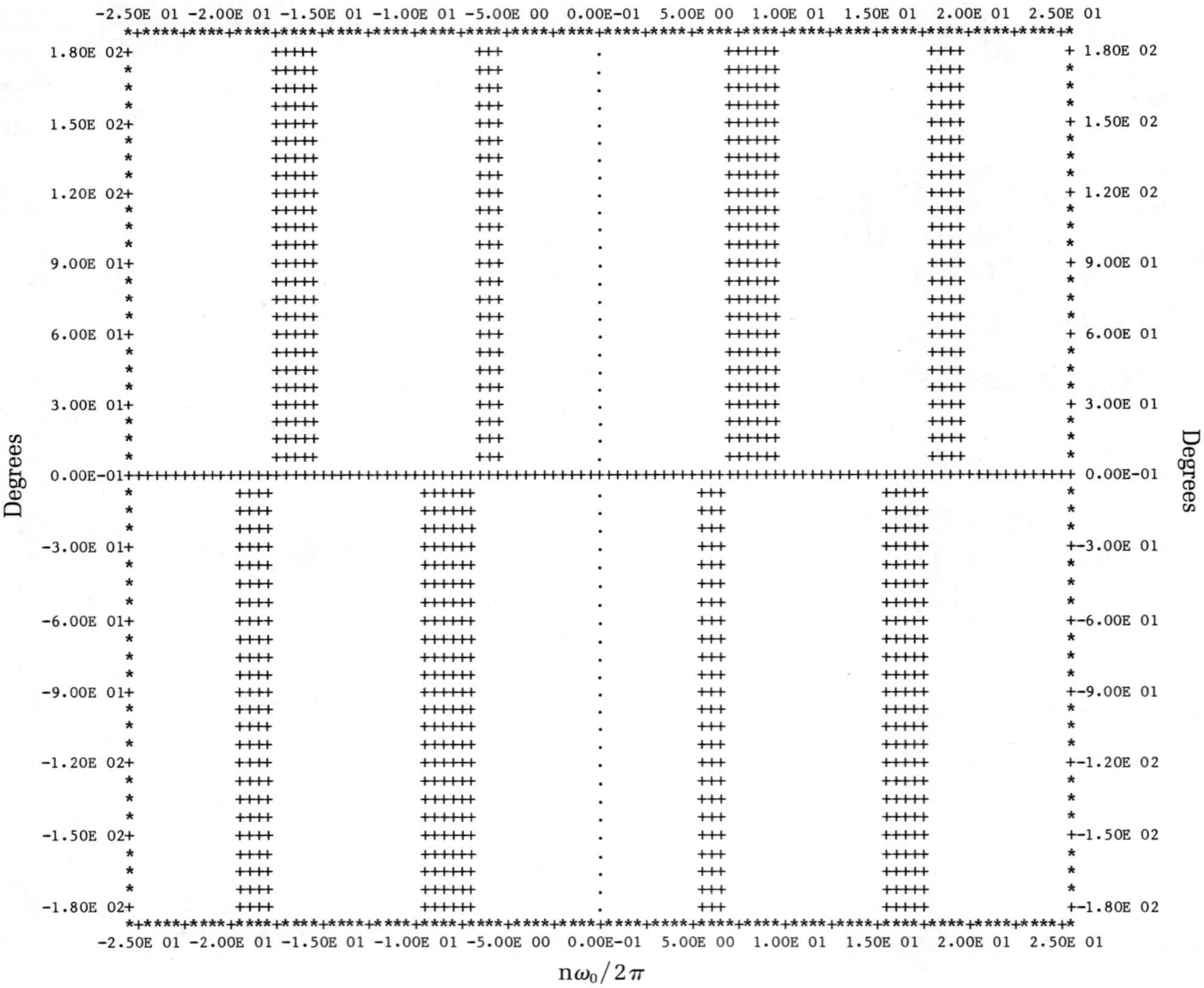

Degrees

$n\omega_0/2\pi$

(c)

Figure 9.12

(a) Rectangular pulse approximation resulting from truncated Fourier series of 11 terms in Equation (9.106). (b) Rectangular pulse approximation resulting from truncated Fourier series of 19 terms in Eq. (9.106). (c) Rectangular pulse approximation resulting from truncated Fourier series of 43 terms in Eq. (9.106). (d) Rectangular pulse approximation resulting from truncated Fourier series of 99 terms in Eq. (9.106).

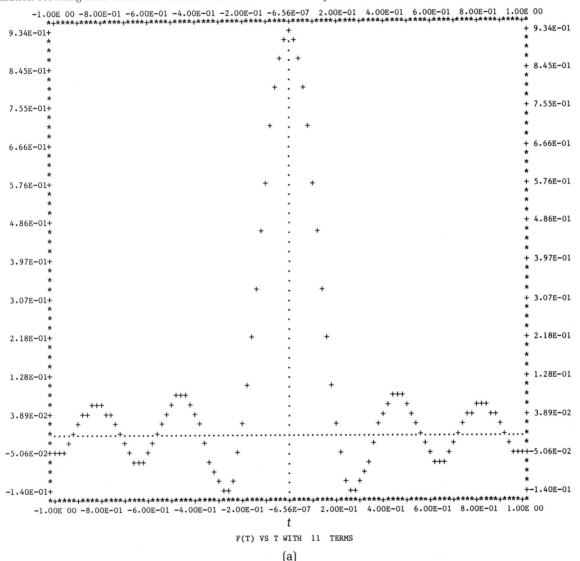

F(T) VS T WITH 11 TERMS

(a)

Figure 9.12 (continued)

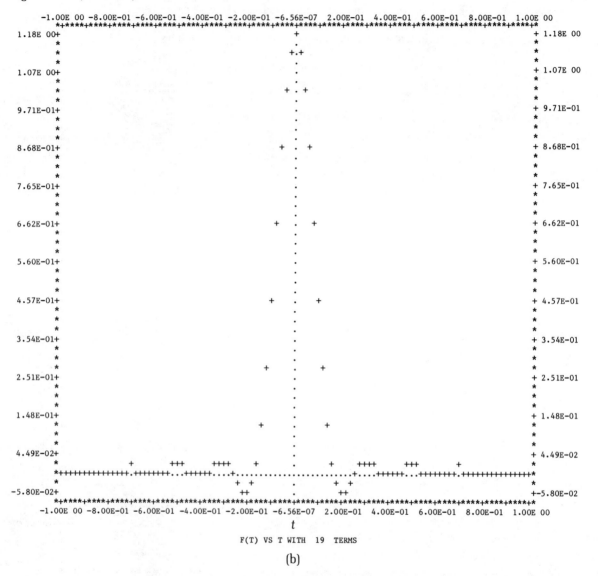

F(T) VS T WITH 19 TERMS

(b)

$$\hat{c}_1 = \cdots + \tilde{c}_{-999} + \tilde{c}_{-499} + \tilde{c}_1 + \tilde{c}_{501} + \tilde{c}_{1001} + \cdots$$

$$= \cdots + \frac{\sin(-99.9\pi)}{-999\pi} + \frac{\sin(-49.9\pi)}{-499\pi} + \frac{\sin 0.1\pi}{\pi}$$

$$+ \frac{\sin 50.1\pi}{501\pi} + \frac{\sin 100.1\pi}{1001\pi} + \cdots \qquad \textbf{(9.109b)}$$

Figure 9.12 (continued)

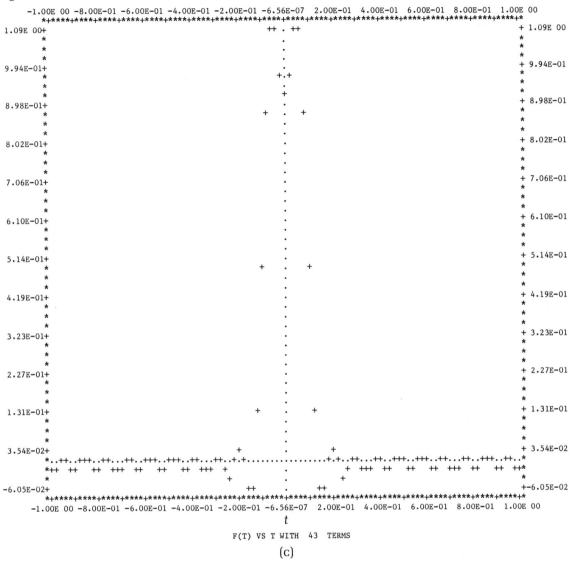

F(T) VS T WITH 43 TERMS

(c)

$$\hat{c}_2 = \cdots + \tilde{c}_{-998} + \tilde{c}_{-498} + \tilde{c}_2 + \tilde{c}_{502} + \tilde{c}_{1002} + \cdots$$

$$= \cdots + \frac{\sin(-99.8\pi)}{-998\pi} + \frac{\sin(-49.8\pi)}{-498\pi} + \frac{\sin 0.2\pi}{2\pi}$$

$$+ \frac{\sin 50.2\pi}{502\pi} + \frac{\sin 100.2\pi}{1002\pi} + \cdots \qquad\qquad \textbf{(9.109c)}$$

Figure 9.12   (*continued*)

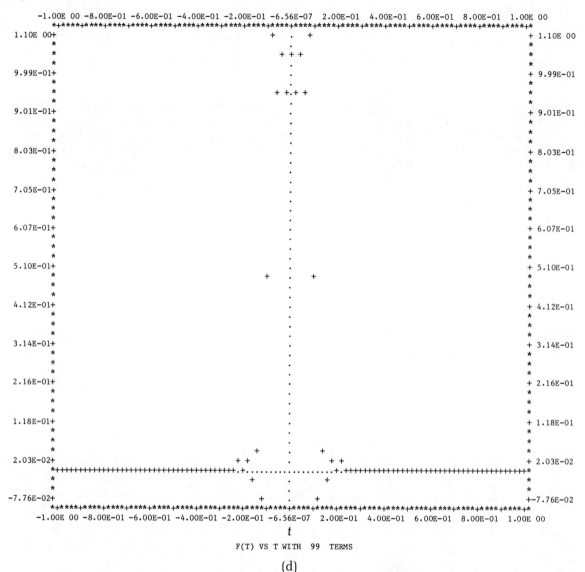

F(T) VS T WITH 99 TERMS

(d)

EXAMPLE 9.4

Consider the square waveform $f(t)$ of Figure 9.14, having a Fourier series found from (8.54) as

$$f(t) = -j\frac{2}{\pi} \sum_{n=-\infty}^{\infty} \frac{1}{2n-1} e^{j(2n-1)\pi t} \tag{9.110}$$

Figure 9.13

Listing of the partial sums of 99 terms of the truncated Fourier series representation of the rectangular pulse of Figure 9.9 for t from -1 to 1 with increment of 0.02 sec.

```
F(T) WITH  99 TERMS
T= -0.10000000D 01     F(T)= -0.974707958740D-03     T=  0.80000000D-01     F(T)=  0.109668139867D 01
T= -0.98000000D 00     F(T)=  0.975694125678D-03     T=  0.10000000D 00     F(T)=  0.490548311321D 00
T= -0.96000000D 00     F(T)= -0.978660716711D-03     T=  0.12000000D 00     F(T)= -0.775823196916D-01
T= -0.94000000D 00     F(T)=  0.983632144078D-03     T=  0.14000000D 00     F(T)=  0.390862518599D-01
T= -0.92000000D 00     F(T)= -0.990649576215D-03     T=  0.16000000D 00     F(T)= -0.247756608426D-01
T= -0.90000000D 00     F(T)=  0.999771675474D-03     T=  0.18000000D 00     F(T)=  0.175296294254D-01
T= -0.88000000D 00     F(T)= -0.101107566298D-02     T=  0.20000000D 00     F(T)= -0.132405046984D-01
T= -0.86000000D 00     F(T)=  0.102465874456D-02     T=  0.22000000D 00     F(T)=  0.104484494605D-01
T= -0.84000000D 00     F(T)= -0.104063994422D-02     T=  0.24000000D 00     F(T)= -0.851041191765D-02
T= -0.82000000D 00     F(T)=  0.105916240632D-02     T=  0.26000000D 00     F(T)=  0.710127594026D-02
T= -0.80000000D 00     F(T)= -0.108039624606D-02     T=  0.28000000D 00     F(T)= -0.604000992451D-02
T= -0.78000000D 00     F(T)=  0.110454204989D-02     T=  0.30000000D 00     F(T)=  0.521833269006D-02
T= -0.76000000D 00     F(T)= -0.113183515613D-02     T=  0.32000000D 00     F(T)= -0.456779728488D-02
T= -0.74000000D 00     F(T)=  0.116255088168D-02     T=  0.34000000D 00     F(T)=  0.404318990416D-02
T= -0.72000000D 00     F(T)= -0.119701090746D-02     T=  0.36000000D 00     F(T)= -0.361355161775D-02
T= -0.70000000D 00     F(T)=  0.123559109526D-02     T=  0.38000000D 00     F(T)=  0.325704103975D-02
T= -0.68000000D 00     F(T)= -0.127873108798D-02     T=  0.40000000D 00     F(T)= -0.295785477535D-02
T= -0.66000000D 00     F(T)=  0.132694615071D-02     T=  0.42000000D 00     F(T)=  0.270430784279D-02
T= -0.64000000D 00     F(T)= -0.138084185028D-02     T=  0.44000000D 00     F(T)= -0.248759663707D-02
T= -0.62000000D 00     F(T)=  0.144113236197D-02     T=  0.46000000D 00     F(T)=  0.230097851684D-02
T= -0.60000000D 00     F(T)= -0.150866345170D-02     T=  0.48000000D 00     F(T)= -0.213921390281D-02
T= -0.58000000D 00     F(T)=  0.158444154157D-02     T=  0.50000000D 00     F(T)=  0.199817843940D-02
T= -0.56000000D 00     F(T)= -0.166967076902D-02     T=  0.52000000D 00     F(T)= -0.187458806032D-02
T= -0.54000000D 00     F(T)=  0.176580066050D-02     T=  0.54000000D 00     F(T)=  0.176580066050D-02
T= -0.52000000D 00     F(T)= -0.187458806032D-02     T=  0.56000000D 00     F(T)= -0.166967076902D-02
T= -0.50000000D 00     F(T)=  0.199817843940D-02     T=  0.58000000D 00     F(T)=  0.158444154157D-02
T= -0.48000000D 00     F(T)= -0.213921390281D-02     T=  0.60000000D 00     F(T)= -0.150866345170D-02
T= -0.46000000D 00     F(T)=  0.230097851684D-02     T=  0.62000000D 00     F(T)=  0.144113236197D-02
T= -0.44000000D 00     F(T)= -0.248759663707D-02     T=  0.64000000D 00     F(T)= -0.138084185028D-02
T= -0.42000000D 00     F(T)=  0.270430784279D-02     T=  0.66000000D 00     F(T)=  0.132694615071D-02
T= -0.40000000D 00     F(T)= -0.295785477535D-02     T=  0.68000000D 00     F(T)= -0.127873108798D-02
T= -0.38000000D 00     F(T)=  0.325704103975D-02     T=  0.70000000D 00     F(T)=  0.123559109526D-02
T= -0.36000000D 00     F(T)= -0.361355161775D-02     T=  0.72000000D 00     F(T)= -0.119701090746D-02
T= -0.34000000D 00     F(T)=  0.404318990416D-02     T=  0.74000000D 00     F(T)=  0.116255088168D-02
T= -0.32000000D 00     F(T)= -0.456779728488D-02     T=  0.76000000D 00     F(T)= -0.113183515613D-02
T= -0.30000000D 00     F(T)=  0.521833269006D-02     T=  0.78000000D 00     F(T)=  0.110454204989D-02
T= -0.28000000D 00     F(T)= -0.604000992451D-02     T=  0.80000000D 00     F(T)= -0.108039624606D-02
T= -0.26000000D 00     F(T)=  0.710127594026D-02     T=  0.82000000D 00     F(T)=  0.105916240632D-02
T= -0.24000000D 00     F(T)= -0.851041191765D-02     T=  0.84000000D 00     F(T)= -0.104063994422D-02
T= -0.22000000D 00     F(T)=  0.104484494605D-01     T=  0.86000000D 00     F(T)=  0.102465874456D-02
T= -0.20000000D 00     F(T)= -0.132405046984D-01     T=  0.88000000D 00     F(T)= -0.101107566298D-02
T= -0.18000000D 00     F(T)=  0.175296294254D-01     T=  0.90000000D 00     F(T)=  0.999771675474D-03
T= -0.16000000D 00     F(T)= -0.247756608426D-01     T=  0.92000000D 00     F(T)= -0.990649576215D-03
T= -0.14000000D 00     F(T)=  0.390862518599D-01     T=  0.94000000D 00     F(T)=  0.983632144078D-03
T= -0.12000000D 00     F(T)= -0.775823196916D-01     T=  0.96000000D 00     F(T)= -0.978660716711D-03
T= -0.10000000D 00     F(T)=  0.490548311321D 00     T=  0.98000000D 00     F(T)=  0.975694125678D-03
T= -0.80000000D-01     F(T)=  0.109668139867D 01     T=  0.10000000D 01     F(T)= -0.974707958740D-03
T= -0.60000000D-01     F(T)=  0.941204714501D 00
T= -0.40000000D-01     F(T)=  0.104558886981D 01
T= -0.20000000D-01     F(T)=  0.959895959949 5D 00
T=  0.15612511D-15     F(T)=  0.103854954653D 01
T=  0.20000000D-01     F(T)=  0.959895959949 5D 00
T=  0.40000000D-01     F(T)=  0.104558886981D 01
T=  0.60000000D-01     F(T)=  0.941204714501D 00
```

Suppose that we divide the period T into 100 subintervals. Then $N = 100$. Using the subroutine FOURI2, the aliased coefficients $\hat{c}_n$ are shown in Figure 9.15. For comparison we compute several values of $\tilde{c}_n$ from (9.110) together with the aliasing errors. The results are presented in Table 9.2, where $\tilde{c}_{-n} = \overline{\tilde{c}}_n$ and $\hat{c}_{-n} = \overline{\hat{c}}_n$. Figure 9.16 is a plot of the aliasing error in percentile

Figure 9.14
A periodic square
waveform.

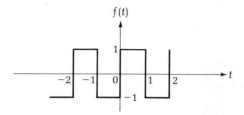

Figure 9.15
The aliased coefficients of
the square waveform of
Figure 9.14.

```
C( -25)= -0.282639776605D-14        0.200000000000D-01
C( -24)= -0.210248485288D-15        0.141692213518D-15
C( -23)= -0.326225157998D-14        0.226855469851D-01
C( -22)= -0.707767178199D-16       -0.999200722163D-17
C( -21)= -0.355063201063D-14        0.257838446357D-01
C( -20)= -0.513478148889D-17       -0.917321774097D-16
C( -19)= -0.353023166255D-14        0.294291063164D-01
C( -18)=  0.299066327258D-15       -0.196648253237D-15
C( -17)= -0.336508598764D-14        0.338181531157D-01
C( -16)= -0.387606613472D-15        0.252992071736D-15
C( -15)= -0.386940479657D-14        0.392522101101D-01
C( -14)= -0.832667268469D-18        0.260902410787D-16
C( -13)= -0.374283937177D-14        0.462172730776D-01
C( -12)= -0.882627304577D-16       -0.840993941154D-16
C( -11)= -0.365124597224D-14        0.555521370783D-01
C( -10)= -0.288657986403D-16       -0.294902990916D-15
C(  -9)= -0.360753094064D-14        0.688404515334D-01
C(  -8)= -0.411892742136D-15        0.685562717706D-16
C(  -7)= -0.363695185079D-14        0.894748565842D-01
C(  -6)= -0.111854969731D-15        0.516253706451D-16
C(  -5)= -0.374963948779D-14        0.126275030294D 00
C(  -4)=  0.123928645124D-15        0.260902410787D-16
C(  -3)= -0.391034427061D-14        0.211577899868D 00
C(  -2)=  0.860422844084D-17        0.152655665886D-16
C(  -1)= -0.415029122180D-14        0.636410319075D 00
C(   0)=  0.000000000000D 00        0.000000000000D 00
C(   1)= -0.415029122180D-14       -0.636410319075D 00
C(   2)=  0.860422844084D-17       -0.152655665886D-16
C(   3)= -0.391034427061D-14       -0.211577899868D 00
C(   4)=  0.123928645124D-15       -0.260902410787D-16
C(   5)= -0.374963948779D-14       -0.126275030294D 00
C(   6)= -0.111854969731D-15       -0.516253706451D-16
C(   7)= -0.363695185079D-14       -0.894748565842D-01
C(   8)= -0.411892742136D-15       -0.685562717706D-16
C(   9)= -0.360753094064D-14       -0.688404515334D-01
C(  10)= -0.288657986403D-16        0.294902990916D-15
C(  11)= -0.365124597224D-14       -0.555521370783D-01
C(  12)= -0.882627304577D-16        0.840993941154D-16
C(  13)= -0.374283937177D-14       -0.462172730776D-01
C(  14)= -0.832667268469D-18       -0.260902410787D-16
C(  15)= -0.386940479657D-14       -0.392522101101D-01
C(  16)= -0.387606613472D-15       -0.252992071736D-15
C(  17)= -0.336508598764D-14       -0.338181531157D-01
C(  18)=  0.299066327258D-15        0.196648253237D-15
C(  19)= -0.353023166255D-14       -0.294291063164D-01
C(  20)= -0.513478148889D-17        0.917321774097D-16
C(  21)= -0.355063201063D-14       -0.257838446357D-01
C(  22)= -0.707767178199D-16        0.999200722163D-17
C(  23)= -0.326225157998D-14       -0.226855469851D-01
C(  24)= -0.210248485288D-15       -0.141692213518D-15
C(  25)= -0.282639776605D-14       -0.200000000000D-01
```

Table 9.2

n	$\hat{c}_n$	$\tilde{c}_n$	$\hat{\epsilon}_n = \hat{c}_n - \tilde{c}_n$
1	$-j0.6366197724$	$-j0.6364103190$	-0.0002094534
5	$-j0.1273239545$	$-j0.1262750302$	-0.0010489243
9	$-j0.0707355303$	$-j0.0688404515$	-0.0018950788
13	$-j0.0489707517$	$-j0.0462172730$	-0.0027534787
17	$-j0.0374482219$	$-j0.0338181531$	-0.0036300688
21	$-j0.0303152273$	$-j0.0257838446$	-0.0045313827
25	$-j0.0254647909$	$-j0.0200000000$	-0.0054647909

Figure 9.16
Plot of the aliasing error in percentile as a function of n.

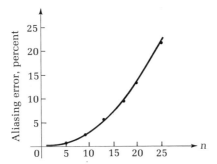

as a function of n. It is seen that the error increases with increasing value of n, reaching to approximately 21.5 percent for 25th harmonic. This is because $\tilde{c}_n$ are not negligible for $|n| > N/2 = 50$. For instance, $\tilde{c}_{51} = -j0.01248$, which is not negligible in comparison with the desired harmonics.

The amplitude and phase spectra are presented in Figure 9.17. The partial sums for $f(t)$ are plotted in Figure 9.18 for 11, 27, and 43 terms of (9.80) with $t = mT_0 = 0.02m$. To observe the Gibbs phenomenon, we dis-play the partial sum $s_{21}(mT_0)$ in Figure 9.19. As can be seen, the maximum deviation occurs approximately at $mT_0 = 0.04, 0.96, 1.04$, and 1.96. At these points $s_{21}(mT_0) = \pm1.1245$ or ±12.45 percent.

EXAMPLE 9.5
Consider the trigonometric polynomial

$$32 \cos^6 20\pi t = 0.5e^{-j120\pi t} + 3e^{-j80\pi t} + 7.5e^{-j40\pi t} + 10$$
$$+ 7.5e^{j40\pi t} + 3e^{j80\pi t} + 0.5e^{j120\pi t} \qquad (9.111)$$

of order $M = 3$. Then

$$\tilde{c}_0 = 10, \qquad \tilde{c}_{-2} = \tilde{c}_2 = 3$$
$$\tilde{c}_{-1} = \tilde{c}_1 = 7.5, \qquad \tilde{c}_{-3} = \tilde{c}_3 = 0.5 \qquad (9.112)$$

Figure 9.17

(a) The amplitude spectrum of the periodic square waveform of Figure 9.14. (b) The phase spectrum of the periodic square waveform of Figure 9.14.

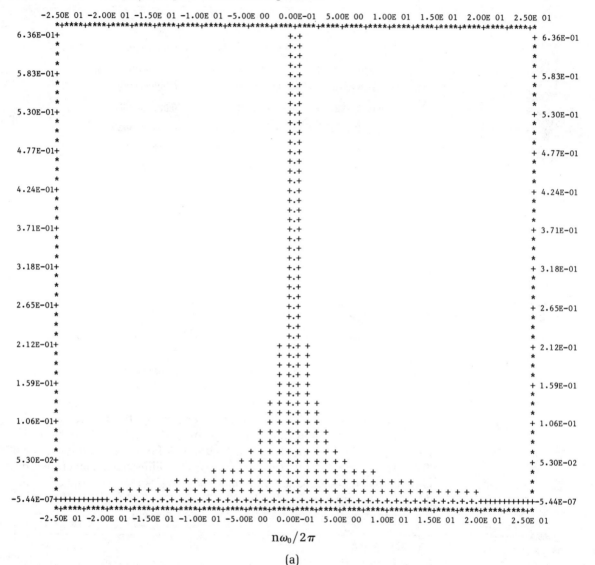

$$n\omega_0/2\pi$$

(a)

and $\tilde{c}_n = 0$ for $|n| \geq 4$. The aliased coefficients $\hat{c}_n$ are obtained from (9.91) as follows. Let $N = 7$. Then

$$\hat{c}_0 = \cdots + \tilde{c}_{-7} + \tilde{c}_0 + \tilde{c}_7 + \cdots = \tilde{c}_0 = 10 \tag{9.113a}$$

$$\hat{c}_1 = \cdots + \tilde{c}_{-6} + \tilde{c}_1 + \tilde{c}_8 + \cdots = \tilde{c}_1 = 7.5 \tag{9.113b}$$

$$\hat{c}_2 = \cdots + \tilde{c}_{-5} + \tilde{c}_2 + \tilde{c}_9 + \cdots = \tilde{c}_2 = 3 \tag{9.113c}$$

Figure 9.17   (continued)

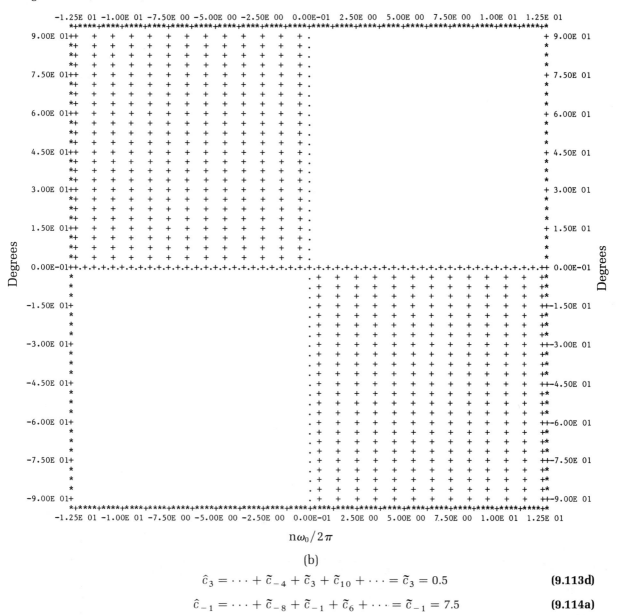

$$n\omega_0/2\pi$$

(b)

$$\hat{c}_3 = \cdots + \tilde{c}_{-4} + \tilde{c}_3 + \tilde{c}_{10} + \cdots = \tilde{c}_3 = 0.5 \tag{9.113d}$$

$$\hat{c}_{-1} = \cdots + \tilde{c}_{-8} + \tilde{c}_{-1} + \tilde{c}_6 + \cdots = \tilde{c}_{-1} = 7.5 \tag{9.114a}$$

$$\hat{c}_{-2} = \cdots + \tilde{c}_{-9} + \tilde{c}_{-2} + \tilde{c}_5 + \cdots = \tilde{c}_{-2} = 3 \tag{9.114b}$$

$$\hat{c}_{-3} = \cdots + \tilde{c}_{-10} + \tilde{c}_{-3} + \tilde{c}_4 + \cdots = \tilde{c}_{-3} = 0.5 \tag{9.114c}$$

This shows that if seven or more terms are used in (9.85) for computing the Fourier coefficients $\tilde{c}_n$, the resulting solution will be exact, the aliasing error being zero.

Figure 9.18
(a) Square-wave approximation resulting from truncated Fourier series of 11 terms in
Eq. (9.110). (b) Square-wave approximation resulting from truncated Fourier series of
27 terms in Eq. (9.110). (c) Square-wave approximation resulting from truncated Fourier
series of 43 terms in Eq. (9.110).

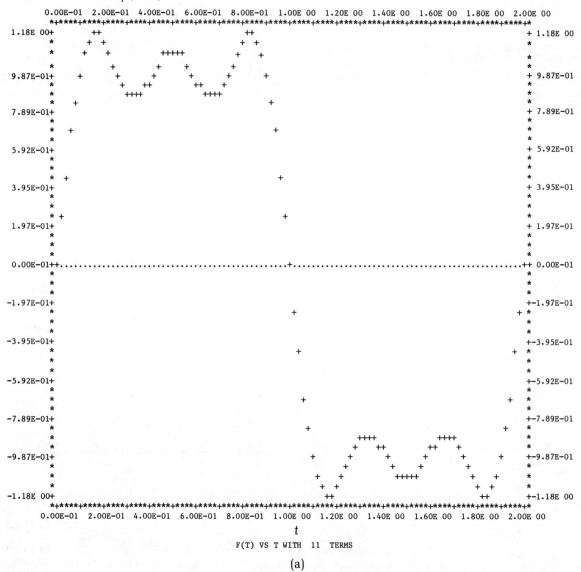

F(T) VS T WITH 11 TERMS

(a)

Figure 9.18   (continued)

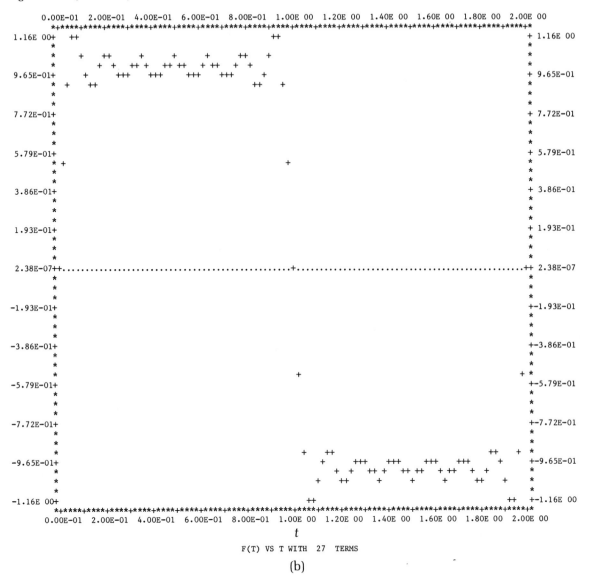

F(T) VS T WITH 27 TERMS

(b)

Unlike $\tilde{c}_n$, the aliased coefficient $\hat{c}_n$ is a periodic sequence of numbers with period N, because from (9.91)

$$\hat{c}_{n+N} = \sum_{k=-\infty}^{\infty} \tilde{c}_{n+N+kN} = \sum_{x=-\infty}^{\infty} \tilde{c}_{n+xN} = \hat{c}_n \qquad (9.115)$$

Figure 9.18   (*continued*)

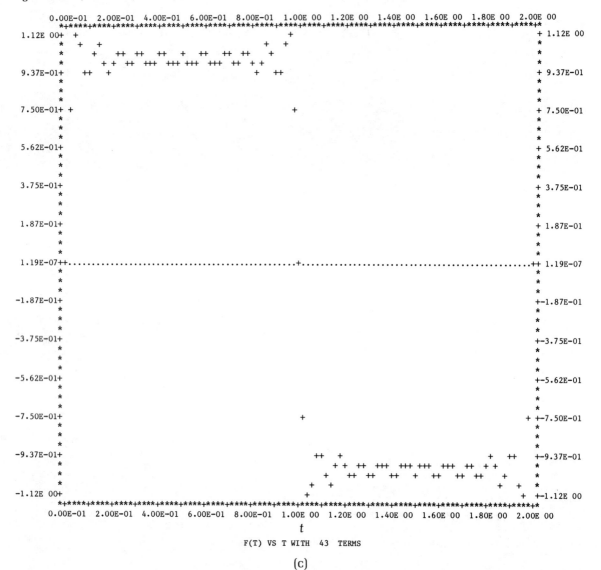

F(T) VS T WITH 43 TERMS

(c)

Thus it suffices to specify $\hat{c}_n$ for $n = 0, 1, 2, \ldots, N-1$ only. Since $\tilde{c}_{-n} = \overline{\tilde{c}}_n$,

$$\hat{c}_{-n} = \overline{\tilde{c}}_n \tag{9.116}$$

This leads to

$$\hat{c}_{N-n} = \overline{\tilde{c}}_n \tag{9.117}$$

For instance, in Example 9.5 again let $N = 7$. The other aliased coefficients are given below:

Figure 9.19

Listing of the partial sums of 43 terms of the truncated Fourier series representation
of the square waveform of Figure 9.14 for t from 0 to 2 with increment of 0.02 sec.

F(T) WITH 43 TERMS

T	F(T)	T	F(T)
0.00000000D 00	-0.827340973508D-13	0.10200000D 01	-0.752876416388D 00
0.20000000D-01	0.752877544963D 00	0.10400000D 01	-0.112449819222D 01
0.40000000D-01	0.112449845116D 01	0.10600000D 01	-0.109312338722D 01
0.60000000D-01	0.109312311479D 01	0.10800000D 01	-0.947952344641D 00
0.80000000D-01	0.947952156759D 00	0.11000000D 01	-0.930942013145D 00
0.10000000D 00	0.930942124084D 00	0.11200000D 01	-0.101814296371D 01
0.12000000D 00	0.101814312194D 01	0.11400000D 01	-0.105562028780D 01
0.14000000D 00	0.105562025663D 01	0.11600000D 01	-0.100250569801D 01
0.16000000D 00	0.100250556759D 01	0.11800000D 01	-0.956394371325D 00
0.18000000D 00	0.956394353272D 00	0.12000000D 01	-0.984076784084D 00
0.20000000D 00	0.984076884762D 00	0.12200000D 01	-0.103155316752D 01
0.22000000D 00	0.103155321684D 01	0.12400000D 01	-0.102415733601D 01
0.24000000D 00	0.102415726659D 01	0.12600000D 01	-0.980505894805D 00
0.26000000D 00	0.980505827604D 00	0.12800000D 01	-0.971870689768D 00
0.28000000D 00	0.971870727861D 00	0.13000000D 01	-0.100788679986D 01
0.30000000D 00	0.100788687384D 01	0.13200000D 01	-0.102842948893D 01
0.32000000D 00	0.102842948049D 01	0.13400000D 01	-0.100268272102D 01
0.34000000D 00	0.100268264972D 01	0.13600000D 01	-0.974407666118D 00
0.36000000D 00	0.974407648318D 00	0.13800000D 01	-0.988373922553D 00
0.38000000D 00	0.988373983298D 00	0.14000000D 01	-0.102019795558D 01
0.40000000D 00	0.102019799461D 01	0.14200000D 01	-0.101842373277D 01
0.42000000D 00	0.101842368873D 01	0.14400000D 01	-0.987097668736D 00
0.44000000D 00	0.987097614783D 00	0.14600000D 01	-0.977327505799D 00
0.46000000D 00	0.977327528899D 00	0.14800000D 01	-0.100443304425D 01
0.48000000D 00	0.100443310590D 01	0.15000000D 01	-0.102411772355D 01
0.50000000D 00	0.102411772355D 01	0.15200000D 01	-0.100443322921D 01
0.52000000D 00	0.100443316756D 01	0.15400000D 01	-0.977327575101D 00
0.54000000D 00	0.977327552000D 00	0.15600000D 01	-0.987097506877D 00
0.56000000D 00	0.987097560830D 00	0.15800000D 01	-0.101842360065D 01
0.58000000D 00	0.101842364469D 01	0.16000000D 01	-0.102019807266D 01
0.60000000D 00	0.102019803364D 01	0.16200000D 01	-0.988374104789D 00
0.62000000D 00	0.988374044043D 00	0.16400000D 01	-0.974407612719D 00
0.64000000D 00	0.974407630518D 00	0.16600000D 01	-0.100268250712D 01
0.66000000D 00	0.100268257842D 01	0.16800000D 01	-0.102842946362D 01
0.68000000D 00	0.102842947205D 01	0.17000000D 01	-0.100788702180D 01
0.70000000D 00	0.100788694782D 01	0.17200000D 01	-0.971870804048D 00
0.72000000D 00	0.971870765955D 00	0.17400000D 01	-0.980505693202D 00
0.74000000D 00	0.980505760403D 00	0.17600000D 01	-0.102415712775D 01
0.76000000D 00	0.102415719717D 01	0.17800000D 01	-0.103155331548D 01
0.78000000D 00	0.103155326616D 01	0.18000000D 01	-0.984077086119D 00
0.80000000D 00	0.984076985440D 00	0.18200000D 01	-0.956394317165D 00
0.82000000D 00	0.956394335218D 00	0.18400000D 01	-0.100250530674D 01
0.84000000D 00	0.100250543717D 01	0.18600000D 01	-0.105562019431D 01
0.86000000D 00	0.105562022547D 01	0.18800000D 01	-0.101814343839D 01
0.88000000D 00	0.101814328017D 01	0.19000000D 01	-0.930942345964D 00
0.90000000D 00	0.930942235024D 00	0.19200000D 01	-0.947951780997D 00
0.92000000D 00	0.947951968877D 00	0.19400000D 01	-0.109312256993D 01
0.94000000D 00	0.109312284236D 01	0.19600000D 01	-0.112449896903D 01
0.96000000D 00	0.112449871009D 01	0.19800000D 01	-0.752879802110D 00
0.98000000D 00	0.752878673537D 00	0.20000000D 01	-0.311864983313D-05
0.10000000D 01	0.155932475400D-05		

$$\hat{c}_4 = \tilde{c}_{-3} = 0.5 = \bar{\tilde{c}}_3 \qquad \hat{c}_{-4} = \tilde{c}_3 = 0.5 = \bar{\tilde{c}}_3$$

$$\hat{c}_5 = \tilde{c}_{-2} = 3 = \bar{\tilde{c}}_2 \qquad \hat{c}_{-5} = \tilde{c}_2 = 3 = \bar{\tilde{c}}_2$$

$$\hat{c}_6 = \tilde{c}_{-1} = 7.5 = \bar{\tilde{c}}_1 \qquad \hat{c}_{-6} = \tilde{c}_1 = 7.5 = \bar{\tilde{c}}_1$$

$$\hat{c}_7 = \tilde{c}_0 = 10 = \bar{\tilde{c}}_0 \qquad \hat{c}_{-7} = \tilde{c}_0 = 10 = \bar{\tilde{c}}_0$$

$$\hat{c}_8 = \tilde{c}_1 = 7.5 = \bar{\tilde{c}}_{-1} \qquad \hat{c}_{-8} = \tilde{c}_{-1} = 7.5 = \bar{\tilde{c}}_1$$

$$\hat{c}_9 = \tilde{c}_2 = 3 = \bar{\tilde{c}}_{-2} \qquad \hat{c}_{-9} = \tilde{c}_{-2} = 3 = \bar{\tilde{c}}_2$$

$$\vdots \qquad\qquad\qquad \vdots \qquad\qquad\qquad\qquad (9.118)$$

confirming (9.116) and (9.117).

In the foregoing we showed how to determine the aliasing coefficients $\hat{c}_n$ from the Fourier coefficients $\tilde{c}_n$. The inverse problem of expressing $\tilde{c}_n$ in terms of $\hat{c}_n$ is not generally possible because there are infinitely many $\tilde{c}_n$. If, however, $f(t)$ is a trigonometric polynomial of order M less than $N/2$,

$$f(t) = \sum_{n=-M}^{M} \tilde{c}_n e^{jn\omega_0 t}, \qquad M < \frac{N}{2} \qquad\qquad (9.119)$$

as in Example 9.5 where $M = 3$, then we can express $\tilde{c}_n$ in terms of $\hat{c}_n$, as follows. Since $\tilde{c}_n = 0$ for $|n| > M$, from (9.103) we have

$$\tilde{c}_n = \hat{c}_n \quad \text{for} \quad |n| \leq M \qquad\qquad (9.120)$$

In practice, however, $\hat{c}_n$ is usually specified in the interval $(0, N-1)$ only. In this case we apply (9.115) and obtain the following relations:

$$\begin{aligned}
\tilde{c}_n &= \hat{c}_n, & 0 \leq n \leq M \\
\tilde{c}_n &= \hat{c}_{n+N}, & -M \leq n < 0
\end{aligned} \qquad\qquad (9.121)$$

For example, we can determine the Fourier coefficients $\tilde{c}_n$ of the function $32 \cos^6 20\pi t$ from its aliased coefficients $\hat{c}_n$ by (9.121). Given

$$\begin{aligned}
&\hat{c}_0 = 10, && \hat{c}_1 = 7.5, && \hat{c}_2 = 3, && \hat{c}_3 = 0.5 \\
&\hat{c}_4 = 0.5, && \hat{c}_5 = 3, && \hat{c}_6 = 7.5
\end{aligned} \qquad\qquad (9.122)$$

then since $M = 3$, for $0 \leq n \leq 3$,

$$\tilde{c}_0 = \hat{c}_0 = 10, \qquad \tilde{c}_1 = \hat{c}_1 = 7.5, \qquad \tilde{c}_2 = \hat{c}_2 = 3, \qquad \tilde{c}_3 = \hat{c}_3 = 0.5 \qquad\qquad (9.123)$$

and for $-3 \leq n < 0$,

$$\tilde{c}_{-3} = \hat{c}_4 = 0.5, \qquad \tilde{c}_{-2} = \hat{c}_5 = 3, \qquad \tilde{c}_{-1} = \hat{c}_6 = 7.5 \qquad\qquad (9.124)$$

where $N = 7$. The trigonometric polynomial (9.111) can then be reconstructed. We remark that the numbers given in (9.122) constitute a periodic sequence, as indicated in (9.115). Once they are given over a period, $\hat{c}_n$ are known for all n. This is depicted in Figure 9.20.

Figure 9.20
A periodic sequence of the aliased coefficients (9.122).

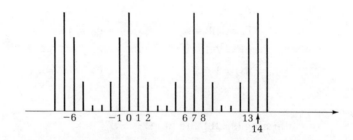

9-4.2 Discrete Fourier Series (DFS)

We have now reached the following conclusion. Given N samples $f(mT_0)$ of a periodic signal $f(t)$ over a period T, its Fourier series at these sampling points can be expressed as

$$f(mT_0) = \sum_{n=0}^{N-1} \hat{c}_n w^{mn}, \qquad m = 0, 1, 2, \ldots, N-1 \tag{9.125}$$

where $T_0 = T/N$ is the time-sampling interval, and

$$\hat{c}_n = \frac{1}{N} \sum_{m=0}^{N-1} f(mT_0) w^{-mn}, \qquad n = 0, 1, 2, \ldots, N-1 \tag{9.126}$$

These equations involve merely two sequences of numbers: the aliased coefficients $\hat{c}_n$ and the samples $f(mT_0)$. If the sequence $\{f(mT_0)\}$ is given, the other sequence $\{\hat{c}_n\}$ can be determined from (9.126). Once $\{\hat{c}_n\}$ is known, $\{f(mT_0)\}$ can be found from (9.125). To stress this fact, in the following we present the analysis directly in terms of the sequences of numbers.

> **Definition 9.1: Discrete Fourier series (DFS)**   *Given a sequence $\{f_n\}$ of N numbers $f_0, f_1, \ldots, f_{N-1}$, real or complex, the discrete Fourier series (DFS) of $\{f_n\}$ is a sequence $\{F_m\}$ of N numbers $F_0, F_1, \ldots, F_{N-1}$ defined by the equation*
>
> $$F_m = \sum_{n=0}^{N-1} f_n w^{-mn}, \qquad m = 0, 1, 2, \ldots, N-1 \tag{9.127}$$
>
> *where $w = \exp(j2\pi/N)$.*

Conversely we can express the numbers f_n in terms of F_m by the *inversion formula*:

$$f_n = \frac{1}{N} \sum_{m=0}^{N-1} F_m w^{mn}, \qquad n = 0, 1, 2, \ldots, N-1 \tag{9.128}$$

To prove this we first change the summation index in (9.127) from n to k and insert the result in the sum of (9.128). This gives

$$\frac{1}{N} \sum_{m=0}^{N-1} F_m w^{mn} = \frac{1}{N} \sum_{m=0}^{N-1} \sum_{k=0}^{N-1} f_k w^{-km} w^{mn}$$

$$= \frac{1}{N} \sum_{k=0}^{N-1} f_k \left[\sum_{m=0}^{N-1} w^{(n-k)m} \right]$$

$$= \frac{1}{N} \sum_{k=0}^{N-1} f_k \frac{1 - w^{(n-k)N}}{1 - w^{n-k}}$$

$$= \frac{1}{N} f_n N = f_n \tag{9.129}$$

In arriving at (9.129), we made use of the relation (9.98).

Thus we have established a one-to-one correspondence between the two sequences $\{f_n\}$ and $\{F_m\}$. We shall say that the two sequences form a *DFS pair*. Because from (9.83),

$$w^{(m+M)n} = w^{mn}, \qquad w^{-(m+N)n} = w^{-mn} \tag{9.130}$$

we conclude that the sequences $\{f_n\}$ and $\{F_m\}$ are periodic with period N:

$$f_{n+N} = f_n, \qquad F_{m+N} = F_m \tag{9.131}$$

As a result the limits of the summations in (9.127) and (9.128) can be changed as long as the total interval is N, or

$$F_m = \sum_{n=n_0}^{n_0+N-1} f_n w^{-mn} \tag{9.132}$$

$$f_n = \frac{1}{N} \sum_{m=m_0}^{m_0+N-1} F_m w^{mn} \tag{9.133}$$

for any integers n_0 and m_0.

Returning to (9.125) and (9.126), we see that if we take the complex conjugate on both sides of the equations, we get

$$f(mT_0) = \sum_{n=0}^{N-1} \overline{\tilde{c}}_n w^{-mn} \tag{9.134}$$

$$\overline{\tilde{c}}_n = \frac{1}{N} \sum_{m=0}^{N-1} f(mT_0) w^{mn} \tag{9.135}$$

where $\overline{f}(mT_0) = f(mT_0)$ because it is real. Comparing these with (9.127) and (9.128), we conclude that the samples $f(mT_0)$ of a periodic function $f(t)$ over a period and the complex conjugate $\overline{\tilde{c}}_n$ of its aliased coefficients form a DFS pair. Thus the problem of evaluating the Fourier series coefficients $\tilde{c}_n$ of $f(t)$ reduces to that of evaluating the DFS of its samples $f(mT_0)$ subject to the aliasing error $\hat{\epsilon}_n = \hat{c}_n - \tilde{c}_n$.

We illustrate the above results by the following examples.

EXAMPLE 9.6

We shall compute the DFS of the periodic sequence

$$f_n = 1, \qquad 0 \leq n \leq 7 \tag{9.136}$$

In this case $N = 8$ and

$$w = e^{j2\pi/8} = e^{j\pi/4} \tag{9.137}$$

From (9.127) we obtain

$$F_m = \sum_{n=0}^{7} e^{-jmn\pi/4} = \frac{1 - w^{-8m}}{1 - w^{-m}}$$

$$= \begin{cases} 8, & m = 0 \\ 0, & 1 \leq m \leq 7 \end{cases} \tag{9.138}$$

Figure 9.21
A periodic sequence (a) and its DFS representation (b).

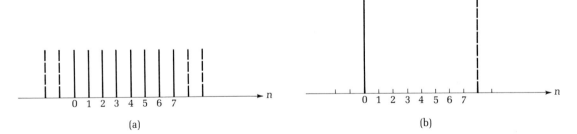

(a)

(b)

The two periodic sequences are displayed in Figure 9.21. Conversely if the sequence $\{8, 0, 0, 0, 0, 0, 0, 0\}$ is given, its inverse DFS can be determined from (9.128):

$$f_n = \frac{1}{8} \sum_{m=0}^{7} F_m w^{mn} = 1, \qquad 0 \leq n \leq 7 \tag{9.139}$$

EXAMPLE 9.7
We determine the DFS representation for a periodic sequence

$$\begin{aligned} f_n &= 1, \qquad 0 \leq n \leq 4 \\ &= 0, \qquad 5 \leq n \leq 9 \end{aligned} \tag{9.140}$$

From (9.127) we have

$$F_m = \sum_{n=0}^{9} f_n w^{-mn} = \sum_{n=0}^{4} e^{-j2mn\pi/10}$$

$$= e^{-j2m\pi/5} \times \frac{\sin(m\pi/2)}{\sin(m\pi/10)} \tag{9.141}$$

giving

$$\begin{aligned}
F_0 &= 5 & F_5 &= 1 \\
F_1 &= 1 - j3.07768 & F_6 &= 0 \\
F_2 &= 0 & F_7 &= 1 + j0.72654 \\
F_3 &= 1 - j0.72654 & F_8 &= 0 \\
F_4 &= 0 & F_9 &= 1 + j3.07768
\end{aligned} \tag{9.142}$$

The two sequences are shown in Figure 9.22, where (a) and (b) are the magnitude spectra of $\{f_n\}$ and $\{F_m\}$ and (c) the phase spectrum of $\{F_m\}$. As can be seen from Figure 9.22(a), the periodic sequence (9.140) gives the samples of a train of rectangular pulses.

Figure 9.22

A periodic sequence (a) and its DFS representation denoted by the magnitude spectrum (b) and the phase spectrum (c).

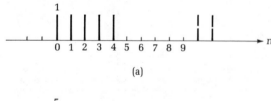

(a)

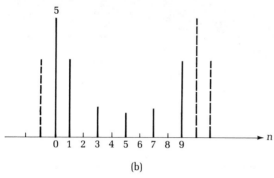

(b)

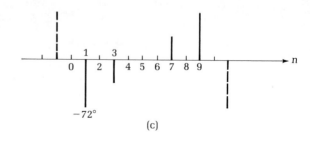

(c)

9-4.3 Properties of the DFS

In this section we present some fundamental properties of the discrete Fourier series that are important in signal processing.

Linearity

Let $\{F_n\}$ and $\{G_n\}$ be the DFS of the two periodic sequences $\{f_n\}$ and $\{g_n\}$ with the same period N. Then the DFS of the combined sequence $\{h_n\}$, written as

$$\{h_n\} = \alpha\{f_n\} + \beta\{g_n\} \tag{9.143}$$

where $h_n = \alpha f_n + \beta g_n$ and α and β are arbitrary constants, is given by

$$\{H_n\} = \alpha\{F_n\} + \beta\{G_n\} \tag{9.144}$$

where $H_n = \alpha F_n + \beta G_n$. All the sequences are periodic with period N.

Shifting

If $\{F_n\}$ is the DFS of a periodic sequence $\{f_n\}$ with period N, then the DFS of the shifted sequence $\{f_{n+m}\}$ is given by the sequence

$$\{H_n\} = \{w^{mn}F_n\} \tag{9.145}$$

where $H_n = w^{mn}F_n$. Likewise the DFS of $\{w^{-mn}f_n\}$ is $\{F_{n+m}\}$.

We remark that any shift that is greater than the period cannot be distinguished in the time domain from a shorter shift $m' = m$ modulo N. This is because all the sequences are periodic with period T. To verify (9.145), we first change the indexes in (9.127) and then apply the result to the sequence $\{f_{n+m}\}$. This yields

$$H_n = \sum_{k=0}^{N-1} f_{k+m} w^{-nk} = \sum_{x=m}^{m+N-1} f_x w^{-(x-m)n}$$

$$= w^{mn} \sum_{x=m}^{m+N-1} f_x w^{-xn} = w^{mn} \sum_{x=0}^{N-1} f_x w^{-xn}$$

$$= w^{mn} F_n \tag{9.146}$$

In arriving at (9.146), we made use of (9.132).

EXAMPLE 9.8

We compute the DFS of the periodic sequence

$$\{g_n\} = \{0, 1, 0, 0, 0, 0, 0, 0\} \tag{9.147}$$

This sequence can be obtained from the periodic sequence

$$\{f_n\} = \{1, 0, 0, 0, 0, 0, 0, 0\} \tag{9.148}$$

by shifting one unit to the right, or $g_n = f_{n-1}$. The DFS of $\{f_n\}$ is found from (9.127):

$$F_n = \sum_{k=0}^{7} f_k w^{-kn} = \sum_{k=0}^{7} f_k e^{-jkn\pi/4} = f_0 = 1 \tag{9.149}$$

giving

$$\{F_n\} = \{1, 1, 1, 1, 1, 1, 1, 1\} \tag{9.150}$$

Thus from (9.145) the DFS of $\{g_n\}$ is given by

$$\{G_n\} = \{1, w^{-1}, w^{-2}, w^{-3}, w^{-4}, w^{-5}, w^{-6}, w^{-7}\}$$

$$= \{1, e^{-j\pi/4}, -j, e^{-j3\pi/4}, -1, e^{-j5\pi/4}, j, e^{-j7\pi/4}\}$$

$$= \left\{1, \frac{1-j}{\sqrt{2}}, -j, \frac{-1-j}{\sqrt{2}}, -1, \frac{-1+j}{\sqrt{2}}, j, \frac{1+j}{\sqrt{2}}\right\} \tag{9.151}$$

Alternatively the sequence $\{G_n\}$ can be computed directly from (9.127):

$$G_n = \sum_{k=0}^{7} g_k w^{-kn} = w^{-n} \tag{9.152}$$

Likewise the DFS of the periodic sequence

$$\{h_n\} = \{0, 0, 1, 0, 0, 0, 0, 0\} \tag{9.153}$$

can be obtained from (9.151) by applying (9.145):

$$\{H_n\} = \{1, -j, -1, j, 1, -j, -1, j\} \tag{9.154}$$

Symmetry

Let $\{F_n\}$ be the DFS of a periodic sequence $\{f_n\}$. Then the DFS of the conjugate sequence $\{\overline{f_n}\}$ is $\{\overline{F}_{-n}\}$ and that of $\{\overline{f}_{-n}\}$ is $\{\overline{F_n}\}$.

To verify we apply (9.127) and obtain the nth coefficient of the DFS of $\{\overline{f_n}\}$ as follows:

$$\sum_{m=0}^{N-1} \overline{f}_m W^{-mn} = \overline{\sum_{m=0}^{N-1} f_m W^{mn}} = \overline{F}_{-n} \tag{9.155}$$

Likewise we can verify the other case, the details being omitted.

For instance, consider the sequence $\{h_n\}$ of (9.153). The conjugate sequence $\{\overline{h}_n\} = \{h_n\}$ is itself. The DFS of $\{h_n\}$ can be obtained from (9.154) by the relation

$$\begin{aligned}
\{\overline{H}_{-n}\} &= \{1, \overline{H}_{-1}, \overline{H}_{-2}, \overline{H}_{-3}, \overline{H}_{-4}, \overline{H}_{-5}, \overline{H}_{-6}, \overline{H}_{-7}\} \\
&= \{1, \overline{j}, -1, -\overline{j}, 1, \overline{j}, -1, -\overline{j}\} \\
&= \{1, -j, -1, j, 1, -j, -1, j\}
\end{aligned} \tag{9.156}$$

which is the same as $\{H_n\}$, as expected.

Next consider the sequence

$$\{\overline{h}_{-n}\} = \{0, 0, 0, 0, 0, 0, 1, 0\} \tag{9.157}$$

the DFS of which is obtained from (9.154) as

$$\begin{aligned}
\{\overline{H}_n\} &= \{\overline{H}_0, \overline{H}_1, \overline{H}_2, \overline{H}_3, \overline{H}_4, \overline{H}_5, \overline{H}_6, \overline{H}_7\} \\
&= \{1, j, -1, -j, 1, j, -1, -j\}
\end{aligned} \tag{9.158}$$

Periodic Convolution

Let $\{F_n\}$ and $\{G_n\}$ be the DFS of two periodic sequences $\{f_n\}$ and $\{g_n\}$ with the same period N. The finite *discrete convolution* $\{h_n\}$ of $\{f_n\}$ and $\{g_n\}$ is defined by the equation

$$h_n = f_n * g_n \triangleq \sum_{m=0}^{N-1} f_m g_{n-m} \tag{9.159}$$

Observe that h_n is obtained by combining f_n and g_n in a manner reminiscent of the convolution integral studied in the preceding chapter. Because $\{f_n\}$ and $\{g_n\}$ are periodic with period N, their discrete convolution $\{h_n\}$ is also periodic with the same period. Therefore we call (9.159) a *periodic convolution of* $\{f_n\}$ and $\{g_n\}$. We next present the periodic convolution theorem as follows.

Property 9.5 Let $\{F_n\}$, $\{G_n\}$, and $\{H_n\}$ be the DFS of the periodic sequences $\{f_n\}$, $\{g_n\}$, and $\{h_n\}$, respectively. If $h_n = f_n * g_n$, then

$$H_n = F_n G_n \tag{9.160}$$

assuming that $\{f_n\}$ and $\{g_n\}$ have the same period N.

To prove, we compute H_n by (9.127), as follows:

$$
\begin{aligned}
H_n &= \sum_{k=0}^{N-1} h_k w^{-kn} = \sum_{k=0}^{N-1}\sum_{x=0}^{N-1} f_x g_{k-x} w^{-kn} = \sum_{x=0}^{N-1}\sum_{k=0}^{N-1} f_x g_{k-x} w^{-kn} \\
&= \sum_{x=0}^{N-1}\sum_{y=-x}^{N-1-x} f_x g_y w^{-(x+y)n} = \sum_{x=0}^{N-1} f_x w^{-xn} \sum_{y=-x}^{N-1-x} g_y w^{-yn} \\
&= \sum_{x=0}^{N-1} f_x w^{-xn} \sum_{y=0}^{N-1} g_y w^{-yn} = F_n G_n
\end{aligned}
\tag{9.161}
$$

as asserted. In arriving at (9.161), use was made of (9.132).

EXAMPLE 9.9
Consider the periodic sequences

$$\{f_n\} = \{1, 0, 0, 0, 0, 0, 0, 0\} \tag{9.162}$$

$$\{g_n\} = \{0, 1, 0, 0, 0, 0, 0, 0\} \tag{9.163}$$

From (9.150) and (9.151), the DFS of $\{f_n\}$ and $\{g_n\}$ are given by

$$\{F_n\} = \{1, 1, 1, 1, 1, 1, 1, 1\} \tag{9.164}$$

$$\{G_n\} = \left\{1, \frac{1-j}{\sqrt{2}}, -j, \frac{-1-j}{\sqrt{2}}, -1, \frac{-1+j}{\sqrt{2}}, j, \frac{1+j}{\sqrt{2}}\right\} \tag{9.165}$$

The periodic convolution of $\{f_n\}$ and $\{g_n\}$ is the periodic sequence

$$\{h_n\} = \{f_n\} * \{g_n\} = \{0, 1, 0, 0, 0, 0, 0, 0\} = \{g_n\} \tag{9.166}$$

having DFS from (9.160)

$$\{H_n\} = \{F_n G_n\} = \{G_n\} \tag{9.167}$$

as in (9.165).

EXAMPLE 9.10
Consider the periodic sequences

$$\{g_n\} = \{0, 1, 0, 0, 0, 0, 0, 0\} \tag{9.168}$$

$$\{h_n\} = \{0, 0, 1, 0, 0, 0, 0, 0\} \tag{9.169}$$

From (9.151) and (9.154), the DFS of $\{g_n\}$ and $\{h_n\}$ are given by

$$\{G_n\} = \left\{1, \frac{1-j}{\sqrt{2}}, -j, \frac{-1-j}{\sqrt{2}}, -1, \frac{-1+j}{\sqrt{2}}, j, \frac{1+j}{\sqrt{2}}\right\} \tag{9.170}$$

$$\{H_n\} = \{1, -j, -1, j, 1, -j, -1, j\} \tag{9.171}$$

The periodic convolution of $\{g_n\}$ and $\{h_n\}$ is the periodic sequence

$$\{k_n\} = \{g_n\} * \{h_n\} = \{0, 0, 0, 1, 0, 0, 0, 0\} \tag{9.172}$$

the DFS of which is from (9.160)

$$\{K_n\} = \{G_n H_n\} = \left\{1, \frac{-1-j}{\sqrt{2}}, j, \frac{1-j}{\sqrt{2}}, -1, \frac{1+j}{\sqrt{2}}, -j, \frac{-1+j}{\sqrt{2}}\right\} \tag{9.173}$$

In Property 9.5 if the roles of time and frequency are interchanged, a similar result is obtained and is stated as follows.

Property 9.6   *Let $\{F_n\}$, $\{G_n\}$, and $\{H_n\}$ be the DFS of the periodic sequences $\{f_n\}$, $\{g_n\}$, and $\{h_n\}$ with the same period N, respectively. If*

$$h_n = f_n g_n \tag{9.174}$$

then H_n equals $1/N$ times the periodic convolution of F_n and G_n:

$$H_n = \frac{1}{N} \sum_{m=0}^{N-1} F_m G_{n-m} \tag{9.175}$$

Proof.   From (9.127) the right-hand side of (9.175) can be expressed as

$$
\begin{aligned}
\frac{1}{N} \sum_{m=0}^{N-1} F_m G_{n-m} &= \frac{1}{N} \sum_{m=0}^{N-1} \sum_{k=0}^{N-1} f_k W^{-km} \sum_{x=0}^{N-1} g_x W^{-x(n-m)} \\
&= \frac{1}{N} \sum_{m=0}^{N-1} \sum_{k=0}^{N-1} \sum_{x=0}^{N-1} f_k g_x W^{(x-k)m - xn} \\
&= \frac{1}{N} \sum_{k=0}^{N-1} \sum_{x=0}^{N-1} f_k g_x W^{-xn} \sum_{m=0}^{N-1} W^{(x-k)m} \\
&= \sum_{k=0}^{N-1} f_k g_k W^{-kn} = H_n \tag{9.176}
\end{aligned}
$$

In arriving at (9.176), we applied the result (9.98).

Since $h_n = f_n g_n = g_n f_n$, we conclude from (9.175) that the operation of a periodic convolution is *commutative*:

$$F_n * G_n = G_n * F_n = \frac{1}{N} \sum_{m=0}^{N-1} F_m G_{n-m} = \frac{1}{N} \sum_{m=0}^{N-1} F_{n-m} G_m \tag{9.177}$$

Likewise since $(f_n g_n)h_n = f_n(g_n h_n)$, we find the operation is also *associative*:

$$(f_n * g_n) * h_n = f_n * (g_n * h_n) \tag{9.178}$$

EXAMPLE 9.11

Consider the periodic sequence

$$\{g_n\} = \{0, 1, 0, 0, 0, 0, 0, 0\} \tag{9.179}$$

the DFS of which is from (9.151)

$$\{G_n\} = \left\{ 1, \frac{1-j}{\sqrt{2}}, -j, \frac{-1-j}{\sqrt{2}}, -1, \frac{-1+j}{\sqrt{2}}, j, \frac{1+j}{\sqrt{2}} \right\} \tag{9.180}$$

Since $g_n = g_n g_n$, we have from (9.175)

$$G_n = \frac{1}{8} \sum_{m=0}^{7} G_m G_{n-m} \tag{9.181}$$

For instance, letting $n = 4$ we obtain

$$
\begin{aligned}
G_4 = \frac{1}{8} \Bigg[& 1 \times (-1) + \left(\frac{1}{\sqrt{2}} - \frac{j}{\sqrt{2}} \right) \times \left(\frac{-1}{\sqrt{2}} - \frac{j}{\sqrt{2}} \right) + (-j) \times (-j) \\
& + \left(\frac{-1}{\sqrt{2}} - \frac{j}{\sqrt{2}} \right) \times \left(\frac{1}{\sqrt{2}} - \frac{j}{\sqrt{2}} \right) + (-1) \times 1 \\
& + \left(\frac{-1}{\sqrt{2}} + \frac{j}{\sqrt{2}} \right) \times \left(\frac{1}{\sqrt{2}} + \frac{j}{\sqrt{2}} \right) + j \times j \\
& + \left(\frac{1}{\sqrt{2}} + \frac{j}{\sqrt{2}} \right) \times \left(\frac{-1}{\sqrt{2}} + \frac{j}{\sqrt{2}} \right) \\
= & -1
\end{aligned}
\tag{9.182}
$$

as given in (9.180). The verification of others is left as an exercise (Problem 9.10).

9-5   FAST FOURIER TRANSFORM

In the preceding section we considered the representation of a periodic sequence by the discrete Fourier series. In the case of a finite-duration sequence, the same representation can be employed if a proper interpretation is provided. Specifically we can represent a finite-duration sequence of length N by a periodic sequence with period N, one period being identical to the finite-duration sequence. For simplicity and without loss of generality, we generally assume that the finite-duration sequence is specified over the interval $0 \leq n \leq N - 1$. This is rather arbitrary, as all of the results can be derived for any interval of N samples. With this interpretation the finite-duration sequence has a unique DFS representation, because we can compute a single period of the periodic sequence and thus the finite-duration sequence from its DFS. The resulting discrete Fourier series representation for the finite-duration sequence is known as the *discrete Fourier transform* (DFT). In this section we shall consider an efficient algorithm that computes the DFS or DFT.

Given a sequence $\{f_n\}$ of length N, its DFT is a sequence $\{F_m\}$ the mth element of which is given by

$$F_m = f_0 + f_1 w^{-m} + \cdots + f_k w^{-km} + \cdots + f_{N-1} w^{-(N-1)m} \qquad (9.183)$$

for $m = 0, 1, \ldots, N - 1$. Thus to find F_m for a fixed m, we need $N - 1$ multiplications and $N - 1$ additions. Thus for m ranging from 0 to $N - 1$, we need $(N - 1)^2$ multiplications and roughly the same number of additions.[†] This number is significant if N is large and particularly if we need to perform many DFT computations, as is the case in many practical situations. For instance, in the case of $N = 2^{12} = 4096$, we need to perform 16,769,025 multiplications and roughly the same number of additions in order to obtain the DFT. In the following we shall discuss an algorithm that reduces the number of multiplications from $(N - 1)^2$ to roughly $(N \log_2 N)/2$. For $N = 4096$, the number of multiplications is reduced from 16,769,025 to roughly 24,576—a great savings! This fast algorithm is called the *fast Fourier transform* (FFT) and finds applications in a variety of problems.

Our objective in this section is to discuss the basic idea used in formulating the FFT, leaving the details of other computational aspects to more advanced texts. This subject has been covered extensively in many books on digital signal processing. For further study we refer to the literature, some of which is listed in the references.

9-5.1 The FFT Theorem

In the course of studying the DFS, we introduced the symbol $w = \exp(j2\pi/N)$ with the understanding that the period N of a periodic sequence is implicitly implied. However, in developing the FFT we are concerned with DFS or DFT of sequences of various periods or lengths. To avoid notational confusion, we insert the subscript N in w to denote the Nth root of unity:

$$w_N = e^{j2\pi/N} \qquad (9.184)$$

giving

$$w_{2N}^2 = w_N, \qquad w_{2N}^N = e^{j\pi} = -1 \qquad (9.185)$$

In the following we derive a theorem that is the cornerstone of the FFT.

Let $\{F_n\}$ be the DFS of a periodic sequence $\{f_n\}$ of period $2N$. For our purposes we form two new sequences $\{g_m\}$ and $\{h_m\}$ of length N from $\{f_n\}$,

---

[†] The exact number of additions required is $N(N - 1)$. For large N, this number is roughly $(N - 1)^2$.

as follows:

$$g_m = f_{2m}, \qquad m = 0, 1, \ldots, N - 1 \tag{9.186}$$

$$h_m = f_{2m+1}, \qquad m = 0, 1, \ldots, N - 1 \tag{9.187}$$

or

$$\{g_m\} = \{f_0, f_2, f_4, \ldots, f_{2N-2}\} \tag{9.188}$$

$$\{h_m\} = \{f_1, f_3, f_5, \ldots, f_{2N-1}\} \tag{9.189}$$

Let $\{G_m\}$ and $\{H_m\}$ be the DFS of $\{g_m\}$ and $\{h_m\}$, respectively—that is,

$$G_m = \sum_{k=0}^{N-1} g_k W_N^{-km}, \qquad m = 0, 1, \ldots, N - 1 \tag{9.190}$$

$$H_m = \sum_{k=0}^{N-1} h_k W_N^{-km}, \qquad m = 0, 1, \ldots, N - 1 \tag{9.191}$$

Note that the nth element of the sequence $\{F_n\}$ is given by

$$F_n = \sum_{k=0}^{2N-1} f_k W_{2N}^{-kn}, \qquad n = 0, 1, \ldots, 2N - 1 \tag{9.192}$$

We shall show the following, which is stated as

The FFT Theorem

$$F_m = G_m + W_{2N}^{-m} H_m, \qquad m = 0, 1, \ldots, 2N - 1 \tag{9.193}$$

Proof.   From (9.192) we have

$$F_m = \sum_{k=0}^{2N-1} f_k W_{2N}^{-km} = \sum_{k=0}^{N-1} f_{2k} W_{2N}^{-2km} + \sum_{k=0}^{N-1} f_{2k+1} W_{2N}^{-(2k+1)m} \tag{9.194}$$

Since from (9.185)

$$W_{2N}^{-2km} = W_N^{-km}, \qquad W_{2N}^{-(2k+1)m} = W_N^{-km} W_{2N}^{-m} \tag{9.195}$$

the two summations on the right-hand side of (9.194) are simplified to

$$\sum_{k=0}^{N-1} f_{2k} W_{2N}^{-2km} = \sum_{k=0}^{N-1} g_k W_N^{-km} = G_m \tag{9.196}$$

$$\sum_{k=0}^{N-1} f_{2k+1} W_{2N}^{-(2k+1)m} = \sum_{k=0}^{N-1} h_k W_N^{-km} W_{2N}^{-m} = W_{2N}^{-m} H_m \tag{9.197}$$

Substituting these in (9.194) yields the desired identity (9.193).

The theorem states that in order to compute the DFS of $\{f_n\}$ of period $2N$, we need only compute the DFS of the sequences $\{g_m\}$ and $\{h_m\}$ of period N, which are combined in accordance with (9.193) to yield the desired DFS of $\{f_n\}$. In doing so the resulting reduction in the number of multiplications is significant. If $\{F_n\}$ is computed directly, the number of multiplications required is $(2N - 1)^2$ or roughly $4N^2$, as noted previously. However, to find $\{F_n\}$ from (9.193) we need only roughly N^2 multiplications for $\{G_m\}$, N^2 multiplications for $\{H_m\}$, and roughly $2N$ multiplications for applying (9.193). Thus the total number of multiplications required in the latter is $2N^2 + 2N$. Furthermore with simple modification an additional reduction of N multiplications is possible, using the following argument. Thus the total number of multiplications required is roughly $2N^2 + N$.

Since $\{G_m\}$ and $\{H_m\}$ are periodic with period N, we have

$$G_{m+N} = G_m, \qquad H_{m+N} = H_m \tag{9.198}$$

Also

$$w_{2N}^{-m+N} = -w_{2N}^{-m} \tag{9.199}$$

Replacing m by $m + N$ in (9.193) and applying (9.198) and (9.199) gives

$$F_{m+N} = G_{m+N} + w_{2N}^{-(m+N)}H_{m+N} = G_m - w_{2N}^{-m}H_m \tag{9.200}$$

Thus to compute F_m from (9.193) for m from N to $2N - 1$, we can use the equations

$$F_{m+N} = G_m - w_{2N}^{-m}H_m, \qquad m = 0, 1, \ldots, N - 1 \tag{9.201}$$

These equations are identical to the last N equations of (9.193). If they are used to compute F_m for m from N to $2N - 1$, the products $w_{2N}^{-m}H_m$ are already computed in (9.193) for m from 0 to $N - 1$. Thus there is a net reduction of the number of multiplications by N.

EXAMPLE 9.12

Given the periodic sequence

$$\{f_n\} = \{1 + j4, j2, j, j2\} \tag{9.202}$$

we determine its DFS by means of (9.193). In this case the sequence is of period $2N = 4$ or $N = 2$.

We form two sequences $\{g_m\}$ and $\{h_m\}$ of period 2 from $\{f_n\}$:

$$\{g_m\} = \{1 + j4, j\} \tag{9.203}$$

$$\{h_m\} = \{j2, j2\} \tag{9.204}$$

the DFS of which are found to be

$$\{G_m\} = \{1 + j5, 1 + j3\} \tag{9.205}$$

$$\{H_m\} = \{j4, 0\} \tag{9.206}$$

From (9.193) we obtain the DFS of $\{f_n\}$, as follows:

$$F_0 = G_0 + H_0 = 1 + j5 + j4 = 1 + j9 \tag{9.207a}$$

$$F_1 = G_1 + w_4^{-1}H_1 = 1 + j3 + e^{-j\pi/2} \times 0 = 1 + j3 \tag{9.207b}$$

$$F_2 = G_2 + w_4^{-2}H_2 = 1 + j5 + e^{-j\pi} \times j4 = 1 + j \qquad \text{(9.207c)}$$

$$F_3 = G_3 + w_4^{-3}H_3 = 1 + j3 + e^{-j6\pi/4} \times 0 = 1 + j3 \qquad \text{(9.207d)}$$

Alternatively F_2 and F_3 can be determined from (9.201):

$$F_2 = G_0 - H_0 = 1 + j5 - j4 = 1 + j \qquad \text{(9.208a)}$$

$$F_3 = G_1 - w_4^{-1}H_1 = 1 + j3 - e^{-j\pi/2} \times 0 = 1 + j3 \qquad \text{(9.208b)}$$

This gives

$$\{F_n\} = \{1 + j9, 1 + j3, 1 + j, 1 + j3\} \qquad \text{(9.209)}$$

9-5.2   The FFT Algorithm

The fast Fourier transform algorithm is based on the formulas (9.193) and (9.201) for computing the DFS or DFT of a sequence $\{f_n\}$ of length $2N$:

$$F_m = G_m + w_{2N}^{-m}H_m, \qquad m = 0, 1, \ldots, N - 1 \qquad \text{(9.210)}$$

$$F_{m+N} = G_m - w_{2N}^{-m}H_m, \qquad m = 0, 1, \ldots, N - 1 \qquad \text{(9.211)}$$

Thus to compute the DFS of $\{f_n\}$ of period $2N$, we need only determine the DFS of the shorter sequences $\{g_m\}$ and $\{h_m\}$, which are of period N. If N is even, the same process can be repeated and can continue until the final sequences are of period 2 if $N = 2^p$ for some positive integer p. We clarify this procedure by considering the special case where $2N = 8$. In order to keep track of changing periods, we introduce the following notation:

$$F_{m,k}(N), \qquad f_{m,k}(N) \qquad \text{(9.212)}$$

$f_{m,k}(N)$ is the mth element of the kth sequence of period N. This sequence is denoted by $\{f_{m,k}(N)\}$. Similarly $F_{m,k}(N)$ represents the mth element of the DFS of the kth sequence $\{f_{m,k}(N)\}$, which is of period N. This kth DFS is denoted by $\{F_{m,k}(N)\}$. In the case where there is only one sequence of period N, the second subscript k together with N will be dropped for simplicity: f_m and F_m.

Consider a sequence $\{f_m\}$ of period 8, written explicitly as

$$\{f_m\} = \{f_0, f_1, f_2, f_3, f_4, f_5, f_6, f_7\} \qquad \text{(9.213)}$$

the DFS of which is denoted by

$$\{F_m\} = \{F_0, F_1, F_2, F_3, F_4, F_5, F_6, F_7\} \qquad \text{(9.214)}$$

We first form two sequences of length 4, using the above notation:

$$\{f_{m,1}(4)\} = \{f_0, f_2, f_4, f_6\} \qquad \text{(9.215)}$$

$$\{f_{m,2}(4)\} = \{f_1, f_3, f_5, f_7\} \qquad \text{(9.216)}$$

Their DFS are written as

$$\{F_{m,1}(4)\} = \{F_{0,1}(4), F_{1,1}(4), F_{2,1}(4), F_{3,1}(4)\} \tag{9.217}$$

$$\{F_{m,2}(4)\} = \{F_{0,2}(4), F_{1,2}(4), F_{2,2}(4), F_{3,2}(4)\} \tag{9.218}$$

The elements of $\{F_m\}$ can be obtained from those of $\{F_{m,1}(4)\}$ and $\{F_{m,2}(4)\}$ by (9.210) and (9.211), as follows:

$$F_m = F_{m,1}(4) + w_8^{-m}F_{m,2}(4), \qquad m = 0, 1, 2, 3 \tag{9.219}$$

$$F_{m+4} = F_{m,1}(4) - w_8^{-m}F_{m,2}(4), \qquad m = 0, 1, 2, 3 \tag{9.220}$$

To compute $\{F_{m,1}(4)\}$ and $\{F_{m,2}(4)\}$, we form new sequences from (9.215) and (9.216), as follows:

$$\{f_{m,1}(2)\} = \{f_0, f_4\} \tag{9.221a}$$

$$\{f_{m,2}(2)\} = \{f_2, f_6\} \tag{9.221b}$$

$$\{f_{m,3}(2)\} = \{f_1, f_5\} \tag{9.222a}$$

$$\{f_{m,4}(2)\} = \{f_3, f_7\} \tag{9.222b}$$

Their DFS are expressed as

$$\{F_{m,1}(2)\} = \{F_{0,1}(2), F_{1,1}(2)\} \tag{9.223a}$$

$$\{F_{m,2}(2)\} = \{F_{0,2}(2), F_{1,2}(2)\} \tag{9.223b}$$

$$\{F_{m,3}(2)\} = \{F_{0,3}(2), F_{1,3}(2)\} \tag{9.224a}$$

$$\{F_{m,4}(2)\} = \{F_{0,4}(2), F_{1,4}(2)\} \tag{9.224b}$$

respectively. Appealing once more to (9.210) and (9.211), the elements of $\{F_{m,1}(4)\}$ and $\{F_{m,2}(4)\}$ can be obtained from those of $\{F_{m,k}(2)\}$, $k = 1, 2, 3, 4$, as follows:

$$F_{m,1}(4) = F_{m,1}(2) + w_4^{-m}F_{m,2}(2), \qquad m = 0, 1 \tag{9.225a}$$

$$F_{m+2,1}(4) = F_{m,1}(2) - w_4^{-m}F_{m,2}(2), \qquad m = 0, 1 \tag{9.225b}$$

$$F_{m,2}(4) = F_{m,3}(2) + w_4^{-m}F_{m,4}(2), \qquad m = 0, 1 \tag{9.226a}$$

$$F_{m+2,2}(4) = F_{m,3}(2) - w_4^{-m}F_{m,4}(2), \qquad m = 0, 1 \tag{9.226b}$$

Finally, to find the elements $F_{m,k}(2)$ we repeat the above process by forming new sequences of period 1 from (9.221) and (9.222):

$$f_{0,1}(1) = f_0 \tag{9.227a}$$

$$f_{0,2}(1) = f_4 \tag{9.227b}$$

$$f_{0,3}(1) = f_2 \tag{9.228a}$$

$$f_{0,4}(1) = f_6 \tag{9.228b}$$

$$f_{0,5}(1) = f_1 \tag{9.229a}$$

$$f_{0,6}(1) = f_5 \tag{9.229b}$$

$$f_{0,7}(1) = f_3 \tag{9.230a}$$

$$f_{0,8}(1) = f_7 \tag{9.230b}$$

Their DFS are simply themselves or

$$F_{0,i}(1) = f_{0,i}(1), \qquad i = 1, 2, \ldots, 8 \tag{9.231}$$

Knowing $F_{0,i}(1)$, the elements of $F_{m,k}(2)$ are obtained again from (9.210) and (9.211), as follows:

$$F_{0,1}(2) = F_{0,1}(1) + F_{0,2}(1) = f_0 + f_4 \tag{9.232a}$$

$$F_{1,1}(2) = F_{0,1}(1) - F_{0,2}(1) = f_0 - f_4 \tag{9.232b}$$

$$F_{0,2}(2) = F_{0,3}(1) + F_{0,4}(1) = f_2 + f_6 \tag{9.233a}$$

$$F_{1,2}(2) = F_{0,3}(1) - F_{0,4}(1) = f_2 - f_6 \tag{9.233b}$$

$$F_{0,3}(2) = F_{0,5}(1) + F_{0,6}(1) = f_1 + f_5 \tag{9.234a}$$

$$F_{1,3}(2) = F_{0,5}(1) - F_{0,6}(1) = f_1 - f_5 \tag{9.234b}$$

$$F_{0,4}(2) = F_{0,7}(1) + F_{0,8}(1) = f_3 + f_7 \tag{9.235a}$$

$$F_{1,4}(2) = F_{0,7}(1) - F_{0,8}(1) = f_3 - f_7 \tag{9.235b}$$

Therefore from the given sequence $\{f_m\}$ we can determine $F_{m,k}(2)$, which are in turn used to find $F_{m,l}(4)$. Knowing $F_{m,l}(4)$, the DFS of $\{f_m\}$ are computed from (9.219).

Observe that the data in $\{f_m\}$ appear in their natural order

$$f_0, f_1, f_2, f_3, f_4, f_5, f_6, f_7 \tag{9.236}$$

whereas in (9.227) to (9.230) they appear in the order

$$f_0, f_4, f_2, f_6, f_1, f_5, f_3, f_7 \tag{9.237}$$

The scrambled order in (9.237) can be determined directly as follows. First we write the natural order n in a three-digit binary form as indicated below:

Decimal	0	1	2	3	4	5	6	7
Binary	000	001	010	011	100	101	110	111

The binary form of n is next written in reverse order, yielding the *reversed bit order*:

Binary	000	100	010	110	001	101	011	111
Decimal	0	4	2	6	1	5	3	7

This is the desired order of (9.237).

Figure 9.23
A graphical display of
Eqs. (9.238) and (9.239)
under which the FFT
algorithm is formulated.

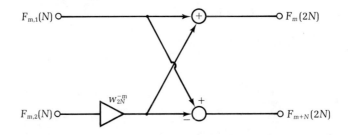

Figure 9.24
A flow chart describing
the operations of the FFT
algorithm for 2N = 8.

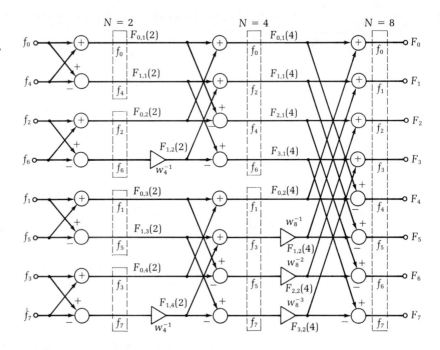

The above procedure for $2N = 8$ is valid in general. In fact, using the notation of (9.212), Equations (9.210) and (9.211) take the form

$$F_m(2N) = F_{m,1}(N) + w_{2N}^{-m}F_{m,2}(N), \qquad m = 0, 1, \ldots, N-1 \qquad \textbf{(9.238)}$$

$$F_{m+N}(2N) = F_{m,1}(N) - w_{2N}^{-m}F_{m,2}(N), \qquad m = 0, 1, \ldots, N-1 \qquad \textbf{(9.239)}$$

These equations are displayed graphically as in Figure 9.23 in a configuration called *butterfly*. The operations described above for $2N = 8$ are presented in Figure 9.24 as a flow chart. The chart consists of $\log_2 2^p = \log_2 8 = 3$ columns, each having $2^{p-1} = 2^{3-1} = 4$ butterflies. Each butterfly requires one multiplication. The quantities associated with the input and output terminals of each butterfly are shown in the figure. The capitals are the DFS of the numbers shown inside each box and are computed from the preceding butterflies as indicated. The procedure is valid only for sequences of length 2^p and is known as the *fast Fourier transform* algorithm (FFT).

To estimate the number of arithmetic operations required in an FFT for a sequence of length 2^p, we see from Figure 9.24 that there are in general $\log_2 2^p = p$ columns. Since each column contains 2^{p-1} butterflies, the total number of butterflies needed is therefore

$$Q(2^p) = 2^{p-1} \times p \qquad\qquad (9.240)$$

As previously noted, each butterfly requires one multiplication. The total number of multiplications needed for an FFT equals $Q(2^p)$. If the DFS of $\{f_m\}$ is computed directly from (9.127), it requires roughly $(2^p)^2 = 2^{2p}$ multiplications. As a comparison, the functions $Q(2^p)$ and 2^{2p} are tabulated in Table 9.3 and plotted in Figure 9.25 for p from 1 to 12. It is seen that for

Table 9.3

p	2^p	$Q(2^p)$	2^{2p}	$2^{2p}/Q(2^p)$
1	2	1	4	4.0
2	4	4	16	4.0
3	8	12	64	5.3
4	16	32	256	8.0
5	32	80	1024	12.8
6	64	192	4096	21.3
7	128	448	16384	36.6
8	256	1024	65536	64.0
9	512	2304	262144	113.8
10	1024	5120	1048576	204.8
11	2048	11264	4194304	372.4
12	4096	24576	16777216	682.7

Figure 9.25

Plots of the numbers of multiplications needed for FFT and those for DFT computed directly from Eq. (9.127) for sequences of lengths 2^p, as p varies from 1 to 12.

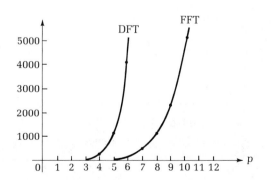

$p \geq 9$ there is a reduction in the number of operations by at least two orders of magnitude. Many practical problems in digital signal processing using DFT can now be solved that were deemed impractical prior to the FFT.

We remark that in the above estimations, we did not include the number of additions while treating each product as a single multiplication. In reality each butterfly involves one complex multiplication and two complex additions. Since

$$(u + jv)(x + jy) = (ux - vy) + j(vx + uy) \tag{9.241}$$

$$(u + jv) + (x + jy) = (u + x) + j(v + y) \tag{9.242}$$

a complex multiplication requires four real multiplications and two real additions, whereas a complex addition requires two real additions. Therefore the total number of real operations required for a butterfly is four multiplications and six additions. If we multiply this number by the number of butterflies of (9.240), the total number of arithmetic operations needed in an FFT for a sequence of length 2^p is given by

$$10 \times 2^{p-1} \times p = 5p \times 2^p \tag{9.243}$$

We illustrate the FFT algorithm by the following numerical example.

EXAMPLE 9.13

We compute the DFT of the sequence $\{f_m\}$ the elements of which are listed below:

$f_0 = 1$ 　　　　　　　 $f_4 = 0.2268 + j0.7776$

$f_1 = 0.9 + j0.3$ 　　　 $f_5 = -0.02916 + j0.76788$

$f_2 = 0.72 + j0.54$ 　　 $f_6 = -0.256608 + j0.682344$

$f_3 = 0.486 + j0.702$ 　 $f_7 = -0.4356504 + j0.5371272$ 　 (9.244)

In this case we have $N = 8$. The elements of $\{f_m\}$ are first scrambled so that they appear in the order of (9.237). Then we compute the DFT of the four sequences of length 2, using (9.232) to (9.235):

$F_{0,1}(2) = 1.2268 + j0.7776$

$F_{1,1}(2) = 0.7732 - j0.7776$

$F_{0,2}(2) = 0.463392 + j1.222344$

$F_{1,2}(2) = 0.976608 - j0.142344$

$F_{0,3}(2) = 0.87084 + j1.06788$

$F_{1,3}(2) = 0.92916 - j0.46788$

$F_{0,4}(2) = 0.0503496 + j1.2391272$

$F_{1,4}(2) = 0.9216504 + j0.1648728$ 　 (9.245)

Knowing $F_{m,k}(2)$, the DFT of the two sequences of length 4 as given in (9.217) and (9.218) are found from (9.225) and (9.226), as follows:

$$F_{0,1}(4) = F_{0,1}(2) + F_{0,2}(2) = 1.690192 + j1.999944$$

$$F_{1,1}(4) = F_{1,1}(2) + w_4^{-1}F_{1,2}(2) = 0.630856 - j1.754208$$

$$F_{2,1}(4) = F_{0,1}(2) - F_{0,2}(2) = 0.763408 - j0.444744$$

$$F_{3,1}(4) = F_{1,1}(2) - w_4^{-1}F_{1,2}(2) = 0.915544 + j0.199008$$

$$F_{0,2}(4) = F_{0,3}(2) + F_{0,4}(2) = 0.9211896 + j2.3070072$$

$$F_{1,2}(4) = F_{1,3}(2) + w_4^{-1}F_{1,4}(2) = 1.0940328 - j1.3895304$$

$$F_{2,2}(4) = F_{0,3}(2) - F_{0,4}(2) = 0.8204904 - j0.1712472$$

$$F_{3,2}(4) = F_{1,3}(2) - w_4^{-1}F_{1,4}(2) = 0.7642872 + j0.4537704 \tag{9.246}$$

Finally, the elements of the above sequences are combined in accordance with (9.219) and (9.220) to yield the DFT of $\{f_m\}$, as follows:

$$F_0 = F_{0,1}(4) + F_{0,2}(4) = 2.6113816 + j4.3069512$$

$$F_1 = F_{1,1}(4) + w_8^{-1}F_{1,2}(4) = 0.4219076432 - j3.51035238$$

$$F_2 = F_{2,1}(4) + w_8^{-2}F_{2,2}(4) = 0.5921608 - j1.2652344$$

$$F_3 = F_{3,1}(4) + w_8^{-3}F_{3,2}(4) = 0.695975465 - j0.6622887888$$

$$F_4 = F_{0,1}(4) - F_{0,2}(4) = 0.7690024 - j0.3070632$$

$$F_5 = F_{1,1}(4) - w_8^{-1}F_{1,2}(4) = 0.8398043568 + j0.0019363802$$

$$F_6 = F_{2,1}(4) - w_8^{-2}F_{2,2}(4) = 0.9346552 + j0.3757464$$

$$F_7 = F_{3,1}(4) - w_8^{-3}F_{3,2}(4) = 1.135112535 + j1.060304789 \tag{9.247}$$

These values are to be compared with the known results for F_m, which are found to be

$$F_m = \frac{1 - f_1^N}{1 - f_1 w_N^{-m}}, \qquad m = 0, 1, \ldots, N - 1 \tag{9.248}$$

where $N = 8$, yielding, of course, the same results.

9-6   COMPUTER SOLUTIONS AND PROGRAMS

In this section we present computer programs in WATFIV that evaluate the DFS or DFT using the standard formulas and the FFT.

9-6.1   Subroutine DFT

1. Purpose. The program, as listed in Figure 9.26, can be used to evaluate the discrete Fourier series or transform (DFS or DFT) of an N-point sequence of real or complex numbers. It can also evaluate the inverse DFS or DFT.

2. Method. The program is based on the definition of the DFS or DFT and its inverse, as given by (9.127) and (9.128).

Figure 9.26
Subroutine DFT used to
evaluate the discrete
Fourier transform and its
inverse of an N point
sequence of complex
numbers.

```
C      SUBROUTINE DFT
C
C      PURPOSE
C         THE PROGRAM CAN BE USED TO EVALUATE THE DISCRETE FOURIER
C            TRANSFORM (DFT) OF AN N POINT SEQUENCE OF COMPLEX
C            NUMBERS.  IT CAN ALSO EVALUATE THE INVERSE DFT.
C
C      METHOD
C         THE PROGRAM IS BASED ON THE DEFINITION OF THE DFT AND ITS
C            INVERSE.
C
C      USAGE
C         CALL DFT (N,NCHOCE,X1,X2)
C
C         N           -THE NUMBER OF GIVEN COMPLEX NUMBERS
C         NCHOCE      -NCHOCE = -1 TO COMPUTE THE DFT, AND NCHOCE = 1
C                         TO COMPUTE THE INVERSE DFT
C         X1          -THE NAME OF THE INPUT ARRAY WHICH HOLDS THE
C                         SEQUENCE TO BE TRANSFORMED
C         X2          -THE NAME OF THE OUTPUT ARRAY WHICH HOLDS THE
C                         SEQUENCE OF THE TRANSFORM
C
C      SUBROUTINES AND FUNCTION SUBPROGRAMS REQUIRED
C         NONE
C
C      REMARKS
C         DOUBLE PRECISION IS USED IN ALL THE COMPUTATIONS.
C         THE INPUT DATA ARE N, NCHOCE AND X1.  THE OUTPUT IS X2.
       SUBROUTINE DFT (N,NCHOCE,X1,X2)
       COMPLEX*16 X1(N),X2(N),W,CD,CDEXP
       W=(0.0,1.0)*6.2831853071796/DFLOAT(N)
       DO 2 K=1,N
       X2(K)=(0.0,0.0)
       DO 1 I=1,N
    1  X2(K)=X2(K)+X1(I)*CDEXP(NCHOCE*W*(I-1)*(K-1))
       IF (NCHOCE.EQ.1) X2(K)=X2(K)/DFLOAT(N)
    2  CONTINUE
       RETURN
       END
```

3. Usage.   The program consists of a subroutine DFT called by

$$\text{CALL DFT (N, NCHOCE, X1, X2)}$$

N	is the length of a given sequence
NCHOCE	$= -1$, compute the DFS or DFT
	$= 1$, compute the inverse DFS or DFT
X1	is the name of the input array that holds the sequence to be transformed
X2	is the name of the output array that holds the sequence of the transform

4. Remarks.   Double precision is used in all the computations. The input data are N, NCHOCE, and X1. The output is X2.

5. Subroutines and Function Subprograms Required.   None

EXAMPLE 9.14
As a test to the program, we compute the DFT of the sequence $\{f_m\}$, having elements defined by

$$f_m = (0.9 + j0.3)^m, \qquad m = 0. 1, 2, \ldots, 31 \tag{9.249}$$

In this case $N = 32$.

Figure 9.27

The main program used to compute the discrete Fourier transform of an N point sequence of complex numbers, using subroutine DFT.

```
        COMPLEX*16 X1(100),X2(100),W,Q
        READ,NP,Q
        N=2**NP
        W=CDEXP(-(0.0,1.0)*6.2831853071796/DFLOAT(N))
        DO 1 I=1,N
        X1(I)=Q**(I-1)
1       X2(I)=(1.0-Q**N)/(1.0-Q*W**(I-1))
        PRINT 2,(I,X1(I),I=1,N)
2       FORMAT (2(4H X1(,I3,2H)=,2D20.12,10X))
        PRINT 3,(I,X2(I),I=1,N)
3       FORMAT (2(4H X2(,I3,2H)=,2D20.12,10X))
        CALL DFT (N,-1,X1,X2)
        PRINT 4
4       FORMAT (36H-VALUES OBTAINED WITH SUBROUTINE DFT)
        PRINT 3,(I,X2(I),I=1,N)
        CALL DFT (N,1,X2,X1)
        PRINT 2,(I,X1(I),I=1,N)
        STOP
        END
```

The main program, as listed in Figure 9.27, computes and prints out $\{f_m\}$, using (9.249), and its transform $\{F_m\}$, using the subroutine DFT. For our purposes we also compute and print out $\{F_m\}$, using (9.248) with $N = 32$ and $\{f_m\}$ as the inverse DFT of (9.248), using the subroutine DFT. Figures 9.28 and 9.29 show the values of $\{f_m\} = \{X1(m)\}$ and $\{F_m\} = \{X2(m)\}$, using (9.249) and (9.248), and their transforms, using the subroutine DFT.

Figure 9.28

Listing of the values of $\{f_m\} = \{X1(m)\}$ and $\{F_m\} = \{X2(m)\}$, using Eqs. (9.249) and (9.248).

X1(1)=	0.100000000000D 01	0.000000000000D 00		X1(2)=	0.900000000000D 00	0.300000000000D 00
X1(3)=	0.720000000000D 00	0.540000000000D 00		X1(4)=	0.486000000000D 00	0.702000000000D 00
X1(5)=	0.226800000000D 00	0.777600000000D 00		X1(6)=	-0.291600000000D-01	0.767880000000D 00
X1(7)=	-0.256608000000D 00	0.682344000000D 00		X1(8)=	-0.435650400000D 00	0.537127200000D 00
X1(9)=	-0.553223520000D 00	0.352719360000D 00		X1(10)=	-0.603716976000D 00	0.151480368000D 00
X1(11)=	-0.588789388800D 00	-0.447827616000D-01		X1(12)=	-0.516475621440D 00	-0.216941302080D 00
X1(13)=	-0.399745668672D 00	-0.350189858304D 00		X1(14)=	-0.254714144314D 00	-0.435094573075D 00
X1(15)=	-0.987143579597D-01	-0.467999359062D 00		X1(16)=	0.515568855548D-01	-0.450813730543D 00
X1(17)=	0.181645316162D 00	-0.390265291823D 00		X1(18)=	0.280560372093D 00	-0.296745167792D 00
X1(19)=	0.341527885221D 00	-0.182902539385D 00		X1(20)=	0.362245858514D 00	-0.621539198798D-01
X1(21)=	0.344667448627D 00	0.527352296625D-01		X1(22)=	0.294380134866D 00	0.150861941284D 00
X1(23)=	0.219683538994D 00	0.224089787616D 00		X1(24)=	0.130488248810D 00	0.267585870552D 00
X1(25)=	0.371636627630D-01	0.279973758140D 00		X1(26)=	-0.505448309552D-01	0.263125481155D 00
X1(27)=	-0.124427992206D 00	0.221649483753D 00		X1(28)=	-0.178480038111D 00	0.162156137716D 00
X1(29)=	-0.209278875615D 00	0.923965125106D-01		X1(30)=	-0.216069941807D 00	0.203731985751D-01
X1(31)=	-0.200574907198D 00	-0.464851038243D-01		X1(32)=	-0.166571885331D 00	-0.102009065601D 00
X2(1)=	0.693972803196D 00	0.349971565599D 01		X2(2)=	0.279226785765D 01	0.805045572144D 01
X2(3)=	0.940296460791D 01	-0.913501355503D 01		X2(4)=	0.186644546746D 01	-0.383383276264D 01
X2(5)=	0.113182268948D 01	-0.223415734713D 01		X2(6)=	0.904793922868D 00	-0.153462930788D 01
X2(7)=	0.799557206779D 00	-0.113960935783D 01		X2(8)=	0.739605630813D 00	-0.882314367551D 00
X2(9)=	0.700861643199D 00	-0.698565363198D 00		X2(10)=	0.673575789604D 00	-0.558478478082D 00
X2(11)=	0.653109437429D 00	-0.446244996656D 00		X2(12)=	0.636991253015D 00	-0.352689135212D 00
X2(13)=	0.623788380776D 00	-0.272085968297D 00		X2(14)=	0.612612873742D 00	-0.200641851239D 00
X2(15)=	0.602883340189D 00	-0.135703205873D 00		X2(16)=	0.594200434347D 00	-0.753136707132D-01
X2(17)=	0.586277479437D 00	-0.179492206265D-01		X2(18)=	0.578899608820D 00	0.376517234751D-01
X2(19)=	0.571898466306D 00	0.926069535069D-01		X2(20)=	0.565135772013D 00	0.147983310050D 00
X2(21)=	0.558492135769D 00	0.204880771268D 00		X2(22)=	0.551859131244D 00	0.264522208758D 00
X2(23)=	0.545133643746D 00	0.328364940349D 00		X2(24)=	0.538214362209D 00	0.398257131749D 00
X2(25)=	0.531001527231D 00	0.476677768576D 00		X2(26)=	0.523403723684D 00	0.567132338630D 00
X2(27)=	0.515362483773D 00	0.674849986673D 00		X2(28)=	0.506925762335D 00	0.808101482253D 00
X2(29)=	0.498467012317D 00	0.980906313952D 00		X2(30)=	0.491389377971D 00	0.121920744159D 01
X2(31)=	0.490732201059D 00	0.157708195516D 01		X2(32)=	0.517353973625D 00	0.218883288454D 01

Figure 9.29

Listing of the values of $\{f_m\} = \{X1(m)\}$ and $\{F_m\} = \{X2(m)\}$, using subroutine DFT.

```
VALUES OBTAINED WITH SUBROUTINE DFT
X2(  1)=  0.693972803196D 00  0.349971565599D 01      X2(  2)=  0.279226785765D 01  0.805045572144D 01
X2(  3)=  0.940296460791D 01 -0.913501355503D 01      X2(  4)=  0.186644546746D 01 -0.383383276264D 01
X2(  5)=  0.113182268948D 01 -0.223415734713D 01      X2(  6)=  0.904793922868D 00 -0.153462930788D 01
X2(  7)=  0.799557206779D 00 -0.113960935783D 01      X2(  8)=  0.739605630813D 00 -0.882314367551D 00
X2(  9)=  0.700861643199D 00 -0.698565363198D 00      X2( 10)=  0.673575789604D 00 -0.558478478082D 00
X2( 11)=  0.653109437429D 00 -0.446244996656D 00      X2( 12)=  0.636991253015D 00 -0.352689135212D 00
X2( 13)=  0.623788380776D 00 -0.272085968297D 00      X2( 14)=  0.612612873742D 00 -0.200641851239D 00
X2( 15)=  0.602883340189D 00 -0.135703205873D 00      X2( 16)=  0.594200434347D 00 -0.753136707132D-01
X2( 17)=  0.586277479437D 00 -0.179492206265D-01      X2( 18)=  0.578899608820D 00  0.376517234751D-01
X2( 19)=  0.571898466306D 00  0.926069535069D-01      X2( 20)=  0.565135772013D 00  0.147983310050D 00
X2( 21)=  0.558492135769D 00  0.204880771268D 00      X2( 22)=  0.551859131244D 00  0.264522208758D 00
X2( 23)=  0.545133643746D 00  0.328364940349D 00      X2( 24)=  0.538214362209D 00  0.398257131749D 00
X2( 25)=  0.531001527231D 00  0.476677768576D 00      X2( 26)=  0.523403723684D 00  0.567132338630D 00
X2( 27)=  0.515362483773D 00  0.674849986673D 00      X2( 28)=  0.506925762335D 00  0.808101482253D 00
X2( 29)=  0.498467012317D 00  0.980906313952D 00      X2( 30)=  0.491389377971D 00  0.121920744159D 01
X2( 31)=  0.490732201060D 00  0.157708195516D 01      X2( 32)=  0.517353973625D 00  0.218883288454D 01
X1(  1)=  0.100000000000D 01  0.164451785523D-14      X1(  2)=  0.900000000000D 00  0.300000000000D 00
X1(  3)=  0.720000000000D 00  0.540000000000D 00      X1(  4)=  0.486000000000D 00  0.702000000000D 00
X1(  5)=  0.226800000000D 00  0.777600000000D 00      X1(  6)= -0.291600000000D-01  0.767880000000D 00
X1(  7)= -0.256608000000D 00  0.682344000000D 00      X1(  8)= -0.435650400000D 00  0.537127200000D 00
X1(  9)= -0.553223520000D 00  0.352719360000D 00      X1( 10)= -0.603716976000D 00  0.151480368000D 00
X1( 11)= -0.588789388800D 00 -0.447827616000D-01      X1( 12)= -0.516475621440D 00 -0.216941302080D 00
X1( 13)= -0.399745668672D 00 -0.350189858304D 00      X1( 14)= -0.254714144314D 00 -0.435094573075D 00
X1( 15)= -0.987143579597D-01 -0.467999359062D 00      X1( 16)=  0.515568855548D-01 -0.450813730543D 00
X1( 17)=  0.181645316162D 00 -0.390265291823D 00      X1( 18)=  0.280560372093D 00 -0.296745167792D 00
X1( 19)=  0.341527885221D 00 -0.182902539385D 00      X1( 20)=  0.362245858514D 00 -0.621539198798D-01
X1( 21)=  0.344667448627D 00  0.527352296625D-01      X1( 22)=  0.294380134866D 00  0.150861941284D 00
X1( 23)=  0.219683538994D 00  0.224089787616D 00      X1( 24)=  0.130488248810D 00  0.267585870552D 00
X1( 25)=  0.371636627631D-01  0.279973758140D 00      X1( 26)= -0.505448309552D-01  0.263125481155D 00
X1( 27)= -0.124427992206D 00  0.221649483753D 00      X1( 28)= -0.178480038111D 00  0.162156137716D 00
X1( 29)= -0.209278875615D 00  0.923965125106D-01      X1( 30)= -0.216069941807D 00  0.203731985751D-01
X1( 31)= -0.200574907198D 00 -0.464851038244D-01      X1( 32)= -0.166571885331D 00 -0.102009065601D 00
```

The computer used was an IBM 370 Model 158, with total execution time of 1.57 sec.

9-6.2   Subroutine FFT

1. Purpose.   The program, as listed in Figure 9.30, can be used to evaluate the DFS or DFT of an N-point sequence of real or complex numbers, where N must be a power of two. It can also evaluate the inverse DFS or DFT.

2. Method.   The program is an implementation of the radix-2 Cooley-Turkey fast Fourier transform algorithm (FFT), as described in Section 9-5.

3. Usage.   The program consists of a subroutine, FFT, called by

CALL FFT (N, NCHOCE, A1, A2)

N   is the length of the sequence being transformed, which must be a power of 2

NCHOCE   = -1, compute the DFS or DFT
             = 1, compute the inverse DFS or DFT

Figure 9.30
Subroutine FFT used to evaluate the discrete Fourier transform and its inverse of an N point sequence of complex numbers, where N must be a power of 2.

```
C
C
C       SUBROUTINE FFT
C
C       PURPOSE
C           THE PROGRAM CAN BE USED TO EVALUATE THE DISCRET FOURIER
C               TRANSFORM (DFT) OF AN N POINT SEQUENCE OF COMPLEX NUMBERS,
C               WHERE N MUST BE A POWER OF TWO.  IT CAN ALSO EVALUATE THE
C               INVERSE DFT.
C
C       METHOD
C           THE PROGRAM IS AN IMPLEMENTATION OF THE RADIX-2 COOLEY-TURKEY
C               FAST FOURIER TRANSFORM ALGORITHM (FFT).
C
C       USAGE
C           CALL FFT (N,NCHOCE,A1,A2)
C
C           N       -THE LENGTH OF THE SEQUENCE BEING TRANSFORMED WHICH
C                       MUST BE A POWER OF TWO
C           NCHOCE  -NCHOCE = -1 TO COMPUTE THE DFT, AND NCHOCE = 1 TO
C                       COMPUTE THE INVERSE DFT
C           A1      -THE NAME OF THE INPUT ARRAY WHICH HOLDS THE SEQUENCE
C                       TO BE TRANSFORMED
C           A2      -THE NAME OF THE OUTPUT ARRAY WHICH HOLDS THE SEQUENCE
C                       OF THE TRANSFORM
C
C       SUBROUTINES AND FUNCTION SUBPROGRAMS REQUIRED
C           NONE
C
C       REMARKS
C           DOUBLE PRECISION IS USED IN ALL THE COMPUTATIONS.
C           THE INPUT DATA ARE N, NCHOCE AND A1.  THE OUTPUT IS A2.
        SUBROUTINE FFT (N,NCHOCE,A1,A2)
        COMPLEX*16 A1(N),A2(N),W,CDEXP
        IF (N-2) 1,1,2
1       A2(1)=A1(1)+A1(2)
        A2(2)=A1(1)+A1(2)*CDEXP(NCHOCE*(0.0,1.0)*6.2831853071796
C       /DFLOAT(N))
        GO TO 14
2       J=1
        DO 8 I=1,N
        IF (I-J) 3,4,4
3       A2(1)=A1(J)
        A1(J)=A1(I)
        A1(I)=A2(1)
4       NH=N/2
5       IF (J-NH) 7,7,6
6       J=J-NH
        NH=(NH+1)/2
        GO TO 5
7       J=J+NH
8       CONTINUE
        DO 9 I=2,N,2
        A2(I-1)=A1(I-1)+A1(I)
9       A2(I)=A1(I-1)-A1(I)
        K=2
10      DO 11 I=1,N
11      A1(I)=A2(I)
        W=CDEXP(NCHOCE*(0.0,1.0)*3.14159 26535898/DFLOAT(K))
        MF=N/K/2
        DO 13 M=1,MF
        DO 12 I=1,K
        A2(I+2*K*(M-1))=A1(I+2*K*(M-1))+A1(I+(2*M-1)*K)*W**(I-1)
12      A2(I+(2*M-1)*K)=A1(I+2*K*(M-1))-A1(I+(2*M-1)*K)*W**(I-1)
13      CONTINUE
        K=K*2
        IF (K-N/2) 10,10,14
14      IF (NCHOCE.EQ.-1) GO TO 16
        DO 15 I=1,N
15      A2(I)=A2(I)/DFLOAT(N)
16      RETURN
        END
```

471

A1   is the name of the input array that holds the sequence to be transformed

A2   is the name of the output array that holds the sequence of the transform

4. Remarks. Double precision is used in all the computations. The input data are N, NCHOCE, and A1. The output is A2.

5. Subroutines and Function Subprograms Required.   None.

EXAMPLE 9.15
We repeat the problem considered in Example 9.14 using the subroutine FFT.

The main program and the printout are shown in Figures 9.31 and 9.32. The total time required for an IBM 370 Model 158 computer was 0.4 sec, whereas for the subroutine DFT the time required was 1.57 sec.

Figure 9.31

The main program used to compute the discrete Fourier transform of an N point sequence of complex numbers, using subroutine FFT.

```
      COMPLEX*16 X1(100),X2(100),W,Q
      READ,NP,Q
      N=2**NP
      W=CDEXP(-(0.0,1.0)*6.2831853071796/DFLOAT(N))
      DO 1 I=1,N
      X1(I)=Q**(I-1)
1     X2(I)=(1.0-Q**N)/(1.0-Q*W**(I-1))
      PRINT 2,(I,X1(I),I=1,N)
2     FORMAT (2(4H X1(,I3,2H)=,2D20.12,10X))
      PRINT 3,(I,X2(I),I=1,N)
3     FORMAT (2(4H X2(,I3,2H)=,2D20.12,10X))
      CALL FFT (N,-1,X1,X2)
      PRINT 4
4     FORMAT (36H-VALUES OBTAINED WITH SUBROUTINE FFT)
      PRINT 3,(I,X2(I),I=1,N)
      CALL FFT (N,1,X2,X1)
      PRINT 2,(I,X1(I),I=1,N)
      STOP
      END
```

Figure 9.32

Listing of the values (a) of $\{f_m\} = \{X1(m)\}$ and $\{F_m\} = \{X2(m)\}$ using Eqs. (9.249) and (9.248) and their transforms (b), using the subroutine FFT.

```
X1(  1)=  0.100000000000D 01  0.000000000000D 00      X1(  2)=  0.900000000000D 00  0.300000000000D 00
X1(  3)=  0.720000000000D 00  0.540000000000D 00      X1(  4)=  0.486000000000D 00  0.702000000000D 00
X1(  5)=  0.226800000000D 00  0.777600000000D 00      X1(  6)= -0.291600000000D-01  0.767880000000D 00
X1(  7)= -0.256608000000D 00  0.682344000000D 00      X1(  8)= -0.435650400000D 00  0.537127200000D 00
X1(  9)= -0.553223520000D 00  0.352719360000D 00      X1( 10)= -0.603716976000D 00  0.151480368000D 00
X1( 11)= -0.588789388800D 00 -0.447827616000D-01      X1( 12)= -0.516475621440D 00 -0.216941302080D 00
X1( 13)= -0.399745668672D 00 -0.350189858304D 00      X1( 14)= -0.254714144314D 00 -0.435094573075D 00
X1( 15)= -0.987143579597D-01 -0.467999359062D 00      X1( 16)=  0.515568855548D-01 -0.450813730543D 00
X1( 17)=  0.181645316162D 00 -0.390265291823D 00      X1( 18)=  0.280560372093D 00 -0.296745167792D 00
X1( 19)=  0.341527885221D 00 -0.182902539385D 00      X1( 20)=  0.362245858514D 00 -0.621539198798D-01
X1( 21)=  0.344667448627D 00  0.527352296625D-01      X1( 22)=  0.294380134866D 00  0.150861941284D 00
X1( 23)=  0.219683538994D 00  0.224089787616D 00      X1( 24)=  0.130488248810D 00  0.267585870552D 00
X1( 25)=  0.371636627630D-01  0.279973758140D 00      X1( 26)= -0.505448309552D-01  0.263125481155D 00
X1( 27)= -0.124427992206D 00  0.221649483753D 00      X1( 28)= -0.178480038111D 00  0.162156137716D 00
X1( 29)= -0.209278875615D 00  0.923965125106D-01      X1( 30)= -0.216069941807D 00  0.203731985751D-01
X1( 31)= -0.200574907198D 00 -0.464851038243D-01      X1( 32)= -0.166571885331D 00 -0.102009065601D 00
```

$$\{f_m\}$$

```
X2(  1)=  0.693972803196D 00  0.349971565599D 01      X2(  2)=  0.279226785765D 01  0.805045572144D 01
X2(  3)=  0.940296460791D 01 -0.913501355503D 01      X2(  4)=  0.186644546746D 01 -0.383383276264D 01
X2(  5)=  0.113182268948D 01 -0.223415734713D 01      X2(  6)=  0.904793922868D 00 -0.153462930788D 01
X2(  7)=  0.799557206779D 00 -0.113960935783D 01      X2(  8)=  0.739605630813D 00 -0.882314367551D 00
X2(  9)=  0.700861643199D 00 -0.698565363198D 00      X2( 10)=  0.673575789604D 00 -0.558478478082D 00
X2( 11)=  0.653109437429D 00 -0.446244996656D 00      X2( 12)=  0.636991253015D 00 -0.352689135212D 00
X2( 13)=  0.623788380776D 00 -0.272085968297D 00      X2( 14)=  0.612612873742D 00 -0.200641851239D 00
X2( 15)=  0.602883340189D 00 -0.135703205873D 00      X2( 16)=  0.594200434347D 00 -0.753136707132D-01
X2( 17)=  0.586277479437D 00 -0.179492206265D-01      X2( 18)=  0.578899608820D 00  0.376517234751D-01
X2( 19)=  0.571898466306D 00  0.926069535069D-01      X2( 20)=  0.565135772013D 00  0.147983310050D 00
X2( 21)=  0.558492135769D 00  0.204880771268D 00      X2( 22)=  0.551859131244D 00  0.264522208758D 00
X2( 23)=  0.545133643746D 00  0.328364940349D 00      X2( 24)=  0.538214362209D 00  0.398257131749D 00
X2( 25)=  0.531001527231D 00  0.476677768576D 00      X2( 26)=  0.523403723684D 00  0.567132338630D 00
X2( 27)=  0.515362483773D 00  0.674849986673D 00      X2( 28)=  0.506925762335D 00  0.808101482253D 00
X2( 29)=  0.498467012317D 00  0.980906313952D 00      X2( 30)=  0.491389377971D 00  0.121920744159D 01
X2( 31)=  0.490732201059D 00  0.157708195516D 01      X2( 32)=  0.517353973625D 00  0.218883288454D 01
```

$$\{F_m\}$$

(a)

```
     VALUES OBTAINED WITH SUBROUTINE FFT
X2(  1)=  0.693972803196D 00  0.349971565599D 01      X2(  2)=  0.279226785765D 01  0.805045572144D 01
X2(  3)=  0.940296460791D 01 -0.913501355503D 01      X2(  4)=  0.186644546746D 01 -0.383383276264D 01
X2(  5)=  0.113182268948D 01 -0.223415734713D 01      X2(  6)=  0.904793922868D 00 -0.153462930788D 01
X2(  7)=  0.799557206779D 00 -0.113960935783D 01      X2(  8)=  0.739605630813D 00 -0.882314367551D 00
X2(  9)=  0.700861643199D 00 -0.698565363198D 00      X2( 10)=  0.673575789604D 00 -0.558478478082D 00
X2( 11)=  0.653109437429D 00 -0.446244996656D 00      X2( 12)=  0.636991253015D 00 -0.352689135212D 00
X2( 13)=  0.623788380776D 00 -0.272085968297D 00      X2( 14)=  0.612612873742D 00 -0.200641851239D 00
X2( 15)=  0.602883340189D 00 -0.135703205873D 00      X2( 16)=  0.594200434347D 00 -0.753136707132D-01
X2( 17)=  0.586277479437D 00 -0.179492206265D-01      X2( 18)=  0.578899608820D 00  0.376517234751D-01
X2( 19)=  0.571898466306D 00  0.926069535069D-01      X2( 20)=  0.565135772013D 00  0.147983310050D 00
X2( 21)=  0.558492135769D 00  0.204880771268D 00      X2( 22)=  0.551859131244D 00  0.264522208758D 00
X2( 23)=  0.545133643746D 00  0.328364940349D 00      X2( 24)=  0.538214362209D 00  0.398257131749D 00
X2( 25)=  0.531001527231D 00  0.476677768576D 00      X2( 26)=  0.523403723684D 00  0.567132338630D 00
X2( 27)=  0.515362483773D 00  0.674849986673D 00      X2( 28)=  0.506925762335D 00  0.808101482253D 00
X2( 29)=  0.498467012317D 00  0.980906313952D 00      X2( 30)=  0.491389377971D 00  0.121920744159D 01
X2( 31)=  0.490732201059D 00  0.157708195516D 01      X2( 32)=  0.517353973625D 00  0.218883288454D 01
```

$$\{F_m\}$$

```
X1(  1)=  0.100000000000D 01  0.346944695195D-16      X1(  2)=  0.900000000000D 00  0.300000000000D 00
X1(  3)=  0.720000000000D 00  0.540000000000D 00      X1(  4)=  0.486000000000D 00  0.702000000000D 00
X1(  5)=  0.226800000000D 00  0.777600000000D 00      X1(  6)= -0.291600000000D-01  0.767880000000D 00
X1(  7)= -0.256608000000D 00  0.682344000000D 00      X1(  8)= -0.435650400000D 00  0.537127200000D 00
X1(  9)= -0.553223520000D 00  0.352719360000D 00      X1( 10)= -0.603716976000D 00  0.151480368000D 00
X1( 11)= -0.588789388800D 00 -0.447827616000D-01      X1( 12)= -0.516475621440D 00 -0.216941302080D 00
X1( 13)= -0.399745668672D 00 -0.350189858304D 00      X1( 14)= -0.254714144314D 00 -0.435094573075D 00
X1( 15)= -0.987143579597D-01 -0.467999359062D 00      X1( 16)=  0.515568855548D-01 -0.450813730543D 00
X1( 17)=  0.181645316162D 00 -0.390265291823D 00      X1( 18)=  0.280560372093D 00 -0.296745167792D 00
X1( 19)=  0.341527885221D 00 -0.182902539385D 00      X1( 20)=  0.362245858514D 00 -0.621539198798D-01
X1( 21)=  0.344667448627D 00  0.527352296625D-01      X1( 22)=  0.294380134866D 00  0.150861941284D 00
X1( 23)=  0.219683538994D 00  0.224089787616D 00      X1( 24)=  0.130488248810D 00  0.267585870552D 00
X1( 25)=  0.371636627630D-01  0.279973758140D 00      X1( 26)= -0.505448309552D-01  0.263125481155D 00
X1( 27)= -0.124427992206D 00  0.221649483753D 00      X1( 28)= -0.178480038111D 00  0.162156137716D 00
X1( 29)= -0.209278875615D 00  0.923965125106D-01      X1( 30)= -0.216069941807D 00  0.203731985751D-01
X1( 31)= -0.200574907198D 00 -0.464851038243D-01      X1( 32)= -0.166571885331D 00 -0.102009065601D 00
```

$$\{f_m\}$$

(b)

Table 9.4

p	$N = 2^p$	Subroutine DFT	Subroutine FFT
2	4	0.05 sec	0.05 sec
3	8	0.16	0.08
4	16	0.50	0.13
5	32	1.57	0.40
6	64	5.75	0.69
7	128	22.62	1.50
8	256	84.80	3.30
9	512	338.04	6.98
10	1024	1116.09	15.49

Figure 9.33
Comparison of relative speed of the subroutines DFT and FFT for the problem treated in Example 9.14 with $N = 2^2, 2^3, \ldots, 2^{10}$, showing a gain in speed for FFT at least one order of magnitude for $N \geq 128$.

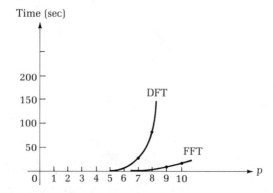

To compare the relative speed of the subroutines DFT and FFT, the problem of Example 9.14 was repeated for $N = 2^2, 2^3, \ldots, 2^{10}$. The results are presented in Table 9.4 and plotted in Figure 9.33, showing a gain in speed for FFT at least one order of magnitude for $N \geq 128$. We remark that the time shown in the table includes that for computing (9.248), (9.249), and their DFS.

9-7 SUMMARY

This chapter continued the study of the signal representation by a Fourier series and its applications. We considered the *steady-state response* of a system to a periodic signal input. We showed that if the input to a system is periodic, the resulting output is also a periodic signal having Fourier coefficients equal to the products of the system transfer function and the corresponding Fourier coefficients of the input signal. The effective or rms value of a nonsinusoidal but periodic signal was shown to equal the square root of the sum of the squares of the effective values of the harmonic components of the signal, dc included. The total average power for a periodic signal equals the sum of the average power of its harmonic components. These are displayed by constructing an average power spectrum.

We studied the *circular convolution* and the *circular convolution theorem*, which states that the Fourier coefficients of the circular convolution of two periodic signals with the same period equals the period times the product of the corresponding Fourier coefficients of the component signals. Using this we derived the *Parseval's formula* and the *energy theorem*.

We next examined the computational aspects of the formulas for evaluating the Fourier series and its coefficients. For this we first grouped together the Fourier coefficients to form the *aliased coefficients* and showed that the *aliasing error* for the nth Fourier coefficient equals the difference between the nth aliased coefficient and the nth Fourier coefficient. Thus the problem of evaluating the Fourier coefficients of a periodic signal is equivalent to the problem of evaluating the aliased coefficients of its samples subject to the aliasing error. This process involves merely two sequences of numbers: the aliased coefficients and the samples of the given periodic signal. To stress this fact, we introduced the *discrete Fourier series* (*DFS*) and discussed some of its fundamental properties. Specifically, they are *linearity*, *shifting*, *symmetry*, and the *periodic convolution*. We showed that the DFS of the product of two periodic sequences is proportional to the periodic convolution of the DFS of the component sequences.

For a finite-duration sequence, with proper interpretation the DFS is called the *discrete Fourier transform* (*DFT*). We presented an efficient algorithm for computing the DFS or DFT. This algorithm is known as the *fast Fourier transform* (*FFT*) and finds applications in a variety of problems. We showed that the DFS of a sequence of length $2N$ can be determined from those of length N. This is the underlying idea for the FFT algorithm. Finally, computer programs using both the standard formulas and the FFT were presented in WATFIV. Illustrative examples were shown and discussed.

REFERENCES AND SUGGESTED READING

Bracewell, R. N. *The Fourier Transform and Its Applications.* 2d ed. New York: McGraw-Hill, 1978, Chapter 18.

Brigham, E. O. *The Fast Fourier Transform.* Englewood Cliffs, N.J.: Prentice-Hall, 1974, Chapters 5–10.

Cooley, J. W., and J. W. Tukey. "An Algorithm for the Machine Computation of Complex Fourier Series." *Math. Computation*, 19 (1965), pp. 297–301.

McGillem, C. D., and G. R. Cooper. *Continuous and Discrete Signal and System Analysis.* New York: Holt, Rinehart & Winston, 1974, Chapter 5.

Oppenheim, A. V., and R. W. Schafer. *Digital Signal Processing.* Englewood Cliffs, N.J.: Prentice-Hall, 1975, Chapters 3 and 6.

Papoulis, A. *Circuits and Systems: A Modern Approach.* New York: Holt, Rinehart & Winston, 1980, Chapter 7.

Rabiner, L. R., and B. Gold. *Theory and Application of Digital Signal Processing.* Englewood Cliffs, N.J.: Prentice-Hall, 1975, Chapters 2 and 6.

Van Valkenburg, M. E. *Network Analysis.* 3d ed. Englewood Cliffs, N.J.: Prentice-Hall, 1974, Chapter 15.

PROBLEMS

9.1 Show that

$$\sum_{n=-M}^{M} e^{jn\omega_0 t} = \frac{\sin (M + \frac{1}{2})\omega_0 t}{\sin \frac{1}{2}\omega_0 t} \qquad (9.250)$$

9.2 Verify that the Fourier series for the triangular voltage waveform v_g of Figure 9.4(a) is that shown in (9.11).

9.3 The full-wave rectified sine voltage waveform of Figure 9.6(a) is applied to the RC network of Figure 9.4(b). Determine the steady-state voltage $v(t)$ across the resistor and plot the power spectrum, assuming $T = 1$.

9.4 Repeat Problem 9.3 for a train of rectangular pulses of Figure 9.9.

9.5 Determine the aliased coefficients $\hat{c}_n$ of the following periodic signals:
   (a) $\cos^3 20\pi t$
   (b) $\sin^4 10\pi t$
   (c) $\sin^3 20\pi t + \cos^4 20\pi t$

9.6 Verify that the Fourier series representation for a train of rectangular pulses of Figure 9.9 is given by (9.106).

9.7 Consider the following periodic sequences
   (a) $\{1, 1, 0, 0\}$
   (b) $\{1, 2, 3, 4\}$
   (c) $\{0, 1 + j, 0, -1 - j\}$
   Verify that the sequences have the respective DFS
   (a) $\{2, 1 - j, 0, 1 + j\}$
   (b) $\{10, -2 + j2, -2, -2 - j2\}$
   (c) $\{0, 2 - j2, 0, -2 - j2\}$

9.8 Find the DFT for the sequence $\{0, 1, 0, 0, 1, 0, 0, 0\}$.

9.9 Using the subrouting DFT, compute the DFS of the periodic signal of Figure 9.6(a) with $T = 1$.

9.10 Using (9.181), verify G_n of (9.180) for all $n \neq 4$.

9.11 Using the subroutine FFT, compute the DFS of the periodic signal of Figure 9.6(a) with $T = 1$.

9.12 Let $\{F_m\}$ be the DFS of a periodic sequence $\{f_n\}$ of period 2N. Show that if $f_{n+N} = f_n$, then $F_{2m+1} = 0$ for all m.

9.13 Let $\{F_m\}$ be the DFS of a periodic sequence $\{f_n\}$ of period 2N. Show that if $F_{m+N} = F_m$ then $f_{2n+1} = 0$ for all n.

9.14 Find the DFT of the sequence $\{f_n\}$ of length 5 with $f_n = n^2$.

9.15 Repeat Example 9.14 for $N = 16$.

9.16 Repeat Example 9.15 for $N = 16$.

9.17 Repeat Example 9.14 for $N = 64$.

9.18 Repeat Example 9.15 for $N = 64$.

9.19 Applying the FFT algorithm, find the DFT for the sequence

$$\{1 + j4, 1 + j3, -1 + j, j\} \qquad (9.251)$$

9.20 For the periodic sequence $\{1, 1, 1, 1, 1, 1, 0, 0\}$, find its DFS and plot its magnitude and phase spectra.

9.21 Find the periodic convolution of the periodic sequences $\{1, 1, 0, 0\}$ and $\{1, 2, 3, 4\}$ and its DFS.

9.22 Find the circular convolution of the triangular waveform of Figure 9.4(a) and the full-wave rectified waveform of Figure 9.6(a) with $T = 1$.

9.23 Let $\{f_n\}$ be a sequence of length N. Find its DFT for $N = 8$ for $f_n = (1 + j)^n$, $n = 0, 1, \ldots, N - 1$.

9.24 Repeat Problem 9.23, using the subroutine FFT for $N = 64$.

9.25 Show that if $\{F_m\}$ is the DFT of an N-point sequence $\{f_n\}$, then

$$\sum_{n=0}^{N-1} |f_n|^2 = \frac{1}{N} \sum_{m=0}^{N-1} |F_m|^2 \tag{9.252}$$

9.26 Let $\{F_m\}$ be the DFT of an N-point sequence $\{f_n\}$. Show that if

$$f_n = -f_{N-n-1} \tag{9.253}$$

then

$$F_0 = 0 \tag{9.254}$$

9.27 Let $\{F_m\}$ be the DFT of a 2N-point sequence $\{f_n\}$. Show that if

$$f_n = f_{2N-n-1} \tag{9.255}$$

then

$$F_N = 0 \tag{9.256}$$

9.28 Using (9.250), verify that

$$\sum_{n=-2M}^{2M} \left(1 - \frac{|n|}{2M + 1}\right) e^{jn\omega_0 t} = \frac{\sin^2 [(M + \frac{1}{2}) \omega_0 t]}{(2M + 1) \sin^2 \frac{1}{2}\omega_0 t} \tag{9.257}$$

CHAPTER TEN

THE FOURIER TRANSFORM
AND CONTINUOUS SPECTRA

In this chapter we introduce the Fourier integral and transform to extend the Fourier methods to include situations where the signals are not periodic and occur only once in some finite time interval. We consider the Fourier integral as a limit of the Fourier series by letting the period become very large. The line spectra thus approach continuous spectra. We shall study the properties of the Fourier transform and its relations to the Fourier series and the discrete Fourier series (DFS) in detail.

10-1 EXPANSION OF NONPERIODIC SIGNALS INTO THE FOURIER SERIES

In the preceding two chapters, we showed that a periodic signal $f(t)$ can be expressed as an infinite sum of trigonometric or exponential harmonics. The resulting expansion was referred to as the Fourier series with the form

$$f(t) = \sum_{n=-\infty}^{\infty} \tilde{c}_n e^{jn\omega_0 t} \tag{10.1}$$

which is valid for every t in the interval $-\infty < t < \infty$. The fundamental frequency ω_0 is unique and is determined by the period T—namely, $\omega_0 = 2\pi/T$. The Fourier coefficients $\tilde{c}_n$ are also uniquely determined by $f(t)$.

In the present section we consider the problem of expressing an arbitrary signal over a finite time interval as a sum of exponentials. As we shall see, the expansion is not unique and cannot hold for every t. The result, however, provides a link between the Fourier series and the Fourier transform.

Suppose that $f(t)$ is a signal defined for every t in the finite interval $0 < t < \gamma$ or in any other interval of length γ. We shall consider the expansion of $f(t)$ in the form of (10.1) for every t in $0 < t < \gamma$. Outside the time interval, the expansion need not be valid: the function $f(t)$ might be undefined out-

Figure 10.1
A nonperiodic signal
$f(t) = t(1 - t)$ that is
defined only over the
interval $t = 0$ to $t = 1$.

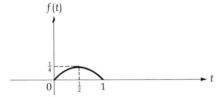

side the interval. To emphasize this we write

$$f(t) = \sum_{n=-\infty}^{\infty} \tilde{c}_n e^{jn\omega_0 t}, \qquad 0 < t < \gamma \tag{10.2}$$

Observe that the sum on the right-hand side of (10.2) is periodic with period $T = 2\pi/\omega_0$, whereas the signal $f(t)$ is arbitrary. Therefore for the equality to hold in (10.2) over an interval of length γ, it is necessary that $T \geq \gamma$ or

$$\omega_0 = \frac{2\pi}{T} \leq \frac{2\pi}{\gamma} \tag{10.3}$$

In other words, if an arbitrary signal is to be expanded into a Fourier series in a time interval of length γ, the fundamental frequency of the resulting expansion cannot exceed $2\pi/\gamma$; however, any value smaller than this is allowable.

As an illustration consider the following signal $f(t)$ defined over an interval of length $\gamma = 1$:

$$f(t) = t(1 - t), \qquad 0 < t < 1 \tag{10.4}$$

Outside the interval the function is not defined. A plot of $f(t)$ over the interval is shown in Figure 10.1. To express $f(t)$ in the form of (10.2), we first introduce a periodic signal $g(t)$ that coincides with $f(t)$ for every t in the interval $0 < t < 1$. There are many ways to construct this $g(t)$; three of them will be exhibited below.

One possibility is to construct a periodic function $g_1(t)$ with period $T = 1$, the first period waveform of which is defined by $f(t)$, as depicted in Figure 10.2. Thus $\omega_0 = 2\pi$ and from (8.88) the Fourier coefficients are

Figure 10.2
A periodic function that is
the same as $f(t)$ of Figure
10.1 over the interval from
$t = 0$ to $t = 1$.

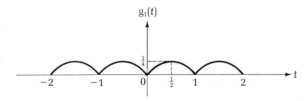

determined as

$$\tilde{c}_n = \frac{1}{1}\int_0^1 t(1-t)e^{-j2n\pi t}\,dt$$

$$= \left[\frac{e^{-j2n\pi t}}{4n^2\pi^2}(j2n\pi t + 1) + \frac{e^{-j2n\pi t}}{j2n\pi}\left(t^2 + \frac{t}{jn\pi} - \frac{1}{2n^2\pi^2}\right)\right]_0^1$$

$$= \frac{-1}{2n^2\pi^2}, \qquad n \neq 0 \tag{10.5a}$$

$$= \frac{1}{6}, \qquad n = 0 \tag{10.5b}$$

Hence it is possible to represent $f(t) = t(1-t)$ for $0 < t < 1$ by the infinite series of exponentials:

$$f(t) = \frac{1}{6} - \frac{1}{2\pi^2}\sum_{\substack{n=-\infty\\n\neq 0}}^{\infty}\frac{1}{n^2}e^{j2n\pi t}, \qquad 0 < t < 1 \tag{10.6}$$

A second possibility is to extend the given function $f(t)$ to the interval $-1 < t < 0$ by reflection in the origin and then to extend periodically the function thus defined over $-1 < t < 1$, as depicted in Figure 10.3(a). The resulting function $g_2(t)$ is periodic with period $T = 2$. Thus $\omega_0 = 2\pi/2 = \pi$,

Figure 10.3

(a) An alternate periodic function that is the same as $f(t)$ of Figure 10.1 over the interval from $t = 0$ to $t = 1$. (b) A third periodic function that is the same as $f(t) = t(1-t)$ over the interval from $t = 0$ to $t = 1$.

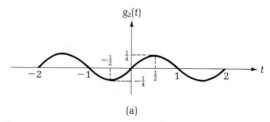

(a)

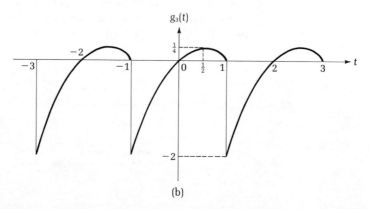

(b)

and from (8.88) the Fourier coefficients are

$$\tilde{c}_n = \frac{1}{2} \left[\int_{-1}^0 t(1 + t)e^{-jn\pi t}\, dt + \int_0^1 t(1 - t)e^{-jn\pi t}\, dt \right]$$

$$= \frac{1}{2} \left[\int_0^1 -t(1 - t)e^{jn\pi t}\, dt + \int_0^1 t(1 - t)e^{-jn\pi t}\, dt \right]$$

$$= \frac{1}{2} \int_0^1 -2jt(1 - t)\sin n\pi t\, dt$$

$$= -j\left[\left(\frac{1}{n^2\pi^2}\sin n\pi t - \frac{t}{n\pi}\cos n\pi t \right) \right.$$

$$\left. - \left(\frac{2t}{n^2\pi^2}\sin n\pi t + \frac{2}{n^3\pi^3}\cos n\pi t - \frac{t^2}{n\pi}\cos n\pi t \right) \right]_0^1$$

$$= -j2\,\frac{1 - \cos n\pi}{n^3\pi^3}$$

$$= \begin{cases} -\dfrac{j4}{n^3\pi^3}, & n \text{ odd} \\[2mm] 0, & n \text{ even} \end{cases} \tag{10.7}$$

Hence the signal $f(t)$ can also be represented for $0 < t < 1$ by the infinite series

$$f(t) = -\frac{j4}{\pi^3} \sum_{k=-\infty}^{\infty} \frac{1}{(2k + 1)^3}\, e^{j(2k+1)\pi t}, \qquad 0 < t < 1 \tag{10.8}$$

Series (10.6) and (10.7) are by no means the only series that will represent $f(t)$ over the interval $0 < t < 1$. In fact, for every possible extension of $f(t)$ from 0 to -1 there corresponds a series representation of $f(t)$ for $0 < t < 1$. For instance, a third possibility might be obtained by letting the extension simply be the function $f(t) = t(1 - t)$ itself for $-1 < t < 0$, as illustrated in Figure 10.3(b). In this case the resulting periodic function $g_3(t)$ is of period $T = 2$ with $\omega_0 = \pi$, and the corresponding Fourier coefficients are obtained as

$$\tilde{c}_n = \frac{1}{2} \int_{-1}^1 t(1 - t)e^{-jn\pi t}\, dt$$

$$= -\frac{2\cos n\pi}{n^2\pi^2} + j\,\frac{\cos n\pi}{n\pi}$$

$$= \frac{(-1)^n}{n\pi}\left(-\frac{2}{n\pi} + j \right), \qquad n \neq 0 \tag{10.9a}$$

$$\tilde{c}_0 = -\frac{1}{3} \tag{10.9b}$$

Therefore for $0 < t < 1$ the function $f(t)$ can also be represented by the series

$$f(t) = -\frac{1}{3} + \sum_{\substack{n=-\infty \\ n \neq 0}}^{\infty} \frac{(-1)^n(-2 + jn\pi)}{n^2\pi^2} e^{jn\pi t}, \qquad 0 < t < 1 \qquad \textbf{(10.10)}$$

It is significant to observe that the function $g_3(t)$ of Figure 10.3(b) is not continuous and has discontinuities at the points $t = \pm1, \pm3, \pm5, \ldots$. In the corresponding series representation (10.10), the Fourier coefficients decrease at least as rapidly as $1/n$. The function $g_1(t)$ of Figure 10.2, on the other hand, is everywhere continuous except at the points $t = \pm1, \pm2, \pm3, \ldots$, where the tangent changes direction discontinuously. In the corresponding series (10.6), the coefficients decrease with n at a rate of $1/n^2$. Finally, in Figure 10.3(a) the function $g_2(t)$ and its derivative are everywhere continuous. This smoother behavior of $g_2(t)$ is reflected in the coefficients in the corresponding series representation (10.8) in that they approach zero at a rate that is proportional to $1/n^3$. Thus the smoother the curves, the faster the series converges—an observation that was pointed out in Section 8-7. Normalized amplitude spectra $|\tilde{c}_n/\tilde{c}_1|$ of (10.10), (10.6), and (10.8) are exhibited in Figure 10.4 (a), (b), and (c), respectively.

Figure 10.4
Normalized amplitude spectra $|\tilde{c}_n/\tilde{c}_1|$ of
(a) Eq. (10.10),
(b) Eq. (10.6), and
(c) Eq. (10.8).

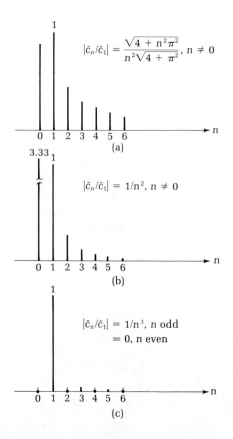

Our conclusion is that corresponding to each extension there is a Fourier series representation that converges to the periodic extension. The rate of convergence depends on the smoothness of the extension, but irrespective of the extension the series represents the given signal over a specified interval length, as desired.

10-2   FOURIER INTEGRAL AS THE LIMIT OF A FOURIER SERIES

Assume that a finite duration signal $f(t)$ is defined for every t in the interval $|t| < \frac{1}{2}\gamma$ and zero outside, as depicted in Figure 10.5:

$$f(t) = 0, \qquad |t| > \tfrac{1}{2}\gamma \tag{10.11}$$

This assumption is not to be deemed restrictive because the interval length γ chosen can be arbitrarily large. In the limit, as γ approaches infinity, signals of infinite duration are included. In (10.11) the origin is chosen at the center of the interval under consideration. This choice is not necessary but is convenient for our discussion.

As demonstrated in the preceding section, the signal $f(t)$ can be represented by a Fourier series in any interval of length T:

$$f(t) = \sum_{n=-\infty}^{\infty} \tilde{c}_n e^{jn\omega_0 t}, \qquad |t| < \tfrac{1}{2}T \tag{10.12}$$

where T is a preassigned period not less than γ and $\omega_0 = 2\pi/T$. Since $f(t)$ is zero for $|t| > \frac{1}{2}\gamma$, the Fourier coefficients are given by

$$\tilde{c}_n = \frac{1}{T} \int_{-\gamma/2}^{\gamma/2} f(t) e^{-jn\omega_0 t}\, dt$$

$$= \frac{1}{T} F(jn\omega_0) = \frac{\omega_0}{2\pi} F(jn\omega_0) \tag{10.13}$$

where

$$F(jn\omega_0) = \int_{-T/2}^{T/2} f(t) e^{-jn\omega_0 t}\, dt \tag{10.14}$$

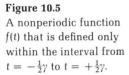

Figure 10.5
A nonperiodic function $f(t)$ that is defined only within the interval from $t = -\frac{1}{2}\gamma$ to $t = +\frac{1}{2}\gamma$.

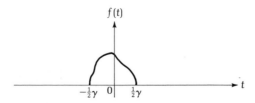

Inserting (10.13) into (10.12) yields the infinite series representation

$$f(t) = \frac{1}{2\pi} \sum_{n=-\infty}^{\infty} F(jn\omega_0)e^{jn\omega_0 t}\omega_0, \qquad |t| < \tfrac{1}{2}T \tag{10.15}$$

The two sides of (10.15) are equal for $|t| < \tfrac{1}{2}T$ only. Outside the interval, $f(t)$ is zero but the sum on the right-hand side is periodic. For $T = \gamma$, $T = 2\gamma$, and $T = 3\gamma$ the periodic sums are shown in Figure 10.6. Clearly as T approaches infinity, (10.15) becomes an identity for all t.

Now as $T \to \infty$, $\omega_0 = 2\pi/T$ tends to zero. Let us concentrate on a particular frequency $\omega = n\omega_0$. If ω_0 has a decreased value, n must be increased proportionately so that the product $n\omega_0$ will stay the same. In the limit as T

Figure 10.6
Periodic function representation of the nonperiodic function $f(t)$ given in Figure 10.5 over the interval from $t = -\tfrac{1}{2}\gamma$ to $t = +\tfrac{1}{2}\gamma$ for three different periods $T = \gamma$, 2γ, and 3γ.

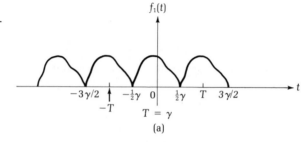

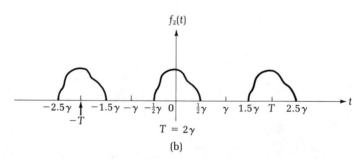

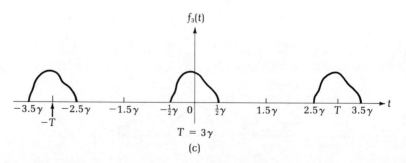

approaches infinity, (10.15) becomes

$$f(t) = \frac{1}{2\pi} \lim_{T \to \infty} \sum_{n=-\infty}^{\infty} F(jn\omega_0)e^{jn\omega_0 t}\omega_0$$

$$= \frac{1}{2\pi} \int_{-\infty}^{\infty} F(j\omega)e^{j\omega t}\,d\omega \tag{10.16}$$

The function $F(j\omega)$ is found directly from (10.14) by making the same substitutions that were made in deriving (10.16):

$$\lim_{T \to \infty} \int_{-T/2}^{T/2} f(t)e^{-jn\omega_0 t}\,dt = \int_{-\infty}^{\infty} f(t)e^{-j\omega t}\,dt$$

$$= \lim_{T \to \infty} F(jn\omega_0) = F(j\omega) \tag{10.17}$$

subject to the sufficient but not necessary condition

$$\int_{-\infty}^{\infty} |f(t)|\,dt < \infty \tag{10.18}$$

Equations (10.17) and (10.16), rewritten as

$$F(j\omega) = \int_{-\infty}^{\infty} f(t)e^{-j\omega t}\,dt \tag{10.19}$$

$$f(t) = \frac{1}{2\pi} \int_{-\infty}^{\infty} F(j\omega)e^{j\omega t}\,d\omega \tag{10.20}$$

constitute the *Fourier transform pair* in the same sense that $F(s)$ and $f(t)$ comprise the Laplace transform pair introduced in Chapter 5. $F(j\omega)$ is called the *Fourier transform* of the time function $f(t)$, whereas $f(t)$ is the *inverse Fourier transform* of $F(j\omega)$. The integral as defined by the right-hand side of (10.19) is known as the *Fourier integral* of $f(t)$. In other words, the Fourier integral is also called the Fourier transform.

By comparing (10.19) with (7.104), we see that $F(j\omega)$ equals the bilateral Laplace transform $F(s)$ of $f(t)$ evaluated for $s = j\omega$.[†] Therefore the Fourier transform $F(j\omega)$ of $f(t)$ is a special case of the more general bilateral Laplace transform $F(s)$ of $f(t)$ for $s = j\omega$.

EXAMPLE 10.1
A rectangular pulse of magnitude V_0 and duration a is shown in Figure 10.7. For simplicity the coordinates are chosen so that

$$\begin{aligned} f(t) &= V_0, & |t| &< \tfrac{1}{2}a \\ &= 0, & |t| &> \tfrac{1}{2}a \end{aligned} \tag{10.21}$$

We compute the Fourier transform of this signal.

[†] The real part of the complex frequency $s = \sigma + j\omega$ is omitted, as we are not here concerned with attenuation concepts.

Figure 10.7
A nonperiodic pulse of
magnitude V_0 and
duration a.

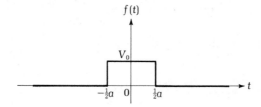

Figure 10.8
Plot of the sinc
function $(\sin x)/x$ that
arises from the Fourier
transform of a pulse
function in the time
domain.

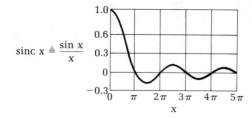

Figure 10.9
Plot in the frequency
domain of the Fourier
transform of the single
time-domain pulse of
Figure 10.7.

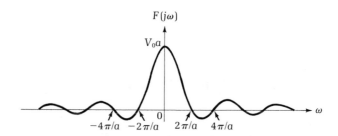

From (10.19) the Fourier transform of $f(t)$ is found to be

$$F(j\omega) = \int_{-a/2}^{a/2} V_0 e^{-j\omega t}\, dt = -\frac{V_0}{j\omega}(e^{-j\omega a/2} - e^{j\omega a/2})$$

$$= V_0 a\, \frac{\sin \frac{1}{2}\omega a}{\frac{1}{2}\omega a} \qquad\qquad\qquad (10.22)$$

using Euler's identity. If we let $x = \frac{1}{2}\omega a$, (10.22) is of the form $(\sin x)/x$
known as the *sinc function*, $\operatorname{sinc} x \triangleq (\sin x)/x$, which is plotted in Figure
10.8. Note that the function $(\sin x)/x$ is even, being symmetric with respect
to the vertical axis, and has the value 1 when $x = 0$. A plot of $F(j\omega)$ as a
function of ω is presented in Figure 10.9.

We next consider the relation between the Fourier transform (10.22) of
a rectangular pulse and the Fourier series of a periodic rectangular pulse
train, as shown in Figure 10.10. The Fourier coefficients for this pulse
train may be determined using (8.89) for all n as

$$\tilde{c}_n = \frac{1}{T}\int_{-a/2}^{a/2} V_0 e^{-jn\omega_0 t}\, dt = \frac{V_0 a}{T}\frac{\sin \frac{1}{2}n\omega_0 a}{\frac{1}{2}n\omega_0 a} \qquad\qquad (10.23)$$

Figure 10.10
Illustration of a periodic pulse train that is equivalent to the single pulse of Figure 10.7 within the interval from $t = -\frac{1}{2}a$ to $t = +\frac{1}{2}a$.

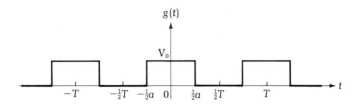

where $\omega_0 = 2\pi/T$, or

$$\tilde{c}_n T = V_0 a \, \frac{\sin \frac{1}{2}n\omega_0 a}{\frac{1}{2}n\omega_0 a} \tag{10.24}$$

Observe that $F(j\omega)$ of (10.22) and $\tilde{c}_n T$ of (10.24) are identical in form except that the latter has values only for the discrete frequencies $n\omega_0$; these are the values of frequencies at which there will be lines in the magnitude spectrum for $\tilde{c}_n$. The *envelope* of the magnitudes of $\tilde{c}_n T$ is a continuous function found by taking the magnitude of the Fourier transform $F(j\omega)$. This leads to the interpretation of $|F(j\omega)|$ as the *continuous spectrum* of $f(t)$ as opposed to the discrete or line spectrum in the context of Fourier series analysis. The term "spectrum" is used to mean the sinusoidal component of a signal.

We next examine $\tilde{c}_n T$ as the ratio a/T changes. We do this for $a/T = \frac{1}{2}$ and $a/T = \frac{1}{4}$. In either case the envelope for $|\tilde{c}_n T|$ is from (10.22)

$$|F(j\omega)| = V_0 a \, \frac{\sin \frac{1}{2}\omega a}{\frac{1}{2}\omega a} \tag{10.25}$$

which is plotted in Figure 10.11. Observe that this envelope is independent of the ratio a/T. In fact, this property holds for all $f(t)$ that are nonrecurring. The zeros or nodes of the envelope occur when

$$\tfrac{1}{2}\omega a = \pm\pi, \ \pm 2\pi, \ \pm 3\pi, \ldots \tag{10.26}$$

Therefore the frequencies at the zeros are

$$\omega = \pm\frac{2\pi}{a}, \ \pm\frac{4\pi}{a}, \ \pm\frac{6\pi}{a}, \cdots \tag{10.27}$$

Figure 10.11
Frequency spectrum of a periodic pulse train (magnitude V_0) for duration to period ratio (a/T) of $\frac{1}{2}$.

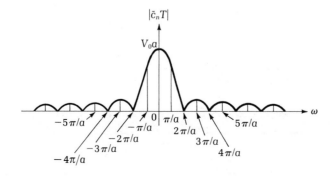

For $T = 2a$ the amplitude spectrum of $\tilde{c}_n T$ has lines at the frequencies

$$n\omega_0 = \frac{n\pi}{a} \tag{10.28}$$

where $\omega_0 = 2\pi/T = \pi/a$ with values

$$|\tilde{c}_n T| = V_0 a \frac{|\sin \frac{1}{2}n\pi|}{|\frac{1}{2}n\pi|} \tag{10.29}$$

Comparing this with (10.25) in conjunction with (10.27), we see that the even-ordered harmonics have zero value, except for $\tilde{c}_0 T$, which has a value $V_0 a$. These are shown in Figure 10.11 for the amplitude spectrum of $\tilde{c}_n T$.

Repeating next for $T = 4a$, the amplitude spectrum of $\tilde{c}_n T$ has lines at the frequencies

$$n\omega_0 = \frac{n\pi}{2a} \tag{10.30}$$

with values

$$|\tilde{c}_n T| = V_0 a \frac{|\sin \frac{1}{4}n\pi|}{|\frac{1}{4}n\pi|} \tag{10.31}$$

where $\omega_0 = \pi/2a$. The amplitude spectrum of $\tilde{c}_n T$ is presented in Figure 10.12.

Observe that when T is increased from $2a$ to $4a$, the number of lines in a given frequency interval is doubled and the separation of the lines is cut in half. As T becomes larger and larger, the separation of the lines will become smaller and smaller. In the limit as T approaches infinity, the line spectrum becomes the continuous spectrum represented by the envelope of $|\tilde{c}_n T|$. In terms of the line plot of $|\tilde{c}_n|$, we see that as T is increased, more frequency components are required to make up the signal, but the amplitude of each frequency component, given by $|\tilde{c}_n T|/T$, is smaller. The product $\tilde{c}_n T$, however, will remain the same.

The above interpretation holds in general. For an arbitrary signal $f(t)$ of finite duration $a < T$, centered at the origin, the limits of integration for $\tilde{c}_n$ of (8.89) and $F(j\omega)$ of (10.19) are identical and

$$\tilde{c}_n T = \int_{-a/2}^{a/2} f(t)e^{-jn\omega_0 t}\, dt \tag{10.32}$$

Figure 10.12
Frequency spectrum of a
periodic pulse train
(magnitude V_0) for
duration to period ratio
(a/T) of $\frac{1}{4}$.

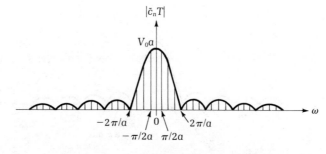

$$F(j\omega) = \int_{-a/2}^{a/2} f(t)e^{-j\omega t} \, dt \tag{10.33}$$

Comparing these two equations, we see that they are identical except that one is a continuous function and the other is defined only for discrete values of frequency. In fact, the product $\tilde{c}_n T$ equals the samples of the Fourier transform $F(j\omega)$ for $\omega = n\omega_0 = 2\pi n/T$. Thus if we write

$$F(j\omega) = |F(j\omega)|e^{j\phi(\omega)} \tag{10.34}$$

$|F(j\omega)|$ is called the *continuous amplitude spectrum* and $\phi(\omega)$ is the *continuous phase spectrum* for a nonrecurring signal $f(t)$, as opposed to the discrete or line amplitude and phase spectra discussed in Chapter 8 in the context of Fourier series analysis of periodic signals.

Our conclusion is that the continuous amplitude and phase spectra for a nonrecurring pulse signal are identical with the envelopes of the line spectra for the corresponding recurring pulse signal, multiplied by a factor equal to the period of the recurring pulse signal. In terms of frequency content, all frequency components are required to make up a nonrecurring pulse signal in the sense that its frequency spectrum is continuous and is defined for all frequencies. A pulse of lightning is an example of a nonrecurring pulse signal. It is a common experience that such a pulse will result in a burst of static on all receivers from low-frequency radio to ultra-high-frequency television.

So far we have demonstrated how to compute the Fourier transform of a nonrecurring signal. An important question at this point is whether a function $f(t)$ may have more than one Fourier transform. Or, putting it differently, can two different functions defined for all t possess the same Fourier transform? From (8.122) we see that if a function is defined as

$$f(t) = \tfrac{1}{2}[f(t+) + f(t-)] \tag{10.35}$$

for all t, there is fortunately one and only one Fourier transform. No two different functions can have the same Fourier transform.

For example, we take the rectangular pulse $f(t)$ of Figure 10.7. If we define

$$f(t) = \begin{cases} V_0, & |t| < \tfrac{1}{2}a \\ \tfrac{1}{2}[f(\pm\tfrac{1}{2}a+) + f(\pm\tfrac{1}{2}a-)] = \tfrac{1}{2}V_0, & |t| = \tfrac{1}{2}a \\ 0, & |t| > \tfrac{1}{2}a \end{cases} \tag{10.36}$$

then there can be one and only one Fourier transform of this function. The essential requirement is that at the points of discontinuities the values of the functions are usually not defined. For the uniqueness of the transform, we choose the average value of $f(t+)$ and $f(t-)$. On the other hand, if the Fourier transform $F(j\omega)$ of a function $f(t)$ is known, there is a unique time function defined by the integral (10.20):

$$\frac{1}{2\pi}\int_{-\infty}^{\infty} F(j\omega)e^{j\omega t} \, d\omega = \tfrac{1}{2}[f(t+) + f(t-)] \tag{10.37}$$

Figure 10.13
Frequency spectrum of an
unknown time-domain
function $f(t)$.

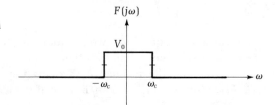

EXAMPLE 10.2

The Fourier transform $F(j\omega)$ of a function is shown in Figure 10.13. We wish
to find the time function $f(t)$.

We apply the inversion formula (10.20) by inserting

$$F(j\omega) = V_0, \qquad |\omega| < \omega_c$$
$$= 0, \qquad |\omega| > \omega_c \qquad \text{(10.38)}$$

in it and obtain

$$f(t) = \frac{1}{2\pi} \int_{-\omega_c}^{\omega_c} V_0 e^{j\omega t}\, d\omega = \frac{V_0}{j2\pi t}(e^{j\omega_c t} - e^{-j\omega_c t})$$

$$= \frac{V_0 \omega_c}{\pi} \cdot \frac{\sin \omega_c t}{\omega_c t} \qquad \text{(10.39)}$$

A plot of (10.39) is shown in Figure 10.14.

If, on the other hand, $f(t)$ of (10.39) is known, its Fourier transform $F(j\omega)$
is found from (10.19) with

$$F(j\omega) = \begin{cases} V_0, & |\omega| < \omega_c \\ \tfrac{1}{2}[F(j\omega+) + F(j\omega-)] = \tfrac{1}{2}V_0, & |\omega| = \omega_c \\ 0, & |\omega| > \omega_c \end{cases} \qquad \text{(10.40)}$$

as indicated in Figure 10.13. Note that in computing the time function $f(t)$
from (10.38), the values of the Fourier transform $F(j\omega)$ at the points of dis-

Figure 10.14
The time-domain function
$f(t)$ the Fourier transform
of which is the $F(j\omega)$ given
in Figure 10.13.

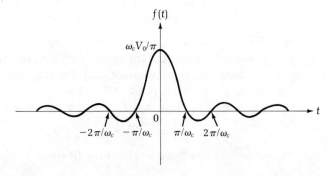

continuities $\omega = \pm \omega_c$ are not required. Once $f(t)$ is known, the Fourier integral (10.19) takes the average values at the points of discontinuities of $F(j\omega)$.

10-3   PROPERTIES OF FOURIER TRANSFORMS

The Fourier transform $F(j\omega)$ is an alternate and equivalent frequency-domain representation of a time-domain signal $f(t)$. Applications to network problems will be taken up later in this chapter. In the present section we study the effect in one domain caused by an operation in the other domain. Not only does this facilitate the transformation of signals from one domain to the other, but also certain physical properties of signals and systems will be revealed that are not otherwise apparent. This elucidation of the duality is enhanced by the symmetry that exists between $f(t)$ and its Fourier transform $F(j\omega)$ by the defining equations repeated below:

$$\mathbf{F}[f(t)] \triangleq F(j\omega) = \int_{-\infty}^{\infty} f(t)e^{-j\omega t}\, dt \tag{10.41}$$

$$\mathbf{F}^{-1}[F(j\omega)] \triangleq f(t) = \frac{1}{2\pi} \int_{-\infty}^{\infty} F(j\omega)e^{j\omega t}\, d\omega \tag{10.42}$$

Observe that if we multiply each integral by $1/\sqrt{2\pi}$ instead of 1 and $1/2\pi$ in (10.41) and (10.42), the symmetry can be made more complete. However, convention dictates that we retain the use of (10.41) and (10.42) as the defining equations for $F(j\omega)$ and $f(t)$.

Elementary Properties

By Euler's formula (4.112), the Fourier integral (10.41) can be rewritten as

$$F(j\omega) = \int_{-\infty}^{\infty} f(t)(\cos \omega t - j \sin \omega t)\, dt \tag{10.43}$$

Writing

$$F(j\omega) = R(\omega) + jX(\omega) \tag{10.44}$$

and equating real and imaginary parts of (10.43), we obtain

$$R(\omega) = \int_{-\infty}^{\infty} f(t) \cos \omega t\, dt \tag{10.45}$$

$$X(\omega) = -\int_{-\infty}^{\infty} f(t) \sin \omega t\, dt \tag{10.46}$$

This shows that since

$$R(-\omega) = R(\omega) \tag{10.47}$$

$$X(-\omega) = -X(\omega) \tag{10.48}$$

$R(\omega)$ is an even function and $X(\omega)$ is an odd function, giving

$$F(-j\omega) = \overline{F}(j\omega) \tag{10.49}$$

where the bar, as before, denotes the complex conjugate.

The implication of (10.49) is that if $F(j\omega)$ is known for $\omega > 0$, then it is also known for $\omega < 0$. For this reason $F(j\omega)$ is usually plotted for $\omega > 0$ only. Note that in this text we consider only algebraically real time signals. Equations (10.45) and (10.46) were derived based on this assumption.

Likewise the inversion formula (10.42) can be rewritten as

$$f(t) = \frac{1}{2\pi} \int_{-\infty}^{\infty} [R(\omega) + jX(\omega)](\cos \omega t + j \sin \omega t) \, d\omega \tag{10.50}$$

Since $f(t)$ is real, the imaginary part of the integral must be zero, so we obtain

$$f(t) = \frac{1}{2\pi} \int_{-\infty}^{\infty} [R(\omega) \cos \omega t - X(\omega) \sin \omega t] \, d\omega \tag{10.51}$$

Observe that the integrand in (10.51) is an even function. As a result the integral from $-\infty$ to ∞ equals twice the integral from 0 to ∞:

$$f(t) = \frac{1}{\pi} \int_{0}^{\infty} [R(\omega) \cos \omega t - X(\omega) \sin \omega t] \, d\omega \tag{10.52}$$

Finally, we mention that if $f(t)$ is even, from (10.46) $X(\omega) = 0$ because the integrand $f(t) \sin \omega t$ is odd. Similarly, if $f(t)$ is odd, then from (10.45) $R(\omega) = 0$. Thus the Fourier transform of an even function is algebraically real, whereas the Fourier transform of an odd function is entirely imaginary:

$$F(j\omega) = R(\omega) \qquad \text{if } f(-t) = f(t) \tag{10.53}$$

$$F(j\omega) = jX(\omega) \qquad \text{if } f(-t) = -f(t) \tag{10.54}$$

The rectangular pulse of Figure 10.7, for instance, is even. This property is reflected in its transform $F(j\omega)$ of (10.22) in that it is not only real but also even, as illustrated in Figure 10.9. For an example of an odd function, we consider the following.

EXAMPLE 10.3
Consider the pulse signal $f(t)$ of Figure 10.15, the Fourier transform of which is found to be

$$F(j\omega) = \int_{-a/2}^{0} V_0 e^{-j\omega t} \, dt - \int_{0}^{a/2} V_0 e^{-j\omega t} \, dt$$

$$= j\frac{2V_0(1 - \cos \frac{1}{2}\omega a)}{\omega} = j4V_0 \frac{\sin^2 \frac{1}{4}\omega a}{\omega} \tag{10.55}$$

Since $f(t)$ is an odd function, not only is its Fourier transform also odd but also the transform is entirely imaginary, as expected from (10.48) and (10.54). A plot of $F(j\omega)$ is presented in Figure 10.16.

Figure 10.15
A nonrepeating signal pulse illustrated in the time domain.

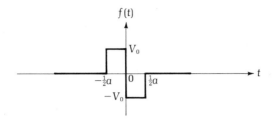

Figure 10.16
The appearance of signal pulse of Figure 10.15 expressed in the frequency domain using the Fourier transform.

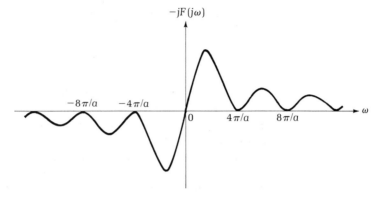

Linearity

If $F_1(j\omega)$ and $F_2(j\omega)$ are the Fourier transforms of $f_1(t)$ and $f_2(t)$, respectively, then for arbitrary constants c_1 and c_2 the Fourier transform of the sum $c_1 f_1(t) + c_2 f_2(t)$ is $c_1 F_1(j\omega) + c_2 F_2(j\omega)$.

The above result follows immediately from the properties that the transforms are integrals of time functions and that integration is a linear operation. Using the notation of (10.41) and (10.42), the linear property of the Fourier transform can be stated as follows: If

$$\mathbf{F}[f_1(t)] = F_1(j\omega), \qquad \mathbf{F}[f_2(t)] = F_2(j\omega) \tag{10.56}$$

then

$$\mathbf{F}[c_1 f_1(t) + c_2 f_2(t)] = c_1 F_1(j\omega) + c_2 F_2(j\omega) \tag{10.57}$$

for arbitrary constants c_1 and c_2. Even though we are primarily concerned with algebraically real signals, (10.57) holds for complex as well as real c_1 and c_2.

Scaling

If

$$\mathbf{F}[f(t)] = F(j\omega) \tag{10.58}$$

then for any real constant a,

$$\mathbf{F}[f(at)] = \frac{1}{|a|} F\left(\frac{j\omega}{a}\right) \tag{10.59}$$

To verify this we appeal to (10.41) and consider the cases of positive and negative values of a separately. For positive a let $x = at$, so that $dx = a\,dt$. Substituting these into the defining integral of $\mathbf{F}[f(at)]$ yields

$$\mathbf{F}[f(at)] = \int_{-\infty}^{\infty} f(at)e^{-j\omega t}\,dt = \frac{1}{a}\int_{-\infty}^{\infty} f(x)e^{-j(\omega/a)x}\,dx$$

$$= \frac{1}{a}F\left(\frac{j\omega}{a}\right) \tag{10.60}$$

When a is negative, the limits on the integral will be reversed when the variable of integration is changed from t to $x = at$. This leads to

$$\mathbf{F}[f(at)] = -\frac{1}{a}F\left(\frac{j\omega}{a}\right) \tag{10.61}$$

Combining these two results gives the desired identity (10.59).

The scaling property (10.59) says that when the time scale of a function is expanded by a factor, its frequency spectrum is contracted by the same factor. Conversely if the time scale of a function is compressed by a factor, its frequency spectrum will be expanded by the same factor. Recall that the waveforms resulting from the effect of time scaling were illustrated in Section 1-3. A similar conclusion can be reached with regard to the waveforms resulting from the effect of frequency scaling.

Symmetry

The symmetry between the Fourier integral (10.41) and its inversion (10.42) can be used to great advantage in finding the frequency-domain Fourier transform of a time-domain signal.

If

$$\mathbf{F}[f(t)] = F(j\omega) \tag{10.62}$$

then

$$\mathbf{F}[F(jt)] = 2\pi f(-\omega) \tag{10.63}$$

Proof.   Because ω in (10.42) is a dummy variable, we can replace it by an alternate symbol x and obtain

$$2\pi f(t) = \int_{-\infty}^{\infty} F(jx)e^{jxt}\,dx \tag{10.64}$$

Now if we replace t by $-\omega$, we have

$$2\pi f(-\omega) = \int_{-\infty}^{\infty} F(jx)e^{-j\omega x}\,dx$$

$$= \int_{-\infty}^{\infty} F(jt)e^{-j\omega t}\,dt = \mathbf{F}[F(jt)] \tag{10.65}$$

where the dummy variable x has been replaced by t.

Figure 10.17
(a) Nonperiodic rectangular pulse in the time domain. (b) Representation in the frequency domain of the rectangular pulse shown in the time domain in (a) above.

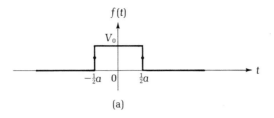

(a)

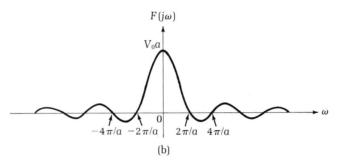

(b)

EXAMPLE 10.4

Consider the rectangular pulse $f(t)$ of Figure 10.17(a), the Fourier transform of which was found in Example 10.1 and is given by

$$F(j\omega) = V_0 a \, \frac{\sin \frac{1}{2}\omega a}{\frac{1}{2}\omega a} \tag{10.66}$$

A plot of $F(j\omega)$ is presented in Figure 10.17(b).

The symmetry property states that the time function

$$F(jt) = V_0 a \, \frac{\sin \frac{1}{2}at}{\frac{1}{2}at} \tag{10.67}$$

of Figure 10.18(a) has the Fourier transform

$$2\pi f(-\omega) = \begin{cases} 2\pi V_0, & |\omega| < \frac{1}{2}a \\ \pi V_0, & |\omega| = \frac{1}{2}a \\ 0, & |\omega| > \frac{1}{2}a \end{cases} \tag{10.68}$$

of Figure 10.18(b), which confirms the result of Example 10.2.

Time-Shifting

If a time signal $f(t)$ is delayed by t_0 sec, the delayed signal is $f(t - t_0)$. If $\mathbf{F}[f(t)] = F(j\omega)$, then

$$\mathbf{F}[f(t - t_0)] = F(j\omega)e^{-j\omega t_0} \tag{10.69}$$

Figure 10.18
Representation (a) of the
sinc function in the time
domain the Fourier
transform of which yields
the rectangular pulse in
the frequency domain as
shown in (b).

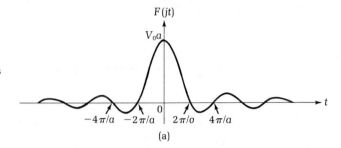

(a)

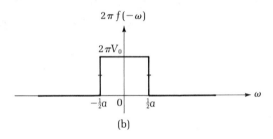

(b)

Proof. From definition (10.41) we have

$$\mathbf{F}[f(t - t_0)] = \int_{-\infty}^{\infty} f(t - t_0)e^{-j\omega t}\,dt = \int_{-\infty}^{\infty} f(x)e^{-j\omega(x + t_0)}\,dx$$

$$= e^{-j\omega t_0}\int_{-\infty}^{\infty} f(x)e^{-j\omega x}\,dx = e^{-j\omega t_0}F(j\omega) \qquad \textbf{(10.70)}$$

in which a change of variable was made with $x = t - t_0$.

Equation (10.69) states that if a signal is delayed by t_0 sec, its spectrum equals the original spectrum multiplied by $e^{-j\omega t_0}$. Thus the amplitude spectrum is not affected, but the phase of each frequency component is shifted by an amount equal to $-\omega t_0$.

EXAMPLE 10.5

We wish to compute the Fourier transform of the double gate function of Figure 10.19. The gate function g(t) can be expressed in terms of the rect-

Figure 10.19
The double gate function
in the time domain.

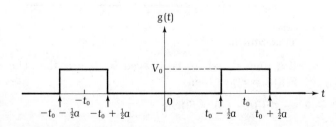

Figure 10.20
Appearance in the frequency domain of the Fourier transform of the double gate function shown in the time domain in Figure 10.19.

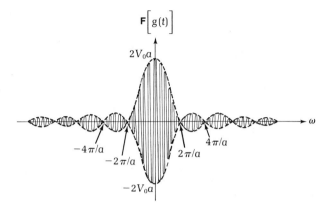

angular pulse $f(t)$ of Figure 10.17(a) by the relation

$$g(t) = f(t - t_0) + f(t + t_0) \tag{10.71}$$

By using the linearity property and applying (10.69) in conjunction with (10.66), we obtain the Fourier transform $G(j\omega)$ of $g(t)$, as follows:

$$\begin{aligned} G(j\omega) &= \mathbf{F}[f(t - t_0)] + \mathbf{F}[f(t + t_0)] \\ &= F(j\omega)e^{-j\omega t_0} + F(j\omega)e^{j\omega t_0} \\ &= V_0 a \frac{\sin \frac{1}{2}\omega a}{\frac{1}{2}\omega a} [e^{-j\omega t_0} + e^{j\omega t_0}] \\ &= 2V_0 a \frac{\sin \frac{1}{2}\omega a}{\frac{1}{2}\omega a} \cos \omega t_0 \end{aligned} \tag{10.72}$$

Equation (10.72) is plotted in Figure 10.20 as a function of ω.

Consider the time-domain signal having a functional form like that of (10.72)—namely,

$$G(jt) = 2V_0 a \frac{\sin \frac{1}{2}at}{\frac{1}{2}at} \cos \omega_0 t \tag{10.73}$$

where $\omega_0 = t_0$, which is obtained from (10.72) by replacing ω by t. The symmetry property (10.63) states that the Fourier transform of $G(jt)$ is given by

$$2\pi g(-\omega) = 2\pi f(-\omega - \omega_0) + 2\pi f(-\omega + \omega_0) \tag{10.74}$$

as depicted in Figure 10.21. This illustrates the symmetry property of the Fourier transform.

Figure 10.21
Fourier transform of the modulated sinc function in the time domain having functional form like that of Figure 10.20 and giving rise to the double gate function in the frequency domain, which illustrates the symmetry property of the Fourier transform.

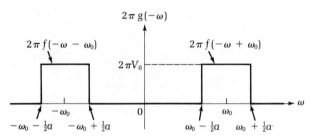

Frequency Shifting—Modulation

Frequency shifting is the displacement of a signal in the frequency domain and is often known as *modulation*. If $\mathbf{F}[f(t)] = F(j\omega)$, then

$$\mathbf{F}[f(t)e^{j\omega_0 t}] = F(j\omega - j\omega_0) \tag{10.75}$$

Proof.   From definition (10.41) we have

$$\mathbf{F}[f(t)e^{j\omega_0 t}] = \int_{-\infty}^{\infty} f(t)e^{j\omega_0 t}e^{-j\omega t}\, dt$$

$$= \int_{-\infty}^{\infty} f(t)e^{-j(\omega - \omega_0)t}\, dt = F(j\omega - j\omega_0) \tag{10.76}$$

The expression (10.75) is essential for understanding the process of modulation. Consider, for instance, the product signal

$$g(t) = f(t)\cos \omega_0 t \tag{10.77}$$

The signal $f(t)$ is called the *modulating signal* and the sinusoid $\cos \omega_0 t$ is the *carrier* or *modulated signal*. From (10.75), Euler's relations, and linearity, the Fourier transform of $g(t)$ is found to be

$$\begin{aligned}\mathbf{F}[g(t)] = \mathbf{F}[f(t)\cos \omega_0 t] &= \mathbf{F}[\tfrac{1}{2}f(t)e^{j\omega_0 t} + \tfrac{1}{2}f(t)e^{-j\omega_0 t}]\\ &= \tfrac{1}{2}\mathbf{F}[f(t)e^{j\omega_0 t}] + \tfrac{1}{2}\mathbf{F}[f(t)e^{-j\omega_0 t}]\\ &= \tfrac{1}{2}F(j\omega - j\omega_0) + \tfrac{1}{2}F(j\omega + j\omega_0)\end{aligned} \tag{10.78}$$

Relationship (10.78) states, in effect, that by multiplying the time function by $\cos \omega_0 t$, the original information spectrum $F(j\omega)$ is shifted so that half

Figure 10.22
Illustration (a) of a general nonperiodic information signal $f(t)$ and its Fourier transform $F(j\omega)$ in the frequency domain (b).

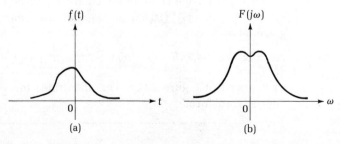

Figure 10.23
(a) Illustration of a carrier cos ωt modulated by the information signal $f(t)$ of Figure 10.22(a). (b) Appearance in the frequency domain of the time-domain modulated carrier signal given in (a).

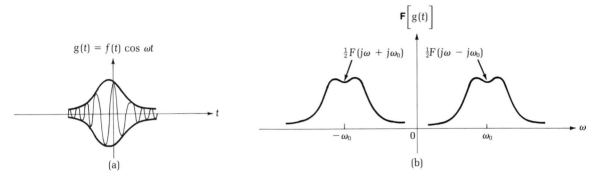

the original spectrum is centered about the carrier frequency ω_0 and the other half is centered about the mirror image $-\omega_0$.

As an example, we consider the time signal $f(t)$ of Figure 10.22(a), the Fourier transform $F(j\omega)$ of which is depicted in Figure 10.22(b). The modulated signal $g(t) = f(t) \cos \omega_0 t$ is shown in Figure 10.23(a) with its spectrum presented in Figure 10.23(b).

EXAMPLE 10.6

Consider the rectangular pulse $f(t)$ of Figure 10.24(a), having the Fourier transform, as computed in (10.22), given by sinc function

$$F(j\omega) = V_0 a \, \frac{\sin \frac{1}{2}\omega a}{\frac{1}{2}\omega a} \tag{10.79}$$

The modulated signal

$$g(t) = f(t) \cos \omega_0 t$$
$$= \begin{cases} \cos \omega_0 t, & |t| < \frac{1}{2}a \\ 0, & |t| > \frac{1}{2}a \end{cases} \tag{10.80}$$

Figure 10.24
Appearance (b) in the time domain of a carrier frequency cos $\omega_0 t$ modulated by the rectangular pulse signal given in (a).

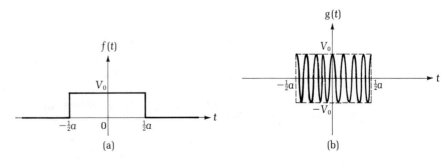

Figure 10.25
Frequency spectrum of the pulse modulated signal given in Figure 10.24(b).

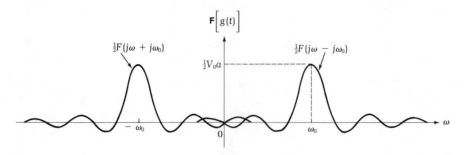

is shown in Figure 10.24(b). From (10.78) the Fourier transform of the modulated signal is

$$\mathbf{F}[g(t)] = \tfrac{1}{2}F(j\omega - j\omega_0) + \tfrac{1}{2}F(j\omega + j\omega_0)$$

$$= \frac{V_0 a}{2}\left[\frac{\sin\tfrac{1}{2}(\omega - \omega_0)a}{\tfrac{1}{2}(\omega - \omega_0)a} + \frac{\sin\tfrac{1}{2}(\omega + \omega_0)a}{\tfrac{1}{2}(\omega + \omega_0)a}\right]$$

$$= \tfrac{1}{2}V_0 a[\text{sinc}\,\tfrac{1}{2}(\omega - \omega_0)a + \text{sinc}\,\tfrac{1}{2}(\omega + \omega_0)a] \qquad \textbf{(10.81)}$$

The pulse spectrum $F(j\omega)$ was plotted earlier in Figure 10.9. Using this pulse spectrum, the amplitude spectrum of the modulated signal is depicted in Figure 10.25.

Time Differentiation and Integration

The Fourier transform is frequently used to solve linear integrodifferential equations arising in linear systems. In such applications transforms of the derivatives and/or integrals of a time function are needed. Thus if $\mathbf{F}[f(t)] = F(j\omega)$, then

$$\mathbf{F}[\dot{f}(t)] = j\omega F(j\omega) \qquad \textbf{(10.82)}$$

where, as before, $\dot{f}(t) = df(t)/dt$.

Proof. Differentiating both sides of (10.42) with respect to t gives

$$\dot{f}(t) = \frac{1}{2\pi} \cdot \frac{d}{dt} \int_{-\infty}^{\infty} F(j\omega)e^{j\omega t}\,d\omega$$

$$= \frac{1}{2\pi} \int_{-\infty}^{\infty} j\omega F(j\omega)e^{j\omega t}\,d\omega \qquad \textbf{(10.83)}$$

where an interchange of differentiation and integration was made. From this equation it is evident that

$$\mathbf{F}[\dot{f}(t)] = j\omega F(j\omega) \qquad \textbf{(10.84)}$$

The result can be extended to the nth derivative by repeated differentiations within the integral of (10.42):

$$\mathbf{F}\left[\frac{d^n f(t)}{dt^n}\right] = (j\omega)^n F(j\omega) \tag{10.85}$$

We next consider the integral of the function $f(t)$ defined by

$$h(t) = \int_{-\infty}^{t} f(x)\, dx \tag{10.86}$$

with Fourier transform $\mathbf{F}[h(t)] = H(j\omega)$. Since

$$f(t) = \dot{h}(t) \tag{10.87}$$

from (10.82) we have

$$F(j\omega) = j\omega H(j\omega) \tag{10.88}$$

giving

$$H(j\omega) = \frac{1}{j\omega} F(j\omega) \tag{10.89}$$

Note that in deriving (10.88) we implicitly assumed the existence of $H(j\omega)$. One sufficient condition is that

$$\lim_{t \to \infty} h(t) = 0 \tag{10.90}$$

or from (10.86),

$$\int_{-\infty}^{\infty} f(t)\, dt = 0 \tag{10.91}$$

This condition is equivalent to

$$F(0) = 0 \tag{10.92}$$

because from (10.41)

$$F(j\omega)\Big|_{\omega=0} = \int_{-\infty}^{\infty} f(t)\, dt \tag{10.93}$$

In the case that $F(0) \neq 0$, an impulse function is included in the transform of $h(t)$:

$$H(j\omega) = \frac{1}{j\omega} F(j\omega) + \pi F(0)\, \delta(\omega) \tag{10.94}$$

EXAMPLE 10.7
The triangular pulse of Figure 10.26(a) is defined by

$$
\begin{aligned}
f(t) &= 1 - \frac{|t|}{a}, & |t| < a \\
&= 0, & |t| > a
\end{aligned}
\tag{10.95}
$$

Figure 10.26

Illustration of a triangular pulse (a) in the time domain and its Fourier transform (b) in the frequency domain.

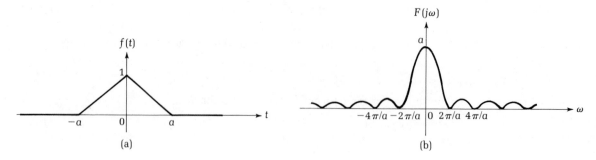

(a)   (b)

Figure 10.27

The derivative of the triangular pulse of Figure 10.26(a).

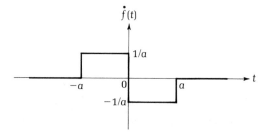

the Fourier transform of which is found to be (Problem 10.5)

$$F(j\omega) = a \, \frac{\sin^2 \frac{1}{2}\omega a}{(\frac{1}{2}\omega a)^2} = a \, \text{sinc}^2 \frac{1}{2}\omega a \qquad (10.96)$$

A plot of $F(j\omega)$ is presented in Figure 10.26(b).

The derivative of $f(t)$ is the rectangular pulse signal of Figure 10.27. Thus from (10.84) its Fourier transform is given by

$$\mathbf{F}[\dot{f}(t)] = j\omega F(j\omega) = j4 \, \frac{\sin^2 \frac{1}{2}\omega a}{\omega a} \qquad (10.97)$$

confirming (10.55) for the pulse signal of Figure 10.15.

Frequency Differentiation and Integration

Instead of taking the derivative of (10.42) with respect to time, we differentiate both sides of (10.41) with respect to the real frequency variable ω. We can show that if $\mathbf{F}[f(t)] = F(j\omega)$, then

$$\mathbf{F}[-jtf(t)] = \frac{dF(j\omega)}{d\omega} \qquad (10.98)$$

or, more generally,

$$\mathbf{F}[(-jt)^n f(t)] = \frac{d^n F(j\omega)}{d\omega^n} \tag{10.99}$$

Likewise within an additive constant (Problem 10.11)

$$\mathbf{F}\left[\frac{f(t)}{-jt}\right] = \int F(j\omega)\, d\omega \tag{10.100}$$

Reversal

We show that when a time function is reflected about the origin, its spectrum in the frequency domain will also be reflected about the origin. Mathematically the reversal property states that if $\mathbf{F}[f(t)] = F(j\omega)$, then

$$\mathbf{F}[f(-t)] = F(-j\omega) \tag{10.101}$$

To prove this we take conjugates on both sides of (10.41) and obtain

$$\bar{F}(j\omega) = F(-j\omega) = \int_{-\infty}^{\infty} f(t)e^{j\omega t}\, dt \tag{10.102}$$

Changing t to $-t$ yields

$$F(-j\omega) = \int_{-\infty}^{\infty} f(-t)e^{-j\omega t}\, dt = \mathbf{F}[f(-t)] \tag{10.103}$$

EXAMPLE 10.8

The Fourier transform of the exponential signal

$$f(t) = e^{-\alpha(t + t_0)}u(t + t_0) \tag{10.104}$$

of Figure 10.28(a) is found to be (Problem 10.15)

$$F(j\omega) = e^{j\omega t_0}\frac{1}{j\omega + \alpha} \tag{10.105}$$

By (10.101) the transform of the signal

$$f(-t) = e^{\alpha(t - t_0)}u(-t + t_0) \tag{10.106}$$

Figure 10.28
Illustration of an exponential function and its reversal (reflection) in the time domain.

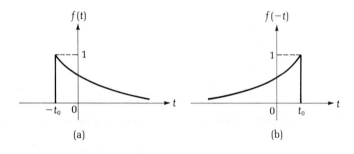

(a)   (b)

of Figure 10.28(b) is therefore

$$\mathbf{F}[f(-t)] = F(-j\omega) = -e^{-j\omega t_0}\frac{1}{j\omega - \alpha} \tag{10.107}$$

Observe that since $f(t)$ is neither even nor odd, its transform is complex, having real and imaginary parts given by

$$R(\omega) = \frac{\alpha \cos \omega t_0 + \omega \sin \omega t_0}{\alpha^2 + \omega^2} \tag{10.108a}$$

$$X(\omega) = \frac{\alpha \sin \omega t_0 - \omega \cos \omega t_0}{\alpha^2 + \omega^2} \tag{10.108b}$$

This is in contrast to other examples, where the transforms are either real or purely imaginary.

10-4   CONVOLUTION

As we showed in Chapter 7, convolution is a powerful means of characterizing the input-output relationship of a linear, time-invariant system. In this section we discuss the convolution integral in the context of the Fourier transform. There are two forms of convolution theorems that we will consider, one in the time domain and the other in the frequency domain.

Time-Domain Convolution

If $\mathbf{F}[f_1(t)] = F_1(j\omega)$ and $\mathbf{F}[f_2(t)] = F_2(j\omega)$, then

$$\mathbf{F}[f_1(t) * f_2(t)] = F_1(j\omega)F_2(j\omega) \tag{10.109}$$

where

$$f_1(t) * f_2(t) \triangleq \int_{-\infty}^{\infty} f_1(\tau)f_2(t-\tau)\,d\tau \tag{10.110}$$

This is called the *time convolution theorem*.

Proof. From the Fourier transform definition (10.41) and the bilateral convolution definition (7.92), we have

$$\mathbf{F}[f_1(t) * f_2(t)] = \int_{-\infty}^{\infty}\left[\int_{-\infty}^{\infty} f_1(\tau)f_2(t-\tau)\,d\tau\right]e^{-j\omega t}\,dt$$

$$= \int_{-\infty}^{\infty} f_1(\tau)\left[\int_{-\infty}^{\infty} f_2(t-\tau)e^{-j\omega t}\,dt\right]d\tau \tag{10.111}$$

Now let $x = t - \tau$. Then $dx = dt$ and $t = x + \tau$, and (10.111) becomes

$$\mathbf{F}[f_1(t) * f_2(t)] = \int_{-\infty}^{\infty} f_1(\tau)e^{-j\omega\tau}\,d\tau \int_{-\infty}^{\infty} f_2(x)e^{-j\omega x}\,dx$$

$$= F_1(j\omega)F_2(j\omega) \tag{10.112}$$

Figure 10.29
Two rectangular pulses the bilateral convolution of which is the triangular pulse given in Figure 10.30.

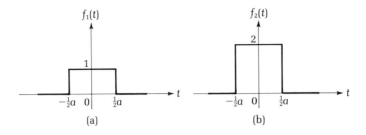

(a)

(b)

Figure 10.30
The triangular pulse that results from the bilateral convolution of the two pulses of Figure 10.29.

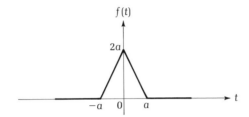

EXAMPLE 10.9

As shown in Example 7.11, the bilateral convolution of the two rectangular pulses of Figure 10.29 is the triangular pulse of Figure 10.30:

$$f(t) = f_1(t) * f_2(t) \tag{10.113}$$

From (10.22) the Fourier transforms of $f_1(t)$ and $f_2(t)$ are

$$F_1(j\omega) = 2 \, \frac{\sin \frac{1}{2}\omega a}{\omega} \tag{10.114}$$

$$F_2(j\omega) = 4 \, \frac{\sin \frac{1}{2}\omega a}{\omega} \tag{10.115}$$

Applying the convolution theorem (10.109), we obtain the Fourier transform of $f(t)$:

$$F(j\omega) = F_1(j\omega)F_2(j\omega) = 8 \, \frac{\sin^2 \frac{1}{2}\omega a}{\omega^2} \tag{10.116}$$

Frequency-Domain Convolution

If $\mathbf{F}[f_1(t)] = F_1(j\omega)$ and $\mathbf{F}[f_2(t)] = F_2(j\omega)$, then

$$\mathbf{F}[f_1(t)f_2(t)] = \frac{1}{2\pi} \, F_1(j\omega) * F_2(j\omega) \tag{10.117}$$

where

$$F_1(j\omega) * F_2(j\omega) \triangleq \int_{-\infty}^{-\infty} F_1(jy)F_2(j\omega - jy) \, dy \tag{10.118}$$

This is called the *frequency convolution theorem*.

To verify this we multiply both sides of the equation

$$f_1(t) = \frac{1}{2\pi} \int_{-\infty}^{\infty} F_1(jy)e^{jyt}\, dy \tag{10.119}$$

by $f_2(t)$ to yield

$$f_1(t)f_2(t) = \frac{1}{2\pi} \int_{-\infty}^{\infty} F_1(jy)f_2(t)e^{jyt}\, dy \tag{10.120}$$

Since from (10.75)

$$F_2(j\omega - jy) = \mathbf{F}[f_2(t)e^{jyt}] = \int_{-\infty}^{\infty} f_2(t)e^{-j(\omega - y)t}\, dt \tag{10.121}$$

the Fourier transform of (10.120) becomes

$$\mathbf{F}[f_1(t)f_2(t)] = \frac{1}{2\pi} \int_{-\infty}^{\infty} \left[\int_{-\infty}^{\infty} F_1(jy)f_2(t)e^{jyt}\, dy \right] e^{-j\omega t}\, dt$$

$$= \frac{1}{2\pi} \int_{-\infty}^{\infty} F_1(jy) \left[\int_{-\infty}^{\infty} f_2(t)e^{j(y - \omega)t}\, dt \right] dy$$

$$= \frac{1}{2\pi} \int_{-\infty}^{\infty} F_1(jy)F_2(j\omega - jy)\, dy$$

$$= \frac{1}{2\pi} F_1(j\omega) * F_2(j\omega) \tag{10.122}$$

EXAMPLE 10.10

We wish to express the transform of a modulated signal $g(t) = f(t)\cos \omega_0 t$ in terms of the transforms of $f(t)$ and $\cos \omega_0 t$.

As will be demonstrated in the following section, the Fourier transform of the carrier signal $\cos \omega_0 t$ is given by

$$\mathbf{F}[\cos \omega_0 t] = \pi[\delta(\omega - \omega_0) + \delta(\omega + \omega_0)] \tag{10.123}$$

Appealing to the frequency convolution theorem (10.117), we obtain

$$\mathbf{F}[g(t)] = \frac{1}{2} \int_{-\infty}^{\infty} [\delta(y - \omega_0) + \delta(y + \omega_0)]F(j\omega - jy)\, dy$$

$$= \tfrac{1}{2}F(j\omega - j\omega_0) + \tfrac{1}{2}F(j\omega + j\omega_0) \tag{10.124}$$

where $\mathbf{F}[f(t)] = F(j\omega)$. This also verifies (10.78).

Equation (10.117) is a shorthand notation of

$$\int_{-\infty}^{\infty} f_1(t)f_2(t)e^{-j\omega t}\, dt = \frac{1}{2\pi} \int_{-\infty}^{\infty} F_1(jy)F_2(j\omega - jy)\, dy \tag{10.125}$$

Setting $\omega = 0$ in (10.125) gives the *Parseval's formula*:[†]

$$\int_{-\infty}^{\infty} f_1(t)f_2(t)\, dt = \frac{1}{2\pi} \int_{-\infty}^{\infty} F_1(j\omega)\bar{F}_2(j\omega)\, d\omega \qquad (10.126)$$

where the dummy variable y was changed to ω.

In particular if $f_1(t) = f_2(t) = f(t)$, Parseval's formula reduces to the energy theorem:[‡]

$$\int_{-\infty}^{\infty} f^2(t)\, dt = \frac{1}{2\pi} \int_{-\infty}^{\infty} |F(j\omega)|^2 \, d\omega \qquad (10.127)$$

because the energy associated with a nonrecurring signal is defined by

$$E \triangleq \int_{-\infty}^{\infty} f^2(t)\, dt \qquad (10.128)$$

Equation (10.127) expresses the energy in $f(t)$ in terms of the continuous amplitude spectrum of $f(t)$. The energy of $f(t)$ is therefore given by the area under the curve $|F(j\omega)|^2/2\pi$. For this reason the term $|F(j\omega)|^2$ is referred to as the *energy spectrum* of $f(t)$. Since the function $|F(j\omega)|^2$ is algebraically real, and even for a real $f(t)$, the integral from $-\infty$ to ∞ on the right-hand side of (10.127) equals twice the integral from 0 to ∞:

$$E = \int_{-\infty}^{\infty} f^2(t)\, dt = \frac{1}{\pi} \int_{0}^{\infty} |F(j\omega)|^2 \, d\omega \qquad (10.129)$$

10-5   FOURIER TRANSFORMS OF SINGULARITY FUNCTIONS

The Fourier transform of the unit impulse is readily obtained by making use of the sampling property of the impulse:

$$\mathbf{F}[\delta(t)] = \int_{-\infty}^{\infty} \delta(t)e^{-j\omega t}\, dt = 1 \qquad (10.130)$$

or

$$\mathbf{F}[\delta(t)] = 1 \qquad (10.131)$$

Applying the time-shifting property (10.69), the transform of the shifted impulse $\delta(t - t_0)$ is given by

$$\mathbf{F}[\delta(t - t_0)] = e^{-j\omega t_0} \qquad (10.132)$$

Using the symmetry property (10.63), we have the transform of $\exp(-jtt_0)$:

$$\mathbf{F}[e^{-jtt_0}] = 2\pi\,\delta(-\omega - t_0) = 2\pi\,\delta(\omega + t_0) \qquad (10.133)$$

[†] Named after Marc Antoine Parseval-Deschènes (?–1836).
[‡] It is frequently called *Rayleigh's theorem* because it was first used by Lord Rayleigh in his study of blackbody radiation. This theorem is sometimes also referred to as *Plancherel's theorem* after M. Plancherel.

Figure 10.31
Illustration of the unit impulse (a) and constant (b) Fourier transform pair, which shows that the unit impulse has a uniform energy spectrum over all frequencies.

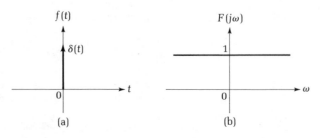

Figure 10.32
The converse of the Fourier transform pair of Figure 10.31, which shows that the frequency spectrum of a constant signal is concentrated at zero frequency.

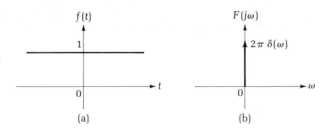

It is convenient to use ω_0 in the place of t_0, giving

$$\mathbf{F}[e^{-j\omega_0 t}] = 2\pi\, \delta(\omega + \omega_0) \tag{10.134}$$

In particular, for $\omega_0 = 0$, (10.134) reduces to

$$\mathbf{F}[1] = 2\pi\, \delta(\omega) \tag{10.135}$$

Thus the transform of the unit impulse is the constant 1, whereas the transform of the constant 1 is an impulse at the origin of area or strength 2π. The transform pairs are depicted in Figures 10.31 and 10.32. The results indicate that the unit impulse has a uniform frequency spectrum over the entire frequency range, in contrast to a dc signal, which has its entire frequency spectrum concentrated at the zero frequency. For an eternal exponential signal $\exp(-j\omega_0 t)$, it is expected from (10.134) that its entire frequency spectrum is concentrated at $\omega = -\omega_0$.

We now apply (10.134) to find the transforms of sinusoids. Since

$$\cos \omega_0 t = \tfrac{1}{2}[e^{j\omega_0 t} + e^{-j\omega_0 t}] \tag{10.136}$$

by linearity we have

$$\mathbf{F}[\cos \omega_0 t] = \pi[\delta(\omega - \omega_0) + \delta(\omega + \omega_0)] \tag{10.137}$$

The result is depicted in Figure 10.33. Likewise the transform of $\sin \omega_0 t$ is found to be (Problem 10.18)

$$\mathbf{F}[\sin \omega_0 t] = j\pi[\delta(\omega + \omega_0) - \delta(\omega - \omega_0)] \tag{10.138}$$

as illustrated in Figure 10.34. Thus the spectrum of a sinusoid is composed of two impulses, each of strength π or $\pm j\pi$ and located at $\pm\omega_0$.

Figure 10.33
Illustration of the cosinusoidal signal (a) in the time domain and its Fourier transform
(b) in the frequency domain.

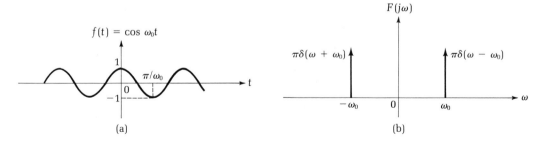

(a)   (b)

Figure 10.34
Illustration of the sinusoidal signal (a) in the time domain and its Fourier transform
(b) in the frequency domain.

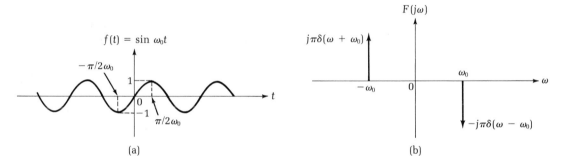

(a)   (b)

Another commonly used function is the *signum function* defined as

$$\operatorname{sgn} t = \begin{cases} -1, & t < 0 \\ 0, & t = 0 \\ 1, & t > 0 \end{cases}$$ **(10.139)**

as shown in Figure 10.35. Its derivative is given by

$$\frac{d}{dt} \operatorname{sgn} t = 2\delta(t)$$ **(10.140)**

Figure 10.35
Illustration of the signum
function.

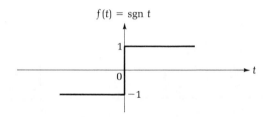

Taking the Fourier transform on both sides of (10.140) and using (10.82) gives

$$j\omega \mathbf{F}[\text{sgn } t] = 2\mathbf{F}[\delta(t)] = 2 \times 1 = 2 \qquad (10.141)$$

or

$$F(j\omega) = \mathbf{F}[\text{sgn } t] = \frac{2}{j\omega} \qquad (10.142)$$

Sketches of the amplitude and phase spectra in the frequency domain for sgn t are presented in Figure 10.36.

The unit step can be expressed in terms of the signum function by

$$u(t) = \tfrac{1}{2} + \tfrac{1}{2} \text{ sgn } t \qquad (10.143)$$

Thus the transform of $u(t)$ is given from (10.135) and (10.142) as

$$\mathbf{F}[u(t)] = \pi\delta(\omega) + \frac{1}{j\omega} \qquad (10.144)$$

Finally, we consider the Fourier transform of a periodic signal $f(t)$ with period T. This periodic signal can be expressed by a Fourier series as

$$f(t) = \sum_{n=-\infty}^{\infty} \tilde{c}_n e^{jn\omega_0 t} \qquad (10.145)$$

where $\omega_0 = 2\pi/T$. Taking the Fourier transform on each side yields

$$\mathbf{F}[f(t)] = \sum_{n=-\infty}^{\infty} \tilde{c}_n \mathbf{F}[e^{jn\omega_0 t}] = \sum_{n=-\infty}^{\infty} 2\pi\tilde{c}_n \,\delta(\omega - n\omega_0) \qquad (10.146)$$

or

$$\mathbf{F}[f(t)] = 2\pi \sum_{n=-\infty}^{\infty} \tilde{c}_n \,\delta(\omega - n\omega_0) \qquad (10.147)$$

In other words, the frequency spectrum of a periodic signal is not continuous but is comprised of discrete impulses located at the harmonic frequencies

Figure 10.36
Illustration in the frequency domain of the amplitude (a) and phase (b) spectra of the signum signal of Figure 10.35.

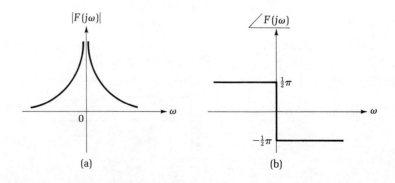

of the periodic signal. The strength of each impulse equals 2π times the corresponding Fourier coefficient. This is nothing more than an alternate representation of the Fourier series and the result hardly needs any further elaboration.

Table 10.1 summarizes many of the Fourier transform pairs used in engineering systems analysis and design.

Table 10.1 Fourier Transform Pairs Frequently Used in Signal Analysis

Time-domain function $f(t)$	Fourier transform $F(j\omega)$
$\delta(t)$	1
1	$2\pi\,\delta(\omega)$
$u(t)$	$\pi\,\delta(\omega) + \dfrac{1}{j\omega}$
sgn t	$\dfrac{2}{j\omega}$
$\cos\omega_0 t$	$\pi[\delta(\omega - \omega_0) + \delta(\omega + \omega_0)]$
$\sin\omega_0 t$	$j\pi[\delta(\omega + \omega_0) - \delta(\omega - \omega_0)]$
$e^{j\omega_0 t}$	$2\pi\,\delta(\omega - \omega_0)$
$e^{-\alpha t}(\cos\omega_0 t)u(t)$	$\dfrac{j\omega + \alpha}{(j\omega + \alpha)^2 + \omega_0^2}$
$e^{-\alpha t}(\sin\omega_0 t)u(t)$	$\dfrac{\omega_0}{(j\omega + \alpha) + \omega_0^2}$
$e^{-\alpha t}u(t)$	$\dfrac{1}{j\omega + \alpha}$
$e^{-\alpha|t|}$	$\dfrac{2\alpha}{\omega^2 + \alpha^2}$
$f(-t)$	$\bar{F}(j\omega)$
$f(at)$	$\dfrac{1}{|a|}F\left(\dfrac{j\omega}{a}\right)$
$f(t)e^{j\omega_0 t}$	$F(j\omega - j\omega_0)$
$2f(t)\cos\omega_0 t$	$F(j\omega - j\omega_0) + F(j\omega + j\omega_0)$
$\displaystyle\sum_{n=-\infty}^{\infty}\delta(t - nT)$	$\dfrac{2\pi}{T}\displaystyle\sum_{n=-\infty}^{\infty}\delta\left(\omega - \dfrac{2n\pi}{T}\right)$
$f_1(t) * f_2(t)$	$F_1(j\omega)F_2(j\omega)$
$f_1(t)f_2(t)$	$\dfrac{1}{2\pi}F_1(j\omega) * F_2(j\omega)$

Table 10.1   (continued)

Time-domain function $f(t)$	Fourier transform $F(j\omega)$
$f(t - t_0)$	$F(j\omega)e^{-j\omega t_0}$
$F(jt)$	$2\pi f(-\omega)$
$\dfrac{d^n f(t)}{dt^n}$	$(j\omega)^n F(j\omega)$
$(-jt)^n f(t)$	$\dfrac{d^n F(j\omega)}{d\omega^n}$
$\displaystyle\int_{-\infty}^{t} f(x)\,dx$	$\dfrac{1}{j\omega} F(j\omega) + \pi F(0)\,\delta(\omega)$
$-\dfrac{f(t)}{jt}$	$\displaystyle\int F(j\omega)\,d\omega$
$\displaystyle\int_{-\infty}^{\infty} f_1(t) f_2(t)\,dt$	$\dfrac{1}{2\pi}\displaystyle\int_{-\infty}^{\infty} F_1(j\omega)\overline{F}_2(j\omega)\,d\omega$
$\displaystyle\int_{-\infty}^{\infty} f^2(t)\,dt$	$\dfrac{1}{2\pi}\displaystyle\int_{-\infty}^{\infty} |F(j\omega)|^2\,d\omega$
$\displaystyle\sum_{n=-\infty}^{\infty} \tilde{c}_n e^{jn\omega_0 t}$	$2\pi \displaystyle\sum_{n=-\infty}^{\infty} \tilde{c}_n\,\delta(\omega - n\omega_0)$
$tu(t)$	$j\pi\,\delta'(\omega) - \dfrac{1}{\omega^2}$
$\dfrac{d^n\,\delta(t)}{dt^n}$	$(j\omega)^n$
t^n	$2\pi j^n \dfrac{d^n\,\delta(\omega)}{d\omega^n}$

10-6   APPLICATION TO SYSTEMS WITH ARBITRARY INPUTS

The Fourier transform is a special case of the more general bilateral Laplace transform (7.104) evaluated for $s = j\omega$. If $R(s)$ and $E(s)$ are the bilateral Laplace transforms of the response $r(t)$ and the excitation $e(t)$ of a system, then they are related by the system transfer function $H(s)$:

$$R(s) = H(s)E(s) \qquad \textbf{(10.148)}$$

Setting $s = j\omega$, we obtain[†]

$$R(j\omega) = H(j\omega)E(j\omega) \qquad \textbf{(10.149)}$$

[†] $R(j\omega)$ should not be confused with $R(\omega)$ of (10.44), which is the real part of $F(j\omega)$.

the inverse Fourier transform of which is found to be

$$r(t) = \frac{1}{2\pi} \int_{-\infty}^{\infty} H(j\omega)E(j\omega)e^{j\omega t}\, d\omega \qquad (10.150)$$

Equation (10.149) states that the response spectrum equals the product of a system-characterizing spectrum and the excitation spectrum. This conceptual interpretation has advantages in engineering applications.

EXAMPLE 10.11

In the RC network of Figure 10.37, let the input current be

$$i_g(t) = e^{-t}u(t) \qquad (10.151)$$

We wish to find the output voltage $v(t)$.

If $V(j\omega)$ and $I_g(j\omega)$ are the transforms of $v(t)$ and $i_g(t)$, their ratio is the transfer impedance given by

$$Z(j\omega) = \frac{V(j\omega)}{I_g(j\omega)} = \frac{j\omega + 1}{(j\omega + 2)(j\omega + 0.5)} \qquad (10.152)$$

giving

$$V(j\omega) = Z(j\omega)I_g(j\omega) = \frac{1}{(j\omega + 2)(j\omega + 0.5)} \qquad (10.153)$$

where from Table 10.1

$$I_g(j\omega) = \frac{1}{j\omega + 1} \qquad (10.154)$$

Using partial-fraction expansion, the inverse Fourier transform of (10.153) is found to be

$$v(t) = \mathbf{F}^{-1}\left[\frac{2}{3(j\omega + 0.5)} - \frac{2}{3(j\omega + 2)} \right]$$

$$= \frac{2}{3}(e^{-0.5t} - e^{-2t})u(t) \qquad (10.155)$$

A sketch of $r(t)$ is shown in Figure 10.38.

Figure 10.37
System transfer network.

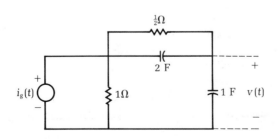

Figure 10.38

The response function of the transfer network of Figure 10.37 to the excitation signal $i_g(t) = e^{-t}u(t)$.

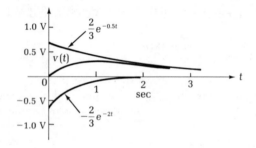

Figure 10.39

Frequency response (b) of an ideal bandpass filter (a).

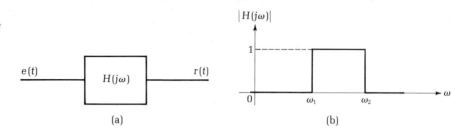

We next express the energy of the response $r(t)$ in terms of the system spectrum and the input spectrum. Applying the energy theorem (10.129) to $r(t)$ and using (10.149), we obtain[†]

$$\int_{-\infty}^{\infty} r^2(t)\, dt = \frac{1}{\pi} \int_0^{\infty} |H(j\omega)|^2\, |E(j\omega)|^2\, d\omega \tag{10.156}$$

Suppose that the system under consideration is an ideal bandpass filter having the frequency response $H(j\omega)$ shown in Figure 10.39. In such case (10.156) becomes

Figure 10.40

Representation of a one-port network with input impedance $Z(j\omega)$.

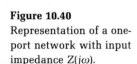

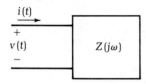

$$\int_{-\infty}^{\infty} r^2(t)\, dt = \frac{1}{\pi} \int_{\omega_1}^{\omega_2} |E(j\omega)|^2\, d\omega \tag{10.157}$$

which states that the energy of the response of the system equals the area under the energy spectrum $|E(j\omega)|^2$ of the excitation $e(t)$ in the frequency band from ω_1 to ω_2. In this sense the energy of $e(t)$ is *localized* on the frequency axis.

As an application of Parseval's formula (10.126), we compute the total energy delivered to a one-port network of Figure 10.40. Assume that the one-port network is characterized by its input impedance $Z(j\omega)$. With zero initial conditions, the transforms of the port voltage $v(t)$ and current $i(t)$ are related by

$$V(j\omega) = Z(j\omega)I(j\omega) \tag{10.158}$$

[†] The symbol $E(j\omega)$ should not be confused with the energy E associated with a nonrecurring signal as defined in (10.128).

From Parseval's formula (10.126) with $f_1(t) = v(t)$ and $f_2(t) = i(t)$, we obtain

$$\int_{-\infty}^{\infty} v(t)i(t)\,dt = \frac{1}{2\pi} \int_{-\infty}^{\infty} Z(j\omega)I(j\omega)\overline{I}(j\omega)\,d\omega$$

$$= \frac{1}{2\pi} \int_{-\infty}^{\infty} |I(j\omega)|^2 \operatorname{Re} Z(j\omega)\,d\omega$$

$$+ j\frac{1}{2\pi} \int_{-\infty}^{\infty} |I(j\omega)|^2 \operatorname{Im} Z(j\omega)\,d\omega$$

$$= \frac{1}{\pi} \int_{0}^{\infty} |I(j\omega)|^2 \operatorname{Re} Z(j\omega)\,d\omega \tag{10.159}$$

The integral involving $\operatorname{Im} Z(j\omega)$ is zero because the integrand is an odd function. The last equality follows because $|I(j\omega)|^2 \operatorname{Re} Z(j\omega)$ is an even function.

Consider, for example, the network of Figure 10.37. The impedance facing the current generator is found to be

$$Z(j\omega) = \frac{2 + j3\omega}{2 - 2\omega^2 + j5\omega} \tag{10.160}$$

For the input signal of (10.151), the total energy delivered to the network is determined from (10.159) as

$$\int_{-\infty}^{\infty} v(t)i(t)\,dt = \frac{1}{\pi} \int_{0}^{\infty} \frac{1}{\omega^2 + 1} \cdot \frac{11\omega^2 + 4}{4\omega^4 + 17\omega^2 + 4}\,d\omega \tag{10.161}$$

where the first term in the integrand is $|I(j\omega)|^2$ and the second term is the real part of $Z(j\omega)$.

10-6.1   Bandwidth and Pulse Duration

An *ideal* or *distortionless* signal transmission system is one in which the output waveform is the same as the input waveform except that the magnitude is scaled by a constant K and the waveform is delayed by t_d seconds, as depicted in Figure 10.41 with

$$r(t) = Ke(t - t_d) \tag{10.162}$$

If $K > 1$, the system is said to have *gain*, and if $K < 1$ the system has *loss* or *attenuation*. From the time-shifting relation (10.69), the transforms of the response and excitation are related by

$$R(j\omega) = KE(j\omega)e^{-j\omega t_d} \tag{10.163}$$

Figure 10.41
Representation of an ideal, distortionless signal transmission system.

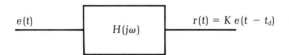

Figure 10.42
The frequency spectrum
of an ideal, distortionless
transmission system that
has a constant amplitude
spectrum (a) and a linear
phase spectrum (b).

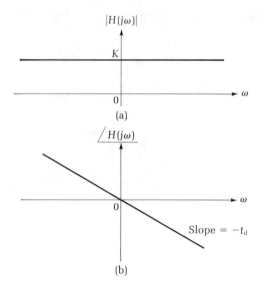

(a)

(b)

As a result the transfer function of such an ideal transmission system is

$$H(j\omega) = \frac{R(j\omega)}{E(j\omega)} = Ke^{-j\omega t_d} \tag{10.164}$$

Thus the amplitude spectrum of the ideal system is a constant and its phase
spectrum, being equal to $-\omega t_d$, is a linear function of ω, as depicted in
Figure 10.42. The *delay* of a transmission system is defined as the negative
of the slope of the phase function. For the ideal system therefore

$$\text{delay} = -\frac{d(-\omega t_d)}{d\omega} = t_d \tag{10.165}$$

showing that the delay is a constant independent of ω, as shown in Figure
10.43.

In practice it is impossible to design an ideal transmission system
because the response of all physical components diminishes at high frequencies. For this reason we attempt to design systems to have distortionless characteristics over a certain frequency band of interest. The range of

Figure 10.43
The delay versus fre-
quency of an ideal signal
transmission system.

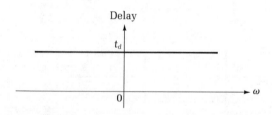

Figure 10.44
The bandwidth defined in terms of the amplitude spectrum of a transmission system.

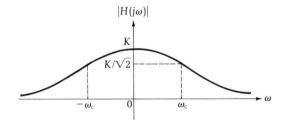

frequencies over which the magnitude of the transfer function $H(j\omega)$ remains within a ratio of $\sqrt{2}$, which corresponds to 3 dB, is called the *bandwidth* of the transmission system, as illustrated in Figure 10.44. This definition is somewhat arbitrary but is a widely accepted criterion of measuring a system's bandwidth. In Figure 10.44 the frequency ω_c is called the *cut-off frequency* and the system's bandwidth is ω_c. Because the amplitude spectrum $|H(j\omega)|$ is even, we generally do not consider $|H(j\omega)|$ for $\omega < 0$ and we plot $|H(j\omega)|$ only for $\omega > 0$.

The transmission characteristic for the network of Figure 10.37 is described by the transfer impedance

$$Z(j\omega) = \frac{V(j\omega)}{I_g(j\omega)} = \frac{j\omega + 1}{(j\omega + 2)(j\omega + 0.5)} \tag{10.166}$$

as given by (10.152), the amplitude and phase spectra of which are presented in Figure 10.45. To ascertain the bandwidth, we set

$$|Z(j\omega_c)|^2 = \tfrac{1}{2} \tag{10.167}$$

yielding

$$\omega_c^4 + 2.25\omega_c^2 - 1 = 0 \tag{10.168}$$

Figure 10.45
Amplitude (a) and phase (b) spectra of the transfer impedance of the transmission network of Figure 10.37.

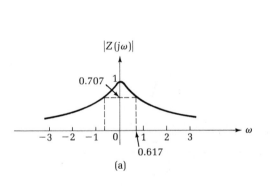

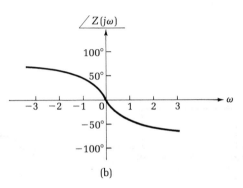

Figure 10.46
Amplitude (a) and frequency (b) spectra of an ideal low-pass filter.

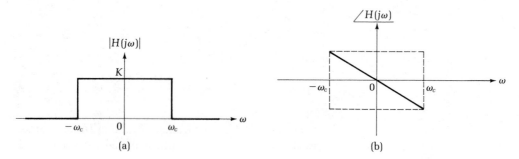

(a)   (b)

the solution of which gives the value of the bandwidth—namely,

$$\omega_c = 0.617 \text{ rad/s} \qquad\qquad (10.169)$$

The band-limited characteristic of a realizable physical system depends upon the particular system. For our purposes of studying the effect of bandwidth on pulse duration, we will consider the ideal *low-pass filter* having the frequency characteristics shown in Figure 10.46. The ideal low-pass filter does not exist in nature any more than the ideal transmission characteristics of Figure 10.42, but it can be used as a measure against which the performance of other physical systems may be compared. In the following we show that the pulses that pass through the ideal filter are distorted. The amount of distortion depends on both the bandwidth of the transmission system and the pulse width of the signal.

Consider the rectangular pulse of Figure 10.47(a), the spectrum of which was found earlier in Example 10.1 and is given by

$$F(j\omega) = V_0 a \, \frac{\sin \frac{1}{2}\omega a}{\frac{1}{2}\omega a} \qquad\qquad (10.170)$$

Figure 10.47
A rectangular pulse (a) of duration a and its amplitude spectrum (b) the first zero of which occurs at $\omega = 2\pi/a$.

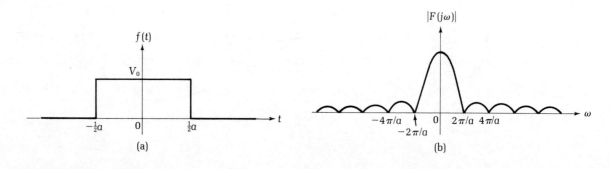

(a)   (b)

The amplitude spectrum is shown in Figure 10.47(b). The first zero of this equation occurs when $\frac{1}{2}\omega a = \pi$ or

$$\omega_1 = \frac{2\pi}{a} \tag{10.171}$$

as indicated in Figure 10.47(b). Thus the width of the main feature in the frequency spectrum is $2\pi/a$. This width is important in that it includes frequency components with relatively large values. Taking $\omega_1 = 2\pi/a$ as the definition of bandwidth, then the product of the bandwidth and the pulse width is

$$\omega_1 a = 2\pi \tag{10.172}$$

a constant—namely, 2π.

We consider next the triangular pulse of Figure 10.48(a). From Example 10.7, the frequency spectrum of this pulse is found to be

$$F(j\omega) = V_0 a \frac{\sin^2 \frac{1}{2}\omega a}{(\frac{1}{2}\omega a)^2} \tag{10.173}$$

Its amplitude spectrum is shown in Figure 10.48(b). The first zero of $F(j\omega)$ occurs when $\frac{1}{2}\omega a = \pi$, or

$$\omega_1 = \frac{2\pi}{a} \tag{10.174}$$

This again shows that the product of the bandwidth and the pulse width is a constant:

$$\omega_1 2a = 4\pi \tag{10.175}$$

As a final example, let us consider the exponential pulse

$$f(t) = V_0 e^{-at} u(t) \tag{10.176}$$

Figure 10.48
The triangular pulse (a) of duration $2a$ and its amplitude spectrum (b) the first zero of which occurs at $\omega = 2\pi/a$.

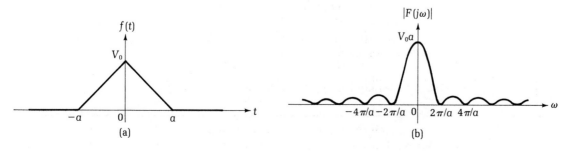

Figure 10.49
An exponential pulse (a) of time constant $1/a$ and its amplitude spectrum (b), which
exhibits a 3 dB bandwidth of a.

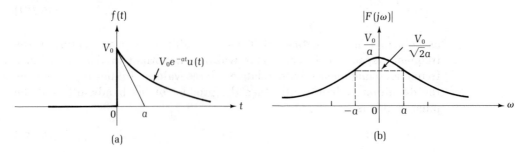

as shown in Figure 10.49(a). From Table 10.1, its transform is given by

$$F(j\omega) = \frac{V_0}{j\omega + a} \tag{10.177}$$

The amplitude spectrum of the signal is presented in Figure 10.49(b). In
this case neither $f(t)$ nor $F(j\omega)$ vanishes. The cut-off frequency ω_c is the
frequency at which $|F(j\omega_c)| = F(0)/\sqrt{2}$, yielding

$$\omega_c = a \tag{10.178}$$

For a measure of the width of the pulse, we use the *time constant*, which
is defined as the time interval for an exponential function to decrease to
37 percent of its initial value or more precisely $1/e$ percent. From (10.176)
we see that the time constant is $1/a$. The product of the pulse width (which
is taken to be $1/a$) and the bandwidth is again a constant:

$$\omega_c \left(\frac{1}{a} \right) = 1 \tag{10.179}$$

From these examples our conclusion is that the bandwidth required for
transmission is inversely proportional to the duration of the pulse being
transmitted. In other words,

(Bandwidth required) × (pulse duration) = constant   (10.180)

We see that relation (10.180) is satisfied even in the limiting case of
transmitting an impulse of zero width. The bandwidth required must be
infinite because the spectrum of the unit impulse is unity, as shown in
Figure 10.31(b), meaning that it contains all frequencies with equal amplitudes and has infinite bandwidth.

10-6.2   Bandwidth and Rise Time

In this section we study the relationship between bandwidth and rise
time. Specifically, we apply a step input to a band-limited system and
observe the rate of rise of the output. We will show that the larger the

bandwidth a system has, the faster the output will rise. This is reasonable because a step input contains a jump discontinuity. This discontinuity is reflected in its spectrum in that it contains very high frequency components. As a result if these high frequency components are suppressed because of small bandwidth, the output will no longer show a jump discontinuity but will exhibit a gradual rise in the output.

Consider again the ideal low-pass filter characteristics of Figure 10.46. As before, let e(t) be the input excitation and r(t) the response. Their Fourier transforms are related by the equation

$$R(j\omega) = H(j\omega)E(j\omega) \tag{10.181}$$

If $e(t) = \delta(t)$, then $E(j\omega) = 1$ from Table 10.1. The impulse response is obtained by computing the corresponding inverse Fourier transform:

$$r_\delta(t) = \frac{1}{2\pi} \int_{-\infty}^{\infty} H(j\omega)e^{j\omega t}\, d\omega \tag{10.182}$$

where from Figure 10.46

$$H(j\omega) = \begin{cases} Ke^{-j\omega t_d}, & |\omega| < \omega_c \\ 0, & |\omega| > \omega_c \end{cases} \tag{10.183}$$

Substituting this in (10.182) yields

$$r_\delta(t) = \frac{K}{2\pi} \int_{-\omega_c}^{\omega_c} e^{j\omega(t - t_d)}\, d\omega = \frac{K\omega_c}{\pi} \cdot \frac{\sin \omega_c(t - t_d)}{\omega_c(t - t_d)} \tag{10.184}$$

The maximum value of this response $r_\delta(t)$ occurs when $\omega_c(t - t_d) = 0$ or $t = t_d$, giving

$$r_\delta(t_d) = \frac{K\omega_c}{\pi} \tag{10.185}$$

A sketch of the impulse response is presented in Figure 10.50. Observe that the peak value of the response, as given by (10.185), is proportional to the cut-off frequency ω_c, and the width of the main feature is $2\pi/\omega_c$. As ω_c approaches infinity, this width approaches zero and the peak value approaches infinity. In other words, the output response approaches the input, an impulse. Also observe that the impulse response is nonzero for

Figure 10.50
Impulse response of the ideal low-pass filter of Figure 10.46.

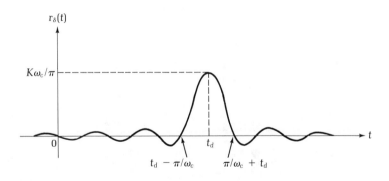

Figure 10.51
The function Si y $\triangleq$
$\int_0^y$ sinc x dx.

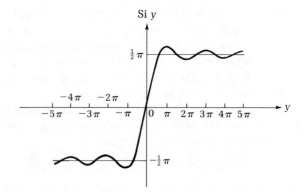

$t < 0$. This is strange in that the system anticipates the input even before the input is applied. The reason is that the ideal low-pass filter characteristic of Figure 10.46 is not realizable by physical components. In real systems causality ensures that effect follows cause.

To find the step response of the ideal low-pass filter, we integrate the impulse response of (10.184) and obtain

$$r_u(t) = \int_{-\infty}^t r_\delta(\tau)\, d\tau = \frac{K\omega_c}{\pi} \int_{-\infty}^t \frac{\sin \omega_c(\tau - t_d)}{\omega_c(\tau - t_d)}\, d\tau$$

$$= \frac{K}{\pi} \int_{-\infty}^{\omega_c(t-t_d)} \frac{\sin x}{x}\, dx \qquad (10.186)$$

The last integral is a tabulated function known as the *sine integral function* and is denoted by the symbol[†]

$$\text{Si } y = \int_0^y \frac{\sin x}{x}\, dx \qquad (10.187)$$

which possesses the following properties:

(i) $\text{Si}(-y) = -\text{Si } y$

(ii) $\text{Si } 0 = 0$

(iii) $\text{Si } \infty = \dfrac{\pi}{2}$ and $\text{Si}(-\infty) = -\dfrac{\pi}{2}$

A sketch of Si y is presented in Figure 10.51. Using this, (10.186) can be written as

$$r_u(t) = \frac{K}{\pi}\left[\int_{-\infty}^0 \frac{\sin x}{x}\, dx + \int_0^{\omega_c(t-t_d)} \frac{\sin x}{x}\, dx \right]$$

$$= \frac{K}{\pi}\left[\frac{\pi}{2} + \text{Si } \omega_c(t - t_d) \right] \qquad (10.188)$$

A plot of the unit step response $r_u(t)$ is shown in Figure 10.52.

[†] See, for example, E. A. Guillemin, *The Mathematics of Circuit Analysis* (New York: Wiley, 1949), pp. 491–96.

Figure 10.52
The response of an ideal low-pass filter to a unit step input. The rise time is t_r.

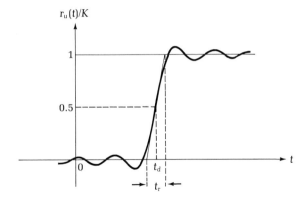

Due to limited bandwidth, we again observe distortion in the step response. Also, the response is nonzero for $t < 0$. As the cut-off frequency ω_c approaches infinity, the response becomes

$$r_u(t) = \begin{cases} \dfrac{K}{\pi}\left(\dfrac{\pi}{2} + \text{Si} \infty\right) = K, & t > t_d \\[3mm] \dfrac{K}{\pi}\left(\dfrac{\pi}{2} + \text{Si}(-\infty)\right) = 0, & t < t_d \end{cases} \qquad \textbf{(10.189)}$$

showing that the step response $r_u(t)$ approaches a delayed step $Ku(t - t_d)$ as the bandwidth ω_c approaches infinity.

For a limited bandwidth, there is a gradual rise of the output as shown in Figure 10.52. The rate of this buildup of the response can be related to cut-off frequency ω_c. Define the *rise time* t_r as the time interval between the intercepts of the tangent at $t = t_d$ with the lines $r_u(t) = 0$ and $r_u(t) = K$, as indicated in Figure 10.52. From (10.188) we have

$$\frac{dr_u(t)}{dt} = \frac{d}{dt}\left[\frac{K}{\pi} \text{Si } \omega_c(t - t_d)\right] = r_\delta(t) \qquad \textbf{(10.190)}$$

which when evaluated at $t = t_d$ gives

$$\frac{dr_u(t)}{dt}\bigg|_{t=t_d} = \frac{K}{t_r} = r_\delta(t_d) \qquad \textbf{(10.191)}$$

Since $(\sin x)/x = 1$ for $x = 0$, we obtain from (10.184) or (8.185)

$$t_r = \frac{\pi}{\omega_c} \qquad \textbf{(10.192)}$$

Rearranging,

$$\omega_c t_r = \pi \qquad \textbf{(10.193)}$$

In other words, the rise time of the response of a band-limited system is inversely proportional to the bandwidth of the system:

$$(\text{Bandwidth}) \times (\text{rise time}) = \text{constant} \qquad \textbf{(10.194)}$$

Figure 10.53
The rise time t_r' of the
unit-step response of an
ideal low-pass filter, which
is defined as the time
required for the response
to rise from 10 percent to
90 percent of the final
value.

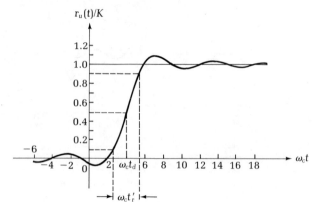

Frequently the *rise time* t_r' is also defined as the time required for the response to rise from 10 percent to 90 percent of the final value. In that case we plot the normalized response $r_u(t)/K$ as a function of the normalized time $\omega_c t$. The result is shown in Figure 10.53, from which we read that

$$\omega_c t_r' = 2.8 \qquad\qquad\qquad (10.195)$$

again showing that the product of bandwidth and rise time is a constant.

10-7 THE SAMPLING THEOREM

In this section we demonstrate that if a signal is band limited, then the complete information about the signal is contained in samples of it spaced uniformly at a distance not greater than a certain value determined by the bandwidth.

The Sampling Theorem *Given a signal f(t) the Fourier transform $F(j\omega)$ of which vanishes for $|\omega|$ larger than some constant ω_c,*

$$\left|F(j\omega)\right| = 0, \qquad |\omega| > \omega_c \qquad\qquad (10.196)$$

then

$$f(t) = \sum_{n=-\infty}^{\infty} f(nT)\,\frac{\sin \omega_c(t - nT)}{\omega_c(t - nT)} \qquad\qquad (10.197)$$

where

$$T = \frac{\pi}{\omega_c} \qquad\qquad\qquad (10.198)$$

Figure 10.54
Illustration of a signal (a)
with a band-limited
spectrum (b).

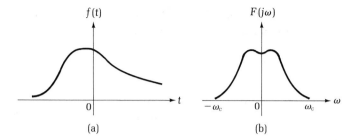

(a)

(b)

Figure 10.55
Illustration of a periodic sequence of unit impulses (a) of period $\hat{T}$ and its Fourier
transform (b).

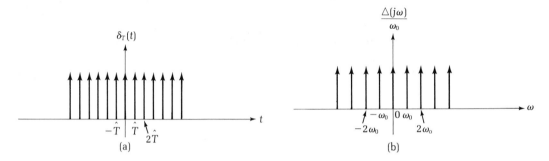

(a)

(b)

The number π/ω_c is known as the *Nyquist interval*,[†] and its reciprocal
ω_c/π is called the *Nyquist rate*.

Proof   The signal $f(t)$, with its band-limited spectrum $F(j\omega)$, is illustrated
in Figure 10.54. This signal will be multiplied by a sequence of unit impulses
located at regular intervals of $\hat{T}$ seconds, as depicted in Figure 10.55(a), to
yield a sequence of samples as input. The pulse train can be written as

$$\delta_T(t) = \sum_{n=-\infty}^{\infty} \delta(t - n\hat{T})$$ (10.199)

which is periodic with period $\hat{T}$. This periodic pulse train can be represented
by a Fourier series the Fourier coefficients of which are all equal:

$$\tilde{c}_n = \frac{1}{\hat{T}} \int_{-\hat{T}/2}^{\hat{T}/2} \delta(t) e^{-jn\omega_0 t}\, dt = \frac{1}{\hat{T}}$$ (10.200)

where $\omega_0 = 2\pi/\hat{T}$, giving

$$\delta_T(t) = \sum_{n=-\infty}^{\infty} \delta(t - n\hat{T}) = \frac{1}{\hat{T}} \sum_{n=-\infty}^{\infty} e^{jn\omega_0 t}$$ (10.201)

[†] Named for the Bell Laboratories electrical engineer Harry Nyquist (1884–   ).

From (10.147) for the Fourier transform of a periodic signal, the Fourier transform of $\delta_T(t)$ is found to be

$$\Delta(j\omega) = \frac{2\pi}{\hat{T}} \sum_{n=-\infty}^{\infty} \delta(\omega - n\omega_0) \tag{10.202}$$

as shown in Figure 10.55(b).

We next consider the product of the functions $f(t)$ and $\delta_T(t)$, as denoted by

$$f_s(t) \triangleq f(t)\,\delta_T(t) = \sum_{n=-\infty}^{\infty} f(n\hat{T})\,\delta(t - n\hat{T}) \tag{10.203}$$

This product signal is a sequence of impulses located at regular intervals of $\hat{T}$ seconds and having strengths equal to values of $f(t)$ at the corresponding instants, as illustrated in Figure 10.56(a). Thus it represents the signal $f(t)$ sampled every $\hat{T}$ seconds. Applying the frequency convolution theorem (10.117) to $f(t)\,\delta_T(t)$, the Fourier transform of $f_s(t)$ is given by

$$F_s(j\omega) = \frac{1}{2\pi} F(j\omega) * \Delta(j\omega) \tag{10.204}$$

Since

$$F(j\omega) * \delta(\omega - n\omega_0) = F(j\omega - jn\omega_0) \tag{10.205}$$

Figure 10.56
The Fourier transform (b) of the sampled signal $f_s(t)$ of (a).

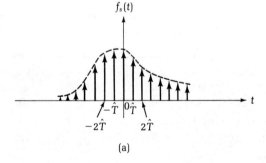

(a)

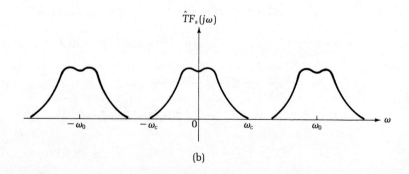

(b)

then using (10.202), (10.204) becomes

$$F_s(j\omega) = \frac{1}{\hat{T}} \sum_{n=-\infty}^{\infty} F(j\omega - jn\omega_0) \tag{10.206}$$

as shown in Figure 10.56(b). Therefore if $|F(j\omega)| = 0$ for $|\omega| > \omega_c$ holds and if $\omega_0 \geqq 2\omega_c$,

$$\hat{T}F_s(j\omega) = F(j\omega), \qquad |\omega| < \omega_c \tag{10.207}$$

In other words, if $\omega_0 \geqq 2\omega_c$, $F(j\omega)$ will repeat in the spectrum $\hat{T}F_s(j\omega)$ without overlap. Or, equivalently, as long as the samples are taken no more than $\hat{T}$ seconds apart, where

$$\hat{T} \leqq \frac{\pi}{\omega_c} \tag{10.208}$$

$F_s(j\omega)$ will be periodic replica of $F(j\omega)$. Notice that there is a factor $1/\hat{T}$ associated with the spectrum $F_s(j\omega)$.

To recover $f(t)$ from its samples $f_s(t)$, we pass $f_s(t)$ through an ideal low-pass filter, the frequency characteristic of which is described by

$$H(j\omega) = \begin{cases} 1, & |\omega| < \omega_c \\ 0, & |\omega| > \omega_c \end{cases} \tag{10.209}$$

Thus for $\omega_0 \geqq 2\omega_c$ the output of the filter is $F(j\omega)/\hat{T}$, which also equals

$$\frac{F(j\omega)}{\hat{T}} = F_s(j\omega)H(j\omega) \tag{10.210}$$

the inverse Fourier transform of which is obtained through the time convolution theorem (10.109):

$$f(t) = \hat{T}f_s(t) * h(t) \tag{10.211}$$

where

$$h(t) = \mathbf{F}^{-1}[H(j\omega)] = \frac{1}{2\pi} \int_{-\omega_c}^{\omega_c} e^{j\omega t}\, d\omega = \frac{\sin \omega_c t}{\pi t} \tag{10.212}$$

is the impulse response of the filter. Substituting (10.203) and (10.212) in (10.211) gives

$$f(t) = \frac{\hat{T}}{\pi} \sum_{n=-\infty}^{\infty} f(n\hat{T})\, \delta(t - n\hat{T}) * \left(\frac{\sin \omega_c t}{t} \right)$$

$$= \hat{T} \sum_{n=-\infty}^{\infty} f(n\hat{T}) \frac{\sin \omega_c(t - n\hat{T})}{\pi(t - n\hat{T})} \tag{10.213}$$

Choosing $\hat{T} = \pi/\omega_c$, (10.213) becomes (10.197).

Our conclusion is that as long as the signal $f(t)$ is sampled at a rate not less than the Nyquist rate, the signal can be completely recovered from its samples. The Nyquist rate, defined as

$$\frac{1}{T} = \frac{\omega_c}{\pi} = \frac{2\pi f_c}{\pi} = 2f_c \tag{10.214}$$

equals twice the highest frequency f_c in Hz present in the spectrum of $f(t)$. In other words, a band-limited signal having no spectral components above a frequency f_c is uniquely determined by its samples taken at uniform intervals at a rate at least twice during each period or cycle of its highest frequency component f_c.

The formula (10.197) can be viewed as a method of interpolation because it allows one to find the signal $f(t)$ by interpolation among the sampled values $f(nT)$. Graphically the result is presented in Figure 10.57(b). For $n = 0$, 1, and 2, the corresponding weighted plots of the terms in the sum-

Figure 10.57

Superposition (b) of the weighted plots (a) of the terms of Eq. (10.197) to yield sample points of the signal $f(t)$ of Figure 10.54(a).

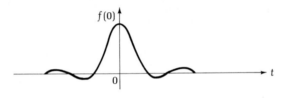

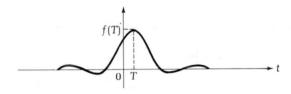

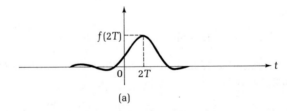

(a)

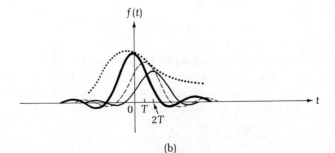

(b)

mation are sketched in Figure 10.57(a). Superimposing these plots yields the signal $f(t)$ as shown in Figure 10.57(b).

We emphasize that the formula is valid only for band-limited signals. It cannot be used in the representation of a finite-duration signal because the spectrum of the latter contains all frequencies. To see this, let $f(t)$ be a finite duration signal with

$$f(t) = 0, \qquad |t| > \tfrac{1}{2}T \tag{10.215}$$

Let

$$g(t) = \begin{cases} 1, & |t| < \tfrac{1}{2}T \\ 0, & |t| > \tfrac{1}{2}T \end{cases} \tag{10.216}$$

Then we can write

$$f(t) = g(t)f(t) \tag{10.217}$$

Applying the frequency-domain convolution (10.117), the spectrum of $f(t)$ is found from (10.22) to be

$$F(j\omega) = \frac{T}{2\pi} \left(\frac{\sin \tfrac{1}{2}\omega T}{\tfrac{1}{2}\omega T} \right) * F(j\omega) \tag{10.218}$$

No matter what the form is for $F(j\omega)$, the convolution on the right-hand side of (10.218) will yield a spectrum $F(j\omega)$ that exists for all ω. Nevertheless the formula can be used as an approximation for a finite-duration signal.

10-8   NUMERICAL COMPUTATION OF FOURIER TRANSFORM

In the present section, we consider the problem of evaluating numerically the Fourier transform $F(j\omega)$ of a given signal $f(t)$. We shall show that the problem can be reduced to the discrete Fourier series problem (DFS) involving the computational economy of the fast Fourier transform (FFT) algorithm discussed in the preceding chapter.

Given a signal $f(t)$ and its Fourier transform $F(j\omega)$, as shown in Figure 10.58, a periodic signal of period T is formed by adding to $f(t)$ all its

Figure 10.58
Illustration of a representative signal (a) and its Fourier transform (b).

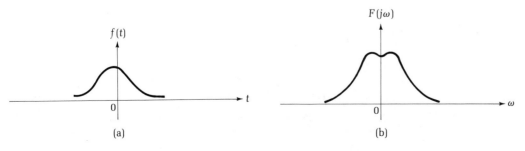

$f(t)$

$F(j\omega)$

0

0

t

ω

(a)

(b)

Figure 10.59
Formation of a periodic
signal of period T by
adding to $f(t)$ of Figure
10.58(a) all of its
displacements.

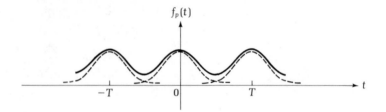

displacements

$$f_p(t) = \sum_{n=-\infty}^{\infty} f(t + nT) \tag{10.219}$$

as depicted in Figure 10.59, where T is an arbitrary number. We first show
that

$$f_p(t) = \sum_{n=-\infty}^{\infty} f(t + nT) = \frac{1}{T} \sum_{n=-\infty}^{\infty} F(jn\omega_0)e^{jn\omega_0 t} \tag{10.220}$$

The identity is known as the *Poisson sum formula*[†] and is important in
Fourier analysis.

To verify this we consider a system with input and output transforms
related by $F(j\omega)$:

$$R(j\omega) = F(j\omega)E(j\omega) \tag{10.221}$$

For the Poisson sum formula, the system need not be physically realizable.
Appealing to the time-domain convolution theorem (10.109), the response of
the system to the excitation $e(t)$ is given by

$$r(t) = f(t) * e(t) \tag{10.222}$$

Thus if a periodic sequence of unit impulses

$$g(t) = \sum_{n=-\infty}^{\infty} \delta(t + nT) \tag{10.223}$$

is applied to the system, the corresponding response is

$$f(t) * g(t) = \sum_{n=-\infty}^{\infty} f(t) * \delta(t + nT) = \sum_{n=-\infty}^{\infty} f(t + nT) \tag{10.224}$$

Since $g(t)$ is periodic with period T, its Fourier series is found to be

$$g(t) = \sum_{n=-\infty}^{\infty} \delta(t + nT) = \frac{1}{T} \sum_{n=-\infty}^{\infty} e^{jn\omega_0 t} \tag{10.225}$$

where $\omega_0 = 2\pi/T$, and the Fourier coefficients are

$$\tilde{c}_n = \frac{1}{T} \int_{-T/2}^{T/2} \delta(t)e^{-jn\omega_0 t}\, dt = \frac{1}{T} \tag{10.226}$$

[†] Named for the French mathematical physicist Siméon Denis Poisson (1781–1840).

Suppose that we use the Fourier series representation of $g(t)$ in (10.224) instead of (10.223). The same response can be expressed as

$$f(t) * g(t) = \sum_{n=-\infty}^{\infty} f(t) * \left(\frac{1}{T} e^{jn\omega_0 t} \right) = \frac{1}{T} \sum_{n=-\infty}^{\infty} F(jn\omega_0) e^{jn\omega_0 t} \qquad (10.227)$$

The last equality is obtained by appealing again to the time-domain convolution theorem (10.109). Equating (10.224) and (10.227) yields (10.220).

We note that if $f(t) = 0$ for $|t| > \frac{1}{2}T$, as depicted in Figure 10.60(a), then the repeated versions of $f(t)$ that make up $f_p(t)$ do not overlap, as indicated in Figure 10.60(b). The signal $f(t)$ is simply the segment of $f_p(t)$ in the time interval $|t| < \frac{1}{2}T$. The right-hand side of (10.220) becomes the Fourier series of $f_p(t)$ because under the stipulated condition the Fourier coefficients are given by

$$\tilde{c}_n = \frac{1}{T} \int_{-T/2}^{T/2} f(t) e^{-jn\omega_0 t} \, dt = \frac{1}{T} F(jn\omega_0) \qquad (10.228)$$

In the case $f(t) \neq 0$ for $|t| > \frac{1}{2}T$, the integral in (10.228) does not hold. Nevertheless formula (10.220) remains valid for general signals, which may either be of finite duration or infinite duration.

Our conclusion is that given a signal $f(t)$, we form a periodic signal $f_p(t)$ as in (10.219) with period T. The Fourier series representation of $f_p(t)$ can be written as

$$f_p(t) = \sum_{n=-\infty}^{\infty} \frac{1}{T} F(jn\omega_0) e^{jn\omega_0 t} \qquad (10.229)$$

Thus the Fourier series coefficients $\tilde{c}_n$ of the signal $f_p(t)$ equals $F(jn\omega_0)/T$ or

$$\tilde{c}_n = \frac{1}{T} F(jn\omega_0) \qquad (10.230)$$

To compute $F(jn\omega_0)$, we form the N samples $f_p(0), f_p(T_0), \ldots, f_p(NT_0 - T_0)$ of $f_p(t)$ in the interval from 0 to T, where $T_0 = T/N$ is the sampling interval. As demonstrated in Section 9-4.2, the samples $f_p(mT_0)$ of $f_p(t)$ and the complex

Figure 10.60
Illustration of a time-limited signal (a) with repeated versions (b) that do not overlap.

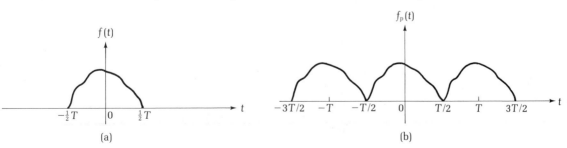

conjugate $\bar{\hat{c}}_n$ of its aliased coefficients

$$\hat{c}_n = \sum_{k=-\infty}^{\infty} \tilde{c}_{n+kN} \tag{10.231}$$

form a DFS pair with $\bar{c}_n$ being the inverse DFS of $f_p(mT_0)$. From (10.230) we have

$$\hat{c}_n = \frac{1}{T} \sum_{k=-\infty}^{\infty} F[j(n+kN)\omega_0] \tag{10.232}$$

with $\omega_0 = 2\pi/T$.

For our purposes we next introduce the new frequency variable

$$\gamma = \frac{\pi}{T_0} = \frac{N\pi}{T} = \frac{N\omega_0}{2} \tag{10.233}$$

Thus $N\omega_0 = 2\gamma$ and (10.232) becomes

$$\hat{c}_n = \frac{1}{T} \sum_{k=-\infty}^{\infty} F(jn\omega_0 + j2k\gamma) \tag{10.234}$$

Define

$$F_p(j\omega) \triangleq \sum_{k=-\infty}^{\infty} F(j\omega + j2k\gamma) \tag{10.235}$$

This function is formed by adding to $F(j\omega)$ all of its displacements, as shown in Figure 10.61, and therefore is periodic with period 2γ. Using this (10.234) can be rewritten as

$$\hat{c}_n = \frac{1}{T} F_p(jn\omega_0) \tag{10.236}$$

showing that the samples of $F_p(j\omega)/T$ equal the aliased coefficients of $f_p(t)$. As a result we conclude that the samples $f_p(mT_0)$ of $f_p(t)$ and the samples $\bar{F}_p(jm\omega_0)/T$ of the periodic function $\bar{F}_p(j\omega)/T$ form a DFS pair. This is the basis for the evaluation of the Fourier integrals with the economy of the FFT algorithm discussed in Section 9-5. The details are described as follows.

Given a signal $f(t)$, choose an appropriate period T and a desired number N of samples $f_p(mT_0)$ of $f_p(t)$. The choice of T is dictated by two conflicting

Figure 10.61
Illustration of the addition to $F(j\omega)$ of all its displacements to form a periodic function $F_p(j\omega)$ with a period 2γ.

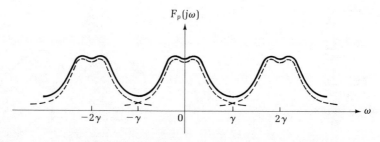

requirements to be discussed later. To find the samples $f_p(mT_0)$, we set
$t = mT_0$, $m = 0, 1, \ldots, N - 1$ in (10.219) and obtain

$$f_p(0) = \cdots + f(-T) + f(0) + f(T) + f(2T) + \cdots$$
$$f_p(T_0) = \cdots + f(T_0 - T) + f(T_0) + f(T_0 + T) + f(T_0 + 2T) + \cdots$$
$$f_p(2T_0) = \cdots + f(2T_0 - T) + f(2T_0) + f(2T_0 + T) + f(2T_0 + 2T) + \cdots$$
$$\vdots$$
$$f_p(NT_0 - T_0) = \cdots + f(NT_0 - T_0) + f(NT_0 - T_0 + T)$$
$$+ f(NT_0 - T_0 + 2T) + \cdots \tag{10.237}$$

Using these samples $f_p(mT_0)$, we use the FFT algorithm to compute the
samples $\overline{F}_p(jm\omega_0)/T$. This involves a DFT of order N. Once we have
$\overline{F}_p(jm\omega_0)/T$, the values of $F_p(j\omega)$ are known at the points $\omega = m\omega_0$. However,
our objective is to find not $F_p(jm\omega_0)$ but the samples $F(jm\omega_0)$ of $F(j\omega)$. This
can be done only by approximation.

If

$$F(j\omega) \approx 0, \qquad |\omega| > \gamma \tag{10.238}$$

then as we can see from Figure 10.61

$$F(j\omega) \approx F_p(j\omega), \qquad |\omega| < \gamma \tag{10.239}$$

or

$$F(jm\omega_0) \approx F_p(jm\omega_0), \qquad |m| < \gamma/\omega_0 = \frac{N}{2} \tag{10.240}$$

In other words, if $F(j\omega)$ is negligible for $|\omega| > \gamma$, then its N samples $F(jm\omega_0)$
in the interval from $-\gamma$ to γ equal approximately $F_p(jm\omega_0)$. The difference
between $F_p(jm\omega_0)$ and $F(jm\omega_0)$,

$$\epsilon_m = F_p(jm\omega_0) - F(jm\omega_0) \tag{10.241}$$

is called the *aliasing error*. This error depends on the value of N, the choice
of T, and the waveform of $F(j\omega)$. The waveform of $F(j\omega)$ is determined by a
given signal $f(t)$ and is therefore beyond our control. The value of N specifies
the order of the pertinent FFT algorithm and hence the complexity of the
computations involved. The choice of T is dictated by two conflicting requirements to be discussed below.

If T is large, the frequency $\gamma = N\pi/T$ is small and the sampling frequency
$\omega_0 = 2\gamma/N$ is small. Therefore the samples $F(jm\omega_0)$ are close and we say
that *resolution* of $F(j\omega)$ is good. But with small γ, the approximation (10.239)
is poor and the aliasing error is large because for ordinary functions $F(j\omega)$
approaches zero as $\omega \to \infty$. On the other hand, for small values of T, the
aliasing error will be small but the resolution of $F(j\omega)$ will be inadequate.
Both of the requirements of small aliasing error and high resolution can
be attained if the number of samples is sufficiently large. But for a large
number of samples, the resulting computational complexity may be pro-
hibitive. A compromise among the various constraints has to be made.

The above procedure for computing the Fourier transform via an FFT algorithm is illustrated in Figure 10.62. In (a) samples of $f(t)$ are made that are used to evaluate the samples $f_p(mT_0)$ as indicated in (b). Once the $f_p(mT_0)$ values are known, the results of the inverse DFS calculations yield $F_p(jm\omega_0)$ as shown in (c) for m from 0 to $N-1$. Finally, the unknown $F(jm\omega_0)$ is obtained approximately from $F_p(jm\omega_0)$ and is sketched in (d).

We note that if $f(t)$ is time limited, then $f_p(mT_0)$ contains only a finite number of terms. For instance, if $f(t) = 0$ for $t < 0$ or $t \geq kT$, the right-hand side of (10.237) has at most k terms that are different from zero. For $k = 2$ we have

$$f_p(mT_0) = f(mT_0) + f(mT_0 + T), \qquad m = 0, 1, \ldots, N - 1 \qquad \textbf{(10.242)}$$

and for $k = 1$,

$$f_p(mT_0) = f(mT_0), \qquad m = 0, 1, \ldots, N - 1 \qquad \textbf{(10.243)}$$

We illustrate the above results by the following example.

Figure 10.62
Illustration of the procedure for computing the Fourier transform of a signal $f(t)$ using a fast Fourier transform (FFT) algorithm.

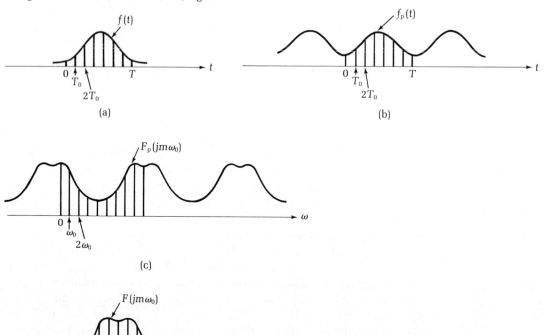

EXAMPLE 10.12

Consider the rectangular pulse $f(t)$ of Figure 10.63(a) described by the equation

$$f(t) = 1, \qquad 0 < t < a$$
$$\quad\; = 0, \qquad t < 0 \quad \text{or} \quad t > a \tag{10.244}$$

For our purposes choose $T = a$. Then $T_0 = T/N = a/N$ and

$$f(mT_0) = 1, \quad m = 0, 1, 2, \ldots, N-1 \tag{10.245}$$

The associated periodic function $f_p(t)$ is a straight line as depicted in Figure 10.63(b) with

$$f_p(mT_0) = f(mT_0) = 1, \qquad m = 0, 1, 2, \ldots, N-1 \tag{10.246}$$

having an inverse DFS, being equal to $\overline{F}_p(jm\omega_0)/T$, found from (9.128) as

$$\overline{F}_p(jm\omega_0)/T = \frac{1}{N} \sum_{n=0}^{N-1} w^{mn} = \frac{1}{N} \cdot \frac{1 - w^{mn}}{1 - w^m}$$

$$= \begin{cases} 1, & m = 0 \\ 0, & m \neq 0 \end{cases} \tag{10.247}$$

Figure 10.63
Illustration of the computation of a discrete Fourier series (DFS) $F_p(jm\omega_0)$ in (c) of a sampled rectangular pulse $f(t)$ in (a) and its associated periodic function $f_p(t)$ in (b).

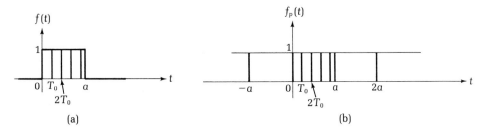

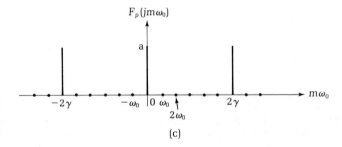

Figure 10.64
The amplitude spectrum
of the signal $f(t)$ of Figure
10.63(a).

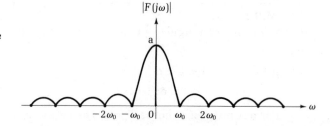

where $\omega_0 = 2\pi/a$ and $w = \exp(j2\pi/N)$. This gives

$$F_p(jm\omega_0) = \begin{cases} a, & m = 0 \\ 0, & m = 1, 2, \ldots, N-1 \end{cases} \qquad (10.248)$$

The sequence $F_p(jm\omega_0)$ is periodic with period $2\gamma = N\omega_0$, and is shown in Figure 10.63(c). For illustrative purposes we compute the Fourier transform $F(j\omega)$ of $f(t)$. From Example 10.1 the desired transform is found by appealing to the time-shifting property (10.69):

$$F(j\omega) = a \frac{\sin \frac{1}{2}\omega a}{\frac{1}{2}\omega a} e^{-j\omega a/2} \qquad (10.249)$$

the amplitude spectrum of which is shown in Figure 10.64 together with its samples $a, 0, 0, \ldots, 0$ at the frequencies $0, \omega_0, 2\omega_0, \ldots, (N-1)\omega_0$.

The periodic function $F_p(j\omega)$ is formed by adding to $F(j\omega)$ all its displacements $F(j\omega + j2k\gamma)$:

$$F_p(j\omega) = a \sum_{k=-\infty}^{\infty} \frac{\sin \frac{1}{2}(\omega + kN\omega_0)a}{\frac{1}{2}(\omega + kN\omega_0)a} e^{-j(\omega + kN\omega_0)a/2} \qquad (10.250)$$

The amplitude spectrum of $F_p(j\omega)$ together with its samples $a, 0, 0, \ldots, 0$ at the frequencies $0, \omega_0, 2\omega_0, \ldots, (N-1)\omega_0$ is sketched and shown in Figure 10.65. The results are consistent with those given in (10.248).

Figure 10.65
Amplitude spectrum of
the periodic function
formed by adding to the
$F(j\omega)$ of Figure 10.64 all of
its displacements.

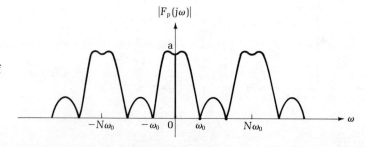

Instead of using $T = a$, suppose that we choose $T = \frac{1}{2}a$. Then $T_0 = a/2N$. The associated periodic function $f_p(t)$ is formed by adding to $f(t)$ all its displacements $f(t + \frac{1}{2}na)$, as depicted in Figure 10.66(a). The resulting function is a constant of value 2 shown in Figure 10.66(b):

$$f_p(t) = \sum_{n=-\infty}^{\infty} f(t + \tfrac{1}{2}na) = 2 \tag{10.251}$$

with samples

$$f_p(mT_0) = 2, \qquad m = 0, 1, 2, \ldots, N - 1 \tag{10.252}$$

The inverse DFS is found to be

$$\frac{\overline{F}_p(jm\omega_0)}{T} = \begin{cases} 2, & m = 0 \\ 0, & m = 1, 2, \ldots, N - 1 \end{cases} \tag{10.253}$$

Figure 10.66

The constant value (b) resulting from adding to the pulse $f(t)$ all of its displacements (a). The time reference is chosen to give an even function in the time domain.

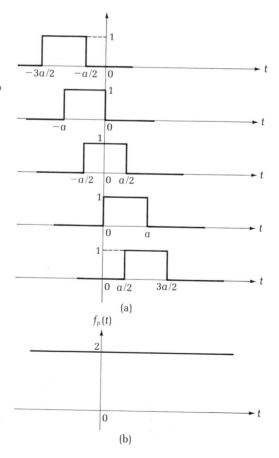

(a)

(b)

Figure 10.67
Amplitude spectrum in which the samples are taken twice as far apart as those illustrated in Figure 10.65.

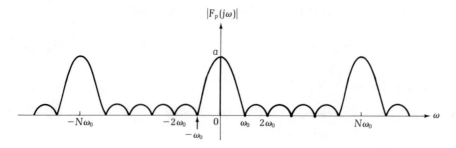

where $\omega_0 = 4\pi/a$, or

$$F_p(jm\omega_0) = \begin{cases} a, & m = 0 \\ 0, & m = 1, 2, \ldots, N - 1 \end{cases} \qquad (10.254)$$

This shows that if the period T is cut in half, the sampling frequency ω_0 is doubled. Therefore the samples $F_p(jm\omega_0)$ are twice as far apart, as indicated in Figure 10.67. As a result the resolution of $F(j\omega)$ is poor.

To improve the resolution of $F(j\omega)$, suppose that we choose $T = 2a$. The associated periodic function $f_p(t)$ is formed by adding to $f(t)$ all its displacements $f(t + 2na)$ and the resulting function is shown in Figure 10.68.

To be specific let $N = 8$. Then $T_0 = T/N = a/4$ and $\omega_0 = 2\pi/T = \pi/a$. The samples of $f_p(t)$ are given by

$$f_p(mT_0) = \begin{cases} 1, & m = 0, 1, 2, 3 \\ 0, & m = 4, 5, 6, 7 \end{cases} \qquad (10.255)$$

the inverse DFS of which is found from (9.128) to be

$$\frac{\overline{F}_p(jm\omega_0)}{T} = \frac{1}{8} \sum_{k=0}^{3} e^{jkm\pi/4}, \qquad m = 0, 1, 2, \ldots, 7 \qquad (10.256)$$

This leads to

$$\overline{F}_p(j0) = a$$
$$\overline{F}_p(j\omega_0) = (0.25 + j0.60355)a = 0.65328ae^{j67.5°}$$

Figure 10.68
The periodic function formed by adding to $f(t)$ all of its displacements for the case in which the period is $T = 2a$.

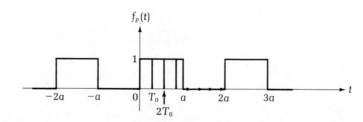

Figure 10.69

Amplitude spectrum corresponding to the periodic pulse function of Figure 10.68 in which the period is $T = 2a$.

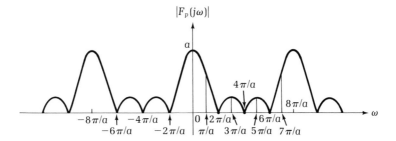

$$\bar{F}_p(j2\omega_0) = 0$$
$$\bar{F}_p(j3\omega_0) = (0.25 + j0.10355)a = 0.27060ae^{j22.5°}$$
$$\bar{F}_p(j4\omega_0) = 0$$
$$\bar{F}_p(j5\omega_0) = (0.25 - j0.10355)a = 0.27060ae^{-j22.5°}$$
$$\bar{F}_p(j6\omega_0) = 0$$
$$\bar{F}_p(j7\omega_0) = (0.25 - j0.60355)a = 0.65328ae^{-j67.5°} \qquad \textbf{(10.257a)}$$

A plot of the amplitudes of $F_p(jm\omega_0)$ is presented in Figure 10.69. For comparison we compute the values of $F(j\omega)$ at these frequencies directly from (10.249) and the results are given by

$$F(j0) = a$$
$$F(j\omega_0) = -j0.63662a$$
$$F(j2\omega_0) = 0$$
$$F(j3\omega_0) = -j0.21221a$$
$$F(j4\omega_0) = 0$$
$$F(j5\omega_0) = -j0.12732a$$
$$F(j6\omega_0) = 0$$
$$F(j7\omega_0) = -j0.09095a \qquad \textbf{(10.257b)}$$

In Figure 10.69 we show also the magnitude function $|F_p(j\omega)|$. As T is increased from a to $2a$, the separation between the displacements of $F(j\omega)$ in $F_p(j\omega)$ is cut in half. As a result the aliasing error is large, but the samples $F(jm\omega_0)$ are closer and the resolution of $F(j\omega)$ is therefore improved. Both small aliasing error and high resolution can be met if N is chosen sufficiently large.

Suppose that we further increase the period of $f_p(t)$ by choosing $T = 16a$ and $N = 64$. Then $T_0 = T/N = a/4$ and $\omega_0 = \pi/8a$. Using the subroutine DFT or FFT developed in Section 9-6, the inverse DFS of the sequence

$$\{1, 1, 1, 1, 0, 0, \ldots, 0\}$$

of the samples of $f_p(t)$ are found to be

$$\overline{F}_p(j0) = (1.00000 + j0.00000)a = a = F_p(j64\omega_0)$$
$$\overline{F}_p(j\omega_0) = (0.98323 + j0.14585)a = 0.99399ae^{j8.44°} = F_p(j63\omega_0)$$
$$\overline{F}_p(j2\omega_0) = (0.93403 + j0.28334)a = 0.97606ae^{j16.88°} = F_p(j62\omega_0)$$
$$\overline{F}_p(j3\omega_0) = (0.85570 + j0.40472)a = 0.94658ae^{j25.31°} = F_p(j61\omega_0)$$
$$\overline{F}_p(j4\omega_0) = (0.75342 + j0.50342)a = 0.90613ae^{j33.75°} = F_p(j60\omega_0)$$
$$\overline{F}_p(j5\omega_0) = (0.63388 + j0.57451)a = 0.85549ae^{j42.19°} = F_p(j59\omega_0)$$
$$\overline{F}_p(j6\omega_0) = (0.50477 + j0.61506)a = 0.79567ae^{j50.62°} = F_p(j58\omega_0)$$
$$\overline{F}_p(j7\omega_0) = (0.37418 + j0.62427)a = 0.72782ae^{j59.06°} = F_p(j57\omega_0)$$
$$\overline{F}_p(j8\omega_0) = (0.25000 + j0.60355)a = 0.65328ae^{j67.50°} = F_p(j56\omega_0)$$
$$\overline{F}_p(j9\omega_0) = (0.13935 + j0.55630)a = 0.57348ae^{j75.94°} = F_p(j55\omega_0)$$
$$\overline{F}_p(j10\omega_0) = (0.04803 + j0.48761)a = 0.48997ae^{j84.37°} = F_p(j54\omega_0)$$
$$\overline{F}_p(j11\omega_0) = (-0.01984 + j0.40384)a = 0.40433ae^{j92.81°} = F_p(j53\omega_0)$$
$$\overline{F}_p(j12\omega_0) = (-0.06208 + j0.31208)a = 0.31819ae^{j101.25°} = F_p(j52\omega_0)$$
$$\overline{F}_p(j13\omega_0) = (-0.07855 + j0.21953)a = 0.23316ae^{j109.69°} = F_p(j51\omega_0)$$
$$\overline{F}_p(j14\omega_0) = (-0.07109 + j0.13300)a = 0.15081ae^{j118.13°} = F_p(j50\omega_0)$$
$$\overline{F}_p(j15\omega_0) = (-0.04326 + j0.05833)a = 0.07263ae^{j126.56°} = F_p(j49\omega_0)$$
$$\overline{F}_p(j16\omega_0) = (0.00000 + j0.00000)a = 0.00000 = F_p(j48\omega_0)$$
$$\overline{F}_p(j17\omega_0) = (0.05287 - j0.03921)a = 0.06582ae^{-j36.56°} = F_p(j47\omega_0)$$
$$\overline{F}_p(j18\omega_0) = (0.10915 - j0.05834)a = 0.12376ae^{-j28.12°} = F_p(j46\omega_0)$$
$$\overline{F}_p(j19\omega_0) = (0.16281 - j0.05826)a = 0.17292ae^{-j19.69°} = F_p(j45\omega_0)$$
$$\overline{F}_p(j20\omega_0) = (0.20852 - j0.04148)a = 0.21261ae^{-j11.25°} = F_p(j44\omega_0)$$
$$\overline{F}_p(j21\omega_0) = (0.24205 - j0.01189)a = 0.24235ae^{-j2.81°} = F_p(j43\omega_0)$$
$$\overline{F}_p(j22\omega_0) = (0.26063 + j0.02567)a = 0.26189ae^{j5.63°} = F_p(j42\omega_0)$$
$$\overline{F}_p(j23\omega_0) = (0.26311 + j0.06591)a = 0.27124ae^{j14.06°} = F_p(j41\omega_0)$$
$$\overline{F}_p(j24\omega_0) = (0.25000 + j0.10355)a = 0.27060ae^{j22.50°} = F_p(j40\omega_0)$$
$$\overline{F}_p(j25\omega_0) = (0.22337 + j0.13388)a = 0.26042ae^{j30.94°} = F_p(j39\omega_0)$$
$$\overline{F}_p(j26\omega_0) = (0.18658 + j0.15312)a = 0.24136ae^{j39.37°} = F_p(j38\omega_0)$$
$$\overline{F}_p(j27\omega_0) = (0.14391 + j0.15878)a = 0.21429ae^{j47.81°} = F_p(j37\omega_0)$$
$$\overline{F}_p(j28\omega_0) = (0.10014 + j0.14986)a = 0.18024ae^{j56.25°} = F_p(j36\omega_0)$$
$$\overline{F}_p(j29\omega_0) = (0.06003 + j0.12693)a = 0.14041ae^{j64.69°} = F_p(j35\omega_0)$$
$$\overline{F}_p(j30\omega_0) = (0.02791 + j0.09199)a = 0.09613ae^{j73.12°} = F_p(j34\omega_0)$$
$$\overline{F}_p(j31\omega_0) = (0.00717 + j0.04830)a = 0.04883ae^{j81.56°} = F_p(j33\omega_0)$$
$$\overline{F}_p(j32\omega_0) = (0.00000 + j0.00000)a = 0.00000$$

A computer plot of the amplitudes of $F_p(jm\omega_0)$ is shown in Figure 10.70. For comparison, a similar plot of (10.257a) is presented in Figure 10.71. Thus,

Figure 10.70

A computer plot of the amplitudes of $F_p(jm\omega_0)$ for a rectangular pulse with $T = 16a$ and $N = 64$.

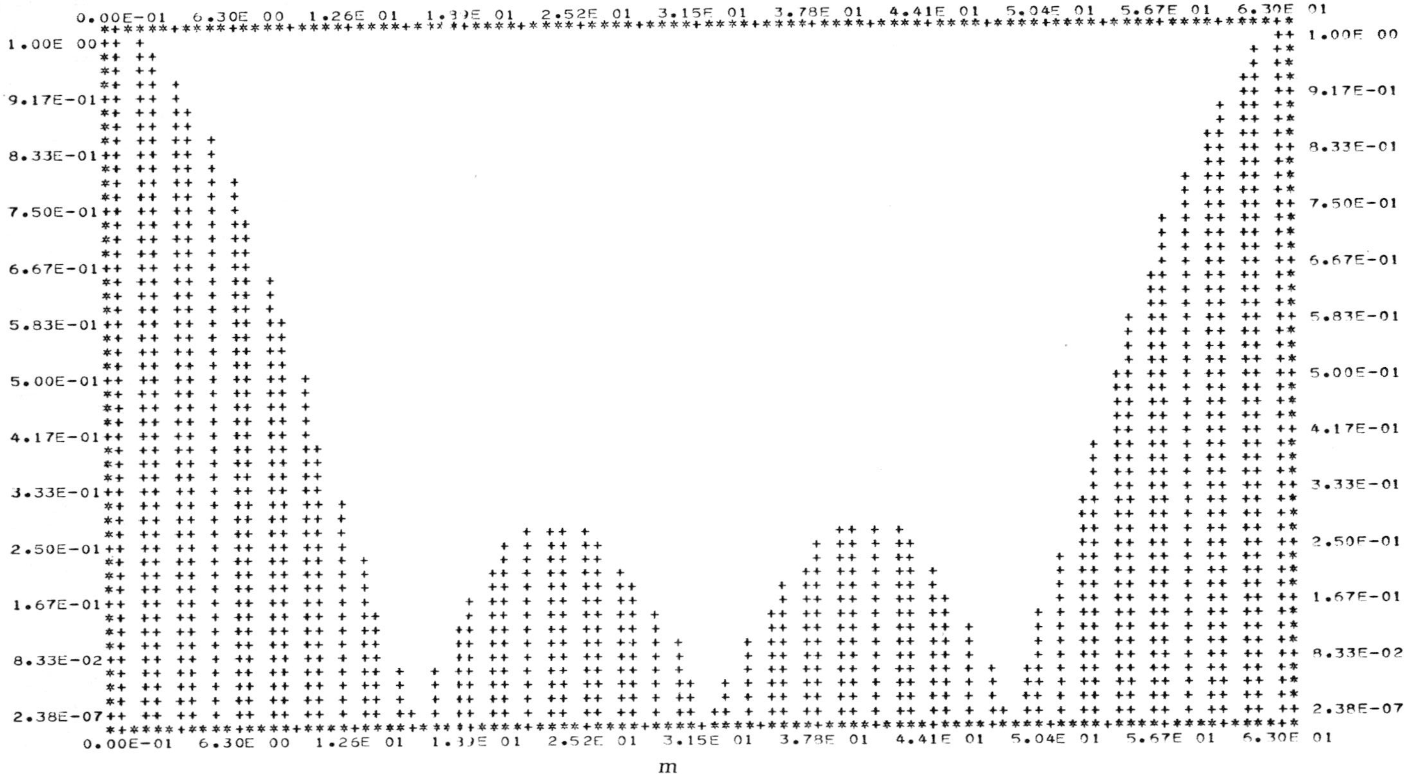

Figure 10.71

A computer plot of the amplitudes of $F_P(jm\omega_0)$ for a rectangular pulse with $T = 2a$ and $N = 8$.

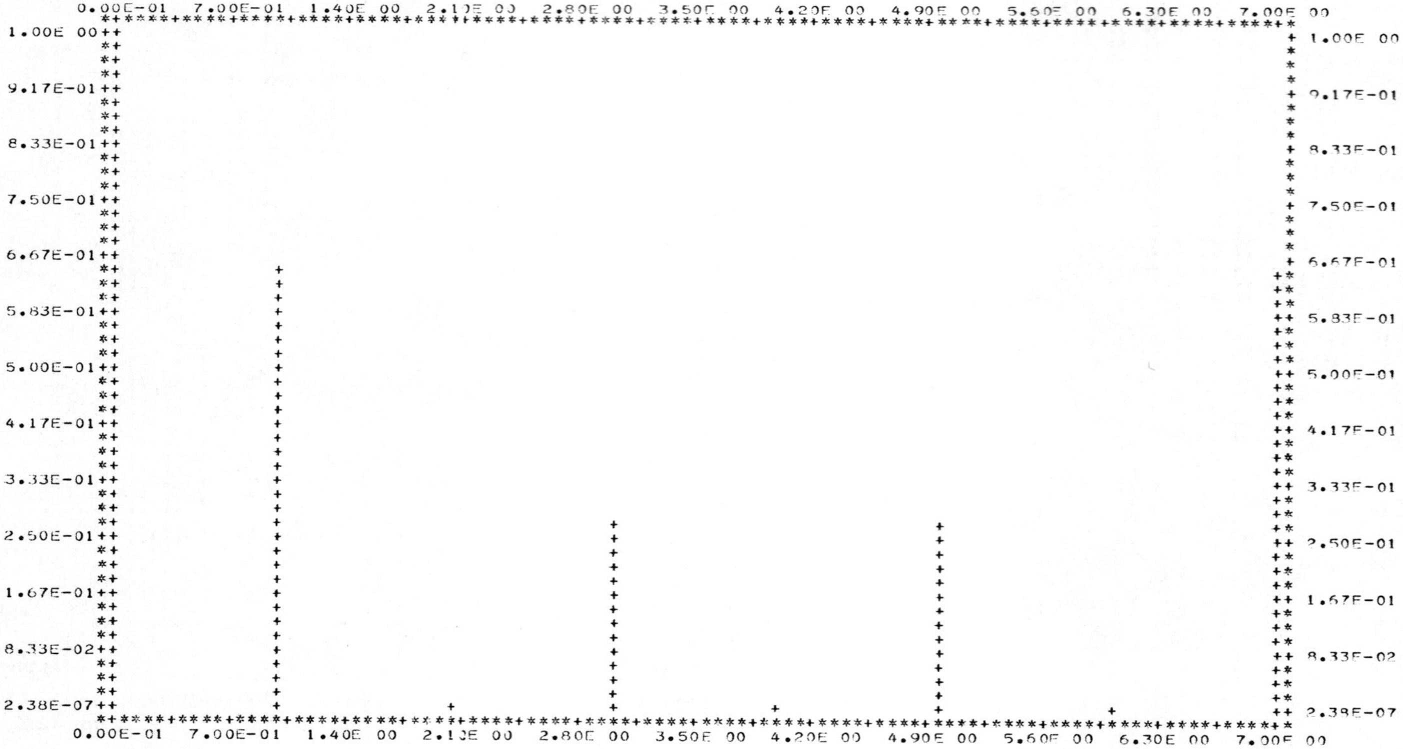

the larger the period we choose for $f_p(t)$, the closer the separation of the samples of $F_p(jm\omega_0)$ will result. For $|m| < \gamma/\omega_0 = N/2 = 32$, we have from (10.240) $F(jm\omega_0) \approx F_p(jm\omega_0)$.

EXAMPLE 10.13

Consider a sinusoidal pulse $f(t)$ described by the equation

$$f(t) = \sin \tfrac{1}{4}\pi t, \qquad 0 \le t \le 4$$
$$= 0, \qquad\qquad t < 0 \quad \text{or} \quad t > 4$$

For our purposes choose $T = 16$ and $N = 16$. Then $T_0 = 1$, $\omega_0 = \pi/8$ and $\gamma = \pi$. Using the subroutine DFT or FFT of Section 9-6, the inverse DFS of the sequence of samples of $f_p(t)$ are found to be

$$\overline{F}_p(j0\omega_0) = 0.60355 + j0.00000 = 0.60355 = F_p(j16\omega_0)$$
$$\overline{F}_p(j1\omega_0) = 0.53275 + j0.22067 = 0.57664e^{j22.50°} = F_p(j15\omega_0)$$
$$\overline{F}_p(j2\omega_0) = 0.35355 + j0.35355 = 0.50000e^{j45°} = F_p(j14\omega_0)$$
$$\overline{F}_p(j3\omega_0) = 0.14745 + j0.35597 = 0.38530e^{j67.50°} = F_p(j13\omega_0)$$
$$\overline{F}_p(j4\omega_0) = 0.00000 + j0.25000 = 0.25000e^{j90°} = F_p(j12\omega_0)$$
$$\overline{F}_p(j5\omega_0) = -0.04389 + j0.10597 = 0.11470e^{j112.50°} = F_p(j11\omega_0)$$
$$\overline{F}_p(j6\omega_0) = 0.00000 + j0.00000 = 0.00000 = F_p(j10\omega_0)$$
$$\overline{F}_p(j7\omega_0) = 0.07081 - j0.02933 = 0.07664e^{-j22.50°} = F_p(j9\omega_0)$$
$$\overline{F}_p(j8\omega_0) = 0.10355 + j0.00000 = 0.10355 \qquad\qquad \textbf{(10.258a)}$$

A computer plot of the amplitudes of $F_p(jm\omega_0)$ is shown in Figure 10.72.

Suppose now that we increase the period from $T = 16$ to $T = 64$ and choose $N = 64$. Then $T_0 = 1$, $\omega_0 = \pi/32$ and $\gamma = \pi$. Using the subroutine DFT or FFT, the inverse DFS of the sequence of samples of $f_p(t)$ are found to be

$$\overline{F}_p(j0\omega_0) = 0.60355 + j0.00000 = 0.60355 = F_p(j64\omega_0)$$
$$\overline{F}_p(j1\omega_0) = 0.59895 + j0.05899 = 0.60185e^{j5.62°} = F_p(j63\omega_0)$$
$$\overline{F}_p(j2\omega_0) = 0.58529 + j0.11642 = 0.59676e^{j11.25°} = F_p(j62\omega_0)$$
$$\overline{F}_p(j3\omega_0) = 0.56300 + j0.17078 = 0.58833e^{j16.87°} = F_p(j61\omega_0)$$
$$\overline{F}_p(j4\omega_0) = 0.53275 + j0.22067 = 0.57664e^{j22.50°} = F_p(j60\omega_0)$$
$$\overline{F}_p(j5\omega_0) = 0.49547 + j0.26483 = 0.56181e^{j28.12°} = F_p(j59\omega_0)$$
$$\overline{F}_p(j6\omega_0) = 0.45229 + j0.30221 = 0.54397e^{j33.75°} = F_p(j58\omega_0)$$
$$\overline{F}_p(j7\omega_0) = 0.40452 + j0.33198 = 0.52330e^{j39.37°} = F_p(j57\omega_0)$$
$$\overline{F}_p(j8\omega_0) = 0.35355 + j0.35355 = 0.50000e^{j45°} = F_p(j56\omega_0)$$
$$\overline{F}_p(j9\omega_0) = 0.30089 + j0.36663 = 0.47429e^{j50.62°} = F_p(j55\omega_0)$$
$$\overline{F}_p(j10\omega_0) = 0.24802 + j0.37119 = 0.44642e^{j56.25°} = F_p(j54\omega_0)$$
$$\overline{F}_p(j11\omega_0) = 0.19641 + j0.36746 = 0.41666e^{j61.88°} = F_p(j53\omega_0)$$
$$\overline{F}_p(j12\omega_0) = 0.14745 + j0.35597 = 0.38530e^{j67.50°} = F_p(j52\omega_0)$$

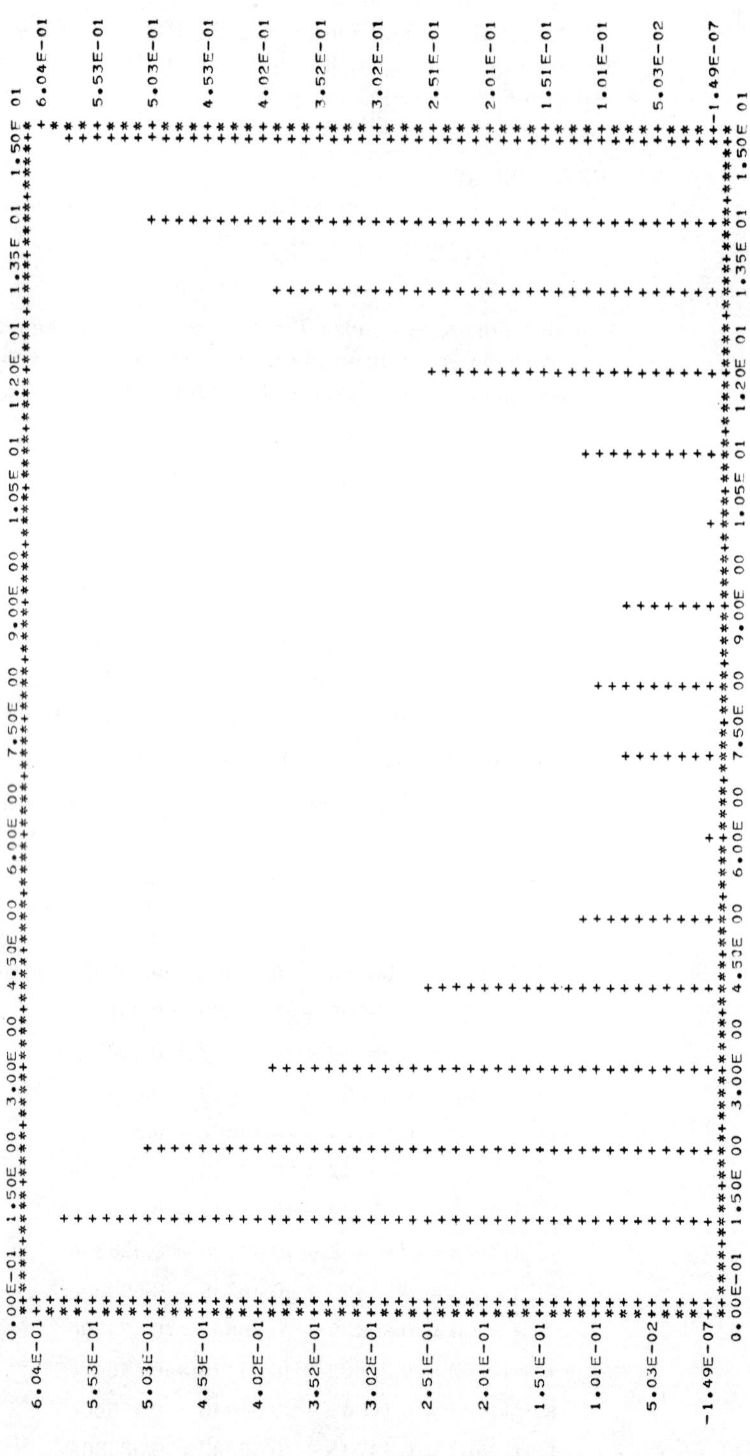

Figure 10.72

A computer plot of the amplitudes of $F_p(jm\omega_0)$ for a sinusoidal pulse with $T = 16$ and $N = 16$.

Figure 10.73

A computer plot of the amplitudes of $F_p(jm\omega_0)$ for a sinusoidal pulse with $T = 64$ and $N = 64$.

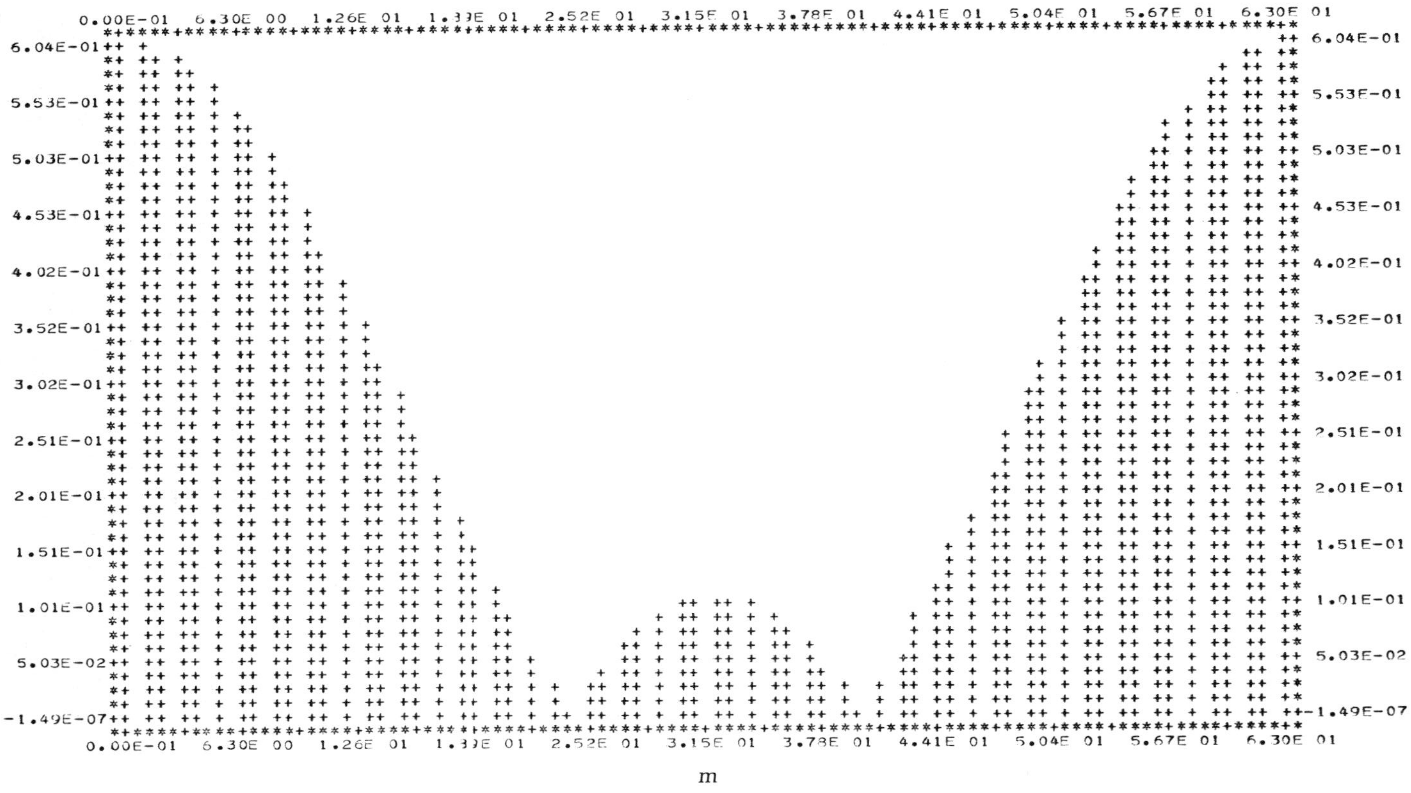

$$\bar{F}_p(j13\omega_0) = 0.10236 + j0.33745 = 0.35263e^{j73.13°} = F_p(j51\omega_0)$$

$$\bar{F}_p(j14\omega_0) = 0.06223 + j0.31285 = 0.31897e^{j78.75°} = F_p(j50\omega_0)$$

$$\bar{F}_p(j15\omega_0) = 0.02790 + j0.28328 = 0.28465e^{j84.38°} = F_p(j49\omega_0)$$

$$\bar{F}_p(j16\omega_0) = 0.00000 + j0.25000 = 0.25000e^{j90°} = F_p(j48\omega_0)$$

$$\bar{F}_p(j17\omega_0) = -0.02111 + j0.21431 = 0.21535e^{j95.63°} = F_p(j47\omega_0)$$

$$\bar{F}_p(j18\omega_0) = -0.03532 + j0.17755 = 0.18103e^{j101.25°} = F_p(j46\omega_0)$$

$$\bar{F}_p(j19\omega_0) = -0.04278 + j0.14102 = 0.14737e^{j106.88°} = F_p(j45\omega_0)$$

$$\bar{F}_p(j20\omega_0) = -0.04389 + j0.10597 = 0.11470e^{j112.50°} = F_p(j44\omega_0)$$

$$\vec{F}_p(j21\omega_0) = -0.03928 + j0.07350 = 0.08334e^{j118.12°} = F_p(j43\omega_0)$$

$$\bar{F}_p(j22\omega_0) = -0.02977 + j0.04455 = 0.05358e^{j123.75°} = F_p(j42\omega_0)$$

$$\bar{F}_p(j23\omega_0) = -0.01631 + j0.01987 = 0.02571e^{j129.38°} = F_p(j41\omega_0)$$

$$\bar{F}_p(j24\omega_0) = 0.00000 + j0.00000 = 0.00000 = F_p(j40\omega_0)$$

$$\bar{F}_p(j25\omega_0) = 0.01801 - j0.01478 = 0.02330e^{-j39.37°} = F_p(j39\omega_0)$$

$$\bar{F}_p(j26\omega_0) = 0.03656 - j0.02443 = 0.04397e^{-j33.75°} = F_p(j38\omega_0)$$

$$\bar{F}_p(j27\omega_0) = 0.05451 - j0.02914 = 0.06181e^{-j28.13°} = F_p(j37\omega_0)$$

$$\bar{F}_p(j28\omega_0) = 0.07081 - j0.02933 = 0.07664e^{-j22.50°} = F_p(j36\omega_0)$$

$$\bar{F}_p(j29\omega_0) = 0.08453 - j0.02564 = 0.08833e^{-j16.87°} = F_p(j35\omega_0)$$

$$\bar{F}_p(j30\omega_0) = 0.09490 - j0.01888 = 0.09676e^{-j11.25°} = F_p(j34\omega_0)$$

$$\bar{F}_p(j31\omega_0) = 0.10136 - j0.00998 = 0.10185e^{-j5.62°} = F_p(j33\omega_0)$$

$$\bar{F}_p(j32\omega_0) = 0.10355 + j0.00000 = 0.10355 \tag{10.258b}$$

A computer plot of the amplitudes of $F_p(jm\omega_0)$ is presented in Figure 10.73. A comparison of this with that of Figure 10.72 again shows that as the period T is increased from 16 to 64, the samples of $F_p(j\omega)$ are taken four times closer. The separation between the displacements of $F(j\omega)$ in $F_p(j\omega)$, however, remains the same with $\gamma = \pi$.

10-9 SAMPLING BAND-LIMITED PERIODIC SIGNALS

In Section 10-7 we showed that the complete information about a band-limited signal is contained in its samples spaced uniformly at a distance not greater than a certain value determined by the bandwidth. In the present section we demonstrate that the same statement can be made for band-limited periodic signals. We do this by means of the discrete Fourier series (DFS), thus affording an accessible application of the Fourier transform.

Let $f(t)$ be a periodic signal with period T satisfying the Dirichlet conditions; then $f(t)$ can be represented by the exponential Fourier series

$$f(t) = \sum_{m=-\infty}^{\infty} \tilde{c}_m e^{jm\omega_0 t} \tag{10.259}$$

where $\omega_0 = 2\pi/T$ and the $\tilde{c}_m$'s are the Fourier coefficients. As before, we divide the period T into N equal subintervals $T_0 = T/N$ and write $w = \exp(j2\pi/N)$. We have from (10.259)

$$f(nT_0) = \sum_{m=-\infty}^{\infty} \tilde{c}_m e^{j2mn\pi/N} = \sum_{m=-\infty}^{\infty} \tilde{c}_m w^{mn} \tag{10.260}$$

for $n = 0, 1, 2, \ldots, N-1$.

As demonstrated in (9.92), $f(nT_0)$ can be expressed in terms of the aliased coefficients $\hat{c}_m$ of $f(t)$ by the relation

$$f(nT_0) = \sum_{m=0}^{N-1} \hat{c}_m w^{mn} \tag{10.261}$$

where

$$\hat{c}_k = \sum_{v=-\infty}^{\infty} \tilde{c}_{k+vN} \tag{10.262}$$

The sequences $\{f(nT_0)\}$ and $\{\hat{c}_k\}$ are each periodic with period N. In fact, from (9.134) and (9.135), we can see that $\{\overline{\hat{c}}_k\}$ and $\{f(nT_0)\}$ form a DFS pair.

The Sampling Theorem   *Given a band-limited periodic signal $f(t)$ with period T, which can be expressed as*

$$f(t) = \sum_{m=-M}^{M} \tilde{c}_m e^{jm\omega_0 t} \tag{10.263}$$

where $\omega_0 = 2\pi/T$, then for any $N \geq 2M+1$ and $T_0 = T/N$, we have

$$f(t) = \frac{1}{N} \sum_{n=0}^{N-1} f(nT_0) \frac{\sin\left[(2M+1)(\omega_0 t/2 - n\pi/N)\right]}{\sin\left(\omega_0 t/2 - n\pi/N\right)} \tag{10.264}$$

Proof.   Observe from (10.263) that $\tilde{c}_m = 0$ for $|m| > M$. As a result, if $N \geq 2M+1$, then

$$\hat{c}_k = \sum_{v=-\infty}^{\infty} \tilde{c}_{k+vN} = \tilde{c}_k, \qquad |k| \leq M \tag{10.265}$$

Taking the complex conjugate on both sides of (10.261), we obtain

$$f(nT_0) = \sum_{m=0}^{N-1} \overline{\hat{c}}_m w^{-mn}, \qquad n = 0, 1, \ldots, N-1 \tag{10.266}$$

the inverse DFS of which is given by

$$\overline{\hat{c}}_m = \frac{1}{N} \sum_{n=0}^{N-1} f(nT_0) w^{mn}, \qquad n = 0, 1, \ldots, N-1 \tag{10.267}$$

Combining (10.263), (10.265), and (10.267) and taking the complex conjugate yield

$$f(t) = \sum_{m=-M}^{M} \bar{\bar{c}}_m e^{-jm\omega_0 t} = \sum_{m=-M}^{M} \bar{\bar{c}}_m e^{-jm\omega_0 t}$$

$$= \frac{1}{N} \sum_{m=-M}^{M} \sum_{n=0}^{N-1} f(nT_0) e^{-jm(\omega_0 t - 2n\pi/N)}$$

$$= \frac{1}{N} \sum_{n=0}^{N-1} f(nT_0) \left[\sum_{m=-M}^{M} e^{-jm(\omega_0 t - 2n\pi/N)} \right] \quad \text{see (9.250)}$$

$$= \frac{1}{N} \sum_{n=0}^{N-1} f(nT_0) \frac{\sin\left[(2M+1)(\omega_0 t/2 - n\pi/N)\right]}{\sin(\omega_0 t/2 - n\pi/N)} \tag{10.268}$$

The sampling theorem states that if a periodic signal $f(t)$ of period T having no spectral components above $M\omega_0 = 2\pi M/T$, then the signal is uniquely determined by its samples taken at uniform intervals at a rate of at least $2M+1$ samples over the period T. The sampling expansion (10.263) can be viewed as a method of interpolation: the value of the signal $f(t)$ between samples is determined as a weighted sum of its sampled values. When $N = 2M + 1$, (10.264) becomes

$$f(t) = \frac{1}{N} \sum_{n=0}^{2M} f(nT_0) \frac{\sin(\pi t/T_0 - n\pi)}{\sin(\pi t/NT_0 - n\pi/N)} \tag{10.269}$$

At the points $t = kT_0$, we have

$$f(kT_0) = \frac{1}{N} \sum_{n=0}^{2M} f(nT_0) \frac{\sin(k-n)\pi}{\sin[(k-n)\pi/N]} \tag{10.270}$$

However,

$$\frac{\sin(k-n)\pi}{\sin[(k-n)\pi/N]} = \begin{cases} N, & k = n \\ 0, & k \neq n \end{cases} \quad 0 \leq k, n \leq 2M \tag{10.271}$$

so the right-hand side of (10.270) reduces to $f(kT_0)$. In this sense the sampled values $f(0), f(T_0), \ldots, f(2MT_0)$ are algebraically independent in that an error made in one of the samples does not affect the values of the others.

In the oversampling case where $N > 2M + 1$, (10.264) is still valid, but an error or change made in a sample will induce changes in the remaining sample values. As a result the sampled values are no longer algebraically independent.

Parseval's Theorem for the DFT

We next derive the Parseval's theorem for the DFT and then use this result to show that if $N = 2M + 1$, the average signal power of $f(t)$ may be recovered directly from samples of the periodic signal.

Refer to Definition 9.1. Let $\{f_n\}$ be a sequence of N numbers, real or complex. The DFS of $\{f_n\}$ is a sequence $\{F_m\}$ of N numbers defined by

$$F_m = \sum_{n=0}^{N-1} f_n w^{-mn}, \quad m = 0, 1, 2, \ldots, N-1 \tag{10.272}$$

the inverse of which is determined by the formula

$$f_n = \frac{1}{N} \sum_{m=0}^{N-1} F_m w^{mn}, \qquad n = 0, 1, 2, \ldots, N-1 \tag{10.273}$$

Taking the complex conjugate on each side of (10.273) yields

$$\bar{f}_n = \frac{1}{N} \sum_{k=0}^{N-1} \bar{F}_k w^{-kn}, \qquad n = 0, 1, 2, \ldots, N-1 \tag{10.274}$$

where a change of the summation index from m to k has been made.

We next form the summation of the products $f_n \bar{f}_n$. Using (10.273) and (10.274), we obtain

$$\sum_{n=0}^{N-1} |f_n|^2 = \frac{1}{N^2} \sum_{n=0}^{N-1} \sum_{m=0}^{N-1} \sum_{k=0}^{N-1} F_m \bar{F}_k w^{(m-k)n}$$

$$= \frac{1}{N^2} \sum_{m=0}^{N-1} \sum_{k=0}^{N-1} F_m \bar{F}_k \left[\sum_{n=0}^{N-1} w^{(m-k)n} \right] \tag{10.275}$$

Since from (9.98) we have

$$\sum_{n=0}^{N-1} w^{(m-k)n} = \begin{cases} 0, & m \neq k \\ N, & m = k \end{cases} \tag{10.276}$$

the summations in (10.275) are simplified to

$$\sum_{n=0}^{N=1} |f_n|^2 = \frac{1}{N} \sum_{m=0}^{N-1} |F_m|^2 \tag{10.277}$$

which is known as *Parseval's theorem* for the DFT.

As pointed out in (9.134) and (9.135), the sequences $\{\bar{c}_n\}$ and $\{f(mT_0)\}$ of N numbers form a DFS pair. Thus (10.277) applies and we have

$$\sum_{n=0}^{N-1} |\hat{c}_n|^2 = \frac{1}{N} \sum_{m=0}^{N-1} |f(mT_0)|^2 \tag{10.278}$$

When $N = 2M + 1$, (10.278) may be rewritten as

$$\frac{1}{2M+1} \sum_{m=0}^{2M} |f(mT_0)|^2 = \sum_{n=0}^{2M} |\hat{c}_n|^2 = \sum_{n=0}^{M} |\hat{c}_n|^2 + \sum_{k=M+1}^{2M} |\hat{c}_k|^2$$

$$= \sum_{n=0}^{M} |\hat{c}_n|^2 + \sum_{m=1}^{M} |\hat{c}_{m+M}|^2$$

$$= \sum_{n=0}^{M} |\hat{c}_n|^2 + \sum_{m=1}^{M} |\hat{c}_{-m}|^2 \qquad \text{see (9.117)}$$

$$= \sum_{k=-M}^{M} |\hat{c}_k|^2 = \sum_{k=-M}^{M} |\tilde{c}_k|^2 \tag{10.279}$$

The last equality follows from (10.265).

The average signal power of $f(t)$ may be calculated from the energy theorem (9.58) and is given by

$$P = \frac{1}{T} \int_{-T/2}^{T/2} f^2(t) \, dt = \sum_{m=-\infty}^{\infty} |\tilde{c}_m|^2 = \sum_{k=-M}^{M} |\tilde{c}_k|^2 \tag{10.280}$$

which, when combined with (10.279), gives

$$P = \frac{1}{2M+1} \sum_{m=0}^{2M} f^2(mT_0) \tag{10.281}$$

The conclusion is that the average power of a periodic signal with period T having no spectral components above $2\pi M/T$ may be recovered directly from the $2M + 1$ samples of the periodic signal, taken at uniform intervals over a period, without the necessity of first calculating the Fourier coefficients.

10-10 SUMMARY

In this chapter we considered situations in which signals are not periodic. We showed that such a signal can be represented as an infinite sum of exponentials. Unlike the Fourier series for periodic signals, the representation for a nonperiodic signal is not unique and holds only over a finite time interval.

The *Fourier integral* was introduced as the limit of a Fourier series by letting the period approach infinity. As a result the line spectrum of a Fourier series becomes the *continuous spectrum* of the *Fourier transform*. Specifically, we showed that the envelope of the magnitudes of the products of the Fourier coefficients and the period is a continuous function found by taking the magnitude of the corresponding Fourier transform. A similar statement can be made with regard to the phase. In other words, the *continuous amplitude* and *phase spectra* for a nonrecurring pulse signal are identical with the envelopes of the line spectra for the corresponding recurring pulse signal, multiplied by a factor equal to the period of the recurring pulse signal. In terms of frequency content, frequency components extending over an infinite frequency range are required to make up a nonrecurring pulse signal as compared with the discrete harmonics that are required to synthesize a recurring pulse.

The Fourier transform of a transformable signal is unique if the average values are taken at the points of discontinuities of the signal. Conversely, if the Fourier transform of a signal is known, there is a unique time function that can be defined. As a result the Fourier transform of a signal may be regarded as an alternate and equivalent *frequency-domain representation* of the *time-domain signal*. For this we studied the effect in one domain caused by an operation in the other domain. Not only does this allow the transformation of signals from one domain to another, but also it reveals certain physical properties of signals and systems that are not otherwise apparent. This is particularly true in view of the symmetry that exists between the

Fourier transform pair. Special properties studied are: *linearity, scaling, symmetry, time-shifting, frequency shifting, time differentiation and integration, frequency differentiation and integration, reversal, time-domain convolution,* and *frequency-domain convolution.* Finally, the Fourier transforms of singularity functions such as the unit impulse and step functions were presented.

Applications of the Fourier transform to systems with arbitrary inputs were discussed. We showed by several examples that the product of the pulse width and the bandwidth of a system is a constant. Thus the bandwidth required for transmission is inversely proportional to the duration of the pulse being transmitted. Likewise the product of the rise time and the bandwidth is a constant. As a result the rise time of the response of a band-limited system is inversely proportional to the bandwidth of the system.

We also demonstrated that if a signal has no spectral components above a certain frequency, then this signal is uniquely recoverable from its samples taken at uniform intervals at a rate of at least two times during each period or cycle of that certain frequency. A similar result can be stated for a band-limited periodic signal. In fact, if the intervals are properly chosen, the average power of a periodic signal may be recovered directly from the samples of the periodic signal without calculating the Fourier coefficients.

Finally, we considered the problem of numerically evaluating the Fourier transform of a signal. We showed that the problem can be reduced to the *discrete Fourier series problem (DFS)* involving the computational economy of the *fast Fourier transform (FFT)* algorithm. Given a signal, we first form a periodic signal by adding to it all its displacements. The samples of this periodic signal are then used to compute the inverse DFS by the FFT algorithm. Once the inverse DFS is known, the samples of the Fourier transform of the original signal are ascertained by approximation. The error between the actual value and the approximation is called the *aliasing error.* The aliasing error depends upon the number of samples, the choice of period, and the waveform of the Fourier transform. The waveform of the Fourier transform is determined by a given signal and is therefore beyond our control. The number of samples specifies the order of the pertinent FFT algorithm and hence the computational complexity. The selection of the period is dictated by the two conflicting requirements of small aliasing error and high resolution, so an engineering compromise has to be made. However, both requirements can be met if the number of samples taken is sufficiently large. Thus we can use the FFT program developed in the preceding chapter for numerically evaluating the Fourier transform of a signal.

REFERENCES AND SUGGESTED READING

Bracewell, R. N. *The Fourier Transform and Its Applications.* 2d ed. New York: McGraw-Hill, 1978, Chapters 6, 7, and 10.

Gabel, R. A., and R. A. Roberts. *Signals and Linear Systems.* New York: Wiley, 1973, Chapter 5.

Lathi, B. P. *Signals, Systems, and Controls.* New York: Intext, 1974, Chapters 3 and 4.

Liu, C. L., and J. W. S. Liu. *Linear Systems Analysis.* New York: McGraw-Hill, 1975, Chapters 7 and 8.

McGillem, C. D., and G. R. Cooper. *Continuous and Discrete Signal and System Analysis.* New York: Holt, Rinehart & Winston, 1974, Chapter 5.

Oppenheim, A. V., and R. W. Schafer. *Digital Signal Processing.* Englewood Cliffs, N.J.: Prentice-Hall, 1975, Chapter 3.

Papoulis, A. *Circuits and Systems: A Modern Approach.* New York: Holt, Rinehart & Winston, 1980, Chapter 8.

Rabiner, L. R., and B. Gold. *Theory and Application of Digital Signal Processing.* Englewood Cliffs, N.J.: Prentice-Hall, 1975, Chapters 2 and 6.

Roden, M. S. *Analog and Digital Communication Systems.* Englewood Cliffs, N.J.: Prentice-Hall, 1978, Chapter 1.

Van Valkenburg, M. E. *Network Analysis.* 3d ed. Englewood Cliffs, N.J.: Prentice-Hall, 1974, Chapter 16.

PROBLEMS

10.1 Consider the following signal defined over an interval of length 1:

$$f(t) = e^{-t}, \qquad 0 < t < 1 \tag{10.282}$$

Express this signal in the form of (10.2) by introducing a periodic signal $g(t)$ that coincides with $f(t)$ for every t in the interval $0 < t < 1$. The periodic signal $g(t)$ must be (a) even, (b) odd, and (c) neither even nor odd.

10.2 Repeat Problem 10.1 for the signal

$$f(t) = 1, \qquad 0 < t < \tfrac{1}{2}$$
$$= 2, \qquad \tfrac{1}{2} < t < 1 \tag{10.283}$$

10.3 Repeat Problem 10.1 for the signal

$$f(t) = \sin t, \qquad 0 < t < \pi \tag{10.284}$$

10.4 The waveforms shown in Figure 10.74 are pulses of different shapes. Verify that the Fourier transforms $F_k(j\omega)$ of these pulses $f_k(t)$ $(k = 1, 2, \text{ and } 3)$ are given, respectively, by

(a) $F_1(j\omega) = \dfrac{1}{j\omega}(1 + e^{-j\omega a} - 2e^{-j2\omega a})$ $\tag{10.285a}$

(b) $F_2(j\omega) = \dfrac{2}{\omega}(2 \sin 2\omega a - \sin \omega a)$ $\tag{10.285b}$

(c) $F_3(j\omega) = a\,\dfrac{\sin \frac{1}{2}\omega a}{\frac{1}{2}\omega a} + a\,\dfrac{\sin \frac{1}{4}\omega a}{\frac{1}{4}\omega a}$ $\tag{10.285c}$

Sketch the continuous amplitude spectrum for each of these waveforms. Also sketch the continuous phase spectrum for the waveform of Figure 10.74(a).

10.5 Verify that the Fourier transform $F(j\omega)$ of the triangular pulse of Figure 10.26(a) is given by (10.96).

10.6 Ascertain the Fourier transforms $F_k(j\omega)$ of the waveforms $f_k(t)$ $(k = 1, 2, 3, \text{ and } 4)$ shown in Figure 10.75. Sketch the amplitude spectrum for each of these waveforms.

10.7 For each of the pulses shown in Figure 10.74, show that the product of the bandwidth and pulse duration is a constant. Ascertain this constant.

Figure 10.74
Waveforms of pulses of
different shapes.

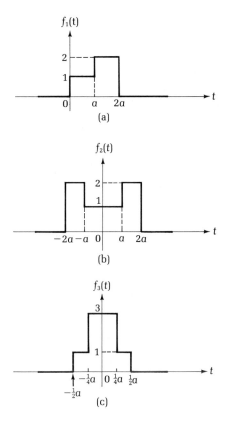

Figure 10.75
Waveforms of four types
of pulses.

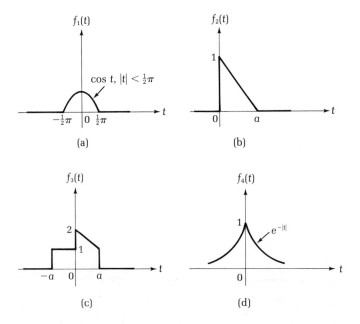

Figure 10.76
Waveform of a triangular-
shaped pulse.

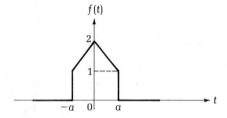

10.8 Show that if $\mathbf{F}[f(t)] = F(j\omega)$, then

$$\mathbf{F}[-jtf(t)] = \frac{dF(j\omega)}{d\omega} \qquad\qquad (10.286)$$

10.9 For each pulse shown in Figure 10.75(b) and (c), demonstrate that the product of the bandwidth and pulse duration is a constant and ascertain this constant.

10.10 Determine the Fourier transforms for the signal waveforms described by the following equations:

(a) $f_1(t) = e^{-at} (\cos \omega_0 t)u(t), \quad$ real $a > 0$ (10.287a)

(b) $f_2(t) = (1 + V_0 \sin \omega_1 t) \cos \omega_0 t$ (10.287b)

(c) $f_3(t) = \exp(-t^2)$ (10.287c)

(d) $f_4(t) = e^{-at} (\sin \omega_0 t)u(t), \quad$ real $a > 0$ (10.287d)

Sketch the continuous amplitude spectrum and the continuous phase spectrum for each of these waveforms.

10.11 Prove the identities (10.94) and (10.100).

10.12 Show that if $\mathbf{F}[f(t)] = F(j\omega)$ and $F(-j\omega) = \bar{F}(j\omega)$, then $f(t)$ is real.

10.13 Determine the Fourier transform for a voltage signal waveform described by the equation

$$v(t) = (\cos t)u(t) \qquad\qquad (10.288)$$

and compute and sketch the amplitude and phase spectra.

10.14 Verify that the Fourier transform for the pulse of Figure 10.76 is given by

$$F(j\omega) = \frac{2}{\omega^2 a} (\omega a \sin \omega a + 2 \sin^2 \tfrac{1}{2}\omega a) \qquad\qquad (10.289)$$

Sketch the amplitude spectrum and show that the product of bandwidth and pulse duration equals 4π.

10.15 Verify that the Fourier transform of the exponential signal of (10.104) is given by (10.105).

10.16 Verify that the Fourier transform for the stepped pulse of Figure 10.77 is given by

$$F(j\omega) = \frac{4}{\omega} \sin \frac{3\omega a}{8} \cos \frac{\omega a}{8} \qquad\qquad (10.290)$$

Sketch the amplitude spectrum and show that the product of bandwidth and pulse duration equals $8\pi/3$.

Figure 10.77
Waveform of a stepped
pulse.

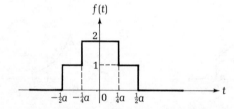

Figure 10.78
An RL network (b)
excited by a triangular
voltage source $v_g(t)$,
the waveform of which is
shown in (a).

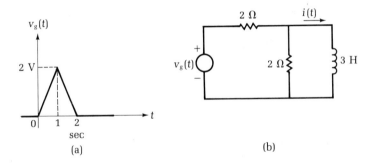

(a)

(b)

10.17 In the network of Figure 10.78(b), the voltage source $v_g(t)$ has the triangular waveform shown in (a). Verify that the Fourier transform $I(j\omega)$ of the inductor current $i(t)$ is given by

$$I(j\omega) = \frac{1}{1 + j3\omega} \cdot \frac{\sin^2 \frac{1}{2}\omega}{(\frac{1}{2}\omega)^2} e^{-j\omega} \qquad (10.291)$$

Sketch the amplitude and phase spectra for the inductor current $i(t)$.

10.18 Prove the identity (10.138).

10.19 In Figure 10.79, the network and its excitation are as shown. Determine the Fourier transforms of the source voltage $v_g(t)$ and inductor current $i(t)$. Sketch the amplitude and phase spectra for $v_g(t)$ and $i(t)$.

10.20 Show that if $\mathbf{F}[f(t)] = F(j\omega)$, then

$$F(j\omega) * e \quad = 2\pi f(t) e^{-j\omega t} \qquad (10.292)$$

10.21 Show that

$$u(t) * \frac{\sin at}{\pi t} = \frac{1}{2} + \frac{1}{\pi} \text{Si } at \qquad (10.293)$$

10.22 Show that if $\mathbf{F}[f(t)] = F(j\omega)$, then for an arbitrary constant T_0

$$T_0 \sum_{n=-\infty}^{\infty} f(nT_0) = \sum_{n=-\infty}^{\infty} F\left(\frac{j2n\pi}{T_0}\right) \qquad (10.294)$$

10.23 Repeat Problem 10.19 for each of the following input excitations:

(a) $v_g(t) = e^{-2t}u(t)$ (10.295a)

(b) $v_g(t) = (\sin 2t)u(t)$ (10.295b)

(c) $v_g(t) = e^t u(-t)$ (10.295c)

Figure 10.79
An RL network (b) excited
by a cosine voltage source
$v_g(t)$, the waveform of
which is shown in (a).

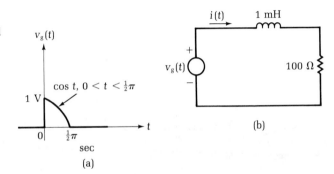

(a)

(b)

10.24 The triangular pulse of Figure 10.75(b) with $a = 0.01$ sec is applied to an ideal low-pass filter having a cutoff frequency of 200 Hz. Ascertain the output waveform of the filter.

10.25 Using the FFT algorithm, compute the amplitude and phase spectra of the signals shown in Figure 10.75 with $a = 1$.

10.26 Find the time function $f(t)$ for each of the frequency spectra $F(j\omega)$ as shown in Figure 10.80.

10.27 Use the energy theorem to evaluate the following integral

$$\int_{-\infty}^{\infty} \frac{\sin^2 t}{t^2}\, dt \tag{10.296}$$

10.28 Use the energy theorem to verify the following integrals:

(a) $\displaystyle\int_{-\infty}^{\infty} \frac{1}{(a^2 + \omega^2)^2}\, d\omega = \frac{\pi}{2|a|^3}$ $\tag{10.297a}$

(b) $\displaystyle\int_{-\infty}^{\infty} \frac{\sin^4 a\omega}{\omega^4}\, d\omega = \frac{2\pi}{3|a|^3}$ $\tag{10.297b}$

10.29 Confirm that if $f(t)$ is band limited, as in (10.196), then for an arbitrary real constant b the sampling theorem (10.197) can be written in any one of the following forms:

(a) $\displaystyle f(t - b) = \sum_{n=-\infty}^{\infty} f(nT - b) \frac{\sin \omega_c(t - nT)}{\omega_c(t - nT)}$ $\tag{10.298a}$

(b) $\displaystyle f(t) = \sum_{n=-\infty}^{\infty} f(nT - b) \frac{\sin \omega_c(t - nT + b)}{\omega_c(t - nT + b)}$ $\tag{10.298b}$

(c) $\displaystyle f(t) = \sin \omega_c t \sum_{n=-\infty}^{\infty} (-1)^n \frac{f(nT)}{\omega_c t - n\pi}$ $\tag{10.298c}$

Figure 10.80
The Fourier transforms of
two unknown time-
domain signals.

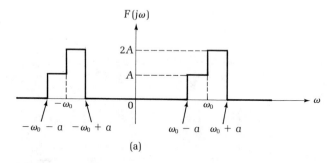

(a)

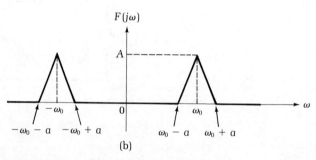

(b)

10.30 A cosine pulse is described by the equation

$$f(t) = (\cos \omega_0 t)[u(t + \tfrac{1}{2}T) - u(t - \tfrac{1}{2}T)]$$ (10.299)

Show that the Fourier transform for this pulse is given by

$$F(j\omega) = \frac{T}{2}\left[\frac{\sin \tfrac{1}{2}(\omega - \omega_0)T}{\tfrac{1}{2}(\omega - \omega_0)T} + \frac{\sin \tfrac{1}{2}(\omega + \omega_0)T}{\tfrac{1}{2}(\omega + \omega_0)T}\right]$$ (10.300)

Sketch the amplitude spectrum for $f(t)$.

10.31 Verify that the Fourier transform of the signal $f(t) = \exp(-a^2 t^2)$ is given by

$$F(j\omega) = \frac{\sqrt{\pi}}{a}\exp\left(\frac{-\omega^2}{4a^2}\right)$$ (10.301)

CHAPTER ELEVEN

STATE EQUATIONS

The primary system of equations of a network having b network branches is a system of $2b$ linear algebraic and differential equations, which is difficult to handle. To simplify the problem, as demonstrated in Chapter 5, we apply the Laplace transform technique to the primary system and obtain a system of $2b$ linear algebraic equations, which is relatively simple to manipulate. The main drawback is that it contains a large number of equations—namely, $2b$ equations. To reduce this number, in Chapter 5 we presented three secondary systems of equations: the nodal system, the cutset system, and the loop system. In nodal or cutset analysis, there are $n - 1$ linearly independent equations corresponding, for example, to the set of f-cutsets with respect to a chosen tree. In loop analysis there are $b - n + 1$ linearly independent equations that can be obtained, for example, with respect to a chosen set of f-circuits. To obtain the final time-domain solution, we must take the inverse Laplace transformation. For most practical networks, the procedure is usually long and complicated and requires an excessive amount of computer time.

As an alternative we can formulate the network equations in the time domain as a system of first-order differential equations which describes the dynamic behavior of the network. Some advantages of representing the network equations in this form are the following. First of all, such a system has been widely studied in mathematics and its solution, both analytical and numerical, is readily known and available. Secondly, the representation can easily and naturally be extended to time-varying and nonlinear networks. In fact, computer-aided solution of time-varying, nonlinear network problems is almost always accomplished using the state-variable approach. Finally, the first-order differential equations can easily be programmed for a digital computer or simulated on an analog computer. Even if it were not for the above reasons, the new approach provides an alternate view of the physical behavior of the network.

The term "state" is an abstract concept that may be represented in many ways. As an analog think of the number 5. This number may be represented by the numeral 5 or by the symbol V. In the binary system it is represented by the symbol 101. We may even picture it as a point in the middle of a real line with abscissa 0 and 10. All of these representations evoke in our minds the concept of the number 5. Thus when we speak of "the state of the system is ... ," we really mean that "the state of the system is represented by ... ."

If we call the set of instantaneous values of all the branch currents and voltages as the "state" of the network, then the knowledge of the instantaneous values of all these variables determines this instantaneous state. However, not all of these instantaneous values are required in order to determine the instantaneous state, because some can be calculated from the others. A set of data qualifies to be called the *state* of a system if it fulfills the following two requirements:

1.   The state at any time, say, t_0 and the input to the system from t_0 on determine uniquely the state at any time $t > t_0$.
2.   The state at time t and the inputs together with some of their derivatives at time t determine uniquely the value of any system variable at the time t.

The state may be regarded as a vector the components of which are the *state variables*. Network variables that are candidates for the state variables are the branch currents and voltages. Our problem is to choose state variables in order to formulate the state equations. In this chapter we first present procedures for systematic writing of network equations in the form of a system of first-order differential equations known as the state equations and then discuss the number of dynamically independent state variables required in the formulation of the state equations. Like the nodal, cutset, or loop system of equations, the state equations are formulated from the primary system of network equations and are considered as another secondary system of equations. In addition, we shall demonstrate how to obtain state equations for systems described by scalar differential equations.

11-1   STATE EQUATIONS IN NORMAL FORM

Consider the network of Figure 11.1(a) the associated directed graph of which is shown in Figure 11.1(b). As before, let $i_k(t)$ and $v_k(t)$ be the current and

Figure 11.1

A network (a) and its associated directed graph (b) used to illustrate the formulation of state equations in normal form.

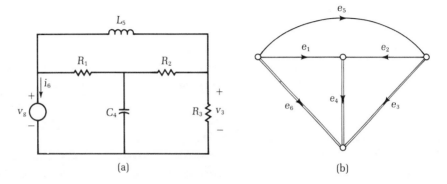

(a)   (b)

voltage of edge e_k $(k = 1, 2, \ldots, 6)$. For simplicity we shall drop the time variable t in all the variables and functions unless it is used for emphasis. We first determine the primary system of equations and then demonstrate the procedure by which the state equations are obtained. A systematic procedure for writing the state equations is presented in Section 11-4.

In Figure 11.1(b) choose the tree $T = e_3e_4e_6$.[†] The KCL and KVL equations corresponding to this tree are obtained as follows:

$$
\mathbf{Q}_f \mathbf{i}(t) =
\begin{array}{c}
\begin{array}{cccccc} e_1 & e_2 & e_5 & e_3 & e_4 & e_6 \end{array} \\
\begin{bmatrix}
0 & 1 & -1 & 1 & 0 & 0 \\
-1 & -1 & 0 & 0 & 1 & 0 \\
1 & 0 & 1 & 0 & 0 & 1
\end{bmatrix}
\end{array}
\begin{bmatrix}
i_1 \\ i_2 \\ i_5 \\ i_3 \\ i_4 \\ i_6
\end{bmatrix}
=
\begin{bmatrix}
0 \\ 0 \\ 0
\end{bmatrix}
\tag{11.1}
$$

$$
\mathbf{B}_f \mathbf{v}(t) =
\begin{bmatrix}
1 & 0 & 0 & 0 & 1 & -1 \\
0 & 1 & 0 & -1 & 1 & 0 \\
0 & 0 & 1 & 1 & 0 & -1
\end{bmatrix}
\begin{bmatrix}
v_1 \\ v_2 \\ v_5 \\ v_3 \\ v_4 \\ v_6
\end{bmatrix}
=
\begin{bmatrix}
0 \\ 0 \\ 0
\end{bmatrix}
\tag{11.2}
$$

The element v-i equations are found to be

$$v_1 = R_1 i_1 \tag{11.3a}$$

$$v_2 = R_2 i_2 \tag{11.3b}$$

$$i_3 = \frac{v_3}{R_3} \tag{11.3c}$$

$$i_4 = C_4 \frac{dv_4}{dt} \tag{11.3d}$$

$$v_5 = L_5 \frac{di_5}{dt} \tag{11.3e}$$

$$v_6 = v_g \tag{11.3f}$$

Our aim is to express dv_4/dt and di_5/dt in terms of v_4, i_5, and the known source v_g. To this end we substitute (11.3) in (11.1) and (11.2) to get

$$i_5 - i_2 = \frac{v_3}{R_3} \tag{11.4a}$$

[†] A tree may be represented by the juxtaposition or "product" of its branch designation symbols.

$$C_4 \frac{dv_4}{dt} = i_1 + i_2 \tag{11.4b}$$

$$i_6 = -i_1 - i_5 \tag{11.4c}$$

$$R_1 i_1 = v_g - v_4 \tag{11.4d}$$

$$R_2 i_2 = -v_4 + v_3 \tag{11.4e}$$

$$L_5 \frac{di_5}{dt} = v_g - v_3 \tag{11.4f}$$

From (11.4a) and (11.4e), we can solve i_2 and v_3 in terms of i_5 and v_4 giving

$$i_2 = \frac{R_3 i_5 - v_4}{R_2 + R_3} \tag{11.5a}$$

$$v_3 = \frac{R_2 R_3 i_5 + R_3 v_4}{R_2 + R_3} \tag{11.5b}$$

Finally, substituting i_1, i_2, and v_3 from (11.4d) and (11.5) in (11.4b) and (11.4f), we get

$$\frac{dv_4}{dt} = -\frac{v_4}{R_1 C_4} - \frac{v_4}{C_4(R_2 + R_3)} + \frac{R_3 i_5}{C_4(R_2 + R_3)} + \frac{v_g}{R_1 C_4} \tag{11.6a}$$

$$\frac{di_5}{dt} = -\frac{R_3 v_4}{L_5(R_2 + R_3)} - \frac{R_2 R_3 i_5}{L_5(R_2 + R_3)} + \frac{v_g}{L_5} \tag{11.6b}$$

or in matrix notation

$$\begin{bmatrix} \dfrac{dv_4}{dt} \\[4mm] \dfrac{di_5}{dt} \end{bmatrix} = \begin{bmatrix} -\dfrac{1}{R_1 C_4} - \dfrac{1}{C_4(R_2 + R_3)} & \dfrac{R_3}{C_4(R_2 + R_3)} \\[4mm] -\dfrac{R_3}{L_5(R_2 + R_3)} & -\dfrac{R_2 R_3}{L_5(R_2 + R_3)} \end{bmatrix} \begin{bmatrix} v_4 \\[4mm] i_5 \end{bmatrix} + \begin{bmatrix} \dfrac{1}{R_1 C_4} \\[4mm] \dfrac{1}{L_5} \end{bmatrix} [v_g] \tag{11.7}$$

The example demonstrates that with appropriate elimination of "un-wanted" variables the primary system of network equations can be reduced to a system of first-order differential equations involving only capacitor voltage v_4, inductor current i_5, and voltage source v_g. We note that in deriving (11.7) we choose the voltage across the capacitor v_4 and the current through the inductor i_5. In so doing we avoid the integral expressions like

$$\frac{1}{L} \int v(t)\, dt \quad \text{and} \quad \frac{1}{C} \int i(t)\, dt$$

for the current through the inductor and for the voltage across the capacitor during the elimination process and obtain two first-order differential equations of the form

$$\begin{bmatrix} \dot{x}_1 \\ \dot{x}_2 \end{bmatrix} = \begin{bmatrix} a_{11} & a_{12} \\ a_{21} & a_{22} \end{bmatrix} \begin{bmatrix} x_1 \\ x_2 \end{bmatrix} + \begin{bmatrix} b_{11} \\ b_{21} \end{bmatrix} [u_1] \tag{11.8}$$

where $\dot{x}_j$ $(j = 1, 2)$ denotes the time derivative of x_j. Excluding degnerate networks, which will be discussed in Section 11-5, if a nondegenerate network has k energy storage elements and h independent sources, the primary system of network equations can be reduced to a system of k first-order differential equations:

$$\dot{x}_i(t) = \sum_{j=1}^{k} a_{ij}x_j(t) + \sum_{j=1}^{h} b_{ij}u_j(t), \qquad (i = 1, 2, \ldots, k) \tag{11.9}$$

or in matrix notation

$$\begin{bmatrix} \dot{x}_1(t) \\ \dot{x}_2(t) \\ \vdots \\ \dot{x}_k(t) \end{bmatrix} = \begin{bmatrix} a_{11} & a_{12} & \cdots & a_{1k} \\ a_{21} & a_{22} & \cdots & a_{2k} \\ \vdots & \vdots & \vdots & \vdots \\ a_{k1} & a_{k2} & \cdots & a_{kk} \end{bmatrix} \begin{bmatrix} x_1(t) \\ x_2(t) \\ \vdots \\ x_k(t) \end{bmatrix}$$

$$+ \begin{bmatrix} b_{11} & b_{12} & \cdots & b_{1h} \\ b_{21} & b_{22} & \cdots & b_{2h} \\ \vdots & \vdots & \vdots & \vdots \\ b_{k1} & b_{k2} & \cdots & b_{kh} \end{bmatrix} \begin{bmatrix} u_1(t) \\ u_2(t) \\ \vdots \\ u_h(t) \end{bmatrix}$$

or more compactly

$$\dot{\mathbf{x}}(t) = \mathbf{A}\mathbf{x}(t) + \mathbf{B}\mathbf{u}(t) \tag{11.10}$$

The real functions $x_1(t)$, $x_2(t)$, $\ldots$, $x_k(t)$ of the time t are called the *state variables*, and the k-vector $\mathbf{x}(t)$ formed by the state variables is known as the *state vector*. The h-vector $\mathbf{u}(t)$ formed by the h known forcing functions or excitations $u_j(t)$ is referred to as the *input vector*. The coefficient matrices $\mathbf{A}$ and $\mathbf{B}$, depending only upon the network parameters, are of orders $k \times k$ and $k \times h$, respectively.[†] Equation (11.10) is usually called the *state equation in normal form*. For the network of Figure 11.1(a), the state equation is given by (11.7). The state variables are the capacitor voltage v_4 and inductor current i_5, and the input vector $\mathbf{u}(t)$ is a vector of one element representing the only excitation v_g.

The state variables x_j may or may not be the desired output variables. We therefore must express the desired output variables in terms of the state variables and excitations. In Figure 11.1(a) suppose that we choose the resistor voltage v_3 and generator current i_6 as the desired output vari-

---

[†] The coefficient matrices $\mathbf{A}$ and $\mathbf{B}$ should not be confused with the basis incidence matrix $\mathbf{A}$ and the basis circuit matrix $\mathbf{B}$. Since they are widely accepted in the literature, we will introduce no new symbols here, as they can be differentiated by their context.

ables. Then from (11.4c) and (11.5b) in conjunction with (11.4d) we obtain

$$v_3 = \frac{R_2 R_3 i_5 + R_3 v_4}{R_2 + R_3} \qquad \text{(11.11a)}$$

$$i_6 = \frac{v_4 - v_g}{R_1} - i_5 \qquad \text{(11.11b)}$$

or in matrix notation

$$\begin{bmatrix} v_3 \\ i_6 \end{bmatrix} = \begin{bmatrix} \dfrac{R_3}{R_2 + R_3} & \dfrac{R_2 R_3}{R_2 + R_3} \\ \dfrac{1}{R_1} & -1 \end{bmatrix} \begin{bmatrix} v_4 \\ i_5 \end{bmatrix} + \begin{bmatrix} 0 \\ -\dfrac{1}{R_1} \end{bmatrix} [v_g] \qquad \text{(11.12)}$$

which expresses the output variables in terms of the state variables and the input excitation. Thus the solution of the state equation (11.7) for the state vector $[v_4, i_5]'$ also gives answer to $[v_3, i_6]'$. Equation (11.12) has the standard form

$$\begin{bmatrix} y_1(t) \\ y_2(t) \end{bmatrix} = \begin{bmatrix} c_{11} & c_{12} \\ c_{21} & c_{22} \end{bmatrix} \begin{bmatrix} x_1 \\ x_2 \end{bmatrix} + \begin{bmatrix} d_{11} \\ d_{21} \end{bmatrix} [v_g] \qquad \text{(11.13)}$$

In general, if there are q output variables $y_j(t)$ ($j = 1, 2, \ldots, q$) and h input excitations, the *output vector* $\mathbf{y}(t)$ formed by the q output variable $y_j(t)$ can be expressed in terms of the state vector $\mathbf{x}(t)$ and the input vector $\mathbf{u}(t)$ by the matrix equation

$$\mathbf{y}(t) = \mathbf{C}\mathbf{x}(t) + \mathbf{D}\mathbf{u}(t) \qquad \text{(11.14)}$$

where the known coefficient matrices $\mathbf{C}$ and $\mathbf{D}$, depending only on the network parameters, are of orders $q \times k$ and $q \times h$, respectively. Equation (11.14) is called the *output equation*. The state equation (11.10) and the output equation (11.14) together are known as the *state equations*.

We illustrate the above discussion by the following example.

EXAMPLE 11.1
We shall write the state equations for the network of Figure 11.2(a) the associated directed graph of which is shown in Figure 11.2(b). There are many trees in the directed graph of Figure 11.2(b), and for our purposes we choose the tree $T = e_2 e_3 e_6$. The primary system of equations becomes

$$\mathbf{Q}_f \mathbf{i}(t) = \begin{array}{cccccc} e_1 & e_4 & e_5 & e_2 & e_3 & e_6 \end{array} \\ \begin{bmatrix} 0 & -1 & 0 & 1 & 0 & 0 \\ 1 & 1 & -1 & 0 & 1 & 0 \\ 0 & -1 & 0 & 0 & 0 & 1 \end{bmatrix} \begin{bmatrix} i_1 \\ i_4 \\ i_5 \\ i_2 \\ i_3 \\ i_6 \end{bmatrix} = \begin{bmatrix} 0 \\ 0 \\ 0 \end{bmatrix} \qquad \text{(11.15)}$$

Figure 11.2
A network (a) and its associated directed graph (b) used to illustrate the formulation of state equations in normal form.

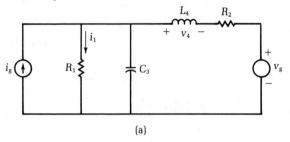

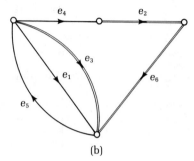

(a)

(b)

$$\mathbf{B}_f \mathbf{v}(t) = \begin{bmatrix} 1 & 0 & 0 & 0 & -1 & 0 \\ 0 & 1 & 0 & 1 & -1 & 1 \\ 0 & 0 & 1 & 0 & 1 & 0 \end{bmatrix} \begin{bmatrix} v_1 \\ v_4 \\ v_5 \\ v_2 \\ v_3 \\ v_6 \end{bmatrix} = \begin{bmatrix} 0 \\ 0 \\ 0 \end{bmatrix} \tag{11.16}$$

$$v_1 = R_1 i_1 \tag{11.17a}$$

$$i_2 = \frac{v_2}{R_2} \tag{11.17b}$$

$$i_3 = C_3 \frac{dv_3}{dt} = C_3 \dot{v}_3 \tag{11.17c}$$

$$v_4 = L_4 \frac{di_4}{dt} = L_4 \dot{i}_4 \tag{11.17d}$$

$$i_5 = i_g \tag{11.17e}$$

$$v_6 = v_g \tag{11.17f}$$

The capacitor voltage v_3 and the inductor current i_4 are chosen as the state variables. Our objective is to eliminate the resistor voltages v_1 and v_2 and the resistor currents i_1 and i_2 among the equations (11.15) to (11.17). To this end we substitute (11.17) in (11.15) and (11.16) and obtain

$$i_4 = i_2 = \frac{v_2}{R_2} \tag{11.18a}$$

$$C_3 \frac{dv_3}{dt} = C_3 \dot{v}_3 = -i_1 - i_4 + i_g \tag{11.18b}$$

$$i_6 = i_4 \tag{11.18c}$$

$$v_3 = v_1 = R_1 i_1 \tag{11.18d}$$

$$L_4 \frac{di_4}{dt} = L_4 \dot{i}_4 = -v_2 + v_3 - v_g \tag{11.18e}$$

$$v_5 = -v_3 \tag{11.18f}$$

To continue our elimination process, we substitute v_2 and i_1 from (11.18a) and (11.18d) in (11.18b) and (11.18e), yielding

$$C_3 \dot{v}_3 = C_3 \frac{dv_3}{dt} = -\frac{v_3}{R_1} - i_4 + i_g \tag{11.19a}$$

$$L_4 \dot{i}_4 = L_4 \frac{di_4}{dt} = -R_2 i_4 + v_3 - v_g \tag{11.19b}$$

which may be put in matrix form as

$$\begin{bmatrix} \dot{v}_3 \\ \dot{i}_4 \end{bmatrix} = \begin{bmatrix} -\dfrac{1}{R_1 C_3} & -\dfrac{1}{C_3} \\ \dfrac{1}{L_4} & -\dfrac{R_2}{L_4} \end{bmatrix} \begin{bmatrix} v_3 \\ i_4 \end{bmatrix} + \begin{bmatrix} \dfrac{1}{C_3} & 0 \\ 0 & -\dfrac{1}{L_4} \end{bmatrix} \begin{bmatrix} i_g \\ v_g \end{bmatrix} \tag{11.20}$$

Equation (11.20) is the desired state equation in normal form. Upon comparison of (11.20) to (11.10), we find that

$$\mathbf{A} = \begin{bmatrix} -\dfrac{1}{R_1 C_3} & -\dfrac{1}{C_3} \\ \dfrac{1}{L_4} & -\dfrac{R_2}{L_4} \end{bmatrix}, \qquad \mathbf{B} = \begin{bmatrix} \dfrac{1}{C_3} & 0 \\ 0 & -\dfrac{1}{L_4} \end{bmatrix} \tag{11.21}$$

$$\mathbf{x}(t) = \begin{bmatrix} v_3 \\ i_4 \end{bmatrix}, \qquad \mathbf{u}(t) = \begin{bmatrix} i_g \\ v_g \end{bmatrix} \tag{11.22}$$

so that

$$\dot{\mathbf{x}}(t) = \begin{bmatrix} \dot{v}_3 \\ \dot{i}_4 \end{bmatrix} \tag{11.23}$$

Suppose that the resistor current i_1 and inductor voltage v_4 are the desired output variables. From (11.17a) and (11.18e) in conjunction with (11.17d), (11.18a), and (11.18d), we get

$$i_1 = \frac{v_3}{R_1} \tag{11.24a}$$

$$v_4 = v_3 - R_2 i_4 - v_g \tag{11.24b}$$

which can be combined into a single matrix equation as

$$\begin{bmatrix} i_1 \\ v_4 \end{bmatrix} = \begin{bmatrix} 1/R_1 & 0 \\ 1 & -R_2 \end{bmatrix} \begin{bmatrix} v_3 \\ i_4 \end{bmatrix} + \begin{bmatrix} 0 & 0 \\ 0 & -1 \end{bmatrix} \begin{bmatrix} i_g \\ v_g \end{bmatrix} \tag{11.25}$$

Equation (11.25) is the output equation in the standard form of (11.14) with the state vector $\mathbf{x}$ and the input vector $\mathbf{u}$ unaltered as shown in (11.22). The output vector $\mathbf{y}$ and the matrices $\mathbf{C}$ and $\mathbf{D}$ are identified as

$$\mathbf{y}(t) = \begin{bmatrix} i_1 \\ v_4 \end{bmatrix}, \qquad \mathbf{C} = \begin{bmatrix} 1/R_1 & 0 \\ 1 & -R_2 \end{bmatrix}, \qquad \mathbf{D} = \begin{bmatrix} 0 & 0 \\ 0 & -1 \end{bmatrix} \qquad \textbf{(11.26)}$$

Equations (11.20) and (11.25) constitute the state equations of the network of Figure 11.2(a).

The preceding example illustrates how to obtain the state equations from the primary systems of network equations. For simple networks such as those given in this section, it is quite straightforward to choose the right combinations of equations to produce the state equations. However, when the network under study contains a large number of elements, the problem becomes difficult. For this reason we need a systematic procedure for writing the state equations. However, before we do this, we first discuss the concept of state and state variables for networks.

11-2   THE CONCEPT OF STATE AND STATE VARIABLES FOR NETWORKS

Our immediate problem is to choose the network variables as the state variables in order to formulate the state equations. If we call the set of instantaneous values of all the branch currents and voltages the "state" of the network, then the knowledge of the instantaneous values of all these variables determines this instantaneous state. However, not all of these instantaneous values are required in order to determine the instantaneous state, since some can be calculated from the others. For example, the instantaneous voltage of a resistor can be obtained from its instantaneous current through Ohm's law. The question arises as to how many instantaneous values of branch voltages and currents are sufficient to determine completely the instantaneous state of the network. This leads to the following definition.

> **Definition 11.1 Complete Set of State Variables**   *In a given network, a minimal set of its branch variables is said to be a complete set of state variables if their instantaneous values are sufficient to determine completely the instantaneous values of all the branch variables.*

We emphasize that the state of a system is unique, but its representation is generally not. In fact, there may be infinite representations for a state.

We illustrate the above concepts by means of a very simple and well-known network problem. Figure 11.3(a) is a simple RC network with pertinent voltages and current as indicated. The input excitation voltage is

Figure 11.3

Plot of the capacitor voltage (c) as a function of t for the RC network (a) with input excitation as shown in (b).

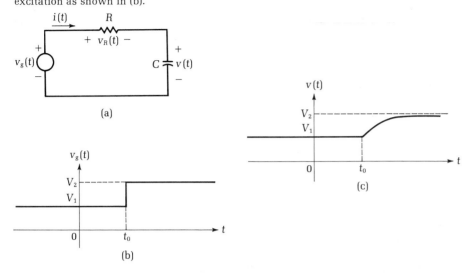

shown in Figure 11.3(b). It is easy to verify that the network is described by the state equation

$$\dot{v}(t) = -\frac{1}{RC} v(t) + \frac{1}{RC} v_g(t) \tag{11.27}$$

the solution of which, as will be shown in the following chapter, is given by

$$v(t) = (V_1 - V_2)e^{-(t-t_0)/RC} + V_2, \qquad t \geq t_0 \tag{11.28}$$

and is plotted in Figure 11.3(c). If we consider t_0 as a time reference, the capacitor voltage $v(t)$ at the time t_0 is the initial voltage across the capacitor, or $v(t_0) = V_1$. For $t \geq t_0$ the present and future input voltage $v_g(t)$ is V_2. Equation (11.28) indicates that regardless of what the input voltage was prior to t_0, the capacitor voltage $v(t)$ for $t \geq t_0$ is completely determined by its initial voltage $v(t_0) = V_1$ and the present and future input voltage $v_g(t) = V_2$. In other words the initial value of $v(t)$ at $t = t_0$ can be regarded as a quantity that describes the complete history of the network up to the time t_0. This value together with the input signal over the interval from t_0 to t is all that is necessary to determine the response at time t.

Once the instantaneous capacitor voltage $v(t)$ is known, the instantaneous generator current, resistor current, and capacitor current are all given by

$$i(t) = C\dot{v}(t)$$

$$= \frac{V_2 - V_1}{R} e^{-(t-t_0)/RC}, \qquad t \geq t_0 \tag{11.29}$$

The instantaneous resistor voltage is obtained as $v_R(t) = Ri(t)$ through Ohm's law. This demonstrates that the instantaneous value of the capacitor voltage $v(t)$ is sufficient to determine completely the instantaneous values of all the branch variables. Thus $v(t)$ constitutes a complete set of the state variable. On the other hand, the knowledge of the instantaneous values of $v(t)$ and $i(t)$ also determines the instantaneous values of all the branch variables or the state of the network. The variables $v(t)$ and $i(t)$, however, do not constitute a complete set of state variables because it is not a minimal collection of branch variables with this property. As an alternative, the current $i(t)$ alone forms a complete set of the state variable because, knowing $i(t)$, $v(t)$ is determined by the equation

$$v(t) = v_g - Ri(t) \tag{11.30}$$

Thus, in general, the complete set of state variables is not unique. In fact, the choice of state variables in a given physical system is somewhat arbitrary. In network problems the state variables are chosen to be the voltages across the capacitors and the currents in the inductors. The main reason for this choice is that the initial conditions may be applied directly to the solution of the problem. This is why in all the illustrative examples the capacitor voltages and inductor currents were chosen as the state variables. In most situations these variables will constitute a complete set of state variables for the network. However, this is not always the case. For example, in a network containing two capacitors connected in parallel, the voltages of these two capacitors cannot be chosen independently, one being determined by the other. Hence they cannot both be the members of a complete set, since any complete set must be minimal. Methods of selecting a complete set of state variables will be discussed in the following sections.

11-3 NORMAL TREE AND ITS SELECTION

As discussed in Chapter 2, we recall that with respect to a chosen tree, the tree-branch voltages determine all other voltages by means of the f-circuits, and the cotree-link currents determine all other currents by way of the f-cutsets. Thus, to develop state equations using capacitor voltages and inductor currents as the state variables, we should place all the voltage sources and as many capacitors as possible in a tree and all the current sources and as many inductors as possible in a cotree. This leads to the following definition.

> **Definition 11.2 Normal Tree**   *In the connected directed graph associated with a network, a normal tree is a tree that contains all the independent-voltage-source edges, the maximum number of capacitive edges, the minimum number of inductive edges, and none of the independent-current-source edges.*

Figure 11.4

A network (a) and its associated directed graph (b) used to illustrate the selection of a normal tree.

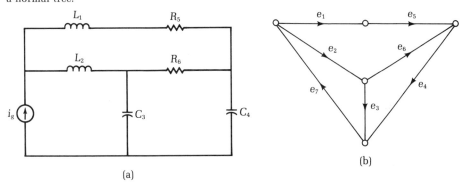

Notice that in this definition we exclude the possibility of having unconnected networks. In the case of unconnected networks, we consider the normal trees of the components individually. In Figure 11.2 the tree $T = e_2 e_3 e_6$ as depicted in Figure 11.2(b) is a normal tree. If the tree-branch voltages v_2, v_3, and v_6 are known, all other voltages v_1, v_4, and v_5 are determined by means of the f-circuits as given in (11.16). Similarly if the cotree-link currents i_1, i_4, and i_5 are specified, all other currents i_2, i_3, and i_6 can be obtained by way of the f-cutsets as shown in (11.15).

The name *normal tree* is used because it is the tree that will enable us to obtain the state equation in normal form. The normal tree may or may not be unique. For instance, the normal tree $T = e_2 e_3 e_6$ in Figure 11.2(b) is unique whereas in Figure 11.1(b) we have two normal trees $e_3 e_4 e_6$ and $e_2 e_4 e_6$. As a more complicated example, consider the network and its associated directed graph G of Figure 11.4. Observe that the independent current-source edge e_7 and the inductive edges e_1 and e_2 constitute a cutset. Therefore at least one of these edges must be contained in every tree of G, since a tree contains all the nodes of G. To construct a normal tree, we must exclude e_7

Figure 11.5

Normal trees (a) and nonnormal trees (b) of the network of Figure 11.4(a).

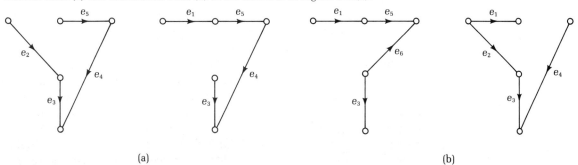

Figure 11.6
A network (a) and its associated directed graph (b) used to illustrate the selection of a normal tree.

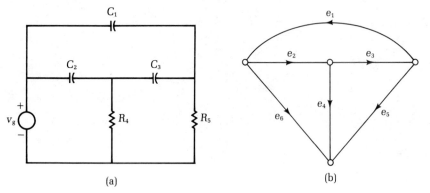

(a)   (b)

Figure 11.7
Normal trees of the
network of Figure 11.6(a).

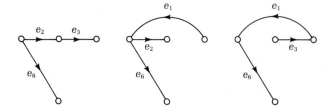

and include e_3 and e_4 as tree branches, yielding two normal trees $e_2e_3e_4e_5$ and $e_1e_3e_4e_5$ as shown in Figure 11.5(a). The two trees $e_1e_3e_5e_6$ and $e_1e_2e_3e_4$ of Figure 11.5(b) are not normal trees because $e_1e_3e_5e_6$ does not contain the maximum number of capacitive edges that can be put in a tree, which is two, and $e_1e_2e_3e_4$ contains one more inductive edge than the required absolute minimum, which is one. On the other hand, for some networks not all the capacitive edges can be made part of a tree. In Figure 11.6, the three capacitive edges e_1, e_2, and e_3 form a circuit, making it impossible to include them all in any tree. Thus the maximum number of capacitive edges that can be included in any tree is two. Figure 11.7 shows the three normal trees associated with the network of Figure 11.6(a).

In general, a subgraph can be made part of a tree if and only if it does not contain any circuit. Likewise a subgraph can be made part of a cotree if and only if it does not contain any cutset of the given graph. In Figure 11.6(b) the subgraphs e_3e_6, $e_1e_2e_6$, and e_1e_3 do not contain any circuit, and therefore each can be made part of a tree as shown in Figure 11.7. The subgraph e_4e_5 does not contain any cutset of Figure 11.6(b), so it can be made part of a cotree, being excluded from a tree. Some of the trees that exclude e_4e_5 are given in Figure 11.7. As another example, consider the directed graph G of Figure 11.4(b). The subgraph formed by e_7 and either e_1 or e_2 contains no cutset of G, and thus can be included in a cotree, as depicted in Figure 11.8.

Figure 11.8
Cotrees of the network
of Figure 11.4(b).

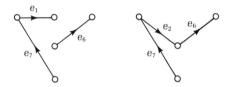

The subgraph $e_3 e_4 e_5$ does not contain any circuit, meaning that there is at least one tree containing $e_3 e_4 e_5$ as a subgraph. Two such trees are found in Figure 11.5(a). In fact, the inclusion of a subgraph g_1 in some tree T_1 and the exclusion of a subgraph g_2 with respect to some tree T_2, as discussed above, can be achieved with respect to the same tree T provided that g_1 and g_2 do not have edges in common. We state this result formally below, a proof of which may be found in Chen.[†]

> **Property 11.1**   *Assume that the subgraphs g_1 and g_2 of a connected directed graph G do not have edges in common. If there exists a tree T_1 containing g_1 as a subgraph and, furthermore, if there exists a tree T_2 and in its complement $\bar{T}_2$, g_2 is contained, then there exists a tree T of G for which g_1 is contained in T and g_2 is contained in its complement $\bar{T}$.*

The result is extremely useful in determining the maximum number of capacitive edges and the minimum number of inductive edges that can be included in a tree. For instance, in Figure 11.9 edges e_1, e_2, and e_7 constitute a cutset and edges e_3, e_4, and e_8 a circuit. Hence at least one of the

---

[†] W. K. Chen, *Applied Graph Theory: Graphs and Electrical Networks*, 2d ed. (Amsterdam: North-Holland, and New York: American Elsevier), 1976, p. 322.

Figure 11.9
A network (a) and its associated directed graph (b) used to illustrate the determination of the maximum number of capacitive edges and the minimum number of inductive edges that can be included in a tree.

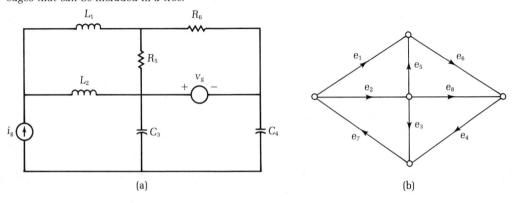

(a)   (b)

Figure 11.10

A tree (a) containing edges e_3 and e_8, and a tree (b) for which e_2e_7 is contained in its complement, showing the existence of a tree T in the directed graph of Figure 11.9(b) for which e_3e_8 is contained in T and e_2e_7 is contained in its complement.

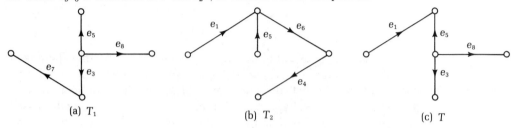

(a) T_1 (b) T_2 (c) T

edges e_1, e_2, and e_7 will be in every tree, and at most two of the edges e_3, e_4, and e_8 can be made part of any tree. Figure 11.10(a) is a tree T_1 containing edges e_3 and e_8, and Figure 11.10(b) shows a tree $T_2 = e_1e_4e_5e_6$ for which e_2e_7 is contained in its complement $\bar{T}_2 = e_2e_3e_7e_8$. According to Property 11.1, there exists a tree T for which e_3e_8 is contained in T and e_2e_7 is contained in its complement $\bar{T}$. Such a tree T is presented in Figure 11.10(c), showing that a normal tree must contain one inductive edge and one capacitive edge.

11-4 SYSTEMATIC PROCEDURE IN WRITING STATE EQUATIONS FOR NETWORKS

In this section we present step by step a systematic procedure for writing the state equation for a network. These steps are a systematic way to eliminate the unwanted variables in the primary system of equations.

Step 1. From a given network, construct a directed graph G representing its interconnection and indicating the voltage and current references of its elements—namely, the directions of the edges.

Step 2. In G select a normal tree T and choose as the state variables the capacitor voltages of T and the inductor currents of the cotree $\bar{T}$, the complement of T in G.

Step 3. Assign each branch of tree T a voltage symbol, and assign each link of cotree $\bar{T}$ a current symbol.

Step 4. Using KCL express each tree-branch current as a sum of cotree-link currents by way of the f-cutsets defined with respect to T, and indicate it in G if necessary.

Step 5. Using KVL express each cotree-link voltage as a sum of tree–branch voltages by means of the f-circuits defined with respect to T, and indicate it in G if necessary.

Step 6. Write the element v-i equations for the passive elements and separate these equations into two groups:

    (a)   those element v-i equations for the tree-branch capacitors and the cotree-link inductors

    (b)   those element v-i equations for all other passive elements

Step 7. Eliminate the nonstate variables among the equations obtained in the preceding step. *Nonstate variables* are defined as those variables that are neither state variables nor known independent sources.

Step 8. Rearrange the terms and write the resulting equations in normal form.

We shall illustrate the above steps by the following examples.

EXAMPLE 11.2

Let us consider again the network N of Figure 11.2(a) discussed in Example 11.1. We shall determine its state equation by the eight steps outlined above.

Step 1. The associated directed graph G is given in Figure 11.2(b).

Step 2. The normal tree consisting of edges e_2, e_3, and e_6 is selected. The normal tree T is represented by the hollow lines in Figure 11.2(b).

Step 3. As indicated in Figure 11.11, the tree branches e_2, e_3, and e_6 are assigned the voltage symbols v_2, v_3, and v_g and the cotree links e_1, e_4, and e_5 are assigned the current symbols i_1, i_4, and i_g, respectively.

Step 4. By writing KCL equations for the f-cutsets defined by T, the branch currents i_2, i_3, and i_6 can each be expressed as the sums of cotree-link currents:

$$i_2 = i_4 \tag{11.31a}$$

$$i_3 = i_g - i_1 - i_4 \tag{11.31b}$$

$$i_6 = i_4 \tag{11.31c}$$

The branch currents are also indicated in Figure 11.12.

Step 5. By writing KVL equations for the f-circuits defined with respect to T, the link voltages v_1, v_4, and v_5 can each be expressed as the sums of

Figure 11.11

The assignment of voltage symbols v_2, v_3, and v_g to the tree branches and the current symbols i_1, i_4, and i_g to the cotree links.

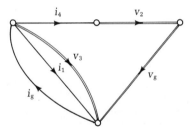

Figure 11.12
Expressions for the tree-branch currents in terms of cotree-link currents and the cotree-link voltages in terms of the tree-branch voltages.

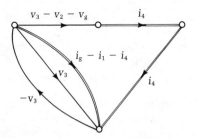

tree-branch voltages:

$$v_1 = v_3 \tag{11.32a}$$

$$v_4 = v_3 - v_2 - v_g \tag{11.32b}$$

$$v_5 = -v_3 \tag{11.32c}$$

They are also indicated in Figure 11.12.

Step 6. The element v-i equations for the tree-branch capacitor and the cotree-link inductor are found to be

$$C_3\dot{v}_3 = i_3 = i_g - i_1 - i_4 \tag{11.33a}$$

$$L_4\dot{i}_4 = v_4 = v_3 - v_2 - v_g \tag{11.33b}$$

Likewise the element v-i equations for other passive elements are obtained as

$$R_1 i_1 = v_1 = v_3 \tag{11.34a}$$

$$v_2 = R_2 i_2 = R_2 i_4 \tag{11.34b}$$

Step 7. The state variables are the capacitor voltage v_3 and inductor current i_4 and the known independent sources are i_g and v_g. To obtain the state equation, we must eliminate the nonstate variables i_1 and v_2 in (11.33). For this we substitute i_1 and v_2 from (11.34) in (11.33) and obtain

$$C_3\dot{v}_3 = i_g - \frac{v_3}{R_1} - i_4 \tag{11.35a}$$

$$L_4\dot{i}_4 = v_3 - R_2 i_4 - v_g \tag{11.35b}$$

Step 8. Equations (11.35) are written in matrix form as

$$\begin{bmatrix} \dot{v}_3 \\ \dot{i}_4 \end{bmatrix} = \begin{bmatrix} -\dfrac{1}{R_1 C_3} & -\dfrac{1}{C_3} \\ \dfrac{1}{L_4} & -\dfrac{R_2}{L_4} \end{bmatrix} \begin{bmatrix} v_3 \\ i_4 \end{bmatrix} + \begin{bmatrix} \dfrac{1}{C_3} & 0 \\ 0 & -\dfrac{1}{L_4} \end{bmatrix} \begin{bmatrix} i_g \\ v_g \end{bmatrix} \tag{11.36}$$

confirming (11.20).

Figure 11.13

An active network (a) and its associated directed graph (b) used to illustrate the procedure for writing the state equation in normal form.

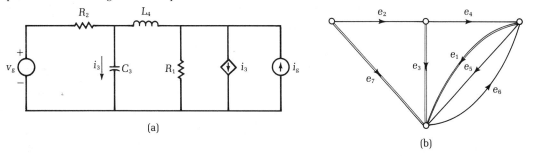

(a)

(b)

EXAMPLE 11.3

We shall write the state equation for the network of Figure 11.13, which contains a current-controlled current source.

Step 1.   The associated directed graph G of the active network is presented in Figure 11.13(b).

Step 2.   Select a normal tree T consisting of the edges e_1, e_3, and e_7. Edge e_5 is excluded from the tree because it is usually simpler to put controlled current sources in the cotree. The subgraph $e_3e_5e_7$, nevertheless, is also a normal tree. The normal tree $T = e_1e_3e_5$ is exhibited explicitly in Figure 11.13(b) by the hollow lines.

Step 3.   As illustrated in Figure 11.14(a), the tree-branches e_1, e_3, and e_7 are assigned the voltage symbols v_1, v_3, and v_g; and the cotree-links e_2, e_4, e_5, and e_6 are assigned the current symbols i_2, i_4, i_3, and i_g respectively. Edge e_5 is given the symbol i_3 because its current is controlled by the current

Figure 11.14

The assignment (a) of the voltage symbols to the tree branches and the current symbols to the cotree links, and the expressions (b) for the tree-branch currents in terms of cotree-link currents and the cotree-link voltages in terms of the tree-branch voltages.

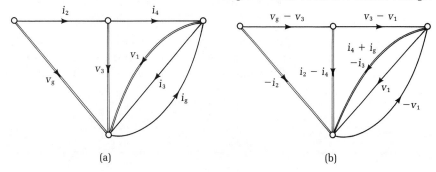

(a)

(b)

of e_3, which is i_3. Edges e_6 and e_7 are assigned i_g and v_g because they represent known independent sources.

Step 4. By writing KCL equations for the f-cutset defined by T, the branch current i_1, i_3, and i_7 are found to be

$$i_1 = i_4 + i_g - i_3 \tag{11.37a}$$

$$i_3 = i_2 - i_4 \tag{11.37b}$$

$$i_7 = -i_2 \tag{11.37c}$$

which are indicated in Figure 11.14(b).

Step 5. By writing KVL equations for the f-circuits defined with respect to T, the link voltages v_2, v_4, v_5, and v_6 are obtained as

$$v_2 = v_g - v_3 \tag{11.38a}$$

$$v_4 = v_3 - v_1 \tag{11.38b}$$

$$v_5 = v_1 \tag{11.38c}$$

$$v_6 = -v_1 \tag{11.38d}$$

which are also indicated in Figure 11.14(b).

Step 6. The element v-i equations for the tree-branch capacitor and the cotree-link inductor are given by

$$C_3\dot{v}_3 = i_3 = i_2 - i_4 \tag{11.39a}$$

$$L_4\dot{i}_4 = v_4 = v_3 - v_1 \tag{11.39b}$$

For the other passive elements, we have

$$v_1 = R_1 i_1 = R_1(i_4 + i_g - i_3) \tag{11.40a}$$

$$i_2 = \frac{v_2}{R_2} = \frac{v_g - v_3}{R_2} \tag{11.40b}$$

In fact, these equations can be obtained directly from Figure 11.14 by writing the v-i equations for the voltage and current variables of the corresponding edges of the two directed graphs, depending of course on the character of the individual elements.

Step 7. The state variables are v_3 and i_4 and the nonstate variables to be eliminated in (11.39) are i_2 and v_1. For this we express i_2 and v_1 in terms of the state variables. From (11.37b) and (11.40), we get

$$v_1 = R_1\left(2i_4 + i_g + \frac{v_3}{R_2} - \frac{v_g}{R_2}\right) \tag{11.41a}$$

$$i_2 = \frac{v_g - v_3}{R_2} \tag{11.41b}$$

Substituting these in (11.39) yields

$$C_3 \dot{v}_3 = \frac{v_g - v_3}{R_2} - i_4 \qquad\qquad\qquad\qquad\qquad\text{(11.41c)}$$

$$L_4 \dot{i}_4 = \left(1 - \frac{R_1}{R_2}\right) v_3 - 2R_1 i_4 - R_1 i_g + \frac{R_1 v_g}{R_2} \qquad\text{(11.41d)}$$

Step 8.   Writing these in matrix form gives the state equation in normal form for the active network of Figure 11.13(a):

$$\begin{bmatrix} \dot{v}_3 \\ \dot{i}_4 \end{bmatrix} = \begin{bmatrix} -\dfrac{1}{R_2 C_3} & -\dfrac{1}{C_3} \\[2ex] \dfrac{1}{L_4} - \dfrac{R_1}{R_2 L_4} & -\dfrac{2R_1}{L_4} \end{bmatrix} \begin{bmatrix} v_3 \\ i_4 \end{bmatrix} + \begin{bmatrix} \dfrac{1}{R_2 C_3} & 0 \\[2ex] \dfrac{R_1}{R_2 L_4} & -\dfrac{R_1}{L_4} \end{bmatrix} \begin{bmatrix} v_g \\ i_g \end{bmatrix} \qquad\text{(11.42)}$$

EXAMPLE 11.4

We shall write the state equation for the network N of Figure 11.15(a), which contains two mutually coupled inductors with mutual inductance M. Again we follow the eight steps with very brief comments.

Step 1.   The associated directed graph G is shown in Figure 11.15(b).

Step 2.   Select a normal tree T in Figure 11.15(b) as shown by the hollow lines.

Step 3.   The tree-branches are assigned the voltages v_3, v_4, and v_g; and the cotree-links the currents i_1, i_2, αi_3, and i_5. These are indicated by the first elements of the pairs associated with the edges of G in Figure 11.15(b).

Steps 4 and 5.   By writing KCL and KVL equations for the f-cutsets and f-circuits defined for T, the branch currents are expressed in terms of the

Figure 11.15
A network (a) containing two mutually coupled inductors and its associated directed graph (b) used to illustrate the procedure for writing the state equation in normal form.

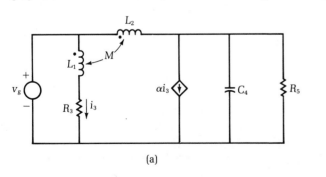

(a)

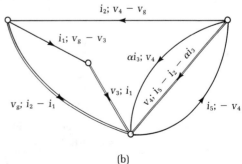

(b)

link currents and the link voltages are expressed in terms of the branch voltages. These are shown as the second elements of the pairs associated with the edges of G in Figure 11.15(b).

Step 6. The element v-i equations can now be written down directly from Figure 11.15 together with the fact that $i_3 = i_1$. We obtain

$$C_4 \dot{v}_4 = i_5 - i_2 - \alpha i_1 \tag{11.43a}$$

$$L_1 \dot{i}_1 - M \dot{i}_2 = v_g - v_3 \tag{11.43b}$$

$$L_2 \dot{i}_2 - M \dot{i}_1 = v_4 - v_g \tag{11.43c}$$

$$v_3 = R_3 i_1 \tag{11.44a}$$

$$i_5 = -\frac{v_4}{R_5} \tag{11.44b}$$

Step 7. The state variables are v_4, i_1, and i_2 and the nonstate variables to be eliminated in (11.43) are v_3 and i_5. Substituting (11.44) in (11.43), we obtain three equations involving only the state variables and the known source v_g. Using matrix notation these three equations can be written as

$$
\begin{bmatrix} C_4 & 0 & 0 \\ 0 & L_1 & -M \\ 0 & -M & L_2 \end{bmatrix}
\begin{bmatrix} \dot{v}_4 \\ \dot{i}_1 \\ \dot{i}_2 \end{bmatrix}
=
\begin{bmatrix} -1/R_5 & -\alpha & -1 \\ 0 & -R_3 & 0 \\ 1 & 0 & 0 \end{bmatrix}
\begin{bmatrix} v_4 \\ i_1 \\ i_2 \end{bmatrix}
+
\begin{bmatrix} 0 \\ v_g \\ -v_g \end{bmatrix}
$$

Step 8. Premultiplying both sides of (11.45) by the matrix

$$
\begin{bmatrix} C_4 & 0 & 0 \\ 0 & L_1 & -M \\ 0 & -M & L_2 \end{bmatrix}^{-1}
= \frac{1}{\Delta}
\begin{bmatrix} \Delta/C_4 & 0 & 0 \\ 0 & L_2 & M \\ 0 & M & L_1 \end{bmatrix}
\tag{11.46}
$$

where $\Delta = L_1 L_2 - M^2 \neq 0$ by assuming that the inductors are not perfectly coupled, we obtain the state equation for the network of Figure 11.15(a):

$$
\begin{bmatrix} \dot{v}_4 \\ \dot{i}_1 \\ \dot{i}_2 \end{bmatrix}
=
\begin{bmatrix}
-\dfrac{1}{C_4 R_5} & -\dfrac{\alpha}{C_4} & -\dfrac{1}{C_4} \\[2mm]
\dfrac{M}{\Delta} & -\dfrac{L_2 R_3}{\Delta} & 0 \\[2mm]
\dfrac{L_1}{\Delta} & -\dfrac{M R_3}{\Delta} & 0
\end{bmatrix}
\begin{bmatrix} v_4 \\ i_1 \\ i_2 \end{bmatrix}
+
\begin{bmatrix} 0 \\[2mm] \dfrac{L_2 - M}{\Delta} \\[2mm] \dfrac{M - L_1}{\Delta} \end{bmatrix}
[v_g]
\tag{11.47}
$$

Finally, we note that if the output variables are specified, they can be expressed in terms of the state variables and the known source through the weighted directed graph of Figure 11.15(b) and (11.43).

11-5   STATE EQUATIONS FOR DEGENERATE NETWORKS

A network is said to be *degenerate* if one or both of the following conditions are met:

1. It contains a circuit composed only of capacitors and/or independent or dependent voltage sources.
2. It contains a cutset composed only of inductors and/or independent or dependent current sources.

In the present section we demonstrate by simple examples that a degenerate network may possess any one of the following attributes:

1. It has no solution.
2. It has a solution but has no unique solution.
3. If capacitor voltages and inductor currents are routinely chosen as the state variables, its state equation cannot be put in the normal form of (11.10).

As is intuitively true, a linear physical network always has a solution— a unique solution. However, the networks that we are dealing with are models representing the physical networks. The models are made up of the idealized elements such as ideal resistors, capacitors, inductors, and independent or dependent sources. As a result we cannot always assume the existence and uniqueness of their solutions. For example, in a degenerate network if there is a circuit consisting only of independent or dependent voltage sources, then for the network to have a solution, KVL requires that the algebraic sum of these voltages be zero. Otherwise the network will have no solution at all. Even the existence of a solution in such a network is not unique. For instance, in Figure 11.16 a resistor R is connected in parallel with two identical batteries. There is a circuit composed only of independent voltage sources v_1 and v_2. It is straightforward to verify that the complete solution of the network is given by

$$\mathbf{v} = \begin{bmatrix} v_1 \\ v_2 \\ v_R \end{bmatrix} = \begin{bmatrix} V \\ V \\ V \end{bmatrix} \tag{11.48a}$$

$$\mathbf{i} = \begin{bmatrix} i_1 \\ i_2 \\ i_R \end{bmatrix} = \begin{bmatrix} \gamma \\ V/R - \gamma \\ V/R \end{bmatrix} \tag{11.48b}$$

Figure 11.16
A simple network that does not possess a unique solution.

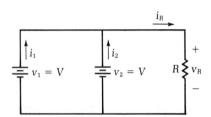

where γ is an arbitrary real constant. Thus the network possesses infinitely many solutions.

Likewise if a degenerate network contains a cutset comprised only of independent or dependent current sources, such a network either has no solution or no unique solution if one exists. For our purposes we shall exclude these networks from our consideration because they cannot be adequately represented by the state equations.

The network of Figure 11.9(a) is degenerate because there is a circuit composed of the capacitors C_3 and C_4 and the independent voltage source v_g. In fact, it also contains a cutset composed of the inductors L_1 and L_2 and the independent current source i_g. Such a network cannot be described by the normal-form state equation (11.10) if capacitor voltages and inductor currents are used as the state variables. In the following we shall demonstrate this by a simple example and show how to select the state variables so that the resulting equations can be put in normal form.

EXAMPLE 11.5
We use the eight steps outlined in the preceding section to obtain the state equation for the degenerate network of Figure 11.9(a).

Step 1.   The associated directed graph G of the network is shown in Figure 11.9(b).

Step 2.   Select a normal tree $T = e_1 e_3 e_5 e_8$, which is exhibited explicitly in Figure 11.17 by the hollow lines.

Step 3.   The tree-branches e_1, e_3, e_5, and e_8 are assigned the voltages v_1, v_3, v_5, and v_g; and the cotree-links e_2, e_4, e_6, and e_7 the currents i_2, i_4, i_6, and i_g, as indicated in Figure 11.17 (the first elements of the pairs).

Step 4.   The branch currents i_1, i_3, i_5, and i_8 are expressed in terms of the link currents by

$$i_1 = i_g - i_2 \tag{11.49a}$$

$$i_3 = i_g - i_4 \tag{11.49b}$$

$$i_5 = i_2 + i_6 - i_g \tag{11.49c}$$

$$i_8 = i_4 - i_6 \tag{11.49d}$$

Step 5.   The cotree-link voltages v_2, v_4, v_6, and v_7 are expressed in terms of the branch voltages by

$$v_2 = v_1 - v_5 \tag{11.50a}$$

$$v_4 = v_3 - v_g \tag{11.50b}$$

$$v_6 = v_g - v_5 \tag{11.50c}$$

$$v_7 = v_5 - v_1 - v_3 \tag{11.50d}$$

Figure 11.17
The associated directed
graph of the degenerate
network of Figure 11.9(a)
with a selected normal
tree exhibited explicitly.

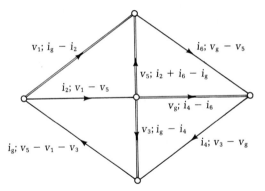

Step 6.   The element v-i equations are found to be

$$L_2\dot{i}_2 = v_1 - v_5 \tag{11.51a}$$

$$C_3\dot{v}_3 = i_g - i_4 \tag{11.51b}$$

$$v_5 = R_5(i_2 + i_6 - i_g) \tag{11.52a}$$

$$i_6 = \frac{v_g - v_5}{R_6} \tag{11.52b}$$

$$L_1(\dot{i}_g - \dot{i}_2) = v_1 \tag{11.52c}$$

$$C_4(\dot{v}_3 - \dot{v}_g) = i_4 \tag{11.52d}$$

Step 7.   The state variables are i_2 and v_3. Using (11.52) we can eliminate
the nonstate variables v_1, i_4, and v_5 in (11.51), giving

$$(L_1 + L_2)\dot{i}_2 = L_1\dot{i}_g - \frac{R_5R_6}{R_5 + R_6}\left(i_2 - i_g + \frac{v_g}{R_6}\right) \tag{11.53a}$$

$$(C_3 + C_4)\dot{v}_3 = i_g + C_4\dot{v}_g \tag{11.53b}$$

Observe that the two equations involve not only i_2, v_3, i_g, and v_g but also
the derivatives of i_g and v_g, which may not exist.

Step 8.   Writing (11.53) in matrix notation yields

$$\begin{bmatrix} \dot{i}_2 \\ \dot{v}_3 \end{bmatrix} = \begin{bmatrix} -\dfrac{R_0}{L_1 + L_2} & 0 \\ 0 & 0 \end{bmatrix}\begin{bmatrix} i_2 \\ v_3 \end{bmatrix} + \begin{bmatrix} \dfrac{R_0}{L_1 + L_2} & -\dfrac{R_0}{R_6(L_1 + L_2)} \\ \dfrac{1}{C_3 + C_4} & 0 \end{bmatrix}\begin{bmatrix} i_g \\ v_g \end{bmatrix}$$

$$+ \begin{bmatrix} \dfrac{L_1}{L_1 + L_2} & 0 \\ 0 & \dfrac{C_4}{C_3 + C_4} \end{bmatrix}\begin{bmatrix} \dot{i}_g \\ \dot{v}_g \end{bmatrix} \tag{11.54}$$

where $R_0 = R_5R_6/(R_5 + R_6)$ and $\dot{i}_g$ and $\dot{v}_g$ are assumed to exist.

The example shows that if the state equation for a degenerate network exists, it takes the form

$$\dot{\mathbf{x}}(t) = \mathbf{A}\mathbf{x}(t) + \mathbf{B}_1\mathbf{u}(t) + \mathbf{B}_2\dot{\mathbf{u}}(t) \qquad (11.55)$$

and the output equation becomes[†]

$$\mathbf{y}(t) = \mathbf{C}\mathbf{x}(t) + \mathbf{D}_1\mathbf{u}(t) + \mathbf{D}_2\dot{\mathbf{u}}(t) \qquad (11.56)$$

where $\mathbf{B}_1$, $\mathbf{B}_2$ and $\mathbf{D}_1$, $\mathbf{D}_2$ are the known parameter matrices of orders $k \times h$ and $q \times h$, respectively. Equation (11.55) may be regarded as the *modified normal form* of the state equation since it can be transformed to

$$\dot{\mathbf{x}}(t) = \mathbf{A}\mathbf{x}(t) + \mathbf{F}\mathbf{u}(t) \qquad (11.57)$$

where

$$\mathbf{F} = \mathbf{B}_1 + \mathbf{B}_2 p \qquad (11.58)$$

and p, as in (2.64), being the *linear differential operator* for any vector function of time t,

$$p\mathbf{u}(t) = \frac{d\mathbf{u}(t)}{dt} = \dot{\mathbf{u}}(t) \qquad (11.59)$$

If, however, one would insist on considering (11.10) as the only acceptable normal form of the state equation, the derivative of the input vector can be removed by the transformation

$$\mathbf{x}(t) \longrightarrow \mathbf{x}(t) + \mathbf{B}_2\mathbf{u}(t) \qquad (11.60)$$

which, when substituted in (11.55), yields

$$\dot{\mathbf{x}}(t) = \mathbf{A}\mathbf{x}(t) + \mathbf{B}\mathbf{u}(t) \qquad (11.61)$$

where

$$\mathbf{B} = \mathbf{B}_1 + \mathbf{A}\mathbf{B}_2 \qquad (11.62)$$

The price we paid in achieving this is that, instead of using $\mathbf{x}(t)$ as the state vector, we use $\mathbf{x}(t) + \mathbf{B}_2\mathbf{u}(t)$ as the new state vector.

We illustrate the above results by the following example.

EXAMPLE 11.6

We transform the equation (11.54) to the forms of (11.57) and (11.61). Clearly (11.54) can be rewritten as

$$\begin{bmatrix} \dot{i}_2 \\ \dot{v}_3 \end{bmatrix} = \begin{bmatrix} -\dfrac{R_0}{L_1 + L_2} & 0 \\ 0 & 0 \end{bmatrix} \begin{bmatrix} i_2 \\ v_3 \end{bmatrix} + \begin{bmatrix} \dfrac{R_0 + L_1 p}{L_1 + L_2} & -\dfrac{R_0}{R_6(L_1 + L_2)} \\ \dfrac{1}{C_3 + C_4} & \dfrac{C_4 p}{C_3 + C_4} \end{bmatrix} \begin{bmatrix} i_g \\ v_g \end{bmatrix} \qquad (11.63)$$

[†] In some situations the output equation (11.56) may also involve the second- and higher-order derivatives of $\mathbf{u}(t)$.

which is in the form of (11.57). Thus (11.63) is the state equation in modified normal form for the network of Figure 11.9(a). Observe that the state variables are still the inductor current i_2 and the capacitor voltage v_3.

Alternatively, to convert (11.54) to the form of (11.61) we compute first the matrix **B** from (11.62), giving

$$\mathbf{B} = \mathbf{B}_1 + \mathbf{A}\mathbf{B}_2 = \begin{bmatrix} \dfrac{R_0 L_2}{(L_1 + L_2)^2} & -\dfrac{R_0}{R_6(L_1 + L_2)} \\[3mm] \dfrac{1}{C_3 + C_4} & 0 \end{bmatrix} \tag{11.64}$$

The new state vector is determined from (11.60) as

$$\mathbf{x}(t) = \begin{bmatrix} x_1 \\ x_2 \end{bmatrix} = \begin{bmatrix} i_2 \\ v_3 \end{bmatrix} + \begin{bmatrix} \dfrac{L_1 i_g}{L_1 + L_2} \\[3mm] \dfrac{C_4 v_g}{C_3 + C_4} \end{bmatrix} = \begin{bmatrix} i_2 + \dfrac{L_1 i_g}{L_1 + L_2} \\[3mm] v_3 + \dfrac{C_4 v_g}{C_3 + C_4} \end{bmatrix} \tag{11.65}$$

Substituting these in (11.61) yields the state equation in normal form for the network of Figure 11.9(a):

$$\begin{bmatrix} \dot{x}_1 \\ \dot{x}_2 \end{bmatrix} = \begin{bmatrix} -\dfrac{R_0}{L_1 + L_2} & 0 \\[3mm] 0 & 0 \end{bmatrix} \begin{bmatrix} x_1 \\ x_2 \end{bmatrix} + \begin{bmatrix} \dfrac{R_0 L_2}{(L_1 + L_2)^2} & -\dfrac{R_0}{R_6(L_1 + L_2)} \\[3mm] \dfrac{1}{C_3 + C_4} & 0 \end{bmatrix} \begin{bmatrix} i_g \\ v_g \end{bmatrix}$$

$$\tag{11.66}$$

Observe that the state variables in (11.66) are no longer the inductor current i_2 and the capacitor voltage v_3. Instead they are the linear combinations of i_2 and i_g and v_3 and v_g, respectively, as defined in (11.65).

11-6   ORDER OF COMPLEXITY

Once the state variables are determined, the remaining network variables can be found purely algebraically as can be seen from the output equation (11.14). Since the state variables that we choose are the capacitor voltages and inductor currents, each of their element v-i equations contains a derivative. Thus, unlike the variables associated with the resistive elements, which are *static*, these variables are *dynamic*. The question arises as to how many *dynamically independent* state variables there are, the instantaneous values of which are sufficient to determine completely the instantaneous state of the network. This leads to the concept of the *order of complexity* of a network. In systems theory this is known as the *order* of the system. Another commonly used name is called the *degrees of freedom*, which originates in the study of mechanical systems where one can attach significance to the word *position* or *configuration*. In this section we shall first define precisely the term "order of complexity" and then show how it can be determined by inspection for RLC networks. Since the initial conditions of a network do not affect our argument to be presented below, for simplicity and without loss of generality we assume that all the initial capacitor voltages and inductor currents have been set to zero.

Our starting point is the primary system of network equations (2.65), which is repeated below:

$$\mathbf{Q}\mathbf{i}(t) = \mathbf{0} \tag{11.67a}$$

$$\mathbf{B}\mathbf{v}(t) = \mathbf{0} \tag{11.67b}$$

$$\tilde{\mathbf{H}}(p)\tilde{\mathbf{u}}(t) = \tilde{\mathbf{y}}(t) \tag{11.67c}$$

Recall that $\tilde{\mathbf{u}}(t)$ and $\tilde{\mathbf{y}}(t)$ together represent the $2b$ branch voltages and currents, b being the number of network branches. Equation (11.67) can be rewritten as a single matrix equation

$$\mathbf{H}(p)\begin{bmatrix} \mathbf{i}(t) \\ \mathbf{v}(t) \end{bmatrix} = \mathbf{0} \tag{11.68}$$

We remark that $\mathbf{H}(p)$ contains the linear differential operator p and must be handled with care. For instance, do not permit the variables to move across the sign of differentiation by a careless interchange of the order of factors containing variable coefficients and the operator p.

With these preliminaries, we now proceed to define precisely the term *natural frequency* of a general network, which in turn is used to define the order of complexity.

Definition 11.3 Natural Frequency *The natural frequencies of a network are defined as the roots of the determinantal polynomial of the operator matrix of the network equations when these are framed as a set of first-order differential and/or algebraic equations for the currents and voltages of the network elements.*

Definition 11.4 Order of Complexity *The number of natural frequencies of a network is called the order of complexity of the network, counting each frequency according to its multiplicity.*

Thus the roots of the polynomial det $\mathbf{H}(s)$ obtained from the operator matrix $\mathbf{H}(p)$ of (11.68) with p being replaced by s are the natural frequencies,[†] the number of which is the order of complexity. As discussed earlier, in state-variable formulation of network equations the order of complexity is the minimal number of dynamically independent branch currents and voltages the instantaneous values of which are sufficient to determine completely

---

[†] Alternatively, det $\mathbf{H}(p)$ may be regarded as a polynomial in p.

the instantaneous state of the network, and hence is the dimension of the *state-space*, a space spanned by the k linearly independent state vectors $\mathbf{x}(t)$. In mathematics it is the number of arbitrary constants appearing in the general solution of the network equations. Physically, it is the maximum number of independent energy-storing elements in the network or equivalently the maximum number of independent initial conditions that can be specified for the network. These are various ways of stating the same thing. For instance, in the network of Figure 11.9(a) if the inductor current i_2 and capacitor voltage v_3 are chosen as the state variables, the initial current $i_2(t_0)$ and initial voltage $v_3(t_0)$ can be specified independently. The initial conditions for the other two energy-storing elements are uniquely determined from (11.49a) and (11.50b) by

$$i_1(t_0) = i_g(t_0) - i_2(t_0) \tag{11.69a}$$

$$v_4(t_0) = v_3(t_0) - v_g(t_0) \tag{11.69b}$$

Thus the maximum number of linearly independent initial conditions that can be specified for this network is 2, showing that its order of complexity is 2.

The following example illustrates some of the points discussed in the foregoing.

EXAMPLE 11.7

Consider the active network N of Figure 11.18(a) together with its associated directed graph G of Figure 11.18(b). Choose $T = e_2 e_3$ as the normal tree. The primary system of network equations becomes

$$\mathbf{Q}\mathbf{i}(t) = \begin{bmatrix} \overset{e_1}{-1} & \overset{e_2}{0} & \overset{e_3}{1} & \overset{e_4}{-1} \\ 1 & 1 & 0 & 1 \end{bmatrix} \begin{bmatrix} i_1 \\ i_2 \\ i_3 \\ i_4 \end{bmatrix} = \begin{bmatrix} 0 \\ 0 \end{bmatrix} \tag{11.70a}$$

$$\mathbf{B}\mathbf{v}(t) = \begin{bmatrix} 1 & -1 & 1 & 0 \\ 0 & -1 & 1 & 1 \end{bmatrix} \begin{bmatrix} v_1 \\ v_2 \\ v_3 \\ v_4 \end{bmatrix} = \begin{bmatrix} 0 \\ 0 \end{bmatrix} \tag{11.70b}$$

Figure 11.18
An active network (a) and its associated directed graph (b) used to illustrate the concept of the order of complexity.

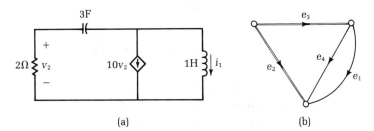

(a)   (b)

$$\tilde{H}(p)\tilde{u}(t) = \begin{bmatrix} p & 0 & 0 & 0 \\ 0 & \frac{1}{2} & 0 & 0 \\ 0 & 0 & 3p & 0 \\ 0 & 10 & 0 & 0 \end{bmatrix} \begin{bmatrix} i_1 \\ v_2 \\ v_3 \\ v_4 \end{bmatrix} = \begin{bmatrix} v_1 \\ i_2 \\ i_3 \\ i_4 \end{bmatrix} = \tilde{y}(t) \qquad (11.70\text{c})$$

which can be combined to give

$$H(p)\begin{bmatrix} i(t) \\ v(t) \end{bmatrix} = \begin{bmatrix} -1 & 0 & 1 & -1 & 0 & 0 & 0 & 0 \\ 1 & 1 & 0 & 1 & 0 & 0 & 0 & 0 \\ 0 & 0 & 0 & 0 & 1 & -1 & 1 & 0 \\ 0 & 0 & 0 & 0 & 0 & -1 & 1 & 1 \\ -p & 0 & 0 & 0 & 1 & 0 & 0 & 0 \\ 0 & -2 & 0 & 0 & 0 & 1 & 0 & 0 \\ 0 & 0 & -1 & 0 & 0 & 0 & 3p & 0 \\ 0 & 0 & 0 & -1 & 0 & 10 & 0 & 0 \end{bmatrix} \begin{bmatrix} i_1 \\ i_2 \\ i_3 \\ i_4 \\ v_1 \\ v_2 \\ v_3 \\ v_4 \end{bmatrix} = 0$$

$$(11.71)$$

Thus the natural frequencies are the roots of the polynomial $\det H(s)$:

$$\det H(s) = -63s^2 - 6s - 1 \qquad (11.72)$$

Setting it to zero and solving for s yield the natural frequencies at

$$s_1, s_2 = -0.0476 \pm j0.1166 \qquad (11.73)$$

The order of complexity of the network is therefore 2, meaning that we can specify two initial conditions independently, the initial capacitor voltage and the initial inductor current. This is also the maximum number of initial conditions that one can independently specify for this network or equivalently, the network has two natural frequencies.

An *RLC network* is a network composed only of resistors, inductors, capacitors, mutual inductances, and independent sources. For such a network, its order of complexity is known and depends only on its topology. For an active network this number, however, is not decided in general purely by the topological structure; it also depends on the values of the parameters of the network. This complicates the matter considerably. In the following we present formulas for determining the order of complexity of an *RLC* network. Before we do this, we introduce the following.

Definition 11.5 C-cutset and L-cutset   *A C-cutset (L-cutset) of a network is a cutset composed only of capacitors (inductors) and, possibly, independent current sources.*

Definition 11.6 C-circuit and L-circuit   *A C-circuit (L-circuit) of a network is a circuit composed only of capacitors (inductors) and, possibly, independent voltage sources.*

Figure 11.19

A network (a) and its associated directed graph (b) with three L-cutsets shown by drawing a dashed line across each L-cutset.

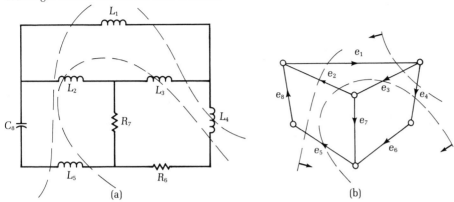

The order of complexity of an RLC network can be stated in terms of L-cutsets and C-circuits as the following.

Property 11.2[†]   *The order of complexity of an RLC network is given by the number of energy storage elements less the number of linearly independent L-cutsets and the number of linearly independent C-circuits.*

By linearly independent cutsets or circuits, we mean that the rows corresponding to these cutsets or circuits in the complete cutset or circuit matrix are linearly independent. For instance, there are three L-cutsets $L_1 L_2 L_5$, $L_1 L_3 L_4$, and $L_2 L_3 L_4 L_5$ in the network N of Figure 11.19(a), the associated directed graph G of which is shown in Figure 11.19(b). These L-cutsets are depicted pictorially in Figure 11.19. They are not linearly independent because the cutset matrix formed by these cutsets, as given below,

$$
\begin{array}{c}
\begin{array}{cccccccc}
\quad e_1 & e_2 & e_3 & e_4 & e_5 & e_6 & e_7 & e_8
\end{array} \\
\begin{array}{c}
\text{cutset } L_1 L_2 L_5 \\
\text{cutset } L_1 L_3 L_4 \\
\text{cutset } L_2 L_3 L_4 L_5
\end{array}
\left[
\begin{array}{cccccccc}
1 & -1 & 0 & 0 & -1 & 0 & 0 & 0 \\
-1 & 0 & 1 & 1 & 0 & 0 & 0 & 0 \\
0 & -1 & 1 & 1 & -1 & 0 & 0 & 0
\end{array}
\right]
\end{array}
$$

(11.74)

is of rank 2. In fact, the third row is the sum of the first two. Thus only two of the three L-cutsets are linearly independent. According to Property 11.2,

[†] A proof of this property may be found in Chen, *Applied Graph Theory*, p. 508.

Figure 11.20

A network (a) and its associated directed graph (b) with three C-circuits shown explicitly.

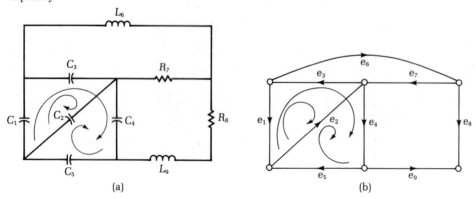

(a)   (b)

the order of complexity of the network is 4 ($=6-2$), 6 being the number of energy storage elements, which means that four dynamically-independent branch currents and voltages have instantaneous values sufficient to determine completely the instantaneous state of the network. As an example using C-circuits, consider the network and its associated directed graph of Figure 11.20. The three C-circuits $C_1C_2C_3$, $C_2C_4C_5$, and $C_1C_3C_4C_5$ are not linearly independent because their associated circuit matrix

$$
\begin{array}{c}
\\
\text{circuit } C_1C_2C_3 \\
\text{circuit } C_2C_4C_5 \\
\text{circuit } C_1C_3C_4C_5
\end{array}
\begin{array}{ccccccccc}
e_1 & e_2 & e_3 & e_4 & e_5 & e_6 & e_7 & e_8 & e_9 \\
\left[\begin{array}{ccccccccc}
-1 & -1 & -1 & 0 & 0 & 0 & 0 & 0 & 0 \\
0 & 1 & 0 & 1 & 1 & 0 & 0 & 0 & 0 \\
-1 & 0 & -1 & 1 & 1 & 0 & 0 & 0 & 0
\end{array}\right]
\end{array}
$$

$$\text{(11.75)}$$

is of rank 2, the third row being the sum of the first two. Therefore there are only two linearly independent C-circuits, and the order of complexity of the network becomes 5 ($=7-2$).

The network of Figure 11.9(a) has one independent L-cutset $L_1L_2i_g$ and one independent C-circuit $C_3C_4v_g$. There are total four energy storage elements: two capacitors and two inductors. Thus, according to Property 11.2, the order of complexity of the network is 2 ($=4-2$). Alternatively we can first remove all the independent sources by open-circuiting the current source i_g and short-circuiting the voltage source v_g. The resulting network is shown in Figure 11.21, which as in Figure 11.9(a) has one independent L-cutset L_1L_2 and one independent C-circuit C_3C_4, giving the same order of complexity.

The natural frequencies, as stated in Definition 11.3, include the zero frequency, which is of limited physical significance. An important question is that of determining the number of nonzero natural frequencies. Remember

Figure 11.21
Network resulting from that of Figure 11.9(a) by removing all the independent sources by open-circuiting the current source and short-circuiting the voltage source.

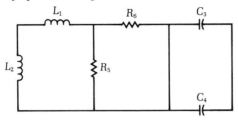

that even the zero frequency is counted according to its multiplicity. To answer this question, we recognize that physically the natural frequencies at the origin correspond to the constant currents that circulate around the circuits comprised only of inductors or the constant voltages that apply across the cutsets comprised only of capacitors. These constant currents and voltages do not give rise to any additional algebraic constraints among the currents in the L-circuits or the voltages across the C-cutsets, only on their rates of changes. For example, an L-circuit with k inductances L_1, $L_2, \ldots, L_k$ gives rise to a differential relation

$$L_1 \dot{i}_1 + L_2 \dot{i}_2 + \cdots + L_k \dot{i}_k = 0 \tag{11.76}$$

Integration of this expression from 0 to t leads to

$$L_1 i_1 + L_2 i_2 + \cdots + L_k i_k = \text{constant} \tag{11.77}$$

It might appear that this also represents a constraint on the inductor currents. However, the constant term on the right-hand side is not specified; its determination requires an independent relationship. A similar statement can be made about voltages across the network elements of a C-cutset. Thus we can state the following.

> **Property 11.3**   *The number of nonzero natural frequencies of an RLC network is equal to its order of complexity less the number of linearly independent L-circuits and the number of linearly independent C-cutsets.*

In Figure 11.21, with an order of complexity of 2, the network has one C-cutset $C_3 C_4$ and no L-circuit. According to Property 11.3, the network has only 1 $(= 2 - 1)$ nonzero natural frequency. The other is the zero frequency corresponding to the independent C-cutset $C_3 C_4$. For the network of Figure 11.19(a), there is one independent L-circuit $L_1 L_2 L_3$. The order of complexity or the number of natural frequencies of the network is known

Figure 11.22
A network used
to illustrate the
determination of its
order of complexity and
the number of nonzero
natural frequencies.

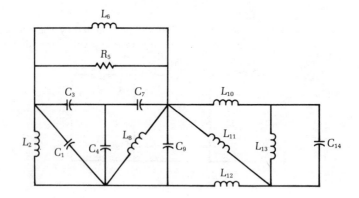

to be 4. Of these natural frequencies, one is zero, corresponding to the L-circuit $L_1 L_2 L_3$. Thus there are 3 nonzero natural frequencies. Finally, for the network of Figure 11.20(a) there is one independent C-cutset $C_1 C_2 C_5$. The number of natural frequencies of the network was determined earlier to be 5. Thus there are 4 ($=5-1$) nonzero natural frequencies.

EXAMPLE 11.8

We shall determine the order of complexity and the number of nonzero natural frequencies of the network of Figure 11.22. For this network we have the following attributes:

> C-circuits: $C_1 C_3 C_4$, $C_4 C_7 C_9$, $C_1 C_3 C_7 C_9$
> C-cutset: $C_3 C_4 C_7$
> L-circuits: $L_2 L_6 L_8$, $L_{10} L_{11} L_{13}$, $L_8 L_{10} L_{12} L_{13}$, $L_8 L_{11} L_{12}$, $L_2 L_6 L_{10} L_{12} L_{13}$,
> $L_2 L_6 L_{11} L_{12}$
>
> L-cutset: $L_{10} L_{11} L_{12}$

Of the three C-circuits, only two are linearly independent. Of the six L-circuits, only three are linearly independent. Since there are 13 inductors and capacitors, according to Property 11.2 the order of complexity or the number of natural frequencies is 10 ($=13-2-1$). Of the 10 natural frequencies, four are zero, corresponding to one C-cutset $C_3 C_4 C_7$ and three independent L-circuits, say, $L_2 L_6 L_8$, $L_8 L_{11} L_{12}$, and $L_{10} L_{11} L_{13}$. Thus there are 6 ($=10-4$) nonzero natural frequencies.

We emphasize that Properties 11.2 and 11.3 are valid only for RLC networks, and the assignment of directions in the associated graphs are arbitrary. For a network containing controlled sources, the order of complexity cannot be decided purely from its topology alone; it also depends on the network parameters. We use the following example to illustrate this point.

Figure 11.23
An active network (a) and its associated directed graph (b) used to demonstrate the invalidity of Properties 11.2 and 11.3 for active networks.

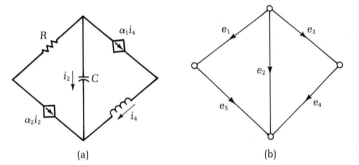

(a)   (b)

EXAMPLE 11.9

Consider the active network of Figure 11.23(a), the associated directed graph of which is shown in Figure 11.23(b). The primary system of network equations can be computed in a straightforward manner and is given by

$$
\mathbf{H}(p)\begin{bmatrix} \mathbf{i}(t) \\ \mathbf{v}(t) \end{bmatrix} =
\begin{bmatrix}
1 & 1 & 1 & 0 & 0 & 0 & 0 & 0 & 0 & 0 \\
1 & 0 & 0 & 0 & -1 & 0 & 0 & 0 & 0 & 0 \\
0 & 0 & 1 & -1 & 0 & 0 & 0 & 0 & 0 & 0 \\
0 & 0 & 0 & 0 & 0 & 1 & -1 & 0 & 0 & 1 \\
0 & 0 & 0 & 0 & 0 & 0 & -1 & 1 & 1 & 0 \\
R & 0 & 0 & 0 & 0 & -1 & 0 & 0 & 0 & 0 \\
0 & -1 & 0 & 0 & 0 & 0 & Cp & 0 & 0 & 0 \\
0 & 0 & -1 & \alpha_1 & 0 & 0 & 0 & 0 & 0 & 0 \\
0 & 0 & 0 & Lp & 0 & 0 & 0 & 0 & -1 & 0 \\
0 & \alpha_2 & 0 & 0 & -1 & 0 & 0 & 0 & 0 & 0
\end{bmatrix}
\begin{bmatrix}
i_1 \\ i_2 \\ i_3 \\ i_4 \\ i_5 \\ v_1 \\ v_2 \\ v_3 \\ v_4 \\ v_5
\end{bmatrix} = \mathbf{0}
$$

(11.78)

After replacing p by s, the determinant of the coefficient matrix is found to be (Problem 11.18)

$$\det \mathbf{H}(s) = C(1 - \alpha_1)(1 + \alpha_2)s \tag{11.79}$$

showing that for $\alpha_1 \neq 1$ and $\alpha_2 \neq -1$ the active network possesses only one natural frequency at the origin. For $\alpha_1 = 1$ or $\alpha_2 = -1$, the coefficient matrix $\mathbf{H}(s)$ becomes singular, implying that the network solution is no longer unique. In fact, under the stipulated conditions the network possesses infinitely many solutions.

On the other hand, if we apply Properties 11.2 and 11.3, the network would have 2 natural frequencies, none of these being zero. Hence we should be careful in applying these formulas.

Figure 11.24
A simple RLC series
network with a voltage
source.

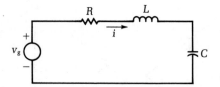

11-7 STATE EQUATIONS FOR SYSTEMS DESCRIBED BY SCALAR DIFFERENTIAL EQUATIONS

In many situations we are faced with systems that are described by scalar differential equations of orders higher than one. Our purpose here is to show that these systems can also be represented by the state equations in normal form.

To motivate our discussion, consider a very simple RLC series network of Figure 11.24. The KCL equation requires that

$$Ri + L\frac{di}{dt} + \frac{1}{C}\int i\,dt = v_g \tag{11.80}$$

at all times. This is an integrodifferential equation, which may be changed to a differential equation of order 2 by differentiation to give

$$L\frac{d^2 i}{dt^2} + R\frac{di}{dt} + \frac{1}{C}i = \frac{dv_g}{dt} \tag{11.81}$$

where the derivatives have been arranged in descending order. In this section we demonstrate that equations of this type can be converted to state equations in normal form.

Consider first the nth-order differential equation

$$\frac{d^n y}{dt^n} + a_1\frac{d^{n-1}y}{dt^{n-1}} + a_2\frac{d^{n-2}y}{dt^{n-2}} + \cdots + a_{n-1}\frac{dy}{dt} + a_n y = bu \tag{11.82}$$

If a system is described by the above equation, its state equation can be obtained by defining

$$
\begin{aligned}
x_1 &= y \\
x_2 &= \dot{x}_1 \\
x_3 &= \dot{x}_2 \\
&\ \ \vdots \\
x_n &= \dot{x}_{n-1}
\end{aligned}
\tag{11.83}
$$

Then (11.82) is equivalent to

$$
\begin{aligned}
\dot{x}_1 &= x_2 \\
\dot{x}_2 &= x_3 \\
&\ \ \vdots \\
\dot{x}_{n-1} &= x_n \\
\dot{x}_n &= -a_n x_1 - a_{n-1}x_2 - \cdots - a_2 x_{n-1} - a_1 x_n + bu
\end{aligned}
\tag{11.84}
$$

which can be written in matrix form as

$$
\begin{bmatrix} \dot{x}_1 \\ \dot{x}_2 \\ \vdots \\ \dot{x}_{n-1} \\ \dot{x}_n \end{bmatrix} = \begin{bmatrix} 0 & 1 & 0 & \cdots & 0 \\ 0 & 0 & 1 & \cdots & 0 \\ \vdots & & & & \\ 0 & 0 & 0 & \cdots & 1 \\ -a_n & -a_{n-1} & -a_{n-2} & \cdots & -a_1 \end{bmatrix} \begin{bmatrix} x_1 \\ x_2 \\ \vdots \\ x_{n-1} \\ x_n \end{bmatrix} + \begin{bmatrix} 0 \\ 0 \\ \vdots \\ 0 \\ b \end{bmatrix} [u]
$$

(11.85)

or more compactly as

$$\dot{\mathbf{x}}(t) = \mathbf{A}\mathbf{x}(t) + \mathbf{B}u(t)$$

(11.86)

The coefficient matrix $\mathbf{A}$ is called the *companion matrix* of (11.82). Thus (11.85) is the state-equation representation of the system described by the differential equation (11.82).

We illustrate the above procedure by the following example.

EXAMPLE 11.10

Consider a system described by the third-order differential equation

$$\frac{d^3y}{dt^3} + 5\frac{d^2y}{dt^2} + 4\frac{dy}{dt} + 2y = 3u$$

(11.87)

with the initial conditions

$$y(0) = 2, \qquad \frac{dy}{dt}\bigg|_{t=0} = 4, \qquad \frac{d^2y}{dt^2}\bigg|_{t=0} = 1$$

(11.88)

Using (11.85), the system can be described by the state equation

$$
\begin{bmatrix} \dot{x}_1 \\ \dot{x}_2 \\ \dot{x}_3 \end{bmatrix} = \begin{bmatrix} 0 & 1 & 0 \\ 0 & 0 & 1 \\ -2 & -4 & -5 \end{bmatrix} \begin{bmatrix} x_1 \\ x_2 \\ x_3 \end{bmatrix} + \begin{bmatrix} 0 \\ 0 \\ 3 \end{bmatrix} [u]
$$

(11.89)

with

$$
\mathbf{x}(t) = \begin{bmatrix} x_1 \\ x_2 \\ x_3 \end{bmatrix} = \begin{bmatrix} y \\ \dfrac{dy}{dt} \\ \dfrac{d^2y}{dt^2} \end{bmatrix} = \begin{bmatrix} y \\ \dot{x}_1 \\ \dot{x}_2 \end{bmatrix}, \qquad \mathbf{x}(0) = \begin{bmatrix} x_1(0) \\ x_2(0) \\ x_3(0) \end{bmatrix} = \begin{bmatrix} 2 \\ 4 \\ 1 \end{bmatrix}
$$

(11.90)

Let us now consider the more general situation where the right-hand side of (11.82) includes derivatives of the input excitation u. In this case the differential equation takes the general form

$$\frac{d^n y}{dt^n} + a_1 \frac{d^{n-1} y}{dt^{n-1}} + a_2 \frac{d^{n-2} y}{dt^{n-2}} + \cdots + a_{n-1} \frac{dy}{dt} + a_n y$$

$$= b_0 \frac{d^n u}{dt^n} + b_1 \frac{d^{n-1} u}{dt^{n-1}} + \cdots + b_{n-1} \frac{du}{dt} + b_n u \tag{11.91}$$

Before we proceed to underline the necessary steps for the determination of the state-equation representation of (11.91), we shall first demonstrate the technique by applying it to the second-order differential equation

$$\frac{d^2 y}{dt^2} + a_1 \frac{dy}{dt} + a_2 y = b_0 \frac{d^2 u}{dt^2} + b_1 \frac{du}{dt} + b_2 u \tag{11.92}$$

Let

$$x_1 = y - c_0 u \tag{11.93a}$$

$$x_2 = \dot{x}_1 - c_1 u \tag{11.93b}$$

where c_0 and c_1 are constants to be chosen. From (11.93) we have

$$\frac{dy}{dt} = \frac{dx_1}{dt} + c_0 \frac{du}{dt} = x_2 + c_1 u + c_0 \frac{du}{dt} \tag{11.94a}$$

$$\frac{d^2 y}{dt^2} = \frac{dx_2}{dt} + c_1 \frac{du}{dt} + c_0 \frac{d^2 u}{dt^2} \tag{11.94b}$$

Substituting these in (11.92) and simplifying, we get

$$\frac{dx_2}{dt} + a_1 x_2 + a_2 x_1 = (b_0 - c_0) \frac{d^2 u}{dt^2} + (b_1 - c_1 - a_1 c_0) \frac{du}{dt}$$

$$+ (b_2 - a_1 c_1 - a_2 c_0) u \tag{11.95}$$

Our objective is to write a system of first-order differential equations involving only the forcing function u, not its first or higher-order derivatives. To this end we set the coefficients of $d^2 u/dt^2$ and du/dt on the right-hand side of (11.95) to zero, yielding

$$c_0 = b_0 \tag{11.96a}$$

$$c_1 = b_1 - a_1 b_0 \tag{11.96b}$$

The two state equations for (11.92) are now given by (11.93b) and (11.95) after the insertion of c_0 and c_1 from (11.96), yielding in matrix notation

$$\begin{bmatrix} \dot{x}_1 \\ \dot{x}_2 \end{bmatrix} = \begin{bmatrix} 0 & 1 \\ -a_2 & -a_1 \end{bmatrix} \begin{bmatrix} x_1 \\ x_2 \end{bmatrix} + \begin{bmatrix} b_1 - a_1 b_0 \\ (b_2 - a_2 b_0) - a_1 (b_1 - a_1 b_0) \end{bmatrix} [u] \tag{11.97}$$

The procedure described above can be extended to higher-order differential equations such as the one shown in (11.91). For instance, when $n = 3$, the state variables may be defined as

$$x_1 = y - c_0 u \qquad\qquad\qquad \textbf{(11.98a)}$$

$$x_2 = \dot{x}_1 - c_1 u \qquad\qquad\qquad \textbf{(11.98b)}$$

$$x_3 = \dot{x}_2 - c_2 u \qquad\qquad\qquad \textbf{(11.98c)}$$

where, as before, c_0, c_1, and c_2 are constants to be chosen. To determine these constants, we differentiate (11.98) to give

$$\frac{dy}{dt} = \frac{dx_1}{dt} + c_0 \frac{du}{dt} = x_2 + c_1 u + c_0 \frac{du}{dt} \qquad\qquad \textbf{(11.99a)}$$

$$\frac{d^2 y}{dt^2} = x_3 + c_2 u + c_1 \frac{du}{dt} + c_0 \frac{d^2 u}{dt^2} \qquad\qquad \textbf{(11.99b)}$$

$$\frac{d^3 y}{dt^3} = \frac{dx_3}{dt} + c_2 \frac{du}{dt} + c_1 \frac{d^2 u}{dt^2} + c_0 \frac{d^3 u}{dt^3} \qquad\qquad \textbf{(11.99c)}$$

After substituting these in (11.91) for $n = 3$ and collecting terms, we get

$$\frac{dx_3}{dt} + a_1 x_3 + a_2 x_2 + a_3 x_1 = (b_0 - c_0) \frac{d^3 u}{dt^3}$$

$$+ (b_1 - c_1 - a_1 c_0) \frac{d^2 u}{dt^2}$$

$$+ (b_2 - c_2 - a_1 c_1 - a_2 c_0) \frac{du}{dt}$$

$$+ (b_3 - a_3 c_0 - a_2 c_1 - a_1 c_2) u \qquad \textbf{(11.100)}$$

As before we set the coefficients of $d^k u / dt^k$ ($k = 1, 2, 3$) to zero and get

$$c_0 = b_0 \qquad\qquad\qquad \textbf{(11.101a)}$$

$$c_1 = b_1 - a_1 c_0 \qquad\qquad\qquad \textbf{(11.101b)}$$

$$c_2 = b_2 - a_2 c_0 - a_1 c_1 \qquad\qquad\qquad \textbf{(11.101c)}$$

The state equation becomes

$$\begin{bmatrix} \dot{x}_1 \\ \dot{x}_2 \\ \dot{x}_3 \end{bmatrix} = \begin{bmatrix} 0 & 1 & 0 \\ 0 & 0 & 1 \\ -a_3 & -a_2 & -a_1 \end{bmatrix} \begin{bmatrix} x_1 \\ x_2 \\ x_3 \end{bmatrix} + \begin{bmatrix} c_1 \\ c_2 \\ c_3 \end{bmatrix} [u] \qquad \textbf{(11.102)}$$

where

$$c_3 = (b_3 - a_3 c_0) - a_2 c_1 - a_1 c_2 \qquad\qquad \textbf{(11.103)}$$

Observe that the elements in the coefficient matrix of [u] in (11.97) are essentially c_1 and c_2 in (11.102). It is not difficult to see that the patterns established for the second- and third-order differential equations can be extended to the nth-order differential equations of (11.91) by defining the state variables as

$$x_1 = y - c_0 u$$
$$x_2 = \dot{x}_1 - c_1 u$$
$$\vdots$$
$$x_n = \dot{x}_{n-1} - c_{n-1} u \tag{11.104}$$

The general state equation becomes

$$
\begin{bmatrix} \dot{x}_1 \\ \dot{x}_2 \\ \vdots \\ \dot{x}_{n-1} \\ \dot{x}_n \end{bmatrix}
=
\begin{bmatrix}
0 & 1 & 0 & \cdots & 0 \\
0 & 0 & 1 & \cdots & 0 \\
\vdots & \vdots & \vdots & \vdots & \vdots \\
0 & 0 & 0 & \cdots & 1 \\
-a_n & -a_{n-1} & -a_{n-2} & \cdots & -a_1
\end{bmatrix}
\begin{bmatrix} x_1 \\ x_2 \\ \vdots \\ x_{n-1} \\ x_n \end{bmatrix}
+
\begin{bmatrix} c_1 \\ c_2 \\ \vdots \\ c_{n-1} \\ c_n \end{bmatrix} [u] \tag{11.105}
$$

where $n > 1$

$$c_1 = b_1 - a_1 b_0$$
$$c_2 = (b_2 - a_2 b_0) - a_1 c_1$$
$$c_3 = (b_3 - a_3 b_0) - a_2 c_1 - a_1 c_2$$
$$\vdots$$
$$c_n = (b_n - a_n b_0) - a_{n-1} c_1 - a_{n-2} c_2 - \cdots - a_2 c_{n-2} - a_1 c_{n-1} \tag{11.106}$$

and

$$x_1 = y - b_0 u \tag{11.107}$$

Finally, if y is the output variable, the output equation becomes

$$y(t) = [1 \quad 0 \quad 0 \quad \cdots \quad 0] \begin{bmatrix} x_1 \\ x_2 \\ \vdots \\ x_n \end{bmatrix} + [b_0][u] \tag{11.108}$$

EXAMPLE 11.11

A system is described by the third-order differential equation

$$\frac{d^3 y}{dt^3} + 5 \frac{d^2 y}{dt^2} + 4 \frac{dy}{dt} + 2y = 2 \frac{d^2 u}{dt^2} + \frac{du}{dt} + 3u \tag{11.109}$$

with the initial conditions

$$y(0) = 2, \qquad \left.\frac{dy}{dt}\right|_{t=0} = 4, \qquad \left.\frac{d^2 y}{dt^2}\right|_{t=0} = 1 \tag{11.110a}$$

$$u(0) = 1, \qquad \frac{du}{dt}\bigg|_{t=0} = 3, \qquad \frac{d^2u}{dt^2}\bigg|_{t=0} = 1 \qquad \textbf{(11.110b)}$$

We wish to determine its state equations.

Using (11.105) to (11.107) for $n = 3$, we obtain the state equation

$$\begin{bmatrix} \dot{x}_1 \\ \dot{x}_2 \\ \dot{x}_3 \end{bmatrix} = \begin{bmatrix} 0 & 1 & 0 \\ 0 & 0 & 1 \\ -2 & -4 & -5 \end{bmatrix} \begin{bmatrix} x_1 \\ x_2 \\ x_3 \end{bmatrix} + \begin{bmatrix} 2 \\ -9 \\ 40 \end{bmatrix} [u] \qquad \textbf{(11.111)}$$

The initial conditions for the state variables are obtained from (11.104), as follows:

$$x_1(0) = y(0) - c_0 u(0) = y(0) - b_0 u(0) = 2 \qquad \textbf{(11.112a)}$$

$$x_2(0) = \dot{x}_1(0) - c_1 u(0) = \dot{y}(0) - b_0 \dot{u}(0) - c_1 u(0) = 2 \qquad \textbf{(11.112b)}$$

$$x_3(0) = \dot{x}_2(0) - c_2 u(0)$$

$$= \frac{d^2y}{dt^2}\bigg|_{t=0} - b_0 \frac{d^2u}{dt^2}\bigg|_{t=0} - c_1 \frac{du}{dt}\bigg|_{t=0} - c_2 u(0) = 4 \qquad \textbf{(11.112c)}$$

Finally, the output equation is found to be

$$y(t) = \begin{bmatrix} 1 & 0 & 0 \end{bmatrix} \begin{bmatrix} x_1 \\ x_2 \\ x_3 \end{bmatrix} + [0][u] \qquad \textbf{(11.113)}$$

EXAMPLE 11.12

The network shown in Figure 11.25(a) contains a new device D with terminal voltage $v_4(t)$ and current $i_4(t)$ related by the differential equation

$$i_4 = \frac{d^3v_4}{dt^3} + 2\frac{d^2v_4}{dt^2} + 4\frac{dv_4}{dt} \qquad \textbf{(11.114)}$$

Figure 11.25
A network (a) containing a new device D having terminal voltage and current related by the differential equation (11.114), and its associated directed graph (b).

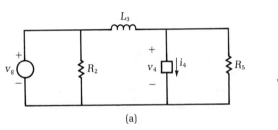
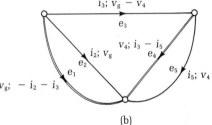

(a)   (b)

We wish to determine the state equation in normal form for an appropriately chosen set of state variables. We do this by following the eight steps outlined in Section 11-4.

The associated directed graph G of the network is shown in Figure 11.25(b). In G the normal tree $T = e_1 e_4$, the voltages, and currents of the edges are as indicated. The element v-i equations are found to be

$$L_3 \frac{di_3}{dt} = v_g - v_4 \tag{11.115a}$$

$$i_3 - i_5 = \frac{d^3 v_4}{dt^3} + 2 \frac{d^2 v_4}{dt^2} + 4 \frac{dv_4}{dt} \tag{11.115b}$$

$$i_2 = \frac{v_g}{R_2} \tag{11.116a}$$

$$i_5 = \frac{v_4}{R_5} \tag{11.116b}$$

Substituting (11.116b) in (11.115b) yields

$$\frac{d^3 v_4}{dt^3} + 2 \frac{d^2 v_4}{dt^2} + 4 \frac{dv_4}{dt} + \frac{v_4}{R_5} = i_3 \tag{11.117}$$

which can be written in the form of (11.105) by defining the state variables as

$$x_1 = v_4 - c_0 i_3 \tag{11.118a}$$

$$x_2 = \dot{x}_1 - c_1 i_3 \tag{11.118b}$$

$$x_3 = \dot{x}_2 - c_2 i_3 \tag{11.118c}$$

giving from (11.106)

$$\begin{bmatrix} \dot{x}_1 \\ \dot{x}_2 \\ \dot{x}_3 \end{bmatrix} = \begin{bmatrix} 0 & 1 & 0 \\ 0 & 0 & 1 \\ -1/R_5 & -4 & -2 \end{bmatrix} \begin{bmatrix} x_1 \\ x_2 \\ x_3 \end{bmatrix} + \begin{bmatrix} 0 \\ 0 \\ 1 \end{bmatrix} [i_3] \tag{11.119}$$

Combining this with (11.115a) and writing them in matrix notation, we get

$$\begin{bmatrix} \dot{x}_1 \\ \dot{x}_2 \\ \dot{x}_3 \\ \dot{i}_3 \end{bmatrix} = \begin{bmatrix} 0 & 1 & 0 & 0 \\ 0 & 0 & 1 & 0 \\ -1/R_5 & -4 & -2 & 1 \\ -1/L_3 & 0 & 0 & 0 \end{bmatrix} \begin{bmatrix} x_1 \\ x_2 \\ x_3 \\ i_3 \end{bmatrix} + \begin{bmatrix} 0 \\ 0 \\ 0 \\ 1/L_3 \end{bmatrix} [v_g] \tag{11.120}$$

where $x_1 = v_4$, $x_2 = \dot{x}_1$, and $x_3 = \dot{x}_2 - i_3$. Equation (11.120) is the desired state equation for the network of Figure 11.25(a).

11-8 SUMMARY

In this chapter we introduced the concept of *state*. For electric networks the set of instantaneous values of all the branch currents and voltages is called the *state of the network*. A minimal set of element variables the instantaneous values of which are sufficient to determine the state of a network is termed a *complete set* of state variables. The inductor currents and capacitor voltages are usually chosen as the *state variables*, since the equations associated with these variables are dynamical. Our objective was to frame the network equations as a system of first-order differential equations involving only the state variables and independent sources. Such a system, when written in the form of (11.10), is called the *state equation in normal form*. Apart from the state equation, the output variables can be expressed algebraically in terms of the state variables and independent sources, resulting in a linear system of equations called the *output equation*. The state equation and the output equation together are referred to as the *state equations*.

To obtain the state equations, we presented an eight-step procedure that amounts to a systematic elimination of nonstate variables among the primary system of network equations. In the case of *degenerate networks* there exists either a circuit composed only of capacitors and/or independent or dependent voltage sources or a cutset composed only of inductors and/or independent or dependent current sources. These networks either have no solution or no unique solution, or their state equations cannot be put in the normal form of (11.10) if capacitor voltages and inductor currents are routinely chosen as the state variables. Instead we must choose the linear combinations of the capacitor voltages, inductor currents, and the independent sources as the state variables.

A problem of considerable physical significance is that of determining the number of natural frequencies of a network. This number, termed the *order of complexity*, is the minimal number of dynamically independent branch currents and voltages having instantaneous values sufficient to determine completely the instantaneous state of the network. In mathematics it is the number of arbitrary constants appearing in the general solution of the network equations. Physically, it is the maximum number of linearly independent initial conditions that can be specified for the network. These are the various ways of stating the same thing. For general active networks, their orders of complexity cannot be determined purely from their topologies alone; network parameters are also involved. For RLC networks, on the other hand, the network structures are sufficient. To this end we presented formulas that enable us to determine the number of natural frequencies and the number of nonzero natural frequencies in an RLC network.

Finally, we indicated that systems which are described by scalar differential equations can be represented by the equivalent state equations. First, we treated the simpler situation where the derivatives of the forcing function are not involved. Then we removed this restriction by considering the general nth-order differential equations.

Once a system has been characterized by its state equations, we can apply the known techniques in mathematics to obtain the solution. This will be the subject of the following chapter.

REFERENCES AND SUGGESTED READING

Balabanian, N., and T. A. Bickart. *Electrical Network Theory.* New York: Wiley, 1969, Chapter 4.

Chan, S. P., S. Y. Chan, and S. G. Chan. *Analysis of Linear Networks and Systems.* Reading, Mass.: Addison-Wesley, 1972, Chapter 7.

Chen, W. K. *Applied Graph Theory: Graphs and Electrical Networks,* 2d ed. Amsterdam: North-Holland, and New York: American Elsevier, 1976, Chapter 7.

Chua, L. O., and P. M. Lin. *Computer-Aided Analysis of Electronics Circuits: Algorithms & Computational Techniques.* Englewood Cliffs, N.J.: Prentice-Hall, 1975, Chapter 8.

Jensen, R. W., and B. O. Watkins. *Network Analysis: Theory and Computer Methods.* Englewood Cliffs, N.J.: Prentice-Hall, 1974, Chapters 10 and 11.

Karni, S. *Intermediate Network Analysis.* Boston, Mass.: Allyn & Bacon, 1971, Chapter 10.

Kuo, B. C. *Linear Networks and Systems,* New York: McGraw-Hill, 1967, Chapters 5 and 6.

Lago, G. V., and L. M. Benningfield. *Circuit and System Theory.* New York: Wiley, 1979, Chapter 13.

Wing, O. *Circuit Theory with Computer Methods.* New York: Holt, Rinehart & Winston, 1972, Chapter 9.

PROBLEMS

11.1 Write the state equation in normal form for each of the networks shown in Figure 11.26.

11.2 Verify that the state equation for the network of Figure 11.27 is given by

$$
\begin{bmatrix} \dot{v}_3 \\ \dot{v}_4 \\ \dot{i}_5 \end{bmatrix} = \begin{bmatrix} -1/C_3R_6 & -1/C_3R_6 & 0 \\ -\beta & -\beta & -1/C_4 \\ 0 & 1/L_5 & 0 \end{bmatrix} \begin{bmatrix} v_3 \\ v_4 \\ i_5 \end{bmatrix} + \begin{bmatrix} 1/C_3 & 0 \\ 1/C_4 & 0 \\ 0 & -1/L_5 \end{bmatrix} \begin{bmatrix} i_g \\ v_g \end{bmatrix}
\tag{11.121}
$$

where $\beta = (1 + \alpha R_6)/C_4R_6$. Also determine the output equation if resistor voltage v_6 and resistor current i_7 are selected as the output variables.

11.3 The state equation for the network of Figure 11.28 is shown below.

$$
\begin{bmatrix} \dot{v}_1 \\ \dot{v}_2 \end{bmatrix} = \begin{bmatrix} -1 & -1 \\ -0.5 & -1.5 \end{bmatrix} \begin{bmatrix} v_1 \\ v_2 \end{bmatrix} + \begin{bmatrix} 1 \\ 0.5 \end{bmatrix} [v_g]
\tag{11.122}
$$

Determine the value of the resistance R in the network.

11.4 Demonstrate that the network of Figure 11.29 has no unique solution.

Figure 11.26
Networks the state equations of which are to be ascertained.

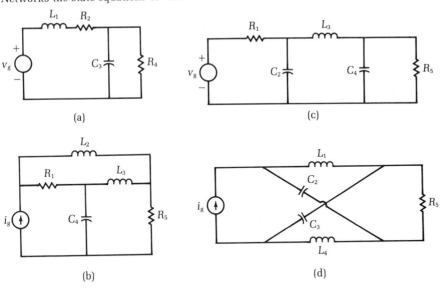

(a)

(b)

(c)

(d)

Figure 11.27
An active network the
state equation of which
is given by (11.121).

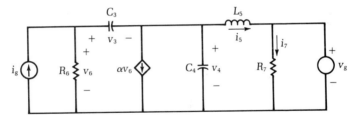

Figure 11.28
A network that can be
characterized by the state
equation (11.122) by an
appropriate choice of the
resistance R.

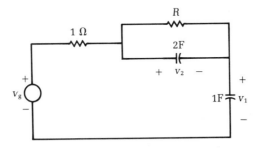

Figure 11.29
A simple network that
does not possess a unique
solution.

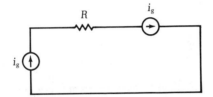

Figure 11.30
A network the state
equation of which is to
be ascertained.

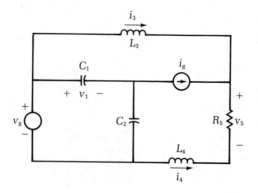

Figure 11.31
A network the state
equation of which is
given by (11.124).

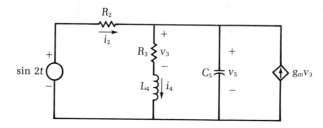

Figure 11.32
An active network the
state equation of which is
to be ascertained.

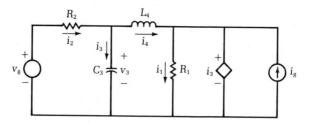

11.5 For the network shown in Figure 11.30, choose

$$x_1 = v_1 + \beta_1 v_g + \beta_2 i_g \qquad\qquad (11.123a)$$
$$x_2 = i_3 + \beta_3 v_g + \beta_4 i_g \qquad\qquad (11.123b)$$

as the state variables. Determine the constants β_j ($j = 1, 2, 3, 4$) and the state equations for the network with output variables i_4 and v_5.

11.6 Verify that the state equation for the network of Figure 11.31 is given by

$$
\begin{bmatrix} \dot{i}_4 \\ \dot{v}_5 \end{bmatrix} =
\begin{bmatrix} -\dfrac{R_3}{L_4} & \dfrac{1}{L_4} \\[2ex] \dfrac{g_m R_3 - 1}{C_5} & -\dfrac{1}{R_2 C_5} \end{bmatrix}
\begin{bmatrix} i_4 \\ v_5 \end{bmatrix} +
\begin{bmatrix} 0 \\[2ex] \dfrac{1}{R_2 C_5} \end{bmatrix} [\sin 2t] \qquad\qquad (11.124)
$$

Obtain the output equation if resistor voltage v_3 and resistor current i_2 are chosen as the output variables.

Figure 11.33
An active network
containing two mutually
coupled inductors that
can be characterized by
the state equation (11.126).

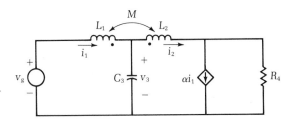

11.7 Choose a normal tree and write the normal-form state equations for the network of Figure 11.32 where the resistor currents i_1 and i_2 are the output variables. Verify that for $R_1 = 20\ \Omega$, $R_2 = 10\ \Omega$, $C_3 = 0.005$ F, and $L_4 = 2$ H, the state equation reduces to

$$\begin{bmatrix} \dot{v}_3 \\ \dot{i}_4 \end{bmatrix} = \begin{bmatrix} -20 & -200 \\ 0.55 & 0.5 \end{bmatrix} \begin{bmatrix} v_3 \\ i_4 \end{bmatrix} + \begin{bmatrix} 0 & 20 \\ 0 & -0.05 \end{bmatrix} \begin{bmatrix} i_g \\ v_g \end{bmatrix} \tag{11.125}$$

11.8 Show that the state equation for the network of Figure 11.33 can be written as

$$\begin{bmatrix} \dot{i}_1 \\ \dot{i}_2 \\ \dot{v}_3 \end{bmatrix} = \begin{bmatrix} \alpha M R_4/\Delta & -MR_4/\Delta & (M - L_2)/\Delta \\ \alpha L_1 R_4/\Delta & -L_1 R_4/\Delta & (L_1 - M)/\Delta \\ 1/C_3 & -1/C_3 & 0 \end{bmatrix} \begin{bmatrix} i_1 \\ i_2 \\ v_3 \end{bmatrix} + \begin{bmatrix} L_2/\Delta \\ M/\Delta \\ 0 \end{bmatrix} [v_g] \tag{11.126}$$

where $\Delta = L_1 L_2 - M^2 \neq 0$.

11.9 Determine the state equation for a system described by the following third-order differential equation

$$3\frac{d^3y}{dt^3} + 9\frac{d^2y}{dt^2} + \frac{dy}{dt} + 2y = 5\frac{d^2u}{dt^2} + u \tag{11.127}$$

with initial conditions as given in (11.110).

11.10 Determine the order of complexity and the number of nonzero natural frequencies for the network of Figure 11.22 when inductor L_{11} is replaced by a resistor.

11.11 Repeat Problem 11.10 when the positions of resistor R_5 and inductor L_8 are interchanged.

11.12 The network shown in Figure 11.34 contains a new kind of device D the terminal voltage of which is equal to the third derivative of its terminal current. Show that the state equation of the network can be written as

$$\begin{bmatrix} \dot{v}_3 \\ \dot{i}_4 \\ \dot{x}_1 \\ \dot{x}_2 \end{bmatrix} = \begin{bmatrix} -1/C_3 R_5 & 1/C_3 & 0 & 0 \\ 0 & 0 & 1 & 0 \\ 0 & 0 & 0 & 1 \\ -1 & -R_2 & 0 & 0 \end{bmatrix} \begin{bmatrix} v_3 \\ i_4 \\ x_1 \\ x_2 \end{bmatrix} + \begin{bmatrix} 0 \\ 0 \\ 0 \\ R_2 \end{bmatrix} [i_g] \tag{11.128}$$

where $x_1 = \dot{i}_4$ and $x_2 = \dot{x}_1$.

11.13 Determine the state equation of the network of Figure 11.34 if the device is characterized by the equation

$$v_4 = \frac{d^3 i_4}{dt^3} + 2\frac{d^2 i_4}{dt^2} + \frac{di_4}{dt} \tag{11.129}$$

Figure 11.34
A network containing a
new device D having
terminal voltage equal to
the third derivative of its
terminal current.

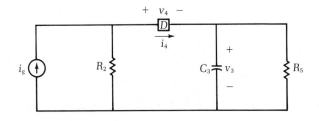

Figure 11.35
Networks the state
equations of which are
to be ascertained.

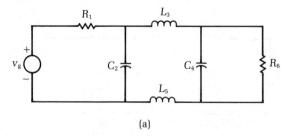

(a)

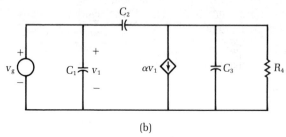

(b)

Figure 11.36
An active network the
state equation of which
is to be ascertained.

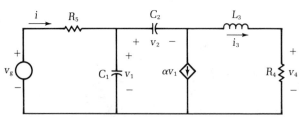

11.14 A system is characterized by the following third-order differential equation. Determine its state-equation representation.

$$\frac{d^3y}{dt^3} + 7\frac{d^2y}{dt^2} + 11\frac{dy}{dt} + 8y = 2\frac{d^3u}{dt^3} + 8\frac{d^2u}{dt^2} + \frac{du}{dt} + u \qquad (11.130)$$

11.15 Write the state equation in normal form for each of the networks shown in Figure 11.35.

11.16 Obtain the state equations for the network of Figure 11.36. The output variables are the input current i and output voltage v_4. Verify that, for $R_4 = 4\ \Omega$, $R_5 = 2\ \Omega$, $C_1 = 1$ F, $C_2 = 3$ F, $L_3 = 5$ H, and $\alpha = 6$, the state equation reduces to

$$\begin{bmatrix} \dot{v}_1 \\ \dot{v}_2 \\ \dot{i}_3 \end{bmatrix} = \begin{bmatrix} -6.5 & 0 & -1 \\ 2 & 0 & 1/3 \\ 0.2 & -0.2 & -0.8 \end{bmatrix} \begin{bmatrix} v_1 \\ v_2 \\ i_3 \end{bmatrix} + \begin{bmatrix} 0.5 \\ 0 \\ 0 \end{bmatrix} [v_g] \qquad (11.131)$$

11.17 Show that the state equation for the network of Figure 11.37 can be written as

$$\begin{bmatrix} \dot{i}_1 \\ \dot{i}_2 \\ \dot{v}_3 \end{bmatrix} = \frac{1}{\Delta}\begin{bmatrix} 0 & -MR_4 & M-L_2 \\ 0 & -L_1R_4 & L_1-M \\ \Delta/C_3 & -\Delta/C_3 & 0 \end{bmatrix} \begin{bmatrix} i_1 \\ i_2 \\ v_3 \end{bmatrix} + \frac{1}{\Delta}\begin{bmatrix} L_2-M & -R_4M \\ M-L_1 & -L_1R_4 \\ 0 & 0 \end{bmatrix}\begin{bmatrix} v_g \\ i_g \end{bmatrix} \qquad (11.132)$$

where $\Delta = L_1L_2 - M^2 \neq 0$.

11.18 Verify that the determinant of the coefficient matrix of (11.78) with s replacing p is given by (11.79).

Figure 11.37
A network containing two mutually coupled inductors the state equation of which is given by (11.132).

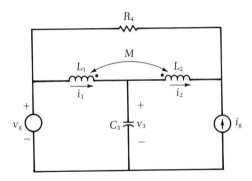

Figure 11.38
A network containing two mutually coupled inductors the state equation of which is given by (11.133).

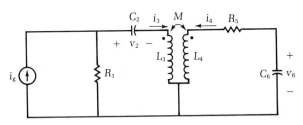

Figure 11.39
An active network the state equation of which is given by (11.134).

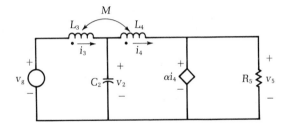

11.19 Confirm that the state equation for the network of Figure 11.38 is given by

$$
\begin{bmatrix} \dot{v}_2 \\ \dot{i}_3 \\ \dot{i}_4 \\ \dot{v}_6 \end{bmatrix} = \begin{bmatrix} 0 & 1/C_2 & 0 & 0 \\ -L_4/\Delta & -L_4R_1/\Delta & MR_5/\Delta & -M/\Delta \\ M/\Delta & MR_1/\Delta & -L_3R_5/\Delta & L_3/\Delta \\ 0 & 0 & -1/C_6 & 0 \end{bmatrix} \begin{bmatrix} v_2 \\ i_3 \\ i_4 \\ v_6 \end{bmatrix} + \begin{bmatrix} 0 \\ L_4R_1/\Delta \\ -MR_1/\Delta \\ 0 \end{bmatrix} [i_g]
$$

(11.133)

where $\Delta = L_3L_4 - M^2 \neq 0$.

11.20 Write the primary system of network equations in the form of (11.68) for the network of Figure 11.20. Using Definitions 11.3 and 11.4, determine the natural frequencies and the order of complexity of the network.

11.21 Show that the state equation for the network of Figure 11.39 can be written as

$$
\begin{bmatrix} \dot{v}_2 \\ \dot{i}_3 \\ \dot{i}_4 \end{bmatrix} = \frac{1}{\Delta} \begin{bmatrix} 0 & \Delta/C_2 & -\Delta/C_2 \\ -M - L_4 & 0 & \alpha M \\ M + L_3 & 0 & -\alpha L_3 \end{bmatrix} \begin{bmatrix} v_2 \\ i_3 \\ i_4 \end{bmatrix} + \frac{1}{\Delta} \begin{bmatrix} 0 \\ L_4 \\ -M \end{bmatrix} [v_g]
$$

(11.134)

where $\Delta = L_3L_4 - M^2 \neq 0$.

CHAPTER TWELVE

<hr>

SOLUTION OF STATE EQUATIONS

In the preceding chapter, we demonstrated that a linear time-invariant system can be described by its state equations, which take the general form[†]

$$\dot{\mathbf{x}}(t) = \mathbf{A}\mathbf{x}(t) + \mathbf{B}\mathbf{u}(t) \tag{12.1}$$

$$\mathbf{y}(t) = \mathbf{C}\mathbf{x}(t) + \mathbf{D}_1\mathbf{u}(t) + \mathbf{D}_2\dot{\mathbf{u}}(t) \tag{12.2}$$

The first of these two equations is a system of first-order differential equations. Once it is solved for the state vector $\mathbf{x}(t)$, the output vector $\mathbf{y}(t)$ is determined algebraically from the second one. Thus the crux of the problem is to solve (12.1). This has been done in mathematics both analytically and numerically. In the present chapter, we shall show how some of these elegant techniques are applied to obtain the solution of (12.1). We first consider explicitly the first-order system and then indicate how the results can be extended to higher-order systems. Both time-domain and frequency-domain solutions are presented, which are amenable to machine computation. In fact, this is one important reason for choosing the state equations to describe a system in general and the network problem in particular.

12-1 FIRST-ORDER SYSTEMS AND TIME-CONSTANT CONCEPTS

To motivate our discussion, in the present section we consider the first-order systems described by the scalar state equation

$$\dot{x}(t) = ax(t) + bu(t) \tag{12.3}$$

and discuss its solution. This solution can also be obtained by using the time-constant concepts.

Recall that in solving a linear differential equation, we usually first assume a general form of its solution and then substitute this solution into

<hr>

[†] In some situations the output equation (12.2) may also involve the second- and higher-order derivatives of $\mathbf{u}(t)$. It is included to show the general form of the output equation.

the original equation to identify specific solutions. For the state equation (12.3), there is no exception. The procedure to be described below is known as the method of *variation of parameters*. Let

$$x(t) = m(t)x_1(t) \tag{12.4}$$

where the function $m(t)$ is assumed to be nonzero for all finite $t \geq t_0$. The functions $m(t)$ and $x_1(t)$ are arbitrary at this point and will be determined shortly. Inserting (12.4) in (12.3) and collecting terms, we get

$$(\dot{m} - am)x_1 + m\dot{x}_1 - bu = 0 \tag{12.5}$$

Note that the time variable t has been dropped for simplicity, unless it is used for emphasis. Observe that in (12.5) if we select m in such a way that the quantity in parentheses is identically zero, the solution of (12.5) is greatly simplified. Proceeding in this way by setting the quantity in parentheses equal to zero, (12.5) reduces to

$$\dot{m}(t) = am(t) \tag{12.6}$$

$$\dot{x}_1(t) = m^{-1}(t)bu(t) \tag{12.7}$$

After (12.6) is solved, m can be inserted in (12.7). The resulting equation can then be directly integrated to find x_1. After both m and x_1 are found, x is determined from (12.4). For the procedure to succeed, we must solve (12.6).

12-1.1 Solution of the Homogeneous Equation

Consider the first-order homogeneous differential equation

$$\dot{m}(t) = am(t) \tag{12.8}$$

with initial condition $m(t_0)$. It is well known that the most general solution can be written as

$$m(t) = ke^{\alpha t} \tag{12.9}$$

where k and α are constants to be determined. Substituting this in (12.8) and simplifying, we get

$$\alpha = a \tag{12.10}$$

The initial condition requires that

$$m(t_0) = ke^{at_0} \tag{12.11}$$

giving

$$k = m(t_0)e^{-at_0} \tag{12.12}$$

Therefore the solution of (12.8) that satisfies the initial condition $m(t_0)$ is given by

$$m(t) = m(t_0)e^{a(t-t_0)} \tag{12.13}$$

Recall that our objective is to solve (12.3). For this we have freedom to specify the initial condition $m(t_0)$. To simplify the problem, we let

$$m(t_0) = 1 \qquad (12.14)$$

The solution (12.9) becomes

$$m(t) = e^{a(t-t_0)} \qquad (12.15)$$

Observe that $m(t)$ is nonzero for all finite $t \geq t_0$, thus satisfying the requirement stated in (12.4)—namely, $m^{-1}(t)$ exists for all finite $t \geq t_0$.

12-1.2 General Solution of the State Equation

With $m(t)$ determined as in (12.15), x_1 can be found by directly integrating (12.7) from t_0 to t, giving

$$\int_{t_0}^{t} \dot{x}_1(\tau)\, d\tau = \int_{t_0}^{t} m^{-1}(\tau) bu(\tau)\, d\tau$$

$$= \int_{t_0}^{t} e^{-a(\tau-t_0)} bu(\tau)\, d\tau \qquad (12.16)$$

or

$$x_1(t) = x_1(t_0) + \int_{t_0}^{t} e^{-a(\tau-t_0)} bu(\tau)\, d\tau \qquad (12.17)$$

To determine the initial condition $x_1(t_0)$, we set $t = t_0$ in (12.4) and obtain from (12.14)

$$x(t_0) = m(t_0)x_1(t_0) = x_1(t_0) \qquad (12.18)$$

Therefore the solution of (12.3) with initial condition $x(t_0)$ is obtained by substituting (12.15) and (12.17) in (12.4) in conjunction with (12.18):

$$x(t) = e^{a(t-t_0)}x(t_0) + \int_{t_0}^{t} e^{a(t-\tau)} bu(\tau)\, d\tau \qquad (12.19)$$

The state variable $x(t)$ at the time $t = t_0$ is called the *initial state*, which describes the complete history of the system for $-\infty < t \leq t_0$. Equation (12.19) indicates that regardless of the input signals prior to t_0, the state $x(t)$ of the system at any time $t > t_0$ can be determined from the information about the initial state $x(t_0)$ and the input excitation from t_0 to t. Thus the evaluation of the response of a system from one state to another through time may be visualized simply as a process of *state transition*.

Finally, the output equation can be written as

$$y(t) = cx(t) + du(t) \qquad (12.20)$$

or when $x(t)$ is substituted from (12.19) we have

$$y(t) = ce^{a(t-t_0)}x(t_0) + \left[c \int_{t_0}^{t} e^{a(t-\tau)} bu(\tau)\, d\tau + du(t) \right] \qquad (12.21)$$

The first term on the right-hand side of (12.21) is the natural response or the zero-input response; the second term is the forced response or the zero-state response. In other words by setting $u(t) = 0$ we obtain the natural response or the zero-input response. Likewise setting $x(t_0) = 0$ results in the forced response or the zero-state response. Together they constitute the complete response of a system.

EXAMPLE 12.1

Consider the simple RC network of Figure 12.1, the state equation of which is given by

$$\dot{v}(t) = -\frac{1}{RC}\, v(t) + \frac{1}{RC}\, v_g \tag{12.22}$$

The input excitation is the step function defined by (Figure 12.2)

$$
\begin{aligned}
v_g &= 0, && t < 0 \\
&= V, && t \geq 0
\end{aligned}
\tag{12.23}
$$

Thus we can make the following identifications: $a = -1/RC$, $b = 1/RC$, $x(t) = v(t)$, so that $\dot{x}(t) = \dot{v}(t)$. Choose $t_0 = 0$. Then $v(t_0) = v(0) = 0$. Applying (12.19) yields the capacitor voltage

$$v(t) = \frac{V}{RC} \int_0^t e^{-(t-\tau)/RC}\, d\tau = \frac{V}{RC}\, e^{-t/RC} \int_0^t e^{\tau/RC}\, d\tau$$

$$= V(1 - e^{-t/RC}), \qquad t \geq 0 \tag{12.24}$$

Figure 12.1
A simple RC network with a voltage excitation.

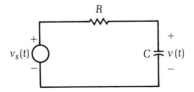

Figure 12.2
Plots of the input voltage $v_g(t)$ and the capacitor voltage $v(t)$ as functions of t for the network of Figure 12.1.

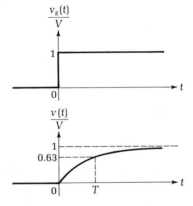

Plots of the input voltage $v_g(t)$ and the capacitor voltage $v(t)$ as functions of the time t are presented in Figure 12.2. The constant $T = RC$ is the *time constant* of the network. The physical significance attached to the time constant is of great importance and will be elaborated later.

EXAMPLE 12.2

Suppose that a square pulse of magnitude V and width T_1 as shown in Figure 12.3 is applied to the network of Figure 12.1.

$$
\begin{aligned}
v_g &= 0, & t &< 0 \\
&= V, & 0 &\leq t \leq T_1 \\
&= 0, & t &> T_1
\end{aligned}
\tag{12.25}
$$

Choose the initial time $t_0 = 0$. Then $v(0) = 0$ and from (12.19) the capacitor voltage can be expressed as

$$
v(t) = \frac{1}{RC} e^{-t/RC} \int_0^t e^{\tau/RC} v_g(\tau)\, d\tau
\tag{12.26}
$$

Since v_g is discontinuous, the integration is performed over two intervals. For $0 \leq t \leq T_1$, $v_g = V$, and we have

$$
\begin{aligned}
v(t) &= \frac{V}{RC} e^{-t/RC} \int_0^t e^{\tau/RC}\, d\tau \\
&= V(1 - e^{-t/RC}), & 0 \leq t \leq T_1
\end{aligned}
\tag{12.27}
$$

For $t > T_1$, (12.26) can be written as sum of two integrals:

$$
\begin{aligned}
v(t) &= \frac{V}{RC} e^{-t/RC} \int_0^{T_1} e^{\tau/RC}\, d\tau + \frac{1}{RC} e^{-t/RC} \int_{T_1}^t e^{\tau/RC} 0\, d\tau \\
&= V(1 - e^{-T_1/RC}) e^{-(t-T_1)/RC}, & t > T_1
\end{aligned}
\tag{12.28}
$$

Observe that the responses over the different intervals match at the boundaries. Alternatively (12.28) can be deduced from (12.27) by the process of state transition. For instance, we may consider the response $v(t)$ at time $t = T_1$ as the initial state, which from (12.27) is

$$
v(T_1) = V(1 - e^{-T_1/RC})
\tag{12.29}
$$

Knowing the initial state $v(T_1)$ and the input excitation from T_1 to t, which is zero, the state of the network at any time $t \geq T_1$ can be written down

Figure 12.3
A square pulse voltage source.

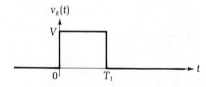

Figure 12.4
Plots of the three typical
responses of the network
of Figure 12.1 due to the
square pulse voltage input
of Figure 12.3.

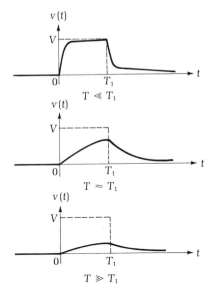

directly from (12.19), as follows:

$$v(t) = e^{-(t-T_1)/RC} v(T_1) + \frac{1}{RC} \int_{T_1}^{t} e^{-(t-\tau)/RC} 0 \, d\tau$$

$$= V(1 - e^{-T_1/RC}) e^{-(t-T_1)/RC}, \qquad t \geq T_1 \qquad \textbf{(12.30)}$$

confirming (12.28).

Depending on the relative values of the time constant $T = RC$ and pulse width T_1, three typical responses of $v(t)$ are plotted in Figure 12.4.

The following example illustrates the decomposition of the complete response into the sum of the zero-input response and the zero-state response.

EXAMPLE 12.3
Consider the simple network of Figure 12.5. For $t < 0$, switch S_1 is on terminal 2, switch S_2 is open, and the inductor is supplied with a constant

Figure 12.5
A network with two
switches activated at
$t = 0$.

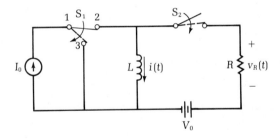

Solution of State Equations

current I_0. At $t = 0$ switch S_1 is flipped to terminal 3 and switch S_2 is closed. Thus for $t \geq 0$ the inductor with initial current I_0 is connected in series with a resistor of resistance R and a battery of voltage V_0. The state equation for the network is found to be

$$\dot{i}(t) = -\frac{R}{L} i(t) + \frac{1}{L} V_0, \qquad t \geq 0 \tag{12.31}$$

with initial state $i(0) = I_0$. Appealing to (12.19), the inductor current is obtained as

$$i(t) = e^{-Rt/L} i(0) + \frac{V_0}{L} \int_0^t e^{-R(t-\tau)/L} \, d\tau$$

$$= I_0 e^{-Rt/L} + \frac{V_0}{R} (1 - e^{-Rt/L}), \qquad t \geq 0 \tag{12.32}$$

If the voltage across the resistor is the output variable, the output equation becomes

$$v_R(t) = -Ri(t) = -RI_0 e^{-Rt/L} - V_0(1 - e^{-Rt/L}), \qquad t \geq 0 \tag{12.33}$$

Equation (12.33) is the complete response that is decomposed into the zero-input response and the zero-state response corresponding to the first and second terms on the right-hand side of the equation. The zero-state response is due to input excitation only, whereas the zero-input response is due to the initial conditions only.

12-1.3 The Time-Constant Concepts

The responses given by (12.24), (12.27), (12.28), and (12.32) may be written in one of the following two general forms:

$$\frac{v}{V_1} = e^{-t/T} \tag{12.34}$$

$$\frac{v}{V_1} = 1 - e^{-t/T} \tag{12.35}$$

where V_1 is the initial value of voltage at $t = 0$ or T_1 as in (12.28), and T is the *time constant* of the system. The values of V_1 and T are different for different problems. The physical significance attached to the time constant is important in that when $t = T$, by (12.34) and (12.35),

$$\frac{v}{V_1} = e^{-1} \approx 0.37 \tag{12.36a}$$

$$\frac{v}{V_1} = 1 - e^{-1} \approx 0.63 \tag{12.36b}$$

For (12.36a) it means that the voltage decreases to 37 percent of its initial value in one time constant; and for (12.36b) it means that the voltage reaches 63 percent of its final value in one time constant. By a similar computation, it can be shown that the voltage decreases to approximately 2 percent of its initial value or reaches 98 percent of its final value in four time constants. Because the response theoretically never disappears or reaches its steady state in finite time, the time interval for an exponential function to decrease to 37 percent of its initial value or to increase to 63 percent of its final value is a convenient value chosen as a standard for comparison. For instance, the time constant for the network of Figure 12.1 is $T = RC$. Suppose that R has the value of 200 Ω and $C = 100$ pF, then $T = 0.02$ μsec. However, if $R = 2$ MΩ and $C = 10$ μF, then $T = 20$ sec. For one combination the voltage $v(t)$, as given in (12.24), would reach 63 percent of its final value V in the small time of 0.02 μsec; for the other the voltage would require 20 sec to reach to 63 percent of the final value.

Another important observation is that the tangents to the equations (12.34) and (12.35) at $t = 0$, given as $-1/T$ and $1/T$, intersect the $v = 0$ and $v = V_1$ lines, respectively, at $t = T$. These two lines are shown as dashed lines in Figure 12.6. Thus if the voltage decreases or increases at the initial rate, it would be reduced to zero or increased to its final value in one time constant.

In Example 12.3 we demonstrated the decomposition of the complete response into the zero-input response and the zero-state response. A different kind of decomposition is also possible. Rewrite (12.33) as

$$v_R(t) = (V_0 - RI_0)e^{-Rt/L} - V_0, \qquad t \geq 0 \tag{12.37}$$

The first term is a decaying exponential and the second term a constant V_0. For very large t, the first term is negligible and the second term becomes dominant. Therefore the first term is the familiar transient response and the second term the steady-state response. In this example it is clear that the transient response depends on both the zero-input response and the zero-state response, whereas the steady-state response is contributed only by the zero-state response. Physically the transient response is due to the initial conditions in the network and the sudden application of the input. If the network is well behaved as time goes on (such as in a stable network), the transient part eventually dies out. The steady-state response is due only

Figure 12.6

Plots of Eqs. (12.34) and (12.35) as functions of the normalized time t/T.

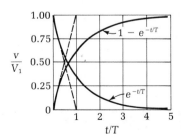

to the input excitation and in many situations has a waveform closely related to that of the input signal. For instance, if the input is a sinusoid of a fixed angular frequency, the steady-state response is also a sinusoid of the same frequency. In Example 12.3 the input excitation is a constant voltage; the steady-state response is therefore also a constant.

In the following we show that (12.19) can be partitioned into the transient response and the steady-state response. Each of these responses can be determined physically, thus avoiding the necessity of using (12.19).

Let

$$f(t) = \int e^{-a\tau} b u(\tau)\, d\tau \tag{12.38}$$

The general solution (12.19) of the state equation can be rewritten as

$$\begin{aligned}
x(t) &= e^{a(t-t_0)}x(t_0) + e^{at}[f(t) - f(t_0)]\\
&= [x(t_0) - f(t_0)e^{at_0}]e^{a(t-t_0)} + f(t)e^{at}\\
&= \gamma e^{-(t-t_0)/T} + x_{ss}(t), \qquad t \geq t_0
\end{aligned} \tag{12.39}$$

where

$$\gamma = x(t_0) - f(t_0)e^{at_0} \tag{12.40a}$$

$$T = -\frac{1}{a} \tag{12.40b}$$

$$x_{ss}(t) = f(t)e^{at} \tag{12.40c}$$

The first term on the right-hand side of (12.39) is the transient response, and the second term the steady-state response. The parameter T is the time constant.

The steady-state response x_{ss} can be computed by any means available. If the inductor current or capacitor voltage is chosen as the state variable, the coefficient γ may be determined from (12.39) by setting $t = t_0$, giving

$$\gamma = x(t_0-) - x_{ss}(t_0+) \tag{12.41}$$

where we have employed the fact that $x(t_0-) = x(t_0+)$.

We illustrate the above procedure by the following examples.

EXAMPLE 12.4

Consider again the network of Figure 12.5, as discussed in Example 12.3. For $t \geq 0$ the network is described by the state equation

$$\dot{i}(t) = -\frac{R}{L} i(t) + \frac{1}{L} V_0, \qquad t \geq 0 \tag{12.42}$$

with initial inductor current $i(0) = I_0$. The time constant of the network is found to be $T = -1/a = L/R$.

For very large t, the steady-state inductor current i_{ss}, as seen from Figure 12.5, is evidently V_0/R. Using this in (12.41) yields

$$\gamma = i(0-) - i_{ss}(0+) = I_0 - \frac{V_0}{R} \tag{12.43}$$

The inductor current for $t \geq 0$ is obtained from (12.39) as

$$i(t) = \gamma e^{-t/T} + i_{ss}(t) = (I_0 - V_0/R)e^{-Rt/L} + \frac{V_0}{R} \tag{12.44}$$

confirming (12.32).

EXAMPLE 12.5

We use the time-constant concept to compute the capacitor voltage of the network of Figure 12.7 due to the input excitation

$$\begin{aligned} v_g &= 0, & t &< 0 \\ &= \tfrac{1}{2} \cos 10^5 t, & t &\geq 0 \end{aligned} \tag{12.45}$$

The state equation of network is found to be

$$\dot{v}(t) = -10^5 \, v(t) + 10^5 \, v_g(t) \tag{12.46}$$

The initial capacitor voltage is $v(0-) = v(0+) = 0$, and the time constant is identified from (12.46) as $T = -1/a = 10^{-5}$ sec.

The steady-state capacitor voltage v_{ss} is the voltage across the capacitor due to the sinusoidal steady-state input voltage $v_g = \tfrac{1}{2} \cos 10^5 t$. This can easily be computed by using the familiar phasor technique. The voltage phasor $V_{ss}(j10^5)$ across the capacitor is found to be

$$V_{ss}(j10^5) = \frac{\dfrac{1}{2} \cdot \dfrac{1}{j10^5 \times 10^{-8}}}{10^3 + \dfrac{1}{j10^5 \times 10^{-8}}} = \frac{1}{2 + j2} = \frac{1}{2\sqrt{2}} \underline{/-\pi/4} \tag{12.47}$$

giving

$$v_{ss}(t) = \frac{1}{4}\sqrt{2} \cos\left(10^5 t - \frac{\pi}{4}\right) \tag{12.48}$$

Using this in (12.41), we get

$$\gamma = v(0-) - v_{ss}(0+) = 0 - \frac{1}{4}\sqrt{2} \cos\left(-\frac{\pi}{4}\right) = -\frac{1}{4} \tag{12.49}$$

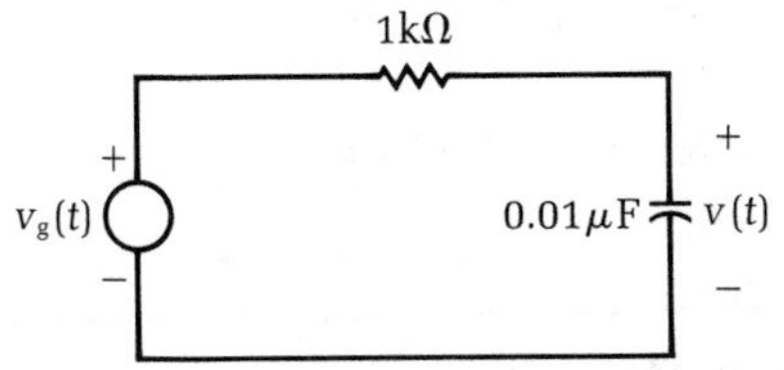

Figure 12.7
A network used to illustrate the time-constant concept in the computation of network response.

Finally, to obtain $v(t)$ we substitute (12.48) and (12.49) in (12.39) and obtain

$$v(t) = \gamma e^{-t/T} + v_{ss}(t) = -\frac{1}{4} e^{-10^5 t} + \frac{1}{4} \sqrt{2} \cos\left(10^5 t - \frac{\pi}{4}\right) \tag{12.50}$$

This can be verified directly by appealing to (12.19), as follows:

$$v(t) = e^{-10^5 t} v(0) + 5 \times 10^4 e^{-10^5 t} \int_0^t e^{10^5 \tau} \cos 10^5 \tau \, d\tau$$

$$= 0 + \frac{1}{4} e^{-10^5 t} \left[(\cos 10^5 \tau + \sin 10^5 \tau) e^{10^5 t} \right]_0^t$$

$$= -\frac{1}{4} e^{-10^5 t} + \frac{1}{4} \sqrt{2} \cos\left(10^5 t - \frac{\pi}{4}\right) \tag{12.51}$$

The first term is the transient response and the second term the steady-state response. Together they constitute the complete response of the network when capacitor voltage is selected as the output variable.

12-2   HIGHER-ORDER SYSTEMS

In the preceding section, we discussed the general solution of the state equation associated with the first-order system. In the present section, we shall extend the previous result by considering the solution of the state equation associated with a general system. The approach is very similar except that now we are dealing with the matrix equation. The scalar quantities are replaced by their matrix counterparts. Since matrix multiplication is in general not commutative, care should be exercised in keeping various matrices in their proper order. For instance, in the scalar equation we can write its solution as either $x = mx_1$ or $x = x_1 m$, whereas in the matrix equation we can only write $\mathbf{x} = \mathbf{M}\mathbf{x}_1$, not $\mathbf{x}_1\mathbf{M}$.

Our objective is to solve the state equation

$$\dot{\mathbf{x}}(t) = \mathbf{A}\mathbf{x}(t) + \mathbf{B}\mathbf{u}(t) \tag{12.52}$$

To obtain the solution to (12.52) in a manner analogous to that described in the foregoing, we let

$$\mathbf{x}(t) = \mathbf{M}(t)\mathbf{x}_1(t) \tag{12.53}$$

where $\mathbf{M}(t)$ is a square matrix of order n that is assumed to be nonsingular for all finite $t \geq t_0$, and $\mathbf{x}_1(t)$ is an n-vector, n being the number of elements of the state vector $\mathbf{x}(t)$. Inserting (12.53) in (12.52) and applying the identity

$$\frac{d}{dt} \mathbf{M}(t)\mathbf{x}_1(t) = \mathbf{M}(t)\dot{\mathbf{x}}_1(t) + \dot{\mathbf{M}}(t)\mathbf{x}_1(t) \tag{12.54}$$

we obtain

$$(\dot{\mathbf{M}} - \mathbf{A}\mathbf{M})\mathbf{x}_1 + \mathbf{M}\dot{\mathbf{x}}_1 - \mathbf{B}\mathbf{u} = 0 \tag{12.55}$$

As in (12.5), if we select the matrix $\mathbf{M}$ to make the quantity in parentheses identically zero, the solution of (12.55) is greatly simplified. Proceeding in this way, (12.55) reduces to

$$\dot{\mathbf{M}}(t) = \mathbf{A}\mathbf{M}(t) \tag{12.56}$$

$$\dot{\mathbf{x}}_1(t) = \mathbf{M}^{-1}(t)\mathbf{B}\mathbf{u}(t) \tag{12.57}$$

The second equation comes from premultiplying by the inverse of $\mathbf{M}(t)$, which exists because $\mathbf{M}(t)$ was assumed to be nonsingular for all finite $t \geqq t_0$. For the procedure to succeed, we must find such an $\mathbf{M}(t)$.

12-2.1   Solution of the Homogeneous Equation

We first demonstrate that solving (12.56) is equivalent to finding n linearly independent solutions satisfying the following homogeneous differential equation

$$\dot{\mathbf{x}}_2(t) = \mathbf{A}\mathbf{x}_2(t) \tag{12.58}$$

where $\mathbf{x}_2(t)$ is an n-vector. To see this let $\mathbf{M}(t)$ be represented by its columns as

$$\mathbf{M}(t) = [\mathbf{M}_1, \mathbf{M}_2, \dots, \mathbf{M}_n] \tag{12.59}$$

Substituting this in (12.56) yields in a different form

$$\dot{\mathbf{M}}_j = \mathbf{A}\mathbf{M}_j, \qquad j = 1, 2, \dots, n \tag{12.60}$$

showing that $\mathbf{M}_j$ are solutions of (12.58). This means that if we can determine n linearly independent solutions satisfying (12.58), the square matrix formed by these solutions will be the desired matrix $\mathbf{M}(t)$.

Since the form of the matrix equation (12.58) is very similar to that of the scalar equation (12.8), it is tempting to seek an exponential solution

$$\mathbf{x}_2(t) = e^{\mathbf{A}t}\mathbf{K} \tag{12.61}$$

where $\mathbf{K}$ is an n-vector. The trouble is that we do not know the meaning of an exponential with a matrix in the exponent. To circumvent this difficulty, we use the series expansion of the exponential function to define the exponential of a matrix. It is well known that e^{at} can be expanded in an infinite series as

$$e^{at} = 1 + at + \frac{a^2 t^2}{2!} + \cdots + \frac{a^k t^k}{k!} + \cdots = \sum_{k=0}^{\infty} \frac{a^k t^k}{k!} \tag{12.62}$$

We define

$$e^{\mathbf{A}t} = \mathbf{1}_n + \mathbf{A}t + \frac{t^2}{2!}\mathbf{A}^2 + \cdots + \frac{t^k}{k!}\mathbf{A}^k + \cdots$$

$$= \sum_{k=0}^{\infty} \frac{t^k}{k!}\mathbf{A}^k \tag{12.63}$$

Solution of State Equations

Since $\mathbf{A}$ is a square matrix of order n and $\mathbf{A}^k$ is well defined, $e^{\mathbf{A}t}$ is also a square matrix. It can be shown that each element of $e^{\mathbf{A}t}$, being an infinite series, converges for all values of t, thus rendering $e^{\mathbf{A}t}$ a well-defined matrix.[†]

EXAMPLE 12.6

Consider the matrix

$$\mathbf{A} = \begin{bmatrix} -1 & -1 \\ -0.5 & -1.5 \end{bmatrix} \tag{12.64}$$

Then we have

$$e^{\mathbf{A}t} = \begin{bmatrix} 1 & 0 \\ 0 & 1 \end{bmatrix} + t\begin{bmatrix} -1 & -1 \\ -0.5 & -1.5 \end{bmatrix} + \frac{t^2}{2!}\begin{bmatrix} -1 & -1 \\ -0.5 & -1.5 \end{bmatrix}^2$$

$$+ \frac{t^3}{3!}\begin{bmatrix} -1 & -1 \\ -0.5 & -1.5 \end{bmatrix}^3 + \cdots$$

$$= \begin{bmatrix} \begin{array}{l} 1 - t + 0.75t^2 - 0.458333t^3 \\ \quad + 0.442708t^4 + \cdots \end{array} & \begin{array}{l} -t + 1.25t^2 - 0.875t^3 \\ \quad + 0.223958t^4 + \cdots \end{array} \\ \begin{array}{l} -0.5t + 0.625t^2 - 0.4375t^3 \\ \quad + 0.221354t^4 + \cdots \end{array} & \begin{array}{l} 1 - 1.5t + 1.375t^2 - 0.89583t^3 \\ \quad + 0.445313t^4 + \cdots \end{array} \end{bmatrix} \tag{12.65}$$

A matrix of infinite series is said to *converge* if each of its elements converges. As will be demonstrated later, the matrix (12.65) converges to

$$e^{\mathbf{A}t} = \frac{1}{3}\begin{bmatrix} e^{-2t} + 2e^{-0.5t} & 2e^{-2t} - 2e^{-0.5t} \\ e^{-2t} - e^{-0.5t} & 2e^{-2t} + e^{-0.5t} \end{bmatrix} \tag{12.66}$$

In general, it is very difficult to obtain the elements of $e^{\mathbf{A}t}$ in closed form by summing up their terms.

As $e^{\mathbf{A}t}$ converges absolutely for any finite t, the term-by-term differentiation of the infinite series (12.63) is permitted, giving

$$\frac{d}{dt}e^{\mathbf{A}t} = \mathbf{A} + t\mathbf{A}^2 + \frac{t^2}{2!}\mathbf{A}^3 + \frac{t^3}{3!}\mathbf{A}^4 + \cdots$$

$$= \mathbf{A}\left[\mathbf{1}_n + t\mathbf{A} + \frac{t^2}{2!}\mathbf{A}^2 + \frac{t^3}{3!}\mathbf{A}^3 + \cdots\right]$$

$$= \mathbf{A}e^{\mathbf{A}t} \tag{12.67}$$

showing that the derivative of a matrix exponential is the same as it is for a scalar exponential. Comparing this equation with (12.56), we recognize

---

[†] See, for example, B. Noble, *Applied Linear Algebra* (Englewood Cliffs, N.J.: Prentice-Hall, 1969).

immediately that $e^{\mathbf{A}t}$ satisfies (12.56) or it amounts to the same thing that each column of $e^{\mathbf{A}t}$ is a solution of (12.58). Thus we let

$$\mathbf{M}(t) = e^{\mathbf{A}t} \tag{12.68}$$

To complete our solution process, we must show that the matrix of (12.68) is nonsingular for all finite $t \geq t_0$. To this end consider

$$e^{-\mathbf{A}t} = \mathbf{l}_n - t\mathbf{A} + \frac{t^2}{2!}\mathbf{A}^2 - \frac{t^3}{3!}\mathbf{A}^3 + \cdots \tag{12.69}$$

Because this matrix and $e^{\mathbf{A}t}$ converge absolutely for all finite t, the term-by-term multiplication of the product of these two matrices is permitted. It is straightforward to verify that

$$e^{\mathbf{A}t}e^{-\mathbf{A}t} = e^{-\mathbf{A}t}e^{\mathbf{A}t} = \mathbf{l}_n \tag{12.70}$$

All other terms cancel. By definition $e^{-\mathbf{A}t}$ is the inverse of $e^{\mathbf{A}t}$. Hence $e^{\mathbf{A}t}$ is nonsingular for all finite $t \geq t_0$, and the most general solution of the homogeneous differential equation (12.58) is found to be

$$\mathbf{x}_2(t) = e^{\mathbf{A}t}\mathbf{K} \tag{12.71}$$

as previously indicated in (12.61).

Our conclusion is that $\mathbf{M}(t) = e^{\mathbf{A}t}$ is a desired solution of (12.56). At this point one might be wondering if there are other solutions. The answer to this question is given below.

Property 12.1 *The system $\dot{\mathbf{M}}(t) = \mathbf{A}\mathbf{M}(t)$ has a unique solution equal to $\mathbf{M}(t_0)$ at time t_0 for all finite $t \geq t_0$.*

In other words, the solution is unique if the initial condition $\mathbf{M}(t_0)$ is specified. A proof of this property is left as an exercise (Problem 12.5). For our purposes we choose the initial condition

$$\mathbf{M}(t_0) = \mathbf{l}_n \tag{12.72}$$

Then the unique solution of the system $\dot{\mathbf{M}} = \mathbf{A}\mathbf{M}$ satisfying the initial condition $\mathbf{M}(t_0) = \mathbf{l}_n$ at time t_0 is given by

$$\mathbf{M}(t) = e^{\mathbf{A}(t-t_0)} \tag{12.73}$$

It is instructive to compare this with (12.15). We note that they are identical in form except for a change from scalar to matrix notation.

12-2.2 General Solution of the State Equation

The procedure for obtaining the solution of the general state equation again parallels that discussed in Section 12-1.2 for the scalar differential equation. Once $\mathbf{M}(t)$ is determined, $\mathbf{x}_1(t)$ can be found by directly integrating (12.57).

For this we define

$$\int_{t_0}^{t} \mathbf{M}(\tau)\, d\tau = \left[\int_{t_0}^{t} m_{ij}(\tau)\, d\tau \right] \tag{12.74}$$

where

$$\mathbf{M}(t) = [m_{ij}(t)] \tag{12.75}$$

Integrating (12.57) from t_0 to t, we get

$$\mathbf{x}_1(t) = \mathbf{x}_1(t_0) + \int_{t_0}^{t} \mathbf{M}^{-1}(\tau)\mathbf{B}\mathbf{u}(\tau)\, d\tau \tag{12.76}$$

To determine the initial condition $\mathbf{x}_1(t_0)$, we set $t = t_0$ in (12.53) and obtain from (12.72)

$$\mathbf{x}(t_0) = \mathbf{M}(t_0)\mathbf{x}_1(t_0) = \mathbf{x}_1(t_0) \tag{12.77}$$

Substituting (12.76) in (12.53) in conjunction with (12.77), and recognizing that the integration is with respect to τ, $\mathbf{M}(t)$ can be taken inside the integral sign, we get

$$\mathbf{x}(t) = \mathbf{M}(t)\mathbf{x}(t_0) + \int_{t_0}^{t} \mathbf{M}(t)\mathbf{M}^{-1}(\tau)\mathbf{B}\mathbf{u}(\tau)\, d\tau \tag{12.78}$$

Finally, inserting (12.73) in (12.78) yields the solution of the general state equation (12.52) with initial state $\mathbf{x}(t_0)$:

$$\mathbf{x}(t) = e^{\mathbf{A}(t-t_0)}\mathbf{x}(t_0) + \int_{t_0}^{t} e^{\mathbf{A}(t-\tau)}\mathbf{B}\mathbf{u}(\tau)\, d\tau \tag{12.79}$$

We remark that in deriving (12.79) we have used the following properties of $e^{\mathbf{A}t}$:

1. $e^{\mathbf{A}t}e^{\mathbf{B}t} = e^{(\mathbf{A}+\mathbf{B})t}$ if and only if $\mathbf{AB} = \mathbf{BA}$. $\tag{12.80a}$

2. $[e^{\mathbf{A}t}]^{-1} = e^{-\mathbf{A}t}$ $\tag{12.80b}$

The output equation of the system is described by

$$\mathbf{y}(t) = \mathbf{C}\mathbf{x}(t) + \mathbf{D}\mathbf{u}(t) \tag{12.81}$$

Putting (12.79) in (12.81) gives

$$\mathbf{y}(t) = \mathbf{C}e^{\mathbf{A}(t-t_0)}\mathbf{x}(t_0) + \left[\mathbf{C} \int_{t_0}^{t} e^{\mathbf{A}(t-\tau)}\mathbf{B}\mathbf{u}(\tau)\, d\tau + \mathbf{D}\mathbf{u}(t) \right] \tag{12.82}$$

As before, the first term on the right-hand side of (12.82) is the natural response or the zero-input response; the second term is the forced response or the zero-state response. Together they constitute the complete response of the system.

Observe from (12.79) and (12.82) that for a given initial state $\mathbf{x}(t_0)$, regardless of the input signals prior to t_0, the state $\mathbf{x}(t)$ and the output $\mathbf{y}(t)$ at any time $t > t_0$ can be determined from the information about the initial state $\mathbf{x}(t_0)$ and the input excitation from t_0 to t. No other information is required. This transition of states from $\mathbf{x}(t_0)$ to $\mathbf{x}(t)$ is governed exclusively by the matrix (12.73), also conveniently written as

$$\boldsymbol{\phi}(t - t_0) \triangleq e^{\mathbf{A}(t - t_0)} \tag{12.83}$$

in the literature. For this reason it is called the *state-transition matrix*.[†] Using this symbol, (12.79) and (12.82) can be rewritten as

$$\mathbf{x}(t) = \boldsymbol{\phi}(t - t_0)\mathbf{x}(t_0) + \int_{t_0}^{t} \boldsymbol{\phi}(t - \tau)\mathbf{B}\mathbf{u}(\tau)\, d\tau \tag{12.84}$$

$$\mathbf{y}(t) = \mathbf{C}\boldsymbol{\phi}(t - t_0)\mathbf{x}(t_0) + \mathbf{C}\int_{t_0}^{t} \boldsymbol{\phi}(t - \tau)\mathbf{B}\mathbf{u}(\tau)\, d\tau + \mathbf{D}\mathbf{u}(t) \tag{12.85}$$

We illustrate the above results by the following examples.

EXAMPLE 12.7

Given the matrix

$$\mathbf{A} = \begin{bmatrix} -1 & -1 \\ -0.5 & -1.5 \end{bmatrix} \tag{12.86}$$

the matrix exponential $e^{\mathbf{A}t}$ was given earlier in Example 12.6 and is repeated below for convenience:

$$e^{\mathbf{A}t} = \frac{1}{3}\begin{bmatrix} e^{-2t} + 2e^{-0.5t} & 2e^{-2t} - 2e^{-0.5t} \\ e^{-2t} - e^{-0.5t} & 2e^{-2t} + e^{-0.5t} \end{bmatrix} \tag{12.87}$$

Differentiating (12.87) with respect to t yields

$$\frac{d}{dt}e^{\mathbf{A}t} = \frac{1}{3}\begin{bmatrix} -2e^{-2t} - e^{-0.5t} & -4e^{-2t} + e^{-0.5t} \\ -2e^{-2t} + \frac{1}{2}e^{-0.5t} & -4e^{-2t} - \frac{1}{2}e^{-0.5t} \end{bmatrix}$$

$$= \mathbf{A}e^{\mathbf{A}t} \tag{12.88}$$

confirming (12.67). The determinant of $e^{\mathbf{A}t}$ is found to be

$$\det e^{\mathbf{A}t} = e^{-2.5t} \tag{12.89}$$

The inverse of $e^{\mathbf{A}t}$ can be computed in two different ways. Using the standard procedure, we have

$$[e^{\mathbf{A}t}]^{-1} = \frac{1}{3\det e^{\mathbf{A}t}}\begin{bmatrix} 2e^{-2t} + e^{-0.5t} & -2e^{-2t} + 2e^{-0.5t} \\ -e^{-2t} + e^{-0.5t} & e^{-2t} + 2e^{-0.5t} \end{bmatrix}$$

$$= \frac{1}{3}\begin{bmatrix} 2e^{0.5t} + e^{2t} & -2e^{0.5t} + 2e^{2t} \\ -e^{0.5t} + e^{2t} & e^{0.5t} + 2e^{2t} \end{bmatrix} \tag{12.90}$$

This matrix is really $e^{\mathbf{A}(-t)} = e^{-\mathbf{A}t}$, verifying (12.70) and (12.80b).

[†] It is also known as the *fundamental matrix* in control theory.

EXAMPLE 12.8

In the network of Figure 12.8, switch S_1 is open and switch S_2 is closed for
$t < 0$. At $t = 0$, switch S_1 is flipped to closed position and switch S_2 is open.
We wish to determine the capacitor voltages $v_1(t)$ and $v_2(t)$ for $t \geq 0$.

At $t = 0$ the initial capacitor voltages are evidently $v_1(0) = 2$ V and
$v_2(0) = 0$. Applying the procedure discussed in the preceding chapter, the
state equation of the network is found to be

$$
\begin{bmatrix} \dot{v}_1 \\ \dot{v}_2 \end{bmatrix} = \begin{bmatrix} -1 & -1 \\ -0.5 & -1.5 \end{bmatrix} \begin{bmatrix} v_1 \\ v_2 \end{bmatrix} + \begin{bmatrix} 1 \\ 0.5 \end{bmatrix} [1], \qquad t \geq 0 \tag{12.91}
$$

The matrix exponential e^{At} is given by (12.87). The state-transition matrix
becomes

$$
\phi(t - \tau) = \frac{1}{3} \begin{bmatrix} e^{-2(t-\tau)} + 2e^{-0.5(t-\tau)} & 2e^{-2(t-\tau)} - 2e^{-0.5(t-\tau)} \\ e^{-2(t-\tau)} - e^{-0.5(t-\tau)} & 2e^{-2(t-\tau)} + e^{-0.5(t-\tau)} \end{bmatrix} \tag{12.92}
$$

Appealing to (12.84) yields the desired solution

$$
\begin{bmatrix} v_1 \\ v_2 \end{bmatrix} = \frac{1}{3} \begin{bmatrix} e^{-2t} + 2e^{-0.5t} & 2e^{-2t} - 2e^{-0.5t} \\ e^{-2t} - e^{-0.5t} & 2e^{-2t} + e^{-0.5t} \end{bmatrix} \begin{bmatrix} 2 \\ 0 \end{bmatrix}
$$

$$
+ \frac{1}{3} \int_0^t \begin{bmatrix} e^{-2(t-\tau)} + 2e^{-0.5(t-\tau)} & 2e^{-2(t-\tau)} - 2e^{-0.5(t-\tau)} \\ e^{-2(t-\tau)} - e^{-0.5(t-\tau)} & 2e^{-2(t-\tau)} + e^{-0.5(t-\tau)} \end{bmatrix} \begin{bmatrix} 1 \\ 0.5 \end{bmatrix} d\tau
$$

$$
= \frac{2}{3} \begin{bmatrix} e^{-2t} + 2e^{-0.5t} \\ e^{-2t} - e^{-0.5t} \end{bmatrix} + \frac{1}{3} \begin{bmatrix} e^{-0.5t} \int_0^t e^{0.5\tau} d\tau + 2e^{-2t} \int_0^t e^{2\tau} d\tau \\ -\dfrac{1}{2} e^{-0.5t} \int_0^t e^{0.5\tau} d\tau + 2e^{-2t} \int_0^t e^{2\tau} d\tau \end{bmatrix}
$$

$$
= \begin{bmatrix} 1 + \dfrac{2}{3} e^{-0.5t} + \dfrac{1}{3} e^{-2t} \\ -\dfrac{1}{3} e^{-0.5t} + \dfrac{1}{3} e^{-2t} \end{bmatrix} \tag{12.93}
$$

Figure 12.8
A network with switch S_1
closed and switch S_2 open
at $t = 0$.

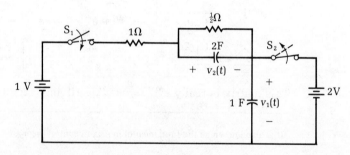

Figure 12.9

Plots of the capacitor voltages as functions of t for the network of Figure 12.8.

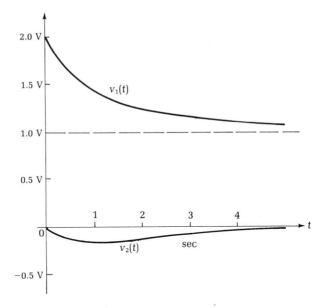

Plots of $v_1(t)$ and $v_2(t)$ as functions of time t are presented in Figure 12.9. The physical significance of the state-transition matrix $\boldsymbol{\phi}(t) = e^{\mathbf{A}t}$ is that it transforms the initial state $[2, 0]'$ at time $t = 0$ to the state (12.93) at any time $t \geqq 0$ after the application of the unit-step voltage at $t = 0$.

It is interesting to point out that if the 2V battery in the network is replaced by a 1V battery, everything else being the same, the capacitor voltages become

$$\begin{bmatrix} v_1(t) \\ v_2(t) \end{bmatrix} = \begin{bmatrix} 1 \\ 0 \end{bmatrix} \tag{12.94}$$

meaning that the state of the network remains the same as its initial state for all $t \geqq 0$ after the application of the unit-step voltage at $t = 0$. The justification is left as an exercise (Problem 12.15).

As in the scalar case, the response (12.93) may be decomposed as

$$\begin{bmatrix} v_1 \\ v_2 \end{bmatrix} = \begin{bmatrix} 1 \\ 0 \end{bmatrix} + \begin{bmatrix} \dfrac{2}{3} e^{-0.5t} + \dfrac{1}{3} e^{-2t} \\[2ex] -\dfrac{1}{3} e^{-0.5t} + \dfrac{1}{3} e^{-2t} \end{bmatrix} \tag{12.95}$$

The first term on the right-hand side is a constant and the second term is a decaying exponential. Therefore the first term is the familiar steady-state response and the second term the transient response. Together they constitute the complete response. Observe that the transient response is contributed by both the zero-input response and the zero-state response, whereas the steady-state response is contributed only by the zero-state response.

EXAMPLE 12.9

A current source

$$i_g(t) = 0, \qquad\qquad t < 0$$
$$\quad\;\; = \cos 2t, \qquad\;\; t \geq 0 \qquad\qquad\qquad\qquad\qquad\qquad\text{(12.96)}$$

and a voltage source

$$v_g(t) = 0, \qquad\qquad\qquad\qquad t < 0$$
$$\quad\;\; = 2\cos\left(2t + \frac{\pi}{4}\right), \qquad t \geq 0 \qquad\qquad\qquad\qquad\text{(12.97)}$$

are applied to the network of Figure 12.10. We compute the inductor current $i(t)$ and capacitor voltage $v(t)$ for $t \geq 0$.

The state equation of the network was computed in Examples 11.1 and 11.2, and for $t \geq 0$ it is given by

$$\begin{bmatrix} \dot{i} \\ \dot{v} \end{bmatrix} = \begin{bmatrix} -5 & 2 \\ -1 & -2 \end{bmatrix}\begin{bmatrix} i \\ v \end{bmatrix} + \begin{bmatrix} -2 & 0 \\ 0 & 1 \end{bmatrix}\begin{bmatrix} 2\cos\left(2t + \dfrac{\pi}{4}\right) \\ \cos 2t \end{bmatrix} \qquad\text{(12.98)}$$

with the initial state $[0\ \ 0]'$. The state-transition matrix is found from (12.63) to be

$$\phi(t) = e^{\mathbf{A}t}$$

$$= \begin{bmatrix} 1 - 5t + 11.5t^2 - 16.8333333t^3 + \cdots & 2t - 7t^2 + 12.3333333t^3 + \cdots \\ -t + 3.5t^2 - 6.1666667t^3 + \cdots & 1 - 2t + t^2 + 1.6666666t^3 + \cdots \end{bmatrix}$$

$$\text{(12.99)}$$

which, as can be shown using a technique developed later, converges to

$$\phi(t) = e^{\mathbf{A}t} = \begin{bmatrix} -e^{-3t} + 2e^{-4t} & 2e^{-3t} - 2e^{-4t} \\ -e^{-3t} + e^{-4t} & 2e^{-3t} - e^{-4t} \end{bmatrix} \qquad\qquad\text{(12.100)}$$

Figure 12.10
A network used to illustrate the computation of the state vector.

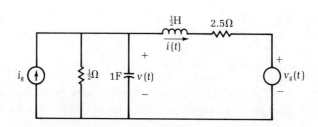

Substituting the appropriate matrices in (12.84), we obtain the state vector

$$\mathbf{x}(t) = \begin{bmatrix} i(t) \\ v(t) \end{bmatrix} = \boldsymbol{\phi}(t) \begin{bmatrix} 0 \\ 0 \end{bmatrix} + e^{\mathbf{A}t} \int_0^t \boldsymbol{\phi}(-\tau) \begin{bmatrix} -4 \cos\left(2\tau + \dfrac{\pi}{4}\right) \\ \\ \cos 2\tau \end{bmatrix} d\tau$$

$$= e^{\mathbf{A}t} \begin{bmatrix} \displaystyle\int_0^t \left\{ 2(e^{3\tau} - e^{4\tau}) \cos 2\tau + 4(e^{3\tau} - 2e^{4\tau}) \cos\left(2\tau + \dfrac{\pi}{4}\right) \right\} d\tau \\ \\ \displaystyle\int_0^t \left\{ (2e^{3\tau} - e^{4\tau}) \cos 2\tau + 4(e^{3\tau} - e^{4\tau}) \cos\left(2\tau + \dfrac{\pi}{4}\right) \right\} d\tau \end{bmatrix}$$

$$= \begin{bmatrix} -1.548e^{-3t} + 2.096e^{-4t} - 0.548 \cos 2t + 0.456 \sin 2t \\ -1.548e^{-3t} + 1.048e^{-4t} + 0.5 \cos 2t + 0.273 \sin 2t \end{bmatrix}$$

$$= \begin{bmatrix} -1.548e^{-3t} + 2.096e^{-4t} \\ -1.548e^{-3t} + 1.048e^{-4t} \end{bmatrix} + \begin{bmatrix} 0.713 \cos (2t - 2.448) \\ 0.57 \cos (2t - 0.5) \end{bmatrix} \qquad \textbf{(12.101)}$$

The first term on the right-hand side is the transient response and the second the steady-state response. The complete responses of the inductor current $i(t)$ and the capacitor voltage $v(t)$ are shown in Figure 12.11.

Now suppose that the voltage source of (12.97) is replaced by

$$v_g(t) = 1, \qquad t \leq 0$$
$$= 0, \qquad t > 0 \qquad\qquad\qquad\qquad \textbf{(12.102)}$$

everything else being the same. The initial state is no longer zero and is given by

$$\mathbf{x}(0) = \begin{bmatrix} i(0) \\ v(0) \end{bmatrix} = \begin{bmatrix} -\frac{1}{3} \\ \frac{1}{6} \end{bmatrix} \qquad\qquad\qquad\qquad \textbf{(12.103)}$$

All other matrices are the same as before except the input vector $\mathbf{u}(t)$ where the first row element is replaced by 0. Appealing once again to (12.84) yields

Figure 12.11
Plots of the complete responses of the inductor current and the capacitor voltage for the network of Figure 12.10.

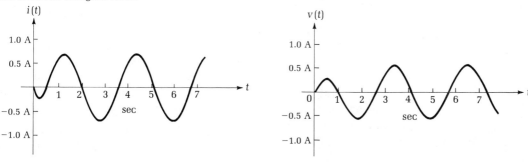

the state vector

$$\begin{bmatrix} i(t) \\ v(t) \end{bmatrix} = \phi(t) \begin{bmatrix} -\dfrac{1}{3} \\[2mm] \dfrac{1}{6} \end{bmatrix} + e^{\mathbf{A}t} \int_0^t \begin{bmatrix} 2(e^{3\tau} - e^{4\tau}) \cos 2\tau \\[1mm] (2e^{3\tau} - e^{4\tau}) \cos 2\tau \end{bmatrix} d\tau$$

$$= \begin{bmatrix} -\dfrac{3}{5} e^{-4t} + \dfrac{8}{39} e^{-3t} + \dfrac{4}{65} \cos 2t + \dfrac{7}{65} \sin 2t \\[4mm] -\dfrac{3}{10} e^{-4t} + \dfrac{8}{39} e^{-3t} + \dfrac{17}{65} \cos 2t + \dfrac{27}{130} \sin 2t \end{bmatrix}$$

$$= \begin{bmatrix} -\dfrac{3}{5} e^{-4t} + \dfrac{8}{39} e^{-3t} + \dfrac{1}{\sqrt{65}} \cos (2t - 1.052) \\[4mm] -\dfrac{3}{10} e^{-4t} + \dfrac{8}{39} e^{-3t} + \dfrac{\sqrt{1885}}{130} \cos (2t - 0.671) \end{bmatrix} \tag{12.104}$$

One might conclude from the above two examples that the computation involved in solving the state equation is long and complicated, but the procedure is readily amenable to computer solutions and is not really meant for hand calculations.

12-3   NATURAL FREQUENCIES AND EIGENVALUES

Recall that in Section 11-6 the natural frequencies of a network are defined as the roots of the determinantal polynomial of the operator matrix of the network equations when these are framed as a set of first-order differential and/or algebraic equations for the currents and voltages of the network elements and when the independent sources and initial conditions are set to zero. The state equation is a secondary system of equations and is obtained from the primary system of equations by the elimination of the algebraic equations and some of the first-order differential equations in the case of degenerate networks. This elimination process is equivalent to the linear combinations of the rows of the operator matrix and therefore will not affect the roots of the determinantal polynomial of the resulting operator matrix. In other words, if we set $\mathbf{u}(t) = \mathbf{0}$ and rewrite the resulting state equation in the form of operator matrix, the roots of the determinantal polynomial of this operator matrix are the natural frequencies of the network. Proceeding in this way, we let $\mathbf{u}(t) = \mathbf{0}$ and write the state equation

$$\dot{\mathbf{x}}(t) = \mathbf{A}\mathbf{x}(t) \tag{12.105}$$

as

$$(p\mathbf{I}_n - \mathbf{A})\mathbf{x}(t) = \mathbf{0} \tag{12.106}$$

The roots of the determinantal polynomial of the operator matrix $p\mathbf{I}_n - \mathbf{A}$ are the natural frequencies, or

$$\det (p\mathbf{I}_n - \mathbf{A}) = (p - \lambda_1)^{r_1}(p - \lambda_2)^{r_2} \cdots (p - \lambda_q)^{r_q} = 0 \tag{12.107}$$

where $r_1 + r_2 + \cdots + r_q = n$, n being the dimension of the state space. In linear algebra the λ_j $(j = 1, 2, \ldots, q)$ are called the *eigenvalues* of the matrix **A**. The eigenvalue λ_j is of multiplicity r_j. Equation (12.107) is known as the *characteristic equation*. Our conclusion is that the eigenvalues of the **A** matrix are the natural frequencies of the given network.

EXAMPLE 12.10

Consider again the problem treated in Example 12.8. The state equation of the network is repeated below:

$$\begin{bmatrix} \dot{v}_1 \\ \dot{v}_2 \end{bmatrix} = \begin{bmatrix} -1 & -1 \\ -0.5 & -1.5 \end{bmatrix}\begin{bmatrix} v_1 \\ v_2 \end{bmatrix} + \begin{bmatrix} 1 \\ 0.5 \end{bmatrix}[1], \qquad t \geq 0 \tag{12.108}$$

The eigenvalues of the **A** matrix are found to be

$$\det (p\mathbf{I}_2 - \mathbf{A}) = \det \begin{bmatrix} p + 1 & 1 \\ 0.5 & p + 1.5 \end{bmatrix} = p^2 + 2.5p + 1 = 0 \tag{12.109}$$

the roots of which are given by

$$\lambda_1 = -2, \qquad \lambda_2 = -0.5 \tag{12.110}$$

Thus the network of Figure 12.8 has two natural frequencies of multiplicity 1 at $\lambda_1 = -2$ and $\lambda_2 = -0.5$.

EXAMPLE 12.11

Consider the network of Figure 12.12. We wish to determine a relation among the elements R, L, and C so that the network possesses a real natural frequency of multiplicity 2.

Figure 12.12
A network used to illustrate the formation of a real natural frequency of multiplicity 2.

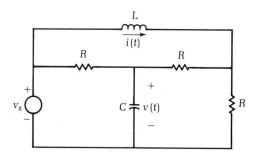

The network was treated before in Section 11-1, and from (11.7) its state equation is given by

$$\begin{bmatrix} \dot{v} \\ \dot{i} \end{bmatrix} = \begin{bmatrix} -3/2RC & 1/2C \\ -1/2L & -R/2L \end{bmatrix} \begin{bmatrix} v \\ i \end{bmatrix} + \begin{bmatrix} 1/RC \\ 1/L \end{bmatrix} [v_g] \qquad (12.111)$$

The natural frequencies are the roots of the characteristic equation

$$\det(p\mathbf{1}_2 - \mathbf{A}) = \det \begin{bmatrix} p + 3/2RC & -1/2C \\ 1/2L & p + R/2L \end{bmatrix}$$

$$= p^2 + \left(\frac{3}{2RC} + \frac{R}{2L} \right) p + \frac{1}{LC} = 0 \qquad (12.112)$$

To produce a real root of multiplicity 2, we set

$$\left(\frac{3}{2RC} + \frac{R}{2L} \right)^2 - \frac{4}{LC} = 0 \qquad (12.113)$$

or

$$(R^2C - 9L)(R^2C - L) = 0 \qquad (12.114)$$

giving a desired relation

$$R^2C = 9L \quad \text{or} \quad R^2C = L \qquad (12.115)$$

Choose $L = R^2C$ and substitute this value for L in (12.112). After simplification we have

$$\det(p\mathbf{1}_2 - \mathbf{A}) = \left(p + \frac{1}{RC} \right)^2 \qquad (12.116)$$

showing that under the stipulated condition the network possesses a real frequency of multiplicity 2 at $-1/RC$.

12-4   THE STATE-TRANSITION MATRIX

The central problem in solving the general state equation is the determination of the state-transition matrix $\boldsymbol{\phi}(t) = e^{\mathbf{A}t}$. As indicated in (12.63), this matrix exponential is defined in terms of an infinite series, and it is usually very difficult to sum up the terms to give a closed-form solution. Once the state-transition matrix is known, the solution can be computed in a straight-forward manner by means of (12.84). The procedure can readily be programmed for a digital computer. In fact, if machine computation is the objective, there is no need to write the state-transition matrix in closed form. In this section we discuss some of the techniques of evaluating the state-transition matrix.

12-4.1   State-Transition Matrix in Closed Form

Using Laplace transformation we present closed-form equivalents of the state-transition matrix. We begin by considering the solution of a linear system described by the homogeneous state equation

$$\dot{\mathbf{x}}(t) = \mathbf{A}\mathbf{x}(t) \tag{12.117}$$

with initial state $\mathbf{x}(0)$ at time $t = 0$. By appealing to (12.84), its solution can be written down directly as

$$\mathbf{x}(t) = \boldsymbol{\phi}(t)\mathbf{x}(0) = e^{\mathbf{A}t}\mathbf{x}(0) \tag{12.118}$$

Taking the Laplace transform of both sides of (12.117) yields

$$s\mathbf{X}(s) - \mathbf{x}(0) = \mathbf{A}\mathbf{X}(s) \tag{12.119}$$

or

$$\mathbf{X}(s) = (s\mathbf{I} - \mathbf{A})^{-1}\mathbf{x}(0) \tag{12.120}$$

where, as before, the capital $\mathbf{X}(s)$ denotes the Laplace transform of the time-domain $\mathbf{x}(t)$. The inverse Laplace transform of (12.120) gives

$$\mathbf{x}(t) = \mathscr{L}^{-1}(s\mathbf{I} - \mathbf{A})^{-1}\mathbf{x}(0) \tag{12.121}$$

Comparing this with (12.118) shows that

$$\boldsymbol{\phi}(t) = e^{\mathbf{A}t} = \mathscr{L}^{-1}(s\mathbf{I} - \mathbf{A})^{-1} \tag{12.122}$$

Thus using (12.122) we can obtain the closed-form equivalents of (12.63).

We remark that the poles (of the elements) of $(s\mathbf{I} - \mathbf{A})^{-1}$ are roots of $\det (s\mathbf{I} - \mathbf{A})$, which is essentially the same as (12.107) except that the linear differential operator p is replaced by the variable s. Therefore the poles of the matrix $(s\mathbf{I} - \mathbf{A})^{-1}$ are natural frequencies of the network. The matrix $(s\mathbf{I} - \mathbf{A})^{-1}$ is called the *resolvent matrix*.

EXAMPLE 12.12

Consider the same problem treated in Examples 12.8 and 12.10. The $\mathbf{A}$ matrix is repeated below.

$$\mathbf{A} = \begin{bmatrix} -1 & -1 \\ -0.5 & -1.5 \end{bmatrix}, \tag{12.123}$$

The resolvent matrix is found to be

$$(s\mathbf{I}_2 - \mathbf{A})^{-1} = \begin{bmatrix} s+1 & 1 \\ 0.5 & s+1.5 \end{bmatrix}^{-1}$$

$$= \frac{1}{(s+0.5)(s+2)} \begin{bmatrix} s+1.5 & -1 \\ -0.5 & s+1 \end{bmatrix} \tag{12.124}$$

Each element of the resolvent matrix (12.124) is divided by $(s + 0.5)(s + 2)$, which is the determinant of $(s\mathbf{1}_2 - \mathbf{A})$. Thus the poles of $(s\mathbf{1}_2 - \mathbf{A})^{-1}$ are natural frequencies of the network. The elements of $(s\mathbf{1}_2 - \mathbf{A})^{-1}$ can each be expanded in partial-fraction expansions and the results are given by

$$(s\mathbf{1}_2 - \mathbf{A})^{-1} = \frac{1}{s + 0.5}\begin{bmatrix} \frac{2}{3} & -\frac{2}{3} \\ -\frac{1}{3} & \frac{1}{3} \end{bmatrix} + \frac{1}{s + 2}\begin{bmatrix} \frac{1}{3} & \frac{2}{3} \\ \frac{1}{3} & \frac{2}{3} \end{bmatrix} \tag{12.125}$$

Finally, performing the inverse Laplace transform of (12.125) yields a closed-form equivalent of $e^{\mathbf{A}t}$:

$$\mathscr{L}^{-1}(s\mathbf{1}_2 - \mathbf{A})^{-1} = e^{-0.5t}\begin{bmatrix} \frac{2}{3} & -\frac{2}{3} \\ -\frac{1}{3} & \frac{1}{3} \end{bmatrix} + e^{-2t}\begin{bmatrix} \frac{1}{3} & \frac{2}{3} \\ \frac{1}{3} & \frac{2}{3} \end{bmatrix}$$

$$= \frac{1}{3}\begin{bmatrix} 2e^{-0.5t} + e^{-2t} & -2e^{-0.5t} + 2e^{-2t} \\ -e^{-0.5t} + e^{-2t} & e^{-0.5t} + 2e^{-2t} \end{bmatrix} \tag{12.126}$$

confirming (12.66).

Let us look back at (12.122). Each element of $(s\mathbf{1} - \mathbf{A})^{-1}$ is a rational function—namely, the ratio of two polynomials. The numerator is a cofactor of an element of $(s\mathbf{1} - \mathbf{A})$ and the denominator is $\det(s\mathbf{1} - \mathbf{A})$. Thus as in (12.125) a partial-fraction expansion of the elements of $(s\mathbf{1} - \mathbf{A})^{-1}$ can be carried out and leads to the general expression

$$(s\mathbf{1} - \mathbf{A})^{-1} = \frac{1}{(s - \lambda_1)}\mathbf{K}_{11} + \frac{1}{(s - \lambda_1)^2}\mathbf{K}_{12} + \cdots + \frac{1}{(s - \lambda_1)^{r_1}}\mathbf{K}_{1r_1}$$

$$+ \frac{1}{(s - \lambda_2)}\mathbf{K}_{21} + \frac{1}{(s - \lambda_2)^2}\mathbf{K}_{22} + \cdots$$

$$+ \frac{1}{(s - \lambda_2)^{r_2}}\mathbf{K}_{2r_2} + \cdots$$

$$+ \frac{1}{(s - \lambda_q)}\mathbf{K}_{q1} + \frac{1}{(s - \lambda_q)^2}\mathbf{K}_{q2} + \cdots + \frac{1}{(s - \lambda_q)^{r_q}}\mathbf{K}_{qr_q}$$

$$= \sum_{i=1}^{q} \sum_{j=1}^{r_i} \frac{1}{(s - \lambda_i)^j}\mathbf{K}_{ij} \tag{12.127}$$

where

$$\det(s\mathbf{1} - \mathbf{A}) = (s - \lambda_1)^{r_1}(s - \lambda_2)^{r_2} \cdots (s - \lambda_q)^{r_q} \tag{12.128}$$

The coefficient matrices $\mathbf{K}_{ij}$ in the partial-fraction expansion of $(s\mathbf{1} - \mathbf{A})$ are called the *constituent matrices*. The λ_j ($j = 1, 2, \ldots, q$) are the eigenvalues of $\mathbf{A}$ or the natural frequencies of the network, as discussed in Section 12-3.

We remark that the Laplace transformation is one way of obtaining the state-transition matrix in closed form. However, if we shall use the Laplace transformation technique, we might as well apply it to the state equation (12.52) and avoid going through all the intervening steps. This of course can be done, but it will defeat the original purpose of solving the network problem in time-domain and using matrix mathematics to its full advantage. Indeed there are other means of finding the state-transition matrix in closed form.

EXAMPLE 12.13

We compute the state-transition matrix for the matrix

$$\mathbf{A} = \frac{1}{2RC}\begin{bmatrix} -3 & R \\ -1/R & -1 \end{bmatrix} \tag{12.129}$$

which was treated in Example 12.11 with $R^2C = L$. The resolvent matrix is found to be

$$(s\mathbf{I}_2 - \mathbf{A})^{-1} = \frac{1}{(s + 1/RC)^2}\begin{bmatrix} s + 1/2RC & 1/2C \\ -1/2R^2C & s + 3/2RC \end{bmatrix} \tag{12.130}$$

which can be expanded in a partial-fraction expansion

$$= \frac{1}{(s + 1/RC)}\begin{bmatrix} 1 & 0 \\ 0 & 1 \end{bmatrix} + \frac{1}{(s + 1/RC)^2}\begin{bmatrix} -1/2RC & 1/2C \\ -1/2R^2C & 1/2RC \end{bmatrix} \tag{12.131}$$

The two constituent matrices are identified as

$$\mathbf{K}_{11} = \begin{bmatrix} 1 & 0 \\ 0 & 1 \end{bmatrix}, \quad \mathbf{K}_{12} = \begin{bmatrix} -1/2RC & 1/2C \\ -1/2R^2C & 1/2RC \end{bmatrix} \tag{12.132}$$

Taking the inverse Laplace transform of (12.131) yields the state-transition matrix

$$\boldsymbol{\phi}(t) = e^{\mathbf{A}t} = \begin{bmatrix} -\dfrac{t}{2RC}e^{-t/RC} + e^{-t/RC} & \dfrac{t}{2C}e^{-t/RC} \\[2ex] -\dfrac{t}{2R^2C}e^{-t/RC} & \dfrac{t}{2RC}e^{-t/RC} + e^{-t/RC} \end{bmatrix} \tag{12.133}$$

EXAMPLE 12.14

The state equation of a network is given by

$$\begin{bmatrix} \dot{v} \\ \dot{i} \end{bmatrix} = \begin{bmatrix} -2 & -\frac{5}{4} \\ 2 & 1 \end{bmatrix}\begin{bmatrix} v \\ i \end{bmatrix} + \begin{bmatrix} 1 \\ -2 \end{bmatrix}[i_g] \tag{12.134}$$

We wish to determine its state-transition matrix in closed form.

The resolvent matrix and its partial-fraction expansion in terms of the constituent matrices K_{11} and K_{21} are found to be

$$(s I_2 - A)^{-1} = \frac{1}{(s + \frac{1}{2} - j\frac{1}{2})(s + \frac{1}{2} + j\frac{1}{2})} \begin{bmatrix} s - 1 & -\frac{5}{4} \\ 2 & s + 2 \end{bmatrix}$$

$$= \frac{\frac{1}{4}}{s + \frac{1}{2} - j\frac{1}{2}} \begin{bmatrix} 2 + j6 & j5 \\ -j8 & 2 - j6 \end{bmatrix}$$

$$+ \frac{\frac{1}{4}}{s + \frac{1}{2} + j\frac{1}{2}} \begin{bmatrix} 2 - j6 & -j5 \\ j8 & 2 + j6 \end{bmatrix} \tag{12.135}$$

where the characteristic equation is given by

$$\det(s I_2 - A) = s^2 + s + 0.5 = (s + \tfrac{1}{2} - j\tfrac{1}{2})(s + \tfrac{1}{2} + j\tfrac{1}{2}) = 0 \tag{12.136}$$

Taking the inverse Laplace transform of (12.135), we get the state-transition matrix

$$\boldsymbol{\phi}(t) = \frac{j}{2} \begin{bmatrix} (3 - j1)e^{\lambda_1 t} - (3 + j1)e^{\lambda_2 t} & 2.5 e^{\lambda_1 t} - 2.5 e^{\lambda_2 t} \\ -4 e^{\lambda_1 t} + 4 e^{\lambda_2 t} & (-3 - j1)e^{\lambda_1 t} + (3 - j1)e^{\lambda_2 t} \end{bmatrix}$$

$$= e^{-t/2} \begin{bmatrix} \cos \frac{1}{2}t - 3 \sin \frac{1}{2}t & -2.5 \sin \frac{1}{2}t \\ 4 \sin \frac{1}{2}t & \cos \frac{1}{2}t + 3 \sin \frac{1}{2}t \end{bmatrix}$$

$$= e^{-t/2} \begin{bmatrix} \sqrt{10} \cos(\frac{1}{2}t + 1.249) & -2.5 \sin \frac{1}{2}t \\ 4 \sin \frac{1}{2}t & \sqrt{10} \cos(\frac{1}{2}t - 1.249) \end{bmatrix} \tag{12.137}$$

where $\lambda_1 = -\frac{1}{2} + j\frac{1}{2}$ and $\lambda_2 = -\frac{1}{2} - j\frac{1}{2}$ are the natural frequencies of the given network.

12-4.2   The Resolvent Matrix Algorithm

The crux of the problem in computing the state-transition matrix, as discussed in the foregoing, is the inversion of the matrix $(sI - A)$, which will be an onerous task for a large system. Once this is done, a partial-fraction expansion of the resolvent matrix $(sI - A)^{-1}$ can be achieved by applying the algorithm outlined in Section 5-4.3. In this section we present an algorithm for rapid calculation of the inverse of the matrix $(sI - A)$. The procedure is known as the *Souriau-Frame algorithm*.[†] A computer program based on this will be presented in Section 12-8.1.

Before we do this, we need the definition of *trace*. The *trace* of a square matrix A, written as tr A, is the sum of the elements on the main diagonal of A. For instance, the trace of the matrix A of (12.129) is $-2/RC$, or tr $A = -2/RC$. Recall that the adjoint of a square matrix A of order n, denoted by

---

[†] J. M. Souriau, "Une méthode pour la Décomposition spectrale à l'inversion des matrices," *Compt. Rend.* 227 (1948): 1010–11 and J. S. Frame, "A Simple Recursion Formula for Inverting a Matrix," *Bull. Am. Math. Soc.* 55 (1949): 1045. The algorithm is also known as the *Leverrier-Faddeeva algorithm*.

adj $\mathbf{A}$, is the $n \times n$ matrix whose ith row and jth column element is the cofactor of the jth row and ith column element of $\mathbf{A}$. Thus the inverse of $(s\mathbf{l}_n - \mathbf{A})$ may be written as

$$(s\mathbf{l}_n - \mathbf{A})^{-1} = \frac{1}{\det(s\mathbf{l}_n - \mathbf{A})} \text{ adj}(s\mathbf{l}_n - \mathbf{A}) \triangleq \frac{1}{\det(s\mathbf{l}_n - \mathbf{A})} \mathbf{P}(s) \qquad (12.138)$$

where $\mathbf{P}(s)$ is a matrix the elements of which are polynomials in s. We now focus our attention on the powers of s by rewriting this matrix as a sum of matrices, one for each power of s. Let

$$\mathbf{P}(s) = \mathbf{P}_0 s^{n-1} + \mathbf{P}_1 s^{n-2} + \cdots + \mathbf{P}_{n-2} s + \mathbf{P}_{n-1} \qquad (12.139)$$

$$\det(s\mathbf{l}_n - \mathbf{A}) = s^n + \alpha_1 s^{n-1} + \cdots + \alpha_{n-1} s + \alpha_n \qquad (12.140)$$

Equation (12.140) is referred to as the *characteristic polynomial*. Premultiplying both sides of (12.138) by $[\det(s\mathbf{l}_n - \mathbf{A})](s\mathbf{l}_n - \mathbf{A})$ gives

$$(s\mathbf{l}_n - \mathbf{A})\mathbf{P}(s) = [\det(s\mathbf{l}_n - \mathbf{A})]\mathbf{l}_n \qquad (12.141)$$

which, upon inserting (12.139) and (12.140), becomes

$$(\mathbf{P}_0 s^n + \mathbf{P}_1 s^{n-1} + \cdots + \mathbf{P}_{n-2} s^2 + \mathbf{P}_{n-1} s) - (\mathbf{A}\mathbf{P}_0 s^{n-1} + \mathbf{A}\mathbf{P}_1 s^{n-2} + \cdots + \mathbf{A}\mathbf{P}_{n-1})$$
$$= \mathbf{l}_n s^n + \mathbf{l}_n \alpha_1 s^{n-1} + \cdots + \mathbf{l}_n \alpha_{n-1} s + \mathbf{l}_n \alpha_n \qquad (12.142)$$

Equating coefficients of like powers of s on both sides, we get

$$\mathbf{P}_0 = \mathbf{l}_n$$
$$\mathbf{P}_1 = \mathbf{A}\mathbf{P}_0 + \alpha_1 \mathbf{l}_n$$
$$\vdots$$
$$\mathbf{P}_k = \mathbf{A}\mathbf{P}_{k-1} + \alpha_k \mathbf{l}_n$$
$$\vdots$$
$$\mathbf{P}_{n-1} = \mathbf{A}\mathbf{P}_{n-2} + \alpha_{n-1} \mathbf{l}_n$$
$$\mathbf{0} = \mathbf{A}\mathbf{P}_{n-1} + \alpha_n \mathbf{l}_n \qquad (12.143)$$

Thus the procedure permits us to compute the matrices $\mathbf{P}_k$ recursively, one at a time. Once this is done, the resolvent matrix can be obtained directly from (12.138). The last equation of (12.143) is used as a check.

We illustrate the above procedure by the following example.

EXAMPLE 12.15

We use the matrix

$$\mathbf{A} = \begin{bmatrix} 0 & -1 & -1 \\ 1 & 0 & 0 \\ 1 & -2 & -2.5 \end{bmatrix} \qquad (12.144)$$

to illustrate the above procedure. The characteristic polynomial is

$$\det(s\mathbf{l}_3 - \mathbf{A}) = s^3 + 2.5s^2 + 2s + 0.5 \qquad (12.145)$$

Our objective is to determine the matrices $\mathbf{P}_0$, $\mathbf{P}_1$, and $\mathbf{P}_2$ in

$$(s\mathbf{I}_3 - \mathbf{A})^{-1} = \frac{1}{\det(s\mathbf{I}_3 - \mathbf{A})} (\mathbf{P}_0 s^2 + \mathbf{P}_1 s + \mathbf{P}_2) \tag{12.146}$$

For this we appeal to (12.143) and obtain

$$\mathbf{P}_0 = \mathbf{I}_3 = \begin{bmatrix} 1 & 0 & 0 \\ 0 & 1 & 0 \\ 0 & 0 & 1 \end{bmatrix} \tag{12.147a}$$

$$\mathbf{P}_1 = \mathbf{A}\mathbf{P}_0 + 2.5\mathbf{I}_3 = \begin{bmatrix} 2.5 & -1 & -1 \\ 1 & 2.5 & 0 \\ 1 & -2 & 0 \end{bmatrix} \tag{12.147b}$$

$$\mathbf{P}_2 = \mathbf{A}\mathbf{P}_1 + 2\mathbf{I}_3 = \begin{bmatrix} 0 & -0.5 & 0 \\ 2.5 & 1 & -1 \\ -2 & -1 & 1 \end{bmatrix} \tag{12.147c}$$

As a check we substitute (12.147c) in $\mathbf{A}\mathbf{P}_2 + 0.5\mathbf{I}_3$ and verify that the sum is identically zero. Finally, putting (12.147) in (12.146) yields the desired inversion:

$$(s\mathbf{I}_3 - \mathbf{A})^{-1} = \frac{1}{(s+1)^2(s+0.5)} \begin{bmatrix} s(s+2.5) & -(s+0.5) & -s \\ s+2.5 & s^2+2.5s+1 & -1 \\ s-2 & -(2s+1) & s^2+1 \end{bmatrix} \tag{12.148}$$

This matrix can be expanded in a partial-fraction expansion in terms of the constituent matrices, as follows:

$$(s\mathbf{I}_3 - \mathbf{A})^{-1} = \frac{1}{(s+1)} \mathbf{K}_{11} + \frac{1}{(s+1)^2} \mathbf{K}_{12} + \frac{1}{(s+0.5)} \mathbf{K}_{21} \tag{12.149}$$

where

$$\mathbf{K}_{11} = \begin{bmatrix} 5 & 0 & -2 \\ -8 & 1 & 4 \\ 10 & 0 & -4 \end{bmatrix}, \quad \mathbf{K}_{12} = \begin{bmatrix} 3 & -1 & -2 \\ -3 & 1 & 2 \\ 6 & -2 & -4 \end{bmatrix}$$

$$\mathbf{K}_{21} = \begin{bmatrix} -4 & 0 & 2 \\ 8 & 0 & -4 \\ -10 & 0 & 5 \end{bmatrix} \tag{12.150}$$

Finally, to compute the state-transition matrix we take the inverse Laplace transform of (12.149) and obtain

$$\boldsymbol{\phi}(t) = e^{\mathbf{A}t} = e^{-t}\mathbf{K}_{11} + te^{-t}\mathbf{K}_{12} + e^{-0.5t}\mathbf{K}_{21} \tag{12.151}$$

The procedure outlined above indicates that in order to compute the matrices $\mathbf{P}_k$, we must first determine the characteristic polynomial. In fact, even this is not necessary. It be can shown[†] that the coefficients α_k of the characteristic polynomial can be determined recursively, as follows:

$$\alpha_1 = -\operatorname{tr}(\mathbf{AP}_0)$$

$$\alpha_2 = -\frac{1}{2}\operatorname{tr}(\mathbf{AP}_1)$$

$$\vdots$$

$$\alpha_k = -\frac{1}{k}\operatorname{tr}(\mathbf{AP}_{k-1})$$

$$\vdots$$

$$\alpha_n = -\frac{1}{n}\operatorname{tr}(\mathbf{AP}_{n-1}) \qquad \textbf{(12.152)}$$

In the case where $\mathbf{A}$ is nonsingular, the Souriau-Frame algorithm may also be used to evaluate the inverse of $\mathbf{A}$. For this we set $s = 0$ in (12.138) to (12.140) and get

$$\mathbf{A}^{-1} = -\frac{1}{\alpha_n}\mathbf{P}_{n-1} \qquad \textbf{(12.153)}$$

Finally, combining (12.143) and (12.152) gives the Souriau-Frame algorithm, as follows:

$$\mathbf{P}_0 = \mathbf{1}_n$$
$$\alpha_1 = -\operatorname{tr}\mathbf{A}$$
$$\mathbf{P}_1 = \mathbf{AP}_0 + \alpha_1\mathbf{1}_n$$
$$\alpha_2 = -\tfrac{1}{2}\operatorname{tr}(\mathbf{AP}_1)$$
$$\vdots$$
$$\alpha_k = -\frac{1}{k}\operatorname{tr}(\mathbf{AP}_{k-1})$$
$$\mathbf{P}_k = \mathbf{AP}_{k-1} + \alpha_k\mathbf{1}_n$$
$$\vdots$$
$$\mathbf{P}_{n-1} = \mathbf{AP}_{n-2} + \alpha_{n-1}\mathbf{1}_n$$
$$\alpha_n = -\frac{1}{n}\operatorname{tr}(\mathbf{AP}_{n-1})$$
$$\mathbf{0} = \mathbf{AP}_{n-1} + \alpha_n\mathbf{1}_n \quad \text{(check)} \qquad \textbf{(12.154)}$$

[†] Souriau, "Une méthode pour la Décomposition spectrale à l'inversion des matrices," and Frame, "A Simple Recursion Formula for Inverting a Matrix."

EXAMPLE 12.16

We repeat the problem treated in Example 12.15 by applying the algorithm (12.154), as follows:

$$\mathbf{P}_0 = \mathbf{I}_3 \tag{12.155a}$$

$$\alpha_1 = -\operatorname{tr}\mathbf{A} = 2.5 \tag{12.155b}$$

$$\mathbf{P}_1 = \mathbf{A}\mathbf{P}_0 + 2.5\mathbf{I}_3 = \begin{bmatrix} 2.5 & -1 & -1 \\ 1 & 2.5 & 0 \\ 1 & -2 & 0 \end{bmatrix} \tag{12.155c}$$

$$\alpha_2 = -\frac{1}{2}\operatorname{tr}(\mathbf{A}\mathbf{P}_1) = 2 \tag{12.155d}$$

$$\mathbf{P}_2 = \mathbf{A}\mathbf{P}_1 + 2\mathbf{I}_3 = \begin{bmatrix} 0 & -0.5 & 0 \\ 2.5 & 1 & -1 \\ -2 & -1 & 1 \end{bmatrix} \tag{12.155e}$$

$$\alpha_3 = -\frac{1}{3}\operatorname{tr}(\mathbf{A}\mathbf{P}_2) = 0.5 \tag{12.155f}$$

$$\mathbf{0} = \mathbf{A}\mathbf{P}_2 + \frac{1}{2}\mathbf{I}_3 \quad \text{(check)} \tag{12.155g}$$

Finally, as a by-product the inverse of $\mathbf{A}$ is found from (12.153) to be

$$\mathbf{A}^{-1} = -\frac{1}{\alpha_3}\mathbf{P}_2 = \begin{bmatrix} 0 & 1 & 0 \\ -5 & -2 & 2 \\ 4 & 2 & -2 \end{bmatrix} \tag{12.156}$$

12-4.3 Estimation of Error

In the preceding section, we showed how the state-transition matrix is obtained in closed form using the Laplace transform as a tool. However, if computer solution is the objective, there is no need to find a closed-form solution. The matrix (12.63) is a very practical way of evaluating the state-transition matrix by truncating the infinite series, say, at the kth term. But in doing so, we must know the error due to truncation. In this section we present a simple formula in estimating this error.

Before we proceed we need the definition of "norm" of a matrix. There are many ways of defining the norm of a matrix. For our purposes the *norm* of a matrix $\mathbf{A}$, written as $\|\mathbf{A}\|$, is defined as

$$\|\mathbf{A}\| = \min(n_1, n_2) \tag{12.157}$$

where

$$n_1 = \max_i \sum_{j=1}^{n} |a_{ij}| \tag{12.158}$$

$$n_2 = \max_j \sum_{i=1}^{n} |a_{ij}| \tag{12.159}$$

$$\mathbf{A} = [a_{ij}]_{n \times n} \tag{12.160}$$

In words n_1 is the largest number among the n numbers, each of which is the row sum of the magnitudes of the elements. Likewise n_2 is the largest number among the n column sums of the magnitudes. For instance, the row sums of the magnitudes of the matrix

$$\mathbf{A} = \begin{bmatrix} 0 & 1 & 0 \\ -5 & -2 & 2 \\ 3 & 2 & -2 \end{bmatrix} \tag{12.161}$$

are 1, 9, and 7 and the column sums of the magnitudes are 8, 5, and 4. Thus $n_1 = 9$, $n_2 = 8$, and $\|\mathbf{A}\| = 8$. Clearly $|a_{ij}| \leq \|\mathbf{A}\|$ for all i and j. It can be shown that[†]

$$\|\mathbf{AB}\| \leq \|\mathbf{A}\| \cdot \|\mathbf{B}\| \tag{12.162}$$

Suppose that we truncate the infinite series (12.63) at the kth term and write for $t = t_1$

$$e^{\mathbf{A}t_1} \approx \mathbf{1}_n + t_1 \mathbf{A} + \frac{t_1^2}{2!} \mathbf{A}^2 + \cdots + \frac{t_1^{k-1}}{(k-1)!} \mathbf{A}^{k-1}$$

$$= \sum_{m=0}^{k-1} \frac{t_1^m}{m!} \mathbf{A}^m \triangleq [w_{ij}] \tag{12.163}$$

The *error matrix* is defined as the difference between $e^{\mathbf{A}t_1}$ and the approximate solution (12.163) or

$$\mathbf{E} = \sum_{m=k}^{\infty} \frac{t_1^m}{m!} \mathbf{A}^m \triangleq [e_{ij}] \tag{12.164}$$

Denote by $q_{ij}^{(m)}$ the ith row and jth column element of $t_1^m \mathbf{A}^m$. Then

$$e_{ij} = \sum_{m=k}^{\infty} \frac{1}{m!} q_{ij}^{(m)} \tag{12.165}$$

[†] See, for example, B. Noble, *Applied Linear Algebra* (Englewood Cliffs, N.J.: Prentice-Hall, 1969).

Appealing to (12.162), which also implies that $\|\mathbf{A}^m\| \le \|\mathbf{A}\|^m$, we get

$$|e_{ij}| \le \sum_{m=k}^{\infty} \frac{1}{m!} |q_{ij}^{(m)}|$$

$$\le \sum_{m=k}^{\infty} \frac{1}{m!} \|t_1^m \mathbf{A}^m\| \le \sum_{m=k}^{\infty} \frac{1}{m!} \|t_1 \mathbf{A}\|^m$$

$$= \frac{1}{k!} \|t_1 \mathbf{A}\|^k \left[1 + \frac{\|t_1 \mathbf{A}\|}{(k+1)} + \frac{\|t_1 \mathbf{A}\|^2}{(k+2)(k+1)} + \cdots \right]$$

$$\le \frac{\|t_1 \mathbf{A}\|^k}{k!} \left[1 + \frac{\|t_1 \mathbf{A}\|}{(k+1)} + \frac{\|t_1 \mathbf{A}\|^2}{(k+1)^2} + \cdots \right]$$

$$= \frac{\|t_1 \mathbf{A}\|^k}{k!} (1 + \epsilon + \epsilon^2 + \epsilon^3 + \cdots) \qquad \text{(12.166)}$$

where

$$\epsilon \triangleq \frac{\|t_1 \mathbf{A}\|}{k+1} \qquad \text{(12.167)}$$

The series inside the parentheses is recognized as the geometric series that converges to $1/(1 - \epsilon)$ for $\epsilon < 1$. Thus we have[†]

$$|e_{ij}| \le \frac{\|t_1 \mathbf{A}\|^k}{k![1 - \|t_1 \mathbf{A}\|/(k+1)]}, \qquad \|t_1 \mathbf{A}\| < k+1 \qquad \text{(12.168)}$$

Equation (12.168) establishes an upper bound of the error when (12.163) is used to approximate the state-transition matrix at time t_1. If each element in $\exp(\mathbf{A}t_1)$ is required within an accuracy of at least d significant digits, in general we must have from (12.163) and (12.164)

$$|e_{ij}| \le 10^{-d} |w_{ij}| \qquad \text{(12.169)}$$

EXAMPLE 12.17

Let

$$\mathbf{A} = \begin{bmatrix} -1 & -1 \\ -0.5 & -1.5 \end{bmatrix} \qquad \text{(12.170)}$$

which was treated in Example 12.6. Let $t_1 = 0.1$ sec. We compute the state-transition matrix $\boldsymbol{\phi}(t_1)$ within two significant digits.

Suppose that we truncate the series (12.63) at the fifth term, giving $k = 4$, and from (12.65)

$$\boldsymbol{\phi}(0.1) = e^{0.1\mathbf{A}} \approx \begin{bmatrix} 0.907064 & -0.088331 \\ -0.044165 & 0.862898 \end{bmatrix} \qquad \text{(12.171)}$$

[†] For machine computation a sharper and more useful bound will be derived in (12.233).

An upper bound error for all the elements is computed as

$$|e_{ij}| \leq \frac{\|0.1\mathbf{A}\|^4}{4!(1 - \|0.1\mathbf{A}\|/5)} = 0.695 \times 10^{-4} \tag{12.172a}$$

where $\|0.1\mathbf{A}\| = 0.20$. Comparing this upper bound error with the elements of (12.171), we see that the largest relative error is approximately given by

$$\frac{0.695 \times 10^{-4}}{0.044165} = 0.157 \times 10^{-2} \tag{12.172b}$$

or 0.16 percent. From (12.169), if the magnitude of the relative error of a number is less than 10^{-d}, then the number is accurate to d digits. Thus the approximate solution (12.171) is accurate to at least two significant figures. In fact, the exact solution, as computed from (12.66), is found to be

$$\boldsymbol{\phi}(0.1) = e^{0.1\mathbf{A}} = \begin{bmatrix} 0.9070632007 & -0.0883324476 \\ -0.0441662238 & 0.8628969769 \end{bmatrix} \tag{12.173}$$

showing that the actual accuracy is to five significant figures using only five terms at $t = 0.1$ sec.

EXAMPLE 12.18
Consider the matrix

$$\mathbf{A} = \begin{bmatrix} 0 & 1 & 0 \\ 0 & 0 & 1 \\ -6 & -11 & -6 \end{bmatrix} \tag{12.174}$$

which is the companion matrix of a third-order differential equation. Choose $t_1 = 0.01$ sec and truncate the series at the seventh term. The approximate solution of the state-transition matrix is found to be

$$\boldsymbol{\phi}(0.01) = e^{0.01\mathbf{A}}$$

$$\approx \begin{bmatrix} 0.9999990149 & 0.0099981915 & 0.0000490103 \\ -0.0002940621 & 0.9994599011 & 0.0097041294 \\ -0.0582247765 & -0.1070394856 & 0.9412351246 \end{bmatrix} \tag{12.175}$$

An upper bound error for all the elements is computed as

$$|e_{ij}| \leq \frac{\|0.01\mathbf{A}\|^6}{6!(1 - \|0.01\mathbf{A}\|/7)} = 4.2195 \times 10^{-9} \tag{12.176}$$

where $\|0.01\mathbf{A}\| = 0.12$. The largest relative error is given by

$$\frac{4.2195 \times 10^{-9}}{4.90103 \times 10^{-5}} = 0.861 \times 10^{-4} \tag{12.177}$$

This means that the approximate solution is accurate to at least four significant figures. This is a very pessimistic estimation. The actual situation is that (12.175) is accurate to at least 10 significant figures. The reader can verify this claim by comparing this solution with the following exact solution: (Problem 12.18)

$$\phi(0.01) = e^{0.01\mathbf{A}}$$

$$= \begin{bmatrix} 3e^{-t} - 3e^{-2t} & 2.5e^{-t} - 4e^{-2t} & 0.5e^{-t} - e^{-2t} \\ \quad + e^{-3t} & \quad + 1.5e^{-3t} & \quad + 0.5e^{-3t} \\ -3e^{-t} + 6e^{-2t} & -2.5e^{-t} + 8e^{-2t} & -0.5e^{-t} + 2e^{-2t} \\ \quad - 3e^{-3t} & \quad - 4.5e^{-3t} & \quad - 1.5e^{-3t} \\ 3e^{-t} - 12e^{-2t} & 2.5e^{-t} - 16e^{-2t} & 0.5e^{-t} - 4e^{-2t} \\ \quad + 9e^{-3t} & \quad + 13.5e^{-3t} & \quad + 4.5e^{-3t} \end{bmatrix}$$

$$(12.178)$$

12-5 THE TRANSFER-FUNCTION MATRIX

In this section we show that the transfer functions of a linear time-invariant system can also be obtained from its state equations

$$\dot{\mathbf{x}}(t) = \mathbf{A}\mathbf{x}(t) + \mathbf{B}\mathbf{u}(t) \tag{12.179}$$

$$\mathbf{y}(t) = \mathbf{C}\mathbf{x}(t) + \mathbf{D}\mathbf{u}(t) \tag{12.180}$$

Taking the Laplace transform of (12.179) and (12.180) yields

$$s\mathbf{X}(s) - \mathbf{x}(0) = \mathbf{A}\mathbf{X}(s) + \mathbf{B}\mathbf{U}(s) \tag{12.181}$$

$$\mathbf{Y}(s) = \mathbf{C}\mathbf{X}(s) + \mathbf{D}\mathbf{U}(s) \tag{12.182}$$

where, as before, the capital letters $\mathbf{X}$, $\mathbf{U}$, and $\mathbf{Y}$ denote the frequency-domain transforms of the corresponding time-domain lowercase $\mathbf{x}$, $\mathbf{u}$, and $\mathbf{y}$. The *transfer-function matrix* $\mathbf{H}(s)$ is a matrix relating the transform of the output vector to the transform of the input vector with all the initial conditions being set to zero:[†]

$$\mathbf{Y}(s) = \mathbf{H}(s)\mathbf{U}(s)\Big|_{\mathbf{x}(0)=\mathbf{0}} \tag{12.183}$$

Thus to express the transfer-function matrix in terms of the coefficient matrices of the state equations, we set $\mathbf{x}(0) = \mathbf{0}$ and eliminate $\mathbf{X}(s)$ in (12.181) and (12.182). From (12.181)

$$(s\mathbf{I} - \mathbf{A})\mathbf{X}(s) = \mathbf{B}\mathbf{U}(s) \tag{12.184}$$

or

$$\mathbf{X}(s) = (s\mathbf{I} - \mathbf{A})^{-1}\mathbf{B}\mathbf{U}(s) \tag{12.185}$$

Substituting (12.185) in (12.182) results in

$$\mathbf{Y}(s) = [\mathbf{C}(s\mathbf{I} - \mathbf{A})^{-1}\mathbf{B} + \mathbf{D}]\mathbf{U}(s) \tag{12.186}$$

[†] The transfer-function matrix $\mathbf{H}(s)$ should not be confused with the operator matrix $\mathbf{H}(p)$ as used in (2.65c) and (11.68). The need for symbols far outstrips the availability.

giving from (12.183)

$$\mathbf{H}(s) = \mathbf{C}(s\mathbf{I} - \mathbf{A})^{-1}\mathbf{B} + \mathbf{D} \tag{12.187}$$

the elements of which are transfer functions of the system. We illustrate this by the following example.

EXAMPLE 12.19

For the network of Figure 12.13, we wish to compute the input admittance facing the voltage source v_g and the transfer voltage ratio between the output and input voltages. The resistances R_1 and R_3 are chosen so that the network has a natural frequency of multiplicity 3 at $s = -1$. All the network element values have been scaled so that they appear in ohms, farads, and henrys.

Applying the procedure of Section 11-4 yields the state equations

$$\begin{bmatrix} \dot{i}_2 \\ \dot{v}_4 \\ \dot{v}_5 \end{bmatrix} = \begin{bmatrix} -R_1 & -1 & -1 \\ 1 & 0 & 0 \\ 1 & -1/R_3 & -1-1/R_3 \end{bmatrix} \begin{bmatrix} i_2 \\ v_4 \\ v_5 \end{bmatrix} + \begin{bmatrix} 1 \\ 0 \\ 0 \end{bmatrix} [v_g] \tag{12.188}$$

$$\begin{bmatrix} i_2 \\ v_6 \end{bmatrix} = \begin{bmatrix} 1 & 0 & 0 \\ 0 & 0 & 1 \end{bmatrix} \begin{bmatrix} i_2 \\ v_4 \\ v_5 \end{bmatrix} + \begin{bmatrix} 0 \\ 0 \end{bmatrix} [v_g] \tag{12.189}$$

where the output variables are the inductor current i_2 and resistor voltage v_6. Equations (12.188) and (12.189) are given in the standard form of (12.179) and (12.180). The characteristic equation of the network is found to be

$$\det(s\mathbf{I}_3 - \mathbf{A}) = s^3 + \left(1 + R_1 + \frac{1}{R_3}\right)s^2 + \left(\frac{R_1}{R_3} + R_1 + 2\right)s + 1 = 0 \tag{12.190}$$

To produce a root of multiplicity 3 at $s = -1$, we set

$$1 + R_1 + \frac{1}{R_3} = 3 \tag{12.191}$$

$$\frac{R_1}{R_3} + R_1 + 2 = 3$$

Figure 12.13
An active network used to
illustrate the Souriau-
Frame algorithm and the
computation of the
transfer-function matrix.

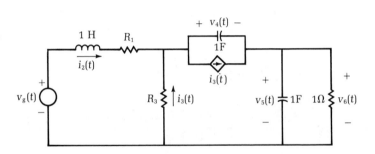

yielding

$$R_1 = 0.382 \ \Omega, \qquad R_3 = 0.618 \ \Omega \tag{12.192}$$

Using these values in (12.188), the $\mathbf{A}$ matrix becomes

$$\mathbf{A} = \begin{bmatrix} -0.382 & -1 & -1 \\ 1 & 0 & 0 \\ 1 & -1.618 & -2.618 \end{bmatrix} \tag{12.193}$$

Applying the Souriau-Frame algorithm of Section 12-4.2 to $\mathbf{A}$, we obtain the resolvent matrix, as follows:

$$\mathbf{P}_0 = \mathbf{I}_3 \tag{12.194a}$$

$$\alpha_1 = -\operatorname{tr} \mathbf{A}\mathbf{P}_0 = 3 \tag{12.194b}$$

$$\mathbf{P}_1 = \mathbf{A}\mathbf{P}_0 + \alpha_1\mathbf{I}_3 = \begin{bmatrix} 2.618 & -1 & -1 \\ 1 & 3 & 0 \\ 1 & -1.618 & 0.382 \end{bmatrix} \tag{12.194c}$$

$$\alpha_2 = -\frac{1}{2}\operatorname{tr} \mathbf{A}\mathbf{P}_1 = 3 \tag{12.194d}$$

$$\mathbf{P}_2 = \mathbf{A}\mathbf{P}_1 + \alpha_2\mathbf{I}_3 = \begin{bmatrix} 0 & -1 & 0 \\ 2.618 & 2 & -1 \\ -1.618 & -1.618 & 1 \end{bmatrix} \tag{12.194e}$$

$$\alpha_3 = -\frac{1}{3}\operatorname{tr} \mathbf{A}\mathbf{P}_2 = 1 \tag{12.194f}$$

$$\mathbf{0} = \mathbf{A}\mathbf{P}_2 + \mathbf{I}_3 \quad \text{(check)} \tag{12.194g}$$

giving

$$(s\mathbf{I}_3 - \mathbf{A})^{-1} = \frac{1}{(s+1)^3}\begin{bmatrix} s^2 + 2.618s & -s-1 & -s \\ s + 2.618 & s^2 + 3s + 2 & -1 \\ s - 1.618 & -1.618(s+1) & s^2 + 0.382s + 1 \end{bmatrix} \tag{12.195}$$

To compute the transfer-function matrix, we insert (12.195) in (12.187) in conjunction with (12.188) and (12.189) and obtain

$$\mathbf{H}(s) = \begin{bmatrix} 1 & 0 & 0 \\ 0 & 0 & 1 \end{bmatrix}(s\mathbf{I}_3 - \mathbf{A})^{-1}\begin{bmatrix} 1 \\ 0 \\ 0 \end{bmatrix} + \begin{bmatrix} 0 \\ 0 \end{bmatrix}$$

$$= \frac{1}{(s+1)^3}\begin{bmatrix} s^2 + 2.618s \\ s - 1.618 \end{bmatrix} = \begin{bmatrix} I_2(s)/V_g(s) \\ V_6(s)/V_g(s) \end{bmatrix} \tag{12.196}$$

where $I_2(s)$, $V_6(s)$, and $V_g(s)$ are the Laplace transforms of $i_2(t)$, $v_6(t)$, and $v_g(t)$, respectively.

Suppose that the network is initially relaxed. We compute the response due to the unit-step voltage input. To this end we first compute the state-transition matrix by a partial-fraction expansion of the resolvent matrix (12.195):

$$(s\mathbf{I}_3 - \mathbf{A})^{-1} = \frac{1}{(s+1)}\mathbf{K}_{11} + \frac{1}{(s+1)^2}\mathbf{K}_{12} + \frac{1}{(s+1)^3}\mathbf{K}_{13} \qquad (12.197)$$

where

$$\mathbf{K}_{11} = \begin{bmatrix} 1 & 0 & 0 \\ 0 & 1 & 0 \\ 0 & 0 & 1 \end{bmatrix}, \qquad \mathbf{K}_{12} = \begin{bmatrix} 0.618 & -1 & -1 \\ 1 & 1 & 0 \\ 1 & -1.618 & -1.618 \end{bmatrix}$$

$$\mathbf{K}_{13} = \begin{bmatrix} -1.618 & 0 & 1 \\ 1.618 & 0 & -1 \\ -2.618 & 0 & 1.618 \end{bmatrix} \qquad (12.198)$$

the inverse Laplace transform of which is the state-transition matrix

$$\boldsymbol{\phi}(t) = e^{\mathbf{A}t} =$$

$$\begin{bmatrix} e^{-t} + 0.618te^{-t} - 0.809t^2 e^{-t} & -te^{-t} & -te^{-t} + 0.5t^2 e^{-t} \\ te^{-t} + 0.809t^2 e^{-t} & e^{-t} + te^{-t} & -0.5t^2 e^{-t} \\ te^{-t} - 1.309t^2 e^{-t} & -1.618te^{-t} & e^{-t} - 1.618te^{-t} + 0.809t^2 e^{-t} \end{bmatrix} \tag{12.199}$$

Appealing to (12.85) yields the desired output

$$\mathbf{y}(t) = \mathbf{C}\int_0^t \boldsymbol{\phi}(t-\tau)\mathbf{B}\mathbf{u}(\tau)\,d\tau + \mathbf{D}\mathbf{u}(t), \qquad t \geq 0 \tag{12.200}$$

or from (12.188), (12.189), and (12.199)

$$\begin{bmatrix} i_2 \\ v_6 \end{bmatrix} = \begin{bmatrix} 1 & 0 & 0 \\ 0 & 0 & 1 \end{bmatrix}\int_0^t \boldsymbol{\phi}(t-\tau)\begin{bmatrix} 1 \\ 0 \\ 0 \end{bmatrix}[1]\,d\tau + \begin{bmatrix} 0 \\ 0 \end{bmatrix}[1]$$

$$= \begin{bmatrix} 1 & 0 & 0 \\ 0 & 0 & 1 \end{bmatrix}e^{\mathbf{A}t}\int_0^t \begin{bmatrix} 1 - 0.618\tau - 0.809\tau^2 \\ -\tau + 0.809\tau^2 \\ -\tau - 1.309\tau^2 \end{bmatrix}e^{\tau}\,d\tau$$

$$= \begin{bmatrix} 1 & 0 & 0 \\ 0 & 0 & 1 \end{bmatrix}\begin{bmatrix} t(1 + 0.809t)e^{-t} \\ 2.618 - (2.618 + 2.618t + 0.809t^2)e^{-t} \\ -1.618 + (1.618 + 1.618t + 1.309t^2)e^{-t} \end{bmatrix}$$

$$= \begin{bmatrix} t(1 + 0.809t)e^{-t} \\ -1.618 + (1.618 + 1.618t + 1.309t^2)e^{-t} \end{bmatrix}, \qquad t \geq 0 \tag{12.201}$$

As a check we compute the inverse Laplace transforms of $I_2(s)$ and $V_6(s)$ for the unit-step input voltage $V_g(s) = 1/s$, as follows. Using the inverse Laplace transform pairs[†]

$$\mathscr{L}^{-1} \frac{s + a}{(s + \alpha)^3} = te^{-\alpha t} + \frac{1}{2}(a - \alpha)t^2 e^{-\alpha t} \tag{12.202}$$

$$\mathscr{L}^{-1} \frac{s + a}{s(s + \alpha)^3} = \frac{a}{\alpha^3}(1 - e^{-\alpha t}) - \frac{a}{\alpha^2} te^{-\alpha t} + \frac{\alpha - a}{2\alpha} t^2 e^{-\alpha t} \tag{12.203}$$

we obtain from (12.196)

$$i_2(t) = \mathscr{L}^{-1} \frac{s + 2.618}{(s + 1)^3} = te^{-t} + 0.809t^2 e^{-t}, \quad t \geq 0 \tag{12.204}$$

$$v_6(t) = \mathscr{L}^{-1} \frac{s - 1.618}{s(s + 1)^3} = -1.618(1 - e^{-t}) + 1.618te^{-t}$$

$$+ 1.309t^2 e^{-t}, \quad t \geq 0 \tag{12.205}$$

confirming (12.201).

12-6   TRAJECTORIES OF THE LINEAR SYSTEMS

So far we have shown that the response of a system described by its state equations can be computed in a straightforward manner once the state-transition matrix is known. No attempt was made to interpret the solutions. In this section we present a physical interpretation of the solutions as trajectories in the state space.

In the network of Figure 12.8, the capacitor voltages $v_1(t)$ and $v_2(t)$ were computed earlier in Example 12.8 and are given by

$$v_1(t) = 1 + \frac{2}{3} e^{-0.5t} + \frac{1}{3} e^{-2t} \tag{12.206a}$$

$$v_2(t) = -\frac{1}{3} e^{-0.5t} + \frac{1}{3} e^{-2t} \tag{12.206b}$$

In Figure 12.9 v_1 and v_2 are plotted as functions of t. These two plots can be described by just one plot as shown in Figure 12.14. In this figure the capacitor voltage v_1 is plotted against the capacitor voltage v_2, using t as an explicit parameter on the curve. The arrowheads on the curve indicate the direction of increasing t. This v_1 versus v_2 plot for a second-order system is generally known as the *phase plane*. The directed curve in the phase plane

---

[†] G. E. Roberts and H. Kaufman, *Table of Laplace Transforms* (Philadelphia: W. B. Saunders, 1966), pp. 182 and 195.

Figure 12.14

A phase-plane trajectory for the equations (12.206).

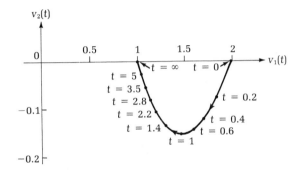

is called a *phase-plane trajectory*. Each point on the phase-plane trajectory corresponds to a state of the network associated with a specific time $t_1 \geq t_0$, the coordinates $[v_1(t_1), v_2(t_1)]$ of which give the values of v_1 and v_2 at time t_1. The *initial state* corresponds to the point $[v_1(t_0), v_2(t_0)]$ that in the present case is $[v_1(0), v_2(0)] = [2, 0]$. When time approaches infinity, the *final state* becomes $[v_1(\infty), v_2(\infty)]$, being equal to $[1, 0]$ in the present situation. As another example consider the network of Figure 12.10. The inductor current $i(t)$ and capacitor voltage $v(t)$ are plotted in Figure 12.11 as functions of t. The phase-plane trajectory is presented in Figure 12.15, which indicates that the steady-state response is sinusoidal.

For higher-order systems the state of a system at any given time t_1 can be represented as a *point* in a multidimensional phase space called the *state space*. A *state-space trajectory* is a directed space curve defined by a solution $\mathbf{x}(t)$ in the state space. Although the state-space trajectory gives a clear picture of the transition of states as time progresses, it is usually very difficult

Figure 12.15

A phase-plane trajectory for the inductor current and capacitor voltage of the network of Figure 12.10

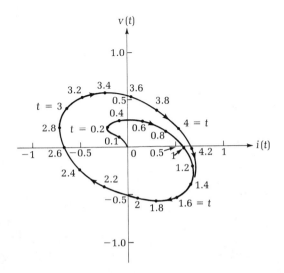

to visualize and to construct such a trajectory for systems with order higher than 2. This is why nearly all of the state-space trajectories are plotted only for the second-order systems.

12-7 CONTROLLABILITY AND OBSERVABILITY

The concepts of controllability and observability of a linear system will be introduced intuitively here and defined in terms of the state.

The term state controllability refers to the ability to get from one state to another with an appropriate input vector. More formally a system described by the state equation

$$\dot{\mathbf{x}}(t) = \mathbf{A}\mathbf{x}(t) + \mathbf{B}\mathbf{u}(t) \tag{12.207}$$

is said to be *completely state controllable* if for any time t_0 it is possible to transfer the system from any initial state $\mathbf{x}(t_0)$ to any other state $\mathbf{x}(t_1)$ in a finite time interval $t_1 - t_0$, $t_1 \geq t_0$, by an appropriate input vector $\mathbf{u}(t)$ defined over $t_0 \leq t \leq t_1$.

As an example, consider the network of Figure 12.1, the state equation of which is given by

$$\dot{v}(t) = -\frac{1}{RC}v(t) + \frac{1}{RC}v_g(t) \tag{12.208}$$

Choose the initial time $t_0 = 0$. We wish to transfer the initial state $v(0)$ to the state $v(t_1)$ at time $t_1 \geq 0$. Our objective is to try to find an appropriate input excitation $v_g(t)$ over the time interval $0 \leq t \leq t_1$, so that when this input is applied to (12.208), which is at the state $v(0)$ at $t = 0$, the capacitor voltage becomes $v(t_1)$ at $t = t_1$.

From (12.79) the solution $v(t_1)$ of (12.208) with initial state $v(0)$ can be written as

$$v(t_1) = e^{-t_1/RC}v(0) + \frac{1}{RC}e^{-t_1/RC}\int_0^{t_1} e^{\tau/RC}v_g(\tau)\,d\tau \tag{12.209}$$

For our purposes let $v_g(t)$ be a constant voltage of unknown magnitude for $0 \leq t \leq t_1$ and write

$$v_g(t) = V, \qquad 0 \leq t \leq t_1 \tag{12.210}$$

Substituting this in (12.209), we get

$$v(t_1) - e^{-t_1/RC}v(0) = e^{-t_1/RC}[e^{t_1/RC} - 1]V \tag{12.211}$$

yielding the required voltage

$$V = \frac{v(t_1) - e^{-t_1/RC}v(0)}{1 - e^{-t_1/RC}} \tag{12.212}$$

Therefore the network of Figure 12.1 is completely state controllable.

Figure 12.16
A network (a) that is not completely state controllable, and its associated directed
graph (b).

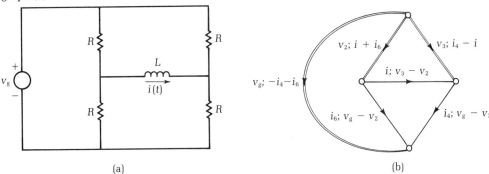

(a)   (b)

On the other hand, the network of Figure 12.16(a) is not completely
state controllable. To see this we first determine its state equation, which
can be derived from the associated directed graph of Figure 12.16(b), resulting in (Problem 12.19)

$$\dot{i}(t) = -\frac{R}{L} i(t) \tag{12.213}$$

Since (12.213) is independent of the input voltage $v_g(t)$, it is impossible to
find any input to transfer the network from one state to another. We conclude
that the network is not completely state controllable. Physically this is
obvious. The network of Figure 12.16(a) is a balanced bridge. The inductor
current $i(t)$ cannot be controlled by any input voltage.

The general characterization of controllability of a linear time-invariant
system described by its state equation is known, and is stated as the
following.[†]

Property 12.2   *A system is completely state controllable if and only if
the composite matrix*

$$[\mathbf{B} \quad \mathbf{AB} \quad \mathbf{A}^2\mathbf{B} \quad \cdots \quad \mathbf{A}^{n-1}\mathbf{B}] \tag{12.214}$$

is of rank n, where n is the order of $\mathbf{A}$.

In the case of a single input, u(t) being a scalar, we have the following.[†]

Property 12.3   *A system with a single input is completely state con-
trollable if and only if there is no cancellation in the matrix* $(s\mathbf{I} - \mathbf{A})^{-1}\mathbf{B}$.

[†] For a proof see, for example, K. Ogata, *State Space Analysis of Control Systems* (Englewood
Cliffs, N.J.: Prentice-Hall, 1967), Chapter 7.

The cancellation of a common factor corresponds to the complete suppression of a natural frequency, thus preventing total control of the state.

EXAMPLE 12.20

The network of Figure 12.8 was treated in Example 12.8, the state equation of which is repeated below for convenience:

$$\begin{bmatrix} \dot{v}_1 \\ \dot{v}_2 \end{bmatrix} = \begin{bmatrix} -1 & -1 \\ -0.5 & -1.5 \end{bmatrix} \begin{bmatrix} v_1 \\ v_2 \end{bmatrix} + \begin{bmatrix} 1 \\ 0.5 \end{bmatrix} [v_g] \tag{12.215}$$

To determine its controllability by means of Property 12.2, we compute the composite matrix

$$[\mathbf{B} \quad \mathbf{AB}] = \begin{bmatrix} 1 & -1.5 \\ 0.5 & -1.25 \end{bmatrix} \tag{12.216}$$

Since this matrix is of rank 2, the network is completely state controllable. On the other hand, if the $\mathbf{B}$ matrix is replaced by $\mathbf{B}' = [1 \quad 1]$ the composite matrix becomes

$$\begin{bmatrix} 1 & -2 \\ 1 & -2 \end{bmatrix} \tag{12.217}$$

which is of rank 1, and the corresponding system is no longer completely. state controllable.

To apply Property 12.3, we must compute the matrix

$$(s\mathbf{I}_2 - \mathbf{A})^{-1}\mathbf{B} = \frac{1}{(s+0.5)(s+2)} \begin{bmatrix} s+1.5 & -1 \\ -0.5 & s+1 \end{bmatrix} \begin{bmatrix} 1 \\ 0.5 \end{bmatrix}$$

$$= \frac{1}{(s+0.5)(s+2)} \begin{bmatrix} s+1 \\ 0.5s \end{bmatrix} \tag{12.218}$$

Because the numerator polynomials $(s+1)$ and $0.5s$ and the denominator polynomial $(s+0.5)(s+2)$ have no common factor, there is no cancellation in $(s\mathbf{I}_2 - \mathbf{A})^{-1}\mathbf{B}$. Therefore the system is completely state controllable.

As in (12.217), if $\mathbf{B}$ is replaced by $\mathbf{B}' = [1 \quad 1]$, the corresponding matrix becomes

$$(s\mathbf{I}_2 - \mathbf{A})^{-1}\mathbf{B} = \frac{1}{(s+0.5)(s+2)} \begin{bmatrix} s+0.5 \\ s+0.5 \end{bmatrix} = \frac{1}{(s+2)} \begin{bmatrix} 1 \\ 1 \end{bmatrix} \tag{12.219}$$

showing the existence of a common factor $(s+0.5)$. Since poles of $(s\mathbf{I}_2 - \mathbf{A})^{-1}$ are natural frequencies, the cancellation of the common factor $(s+0.5)$ corresponds to the suppression of the natural mode $\exp(-\frac{1}{2}t)$. As a result the effect of this mode cannot be controlled by the input, and the system is therefore not completely state controllable.

A concept that is closely related to controllability is observability. Controllability implies that the state of a controllable system can be transferred from one to another with an appropriate input. An observable system is one in which a knowledge of the output vector over a finite time is sufficient to determine its initial state. More formally a system described by the state equations

$$\dot{\mathbf{x}}(t) = \mathbf{A}\mathbf{x}(t) + \mathbf{B}\mathbf{u}(t) \tag{12.220a}$$

$$\mathbf{y}(t) = \mathbf{C}\mathbf{x}(t) \tag{12.220b}$$

is said to be *completely observable*, if with the system initially at the state $\mathbf{x}(t_0)$ at $t = t_0$ it is possible to identify this state $\mathbf{x}(t_0)$ from the knowledge of the output vector $\mathbf{y}(t)$ over the finite time interval $t_0 \leq t \leq t_1$ for some $t_1 > t_0$. Note that in the definition we assume that $\mathbf{D} = \mathbf{0}$. This is not to be deemed a restriction, since the contribution to the output due to the observed input may always be subtracted from the observed output to obtain the output under free motion.

Evidently a system is completely observable only if every transition of its state eventually affects its output. As in the case of controllability, a general test for observability can be formulated in terms of the matrices $\mathbf{A}$ and $\mathbf{C}$. The results are stated as follows.[†]

Property 12.4 *The system (12.220) is completely observable if and only if the composite matrix*

$$[\mathbf{C}' \quad \mathbf{A}'\mathbf{C}' \quad \mathbf{A}'^2\mathbf{C}' \quad \cdots \quad \mathbf{A}'^{n-1}\mathbf{C}'] \tag{12.221}$$

is of rank n, where n is the order of $\mathbf{A}$.

In the case of a single input and a single output, we have the following.[†]

Property 12.5 *The system (12.220) with a single input and a single output is completely observable if and only if there is no cancellation in the matrix* $\mathbf{C}(s\mathbf{I} - \mathbf{A})^{-1}$.

EXAMPLE 12.21

The state equations for the network of Figure 12.17 with $R_1 = 1\ \Omega$, $R_2 = 2\ \Omega$, $R_3 = 3\ \Omega$, $L = 1$ H, and $C = 1$ F are found to be

$$\begin{bmatrix} \dot{v} \\ \dot{i} \end{bmatrix} = \frac{1}{5}\begin{bmatrix} -6 & 3 \\ -3 & -6 \end{bmatrix}\begin{bmatrix} v \\ i \end{bmatrix} + \begin{bmatrix} 1 \\ 1 \end{bmatrix}[v_g] \tag{12.222a}$$

$$\begin{bmatrix} i_2 \\ v_3 \end{bmatrix} = \frac{1}{5}\begin{bmatrix} -1 & 3 \\ 3 & 6 \end{bmatrix}\begin{bmatrix} v \\ i \end{bmatrix} + \begin{bmatrix} 0 \\ 0 \end{bmatrix}[v_g] \tag{12.222b}$$

[†] Ogata, *State Space Analysis of Control Systems*, Chapter 7.

Figure 12.17
A completely observable
network with output
variables i_2 and v_3.

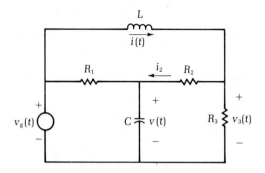

where the output variables are the resistor current i_2 and resistor voltage v_3. To determine observability from the output variables i_2 and v_3, we first compute the composite matrix

$$[\mathbf{C'} \quad \mathbf{A'C'}] = \frac{1}{25}\begin{bmatrix} -5 & 15 & -3 & -36 \\ 15 & 30 & -21 & -27 \end{bmatrix} \qquad (12.223)$$

The rank of this matrix is 2, which is equal to the order of $\mathbf{A}$. Thus by Property 12.4 the network is completely observable.

To apply Property 12.5, we must first determine the resolvent matrix, which is found to be

$$(s\mathbf{I}_2 - \mathbf{A})^{-1} = \frac{1}{5s^2 + 12s + 9}\begin{bmatrix} 5s + 6 & 3 \\ -3 & 5s + 6 \end{bmatrix} \qquad (12.224a)$$

where

$$\det(s\mathbf{I}_2 - \mathbf{A}) = \frac{1}{5}(5s^2 + 12s + 9) = (s + 1.2 + j0.6)(s + 1.2 - j0.6) \qquad (12.224b)$$

Choose $i_2(t)$ as the output variable, giving the output equation

$$[i_2] = [-\tfrac{1}{5} \quad \tfrac{3}{5}]\begin{bmatrix} v \\ i \end{bmatrix} \qquad (12.225)$$

From (12.224a) we have for the system described by (12.222a) and (12.225)

$$\mathbf{C}(s\mathbf{I} - \mathbf{A})^{-1} = \frac{1}{5s^2 + 12s + 9}[-s - 3 \quad 3(s + 1)] \qquad (12.226)$$

Since no cancellation exists, the system is completely observable.

EXAMPLE 12.22
Consider again the network of Figure 12.17 the state equation of which was computed earlier and is given by (11.7). Select resistor current i_2 as the

output variable. The state equations of the network are rewritten as

$$\begin{bmatrix} \dot{i} \\ \dot{v} \end{bmatrix} = \frac{1}{R_2 + R_3} \begin{bmatrix} -R_2 R_3/L & -R_3/L \\ R_3/C & -(R_1 + R_2 + R_3)/CR_1 \end{bmatrix} \begin{bmatrix} i \\ v \end{bmatrix}$$

$$+ \begin{bmatrix} 1/L \\ 1/CR_1 \end{bmatrix} [v_g] \qquad (12.227a)$$

$$[i_2] = \frac{1}{R_2 + R_3} [R_3 \quad -1] \begin{bmatrix} i \\ v \end{bmatrix} + [0][v_g] \qquad (12.227b)$$

We wish to determine a relation among the network elements L, C, R_1, R_2, and R_3 so that the network is *not* completely state controllable and observable. To this end we compute

$$[\mathbf{B} \quad \mathbf{AB}] = \frac{1}{R_2 + R_3} \begin{bmatrix} (R_2 + R_3)/L & -R_2 R_3/L^2 - R_3/LCR_1 \\ (R_2 + R_3)/CR_1 & R_3/LC - (R_1 + R_2 + R_3)/C^2 R_1^2 \end{bmatrix}$$

$$(12.228)$$

$$[\mathbf{C'} \quad \mathbf{A'C'}] = \frac{1}{(R_2 + R_3)^2} \begin{bmatrix} R_3(R_2 + R_3) & -R_3(R_2 R_3/L + 1/C) \\ -R_2 - R_3 & (R_1 + R_2 + R_3)/CR_1 - R_3^2/L \end{bmatrix}$$

$$(12.229)$$

For the network *not* to be completely state controllable and observable, the rank of both matrices (12.228) and (12.229) must be less than 2. Thus we set their determinants to zero and obtain two relations. After simplification these two relations reduce to a single equation

$$R_1 R_3 C = L \qquad (12.230)$$

This is the condition under which the network is not completely state controllable and observable. The result is expected because it is the condition under which the network can be viewed as a balanced bridge.

12-8 COMPUTER SOLUTIONS AND PROGRAMS

In this section we present computer programs in WATFIV that are direct implementation of the Souriau-Frame algorithm for rapid calculation of the resolvent matrix $(s\mathbf{I} - \mathbf{A})^{-1}$, as described in Section 12-4.2, and the power series approximation (12.163) of the state-transition matrix $\boldsymbol{\phi}(t) = e^{\mathbf{A}t}$. The programs are given in the form of subroutines SOUFRA and EXPAT.

In addition, subroutines are provided for the computation of the state vector $\mathbf{x}(t)$. Subroutine STATE is based on Runge-Kutta methods.[†] Subroutine STATE2 was written in accordance with (12.84) using the integration subroutine DQUAD, which takes a five-point formula used over successive

[†] See reference on p. 661.

sets of points as the basis of the quadrature.[†] The subroutine DQUAD is included.

12-8.1   Subroutine SOUFRA

1. Purpose.   The program is for rapid calculation of the resolvent matrix $(sI - A)^{-1}$. It can also be used to evaluate the inverse of a real matrix A. The program is listed in Figure 12.18.

2. Method.   The program is an implementation of the Souriau-Frame algorithm as described in Section 12-4.2. The resolvent matrix $(sI - A)^{-1}$ is expanded as in (12.138) through (12.140).

3. Usage.   The program consists of a subroutine, SOUFRA, called by

$$\text{CALL SOUFRA (N, A, P, AINVRS, ALPHA)}$$

N	is the order of A
A	is the name of a given real matrix A
P	is the name of N − 1 output matrices P_k, $k = 1, 2, \ldots, N - 1$
AINVRS	is the name of the inverse of A
ALPHA	is the name of the coefficients of the characteristic polynomial as defined in (12.140)

4. Remarks.   Double precision is used in all the computations. The order of A cannot exceed 20. Otherwise new dimensional statements are required. P_N is the name of the check matrix which, when printed, should be the zero matrix. The input data are N and elements A(I, J) of A. The outputs are elements P(I, J, K) of P_K, AINVRS (I, J) of AINVRS, and ALPHA (I) of ALPHA.

5. Subroutines and Function Subprograms Required.   None.

EXAMPLE 12.23
We shall use the subroutine SOUFRA to calculate the resolvent matrix $(sI - A)^{-1}$ for the real matrix

$$A = \begin{bmatrix} 0 & -1 & -1 \\ 1 & 0 & 0 \\ 1 & -2 & -2.5 \end{bmatrix} \tag{12.231}$$

considered in Examples 12.15 and 12.16.

The main program and the computer output are presented, respectively, in Figures 12.19 and 12.20. In Figure 12.20 P(i, j, k) denotes the ith row and

[†] See, for example, A. Ralston and H. S. Wilf, *Mathematical Methods for Digital Computers*, vol. II (New York: Wiley, 1967), Chapter 6.

Figure 12.18
Subroutine SOUFRA used
for rapid calculation of
the resolvent matrix.

```
C
C      SUBROUTINE SOUFRA
C
C      PURPOSE
C         THE PROGRAM IS FOR INVERTING MATRICES (S1-A) AND A.
C
C      METHOD
C         THE PROGRAM IS AN IMPLEMENTATION OF THE SOURIAU-FRAME ALGORITHM
C         AS DESCRIBED IN SECTION 12-4.2.  THE RESOLVENT MATRIX
C         IS EXPANDED AS IN (12.138)-(12.140).
C
C      USAGE
C         CALL SOUFRA (N,A,P,AINVRS,ALPHA)
C
C         N      -THE ORDER OF A.
C         A      -A GIVEN REAL MATRIX A.
C         P      -THE NAME OF THE P MATRIX.  THE LAST MATRIX IS USED AS A CHECK
C                   MATRIX, WHICH, WHEN PRINTED, SHOULD BE THE ZERO MATRIX.
C         AINVRS -THE INVERSE OF A, ALSO USED FOR TEMPORARY STORAGE IN
C                   THE PROGRAM
C         ALPHA  -THE COEFFICIENTS OF THE CHARACTERISTIC POLYNOMIAL AS
C                   DEFINED IN (12.140).
C
C      SUBROUTINES AND FUNCTION SUBPROGRAMS REQUIRED
C         NONE
C
C      REMARKS
C         DOUBLE PRECISION IS USED IN ALL THE COMPUTATIONS.
C         THE ORDER OF A CANNOT EXCEED TWENTY.  OTHERWISE, NEW DIMENSIONAL
C            STATEMENTS ARE REQUIRED.
C         THE INPUT DATA ARE N AND ELEMENTS A(I,J) OF A.  THE OUTPUTS ARE
C            ELEMENTS P(I,J,L) OF P(L), AINVRS(I,J) OF AINVRS AND ALPHA(L)
C            OF ALPHA.
       SUBROUTINE SOUFRA (N,A,P,AINVRS,ALPHA)
       DOUBLE PRECISION A(20,20),AINVRS(20,20),P(20,20,20),ALPHA(20)
       DO 1 I=1,N
       DO 1 J=1,N
   1   AINVRS(I,J)=A(I,J)
       DO 4 L=1,N
C
C      COMPUTE CHARACTERISTIC EQUATION COEFFICIENTS ALPHA(L) FROM TRACE
       ALPHA(L)=0.0
       DO 2 K=1,N
   2   ALPHA(L)=ALPHA(L)+AINVRS(K,K)
       ALPHA(L)=-ALPHA(L)/L
C
C      COMPUTE P(L) MATRICES
       DO 3 I=1,N
       DO 3 J=1,N
       P(I,J,L)=AINVRS(I,J)
   3   IF (I.EQ.J) P(I,J,L)=AINVRS(I,J)+ALPHA(L)
       IF (L.EQ.N) GO TO 5
C
C      COMPUTE AP MATRIX
       DO 4 I=1,N
       DO 4 J=1,N
       AINVRS(I,J)=0.0
       DO 4 M=1,N
   4   AINVRS(I,J)=AINVRS(I,J)+A(I,M)*P(M,J,L)
C
C      COMPUTE A INVERSE
   5   LM1=L-1
       DO 6 I=1,N
       DO 6 J=1,N
   6   AINVRS(I,J)=-P(I,J,LM1)/ALPHA(L)
       RETURN
       END
```

Figure 12.19
The main program used
for rapid calculation of
the resolvent matrix, using
subroutine SOUFRA.

```
DOUBLE PRECISION A(20,20),AINVRS(20,20),P(20,20,20),ALPHA(20)
READ,N,((A(I,J),J=1,N),I=1,N)
CALL SOUFRA (N,A,P,AINVRS,ALPHA)
PRINT 1,(((I,J,K,P(I,J,K),J=1,N),I=1,N),K=1,N)
1    FORMAT (2(' P(',I2,',',I2,',',I2,')=',D16.8,4X))
PRINT 2,((I,J,AINVRS(I,J),J=1,N),I=1,N)
2    FORMAT (2(' AINVRS(',I2,',',I2,')=',D16.8,2X))
PRINT 3,(I,ALPHA(I),I=1,N)
3    FORMAT (2(' ALPHA(',I2,')=',D16.8,6X))
STOP
END
```

Figure 12.20
Computer printout of the elements (a) of the P-matrices of the resolvent matrix associated with the real matrix (12.231) and the elements (b) of the inverse of the matrix
(12.231).

```
P( 1, 1, 1)=  0.25000000D 01      P( 1, 2, 1)= -0.10000000D 01
P( 1, 3, 1)= -0.10000000D 01      P( 2, 1, 1)=  0.10000000D 01
P( 2, 2, 1)=  0.25000000D 01      P( 2, 3, 1)=  0.00000000D 00
P( 3, 1, 1)=  0.10000000D 01      P( 3, 2, 1)= -0.20000000D 01
P( 3, 3, 1)=  0.00000000D 00      P( 1, 1, 2)=  0.00000000D 00
P( 1, 2, 2)= -0.50000000D 00      P( 1, 3, 2)=  0.00000000D 00
P( 2, 1, 2)=  0.25000000D 01      P( 2, 2, 2)=  0.10000000D 01
P( 2, 3, 2)= -0.10000000D 01      P( 3, 1, 2)= -0.20000000D 01
P( 3, 2, 2)= -0.10000000D 01      P( 3, 2, 2)=  0.10000000D 01
P( 1, 1, 3)=  0.00000000D 00      P( 1, 2, 3)=  0.00000000D 00
P( 1, 3, 3)=  0.00000000D 00      P( 2, 1, 3)=  0.00000000D 00
P( 2, 2, 3)=  0.00000000D 00      P( 2, 3, 3)=  0.00000000D 00
P( 3, 1, 3)=  0.00000000D 00      P( 3, 2, 3)=  0.00000000D 00
P( 3, 3, 3)=  0.00000000D 00
```

(a)

```
AINVRS( 1, 1)= -0.00000000D 00     AINVRS( 1, 2)=  0.10000000D 01
AINVRS( 1, 3)= -0.00000000D 00     AINVRS( 2, 1)= -0.50000000D 01
AINVRS( 2, 2)= -0.20000000D 01     AINVRS( 2, 3)=  0.20000000D 01
AINVRS( 3, 1)=  0.40000000D 01     AINVRS( 3, 2)=  0.20000000D 01
AINVRS( 3, 3)= -0.20000000D 01
ALPHA( 1)=  0.25000000D 01         ALPHA( 2)=  0.20000000D 01
ALPHA( 3)=  0.50000000D 00
```

(b)

jth column element of the matrix $\mathbf{P}_k$ and AINVRS(i, j) is the ith row and jth
column element of $\mathbf{A}^{-1}$ for $i, j, k = 1, 2, 3$. The computer used was an IBM
370 Model 158 with total execution time 0.09 sec.

EXAMPLE 12.24
Consider the 19×19 real matrix [Figure 12.21(a)]

$$
\mathbf{A} = \begin{bmatrix}
1 & -1 & 0 & 0 & \cdots & 0 & 0 & 0 \\
-1 & 2 & -1 & 0 & \cdots & 0 & 0 & 0 \\
0 & -1 & 2 & -1 & \cdots & 0 & 0 & 0 \\
\vdots & \vdots & \vdots & \vdots & \cdots & \vdots & \vdots & \vdots \\
0 & 0 & 0 & 0 & \cdots & -1 & 2 & -1 \\
0 & 0 & 0 & 0 & \cdots & 0 & -1 & 2
\end{bmatrix}
\tag{12.232}
$$

We use the subroutine SOUFRA to compute its resolvent matrix $(s\mathbf{I} - \mathbf{A})^{-1}$.

Figure 12.21

A given 19 × 19 real matrix (a)† partial listing (b) of the elements of the P-matrices, partial listing (c) of the elements of the inverse of the matrix (a), and the coefficients (d) of the characteristic polynominal of the matrix (a).

(a)

```
19,1.0,-1.0,0.0,0.0,0.0,0.0,0.0,0.0,0.0,0.0,0.0,0.0,0.0,0.0,0.0,0.0,0.0,0.0,0.0
-1.0,2.0,-1.0,0.0,0.0,0.0,0.0,0.0,0.0,0.0,0.0,0.0,0.0,0.0,0.0,0.0,0.0,0.0,0.0
0.0,-1.0,2.0,-1.0,0.0,0.0,0.0,0.0,0.0,0.0,0.0,0.0,0.0,0.0,0.0,0.0,0.0,0.0,0.0
0.0,0.0,-1.0,2.0,-1.0,0.0,0.0,0.0,0.0,0.0,0.0,0.0,0.0,0.0,0.0,0.0,0.0,0.0,0.0
0.0,0.0,0.0,-1.0,2.0,-1.0,0.0,0.0,0.0,0.0,0.0,0.0,0.0,0.0,0.0,0.0,0.0,0.0,0.0
0.0,0.0,0.0,0.0,-1.0,2.0,-1.0,0.0,0.0,0.0,0.0,0.0,0.0,0.0,0.0,0.0,0.0,0.0,0.0
0.0,0.0,0.0,0.0,0.0,-1.0,2.0,-1.0,0.0,0.0,0.0,0.0,0.0,0.0,0.0,0.0,0.0,0.0,0.0
0.0,0.0,0.0,0.0,0.0,0.0,-1.0,2.0,-1.0,0.0,0.0,0.0,0.0,0.0,0.0,0.0,0.0,0.0,0.0
0.0,0.0,0.0,0.0,0.0,0.0,0.0,-1.0,2.0,-1.0,0.0,0.0,0.0,0.0,0.0,0.0,0.0,0.0,0.0
0.0,0.0,0.0,0.0,0.0,0.0,0.0,0.0,-1.0,2.0,-1.0,0.0,0.0,0.0,0.0,0.0,0.0,0.0,0.0
0.0,0.0,0.0,0.0,0.0,0.0,0.0,0.0,0.0,-1.0,2.0,-1.0,0.0,0.0,0.0,0.0,0.0,0.0,0.0
0.0,0.0,0.0,0.0,0.0,0.0,0.0,0.0,0.0,0.0,-1.0,2.0,-1.0,0.0,0.0,0.0,0.0,0.0,0.0
0.0,0.0,0.0,0.0,0.0,0.0,0.0,0.0,0.0,0.0,0.0,-1.0,2.0,-1.0,0.0,0.0,0.0,0.0,0.0
0.0,0.0,0.0,0.0,0.0,0.0,0.0,0.0,0.0,0.0,0.0,0.0,-1.0,2.0,-1.0,0.0,0.0,0.0,0.0
0.0,0.0,0.0,0.0,0.0,0.0,0.0,0.0,0.0,0.0,0.0,0.0,0.0,-1.0,2.0,-1.0,0.0,0.0,0.0
0.0,0.0,0.0,0.0,0.0,0.0,0.0,0.0,0.0,0.0,0.0,0.0,0.0,0.0,-1.0,2.0,-1.0,0.0,0.0
0.0,0.0,0.0,0.0,0.0,0.0,0.0,0.0,0.0,0.0,0.0,0.0,0.0,0.0,0.0,-1.0,2.0,-1.0,0.0
0.0,0.0,0.0,0.0,0.0,0.0,0.0,0.0,0.0,0.0,0.0,0.0,0.0,0.0,0.0,0.0,-1.0,2.0,-1.0
0.0,0.0,0.0,0.0,0.0,0.0,0.0,0.0,0.0,0.0,0.0,0.0,0.0,0.0,0.0,0.0,0.0,-1.0,2.0
```

(b)

```
P( 1, 1, 1)=  -0.36000000D 02     P( 1, 2, 1)=  -0.10000000D 01
P( 1, 3, 1)=   0.00000000D 00     P( 1, 4, 1)=   0.00000000D 00
P( 1, 5, 1)=   0.00000000D 00     P( 1, 6, 1)=   0.00000000D 00
P( 1, 7, 1)=   0.00000000D 00     P( 1, 8, 1)=   0.00000000D 00
P( 1, 9, 1)=   0.00000000D 00     P( 1,10, 1)=   0.00000000D 00
P(12,16, 9)=  -0.34998000D 05     P(12,17, 9)=  -0.75060000D 04
P(12,18, 9)=  -0.13490000D 04     P(12,19, 9)=  -0.19000000D 03
P(13, 1, 9)=   0.00000000D 00     P(13, 2, 9)=  -0.10000000D 01
P(13, 3, 9)=   0.00000000D 00     P(13, 4, 9)=  -0.10000000D 01
P(13, 5, 9)=  -0.19000000D 02     P(13, 6, 9)=  -0.19100000D 03
P(13, 7, 9)=  -0.13490000D 04     P(13, 8, 9)=  -0.75060000D 04
P(13, 9, 9)=  -0.34998000D 05     P(13,10, 9)=  -0.14210200D 06
P(13,11, 9)=  -0.51569800D 06     P(13,12, 9)=  -0.17043770D 07
P(13,13, 9)=  -0.52025230D 07     P(13,14, 9)=  -0.17043770D 07
P(13,15, 9)=  -0.51569800D 06     P(13,16, 9)=  -0.14210200D 06
P(13,17, 9)=  -0.34998000D 05     P(13,18, 9)=  -0.75050000D 04
P(13,19, 9)=  -0.13300000D 04     P(14, 1, 9)=   0.00000000D 00
P(19, 9,19)=   0.00000000D 00     P(19,10,19)=   0.00000000D 00
P(19,11,19)=   0.00000000D 00     P(19,12,19)=   0.00000000D 00
P(19,13,19)=   0.00000000D 00     P(19,14,19)=   0.00000000D 00
P(19,15,19)=   0.00000000D 00     P(19,16,19)=   0.00000000D 00
P(19,17,19)=   0.00000000D 00     P(19,18,19)=   0.00000000D 00
P(19,19,19)=   0.00000000D 00
```

(c)

```
AINVRS( 1, 1)=  0.19000000D 02     AINVRS( 1, 2)=  0.18000000D 02
AINVRS( 1, 3)=  0.17000000D 02     AINVRS( 1, 4)=  0.16000000D 02
AINVRS( 1, 5)=  0.15000000D 02     AINVRS( 1, 6)=  0.14000000D 02
AINVRS( 1, 7)=  0.13000000D 02     AINVRS( 1, 8)=  0.12000000D 02
AINVRS( 1, 9)=  0.11000000D 02     AINVRS( 1,10)=  0.10000000D 02
AINVRS( 1,11)=  0.90000000D 01     AINVRS( 1,12)=  0.80000000D 01
AINVRS( 9,11)=  0.90000000D 01     AINVRS( 9,12)=  0.80000000D 01
AINVRS( 9,13)=  0.70000000D 01     AINVRS( 9,14)=  0.60000000D 01
AINVRS( 9,15)=  0.50000000D 01     AINVRS( 9,16)=  0.40000000D 01
AINVRS( 9,17)=  0.30000000D 01     AINVRS( 9,18)=  0.20000000D 01
AINVRS( 9,19)=  0.10000000D 01     AINVRS(10, 1)=  0.10000000D 02
AINVRS(10, 2)=  0.10000000D 02     AINVRS(10, 3)=  0.10000000D 02
AINVRS(10, 4)=  0.10000000D 02     AINVRS(10, 5)=  0.10000000D 02
AINVRS(10, 6)=  0.10000000D 02     AINVRS(10, 7)=  0.10000000D 02
AINVRS(19, 1)=  0.10000000D 01     AINVRS(19, 2)=  0.10000000D 01
AINVRS(19, 3)=  0.10000000D 01     AINVRS(19, 4)=  0.10000000D 01
AINVRS(19, 5)=  0.10000000D 01     AINVRS(19, 6)=  0.10000000D 01
AINVRS(19, 7)=  0.10000000D 01     AINVRS(19, 8)=  0.10000000D 01
AINVRS(19, 9)=  0.10000000D 01     AINVRS(19,10)=  0.10000000D 01
AINVRS(19,11)=  0.10000000D 01     AINVRS(19,12)=  0.10000000D 01
AINVRS(19,13)=  0.10000000D 01     AINVRS(19,14)=  0.10000000D 01
AINVRS(19,15)=  0.10000000D 01     AINVRS(19,16)=  0.10000000D 01
AINVRS(19,17)=  0.10000000D 01     AINVRS(19,18)=  0.10000000D 01
AINVRS(19,19)=  0.10000000D 01
```

(d)

```
ALPHA( 1)=  -0.37000000D 02     ALPHA( 2)=  0.63000000D 03
ALPHA( 3)=  -0.65450000D 04     ALPHA( 4)=  0.46376000D 05
ALPHA( 5)=  -0.23733600D 06     ALPHA( 6)=  0.90619200D 06
ALPHA( 7)=  -0.26295750D 07     ALPHA( 8)=  0.58529250D 07
ALPHA( 9)=  -0.10015005D 08     ALPHA(10)=  0.13123110D 08
ALPHA(11)=  -0.13037895D 08     ALPHA(12)=  0.96577000D 07
ALPHA(13)=  -0.52003000D 07     ALPHA(14)=  0.19612560D 07
ALPHA(15)=  -0.49031400D 06     ALPHA(16)=  0.74613000D 05
ALPHA(17)=  -0.59850000D 04     ALPHA(18)=  0.19000000D 03
ALPHA(19)=  -0.10000000D 01
```

† The first row and first column element 19 denotes the order of the given matrix **A**, and therefore is not part of the matrix.

The main program is the same as that given in Figure 12.19. A part of the output showing the elements P(i, j, k) of $\mathbf{P}_k$ and AINVRS(i, j) of $\mathbf{A}^{-1}$ is presented in Figure 12.21 for i, j, $k = 1, 2, \ldots, 19$ together with the input data. Figure 12.21(a) shows the input matrix $\mathbf{A}$, and (b), (c), and (d) the elements of $\mathbf{P}_k$, $\mathbf{A}^{-1}$, and α_k, respectively. The total execution time for an IBM 370 Model 158 computer was 31.43 sec.

12-8.2   Subroutine EXPAT

In implementing the power-series method for evaluation of $e^{\mathbf{A}t}$ on a digital computer, two important points should be mentioned.

First of all, the bound for $|e_{ij}|$ given in (12.168) is wider than the information available would make necessary. A considerably sharper bound on $|e_{ij}|$ can be obtained as follows. From (12.166), we have

$$
\begin{aligned}
|e_{ij}| &\leq \frac{1}{k!}\left[\left\|(t_1\mathbf{A})^k\right\| + \frac{\left\|(t_1\mathbf{A})^{k+1}\right\|}{k+1} + \frac{\left\|(t_1\mathbf{A})^{k+2}\right\|}{(k+1)^2} + \cdots \right] \\
&\leqq \frac{\left\|(t_1\mathbf{A})^k\right\|}{k!}\left[1 + \frac{\left\|t_1\mathbf{A}\right\|}{k+1} + \frac{\left\|t_1\mathbf{A}\right\|^2}{(k+1)^2} + \cdots \right] \\
&= \frac{\left\|(t_1\mathbf{A})^k\right\|}{k!}\cdot\frac{1}{(1-\epsilon)} \leqq \frac{\left\|(t_1\mathbf{A})^{k-1}\right\|}{(k-1)!}\cdot\frac{\left\|t_1\mathbf{A}\right\|}{k}\cdot\frac{1}{(1-\epsilon)} \\
&= \left\|\frac{(t_1\mathbf{A})^{k-1}}{(k-1)!}\right\|\cdot\frac{\left\|t_1\mathbf{A}\right\|}{k}\cdot\frac{1}{[1-\left\|t_1\mathbf{A}\right\|/(k+1)]}, \qquad \left\|t_1\mathbf{A}\right\| < k+1
\end{aligned}
$$

$$(12.233)$$

Since the expression $[t_1^{k-1}/(k-1)!]\mathbf{A}^{k-1}$ is computed during iteration, use of its norm instead of $\left\|t_1\mathbf{A}\right\|^k/k!$, as in (12.168), is most convenient.

Secondly, the resulting approximate matrix (12.163) may contain zero elements. For such elements the comparison described in (12.169) of the iterative procedure will never be successful. For practical implementation, it is better to use the the average value of the magnitudes $|w_{ij}|$.

These considerations are included in the WATFIV subroutine EXPAT to be described below.

Subroutine EXPAT

1. Purpose.   The program is for the evaluation of the state-transition matrix $\phi(t) = e^{\mathbf{A}t}$ and is listed in Figure 12.22.

2. Method.   The program is an implementation of the power-series representation of the state-transition matrix, as described in (12.163).

Figure 12.22
Subroutine EXPAT used
for the evaluation of the
state-transition matrix.

```
C       SUBROUTINE EXPAT
C
C       PURPOSE
C           THE PROGRAM IS FOR THE EVALUATION OF THE STATE-TRANSITION
C               MATRIX EXP(AT).
C
C       METHOD
C           THE PROGRAM IS AN IMPLEMENTATION OF THE POWER-SERIES
C               REPRESENTATION OF THE STATE-TRANSITION MATRIX, AS DESCRIBED
C               IN (12.163).
C
C       USAGE
C           CALL EXPAT (N,A,PHI,K,ND,T)
C
C           N      -THE ORDER OF A
C           A      -A GIVEN REAL SQUARE MATRIX A
C           PHI    -THE STATE-TRANSITION MATRIX
C           K      -THE NUMBER OF TRUNCATION TERMS REQUIRED TO REACH AN
C                       ACCURACY OF ND SIGNIFICANT FIGURES
C           ND     -THE NUMBER OF SIGNIFICANT FIGURES REQUIRED IN THE
C                       COMPUTATION OF THE STATE-TRANSITION MATRIX
C           T      -THE TIME
C
C       SUBROUTINES AND FUNCTION SUBPROGRAMS REQUIRED
C
C           FUNCTION DABS (X)              -ABSOLUTE VALUE OF X
C           FUNCTION DMIN1 (X,Y,...)       -SMALLEST VALUE AMONG X,Y,...
C
C       REMARKS
C           DOUBLE PRECISION IS USED IN ALL THE COMPUTATIONS.
C           THE ORDER OF A CANNOT EXCEED TWENTY.  OTHERWISE, NEW
C               DIMENSIONAL STATEMENTS ARE REQUIRED.
C           THE INPUT DATA ARE N, A(I,J) OF A, ND AND T.  THE OUTPUTS ARE K
C               AND PHI(I,J) OF PHI.
        SUBROUTINE EXPAT (N,A,PHI,K,ND,T)
        DOUBLE PRECISION A(20,20),PHI(20,20),AK(20,20),F(20,20),
       CROWSUM(20),COLSUM(20),ROWMAX,COLMAX,ANORM,AKNORM,E,W,DABS,DMAX,
       CDAVG,DMIN1,T
C
C       SET AK AND PHI MATRICES TO THE IDENTITY MATRIX
C       COMPUTE ANORM
        DO 1 I=1,N
        ROWSUM(I)=0.0
        DO 1 J=1,N
        ROWSUM(I)=ROWSUM(I)+DABS(A(I,J))
        AK(I,J)=0.0
        IF (I.EQ.J) AK(I,J)=1.0
   1    PHI(I,J)=AK(I,J)
        ROWMAX=DMAX(N,ROWSUM)
        DO 2 J=1,N
        COLSUM(J)=0.0
        DO 2 I=1,N
   2    COLSUM(J)=COLSUM(J)+DABS(A(I,J))
        COLMAX=DMAX(N,COLSUM)
        ANORM=DMIN1(ROWMAX,COLMAX)*DABS(T)
C
C       COMPUTE PHI MATRIX
        K=1
   3    DO 5 I=1,N
        DO 5 J=1,N
        F(I,J)=0.0
        DO 4 M=1,N
   4    F(I,J)=F(I,J)+AK(I,M)*A(M,J)
        F(I,J)=F(I,J)*T/DFLOAT(K)
   5    PHI(I,J)=PHI(I,J)+F(I,J)
C
C       COMPUTE AKNORM
        DO 6 I=1,N
        DO 6 J=1,N
   6    AK(I,J)=F(I,J)
        K=K+1
        IF (K+1-ANORM) 3,3,7
   7    DO 8 I=1,N
        ROWSUM(I)=0.0
        DO 8 J=1,N
```

Figure 12.22 (continued)

```
      8   ROWSUM(I)=ROWSUM(I)+DABS(AK(I,J))
          ROWMAX=DMAX(N,ROWSUM)
          DO 9 J=1,N
          COLSUM(J)=0.0
          DO 9 I=1,N
      9   COLSUM(J)=COLSUM(J)+DABS(AK(I,J))
          COLMAX=DMAX(N,COLSUM)
          AKNORM=DMIN1(ROWMAX,COLMAX)
      C
      C   COMPUTE E AND W, CHECK THE ACCURACY OF PHI
          E=AKNORM*ANORM/(K*(1.0-ANORM/(K+1)))
          W=DAVG(N,N,PHI)
          IF (E-W*10.0**(-ND)) 10,10,3
     10   CONTINUE
          RETURN
          END
          FUNCTION DMAX (N,X)
      C
      C   SEARCH FOR THE MAXIMUM IN A ONE-DIMENSIONAL ARRAY OF NUMBERS
      C   N         -DIMENSION OF THE ARRAY
      C   X         -THE ONE-DIMENSIONAL ARRAY
          DOUBLE PRECISION X(20) DMAX
          DMAX=0.0
          DO 100 I=1,N
          IF (X(I).GT.DMAX) DMAX=X(I)
    100   CONTINUE
          RETURN
          END
          FUNCTION DAVG (N,M,X)
      C
      C   COMPUTE THE AVERAGE OF THE ABSOLUTE VALUES
      C     IN A TWO-DIMENSIONAL ARRAY OF NUMBERS
      C   N,M       -DIMENSIONS OF THE ARRAY
      C   X         -THE TWO-DIMENSIONAL ARRAY
          DOUBLE PRECISION X(20,20),DABS,DAVG
          DAVG=0.0
          DO 200 I=1,N
          DO 100 J=1,M
          DAVG=DAVG+DABS(X(I,J))
    100   CONTINUE
    200   CONTINUE
          DAVG=DAVG/(N*M)
          RETURN
          END
```

3. Usage. The program consists of a subroutine, EXPAT, called by

$$\text{CALL EXPAT (N, A, PHI, K, ND, T)}$$

N   is the order of **A**

A   is the name of a given real square matrix **A**

PHI is the name of the state-transition matrix $\phi(t)$

K   is the number of truncation terms required to reach an accuracy of ND significant figures

ND   is the number of significant figures required in tne computation of the state-transition matrix $\phi(t)$

T   is the time t

4. Remarks. Double precision is used in all the computations. The order of **A** cannot exceed 20. Otherwise new dimensional statements are required. The input data are N, A(I, J) of A, ND, and T. The outputs are K and PHI(I, J)

Figure 12.23
The main program used for the evaluation of the state-transition matrix, using subroutine EXPAT.

```
      DOUBLE PRECISION A(20,20),PHI(20,20),T
      READ,N,((A(I,J),J=1,N),I=1,N),ND,T
      CALL EXPAT (N,A,PHI,K,ND,T)
      PRINT 1,K,ND,T
1     FORMAT ('-THE SERIES IS TRUNCATED AT THE',I3,'TH TERM WITH AN ',
     C'ACCURACY OF AT LEAST',/I3,' SIGNIFICANT FIGURES AT T=',
     CD15.8,' SEC'/)
      PRINT 2,((I,J,PHI(I,J),J=1,N),I=1,N)
2     FORMAT (2(' PHI(',I2,',',I2,')=',D20.12,10X))
      STOP
      END
```

Figure 12.24
Computer printout of the elements of the state-transition matrix to within an accuracy of 10 significant figures for the companion matrix (12.234).

```
THE SERIES IS TRUNCATED AT THE   7TH TERM WITH AN ACCURACY OF AT LEAST
10 SIGNIFICANT FIGURES AT T= 0.10000000D-01 SEC

PHI( 1, 1)=  0.999999014876D 00        PHI( 1, 2)=  0.999819146867D-02
PHI( 1, 3)=  0.490103420847D-04        PHI( 2, 1)= -0.294062052508D-03
PHI( 2, 2)=  0.999459901113D 00        PHI( 2, 3)=  0.970412941616D-02
PHI( 3, 1)= -0.582247764970D-01        PHI( 3, 2)= -0.107039485630D 00
PHI( 3, 3)=  0.941235124616D 00
```

of PHI. All function subprograms required are provided except those listed below, which are usually available to the user.

5. Subroutine and Function Subprograms Required.

Function DABS (X)—determining the absolute value of X

Function DMINI (X, Y, . . .)—selecting the smallest number among the real numbers X, Y, . . .

EXAMPLE 12.25
Consider the companion matrix

$$\mathbf{A} = \begin{bmatrix} 0 & 1 & 0 \\ 0 & 0 & 1 \\ -6 & -11 & -6 \end{bmatrix} \qquad (12.234)$$

of a third-order differential equation. We shall use the subroutine EXPAT to compute the state-transition matrix $\phi(t)$ at $t = 0.01$ sec to within an accuracy of at least ten significant figures.

The main program and the computer outputs are presented in Figures 12.23 and 12.24, respectively. In Figure 12.24, PHI(i, j) is the ith row and jth column element of $\phi(0.01)$. The desired accuracy is attained after the series (12.163) is truncated at the seventh term. The reader can verify this by comparing this solution with the exact solution of (12.178). The total execution time for an IBM 370 computer was 0.09 sec.

12-8.3   Subroutine STATE

1. Purpose.   The program, as listed in Figure 12.25, is for the solution of a linear state equation in normal form.

Figure 12.25
Subroutine STATE used
for the solution of a linear
state equation in normal
form.

```
C
C
C      SUBROUTINE STATE
C
C      PURPOSE
C          THE PROGRAM IS FOR THE SOLUTION OF A LINEAR STATE EQUATION IN NORMAL
C          FORM.
C
C      METHOD
C          THE PROGRAM IS BASED ON RUNGE-KUTTA METHOD FOR THE SOLUTION OF A SYSTEM
C          OF ORDINARY DIFFERENTIAL EQUATIONS.
C
C      USAGE
C          CALL STATE (A,B,N,NH,XI,X,TI,TF,H)
C
C          A      -THE COEFFICIENT MATRIX A
C          B      -THE COEFFICIENT MATRIX B
C          N      -THE ORDER OF THE MATRIX A
C          NH     -THE DIMENSION OF THE INPUT VECTOR U
C          XI     -THE STATE VECTOR AT TIME TI
C          X      -THE STATE VECTOR AT TIME TF
C          TI     -THE INITIAL TIME
C          TF     -THE FINAL TIME
C          H      -THE STEP SIZE
C
C      SUBROUTINES AND FUNCTION SUBPROGRAMS REQUIRED
C          SUBROUTINE FUNCTU MUST BE PROVIDED BY THE USER.
C
C      REMARKS
C          DOUBLE PRECISION IS USED IN ALL THE COMPUTATIONS.
C
C          THE ORDER OF A AND THE DIMENSION OF U CANNOT EXCEED TWENTY.
C          OTHERWISE, NEW DIMENSIONAL STATEMENTS ARE REQUIRED.
C
C          THE INPUT DATA ARE A, B, N, NH, XI, TI, TF, AND H.  THE OUTPUT IS X.
C
       SUBROUTINE STATE (A,B,N,NH,XI,X,TI,TF,H)
       DOUBLE PRECISION A(20,20),B(20,20),U(20),XI(20),X(20),TI,T,TF,H,
      C  AX(20),BU(20),DKO(20),DK1(20),DK2(20),DK3(20),AK
       T=TI
       CALL FUNCTU(T,U)
       DO 2 I=1,N
       BU(I)=0.0
       DO 1 J=1,NH
 1     BU(I)=BU(I)+B(I,J)*U(J)
 2     X(I)=XI(I)
 3     DO 5 I=1,N
       AX(I)=0.0
       DO 4 J=1,N
 4     AX(I)=AX(I)+A(I,J)*X(J)
 5     DKO(I)=H*(AX(I)+BU(I))
       T=T+H/2.0
       CALL FUNCTU(T,U)
       DO 8 I=1,N
       AK=0.0
       DO 6 J=1,N
 6     AK=AK+A(I,J)*DKO(J)
       BU(I)=0.0
       DO 7 J=1,N
 7     BU(I)=BU(I)+B(I,J)*U(J)
 8     DK1(I)=H*(AX(I)+AK/2.0+BU(I))
       DO 10 I=1,N
       AK=0.0
       DO 9 J=1,N
 9     AK=AK+A(I,J)*DK1(J)
 10    DK2(I)=H*(AX(I)+AK/2.0+BU(I))
       T=T+H/2.0
       CALL FUNCTU(T,U)
       DO 13 I=1,N
       AK=0.0
       DO 11 J=1,N
 11    AK=AK+A(I,J)*DK2(J)
       BU(I)=0.0
       DO 12 J=1,NH
 12    BU(I)=BU(I)+B(I,J)*U(J)
 13    DK3(I)=H*(AX(I)+AK+BU(I))
       DO 14 I=1,N
 14    X(I)=X(I)+(DKO(I)+2.0*DK1(I)+2.0*DK2(I)+DK3(I))/6.0
       TT=(T-TF)/H
       IF (INT(TT)) 3,15,15
 15    TF=T
       RETURN
       END
```

2. Method. The program is based on Runge-Kutta methods[†] for the solution of a system of ordinary differential equations.

3. Usage. The program consists of a subroutine, STATE, called by

$$\text{CALL STATE (A, B, N, NH, XI, X, TI, TF, H)}$$

A   is the name of the coefficient matrix **A**
B   is the name of the coefficient matrix **B**
N   is the order of the matrix **A**
NH   is the dimension of the input vector **u**(t)
XI   is the state vector at time TI
X   is the state vector at time TF
TI   is the initial time
TF   is the final time
H   is the stepsize

4. Remarks. Double precision is used in all the computations. The order of **A** and the dimension of **u**(t) cannot exceed 20. Otherwise new dimensional statements are required. The input data are A, B, N, NH, XI, TI, TF, and H. The output is X.

5. Subroutines and Function Subprograms Required. Subroutine FUNCTU, which gives the input vector **u**(t), must be provided by the user.

EXAMPLE 12.26
We shall use the subroutine STATE to compute the state vector $x(t)$ for the state equation

$$\begin{bmatrix} \dot{i} \\ \dot{v} \end{bmatrix} = \begin{bmatrix} -5 & 2 \\ -1 & -2 \end{bmatrix} \begin{bmatrix} i \\ v \end{bmatrix} + \begin{bmatrix} -2 & 0 \\ 0 & 1 \end{bmatrix} \begin{bmatrix} 2\cos\left(2t + \dfrac{\pi}{4}\right) \\ \cos 2t \end{bmatrix}, \quad t \geq 0 \qquad \textbf{(12.235a)}$$

which were considered in Example 12.9, with the initial state

$$\mathbf{x}(0) = \begin{bmatrix} i(0) \\ v(0) \end{bmatrix} = \begin{bmatrix} 0 \\ 0 \end{bmatrix} \qquad \textbf{(12.235b)}$$

Choose the stepsize H = 0.001 and let $t = 0, 0.05, 0.1, \ldots, 5$. The main program and the required subroutine FUNCTU are shown in Figures 12.26 and 12.27. The computer printout of the inductor current $i(t) = X1$ and capacitor voltage $v(t) = X2$ is presented in Figure 12.28 for only half of the terms that were computed, to save space. The plots of $i(t)$ and $v(t)$ as functions

---

[†] See, for example, D. D. McCracken and W. S. Dorn, *Numerical Methods and FORTRAN Programming* (New York: Wiley, 1964).

Figure 12.26

The main program used
for the solution of a
linear state equation in
normal form, using
subroutine STATE.

```
C
C       THE VARIABLES OF THE PROGRAM ARE DEFINED AS FOLLOWS.
C
C       NCHOCE         0-PLOT THE STATE VECTOR
C                      1-PLOT AND PRINT THE STATE VECTOR
C                      2-PRINT THE STATE VECTOR
C       N             -THE ORDER OF THE A-MATRIX
C       NH            -THE NUMBER OF ROWS OF THE INPUT VECTOR
C       A             -THE NAME OF THE A-MATRIX
C       B             -THE NAME OF THE B-MATRIX
C       XO            -THE NAME OF THE INITIAL VECTOR
C       T             -THE NAME OF A ONE-DIMENSIONAL STORAGE ARRAY OF
C                         DIMENSION 200 FOR TIME
C       TO            -THE INITIAL TIME
C       TF            -THE FINAL TIME
C       DT            -THE TIME INCREMENT
C       H             -THE STEP SIZE
C
        DOUBLE PRECISION A(20,20),B(20,20),XO(20),T(200),X(20,200),TO,TF,
      C DT,H
        DIMENSION JXY(2),XY(200,2)
        READ,NCHOCE,N,NH,((A(I,J),J=1,N),I=1,N),((B(I,J),J=1,NH),I=1,N),
      C (XO(I),I=1,N),TO,TF,DT,H
        IT=1
        T(1)=TO
        DO 1 I=1,N
   1    X(I,1)=XO(I)
   2    IF (NCHOCE-1) 5,3,3
   3    PRINT 4,T(IT),(I,X(I,IT),I=1,N)
   4    FORMAT (3H-T=,D15.8,4H SEC/2(5H     X,I2,2H =,D20.12,6X))
   5    TT=(T(IT)-TF)/DT
        IF (INT(TT)) 6,7,7
   6    ITM1=IT
        IT=IT+1
        T(IT)=T(ITM1)+DT
        CALL STATE (A,B,N,NH,X(1,ITM1),X(1,IT),T(ITM1),T(IT),H)
        GO TO 2
   7    IF(NCHOCE-1) 8,8,14
   8    JXY(1)=1
        JXY(2)=2
        DO 9 J=1,IT
   9    XY(J,1)=T(J)
        DO 11 I=1,N
        DO 10 J=1,IT
  10    XY(J,2)=X(I,J)
        CALL PLOT (XY,JXY,IT,200,1)
  11    PRINT 12,I
  12    FORMAT (1H-,60X,7H CHART ,I3)
        DO 13 I=1,2
        DO 13 J=1,IT
  13    XY(J,I)=X(I,J)
        CALL PLOT (XY,JXY,IT,200,1)
  14    CONTINUE
        STOP
        END
```

Figure 12.27

Subroutine FUNCTU
which gives the input
vector to the main
program of Figure 12.26.

```
        SUBROUTINE FUNCTU (T,U)
C
C       EVALUATE THE INPUT VECTOR
C       T         -TIME
C       U         -THE INPUT VECTOR OF DIMENSION 20
        DOUBLE PRECISION U(20),T,T1,T2
        T2=2*T
        T1=T2+0.78539816339745
        U(1)=2*DCOS(T1)
        U(2)=DCOS(T2)
        RETURN
        END
```

Figure 12.28
Computer printout of the inductor current $i(t) = X1$ and capacitor voltage $v(t) = X2$ for t from 0 to 5 with increments of 0.1 second.

```
T= 0.00000000D 00 SEC
    X 1 =  0.000000000000D 00        X 2 =  0.000000000000D 00
T= 0.10000000D 00 SEC
    X 1 = -0.188310932804D 00        X 2 =  0.100141781935D 00
T= 0.20000000D 00 SEC
    X 1 = -0.234987922879D 00        X 2 =  0.188434077784D 00
T= 0.30000000D 00 SEC
    X 1 = -0.192952459106D 00        X 2 =  0.253383788511D 00
T= 0.40000000D 00 SEC
    X 1 = -0.978642033684D-01        X 2 =  0.289794871477D 00
T= 0.50000000D 00 SEC
    X 1 =  0.257341892296D-01        X 2 =  0.296496669856D 00
T= 0.60000000D 00 SEC
    X 1 =  0.160507476703D 00        X 2 =  0.274912896880D 00
T= 0.70000000D 00 SEC
    X 1 =  0.293878634803D 00        X 2 =  0.228150603232D 00
T= 0.80000000D 00 SEC
    X 1 =  0.416526195132D 00        X 2 =  0.160404607687D 00
T= 0.90000000D 00 SEC
    X 1 =  0.521487270632D 00        X 2 =  0.765488472927D-01
T= 0.10000000D 01 SEC
    X 1 =  0.603640175949D 00        X 2 = -0.181643621673D-01
T= 0.11000000D 01 SEC
    X 1 =  0.659416767297D 00        X 2 = -0.118344024892D 00
T= 0.12000000D 01 SEC
    X 1 =  0.686645400405D 00        X 2 = -0.218665300152D 00
T= 0.13000000D 01 SEC
    X 1 =  0.684459407037D 00        X 2 = -0.314052503637D 00
T= 0.14000000D 01 SEC
    X 1 =  0.653228946824D 00        X 2 = -0.399845145252D 00
T= 0.15000000D 01 SEC
    X 1 =  0.594489768868D 00        X 2 = -0.471947026117D 00
T= 0.16000000D 01 SEC
    X 1 =  0.510853282404D 00        X 2 = -0.526955782290D 00
T= 0.17000000D 01 SEC
    X 1 =  0.405889990816D 00        X 2 = -0.562269113651D 00
T= 0.18000000D 01 SEC
    X 1 =  0.283983821823D 00        X 2 = -0.576163788225D 00
T= 0.19000000D 01 SEC
    X 1 =  0.150158848052D 00        X 2 = -0.567844034664D 00
T= 0.20000000D 01 SEC
    X 1 =  0.988275248086D-02        X 2 = -0.537456896797D 00
T= 0.21000000D 01 SEC
    X 1 = -0.131146587727D 00        X 2 = -0.486073354940D 00
T= 0.22000000D 01 SEC
    X 1 = -0.267223685273D 00        X 2 = -0.415635390485D 00
T= 0.23000000D 01 SEC
    X 1 = -0.392858716279D 00        X 2 = -0.328870580643D 00
T= 0.24000000D 01 SEC
    X 1 = -0.502992745458D 00        X 2 = -0.229177175661D 00
T= 0.25000000D 01 SEC
    X 1 = -0.593196356713D 00        X 2 = -0.120483861398D 00
T= 0.26000000D 01 SEC
    X 1 = -0.659843737481D 00        X 2 = -0.708948848560D-02
T= 0.27000000D 01 SEC
    X 1 = -0.700255213239D 00        X 2 =  0.106511091364D 00
T= 0.28000000D 01 SEC
    X 1 = -0.712802476642D 00        X 2 =  0.215808335985D 00
T= 0.29000000D 01 SEC
    X 1 = -0.696972244456D 00        X 2 =  0.316459397099D 00
T= 0.30000000D 01 SEC
    X 1 = -0.653385739597D 00        X 2 =  0.404462467892D 00
T= 0.31000000D 01 SEC
    X 1 = -0.583773165672D 00        X 2 =  0.476317236516D 00
T= 0.32000000D 01 SEC
    X 1 = -0.490904144912D 00        X 2 =  0.529165120571D 00
T= 0.33000000D 01 SEC
    X 1 = -0.378476854574D 00        X 2 =  0.560903746832D 00
T= 0.34000000D 01 SEC
    X 1 = -0.250970250960D 00        X 2 =  0.570271154191D 00
T= 0.35000000D 01 SEC
    X 1 = -0.113465248131D 00        X 2 =  0.556896394765D 00
```

Figure 12.28   (*continued*)

```
T= 0.36000000D 01 SEC
  X 1 =  0.285580387080D-01        X 2 =  0.521314539964D 00
T= 0.37000000D 01 SEC
  X 1 =  0.169438916705D 00        X 2 =  0.464945511536D 00
T= 0.38000000D 01 SEC
  X 1 =  0.303561901389D 00        X 2 =  0.390037595245D 00
T= 0.39000000D 01 SEC
  X 1 =  0.425580672666D 00        X 2 =  0.299577899471D 00
T= 0.40000000D 01 SEC
  X 1 =  0.530631279099D 00        X 2 =  0.197173336237D 00
T= 0.41000000D 01 SEC
  X 1 =  0.614526095858D 00        X 2 =  0.869068753868D-01
T= 0.42000000D 01 SEC
  X 1 =  0.673920807727D 00        X 2 = -0.268251930104D-01
T= 0.43000000D 01 SEC
  X 1 =  0.706447763056D 00        X 2 = -0.139488496941D 00
T= 0.44000000D 01 SEC
  X 1 =  0.710810384489D 00        X 2 = -0.246591332766D 00
T= 0.45000000D 01 SEC
  X 1 =  0.686834874300D 00        X 2 = -0.343863719815D 00
T= 0.46000000D 01 SEC
  X 1 =  0.635477154261D 00        X 2 = -0.427427619550D 00
T= 0.47000000D 01 SEC
  X 1 =  0.558784764346D 00        X 2 = -0.493951532183D 00
T= 0.48000000D 01 SEC
  X 1 =  0.459815239939D 00        X 2 = -0.540783306715D 00
T= 0.49000000D 01 SEC
  X 1 =  0.342514222149D 00        X 2 = -0.566055869117D 00
T= 0.50000000D 01 SEC
  X 1 =  0.211558160972D 00        X 2 = -0.568761653179D 00
```

of t and also the phase-plane trajectory are displayed in Figure 12.29. They are to be compared with those given in Figures 12.11 and 12.15. The total execution time for an IBM 370 Model 158 computer was 32.47 sec.

12-8.4   Subroutine STATE2

1. Purpose. The program, as listed in Figure 12.30, is for the solution of a linear state equation in normal form.

2. Method. The program is based on the general solution

$$\mathbf{x}(t) = \boldsymbol{\phi}(t - t_0)\mathbf{x}(t_0) + \int_{t_0}^{t} \boldsymbol{\phi}(t - \tau)\mathbf{B}\mathbf{u}(\tau)\, d\tau \qquad (12.236)$$

of the state equation (12.1) using the quadrature integration subroutine DQUAD, which takes a five-point formula used over successive sets of points as the basis of the quadrature.

3. Usage. The program consists of a subroutine, STATE2, called by

CALL STATE2 (N, NH, B, XO, TO, TF, DTAU, PHIT, PHITAU, X)

N	is the order of the matrix **A**
NH	is the dimension of the input vector $\mathbf{u}(t)$
B	is the name of the matrix **B**
XO	is the name of the initial state vector at TO
TO	is the initial time
TF	is the final time

Figure 12.29

(a) Computer plot of the inductor current as a function of t. (b) Computer plot of the capacitor voltage as a function of t. (c) Computer plot of a phase-plane trajectory for the inductor current and capacitor voltage of the state equation (12.235a).

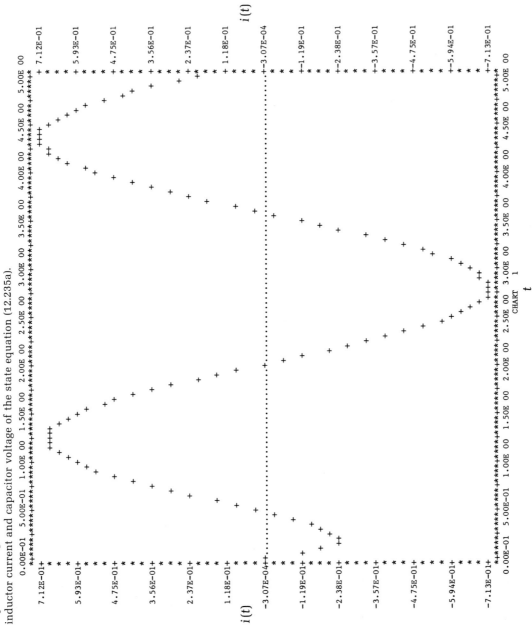

(a)

Figure 12.29 (continued)

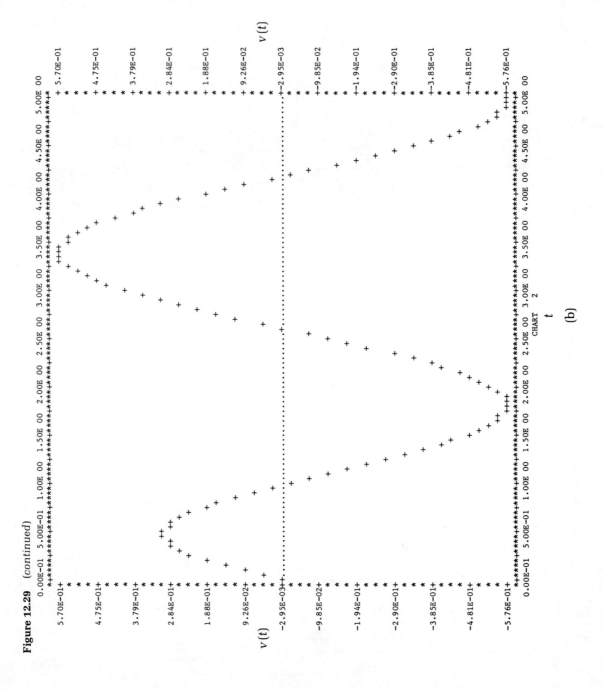

(b)

666

Figure 12.29 *(continued)*

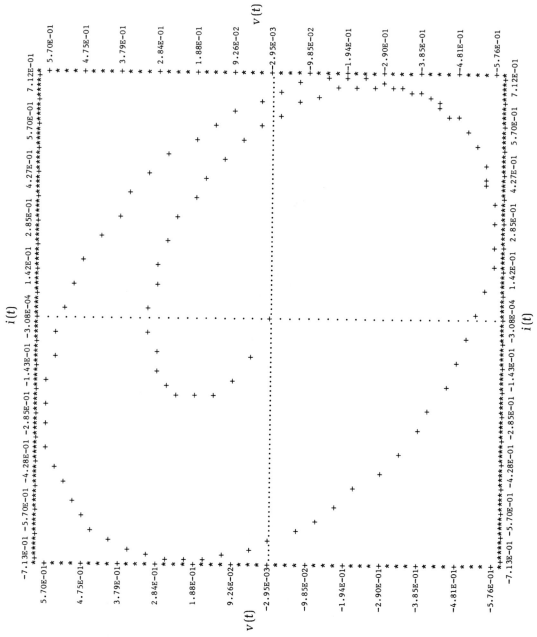

(c)

Figure 12.30
Subroutine STATE2 used
for the solution of a linear
state equation in normal
form.

```
C
C
C       SUBROUTINE STATE2
C
C       PURPOSE
C           THE PROGRAM IS FOR THE SOLUTION OF A LINEAR STATE EQUATION IN
C               NORMAL FORM.
C
C       METHOD
C           THE PROGRAM IS BASED ON THE GENERAL SOLUTION OF THE STATE
C               EQUATION USING THE QUADRATURE INTEGRATION SUBROUTINE DQUAD,
C               WHICH TAKES A FIVE-POINT FORMULA USED OVER SUCCESSIVE SETS
C               OF POINTS AS THE BASIS OF THE QUADRATURE.
C
C       USAGE
C           CALL STATE2 (N,NH,B,XO,TO,TF,DTAU,PHIT,PHITAU,X)
C
C           N       -THE ORDER OF THE A-MATRIX
C           NH      -THE DIMENSION OF THE INPUT VECTOR U
C           B       -THE NAME OF THE B-MATRIX
C           XO      -THE NAME OF THE INITIAL STATE VECTOR AT TO
C           TO      -THE INITIAL TIME
C           TF      -THE FINAL TIME
C           DTAU    -THE INCREMENT FOR INTEGRAGION
C           PHIT    -THE STATE-TRANSITION MATRIX EVALUATED AT (TF-TO)
C           PHITAU  -THE STATE-TRANSITION MATRIX EVALUATED AT -DTAU
C           X       -THE STATE VECTOR AT TF
C
C       SUBROUTINES AND FUNCTION SUBPROGRAMS REQUIRED
C           SUBROUTINE FUNCTU   -EVALUATION OF THE INPUT VECTOR, SUPPLIED
C                               BY USER
C           SUBROUTINE DQUAD    -QUADRATURE INTEGRATION
C
C       REMARKS
C           DOUBLE PRECISION IS USED IN ALL THE COMPUTATIONS.
C           THE ORDER OF A-MATRIX CANNOT EXCEED 20, AND THE NUMBER OF
C               INPUTS CANNOT EXCEED 10.  THE RATIO OF DT TO DTAU CANNOT
C               EXCEED 200.  OTHERWISE, NEW DIMENSIONAL STATEMENTS ARE
C               REQUIRED.
C           THE INPUT DATA ARE N, NH, B, XO, TO, TF, DTAU, PHIT, AND PHITAU.
C               THE OUTPUT IS X.
        SUBROUTINE STATE2 (N,NH,B,XO,TO,TF,DTAU,PHIT,PHITAU,X)
        DOUBLE PRECISION B(20,10),XO(10),TAU(200),PHIT(20,20),PHITAU
     C  (20,20),X(10),PHIB(20,10),PHIBU(200,10),U(10),P(20,10),TO,TF,
     C  DTAU,Z
        L=1
        TAU(1)=TO
        DO 1 I=1,N
        DO 1 J=1,NH
    1   PHIB(I,J)=B(I,J)
    2   CALL FUNCTU(TAU(L),U)
        DO 3 I=1,N
        PHIBU(L,I)=0.0
        DO 3 J=1,NH
    3   PHIBU(L,I)=PHIBU(L,I)+PHIB(I,J)*U(J)
        DO 5 I=1,N
        DO 5 J=1,NH
        P(I,J)=0.0
        DO 4 M=1,N
    4   P(I,J)=P(I,J)+PHITAU(I,M)*PHIB(M,J)
    5   CONTINUE
        DO 6 I=1,N
        DO 6 J=1,NH
    6   PHIB(I,J)=P(I,J)
        L=L+1
        TAU(L)=TAU(L-1)+DTAU
        IF(TAU(L)-TF) 2,2,7
    7   NTAU=L-1
        DO 8 I=1,N
        CALL DQUAD (TAU,PHIBU(1,I),Z,NTAU)
    8   P(I,1)=XO(I)+Z
        DO 9 I=1,N
        X(I)=0.0
        DO 9 J=1,N
    9   X(I)=X(I)+PHIT(I,J)*P(J,1)
        RETURN
        END
```

Figure 12.30   (*continued*)

```
      SUBROUTINE DQUAD (X,Y,Z,N)
C
C     INTEGRATE A TABULATED FUNCTION BASED ON A FIVE-POINT FORMULA, USED OVER
C        SUCCESSIVE SETS OF POINTS AS THE BASIS OF THE QUADRATURE
C
C     USAGE
C        CALL DQUAD (X,Y,Z,N)
C
C        X       -A VECTOR OF LENGTH N CONTAINING THE VALUES OF THE
C                    INDEPENDENT VARIABLE
C        Y       -A VECTOR OF LENGTH N CONTAINING THE VALUES OF THE
C                    FUNCTION TO BE INTEGRATED
C        Z       -THE RESULTANT VALUE OF THE COMPUTED INTEGRAL
C        N       -THE NUMBER OF FUNCTION VALUES TO BE INTEGRATED
C
      DOUBLE PRECISION.X(N),Y(N),Z
      Z=0.00
      IF (N-1) 3,3,1
    1 DO 2 I=2,N
    2 Z=Z+0.500*(X(I)-X(I-1))*(Y(I)+Y(I-1))
    3 RETURN
      END
```

DTAU   is the increment for integration
PHIT   is the state-transition matrix evaluated at (TF − TO)
PHITAU   is the state-transition matrix evaluated at −DTAU
X   is the state vector at TF

4. Remarks.   Double precision is used in all the computations. The order of A-matrix cannot exceed 20, and the number of inputs cannot exceed 10. The ratio of DT to DTAU cannot exceed 200. Otherwise new dimensional statements are required. The input data are N, NH, B, XO, TO, TF, DTAU, PHIT, and PHITAU. The output is X.

5. Subroutines and Function Subprograms Required.   Subroutine FUNCTU for the evaluation of the input vector is supplied by the user. Subroutine DQUAD for quadrature integration is provided.

EXAMPLE 12.27

We use the subroutine STATE2 to compute the state vector $\mathbf{x}(t)$ of the state equation (12.235), as previously considered in Example 12.26.

For our purposes we choose the increment for integration DTAU = 0.001 and let t vary from 0 to 5 with time increment 0.05. We also require that the state-transition matrix $\boldsymbol{\phi}(t)$ be accurate to at least 10 significant figures. The main program is listed in Figure 12.31 and the subroutine FUNCTU was given earlier in Figure 12.27. The computer printout is presented in Figure 12.32, which give the inductor current $i(t) = X1$ and capacitor voltage $v(t) = X2$ as functions of t. The plots of $i(t)$ and $v(t)$ as functions of t and the phase-plane trajectory are the same as those given in Figure 12.29 and are omitted here. The total execution time for an IBM 370 Model 158 computer was 22.66 sec.

Figure 12.31
The main program used for the solution of a linear state equation in normal form using subroutine STATE2.

```
      DOUBLE PRECISION A(20,20),B(20,10),XO(10),T(200),PHIT(20,20),
     C   PHITAU(20,20),X(10,200),TO,TF,DT,DTAU,DMTAU
      DIMENSION JXY(2),AX(200,2)
      READ,NCHOCE,N,NH,((A(I,J),J=1,N),I=1,N),((B(I,J),J=1,NH),I=1,N),
     C   (XO(I),I=1,N),TO,TF,DT,DTAU,ND
      CALL EXPAT (N,A,PHIT,K,ND,DT)
      DMTAU=-DTAU
      CALL EXPAT (N,A,PHITAU,K,ND,DMTAU)
      IT=1
      T(1)=TO
      DO 1 I=1,N
1     X(I,1)=XO(I)
2     IF (NCHOCE-1) 6,3,3
3     PRINT 4,T(IT)
4     FORMAT (3H T=,D15.8,4H SEC)
      PRINT 5,(I,X(I,IT),I=1,N)
5     FORMAT (2(5H   X,I2,2H =,D20.12,10X))
6     IT=IT+1
      T(IT)=T(IT-1)+DT
      IF (T(IT)-TF) 7,7,8
7     CALL STATE2 (N,NH,B,X(1,IT-1),T(IT-1),T(IT),DTAU,PHIT,PHITAU,X(1,
     C   IT))
      GO TO 2
8     IF (NCHOCE-1) 9,9,15
9     NT=IT-1
      JXY(1)=1
      JXY(2)=2
      DO 10 IT=1,NT
10    AX(IT,1)=T(IT)
      DO 12 I=1,N
      DO 11 IT=1,NT
11    AX(IT,2)=X(I,IT)
      CALL PLOT (AX,JXY,NT,200,1)
12    PRINT 13,I
13    FORMAT (1H-,60X,6H CHART,I4)
      DO 14  I=1,2
      DO 14 IT=1,NT
14    AX(IT,I)=X(I,IT)
      CALL PLOT (AX,JXY,NT,200,1)
15    CONTINUE
      STOP
      END
```

Figure 12.32
Computer printout of the inductor current $i(t) = X1$ and capacitor voltage $v(t) = X2$ for t from 0 to 5 with increments of 0.05 sec.

```
T= 0.00000000D 00 SEC
   X 1 =   0.000000000000D 00          X 2 =   0.000000000000D 00
T= 0.50000000D-01 SEC
   X 1 = -0.116072723515D 00          X 2 =   0.505010803453D-01
T= 0.10000000D 00 SEC
   X 1 = -0.188310927410D 00          X 2 =   0.100141722445D 00
T= 0.15000000D 00 SEC
   X 1 = -0.225549695854D 00          X 2 =   0.146669194800D 00
T= 0.20000000D 00 SEC
   X 1 = -0.234987765577D 00          X 2 =   0.188433998972D 00
T= 0.25000000D 00 SEC
   X 1 = -0.222510944541D 00          X 2 =   0.224267305503D 00
T= 0.30000000D 00 SEC
   X 1 = -0.192952062826D 00          X 2 =   0.253383713839D 00
T= 0.35000000D 00 SEC
   X 1 = -0.150299426694D 00          X 2 =   0.275304161219D 00
T= 0.40000000D 00 SEC
   X 1 = -0.978635250559D-01          X 2 =   0.289794813830D 00
T= 0.45000000D 00 SEC
   X 1 = -0.384099277257D-01          X 2 =   0.296818590994D 00
T= 0.50000000D 00 SEC
   X 1 =   0.257351592313D-01          X 2 =   0.296496635419D 00
T= 0.55000000D 00 SEC
   X 1 =   0.926014160719D-01          X 2 =   0.289077578984D 00
T= 0.60000000D 00 SEC
   X 1 =   0.160508722284D 00          X 2 =   0.274912887595D 00
T= 0.65000000D 00 SEC
   X 1 =   0.228015767992D 00          X 2 =   0.254436919885D 00
T= 0.70000000D 00 SEC
   X 1 =   0.293880119919D 00          X 2 =   0.228150618337D 00
T= 0.75000000D 00 SEC
   X 1 =   0.357027425519D 00          X 2 =   0.196607980725D 00
```

Figure 12.32 (*continued*)

```
T= 0.80000000D 00 SEC
   X 1 =  0.416527868581D 00          X 2 =  0.160404644705D 00
T= 0.85000000D 00 SEC
   X 1 =  0.471578339681D 00          X 2 =  0.120168067515D 00
T= 0.90000000D 00 SEC
   X 1 =  0.521489070238D 00          X 2 =  0.765489026615D-01
T= 0.95000000D 00 SEC
   X 1 =  0.565673711809D 00          X 2 =  0.302132718191D-01
T= 0.10000000D 01 SEC
   X 1 =  0.603642032394D 00          X 2 = -0.181642926687D-01
T= 0.10500000D 01 SEC
   X 1 =  0.634994557073D 00          X 2 = -0.679073977065D-01
T= 0.11000000D 01 SEC
   X 1 =  0.659418607743D 00          X 2 = -0.118343945838D 00
T= 0.11500000D 01 SEC
   X 1 =  0.676685301318D 00          X 2 = -0.168811973503D 00
T= 0.12000000D 01 SEC
   X 1 =  0.686647151928D 00          X 2 = -0.218665216242D 00
T= 0.12500000D 01 SEC
   X 1 =  0.689235993724D 00          X 2 = -0.267278422467D 00
T= 0.13000000D 01 SEC
   X 1 =  0.684460999860D 00          X 2 = -0.314052419501D 00
T= 0.13500000D 01 SEC
   X 1 =  0.672406622206D 00          X 2 = -0.358418922257D 00
T= 0.14000000D 01 SEC
   X 1 =  0.653230317296D 00          X 2 = -0.399845065293D 00
T= 0.14500000D 01 SEC
   X 1 =  0.627159958343D 00          X 2 = -0.437837632470D 00
T= 0.15000000D 01 SEC
   X 1 =  0.594490862114D 00          X 2 = -0.471946954360D 00
T= 0.15500000D 01 SEC
   X 1 =  0.555582383892D 00          X 2 = -0.501770441618D 00
T= 0.16000000D 01 SEC
   X 1 =  0.510854054579D 00          X 2 = -0.526955722260D 00
T= 0.16500000D 01 SEC
   X 1 =  0.460781251668D 00          X 2 = -0.547203351898D 00
T= 0.17000000D 01 SEC
   X 1 =  0.405890410886D 00          X 2 = -0.562269068262D 00
T= 0.17500000D 01 SEC
   X 1 =  0.346753798208D 00          X 2 = -0.571965564502D 00
T= 0.18000000D 01 SEC
   X 1 =  0.283983872817D 00          X 2 = -0.576163759690D 00
T= 0.18500000D 01 SEC
   X 1 =  0.218227280888D 00          X 2 = -0.574793549436D 00
T= 0.19000000D 01 SEC
   X 1 =  0.150158527747D 00          X 2 = -0.567844024435D 00
T= 0.19500000D 01 SEC
   X 1 =  0.804733823967D-01          X 2 = -0.555363150059D 00
T= 0.20000000D 01 SEC
   X 1 =  0.988207348718D-02          X 2 = -0.537456905520D 00
T= 0.20500000D 01 SEC
   X 1 = -0.608976601610D-01          X 2 = -0.514287886712D 00
T= 0.21000000D 01 SEC
   X 1 = -0.131147598469D 00          X 2 = -0.486073382452D 00
T= 0.21500000D 01 SEC
   X 1 = -0.200156080517D 00          X 2 = -0.453082939330D 00
T= 0.22000000D 01 SEC
   X 1 = -0.267224987571D 00          X 2 = -0.415635435827D 00
T= 0.22500000D 01 SEC
   X 1 = -0.331676599770D 00          X 2 = -0.374095691510D 00
T= 0.23000000D 01 SEC
   X 1 = -0.392860258297D 00          X 2 = -0.328870642115D 00
T= 0.23500000D 01 SEC
   X 1 = -0.450158766504D 00          X 2 = -0.280405115893D 00
T= 0.24000000D 01 SEC
   X 1 = -0.502994465784D 00          X 2 = -0.229177250892D 00
T= 0.24500000D 01 SEC
   X 1 = -0.550834925096D 00          X 2 = -0.175693596690D 00
T= 0.25000000D 01 SEC
   X 1 = -0.593198186813D 00          X 2 = -0.120483947450D 00
T= 0.25500000D 01 SEC
   X 1 = -0.629657515905D 00          X 2 = -0.640959561231D-01
T= 0.26000000D 01 SEC
   X 1 = -0.659845604433D 00          X 2 = -0.708958197343D-02
```

Figure 12.32 (continued)

```
T= 0.26500000D 01 SEC
    X 1 = -0.683458188742D 00          X 2 =  0.499685745297D-01
T= 0.27000000D 01 SEC
    X 1 = -0.700257042642D 00          X 2 =  0.106510994133D 00
T= 0.27500000D 01 SEC
    X 1 = -0.710072316106D 00          X 2 =  0.161974962481D 00
T= 0.28000000D 01 SEC
    X 1 = -0.712804195587D 00          X 2 =  0.215808238861D 00
T= 0.28500000D 01 SEC
    X 1 = -0.708423868882D 00          X 2 =  0.267474614257D 00
T= 0.29000000D 01 SEC
    X 1 = -0.696973784431D 00          X 2 =  0.316459303935D 00
T= 0.29500000D 01 SEC
    X 1 = -0.678567202054D 00          X 2 =  0.362274121369D 00
T= 0.30000000D 01 SEC
    X 1 = -0.653387039222D 00          X 2 =  0.404462382388D 00
T= 0.30500000D 01 SEC
    X 1 = -0.621684024048D 00          X 2 =  0.442603491080D 00
T= 0.31000000D 01 SEC
    X 1 = -0.583774173144D 00          X 2 =  0.476317162069D 00
T= 0.31500000D 01 SEC
    X 1 = -0.540035619245D 00          X 2 =  0.505267237388D 00
T= 0.32000000D 01 SEC
    X 1 = -0.490904820073D 00          X 2 =  0.529165060142D 00
T= 0.32500000D 01 SEC
    X 1 = -0.436872186072D 00          X 2 =  0.547772371572D 00
T= 0.33000000D 01 SEC
    X 1 = -0.378477170513D 00          X 2 =  0.560903702823D 00
T= 0.33500000D 01 SEC
    X 1 = -0.316302870847D 00          X 2 =  0.568428237749D 00
T= 0.34000000D 01 SEC
    X 1 = -0.250970195087D 00          X 2 =  0.570271128353D 00
T= 0.34500000D 01 SEC
    X 1 = -0.183131651379D 00          X 2 =  0.566414249869D 00
T= 0.35000000D 01 SEC
    X 1 = -0.113464822676D 00          X 2 =  0.556896388123D 00
T= 0.35500000D 01 SEC
    X 1 = -0.426655916185D-01          X 2 =  0.541812857403D 00
T= 0.36000000D 01 SEC
    X 1 =  0.285588167806D-01          X 2 =  0.521314552782D 00
T= 0.36500000D 01 SEC
    X 1 =  0.994969052176D-01          X 2 =  0.495606446474D 00
T= 0.37000000D 01 SEC
    X 1 =  0.169440016374D 00          X 2 =  0.464945543301D 00
T= 0.37500000D 01 SEC
    X 1 =  0.237689416391D 00          X 2 =  0.429638315807D 00
T= 0.38000000D 01 SEC
    X 1 =  0.303563278814D 00          X 2 =  0.390037644688D 00
T= 0.38500000D 01 SEC
    X 1 =  0.366403499286D 00          X 2 =  0.346539295175D 00
T= 0.39000000D 01 SEC
    X 1 =  0.425582272931D 00          X 2 =  0.299577964621D 00
T= 0.39500000D 01 SEC
    X 1 =  0.480508368757D 00          X 2 =  0.249622940821D 00
T= 0.40000000D 01 SEC
    X 1 =  0.530633038406D 00          X 2 =  0.197173414495D 00
T= 0.40500000D 01 SEC
    X 1 =  0.575455500252D 00          X 2 =  0.142753492787D 00
T= 0.41000000D 01 SEC
    X 1 =  0.614527944070D 00          X 2 =  0.869069636330D-01
T= 0.41500000D 01 SEC
    X 1 =  0.647460006291D 00          X 2 =  0.301918633475D-01
T= 0.42000000D 01 SEC
    X 1 =  0.673922671160D 00          X 2 = -0.268250982947D-01
T= 0.42500000D 01 SEC
    X 1 =  0.693651558806D 00          X 2 = -0.835741997403D-01
T= 0.43000000D 01 SEC
    X 1 =  0.706449567420D 00          X 2 = -0.139488399532D 00
T= 0.43500000D 01 SEC
    X 1 =  0.712188843118D 00          X 2 = -0.194009001484D 00
T= 0.44000000D 01 SEC
    X 1 =  0.710812057850D 00          X 2 = -0.246591236548D 00
T= 0.44500000D 01 SEC
    X 1 =  0.702332982567D 00          X 2 = -0.296709705578D 00
```

Figure 12.32 *(continued)*

```
T= 0.45000000D 01 SEC
   X 1 =  0.686836349946D 00          X 2 = -0.343863628624D 00
T= 0.45500000D 01 SEC
   X 1 =  0.664477008048D 00          X 2 = -0.387581848264D 00
T= 0.46000000D 01 SEC
   X 1 =  0.635478373363D 00          X 2 = -0.427427537021D 00
T= 0.46500000D 01 SEC
   X 1 =  0.600130198710D 00          X 2 = -0.463002561775D 00
T= 0.47000000D 01 SEC
   X 1 =  0.558785678301D 00          X 2 = -0.493951461606D 00
T= 0.47500000D 01 SEC
   X 1 =  0.511857918898D 00          X 2 = -0.519964999278D 00
T= 0.48000000D 01 SEC
   X 1 =  0.459815812311D 00          X 2 = -0.540783250904D 00
T= 0.48500000D 01 SEC
   X 1 =  0.403179350509D 00          X 2 = -0.556198202902D 00
T= 0.49000000D 01 SEC
   X 1 =  0.342514430121D 00          X 2 = -0.566055830297D 00
T= 0.49500000D 01 SEC
   X 1 =  0.278427198276D 00          X 2 = -0.570257635603D 00
T= 0.50000000D 01 SEC
   X 1 =  0.211557996251D 00          X 2 = -0.568761632898D 00
```

EXAMPLE 12.28

Consider the state equation

$$\begin{bmatrix} \dot{i} \\ \dot{v} \end{bmatrix} = \begin{bmatrix} -5 & 2 \\ -1 & -2 \end{bmatrix} \begin{bmatrix} i \\ v \end{bmatrix} + \begin{bmatrix} -2 & 0 \\ 0 & 1 \end{bmatrix} \begin{bmatrix} 0 \\ \cos 2t \end{bmatrix} \tag{12.237a}$$

with the initial state

$$\mathbf{x}(0) = \frac{1}{6} \begin{bmatrix} -2 \\ 1 \end{bmatrix} \tag{12.237b}$$

The state equation is the same as in (12.235) except that the input vector and initial conditions have been changed. We use the subroutine STATE2 to compute the state vector $\mathbf{x}(t)$.

Again we choose the increment for integration to be 0.001 and compute the state vector for t from 0 to 5 with time increment 0.05. The main program is the same as that listed in Figure 12.31. However, the subroutine FUNCTU is different and is shown in Figure 12.33. The computer printout and plots of the state variables $i(t) = X1$ and $v(t) = X2$ are presented in Figures 12.34 and 12.35. Figure 12.35(c) is the phase-plane trajectory. The total execution time for the aforementioned computer was 22.27 sec.

Figure 12.33
Subroutine FUNCTU,
which gives the input
vector to the main
program of Figure 12.31.

```
      SUBROUTINE FUNCTU (T,U)
C
C     EVALUATE THE INPUT VECTOR
C     T         -TIME
C     U         -THE INPUT VECTOR OF DIMENSION 10
      DOUBLE PRECISION U(10),T
      U(1)=0.0
      U(2)=DCOS(2*T)
      RETURN
      END
```

Figure 12.34
Computer printout of the state variables $i(t) = X1$ and $v(t) = X2$ for t from 0 to 5 with increments of 0.05 seconds.

```
T= 0.00000000D 00 SEC
    X 1 = -0.333333333333D 00         X 2 =  0.166666666667D 00
T= 0.50000000D-01 SEC
    X 1 = -0.242700703276D 00         X 2 =  0.211902739475D 00
T= 0.10000000D 00 SEC
    X 1 = -0.168522454426D 00         X 2 =  0.248453878149D 00
T= 0.15000000D 00 SEC
    X 1 = -0.107876384683D 00         X 2 =  0.277386510863D 00
T= 0.20000000D 00 SEC
    X 1 = -0.584027244292D-01         X 2 =  0.299550090144D 00
T= 0.25000000D 00 SEC
    X 1 = -0.181965753709D-01         X 2 =  0.315626413177D 00
T= 0.30000000D 00 SEC
    X 1 =  0.142797723979D-01         X 2 =  0.326169486858D 00
T= 0.35000000D 00 SEC
    X 1 =  0.402682869460D-01         X 2 =  0.331637647645D 00
T= 0.40000000D 00 SEC
    X 1 =  0.607733771273D-01         X 2 =  0.332419328819D 00
T= 0.45000000D 00 SEC
    X 1 =  0.766092371602D-01         X 2 =  0.328853609749D 00
T= 0.50000000D 00 SEC
    X 1 =  0.884383328568D-01         X 2 =  0.321246472184D 00
T= 0.55000000D 00 SEC
    X 1 =  0.968026450347D-01         X 2 =  0.309883518524D 00
T= 0.60000000D 00 SEC
    X 1 =  0.102148988521D 00         X 2 =  0.295039769169D 00
T= 0.65000000D 00 SEC
    X 1 =  0.104849482807D 00         X 2 =  0.276987044319D 00
T= 0.70000000D 00 SEC
    X 1 =  0.105218052800D 00         X 2 =  0.255999345280D 00
T= 0.75000000D 00 SEC
    X 1 =  0.103523677048D 00         X 2 =  0.232356577261D 00
T= 0.80000000D 00 SEC
    X 1 =  0.100000969588D 00         X 2 =  0.206346896688D 00
T= 0.85000000D 00 SEC
    X 1 =  0.948585746590D-01         X 2 =  0.178267918467D 00
T= 0.90000000D 00 SEC
    X 1 =  0.882857665667D-01         X 2 =  0.148426980290D 00
T= 0.95000000D 00 SEC
    X 1 =  0.804575761145D-01         X 2 =  0.117140630126D 00
T= 0.10000000D 01 SEC
    X 1 =  0.715387074863D-01         X 2 =  0.847334781566D-01
T= 0.10500000D 01 SEC
    X 1 =  0.616864626017D-01         X 2 =  0.515365342507D-01
T= 0.11000000D 01 SEC
    X 1 =  0.510528519117D-01         X 2 =  0.178851358171D-01
T= 0.11500000D 01 SEC
    X 1 =  0.397860396593D-01         X 2 = -0.158834423643D-01
T= 0.12000000D 01 SEC
    X 1 =  0.280312464757D-01         X 2 = -0.494326155575D-01
T= 0.12500000D 01 SEC
    X 1 =  0.159312117297D-01         X 2 = -0.824292818798D-01
T= 0.13000000D 01 SEC
    X 1 =  0.362630139460D-02         X 2 = -0.114546716122D 00
T= 0.13500000D 01 SEC
    X 1 = -0.874566637519D-02         X 2 = -0.145467506527D 00
T= 0.14000000D 01 SEC
    X 1 = -0.210498168554D-01         X 2 = -0.174886481954D 00
T= 0.14500000D 01 SEC
    X 1 = -0.331547835423D-01         X 2 = -0.202513581717D 00
T= 0.15000000D 01 SEC
    X 1 = -0.449333975986D-01         X 2 = -0.228076624753D 00
T= 0.15500000D 01 SEC
    X 1 = -0.562634636361D-01         X 2 = -0.251323938855D 00
T= 0.16000000D 01 SEC
    X 1 = -0.670285886969D-01         X 2 = -0.272026814551D 00
T= 0.16500000D 01 SEC
    X 1 = -0.771190356602D-01         X 2 = -0.289981751910D 00
T= 0.17000000D 01 SEC
    X 1 = -0.864325760307D-01         X 2 = -0.305012472193D 00
T= 0.17500000D 01 SEC
    X 1 = -0.948753202840D-01         X 2 = -0.316971669854D 00
T= 0.18000000D 01 SEC
    X 1 = -0.102362506768D 00         X 2 = -0.325742483977D 00
```

Figure 12.34 (continued)

```
T= 0.18500000D 01 SEC
   X 1 = -0.108819232682D 00          X 2 = -0.331239671825D 00
T= 0.19000000D 01 SEC
   X 1 = -0.114181112925D 00          X 2 = -0.333410470760D 00
T= 0.19500000D 01 SEC
   X 1 = -0.118394854703D 00          X 2 = -0.332235138398D 00
T= 0.20000000D 01 SEC
   X 1 = -0.121418737740D 00          X 2 = -0.327727164460D 00
T= 0.20500000D 01 SEC
   X 1 = -0.123222991742D 00          X 2 = -0.319933151353D 00
T= 0.21000000D 01 SEC
   X 1 = -0.123790064562D 00          X 2 = -0.308932364059D 00
T= 0.21500000D 01 SEC
   X 1 = -0.123114776111D 00          X 2 = -0.294835953386D 00
T= 0.22000000D 01 SEC
   X 1 = -0.121204354734D 00          X 2 = -0.277785860050D 00
T= 0.22500000D 01 SEC
   X 1 = -0.118078354262D 00          X 2 = -0.257953410322D 00
T= 0.23000000D 01 SEC
   X 1 = -0.113768451481D 00          X 2 = -0.235537617183D 00
T= 0.23500000D 01 SEC
   X 1 = -0.108318125148D 00          X 2 = -0.210763203855D 00
T= 0.24000000D 01 SEC
   X 1 = -0.101782219105D 00          X 2 = -0.183878369488D 00
T= 0.24500000D 01 SEC
   X 1 = -0.942263933030D-01          X 2 = -0.155152319292D 00
T= 0.25000000D 01 SEC
   X 1 = -0.857264678318D-01          X 2 = -0.124872583858D 00
T= 0.25500000D 01 SEC
   X 1 = -0.763676661729D-01          X 2 = -0.933421544822D-01
T= 0.26000000D 01 SEC
   X 1 = -0.662437650085D-01          X 2 = -0.608764631743D-01
T= 0.26500000D 01 SEC
   X 1 = -0.554561588924D-01          X 2 = -0.278002375909D-01
T= 0.27000000D 01 SEC
   X 1 = -0.441128489921D-01          X 2 =  0.555573762839D-02
T= 0.27500000D 01 SEC
   X 1 = -0.323273659036D-01          X 2 =  0.388579206949D-01
T= 0.28000000D 01 SEC
   X 1 = -0.202176372302D-01          X 2 =  0.717733406456D-01
T= 0.28500000D 01 SEC
   X 1 = -0.790481118627D-02          X 2 =  0.103972920151D 00
T= 0.29000000D 01 SEC
   X 1 =  0.448795205600D-02          X 2 =  0.135134759897D 00
T= 0.29500000D 01 SEC
   X 1 =  0.168367092215D-01          X 2 =  0.164947351683D 00
T= 0.30000000D 01 SEC
   X 1 =  0.290179707263D-01          X 2 =  0.193112688079D 00
T= 0.30500000D 01 SEC
   X 1 =  0.409099330600D-01          X 2 =  0.219349237524D 00
T= 0.31000000D 01 SEC
   X 1 =  0.523936944430D-01          X 2 =  0.243394755118D 00
T= 0.31500000D 01 SEC
   X 1 =  0.633544416122D-01          X 2 =  0.265008900970D 00
T= 0.32000000D 01 SEC
   X 1 =  0.736825958722D-01          X 2 =  0.283975639920D 00
T= 0.32500000D 01 SEC
   X 1 =  0.832749069546D-01          X 2 =  0.300105398625D 00
T= 0.33000000D 01 SEC
   X 1 =  0.920354837496D-01          X 2 =  0.313236958434D 00
T= 0.33500000D 01 SEC
   X 1 =  0.998767516028D-01          X 2 =  0.323239065114D 00
T= 0.34000000D 01 SEC
   X 1 =  0.106720326605D 00          X 2 =  0.330011739325D 00
T= 0.34500000D 01 SEC
   X 1 =  0.112497798133D 00          X 2 =  0.333487274729D 00
T= 0.35000000D 01 SEC
   X 1 =  0.117151411811D 00          X 2 =  0.333630913755D 00
T= 0.35500000D 01 SEC
   X 1 =  0.120634646066D 00          X 2 =  0.330441194238D 00
T= 0.36000000D 01 SEC
   X 1 =  0.122912676509D 00          X 2 =  0.323949963470D 00
T= 0.36500000D 01 SEC
   X 1 =  0.123962723493D 00          X 2 =  0.314222059507D 00
```

Figure 12.34   (continued)

```
T= 0.37000000D 01 SEC
   X 1 =  0.123774279370D 00        X 2 =  0.301354662906D 00
T= 0.37500000D 01 SEC
   X 1 =  0.122349213179D 00        X 2 =  0.285476325363D 00
T= 0.38000000D 01 SEC
   X 1 =  0.119701751698D 00        X 2 =  0.266745684949D 00
T= 0.38500000D 01 SEC
   X 1 =  0.115858337065D 00        X 2 =  0.245349880773D 00
T= 0.39000000D 01 SEC
   X 1 =  0.110857362366D 00        X 2 =  0.221502682914D 00
T= 0.39500000D 01 SEC
   X 1 =  0.104748787846D 00        X 2 =  0.195442356291D 00
T= 0.40000000D 01 SEC
   X 1 =  0.975936415705D-01        X 2 =  0.167429279822D 00
T= 0.40500000D 01 SEC
   X 1 =  0.894634095077D-01        X 2 =  0.137743344645D 00
T= 0.41000000D 01 SEC
   X 1 =  0.804393211513D-01        X 2 =  0.106681157408D 00
T= 0.41500000D 01 SEC
   X 1 =  0.706115377954D-01        X 2 =  0.745530765517D-01
T= 0.42000000D 01 SEC
   X 1 =  0.600782515829D-01        X 2 =  0.416801112154D-01
T= 0.42500000D 01 SEC
   X 1 =  0.489447043225D-01        X 2 =  0.839071372851D-02
T= 0.43000000D 01 SEC
   X 1 =  0.373221358793D-01        X 2 = -0.249825022499D-01
T= 0.43500000D 01 SEC
   X 1 =  0.253266726431D-01        X 2 = -0.581060851627D-01
T= 0.44000000D 01 SEC
   X 1 =  0.130781671799D-01        X 2 = -0.906490773483D-01
T= 0.44500000D 01 SEC
   X 1 =  0.699000660233D-03        X 2 = -0.122286321909D 00
T= 0.45000000D 01 SEC
   X 1 = -0.116871399726D-01        X 2 = -0.152701711614D 00
T= 0.45500000D 01 SEC
   X 1 = -0.239564978748D-01        X 2 = -0.181591347372D 00
T= 0.46000000D 01 SEC
   X 1 = -0.359864828694D-01        X 2 = -0.208666574726D 00
T= 0.46500000D 01 SEC
   X 1 = -0.476568963519D-01        X 2 = -0.233656868018D 00
T= 0.47000000D 01 SEC
   X 1 = -0.588511322969D-01        X 2 = -0.256312533412D 00
T= 0.47500000D 01 SEC
   X 1 = -0.694573423661D-01        X 2 = -0.276407203780D 00
T= 0.48000000D 01 SEC
   X 1 = -0.793695534767D-01        X 2 = -0.293740100499D 00
T= 0.48500000D 01 SEC
   X 1 = -0.884887266632D-01        X 2 = -0.308138039580D 00
T= 0.49000000D 01 SEC
   X 1 = -0.967237466547D-01        X 2 = -0.319457162081D 00
T= 0.49500000D 01 SEC
   X 1 = -0.103992332279D 00        X 2 = -0.327584371508D 00
T= 0.50000000D 01 SEC
   X 1 = -0.110221858598D 00        X 2 = -0.332438463845D 00
```

12-9   SUMMARY

We began this chapter by considering the first-order systems described by the scalar state equations and showed how to obtain the general solution of the state equations. We then introduced the *time-constant* concepts and demonstrated how these could be employed to compute the system response.

The previous results were extended to higher-order systems by considering the solution of the associated matrix state equations. We found that the procedure was parallel to that of the scalar case where the scalar quantities were replaced by their matrix counterparts. The physical significance of the general solution is that for a given *initial state* $\mathbf{x}(t_0)$, regardless of the input signals prior to t_0, the state $\mathbf{x}(t)$ and the output $\mathbf{y}(t)$ at any time $t \geqq t_0$

Figure 12.35

(a) Computer plot of the state variable $i(t)$ as a function of t. (b) Computer plot of the state variable $v(t)$ as a function of t. (c) Computer plot of a phase-plane trajectory for the state variables of the state equation (12.237a).

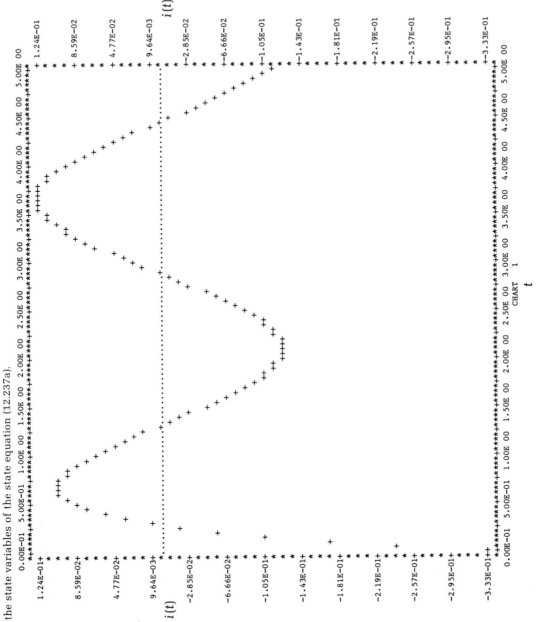

(a)

Figure 12.35 (continued)

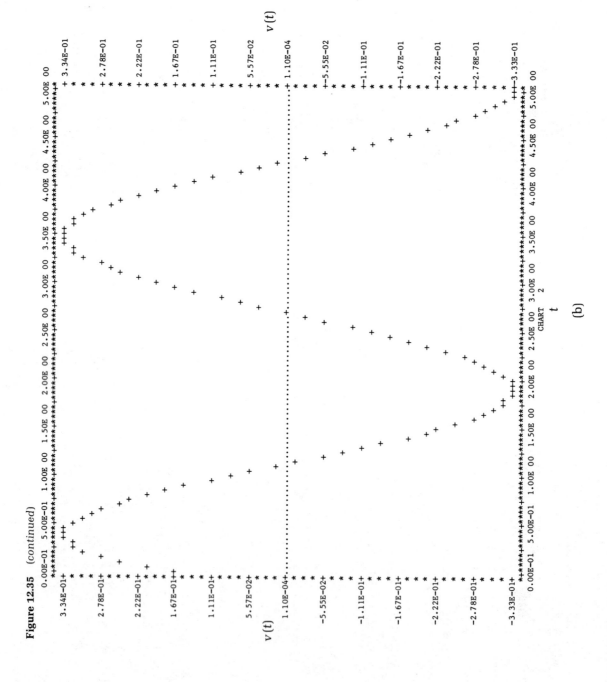

(b)

678

Figure 12.35 (continued)

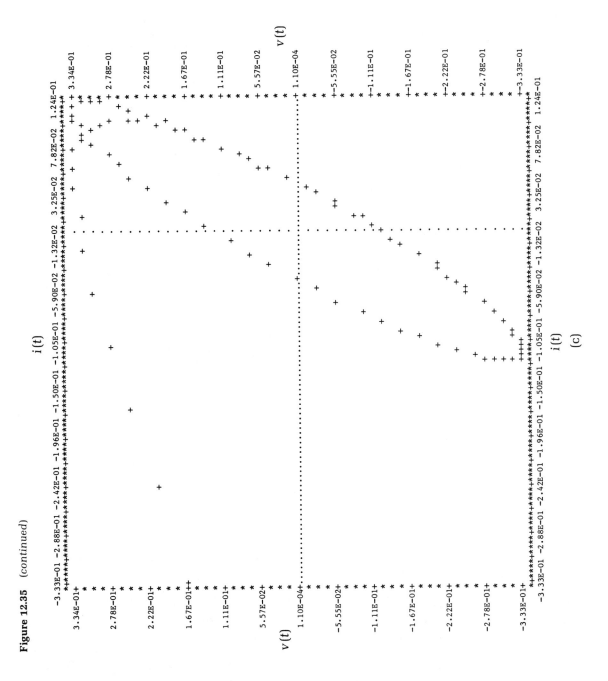

(c)

can be determined from the information about the initial state $\mathbf{x}(t_0)$ and the input excitation from t_0 to t. No other information is required. This *transition of states* from $\mathbf{x}(t_0)$ to $\mathbf{x}(t)$ is governed exclusively by the *state-transition matrix* and the input. Thus once the state-transition matrix is known, the output of a system can be computed in a straightforward manner. The procedure is readily amenable to computer solutions. In addition, we showed that the natural frequencies of a system are the *eigenvalues* of the $\mathbf{A}$ matrix of the state equations.

The central problem in solving the general state equation is the determination of the state-transition matrix, which is defined in terms of an infinite series. If machine computation is the objective, this series representation is convenient and there is no need to obtain a closed-form solution. For this we presented a simple formula in estimating the error when the infinite series is truncated at a finite number of terms. Nevertheless the closed-form equivalents of the state-transition matrix can be obtained. One way to achieve this is by applying the Laplace transform to the state equations. Proceeding in this way, we found that the state-transition matrix is equal to the inverse Laplace transform of the *resolvent matrix* defined as $(s\mathbf{I} - \mathbf{A})^{-1}$. The poles of the resolvent matrix are the *natural frequencies* of the system. A partial-fraction expansion of this matrix about its poles yields constant coefficient matrices called the *constituent matrices*. For large systems the computation of the resolvent matrix is difficult. For this we presented a procedure known as the *Souriau-Frame algorithm*, which permits a rapid calculation of the inverse of $(s\mathbf{I} - \mathbf{A})$. We remark that if we use the Laplace transform technique, we might as well apply it to solve the state equations and avoid the trouble of going through all the intervening steps. This, of course, can be done, but it will defeat the original purpose of solving the system problem in time domain. There are other means of finding the state-transition matrix in closed form. They are not discussed in the text because they will take us far afield into the area of systems theory. Further pursuit of the subject will be left to more advanced treatises.

The *transfer-function matrix* is a matrix relating the transform of the output vector to the transform of the input vector. We showed that this matrix can be expressed in terms of the coefficient matrices of the state equations.

The concepts of state-space trajectory, controllability, and observability were introduced. A *state-space trajectory* is a directed space curve defined by a solution $\mathbf{x}(t)$ in the state space. For the second-order systems, the state-space is called the *phase plane* and the corresponding state-space trajectory is a *phase-plane trajectory*, which gives a clear picture of the transition of states as time progresses. The term *controllability* refers to the ability of a system to get from one state to another with an appropriate input. On the other hand, an *observable* system is one in which a knowledge of the output over a finite time interval is sufficient to determine its *initial state*. General tests for controllability and observability were presented in terms of the composite matrices formed by the coefficient matrices of the state equations.

Finally, to help to obtain the numerical solutions, we presented computer programs for computing the resolvent matrix, the state-transition matrix and the solutions of the state equations. Thus once the state equations are known, their solutions can be calculated routinely.

REFERENCES AND SUGGESTED READING

Balabanian, N., and T. A. Bickart. *Electrical Network Theory.* New York: Wiley, 1969, Chapter 4.

Chan, S. P., S. Y. Chan, and S. G. Chan. *Analysis of Linear Networks and Systems.* Reading, Mass.: Addison-Wesley, 1972, Chapters 7 and 11.

Chen, W. K. *Applied Graph Theory: Graphs and Electrical Networks,* 2d ed. Amsterdam: North-Holland, New York: American Elsevier, 1976, Chapter 7.

Chua, L. O., and P. M. Lin. *Computer-Aided Analysis of Electronics Circuits: Algorithms & Computational Techniques.* Englewood Cliffs, N.J.: Prentice-Hall, 1975, Chapter 9.

DeRusso, P. M., R. J. Roy, and C. M. Close. *State Variables for Engineers.* New York: Wiley, 1965, Chapter 5.

Jensen, R. W., and B. O. Watkins. *Network Analysis: Theory and Computer Methods.* Englewood Cliffs, N.J.: Prentice-Hall, 1974, Chapter 10.

Noble, B. *Applied Linear Algebra.* Englewood Cliffs, N.J.: Prentice-Hall, 1969.

Ogata, K. *State Space Analysis of Control Systems.* Englewood Cliffs, N.J.: Prentice-Hall, 1967, Chapters 6 and 7.

Wing, O. *Circuit Theory with Computer Methods.* New York: Holt, Rinehart & Winston, 1972, Chapter 9.

Zadeh, L. A., and C. A. Desoer. *Linear System Theory.* New York: McGraw-Hill, 1963, Chapters 4 and 5.

PROBLEMS

12.1 For the network of Figure 12.36, use state variable technique to verify the following solutions for the capacitor voltage $v(t)$:

(i) If $v_g(t) = 0$, $t < 0$, and $v_g(t) = 2 \cos 2t$, $t \geqq 0$, then

$$v(t) = -0.98e^{-2.5t} + 1.24 \cos(2t - 0.663) \tag{12.238}$$

(ii) If $v_g(t) = 0$, $t < 0$, and $v_g(t) = t$, $t \geqq 0$, then

$$v(t) = 0.32(e^{-2.5t} - 1) + 0.8t \tag{12.239}$$

Figure 12.36
A network the solutions of which are given by Eqs. (12.238) and (12.239).

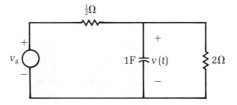

12.2 Repeat Problem 12.1 by using time-constant concepts as in (12.39). Sketch the steady-state response, the transient response, and the complete response.

12.3 Prove the following identities:

$$\boldsymbol{\phi}(t_2 - t_1) = \boldsymbol{\phi}(t_2)\boldsymbol{\phi}(-t_1) \qquad\qquad\qquad (12.240)$$

$$\boldsymbol{\phi}(t_3 - t_2)\boldsymbol{\phi}(t_2 - t_1) = \boldsymbol{\phi}(t_3 - t_1) \qquad\qquad\qquad (12.241)$$

Attach physical significance to these identities.

12.4 In the network of Figure 12.37, let $v_g(t)$ be the unit-step voltage. Applying (12.84), verify that the solution for $t \geq 0$ is given by

$$\begin{bmatrix} i(t) \\ v(t) \end{bmatrix} = \begin{bmatrix} 0.5 + 0.707e^{-0.707t}\cos(0.707t + 5\pi/4) \\ 0.5 + 0.707e^{-0.707t}\cos(0.707t + 3\pi/4) \end{bmatrix} \qquad\qquad (12.242)$$

Plot the phase-plane trajectory.

12.5 Prove Property 12.1. (Hint: Assume that there are two solutions.)

12.6 A voltage source as shown in Figure 12.38 is applied to the network of Figure 12.36. Using both the state-variable technique and the time-constant concepts of (12.39), compute the capacitor voltage $v(t)$ for all t. Sketch the response.

12.7 The switch in the network of Figure 12.39 is closed at $t = 0$. The inductors were initially relaxed. The resistor voltage $v_R(t)$ is the output variable. Using state-variable technique, verify that

$$v_R(t) = 0.6(e^{-6t} - e^{-t}), \quad t \geq 0 \qquad\qquad\qquad (12.243)$$

12.8 Using state-variable technique, compute and sketch the voltage $v_R(t)$ of the network shown in Figure 12.40(b).

Figure 12.37
A network excited by a unit-step voltage, the solution of which is given by Eq. (12.242).

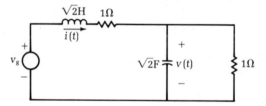

Figure 12.38
A voltage-source waveform.

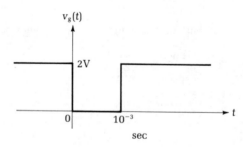

Figure 12.39
A network containing two mutually coupled inductors with the switch closed at $t = 0$.

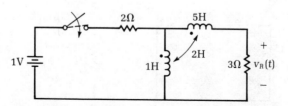

Figure 12.40
A series RL network (b) excited by a voltage source having the waveform (a).

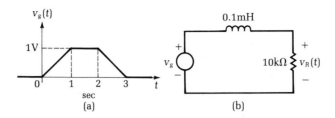

Figure 12.41
A series RC network (b) excited by a voltage source having the waveform (a).

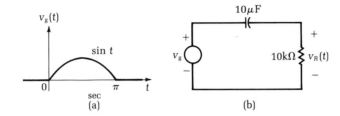

12.9 Using state-variable technique, sketch and verify that the resistor voltage $v_R(t)$ of the network of Figure 12.41(b) is given by

$$v_R(t) = 0.0995 \sin (t + 1.47) - 0.099e^{-10t}, \quad 0 \leq t \leq \pi$$
$$= -0.099(1 + e^{10\pi})e^{-10t}, \quad t > \pi \tag{12.244}$$

12.10 Prove that

$$e^{\mathbf{A}t}e^{\mathbf{B}t} = e^{(\mathbf{A}+\mathbf{B})t} \tag{12.245}$$

if and only if $\mathbf{AB} = \mathbf{BA}$.

12.11 Prove that

$$[e^{\mathbf{A}t}]^{-1} = e^{-\mathbf{A}t} \tag{12.246}$$

12.12 Using the state-variable technique, verify that the resistor current $i_R(t)$ of the network of Figure 12.42 is given by

$$i_R(t) = \frac{1}{82}\left(9 \cos t - \sin t + \frac{1}{9}e^{-t/9}\right), \quad t \geq 0 \tag{12.247}$$

12.13 Sketch and verify that the inductor current $i(t)$ of the network of Figure 12.43(b) is given by

$$i(t) = 2 \text{ mA}, \quad t \leq 1$$
$$= 1 + e^{-10^7(t-1)} \text{ mA}, \quad 1 \leq t \leq 2 \tag{12.248}$$
$$= e^{-10^7(t-1)} + e^{-10^7(t-2)} \text{ mA}, \quad t \geq 2$$

Figure 12.42
A network in which the resistor current is given by (12.247).

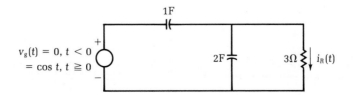

Figure 12.43
An *RL* network (b) excited
by a voltage source having
the waveform (a), in
which the inductor
current is given by
(12.248).

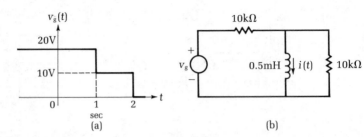

(a)

(b)

12.14 Consider the network of Figure 12.10. Let

$$i_g(t) = 1, \quad t < 2$$

$$\qquad = 2 \cos 3t, \quad t \geq 2 \tag{12.249}$$

$$v_g(t) = 3, \quad t < 2$$

$$\qquad = \cos\left(3t + \frac{\pi}{4}\right), \quad t \geq 2 \tag{12.250}$$

Using (12.84) compute inductor current $i(t)$ and capacitor voltage $v(t)$.

12.15 Repeat Example 12.8 if the 2V battery in the network of Figure 12.8 is replaced by a 1V battery, everything else being the same. Show that the capacitor voltages are given by (12.94). Also justify the result by appealing to physical argument.

12.16 Repeat Problem 12.8 by applying the input voltage of Figure 12.44 instead of that of Figure 12.40(a).

12.17 Repeat Problem 12.9 by applying the input voltage of Figure 12.45 instead of that of Figure 12.41(a).

12.18 Applying the Souriau-Frame algorithm, verify that the state-transition matrix for the **A** matrix of (12.174) is that of (12.178).

12.19 Determine the state equation for the network of Figure 12.16(a).

12.20 Repeat Example 12.17 for the **A** matrix of (12.98).

12.21 Repeat Example 12.18 for the matrix of (12.144) but choose $t_1 = 0.1$ sec.

Figure 12.44
The waveform of a voltage
source.

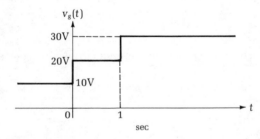

Figure 12.45
The waveform of a voltage
source.

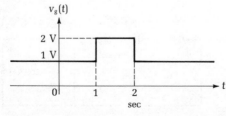

Figure 12.46
A network the state
equation of which is given
by Eq. (12.252).

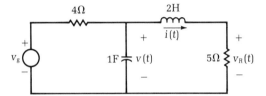

12.22 The state equation of a network is given by

$$\begin{bmatrix} \dot{x}_1 \\ \dot{x}_2 \end{bmatrix} = \begin{bmatrix} -1 & 4 \\ -8 & 0 \end{bmatrix} \begin{bmatrix} x_1 \\ x_2 \end{bmatrix} + \begin{bmatrix} 1 \\ 2 \end{bmatrix} [v_g]$$ (12.251)

Determine the state-transition matrix in closed form and compute the state vector $\mathbf{x}(t)$ with zero initial state.

12.23 The state equation of the network of Figure 12.46 is found to be

$$\begin{bmatrix} \dot{v} \\ \dot{i} \end{bmatrix} = \begin{bmatrix} -0.25 & -1 \\ 0.5 & -2.5 \end{bmatrix} \begin{bmatrix} v \\ i \end{bmatrix} + \begin{bmatrix} 0.25 \\ 0 \end{bmatrix} [v_g]$$ (12.252)

Determine the state-transition matrix. If the input voltage is

$$v_g(t) = 1, \qquad t < 0$$
$$= 2, \qquad t \geq 0$$ (12.253)

verify that the solution of (12.252) for $t \geq 0$ can be written as

$$\begin{bmatrix} v(t) \\ i(t) \end{bmatrix} = \begin{bmatrix} -\frac{36}{63}e^{-0.5t} + \frac{1}{63}e^{-2.25t} + \frac{10}{9} \\ -\frac{1}{7}e^{-0.5t} + \frac{2}{63}e^{-2.25t} + \frac{2}{9} \end{bmatrix}$$ (12.254)

12.24 Sketch the phase-plane trajectory of the solution (12.254).

12.25 In Problem 12.23 select the resistor voltage $v_R(t)$ as the output variable. Determine if the network of Figure 12.46 is completely observable. Is this network completely state controllable?

12.26 A system is described by its state equation

$$\begin{bmatrix} \dot{x}_1 \\ \dot{x}_2 \end{bmatrix} = \begin{bmatrix} -5 & -4 \\ 4 & 3 \end{bmatrix} \begin{bmatrix} x_1 \\ x_2 \end{bmatrix} + \begin{bmatrix} 1 \\ 0 \end{bmatrix} [e^{-t}]$$ (12.255)

with $x_1(0) = 1$ and $x_2(0) = 0$. Verify that the solution of the state equation is given by

$$\mathbf{x}(t) = \begin{bmatrix} e^{-t} - 3te^{-t} - 2t^2e^{-t} \\ 4te^{-t} + 2t^2e^{-t} \end{bmatrix}$$ (12.256)

12.27 Sketch the phase-plane trajectory of the solution (12.256).

12.28 The state equations of the network of Figure 12.47 are given below:

$$\begin{bmatrix} \dot{i} \\ \dot{v} \end{bmatrix} = \begin{bmatrix} -3/2 & 0 \\ 1/6 & -1/12 \end{bmatrix} \begin{bmatrix} i \\ v \end{bmatrix} + \begin{bmatrix} 1/2 \\ 0 \end{bmatrix} [1]$$ (12.257a)

$$[i_R] = [2 \quad -1] \begin{bmatrix} i \\ v \end{bmatrix} + [0][1]$$ (12.257b)

Verify that with $i(0) = 0$ and $v(0) = 1$ the solution of (12.257a) is

$$\begin{bmatrix} i \\ v \end{bmatrix} = \begin{bmatrix} \frac{1}{3}(1 - e^{-1.5t}) \\ \frac{34}{51} + \frac{5}{17}e^{-t/12} + \frac{2}{51}e^{-1.5t} \end{bmatrix}$$ (12.258)

Figure 12.47
A network the state
equations of which are
given by Eq. (12.257).

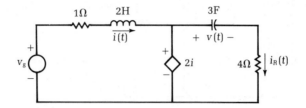

Figure 12.48
An active network (b) excited by a voltage source having the waveform (a).

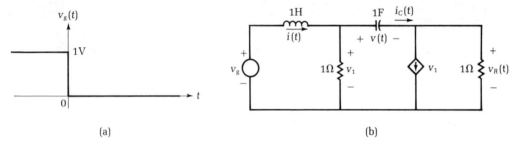

(a)   (b)

12.29 Sketch the phase-plane trajectory of the solution (12.258).

12.30 Determine if the system of (12.257) is completely state controllable. Is this system completely observable?

12.31 The voltage source of Figure 12.48(a) is applied to the network of Figure 12.48(b). The state equations of the network are found to be

$$\begin{bmatrix} i \\ \dot{v} \end{bmatrix} = \frac{1}{3} \begin{bmatrix} -1 & -1 \\ 2 & -1 \end{bmatrix} \begin{bmatrix} i \\ v \end{bmatrix} + \begin{bmatrix} 1 \\ 0 \end{bmatrix} [v_g]$$ (12.259a)

$$\begin{bmatrix} i_C \\ v_R \end{bmatrix} = \frac{1}{3} \begin{bmatrix} 2 & -1 \\ 1 & -2 \end{bmatrix} \begin{bmatrix} i \\ v \end{bmatrix}$$ (12.259b)

    (i) Determine the constituent matrices, transfer-function matrix, and the state-transition matrix.

    (ii) Verify that the solution of (12.259a) is given by

$$\begin{bmatrix} i(t) \\ v(t) \end{bmatrix} = e^{-t/3} \begin{bmatrix} \cos\sqrt{2}t/3 - \sqrt{2}\sin\sqrt{2}t/3 \\ \sqrt{2}\sin\sqrt{2}t/3 + 2\cos\sqrt{2}t/3 \end{bmatrix}, \quad t \geq 0$$ (12.260)

    (iii) Determine if the system of (12.259) is completely state controllable and completely observable.

12.32 Calculate the state-transition matrix of the system of (12.259) by using six terms of the series (12.63) with $t_1 = 0.1$ sec. What is an upper bound of error, as predicted by (12.168)? Compare this result with the closed-form solution obtained by Eq. (12.122). What is an upper bound of error if (12.233) is used?

12.33 Repeat Example 12.23 for the matrix **A** of (12.193).

12.34 Repeat Example 12.25 for the matrix **A** of (12.193).

12.35 Repeat Example 12.26 for the state equation (12.98) with 1 replacing cos 2t. This is equivalent to replacing the current source i_g of (12.96) by a unit-step input current

$$i_g(t) = u(t)$$ (12.261)

APPENDIX A

─────

SUBROUTINE PLOT

1. Purpose. The program is for plotting one or more curves on one plot by a line printer, and is listed in Figure A.

2. Usage. The program consists of a subroutine, PLOT, called by

CALL PLOT (XY, JXY, N, NDIM, NC)

XY   is a two-dimensional array in which each column represents the values of a variable, either dependent or independent.

JXY   is a one-dimensional array of integers. The odd rows of this array are used to specify the independent variables, and the even rows are for the corresponding dependent variables. The columns of XY specified by rows 2K and 2K − 1 of JXY constitute a pair of variables for one curve, where K = 1, 2, . . . , NC. The dimension of JXY is 2NC.

N   is the number of points to be plotted.

NDIM   is the number of rows in XY as declared in the dimensional statement in the main program.

NC   is the number of curves to be plotted on a single plot.

3. Subroutines and Function Subprograms Required.   None

Figure A
Subroutine PLOT used to plot one or more curves on one plot by a line printer.

```
C
C      SUBROUTINE PLOT
C
C      PURPOSE
C         THE PROGRAM IS FOR PLOTTING ONE OR MORE CURVES ON ONE PLOT BY A
C            LINE PRINTER.
C
C      USAGE
C         CALL PLOT (XY,JXY,N,NDIM,NC)
C
C         XY      -A TWO-DIMENSIONAL ARRAY IN WHICH EACH COLUMN REPRESENTS THE
```

Figure A (*continued*)

```
C                         VALUES OF A VARIABLE, EITHER DEPENDENT OR INDEPENDENT.
C            JXY      -A ONE-DIMENSIONAL ARRAY OF INTEGERS.  THE ODD ROWS OF THIS
C                         ARRAY ARE USED TO SPECIFY THE INDEPENDENT VARIABLES, AND THE
C                         EVEN ROWS ARE FOR THE CORRESPONDING DEPENDENT VARIABLES.
C                         THE COLUMNS OF XY SPECIFIED BY ROWS 2K AND (2K-1) OF JXY
C                         CONSTITUTE A PAIR OF VARIABLES FOR ONE CURVE, WHERE K=1,
C                         2,...,NC.  THE DIMENSION OF JXY IS 2NC.
C            N        -THE NUMBER OF POINTS TO BE PLOTTED.
C            NDIM     -THE NUMBER OF ROWS IN XY AS DECLARED IN THE DIMENSIONAL
C                         STATEMENT IN THE MAIN PROGRAM.
C            NC       -THE NUMBER OF CURVES TO BE PLOTTED ON A SINGLE PLOT.
C
C        SUBROUTINES AND FUNCTION SUBPROGRAMS REQUIRED
C            NONE
C
         SUBROUTINE PLOT(XY,JXY,N,NDIM,NC)
         DIMENSION IGRID(101),XS(11),YS(13),ICHAR(7),XY(1),JXY(1)
         DATA ICHAR/1H+,1H*,1H-,1H$,1H=,1H.,1H /
         XMAX=-1.0E20
         XS(1)=-XMAX
         YS(1)=XMAX
         YMIN=XS(1)
         J2=0
         DO 1 J=1,NC
         J2=J2+2
         JIX=(JXY(J2-1)-1)*NDIM
         JIY=(JXY(J2)-1)*NDIM
         DO 1 I=1,N
         IJX=JIX+I
         IJY=JIY+I
         IF(XY(IJX).GT.XMAX) XMAX=XY(IJX)
         IF(XY(IJX).LT.XS(1)) XS(1)=XY(IJX)
         IF(XY(IJY).GT.YS(1)) YS(1)=XY(IJY)
         IF(XY(IJY).LT.YMIN) YMIN=XY(IJY)
1        CONTINUE
         XR=XMAX-XS(1)
         IF(XR.EQ.0.0)XR=1.0E-20
         YR=YS(1)-YMIN
         IF(YR.EQ.0.0) YR=1.0E-20
         XT=XMAX*XS(1)
         YT=YMIN*YS(1)
         IF(XT.LT.0.0) IYAX=100.0*(-XS(1))/XR+1.5
         IF(YT.LE.0.0) IXAX=48.0*YS(1)/YR+1.5
         XMAX=XR/10.0
         DO 2 I=2,11
2        XS(I)=XS(I-1)+XMAX
         XMAX=YR/12.0
         DO 3 I=2,13
3        YS(I)=YS(I-1)-XMAX
         PRINT 4,(XS(I),I=1,11)
4        FORMAT(1H1,1PE15.2,10E10.2/10X,1H*,20(5H+****),2H+*)
         II=1
         KK=0
         DO 14 LINE=1,49
         DO 5 J=1,101
5        IGRID(J)=ICHAR(7)
         IF(YT.GT.0.0) GO TO 7
         IF(LINE.NE.IXAX) GO TO 7
         DO 6 J=1,101
6        IGRID(J)=ICHAR(6)
7        IF(XT.LT.0.0) IGRID(IYAX)=ICHAR(6)
         J2=0
         DO 9 J=1,NC
         J2=J2+2
         JIX=(JXY(J2-1)-1)*NDIM
         JIY=(JXY(J2)-1)*NDIM
         JC=MOD(J,5)
         DO 9 I=1,N
         IJX=JIX+I
         IJY=JIY+I
         IPTY=48.0*(YS(1)-XY(IJY))/YR+1.5
         IF(IPTY.GT.49) IPTY=49
```

Figure A *(continued)*

```
        IF(IPTY.LT.1) IPTY=1
        IF(IPTY.NE.LINE) GO TO 9
        IPTX=100.0*(XY(IJX)-XS(1))/XR+1.5
        IF(IPTX.LT.1)IPTX=1
        IF(IPTX.GT.101) IPTX=101
        IF(JC.NE.0) GO TO 8
        IGRID(IPTX)=ICHAR(5)
        GO TO 9
8       IGRID(IPTX)=ICHAR(JC)
9       CONTINUE
        IF(KK.GT.0) GO TO 11
        PRINT 10,YS(II),(IGRID(I),I=1,101),YS(II)
10      FORMAT(1PE10.2,1H+,101A1,1H+,E9.2)
        II=II+1
        GO TO 13
11      PRINT 12,(IGRID(I),I=1,101)
12      FORMAT(10X,1H*,101A1,1H*)
13      KK=KK+1
        IF(KK.NE.4) GO TO 14
        KK=0
14      CONTINUE
        PRINT 15,(XS(I),I=1,11)
15      FORMAT(10X,1H*,20(5H+****),2H+*/1PE16.2,10E10.2)
        RETURN
        END
```

APPENDIX B

SUBROUTINE PLOT2

1. Purpose.   The program is for plotting a single curve on one plot by a line printer with area beneath the curve filled, and is listed in Figure B.

2. Usage.   The program consists of a subroutine, PLOT2, called by

CALL PLOT2 (XY, N, NDIM)

XY   is a two-dimensional array in which each column represents the values of a variable, either dependent or independent.

N   is the number of points to be plotted.

NDIM is the number of rows in XY as declared in the dimensional statement in the main program.

3. Subroutines and Function Subprograms Required.   None

Figure B

Subroutine PLOT2 used to plot a single curve on one plot by a line printer with area beneath the curve filled.

```
C
C
C       SUBROUTINE PLOT2
C
C       PURPOSE
C          THE PROGRAM IS FOR PLOTTING A SINGLE CURVE ON ONE PLOT BY A
C             LINE PRINTER WITH AREA BENEATH THE CURVE FILLED.
C
C       USAGE
C          CALL PLOT2 (XY,N,NDIM)
C
C          XY      -A TWO-DIMENSIONAL ARRAY IN WHICH EACH COLUMN REPRESENTS
C                      THE VALUES OF A VARIABLE, EITHER DEPENDENT OR
C                      INDEPENDENT.
C          N       -THE NUMBER OF POINTS TO BE PLOTTED.
C          NDIM    -THE NUMBER OF ROWS IN XY AS DECLARED IN THE DIMENSIONAL
C                      STATEMENT IN THE MAIN PROGRAM.
C
C       SUBROUTINES AND FUNCTION SUBPROGRAMS REQUIRED
C          NONE
C
```

Figure B (*continued*)

```
      SUBROUTINE PLOT2(XY,N,NDIM)
      DIMENSION IGRID(101),XS(11),YS(13),ICHAR(3),XY(1)
      DATA ICHAR/1H+,1H.,1H /
      KK=0
      II=1
      JC=1
      LINEO=1
      XMAX=-1.0E20
      XS(1)=-XMAX
      YS(1)=XMAX
      YMIN=XS(1)
      DO 1 I=1,N
      IF(XY(I).GT.XMAX) XMAX=XY(I)
      IF(XY(I).LT.XS(1)) XS(1)=XY(I)
      IF(XY(NDIM+I).GT.YS(1)) YS(1)=XY(NDIM+I)
      IF(XY(NDIM+I).LT.YMIN) YMIN=XY(NDIM+I)
    1 CONTINUE
      XR=XMAX-XS(1)
      IF(XR.EQ.0.0)XR=1.0E-20
      YR=YS(1)-YMIN
      IF(YR.EQ.0.0) YR=1.0E-20
      XT=XMAX*XS(1)
      YT=YMIN*YS(1)
      IF(XT.LT.0.0) IYAX=100.0*(-XS(1))/XR+1.5
      IF(YT.LE.0.0) IXAX=48.0*YS(1)/YR+1.5
      IF(YS(1).LT.0.0) IXAX=0
      IF(YS(1).EQ.0.0) IXAX=1
      IF(YMIN.GT.0.0) IXAX=50
      XMAX=XR/10.0
      DO 2 I=2,11
    2 XS(I)=XS(I-1)+XMAX
      XMAX=YR/12.0
      DO 3 I=2,13
    3 YS(I)=YS(I-1)-XMAX
      PRINT 4,(XS(I),I=1,11)
    4 FORMAT(1H1,1PE15.2,10E10.2/10X,1H*,20(5H+****),2H+*)
      DO 5 J=1,101
    5 IGRID(J)=ICHAR(3)
      IF(YS(1).LT.0.0) LINE=0
      IF(YS(1).LT.0.0) GO TO 18
    6 CONTINUE
      DO 17 LINE=LINEO,49
      IF(LINEO.NE.1.OR.YT.GT.0.0.OR.LINE.NE.IXAX) GO TO 8
      DO 7 J=1,101
    7 IF(IGRID(J).NE.ICHAR(1)) IGRID(J)=ICHAR(2)
    8 CONTINUE
      DO 11 I=1,N
      IPTY=48.0*(YS(1)-XY(NDIM+I))/YR+1.5
      IPTX=100.0*(XY(I)-XS(1))/XR+1.5
      IF(YT.GT.0.0) GO TO 9
      IF(LINE.EQ.IXAX.AND.IPTY.GT.IXAX) IGRID(IPTX)=ICHAR(1)
    9 IF(LINEO.EQ.1.AND.IPTY.NE.LINE) GO TO 10
      IF(LINEO.NE.1.AND.IPTY.NE.(LINE-1)) GO TO 10
      IGRID(IPTX)=ICHAR(JC)
   10 IF(XT.LT.0.0.AND.IGRID(IYAX).NE.ICHAR(1)) IGRID(IYAX)=ICHAR(2)
   11 CONTINUE
      IF(KK.GT.0) GO TO 13
      PRINT 12,YS(II),(IGRID(I),I=1,101),YS(II)
   12 FORMAT(1PE10.2,1H+,101A1,1H+,E9.2)
      II=II+1
      GO TO 15
   13 PRINT 14,(IGRID(I),I=1,101)
   14 FORMAT(10X,1H*,101A1,1H*)
   15 KK=KK+1
      IF(KK.NE.4) GO TO 16
      KK=0
   16 IF(YT.GE.0.0.AND.IXAX.NE.1) GO TO 17
      IF(LINE.EQ.IXAX) GO TO 18
   17 CONTINUE
      GO TO 21
   18 LINEO=LINE+1
      DO 19 J=1,101
```

Figure B *(Continued)*

```
19 IGRID(J)=ICHAR(3)
   DO 20 I=1,N
   IPTY=48.0*(YS(1)-XY(NDIM+I))/YR+1.5
   IPTX=100.0*(XY(I)-XS(1))/XR+1.5
   IF(IPTY.GT.IXAX) IGRID(IPTX)=ICHAR(1)
20 CONTINUE
   JC=3
   GO TO 6
21 PRINT 22,(XS(I),I=1,11)
22 FORMAT(10X,1H*,20(5H+****),2H+*/1PE16.2,10E10.2)
   RETURN
   END
```

INDEX

I
N
D
E
X